Viruses of Microorganisms

https://doi.org/10.21775/9781910190852

Edited by

Paul Hyman

Department of Biology and Toxicology
Ashland University
Ashland, OH
USA

and

Stephen T. Abedon

Department of Microbiology
The Ohio State University
Mansfield, OH
USA

Caister Academic Press

Caister Academic Press
Norfolk, UK

www.caister.com

British Library Cataloguing-in-Publication Data
A catalogue record for this book is available from the British Library

ISBN: 978-1-910190-85-2 (paperback)
ISBN: 978-1-910190-86-9 (ebook)

Cover design adapted from a composite image of living *Chrysochromulina parva* CCMP291 cells (bright-field microscopy; 100x magnification) together with infected cells undergoing lysis (false-coloured TEM thin sections; 12,000x direct magnification). Steven M. Short and Beata Cohan of the University of Toronto Mississauga created the original image. See *Chapter 10* for information on viral infections of *C. parva*.

Contents

Preface

The timing of our taking on the editing of this monograph, *Viruses of Microorganisms*, is of interest as it both was and was not associated with seemingly related events. The first event was our editing of the 2012 monograph, *Bacteriophages in Health and Disease* (CABI Press). It was both well into the editing of that book, and to our knowledge without connection, that Caister Academic Press suggested that we take on the editing of this new volume. We assume that the editing of previous books on viruses and specifically phages by one of us (Abedon) is what motivated the connection, and it clearly was our idea that this be a second joint editing effort, so we accept that no strange coincidences were at work in our being asked to take on (yet) another editing task.

What was strange was the specific subject of this text. For a number of years prior to our being tagged with this new editing task, both of us had been intimately and some might argue excessively involved in the founding of a professional society that ultimately would come to be named the International Society for Viruses of Microorganisms (see ISVM.org). This society, as we were being contacted to write on this very subject, did not yet exist and indeed neither of us had published (at that point) on anything beyond bacteriophages. Because viruses of a specific type of microorganism (bacteria) were our primary research interests (Hyman and Abedon, 2009, 2012a, 2015), there is no reason to assume that either of us knew much virology beyond that of these 'phages'. So here we have what clearly was complete coincidence – ISVM and this monograph – but just as clearly we should assume that the concept of 'Viruses of Microorganisms' must have been in the air.

What followed, in turn, was as not coincidental as can possibly be. Prior to taking on the task of pulling together a book on a subject, one which at best we possessed only tangential familiarity, we decided to write on that very subject. The result, to our knowledge, is the very first in-depth review of the concept of viruses of microorganisms (Hyman and Abedon, 2012b). This article was piloted as a poster at the Viruses of Microbes 2012 meeting, which further convinced us that the article needed to be written. While in terms of timing, it should seem 'obvious' that the editing of this book was our plan all along, as the review was published in 2012, but in fact it was the poster and review that stemmed from the book rather than the other way around. And even worse, in terms of establishing cause and effect, although in the Spring of 2012 we were pondering the writing of a review, on something, we in fact had not yet hit upon just what that something might be. It was only in preparation for going out to eat Chinese food, at Sue Min's (formerly of Wooster, Ohio), that the idea of writing such a review was hatched. Over that dinner it was decided that this indeed would be a good thing to do, and we then spent the month of May, 2012, doing the writing and extracting the tables for the poster.

We, of course, were not alone in building on this theme. In addition to the series of Virus of Microbes meetings which have taken place in Europe (Paris in 2010, Brussels in 2012, Zurich in 2014, and Liverpool in 2016), there has been a recent special issue on viruses of microbes published in the journal, *Viruses* (http://www.mdpi.com/journal/viruses/special_issues/viruses_microbes). Thus, we feel that now is perhaps an ideal moment to take a step back to summarize the totality of the field, i.e. with a multi-authored volume on just this subject, the viruses of microbes or, more formally and reflecting the name of the associated society, the *Viruses of Microorganisms*.

In this volume, which inspired that review, we begin with a chapter that was inspired by that review, that is, considering just what distinguishes viruses of microorganisms from viruses in general. We follow this with a chapter, by Evelien Adriaenssens, that is both comprehensive and basic to what viruses of microorganisms are. This chapter is so comprehensive and basic that we considered it to be what essentially is now a co-introductory chapter. These introductions are then followed by chapters which set out the context of viruses of microorganisms, in terms of their evolution and particularly their ecology as well as environmental microbiology. In other words, what is 'out there'. The next section of the book then examines specific categories of viruses of microorganisms including the viruses of domain Bacteria (a.k.a. bacteriophages or phages, but also bacterial viruses), the viruses of domain Archaea (in the more modern literature usually described as archaeal viruses), and the many categories of viruses of microbial members of domain Eukarya.

While viruses of domain Archaea and Bacteria unambiguously are all viruses of microorganisms, domain Eukarya consists of a combination of microorganisms (especially although not necessarily exclusively single-celled eukaryotes), macroorganisms (animals, plants, many fungi, and a fair number of quite large protists), and organisms that are not necessarily easily categorized as microorganisms versus macroorganisms (especially various fungi and protists that are of intermediate size and/or complexity). Although clearly these lists and distinctions ought to be sufficient, we nonetheless can add as well what can be described as viruses of viruses, that is, viruses whose hosts consist solely of virus-infected cells, here called virophages.

Lastly are various 'application' chapters. First is a chapter about the roles of viruses of microorganisms in biotechnology. Mostly the latter has involved phages, but as we learn more about other viruses this will likely change. This theme is then expanded beyond just phages in terms of the potential for the use of viruses of microorganisms as anti-microorganism agents. Smaller fleas... can be the enemies of our enemies, and therefore viruses of unfriendly microorganisms can be our friends! Lastly is a chapter describing techniques, especially as used to characterize viruses of microorganisms ecologically.

We are hopeful that you will find this book helpful, and that you will come to find the viruses of microorganisms as fascinating as we do.

References

Hyman, P., and Abedon, S.T. (2009). Bacteriophage (overview). In Encyclopaedia of Microbiology, Schaecter, M. ed. (Elsevier, UK), pp. 322–338.

Hyman, P., and Abedon, S.T. (2012a). Bacteriophage (overview). In The Desk Encyclopaedia of Microbiology, 2nd edn, Schaecter, M. ed. (Elsevier, UK), pp. 166–182.

Hyman, P., and Abedon, S.T. (2012b). Smaller fleas: viruses of microorganisms. Scientifica *2012*, 734023. https://doi.org/10.6064/2012/734023.

Hyman, P., and Abedon, S.T. (2015). Bacteriophage: overview. In Reference Module in Biomedical Sciences, 3rd edn, Caplan, M.J. ed. (Elsevier, UK), pp. 322–338.

Part I

Introduction

Viruses of Microorganisms: What are They and Why Care?

1

Paul Hyman[1] and Stephen T. Abedon[2*]

[1]Department of Biology and Toxicology, Ashland University, Ashland, OH, USA.
[2]Department of Microbiology, The Ohio State University, Mansfield, OH, USA.

*Correspondence: phyman@ashland.edu and abedon.1@osu.edu

https://doi.org/10.21775/9781910190852.01

Abstract

Like microorganisms generally, the notion of viruses of microorganisms, or VoMs, that is, viruses that *infect* cellular microorganisms, is an artificial concept, and particularly so from a phylogenetic perspective. Microorganisms are collectively defined by a single, somewhat primitive phenotype – their smallness even at maturity – rather than together making up a monophyletic taxon. There is, therefore, no clade Microorganisms. As individual viruses are more likely to infect organisms which in fact *are* members of the same clade, VoMs too do not collectively form a single viral clade, nor even a collection of distinct, only-microorganism-infecting taxa. VoMs instead consist of viruses of domains Bacteria, Archaea, and Eukarya, with especially the viruses of Eukarya tending to be evolutionarily related between those that infect eukaryotic microorganisms and those that infect eukaryotic macroorganisms. By contrast, all viruses of domains Bacteria and Archaea represent VoMs. All members of those two domains in fact are microorganisms, although even here, these VoMs are far from being collectively monophyletic. The result, ultimately, is a great deal of diversity among both VoMs and their hosts. In this chapter we introduce the concept of viruses of microorganisms. We also offer a first-approximation introduction to what VoMs exist, as considered in great detail in subsequent chapters.

Introduction

Viruses hold a special place within the pantheon of life. They are obligate symbionts of cellular organisms and generally also parasites of these organisms at some point in viral life cycles, versus consistently serving instead as commensalistic or mutualist symbionts. Although viruses are microorganisms themselves, they are not cellular but nevertheless generally exist in a distinct, autonomous, acellular state for at least part of their existence. The existence of this evolved, autonomous state, the virion, contrasts viruses with most other independent genetic entities e.g. such as plasmids. In that extracellular state, viruses do not metabolize on their own but nonetheless they are not always physiologically inert. Nonetheless, like any obligately symbiotic organism, viruses are dependent on their hosts for their reproduction, and also depend on their hosts for expression of the bulk of their physiology during the infection phase of their existence. And like any mobile genetic material that can gain access to the interior of new cells, viruses can carry both novel genes and novel biochemical pathways from one organism to another. Viruses also are ubiquitous, seemingly more so than the organisms they infect, and are even thought to be, collectively, the most numerous semi-autonomous genetic entities on Earth. Viruses, in short, are acellular for at least part of their life cycles, are diverse, and also are commonplace. They are often as well, but not quite always, distinguishable from the non-viral goings on of the cells that they infect.

Aside from being organisms in their own right, we can consider viruses in terms of their impact on other organisms as well as on the environments that they occupy. Since these impacts are a consequence of the basic nature of viruses as organisms themselves, you will see overlap between the following list and the previous paragraph. Nevertheless, ecologically as well as evolutionarily, viruses play three major, direct roles (Abedon, 2011):

1 As exploiters of their cellular hosts – as predators or parasites. Whether a given cellular species is subject to significant viral infection therefore has bearing on its potential to maintain itself within environments at high, i.e. 'winner' densities.
2 As transducers of genetic material between hosts, where the resulting horizontal gene transfer can impact the evolution of both the virus and host, and to some extent cellular host organisms are modified against their 'will'.
3 As solubilizers (lysers, decomposers) of especially microbial hosts, viruses contribute to the densities of bioavailable nutrients within an environment (e.g. dissolved organic carbon).

With regard to horizontal gene transfer, a perhaps hard upper limit on the size of novel blocks of DNA that can be transferred is a function of the lengths of DNA that can be carried by associated viruses. This applies to random genome fragments (generalized transduction) as well as to host genome fragments adapted to be transduced, e.g. pathogenicity islands. Especially among otherwise clonally reproducing organisms, such as prokaryotes generally, it is the transduction of DNA by viruses that can make the difference between a species or strain displaying dynamic adaptability on the one hand and Muller's Ratchet-associated erosion of genomic information on the other (Hendrickson, 2012). See Fig. 1.1 for illustration.

The ecology of VoMs is otherwise covered in a series of four chapters found in this volume. These

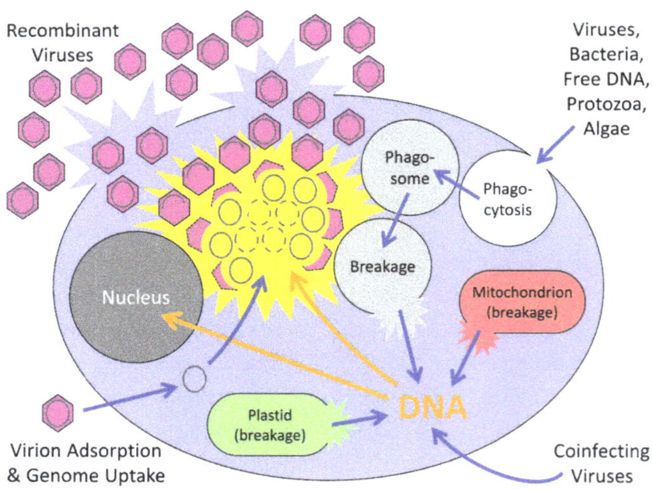

Figure 1.1 Sources of DNA that can be available to viruses for recombination during infection of a eukaryotic cell. In principle, all DNA in a cell has some potential to become incorporated into an infecting virion. Here the site of virion replication, within which such DNA would be incorporated, is a virion factory structure (yellow). This DNA, if virus incorporated, then may be passed on to subsequently infected cells, particularly if the DNA-carrying virus is unable to infect either lytically or cytotoxically. The result is transduction, that is, virus-mediated horizontal gene transfer. Although less dramatic in terms of the potential sources of DNA, VoMs of prokaryotic organisms also can pick up DNA from numerous sources (Abedon, 2009b). This figure is derived from one originally published in Hyman and Abedon (2012).

consider more evolutionary aspects of the ecology of VoMs (Ford *et al.*, Chapter 3), VoMs as they are found in soils (Williamson, Chapter 4), VoMs as they are found in microbiomes, particularly that of corals (Vega Thurber *et al.*, Chapter 5), and VoMs as they are found in aquatic environments (Jacquet *et al.*, Chapter 6).

Viruses thus are important, and perhaps particularly so with regard to the death, gene exchange, and environmental breakdown especially of microorganisms. The concept of microorganism, however, cannot be defined with exceptional precision, so neither can just what constitutes the viruses that infect them. That all cellular organisms possess viruses has been taken as an article of faith at least as long as these authors have been alive (i.e. 50-plus years). This belief, as leading to the search for and discovery of new viruses, tends to serve to further justify that faith. As cellular organisms that can be described as microorganisms thus mostly if not entirely possess viruses, then so too these viruses must collectively define what here we call Viruses of Microorganisms, or VoMs. In this chapter therefore we introduce VoMs as both a concept and as specific types of viruses. For more extensive consideration of VoMs, in what effectively can serve, as well, as an introduction to much of the material found in this monograph, see Hyman and Abedon (2012).

VoM, VoMi, and VoMa

Throughout this monograph we abbreviate viruses of microorganisms as VoMs. We accept that some of the viruses that are described will not necessarily be considered to be VoMs by all, but are confident that the vast majority the viruses considered in this monograph will be. Nonetheless, we still run into an ambiguity as, in principle at least, the viruses of *ma*croorganisms could possess the same acronym, that is, VoM as well. To relieve that ambiguity, we might distinguish VoMi(s) from VoMa(s) (Hyman and Abedon, 2012), but for the sake of aesthetics we will stick with VoM or VoMs to describe viruses of microorganisms as a group and, as necessary, in this chapter use VoMa or VoMas for Viruses of Macroorganisms.

Though complicated, it would be wonderful if we could limit complications to just this naming, or abbreviating, but we cannot. As alluded to in our opening sentence, this is because the distinction between what is a microorganism as what is a macroorganism is not *un*ambiguous. Our first task therefore is to define 'microorganism'. As you will see, this can be accomplished in multiple ways, one of which is simply and 'not macroorganism'. We therefore begin, necessarily, by distinguishing VoMs from VoMas. A summary of what, in our opinion, are the categories of viruses which should be viewed as VoMs (versus VoMas) can be found in Table 1.1.

Table 1.1 Categories of viruses by host type

Cellular host domain		
Bacteria	Archaea	Eukarya
Bacteriophages (i.e. phages, bacterial viruses, or bacterioviruses)	Archaeal viruses (i.e. archaeoviruses; not archaeal phages or archaeophages)	Viruses of unicellular alga (protists)
		Viruses of protozoans (amoeba and microscopic protists)
		Fungal viruses (i.e. mycoviruses; including both unicellular and multicellular fungi)
		Plant viruses (including viroids)
		Animal viruses
Acellular hosts (hosts are cells infected with another virus – the host virus)		
Virophages (specific virus-infected protist host)		
Satellite viruses (specific virus-infected cells of multiple domains)		

Shaded cells indicate VoMs as considered in subsequent chapters. Later sections of this chapter better define these categories, especially the viruses of eukaryotes.

What is a microorganism?

The most basic of questions regarding what are viruses of microorganisms is that of what *is* a microorganism. There are at least three general answers to this question: (1) microorganisms are not macroorganisms; (2) microorganisms to a large extant are basically small organisms (e.g. <<1 mm fully mature); and (3) at least a subset of microorganisms are pathogenic. We thus begin in asking the question of what are VoMs by asking the more basic question of 'What are microorganisms?'. A visual summary of how this question might asked as well as answered can be found in Fig. 1.2. Once we approximately settle that issue of just what constitute microorganisms, we then turn to a brief introduction to those viruses that we feel together make up this concept that we call viruses of microorganisms, or VoMs.

Not macroorganisms

There are two categories of organisms that are *not* microorganisms, although not completely without ambiguity. Nonetheless, these make up a large fraction of macroorganisms, and these are the plants and the animals. Both are mostly large, particularly the plants, and both are exclusively multicellular. The ambiguity comes in, for the animals, in that there are in fact animals which are so small as adults as to be microscopic, or in many cases at least nearly microscopic. As a consequence, we cannot define macroorganisms as consisting solely of organisms that are 'Not microscopic', although perhaps 'Not *very* microscopic' will do (though, alas, there also are bacteria along with single-celled, non-coenocytic protists which also are not *very* microscopic). For plants, contrasting animals, the problem is not so much size, as no plants as adults are so small as to be microscopic, although even here this requires limiting the concept of adults to mature sporophyte individuals. That is, we are not counting the female gametophytes (haploid generation) of angiosperms (flowering plants) among adults – all seven cells of them – even although that designation arguably could be applied. For plants, an additional ambiguity comes from their close evolutionary connection to green algae, which do include unambiguously microorganism members. Indeed, one could make a similar argument for animals and choanoflagellate

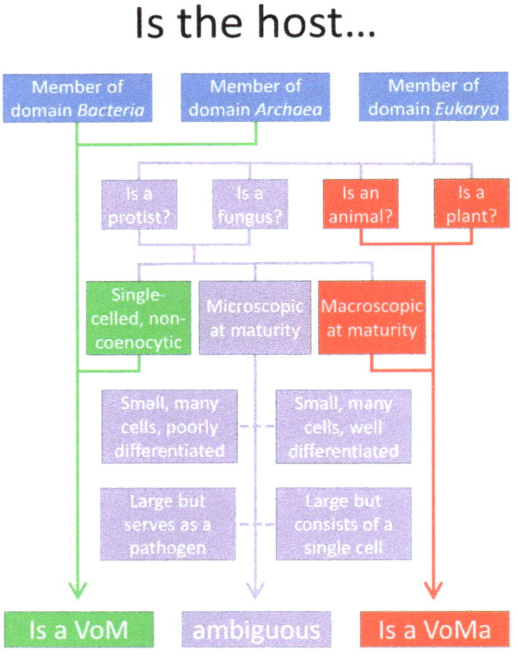

Figure 1.2 Flowchart in determining whether a host organisms is a microorganism (green) or instead a macroorganism (red), and therefore whether its virus is a VoM (a.k.a. 'VoMi') or instead a VoMa, respectively. Uncertainty is represented in the centre, bottom of the figure.

protists, which either represent the direct ancestors of animals or instead are highly degenerate sponges.

These issues for plants and animals and their relationship with unicellular protists leads directly to our next concern: When *is* versus *is not* a protist a microorganism? Although on its surface that can be a difficult question to answer, it becomes even more difficult as one digs deeper. Thus, there are small protists (e.g. most protozoa) and large protists (e.g. seaweed), but there are also protists of intermediate size between microscopic and not microscopic (e.g. large amoeba). There also are both single-celled protists and multicellular protists (with the same examples mostly applying), but also colonial protists. Furthermore, certain algae (again, seaweed) can be unambiguously macroscopic (e.g. the green alga, *Caulerpa*), but nevertheless still single celled. Also among green algae are the volvox, which are relatively large 'specks' of organisms, are arguably multicellular (certainly are multi-celled), but also are described as colonial, although these organisms nevertheless possess both well differentiated if not necessarily particularly complex soma and are fairly close to being microscopic in size. Microorganism or instead macroorganism? Small, unicellular protists thus may be unambiguously designated as microorganisms and large, multicellular protists may be unambiguously designated as macroorganisms, but what about the relatively vast realm in between?

And then there are the fungi. Members of kingdom Fungi range, like protists, from large, multicellular organisms to small unicellular ones. What are known as fleshy or macrofungi, which at a minimum possess macroscopic as well as multicellular fruiting bodies, thus unambiguously represent macroorganisms, e.g. the fungi that give rise to mushrooms. Similarly, yeasts as well as the non-yeast but nonetheless tiny, protozoa-like fungi known as chytrids can be unambiguously designated as microorganisms. But what about coenocytic fungi which can be relatively large but without possessing distinctly separate cells? And what about fungal moulds, which clearly can be manifest as at least seemingly macroscopic phenomena? Thus, among eukaryotes what does serve and does not serve as the interface between microorganisms and macroorganisms is not easily defined, although clearly there exists a large swath of eukaryotic organisms that are microscopic as adults and therefore, at least in that sense, can be unambiguously categorized as microorganisms.

Prokaryotes, unambiguously microorganisms?

Among microorganisms we can include the prokaryotes, and among the prokaryotes we can distinguish between the Bacteria versus the Archaea. Microorganisms thus consist of these two domains plus some subset of domain Eukarya. But are things really that simple? That is, are in fact all prokaryotes microorganisms?

Most prokaryotes, especially as individual cells, certainly are microorganisms. The term prokaryote traditionally referred to bacteria and bacteria-like organisms, such as cyanobacteria. Cyanobacteria from a modern perspective are bacteria, although the idea that cyanobacteria are prokaryotes, that is, versus eukaryotic algae, was not always appreciated. The prokaryotes in any case could and to a large degree still are readily differentiated from eukaryotes, where the cells of the latter possess one or more nuclei whereas prokaryotes, in this same sense, do not. A conceptual upheaval in terms of prokaryote phylogenies, that is, evolutionary relationships, began during the mid to late 1970s, where not only were prokaryotes distinguished into two phylogenetically distinct lineages – which are now known as domains Bacteria and Archaea (Woese *et al.*, 1990) – but one of those lineages (the Archaea) is thought, at least in part, to be more evolutionarily closely related to members of domain Eukarya than domain Bacteria (Dey *et al.*, 2016).

In the previous subsection we considered two characteristics of organisms that together defined a macroorganism, but potentially with ambiguity given deviation by an adult organism from one but not the other of these aspects. These characteristics are large size and multicellularity. Ambiguity is found given near macroscopic dimensions among mature individuals (big enough to be a macroorganism?) or when *individuals* consist of more than one cell, although as a colony rather than as a multicellular individual (complex enough to be a macroorganism?). Just as with protists, large cell size and existing as colonies can give rise to 'microorganisms' that may not quite unambiguously represent microorganisms. In addition, there is the defining aspect of multicellularity, which is cellular differentiation among otherwise clonal aggregations of cells. Should we associate such multicellularity strictly with *ma*croorganisms, rather than microorganisms, or might instead such an association represent simply an historical bias? In particular, is cellular differentiation really something we only see with relatively large organisms, or instead is it that we happen to define cellular differentiation in such a manner that there is a greater tendency to associate it with eukaryotes rather than with prokaryotes?

In terms of cellular differentiation, the most notable example among prokaryotes is the heterocysts seen among certain types of cyanobacteria. These cells are specialized both physiologically and morphologically to fix nitrogen. Other bacteria are able to display distinctive life stages, such as one sees with the bacteria *Myxococcus* or *Caulobacter*, and indeed as seen with spore-forming bacteria. This latter variation, however, occurs over time rather than necessarily from cell to cell among clonal aggregates and therefore is less convincingly a characteristic of something other than a microorganism, that is, since each individual step clearly represents an individual microorganism. Cellular differentiation as seen among certain cyanobacteria thus may be a relatively isolated instance among prokaryotes, as well as one that does not generally raise these organisms too far beyond the concept of colonial (that is, rather than multicellular), and thus beyond the idea that they are microorganisms.

A more general aspect of colonial living among prokaryotes that also may potentially involve cellular differentiation, although less distinctive than that seen with heterocysts, is observed with bacterial microcolonies associated with biofilms. Microcolonies can go through distinct life stages; can consist of cells displaying different positions, physiologies, and/or functions; and also *do* consist of multiple, clonally related cells. Indeed, microcolonies might even effect mechanisms of resistance to viruses which can be seen collectively among their constituent cells rather than being strictly functions of the actions of individual cells (Abedon, 2017a). To a large extent, then, these cellular collectives could be serving as 'macroorganisms' among microorganisms, although few researchers would describe biofilm-associated organisms as not being microorganisms, even in aggregate.

Cellular arrangements – such as diplococci, streptococci, or staphylococci – although not necessarily consisting of as many cells as bacterial microcolonies, nor possessing substantial cellular differentiation, nevertheless do consist of multiple, clonally related cells. Again, however, few would consider cellular arrangements as consisting of anything other than microorganisms, particularly given the small size of these arrangements and their lack of cellular differentiation. Instead they might be viewed as colonial, although not in the typical microbiological sense, e.g. a colony growing in association with solid or semi-solid media, but in the same way that eukaryotic organisms that live in clonal groups also may be described as colonial. But despite their colonial form, these prokaryotes are still microscopic and therefore for all intents should, in our opinion, be viewed as microorganisms.

Generally, then, the most robust distinction between microorganisms and macroorganisms thus continues to be one of relative size, although with the above noted exceptions. An additional exception may be stromatolites. These cyanobacteria-based photosynthetic masses are rare today but were more widespread prior to the evolution of animals, where animals otherwise will tend to serve as stromatolite disruptors. Without animals, stromatolite dimensions can range from centimetres to metres, and even larger in the fossil record. Are stromatolites therefore not microorganisms? Are indeed they even individual organisms, versus simply collections of (microbial) cells? Working against the idea that they are macroorganisms is their relatively simple shape, which does not appear to display substantial differentiation at least on macro scales, that is, other than in terms of layers building upon each other upwards. Of course, one might make the same argument with regard to plasmodial slime moulds, which are non-algal protists that nonetheless can be quite large, and certainly macroscopic as mature individuals. Plasmodial slime moulds, however, are coenocytic, that is, consisting of a single multinucleated cell, with collective parts unquestionably together making up a single individual. In addition, plasmodial slime moulds do display more complex morphologies, especially in the course of their spore-stalk formation. As it seems reasonable to view plasmodial slime moulds as macroorganisms, then perhaps stromatolites too are macroorganisms. But, then again, perhaps they are not.

Pathogens, parasites, but not ectoparasites nor helminths?

Medically as well as historically, a key emphasis of microbiology has been on microorganisms which are pathogenic. Among those pathogens are viruses, and all viruses without ambiguity are considered to be microorganisms, even if not all viruses are necessarily pathogenic. Among bacteria as microorganisms, only a small percentage possess even the potential to be pathogenic against any host organism. Among archaea, also as microorganisms, an even smaller fraction, apparently, seem to serve as pathogens (Abedon, 2017b). Although not necessarily described as pathogens, parasitic protozoa, infestation of which can result in disease, are also unquestionably microorganisms. When an organism can infect, or otherwise infest – living as a parasitic symbiont in association with another, host organism – can that organism therefore always be described as a microorganism?

The short answer to that question is an unambiguous, no! Plenty of organisms are parasitic but not microscopic nor microorganisms, even if they nevertheless might be relatively small. Ectoparasitic arthropods, for example, are animals and not microorganisms. Two categories of parasitic organisms do come closer to the divide between microorganisms and macroorganism, however. Thus, just as we might place plasmodial slime moulds on one side of the microorganism–*ma*croorganism divide (i.e. plasmodial slime moulds are macroorganisms) and stromatolites on the other side (that of microorganisms), so too can similar roles be played by helminths as animals and therefore macroorganisms, on the one hand, and pathogenic fungi on the other? The latter, technically, as the causative agent of mycoses, i.e. diseases caused by fungi, can span this divide. The question, then, is which is more important in defining a microorganism, particularly a non-animal or non-plant organism as a microorganism: its morphology or instead its pathogenicity? Thus, are pathogenic fungi that form complex fruiting bodies, such as smuts which technically are macrofungi, microorganisms or instead are they macroorganisms?

Our bias, and it is only that, a bias, is to lump in general pathogenic fungi under the heading of microorganism, but to not include helminths, as animals, under the same heading. We accept, however, that these designations can be viewed as both arbitrary and to no small degree a product of laziness: 'Let's not worry about worm viruses!' 'Let's not worry about whether a pathogenic fungus is (or is not) truly microscopic!' We encourage others to think about these issues, although perhaps not think about them too hard. Ultimately, without question, there will be viruses of what unambiguously are microorganisms, viruses of what unambiguously are macroorganisms, and then viruses of what some might view as microorganisms and about which others might reasonably disagree (Fig. 1.2).

UOPs and MOPs

Another means by which microorganisms might be distinguished from macroorganisms is in terms of the impact that viruses can have, especially ecologically, on other organisms. In particular, one might posit that the ability of a virus to infect the majority of the cells making up an organism, following a single symbiotic encounter, is a characteristic more readily observed with VoMs versus VoMas. That is, non-colonial macroorganisms presumably generally possess more sophisticated defence systems which to some degree can serve to contain the damage caused by attacking viruses, although not necessarily stave off death of the host organism.

Similarly, virus-induced lysis of whole organisms is more likely for VoMs than it is for VoMas, thereby allowing VoMs to more directly serve as decomposer organisms versus VoMas. There are, however, some exceptions, e.g. baculoviruses which directly effect the dissolving of whole animals, or less virulent VoMs that in fact do not effect lysis but instead are chronically released from host cells. And never mind that technically a 'decomposer' is an organism which acquires and then consumes already dead organisms whereas viruses both start with still-alive organisms and then, also technically and from an ecological standpoint, do not to any large degree actually 'consume' their victims. VoMs are also more likely to infect what might be described in general terms as 'germ line' cells (versus somatic cells), that is, the cells that are charged with giving rise to the next generation, which for unicellular organisms basically is the single cell making up each of these organisms. The result is perhaps a greater potential for VoMs, versus VoMas, to contribute to horizontal gene transfer versus viruses that infect somatic cells in animals or plants that might or might not also infect reproductive cell lines.

Though not 100% equivalently articulated, we have previously expressed the idea that there is a fundamental, especially ecological distinction between VoMs as VoMas. In the introductory chapter to the monograph, *Bacteriophage Ecology* (Abedon, 2008a,b), UOPs versus MOPs are described, that is, unicellular-organism parasites versus multicellular-organism parasites, with the concept of UOPs unquestionably and even deliberately more closely aligned with that of VoMs and MOPs with that of VoMas. Quoting liberally (pp. 3–4):

> 'UOPs' (pronounced ōōps) and 'MOPs' (pronounced mŏps) are the parasites of unicellular organisms (unicellular-organism parasites) and multicellular organisms (multicellular-organism parasites), respectively. I first presented the concepts of UOPs and MOPs as part of the mission statement of the Bacteriophage Ecology Group (BEG; [phage.org])... My intention in describing phages as UOPs was to broaden what one might think of as phage-like when considering the theory of phage ecological interactions...
>
> Included among UOPs, of course, are bacteriophages... which, with the addition of archaeal viruses... are viruses whose hosts are prokaryotes. However, there are also a number of non-bacterial unicellular organisms which can serve as viral hosts, plus bacteria that are (or, at least, appear to be) obligately intracellular parasites of unicellular organisms. Together these non-bacteriophage UOPs include the viruses of protozoa...; the mycoviruses of yeasts...; the viruses of unicellular, eukaryotic algae ...; *Bdellovibrio* spp., and bacterial species that parasitize other bacteria intracellularly... UOPs also include *Legionella* spp., which are bacteria that parasitize protozoa intracellularly... and viruses of multicellular creatures (such as animals) when infecting their quasi-unicellular counterparts grown in the laboratory as tissue cultures...
>
> UOPs may be distinguished from MOPs by a number of criteria. For example, UOPs typically are organisms which (1) are obligate intracellular parasites; (2) employ simpler infection strategies, ones in which within-host growth and intracellular growth are synonymous; and (3) infect morphologically simpler hosts. By contrast: (1) MOPs include obligate intracellular parasites but also [include] many parasites which replicate extracellularly or even extra-organismally. (2) MOPs that are obligate intracellular parasites display four fundamental and differentiable phases: host acquisition (which itself may be broken into two distinct phases, host entry and cell adsorption); an intracellular phase; an extracellular dispersal-within-the-host phase; and finally an extracellular between-host phase. The simpler UOPs, by contrast, exist in just two fundamental and differentiable phases: an extracellular host acquisition phase (i.e. adsorption) and an intracellular phase. (3) MOP hosts tend to be more complex than unicellular organisms.

Most viral UOPs thus are VoMs, although not all UOPs are viruses, and an ambiguity exists because there are coenocytic organisms that are unicellular but nevertheless macroscopic, and thus subject to exploitation by UOPs, but not necessarily by VoMs. By contrast, the distinction between microorganisms and macroorganisms, and therefore between VoMs and VoMas as subsets of MOPs, remains a matter of opinion. Ecologically as considered in the first paragraph of this section, however, there is substantial overlap between UOPs and VoMs and as well between VoMs and the MOPs of nevertheless microscopic cellular assemblages.

What are viruses of microorganisms?

The question of just what is a microorganism is less than fully straightforward and to a degree – perhaps to a large degree – as much philosophical as scientific. Once we have defined what a microorganism is, or at least declared what is and is not one, then the question of what is a virus of a microorganism more or less is the equivalent of stamp collecting. Nonetheless, there still does remain some ambiguity, largely stemming from the question of 'What is a virus?' Again it is fungi, or rather here the viruses of fungi which provide much of that ambiguity, because many of these viruses do not seem to have an extracellular state and instead are transmitted via direct cell-to-cell contact as well vertically, that is from parent to offspring. Viruses of viruses (e.g. virophages) are another interesting deviation from the otherwise standard theme of VoMs being viruses that infect what may be straightforwardly defined as microorganisms. In this section we provide in particular a 'macroscopic' overview of the diversity of viruses of microorganisms, leaving for subsequent chapters of this monograph, particularly section II, to fill in many of the details. As a co-introduction to these viruses, with much more detail and written from the perspective of VoM genomics, see Adriaenssens in Chapter 2.

Viruses of bacteria

The viruses of bacteria are most commonly known as phages, bacteriophages, or bacterial viruses, although bacteriovirus has been proposed as well (Pina *et al.*, 2011). These, at the moment, consist of virion types that include the tailed phages (three basic types), the non-tailed phages with lipid-containing capsids (five basic types, which mostly are icosahedral), and the phages possessing single-stranded genomes (three basic types, which are either icosahedral or filamentous). The majority of bacteriophages are lytic although some chronically infecting phages exist, including particularly the filamentous phages. Latently infecting phages are said to display lysogenic cycles, and the phages themselves are described as temperate. These virus types are considered in detail by Lehman in Chapter 7 of this volume.

Viruses of archaea

The viruses of archaea have been described using a diversity of names although archaeal virus today is the most widely accepted descriptor. The majority of the archaeal virus community appears to have decided that they should not be called phages (Abedon and Murray, 2013). The number of basic virion types among archaeal viruses is at least equal to those seen among bacteriophages, and include tailed viruses, icosahedral viruses, and filamentous viruses as well virus morphological types unlike any seen infecting the other cellular domains. These viruses can be lytic or chronic infectors of archaea, and latently infecting archaeal viruses should be described as proviruses, that is, rather than as prophages. These virus types are considered in detail by Abedon in Chapter 8 of this volume.

Viruses of eukaryotic microorganisms

Eukaryotic microorganisms include especially protozoa, microscopic algae, and yeasts. Mycoviruses, that is, the viruses of fungi, are covered by Vainio and Hantula in Chapter 9 of this volume. As noted above, just what constitutes a microorganism is especially ambiguous among fungi. Algal viruses, that is, viruses of eukaryotic algae, are covered by Short *et al.* in Chapter 10. Giant protozoal viruses, in turn, are covered by Reteno *et al.* in Chapter 11. A representation of the virus factories within which these giant viruses replicate is reprinted here as a figure (Fig. 1.1). Related to these viruses of eukaryotes are virophages, although the latter in fact can also be viewed as viruses of viruses. In addition are the non-giant viruses of protozoa (Hyman and Abedon, 2012), although these have been relatively understudied and are not reviewed here. Generally, VoMs of microscopic eukaryotes are most closely related to VoMas than they are either to bacteriophages or to archaeal viruses.

Viruses of other viruses (virophages)

Viruses of other viruses are actually viruses of virus-infected cells. Those cells may or may not be microorganisms, or may or may not make up multicelled microorganisms, but viruses nevertheless unambiguously are microorganisms. Virophages can be categorized more generally as types of satellite viruses – which include, for example, coliphage P4 – although others make a distinction between virophages of giant protozoal viruses and satellite viruses of other viruses. Other viruses of viruses may include defective interfering (DI) particles, although in this case both the parasitizing virus and the virus being parasitized are derived from the same virus lineage. Brief discussion of these various viruses of other viruses is presented in Hyman and Abedon (2012). See also Fischer in Chapter 12.

VoMs and biotechnology

The final section of this monograph covers 'technologies involving viruses of microorganisms'. These basically represent means by which VoMs can be used for biotechnological purposes or instead technologies that can be used towards VoM characterization. Chapters include: the use of bacteriophages and bacteriophage proteins as components in systems to identify bacteria as well as high-throughput screening of other molecules for binding affinities (Hyman in Chapter 13). The use of bacteriophages and other VoMs as biocontrol agents via their innate ability to kill those hosts that are pathogenic for humans and other species. This process commonly called phage therapy but is not necessarily limited to phages as VoMs against bacteria, as treatment targets has been proposed (Alves *et al.*, in Chapter 14). And lastly, descriptions of the various methods used to identify and characterize VoMs (Jacquet *et al.*, in Chapter 15).

Overview of basic VoM life cycles

Though redundant with some of the information supplied above, in this section we nonetheless introduce the basic VoM life cycles for the sake of reference from the rest of this monograph. We can differentiate the life cycles of VoMs first into virion dissemination and then attachment phases, with the VoM present in both cases in an extracellular state. As noted, some VoMs seem to lack extracellular virions (mainly mycoviruses), so an alternative means of VoM acquisition of new host cells thereby must occur. With the more common extracellular state, the stage of host cell acquisition is often described as one of adsorption, although the term 'extracellular search' can also be used when considering just the pre-host-encounter and thereby pre-attachment portion of virion adsorption.

The second stage is one of infection. Infections can be differentiated into ones which, after Abedon (2009a), may be described as productive, reductive, or destructive. Destructive refers to consequences to the virus rather than to the host and these, as examples of infections which do not work out in the virus' favour, are not of interest here. Rather, productive infections are ones in which new virions are produced, and this occurs in association with virus genome replication. As noted above, some VoMs are replicated within what are known as virus factories (Fig. 1.1), although for most VoMs the replication step, at least visually, is less striking than as depicted in the figure.

Reductive infections leave the virus in a latent state. For VoMs which display lytic cycles, latent cycles generally are referred to as lysogenic cycles. Lysogenic means 'lysis-generating' and refers to the potential of such cycles ultimately to give rise to both lysis of the infected host cell and also of any cells found in the vicinity of that cell which are susceptible to the infecting virus. Technically, for VoMs which display chronic rather than lytic cycles, the term lysogenic therefore really should *not* apply, but nevertheless that term often is used. In any case, with a reductive infection the virus replicates within its host cell, and often does so supported by the replication of its host cell, but no virion progeny are produced until a productive-cycle 'induction' event occurs. In general terms, virions displaying latent cycles may be referred to as proviruses. For latently infecting bacteriophages *only*, the terms prophage may be substituted.

The third aspect of VoM life cycles – with adsorption and infection being the first two – is virion release from VoM infected host cells. For virion release to occur then by definition a productive infection must first also have occurred. For many VoMs those productive infections may be called lytic cycles, and this is particularly the case for tailed phages along with their head-tailed archaeal virus equivalents. For many VoMs, however, virions instead are released chronically, that is, without substantial destruction of the host cell. The latter is why not all latent VoM infections should be described as lysogenic. Consistently, a potential for a virus to be temperate, meaning able to infect latently as well as productively, is not solely the domain of lytic viruses but instead includes chronically released virions as well. For discussion of many of these ideas, particularly as applied to bacteriophages, see Hobbs and Abedon (2016).

Conclusion

The viruses of microorganisms, or VoMs, above all are viruses. They are the viruses especially of microscopic unicellular organisms but also of microscopic colonies of cells. The impact of VoMs differs from that of viruses more generally in being more likely to kill as well as solubilize entire target organisms, or more likely to infect cells that are destined to give rise to the next generation of organisms, with the latter crucial to the evolutionary success of various forms of virus-mediated horizontal gene transfer, i.e. transduction. VoMs thus play important roles in the ecology and evolution of their hosts as well as in ecosystems as a whole, serving in the latter case especially as microorganism decomposers in their conversion of nutrients that are locked within intact organism into nutrients that instead are water soluble. In this form these now dead cells are readily available to other organisms which make a living by consuming dissolved organic matter, which particularly includes the heterotrophic prokaryotes. VoMs also play important roles in biotechnology, in the laboratory in the form of molecular tools as well as outside of the laboratory in the form of highly specific antagonists that may be applied to cellular microorganisms. And particularly in terms of their use of microorganisms as hosts, VoMs by definition are unique among viruses.

References

Abedon, S.T. (2008a). Bacteriophage Ecology: Population Growth, Evolution, and Impact of Bacterial Viruses (Cambridge University Press, UK). https://doi.org/10.1017/CBO9780511541483.

Abedon, S.T. (2008b). Phages, ecology, evolution. In Bacteriophage Ecology, Abedon, S.T. ed. (Cambridge University Press, UK), pp. 1–28. https://doi.org/10.1017/CBO9780511541483.004.

Abedon, S.T. (2009a). Bacteriophage intraspecific cooperation and defection. In Contemporary Trends in Bacteriophage Research, Adams, H.T. ed. (Nova Science Publishers, Hauppauge, New York). pp. 191–215.

Abedon, S.T. (2009b). Phage evolution and ecology. Adv. Appl. Microbiol. *67*, 1–45. https://doi.org/10.1016/S0065-2164(08)01001-0.

Abedon, S.T. (2011). Communication among phages, bacteria, and soil environments. In Biocommunication of Soil Microorganisms, Witzany, G. ed. (New York, Springer), pp. 37–65.

Abedon, S.T. (2017a). Phage 'delay' towards enhancing bacterial escape from biofilms: a more comprehensive way of viewing resistance to bacteriophages. AIMS Microbiol. 3, 186–226.

Abedon, S.T. (2017b). Why archaea are limited in their exploitation of other, living organisms. In Biocommunication of Archaea, Witzany, G., ed. (Springer, Cham), pp. 41-66.

Abedon, S.T., and Murray, K.L. (2013). Archaeal viruses, not archaeal phages: an archaeological dig. Archaea *2013*, 251245. https://doi.org/10.1155/2013/251245.

Dey, G., Thattai, M., and Baum, B. (2016). On the archaeal origins of eukaryotes and the challenges of inferring phenotype from genotype. Trends Cell Biol. *26*, 476–485.

Hendrickson, H. (2012). The lion and the mouse: how bacteriophages create, liberate, and decimate bacterial pathogens. In Bacteriophages in Health and Disease, Hyman, P., and Abedon, S.T. eds. (CABI Press, UK), pp. 61–75.

Hobbs, Z., and Abedon, S.T. (2016). Diversity of phage infection types and associated terminology: the problem with 'lytic or lysogenic'. FEMS Microbiol. Lett. *363*, fnw047. https://doi.org/10.1093/femsle/fnw047.

Hyman, P., and Abedon, S.T. (2012). Smaller fleas: viruses of microorganisms. Scientifica *2012*, 734023. https://doi.org/10.6064/2012/734023.

Pina, M., Bize, A., Forterre, P., and Prangishvili, D. (2011). The archeoviruses. FEMS Microbiol. Rev. *35*, 1035–1054. https://doi.org/10.1111/j.1574-6976.2011.00280.x.

Woese, C.R., Kandler, O., and Wheelis, M.L. (1990). Towards a natural system of organisms: proposal for the domains Archaea, Bacteria, and Eucarya. Proc. Natl. Acad. Sci. U.S.A. *87*, 4576–4579.

Genomics of Viruses of Microorganisms

2

Evelien M. Adriaenssens[1,2*]

[1]Department of Genetics, Centre for Microbial Ecology and Genomics, Genomics Research Institute, University of Pretoria, Pretoria, South Africa.
[2]Institute for Integrative Biology, University of Liverpool, Liverpool, UK.

*Correspondence: evelien.adriaenssens@gmail.com

https://doi.org/10.21775/9781910190852.02

Abstract

Genomics, the study of the sequence and gene set making up the genomes of viruses and of microorganisms, is becoming increasingly important. It allows researchers to access the full sequence diversity of viruses, facilitates the discovery of previously unknown viral types and gives a better insight into evolutionary relationships. In this chapter, I review the current methods used in genomics of viruses of microorganisms, and highlight what is known about the major groups of viruses. I end with the impact genomics research has had on the study and understanding of higher-order genomic relationships.

Introduction

Depending on the dictionary, the word 'genomics' has slightly different definitions, but all agree that it is a field of study which comprises the use of molecular biology techniques to investigate the sequence structure and the function of a set of genes making up the genome of an organism. For viruses of microorganisms this covers a range of genome sizes from 2 kilobases (kb) ssDNA and ssRNA viruses infecting fungi, through (bacterio)phage genomes which can range from 4 kb (ssDNA) to more than 400 kb (dsDNA), to the so-called megaviruses, with genomes which can be over 2 Mb in size, infecting unicellular eukaryotes (NCBI virus genomes database (King *et al.*, 2012)).

When Brüssow and Hendrix first wrote about phage genomics in 2002, telling the world how 'small is beautiful', there were just 100 bacteriophage genomes sequenced (Brüssow and Hendrix, 2002). The archaeal viruses received proportionally less attention, and by 2006, only 23 genomes were published (Prangishvili *et al.*, 2006a). In 2004, with the publication of the 1.2 Gb mimivirus genome, the sequencing of megaviruses also began (Raoult *et al.*, 2004). The development of next-generation-sequencing technologies has significantly increased these numbers (Table 2.1), with the dsDNA bacteriophage genomes receiving the bulk of the sequencing effort at over 1700 genomes published as of early 2016.

In this chapter, I provide an overview of the methods that are currently in use for virus genomics and go into more detail on the different genome types that have been found for viruses of microorganisms, organized according to the Baltimore classification of viruses (Baltimore, 1971). This classification places viruses in one of seven groups based on the type of nucleic acid (RNA or DNA) and the sense of the strand (positive or negative for single-stranded genomes), as well as the method of replication (retro-transcribing or not), i.e. Group I dsDNA viruses, Group II ssDNA viruses, Group III dsRNA viruses, Group IV positive sense ssRNA viruses, Group V negative sense ssRNA viruses, Group VI positive sense ssRNA viruses replicating through a DNA intermediate (retroviruses), and Group VII dsDNA viruses replicating through a ssRNA intermediate.

Table 2.1 Number of genomes per taxon of viruses infecting groups of microorganisms, excluding environmental samples

Virus classification or taxon	Bacteria	Archaea	Algae	Fungi*	Protozoa
dsDNA viruses, no RNA stage	2019	74	28	0	15
Ampullaviridae	0	3	0	0	0
Bicaudaviridae	0	4	0	0	0
Caudovirales	1968	4	0	0	0
Clavaviridae	0	1	0	0	0
Corticoviridae	1	0	0	0	0
Fuselloviridae	0	10	0	0	0
Globuloviridae	0	2	0	0	0
Guttaviridae	0	1	0	0	0
Ligamenvirales	0	15	0	0	0
Marseilleviridae	0	0	0	0	5
Mimiviridae	0	0	0	0	6
Phycodnaviridae	0	0	24	0	0
Plasmaviridae	1	0	0	0	0
Salterprovirus	0	1	0	0	0
Sphaerolipoviridae	0	6	0	0	0
Tectiviridae	5	0	0	0	0
Turriviridae	0	2	0	0	0
Unclassified dsDNA phages	43	0	0	0	0
Unclassified dsDNA viruses	1	25	4	0	4
ssDNA viruses	85	6	1	1	0
Inoviridae	40	0	0	0	0
Microviridae	44	0	0	0	0
Spiraviridae	0	1	0	0	0
Unclassified ssDNA viruses	1	5	1	1	0
Pleolipoviridae	0	8	0	0	0
dsRNA viruses	5	0	1	83	31
Chrysoviridae	0	0	0	7	0
Cystoviridae	5	0	0	0	0
Partitiviridae	0	0	0	23	0
Quadriviridae	0	0	0	1	0
Reoviridae	0	0	1	3	0
Totiviridae	0	0	0	30	31
Unclassified dsRNA viruses	0	0	0	19	0
ssRNA viruses	12	0	8	81	1
Mononegavirales (–)	0	0	0	3	0
Endornaviridae (+)	0	0	0	12	0
Hypoviridae (+)	0	0	0	11	0
Leviviridae (+)	12	0	0	0	0
Narnaviridae (+)	0	0	0	36	1
Picornavirales (+)	0	0	6	0	0
Tymovirales (+)	0	0	0	6	0

Table 2.1 Continued

Virus classification or taxon	Bacteria	Archaea	Algae	Fungi*	Protozoa
Unclassified (–)	0	0	0	3	0
Unclassified (+)	0	0	0	4	0
Unassigned ssRNA viruses	0	0	1	0	0
Unclassified ssRNA viruses	0	0	1	6	0
Unassigned viruses	0	0	5	3	0
Unclassified archaeal viruses	0	3	0	0	0
Unclassified phages	52	0	0	0	0
Unclassified viruses	0	1	0	11	0
Virophages (*Lavidaviridae*)	0	0	1	0	3
Total	1930	89	43	152	55

*For brevity the oomycete *Phytophtora* was included with the fungi.
Data are from the RefSeq Viruses database (Viral Genomes Home, complete genomes only) hosted by NCBI in combination with official taxonomy from ICTV (ICTV Taxonomy: 2015 Release). This information was last updated January 22nd 2017.

Current methods

What sets genomics apart from metagenomics, as described elsewhere in this volume (see Chapter 5), is the need for accurate identification of the virus in question. Depending on the virus–host system, this can be very straightforward, such as for virulent bacteriophages with dsDNA genomes forming easily identifiable plaques on a fast-growing bacterial strain. Isolation of temperate bacteriophages usually involves a lysogen induction step with UV light, or more commonly mitomycin C. Once a pure culture of the bacteriophage is obtained, the genomic content can be relatively easily extracted (Ehara *et al.*, 1997; Payet and Suttle, 2013; Roberts and Devoret, 1983; Williamson *et al.*, 2008). For virus–host systems which can only propagate in aquatic media, such as marine and freshwater cyanobacteria and algae, clonal virus populations are generally isolated using successive extinction-dilution cycles (also called limiting dilution) in which virus concentrates are added in different dilutions to exponentially growing host cultures. In this case, the most dilute concentrate that kills the host cells is assumed to be a single virus thus producing a clonal population in that culture (Chénard *et al.*, 2015; Clasen *et al.*, 2013; Kimura and Tomaru, 2015; Suttle, 1993). When dealing with fungal RNA viruses, the fungal mycelia are usually screened for dsRNA elements representing dsRNA viruses or the replicative form of ssRNA virus genomes, using SDS/phenol extraction and purification through chromatography (Jiang *et al.*, 2015; Morris and Dodds, 1979; Urayama *et al.*, 2010). To test for the presence of ssRNA genomes, a combination of nuclease digestion and RT-PCR amplification can be used (Liu *et al.*, 2014). In many protocols, additional purification steps are added before the extraction of nucleic acids including caesium chloride density gradient centrifugation or anion-exchange chromatography (Adriaenssens *et al.*, 2012; D'Souza *et al.*, 2013; Green and Sambrook, 2012; Kramberger *et al.*, 2010). Extraction of nucleic acids is either performed by standard phenol:chloroform (or phenol) extraction or by using specialized kits, depending on the virus of interest, with additional complementary DNA (cDNA) synthesis steps for RNA viruses.

Most of the currently available sequencing platforms have been used to sequence viruses of microorganisms. While clone library creation and Sanger sequencing has been phased out for the most part after the advent of next-generation sequencing, it can still be cost- and time-effective for small viruses and is still used for mycoviruses, that is, fungal viruses (Li *et al.*, 2014b; Xu *et al.*, 2015). Klumpp and colleagues have found that the majority of bacteriophages can be easily sequenced and the genome assembly completed using 454 sequencing and vendor software (Klumpp *et al.*, 2012). For more difficult phage genomes, and by extension other viruses, they suggest the use of a combination of long read technology for scaffolding (Sanger, PacBio or 454) and short-read, high-throughput platforms for gap filling and error correction (Illumina, Ion Torrent). Roche, however, is phasing out the pyrosequencing platform in 2016–2017, reducing the long read technology options (Buermans and den Dunnen, 2014). Commercial assembly software such as CLC Genomics Workbench (Qiagen) or Seqman NGen (DNAstar), as well as freely available assemblers such as velvet, MIRA, SOAP-denovo or Spades can be used to assemble the reads into complete genomes (Chevreux, 2005; Li *et al.*, 2009; Nurk *et al.*, 2013; Zerbino and Birney, 2008). Success of the assembly, however, will depend on the virus genome in question (genome size, presence of repeats) as well as the sequencing technology.

Correct annotation of complete virus genomes is of paramount importance as incorrect open reading frame (ORF) identification and gene designations tend to get perpetuated and even amplified in databases. Therefore, manual annotations are encouraged, or use of online annotation pipelines followed by manual curation steps. As a first step, the genome termini need to be determined (such as by PCR or primer walking) or in the case of circular or circularly permuted genomes, the genomes need to be collinearized with their closest relative in the database. Then ORFs can be delineated with programs such as ORFfinder (www.ncbi.nlm.nih.gov/orffinder/), Glimmer (Delcher *et al.*, 1999) or GeneMark (Besemer and Borodovsky, 2005). Each ORF which is designated as a coding sequence (CDS) should be preceded by a ribosome-binding site (Shine–Dalgarno sequence) and manual verification followed by reassignment of the start codon might be necessary (Starmer *et al.*, 2006). Assigning predicted functions to the CDSs is traditionally done using BLASTP (or PSI-BLAST) of the translated ORF sequence or, alternatively, BLASTX of the nucleotide sequence against the non-redundant protein database. Limiting the BLAST homology search to the viral database will significantly reduce computation time (Altschul *et al.*, 1990, 1997). The program Prokka, performs gene calling and BLAST-based annotation in one, but does not have a graphical user interface (Seemann, 2014). As an alternative homology search, or for those CDSs which have no BLAST-based homologues, the algorithms known as HHpred or HHblits can be used, which use hidden Markov model profile comparisons with a range of protein databases (Remmert *et al.*, 2012; Söding *et al.*, 2005). Ideally, other genome features such as promoters, terminators and t-RNAs should be predicted as well. Promoters can be found by performing a conserved pattern search in upstream sequences with programs such as MEME/MAST or PHIRE (Bailey *et al.*, 2009; Lavigne *et al.*, 2004), while rho-independent terminators are generally found by looking for potential stem–loop structures in the transcribed RNA by programs such as ARNold or TransTermHP (Gautheret and Lambert, 2001; Kingsford *et al.*, 2007). t-RNAs can be present in certain bacteriophage genomes and can be found with tRNAscan-SE or ARAGORN (Laslett and Canback, 2004; Schattner *et al.*, 2005). Several programs are available, online and/or as stand-alone software, which integrate some of the features described above and can be used as alternatives. The online suite RAST accepts a putative complete genome sequence and will perform ORF identification and BLAST-based annotations with comparisons to the SEED server (Aziz *et al.*, 2008, 2012; Overbeek *et al.*, 2014). For bacteriophages, a separate RAST system was developed as well, phage-RAST or PHAST (http://www.phantome.org/PhageSeed/Phage.cgi?page=phast). The stand-alone tools DNA Master (http://phagesdb.org/DNAMaster/) or Artemis (Carver *et al.*, 2008) can be used to perform the genome annotation in an integrated environment and also provide visualization and comparison tools.

The software described above is by no means an exhaustive list and new programs are being written constantly. Other tools are available from http://molbiol-tools.ca/, a molecular biology tools webpage compiled by Dr Andrew Kropinski, or PhAnToMe (www.phantome.org).

Overview of the genomics of different types of viruses[1]

In the following sections, I review the different genome organizations of viruses of microorganisms

1 Virus genomics is an ever-growing field and has a big effect on taxonomy, with new taxa being created on a yearly basis. For the latest official taxonomy, please visit www.ictvonline.org.

identified so far, grouped according to their Baltimore classification and official taxonomic affiliation.

Negative sense ssRNA viruses

So far only two (−)ssRNA viruses of microorganisms, in this case infecting fungi (see Chapter 9), have been isolated and sequenced. They belong to the order of *Mononegavirales*, viruses with linear, non-segmented, negative-sense RNA (King *et al.*, 2012). Sclerotinia sclerotiorum negative-stranded RNA virus 1 (SsNSRV-1) is the type isolate for the genus *Sclerotimonavirus* in the newly created family *Mymonaviridae* (Afonso *et al.*, 2016). The genomes of SsNSRV-1 and SsNSRV-3 are 10 kb in length and encode six ORFs (Liu *et al.*, 2014). The RNA polymerase domain amino acid sequence of SsNSRV-1 places it in a clade with members of the *Bornaviridae* and *Nyamiviridae* which also contain animal and invertebrate viruses (King *et al.*, 2012; Kuhn *et al.*, 2013; Liu *et al.*, 2014). Other mononegavirus signatures have been found in transcriptome libraries from *Sclerotinia homoeocarpa* and in the genome of a phytopathogenic ascomycete indicating that (−) ssRNA viruses are more widespread in fungi than the two current isolates suggest (Kondo *et al.*, 2013a).

Positive-sense ssRNA viruses

The viruses with positive strand ssRNA genomes form a diverse group, infecting bacteria, algae, and fungi, but no (+)ssRNA viruses infecting archaea or protozoa have been isolated or sequenced yet (Table 2.1). The fungal viruses or mycoviruses make up the largest fraction in this group of viruses of microorganisms.

Mycoviruses

Mycoviruses are summarized and covered in greater detail in Chapter 9. Here and elsewhere in this chapter I focus particularly on the genomic characteristics of these viruses. Mycoviruses can be found as virion-less RNA viruses or encapsidated viruses.

There are multiple groupings of RNA mycoviruses which do not encode a capsid protein and these viruses do not seem to have an extracellular virion form (Ghabrial and Suzuki, 2009). Members of the family *Narnaviridae* are the simplest of (+) ssRNA viruses encoding only an RNA-dependent RNA polymerase (RdRp) (Ghabrial and Suzuki, 2009). The genus *Narnavirus* consists of two linear ssRNA viruses, 20 S RNA and 23 S RNA, found in *Saccharomyces cerevisiae*, while their replicating dsRNA forms are designated W and T, respectively (Esteban *et al.*, 1992). The ssRNA genomes of 2.5 kb (20 S) and 2.8 kb (23 S) encode RdRps which show similarities to those of levivirus replicases (Rodriguez-Cousino *et al.*, 1998). Recently, the first narnavirus infecting a protozoan host was sequenced, PserNV1, infecting a plant protist, comprising a genome of 3782 nt and showing over 80% similarity in RdRp amino acid sequence with the fungal narnaviruses (Akopyants *et al.*, 2016).

Members of the second genus in this family, *Mitovirus*, are generally isolated from the mitochondria of phytopathogenic fungi (King *et al.*, 2012). The genome sizes in this genus range from 2343 nucleotides for Ophiostoma mitovirus 6 (Hong *et al.*, 1999) to 3570 nucleotides for Rhizoctonia solani mitovirus M2 (Lakshman *et al.*, 1998). Their one ORF encodes for an RdRp (Fig. 2.1) which is distinct in its use of the mitochondrial translation code and shows a strong evolutionary relationship between all the members of the genus and a more distant relationship to the narnaviruses 20 S and 23 S (Deng *et al.*, 2003; Hong *et al.*, 1998, 1999; Osaki *et al.*, 2015; Park *et al.*, 2006; Polashock and Hillman, 1994; Tuomivirta and Hantula, 2003a, 2005; Wu *et al.*, 2007). Recent phylogenetic analysis of the RdRps of a large group of mitoviruses suggests the division of the mitoviruses into two distinct clades with viruses of *Ophiostoma* and *Sclerotinia* present in both clades, indicating two separate ancestral species (Doherty *et al.*, 2006; Khalifa and Pearson, 2013; Osaki *et al.*, 2015). Many of the mitoviruses show an enriched A+U content, with certain members having a preferred A or U in the third codon position (Deng *et al.*, 2003; Hong *et al.*, 1999; Khalifa and Pearson, 2013; Lakshman *et al.*, 1998; Osaki *et al.*, 2015; Tuomivirta and Hantula, 2003a, 2005). Using deep sequencing approaches, which can generate massive amounts of RNA sequencing data from cDNA libraries or transcriptomes, complete genomes of new mitoviruses were discovered for the white pine blister rot fungus *Cronartium ribicola*, opening the path for discovery of a new range of mitovirus hosts (Liu *et al.*, 2016).

Viruses belonging to the family *Hypoviridae* seem to have been erroneously classified as dsRNA in NCBI and have been assigned to the (+)ssRNA group by ICTV (2014 Virus Taxonomy Release).

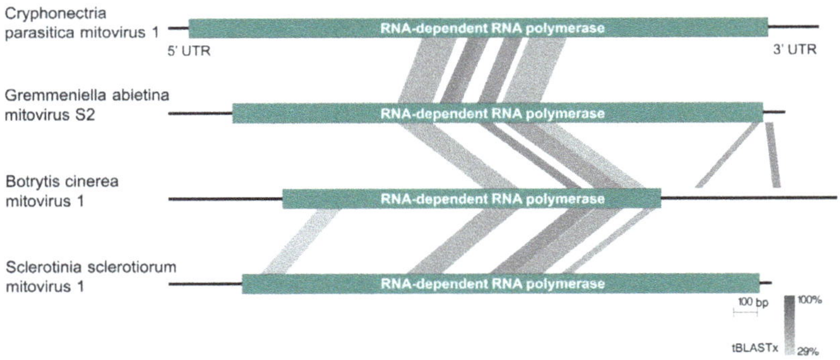

Figure 2.1 *Mitovirus* genome structure. The genomes of representative mitoviruses, including the untranslated regions (UTRs), were compared with tBLASTx and visualized with Easyfig 2.1 (Sullivan *et al.*, 2011). tBLASTx compares translated nucleotide sequences in all six reading frames (comparing putative predicted protein sequences) yielding better detection of distant relationships.

All current members of this family confer hypovirulence to the chestnut blight fungus *Cryphonectria parasitica* and they have a distant phylogenetic relationship with ssRNA potyviruses based on their protease, polymerase and helicase domain sequences (King *et al.*, 2012). The type species, *Cryphonectria hypovirus 1* (CHV1), has a genome of 12,712 nucleotides excluding the poly(A) tail and encodes two overlapping ORFs which have no significant sequence homology outside of the family (Shapira *et al.*, 1991). Three other hypovirus species have been described which show various levels of amino acid similarities, CHV2, CHV3 and CHV4, with CHV2 more similar to CHV1, while the other two virus types only encoded one ORF (Chen and Nuss, 1999; Hillman *et al.*, 1994; Linder-Basso *et al.*, 2005; Smart *et al.*, 1999). The RNA polymerase domain of CHV3 shows some similarity to plant-pathogenic potyviruses, leading to speculation that CHV3 together with its closest potyvirus relative barley yellow mosaic virus cross the family lines between *Hypoviridae* and *Potyviridae* (Smart *et al.*, 1999). What sets CHV3 and CHV4 further apart from CHV1 and CHV2 is the presence of a UDP-glucose/sterol glucosyltransferase (UGT) domain which might have been acquired from the fungal host (Linder-Basso *et al.*, 2005).

Currently, the family *Endornaviridae* is classified with the dsRNA viruses by ICTV, but phylogenetic evidence indicates that the replicative form of these viruses is positive sense ssRNA (Ghabrial and Suzuki, 2009; Gibbs *et al.*, 2000; King *et al.*, 2012). Only a few endornavirus species have microorganism hosts, while the majority infect plants (King *et al.*, 2012). The first one discovered was Phytophtora endornavirus 1 (PEV1) with a genome of 13,883 nucleotides containing one large ORF encoding for a polyprotein of 4548 amino acids (Hacker *et al.*, 2005). This protein had several RdRp motifs and a RNA helicase similar to plant endornaviruses and the UGT domain found in the two hypoviruses described above. A second species, Gremmeniella abietina type B RNA virus XL, with isolates GaBRV-XL1 and GaBRV-XL2, has genomes of 10,375 nucleotides and 10,374 nucleotides respectively, sharing 97% identity. Both viruses encode one ORF with four domains, a methyltransferase, DEAD-like helicase, RNA helicase, and RdRp; the latter showing similarities with other endornaviruses (Tuomivirta *et al.*, 2009). Recently published research challenged some of the classification criteria for endornaviruses, as three *Ceratobasidium* virus genomes were found to encode two ORFs instead of one (Ong *et al.*, 2016).

Sclerotinia sclerotiorum debilitation-associated RNA virus (SsDRV) has a genome of 5419 nucleotides excluding the poly(A) tail containing one ORF and is currently classified as the only species in the genus *Sclerodarnavirus* in the family *Alphaflexiviridae* (Xie *et al.*, 2006). The coding sequence has a conserved methyltransferase domain, as well as helicase and RdRp domains, typical for alphaflexiviruses. But unlike other members, SsDRV does not encode a capsid protein (Xie *et al.*, 2006). Another virus isolated from the same

Sclerotinia sclerotiorum strain, Sclerotinia sclerotiorum RNA virus L (SsRV-L), had a larger genome at 6043 nucleotides and one ORF encoding for the same replication domains, but showed the highest similarity to animal viruses of the genus *Hepevirus* rather than to SsDRV (Liu *et al.*, 2009). Based on phylogenetic analyses, the authors suggest that this virus be classified with the rubiviruses in the family *Togaviridae.*

Lastly among virion-less viruses with a fully sequenced genome is Diaporthe ambigua RNA virus (DaRV). DaRV possesses a genome of 4113 nucleotides encoding two ORFs of which the RdRp domain is similar to that of plant viruses of the family *Tombusviridae* (Preisig *et al.*, 2000).

Encapsidated mycoviruses

Two distinct (+)ssRNA viruses, BVF and BVX, were found in the same isolate of *Botrytis* (Howitt *et al.*, 2001, 2006). Despite their similar genome sizes (6827 nucleotides for BVF and 6966 nucleotides for BVX) and their similar flexible rod-shaped structure, their genome structure was distinct showing only moderate sequence similarity. BVF encoded two ORFs with the first containing replicase domains, while BVX encoded five ORFs with the replicase regions on ORF1. The methyltransferase, helicase and RdRp domains in both viruses show varying degrees of similarity to plant 'tymo-' and 'potex-like' viruses (Howitt *et al.*, 2001, 2006). BVX was more similar to previously isolated viruses and was classified in a new genus *Botrexvirus* in the family *Alphaflexiviridae* which also contains multiple plant-infecting genera, while BVF was classified as the first species in the new family *Gammaflexiviridae* in the genus *Mycoflexivirus* (King *et al.*, 2012).

Several other (+)ssRNA mycovirus genomes have been sequenced, but the viruses could not yet be assigned to an existing taxon. The most peculiar genome structure was found in Sclerophtora macrospora virus A (SmV A) which consists of three capped segments of 2928 nucleotides, 1981 nucleotides and 977 nucleotides encoding two capsid proteins and a RdRp domain, and this genome organization is not found in any other ssRNA mycovirus (Yokoi *et al.*, 2003). A second *Sclerophtora* virus, SmV B, also has a distinct genome structure of 5533 nucleotides with little similarity to other viruses, encoding two ORFs with a RdRp domain, a chymotrypsin-related serine protease domain and a capsid protein, showing similarity to sobomoviruses, barnaviruses and luteoviruses (Yokoi *et al.*, 1999). A third unassigned virus is Fusarium gramineum virus DK21 (FgV-DK21) which has a genome of 6624 nucleotides and encodes five ORFs showing a relatively close relationship with hypoviruses, endornaviruses and partitiviruses in the first ORF while the others show no strong sequence similarity to known proteins (Kwon *et al.*, 2009).

Leviviridae

The first members of the family *Leviviridae* were first discovered as male-specific bacteriophages which could only reproduce on *Escherichia coli* with F-pili (Van Duin and Tsareva, 2006) (see Chapter 7). There are currently two genera defined, *Levivirus* and *Allolevivirus,* with *Enterobacteria* phages MS2 and Qβ the respective type species (King *et al.*, 2012). The genomes of these phages form hairpin loop structures to condense and fit into the virion capsid, prevent formation of a double-stranded structure and to regulate gene expression upon infection of the host (Beekwilder *et al.*, 1996, 1995). The levivirus genome is typically about 3.5 kb long and has four ORFs with the lysis gene overlapping the coat and replicase genes, as exemplified by MS2 (Atkins *et al.*, 1979). Other members of this genus share the typical genome organization with maturation, coat and replicase genes and an overlapping lysis gene location (Inokuchi *et al.*, 1986; Rumnieks and Tars, 2012; Ruokoranta *et al.*, 2006).

The allolevivirus genome, on the other hand, is longer at over 4 kb with a slightly different gene order, encoding a read-through protein downstream from the coat protein and no lysis gene (Inokuchi *et al.*, 1988). Three phages currently classified in the family, Pseudomonas phage PP7, Acinetobacter phage AP205, and Caulobacter phage phiCb5, do not cluster with the levivirus- or allolevivirus clades (Kannoly *et al.*, 2012; Kazaks *et al.*, 2011; Klovins *et al.*, 2002; Olsthoorn *et al.*, 1995; Rumnieks and Tars, 2012). Combining phylogenetic evidence with plasmid specificity information, it was hypothesized that specialization for the F plasmid is recent evolutionarily and that the common ancestors for the two genera independently evolved this plasmid dependence, i.e. that the levivirus and allolevivirus lineages split before pili specialization (Kannoly *et al.*, 2012; Rumnieks and Tars, 2012).

Picornavirales

Viruses belonging to the order *Picornavirales* all have positive-sense RNA genomes encoding polyproteins with a characteristic helicase-proteinase-RNA polymerase block (Hel-Pro-Pol) (King *et al.*, 2012). The viruses of microorganisms in this order are all algal viruses from marine origins (see Chapter 10). The first algal picornavirus to be sequenced, Heterosigma akashiwo RNA virus (HaRNAV), belongs to the genus *Marnavirus* in the family *Marnaviridae* and is currently its only representative (Lang *et al.*, 2004). HaRNAV has a genome of 8587 nucleotides long encoding a single ORF polyprotein, with the separate domains showing similarities to different families within the order *Picornavirales* (Lang *et al.*, 2004). Three diatom-infecting viruses with similar properties were assigned to the genus *Bacillarnavirus* which is unassigned within the order, and has *Rhizosolenia setigera RNA virus* (RsRNAV) as its type species (Shirai *et al.*, 2006). RsRNAV, Chaetoceros tenuissimus RNA virus 01 (CtenRNAV01) and Chaetoceros socialis f. radians RNA virus (CsfrRNAV) have genomes ranging between 8.8 and 9.5 kb and are relatively rich in A + U content (60–63%) (Shirai *et al.*, 2006, 2008; Tomaru *et al.*, 2009). In contrast to the marnavirus HaRNAV, the bacillarnaviruses encode two ORFs, with the replicase polyprotein encoded by the first ORF and a structural polyprotein by ORF2. The genome of the marine fungal protist *Schizochytrium* virus, SssRNAV, was also sequenced, and phylogenetic analysis of the RdRp domain placed it on a branch between the marnaviruses and bacillarnaviruses (Shirai *et al.*, 2008; Takao *et al.*, 2005; Tomaru *et al.*, 2009). When more algal picornaviruses are sequenced it should become clear whether SssRNAV is the first isolate of a new family or whether instead it should be included with the bacillarnaviruses.

dsRNA viruses

The dsRNA viruses of microorganisms are classified into six different families. Unsegmented fungal and protozoan viruses fall within the family *Totiviridae*, while the segmented viruses are grouped in the families *Partitiviridae* (two segments), *Chrysoviridae* (four segments), *Quadriviridae* (four segments) or *Reoviridae* (9–12 segments) (King *et al.*, 2012). One family of bacteriophages has been described thus far, *Cystoviridae*, which groups phages infecting plant-pathogenic *Pseudomonas* bacteria (Ceyssens and Lavigne, 2010) (see Chapters 7, 9, and 10).

Totiviridae

Members of the genus *Totivirus* have linear genomes of 3–5 kb encoding two overlapping ORFs (Fig. 2.2). The genome of the type species *Saccharomyces cerevisiae virus L-A* is 4571 bp long and the two ORFs with a 130 bp overlap create a 180 kDa fusion protein between the coat protein (CP) and the RNA-dependent RNA polymerase (RdRp) (Icho and Wickner, 1989). Correct translation of the second gene is in this case achieved by frameshift translation caused by a stem–loop structure and the −1 reading frame of the second gene, meaning that the tRNAs bound to the stem–loop on the mRNA 'slip' into the correct reading frame for the second ORF (Icho and Wickner, 1989). Three other yeast viruses infecting *Xanthophyllomyces dendrorhous* showed the same characteristics, slippery sites (heptameric sequences downstream from stem–loop structures responsible for the correct expression of the polyprotein) and a −1 frameshift for the second ORF, detected in all three genomes (Baeza *et al.*, 2012). Scheffersomyces segobiensis virus L has a genetic code adapted to that of its host and probably originated from an exogenous totivirus using the standard genetic code (Taylor *et al.*,

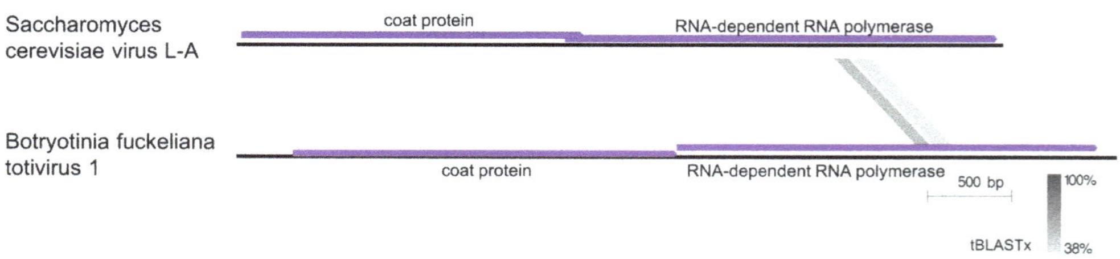

Figure 2.2 *Totivirus* genome structure. The genome of the type isolate Saccharomyces cerevisiae virus L-A (top) was compared with Botryotinia fuckeliana totivirus 1 (bottom) using tBLASTx and visualized using Easyfig 2.1 (Sullivan *et al.*, 2011).

2013). Two other genomes have been sequenced which are larger, Botryotinia fuckeliana totivirus 1 (5261 nucleotides, NC_009224), and smaller, Zygosaccharomyces bailii virus Z (3157 nucleotides, NC_003874), than the other members, but as of this writing no comparative studies have been performed.

The genome of *Helminthosporium victoriae 190S virus* (Hv190SV), the type species of the genus *Victorivirus* consists of similar overlapping ORFs (CP and RdRp) as the totiviruses, but with a shorter overlap of only 16 nucleotides instead of the ~130 nucleotides in totiviruses (Huang and Ghabrial, 1996). Since no heptamer frameshift site (slippery site) was identified, the proposed mechanism for expression of the RdRp (second ORF) is a form of internal initiation, rather than translational frameshifting (Huang and Ghabrial, 1996). Many of the victoriviruses have even shorter overlaps between the two genes, where the stop codon of the CP overlaps with the start codon of the RdRp, witnessed in Coniothyrium minitans RNA virus, Gremmeniella abietina dsRNA virus L1, Magnaporthe oryzae virus 1 and 2, and several *Ustilaginoidea virens* RNA viruses (Cheng *et al.*, 2003; Jiang *et al.*, 2015; Maejima *et al.*, 2008; Tuomivirta and Hantula, 2003b; Yokoi *et al.*, 2007; Zhang *et al.*, 2013; Zhong *et al.*, 2014a,b). While some of the earlier sequenced members of this genus were originally classified as totiviruses or leishmaniaviruses (Nomura *et al.*, 2003; Preisig *et al.*, 1998), recent phylogenetic analysis of the RdRp clustered the victoriviruses together, distinct from other members of the *Totiviridae* family (Chiba *et al.*, 2013a; Herrero and Zabalgogeazcoa, 2011; Park *et al.*, 2005). In this analysis, two separate clusters of victoriviruses were found and a new genus '*Mykitasvirus*' comprising those viruses infecting filamentous fungi was proposed, but has not yet been included in the official taxonomy (Maejima *et al.*, 2008).

Three distinct species of Trichomonas vaginalis virus (TVV1, TVV2 and TVV3) were recently grouped together in a new genus, *Trichomonasvirus*. These protozoan-infecting viruses share characteristics with fungal totiviruses, such as the overlapping ORFs with the presence of a heptameric slippery site indicating translational frameshift, but the CP and RdRp sequences show no or limited sequence homology to other totiviruses (Bessarab *et al.*, 2000, 2011; Tai and Ip, 1995).

The wide range of hosts in this family, from fungi to protozoan animal parasites such as *Leishmania* (genus *Leishmaniavirus*) and *Giardia* (genus *Giardiavirus*), combined with a shared sequence identity of the RdRps, has raised the hypothesis of an ancient origin for the totiviruses, maybe even dating back to before the divergence of animals and fungi (Bruenn, 1993). Evidence of coevolution of totiviruses with yeasts was deduced from the viral avoidance of a modified codon used by one host whereas the codon usage in other exogenous viruses retained standard code, same as in their respective hosts (Taylor *et al.*, 2013). Horizontal gene transfer between totiviruses and their hosts has also given rise to novel eukaryotic gene families from viral origin (Taylor and Bruenn, 2009).

Partitiviridae

Genomes in this family encode similar proteins as the those in the family *Totiviridae*, a coat protein (CP) and a RNA-dependent RNA polymerase (RdRp), but partitivirus genomes consist of two main linear monocistronic dsRNA segments each containing one gene (King *et al.*, 2012). From the presence of only these two genes, the likely minimum gene set for virus propagation was deduced (Nogawa *et al.*, 1996).

Aside from basidiomycete mushroom- and plant-infecting viruses, the genus *Alphapartitivirus* contains four completely sequenced mycoviruses. The cherry chlorotic rust spot-associated partitivirus (CCRS-PV) and Amasya cherry disease-associated partitivirus (ACD-PV) share 98% amino acid identity for their RdRp encoded by segment 1 (2021 and 2006 nucleotides, respectively) and 94% for the CP encoded by segment 2 (1841 and 1839 nucleotides, respectively) and are considered isolates of the same species (Coutts, 2004). Based on RdRp phylogeny, Rosellinia necatrix partitivirus 2 (RnPV2) and Rhizoctonia solani partitivirus 2 were also classified in this genus (Chiba *et al.*, 2013b; Zheng *et al.*, 2014). RnPV2 contained a third segment similar to segment 1, encoding a frameshift-terminated RdRp which was designated as defective interfering RNA (Chiba *et al.*, 2013b). The members of the genus *Betapartitivirus* share the same characteristics such as the conserved motifs encoded by the RdRp and conserved termini between the two segments, but are distinct based on RdRp multiple alignment (Compel *et al.*, 1999; Deng and Boland, 2007; Sasaki *et al.*, 2005; Strauss *et al.*, 2000; Zheng *et al.*, 2014). The type species of this genus is Atkinsonella hypoxolon virus (AhV) containing three segments of 2180, 2135 and 1790 nucleotides of which the latter does not encode an ORF and is considered satellite RNA (Oh and Hillman, 1995).

While the alpha- and betapartitiviruses infect plants and fungi (ascomycetes and basidiomycetes), the members of the genus *Gammapartitivirus* infect only microbes, plant-pathogenic fungi to be specific. The genome segments are generally smaller than those of the alpha- and betapartitiviruses, with segment 1 of the type species *Penicillium stoloniferum virus S* (PsV-S) 1690 nucleotides and segment 2 1523 nucleotides (Kim *et al.*, 2003). The RdRp of PsV-S showed a high degree of similarity to that of two *Discula destructiva* viruses, 1 and 2, which have conserved 5′ and 3′ termini between segments (Rong *et al.*, 2002). In this genus, several viruses have a third or even a fourth dsRNA segment either encoding an ORF of unknown function (PsV-F, DdV1, GaRV-MS1) or encoding no ORF at all (DdV1 dsRNA4) and these segments are likely satellite RNA (Kim *et al.*, 2005; Rong *et al.*, 2002; Tuomivirta and Hantula, 2003b). This hypothesis is strengthened by the discovery of a second *Gremmeniella abietina* RNA virus (GaRV-MS2) which is very similar to GaRV-MS1 (97% at the amino acid level), but lacks the third satellite dsRNA segment (Tuomivirta and Hantula, 2005).

The family *Partitiviridae* currently contains one genus of viruses infecting unicellular protozoa, *Cryspovirus*. This genus is named after its type species, *Cryptosporidium parvum virus 1*, which is distinct from other partitiviruses based on RdRp and CP phylogenies (Khramtsov *et al.*, 1997; Nibert *et al.*, 2009).

Chrysoviridae

Members of the family *Chrysoviridae*, exemplified by the type species *Penicillium chrysogenum virus*, contain four linear dsRNA segments, each encoding a single ORF, with the first segment encoding the RdRp, the second a coat protein and the other two ORFs of unknown function (Jiang and Ghabrial, 2004). The cherry chlorotic rusty spot disease and Amasya cherry disease virus RdRp and CP amino acid sequences place these viruses in this genus (Covelli *et al.*, 2004). Conserved 5′ and 3′ untranslated sequences (UTRs) of the segments were found between the Magnaporthe oryzae chrysoviruses 1-A and 1-B while for the other chrysoviruses conservation of the UTR sequences was only observed between the segments of the same virus (Urayama *et al.*, 2010, 2014; Yu *et al.*, 2011). Phylogenetic analysis of the RdRps showed two separate clades of chrysoviruses, suggesting the creation of two new genera in this family (Darissa *et al.*, 2011; Urayama *et al.*, 2014).

Reoviridae

The family *Reoviridae* consists of viruses with segmented, linear dsRNA genomes with 9–12 segments which show conserved terminal sequences (King *et al.*, 2012). While the majority of the viruses in this family infect vertebrates, insects or plants, members of the genus *Mycoreovirus* in subfamily *Spinareovirinae* infect fungi, and those of the genus *Mimovirus* in subfamily *Sedoreovirinae* have algal hosts.

The type species of the genus *Mycoreovirus* is *Cryphonectria parasitica 9B21 mycoreovirus*, also known as *MYCOREOVIRUS 1* (MYRV-1). The genome is made up of 11 segments from 732 bp to 4127 bp in size, all encoding a single large ORF taking up over 90% of the length of the segment and the segments sharing unique termini (Hillman *et al.*, 2004; Suzuki *et al.*, 2004). The two largest segments show the highest similarity with *MYCOREOVIRUS 3* (Rosellinia necatrix W370 mycoreovirus 3) which is made up of 12 segments, and several other reoviruses such as the coltiviruses Colorado tick fever virus and European Eyach virus (Hillman *et al.*, 2004; Osaki *et al.*, 2002; Wei *et al.*, 2003, 2004). The varying degrees of similarity between MYRV-1, MYRV-3 and more distantly related reoviruses has led to the hypothesis that mycoreoviruses together with coltiviruses evolved from a ancestral virus of 12 segments infecting an Acari arthropod species (Hillman *et al.*, 2004).

The only current member of the genus *Mimovirus* is Micromonas pusilla reovirus (MpRV) with a genome consisting of 11 segments. The first segment, at 5792 bp, is markedly larger than those of other reoviruses (~4000 bp) and it is hypothesized that the protein encoded by this segments provides an extra coat for the viral particle (Attoui *et al.*, 2006).

Quadriviridae

Rosellinia necatrix quadrivirus 1 (RnQV1) is the only species in the genus *Quadrivirus*, which is currently the only genus in this family. As the name suggests, the genome consists of *four* segments of linear dsRNA which each encode a single ORF (Kondo *et al.*, 2013a,b). The RdRp, found on segment 3, showed low sequence homology with other dsRNA mycoviruses, in specific the partitiviruses ACD-PV and CCRS-PV.

Cystoviridae

The family *Cystoviridae* with its only genus, *Cystovirus*, consists of bacteriophages of plant-pathogenic *Pseudomonas* species and contain a lipid envelope around the icosahedral capsid (King *et al.*, 2012; Poranen *et al.*, 2015). The type species genome of *Pseudomonas phage* $\phi6$ is made up of three linear dsRNA segments, named small (S), medium (M) and large (L) according to their sizes, which each encode multiple ORFs. Segment S is 2948 nucleotides long and codes four predicted proteins, the major capsid protein, a morphogenetic protein, the major membrane protein and the lysozyme (McGraw *et al.*, 1986). The second segment (M) at 4057 bp encodes the proteins required for the attachment apparatus of the virus and two membrane-associated proteins (Gottlieb *et al.*, 1988), while the large segment (6374 bp) carries the transcriptase and replicase genes making up the procapsid (Mindich *et al.*, 1988). The S and M segments of phage $\phi8$ show no similarity to those of $\phi6$, but the homology of the predicted RdRps encoded by the L segment places them in the same genus (Hoogstraten *et al.*, 2000).

Phage $\phi12$ is more similar to $\phi6$ with the lysis cassettes showing significant amino acid identities (Gottlieb *et al.*, 2002a). The RdRp of $\phi12$ is similar to those of $\phi6$, $\phi8$ and $\phi13$, which suggests that the evolution of $\phi12$ may have happened through genetic shift (reassortment) of segments from more than one ancestral virus (Gottlieb *et al.*, 2002b). The terminal sequences of the three segments, which are very much conserved in $\phi6$, differ between segments in $\phi13$ and this phage is also able to function with all segments compiled into one RNA molecule, indicating that packaging requirements in this phage are less strict than in $\phi6$ (Qiao *et al.*, 2000). The RdRp of $\phi6$ was found to be similar to that of hepatitis C virus, giving indications that dsRNA viruses and flaviviruses might have originated from a common ancestor (Butcher *et al.*, 2001).

ssDNA viruses

Inoviridae

Members of the genus *Inovirus* are filamentous bacteriophages with ssDNA genomes between 5 and 9 kb in length. They were first discovered as male-specific coliphages and *Escherichia phage M13* was designated as the type species (van Wezenbeek *et al.*, 1980). Three very similar *Escherichia* inoviruses with a genome size of 6.4 kb sharing >95% nucleotide identity, M13, fd and f1, collectively referred to as Ff, each contain eight genes, nine promoters and have a preference for T in the third codon position (Beck *et al.*, 1978; Hill and Petersen, 1982; van Wezenbeek *et al.*, 1980). The slightly larger phage IKe (6.9 kb) has the same gene arrangement as Ff but the ORFs are generally larger and they all likely have evolved from a common ancestor (Peeters *et al.*, 1985). Another related phage, I2–2, shows an identical genome organization but less homology and it is hypothesized to be derived from two different replicons (Stassen *et al.*, 1992). *Pseudomonas*-infecting inoviruses, such as Pf1 and Pf3, generally have the same overall pattern of large and small ORFs following the coat protein and stem–loop structure-forming intergenic regions as the coliphages, but have little sequence homology (Hill *et al.*, 1991; Luiten *et al.*, 1985).

Several of the inoviruses infecting *Vibrio* sp. carry toxins in their genomes, such as CTXΦ which carries cholera toxin and the morphologically indistinguishable fs1 and fs2 phages which encode a new zonula occludens toxin-like protein (Zot) (Davis, 2003; Ehara *et al.*, 1997). Other *Vibrio* phages (Vf12, Vf33, VGJΦ, KSF-1Φ, VCYΦ, VFJΦ) share the same modular structure (replication, structure and assembly, and regulation modules) with around ten homologous ORFs, but they have a distinct region at the 3′ end at the location of cholera toxin in CTXΦ (Campos *et al.*, 2003; Chang *et al.*, 1998; Faruque *et al.*, 2005; Wang *et al.*, 2013; Xue *et al.*, 2012). Homology analyses of the genome modules in these phages revealed that they have a mosaic genome structure which has likely arisen from horizontal gene transfer; for example the structure-assembly module of VCYΦ is 76% similar at the nucleotide level to KSF-1, while the regulatory module is 80% similar to VGJΦ (Xue *et al.*, 2012).

The *Ralstonia* inovirus ΦRSS1 genome (6662 nt, 11 ORFs), while displaying gene synteny with Ff phages, also showed 95% sequence identity with the plasmid pJTPS1 isolated from a *R. solanacearum* strain, which was hypothesized to correspond to the replicative form of a ΦRSS-type phage (Kawasaki *et al.*, 2007). Another *Ralstonia* phage, ΦRSM1, had a much larger genome at 9004 nucleotides, with this greater length caused by gene insertions in the otherwise syntenic genome, while the strongly homologous phage ΦRS603 had a shorter genome but contained a Zot-like protein (Kawasaki *et al.*, 2007; Van *et al.*, 2014). Other inoviruses with a genome organization shared with the previously discussed phages include Propionibacterium phage B5 (5806 nucleotides) and Xanthomonas phage Cf1t (7308 nucleotides) (Chopin *et al.*, 2002; Kuo *et al.*, 1991).

Phages belonging to the genus *Plectrovirus* are rod-shaped, in contrast with the filamentous inoviruses (King *et al.*, 2012). The genome lengths of plectroviruses range between 4.5 and 8.3 kb. The longest genome belongs to Spiroplasma virus SpV1-R8A2 (8273 nucleotides) comprised 12 ORFs and four intergenic regions (Renaudin *et al.*, 1990). The smaller genome size of Spiroplasma virus SVTS2 (6824 nucleotides), encoding 14 ORFs divided in three sections, was due to the absence of certain ORFs compared to SpV1, rather than being due to the absence of a transposase gene (Sha *et al.*, 2000).

Spiraviridae

Perhaps the most unique of the ssDNA viruses is the archaeal virus, Aeropyrum coil-shaped virus (ACV), currently the only member of the family *Spiraviridae*. With its particle structure consisting of a hollow cylinder made from a coiled nucleoprotein filament, there are less restrictions when it comes to genome size, which at 24,893 nucleotides and 57 predicted CDSs is the largest and most complex ssDNA genome found to date (Mochizuki *et al.*, 2012). The large genome size has also allowed ACV to encode genes with auxiliary functions (Mochizuki *et al.*, 2012), a feature more common in large dsDNA viruses.

Microviridae

The family *Microviridae* contains two ICTV-ratified subfamilies, *Bullavirinae* and *Gokushovirinae*, each containing three genera (King *et al.*, 2012; Krupovic *et al.*, 2016). Viruses belonging to this family form small, icosahedral, tailless virions.

The most famous member of the subfamily *Bullavirinae* in genus *Phix174microvirus* is without a doubt *Escherichia* phage ΦX174 at 5386 nucleotides the first fully sequenced DNA genome in history (Sanger *et al.*, 1977). While many similar phages have been screened, 42 distinct genotypes were found and sequenced (Rokyta *et al.*, 2006). All these genomes had homologues of the common 11 microvirus genes, while the variability within groups was caused by insertions and deletions in intergenic regions, e.g. the alpha3-like group had five extra ORFs in the H-A intergenic region (Fig. 2.3). Based on phylogenetic analysis, genome size and gene content, three genera of bullaviruses could be delineated, *Phix174microvirus* (5386–5387 nucleotides), *G4microvirus* (5486–5577 nucleotides) and *Alpha3microvirus* (6061–6259 nucleotides) (Krupovic *et al.*, 2016; Labrie *et al.*, 2014; Rokyta *et al.*, 2006). A distinct feature of these genomes is the presence of overlapping genes which show differing codon usages in the overlapping and non-overlapping sections, leading to the hypothesis that these phages gradually increased in complexity from the ancestral state without overlapping genes to the current state with high gene density (Pavesi, 2006).

Members of the subfamily *Gokushovirinae* generally have smaller genomes, encoding fewer ORFs than the bullaviruses. Spiroplasma virus 4 (SpV4) is currently the only representative of the genus *Spiromicrovirus*, with a genome length of 4421 nucleotides and nine ORFs preceded by a Shine–Dalgarno sequence (Renaudin *et al.*, 1987). Similar to mitoviruses which use an adapted fungal genetic code (described above), TGA is not considered a stop codon in this group of viruses, but expected to code for tryptophan (Paquin *et al.*, 1997; Renaudin *et al.*, 1987). The second genus in this subfamily, *Chlamydiamicrovirus*, consists of a set of *Chlamydia*-infecting phages. The genome of the type species *Chlamydia* phage Chp1 is 10% shorter than ΦX174 at 4877 nt and encodes *five* major ORFs, of which VP1 shows homology to the F protein of ΦX174 and is considered the major capsid protein (Storey

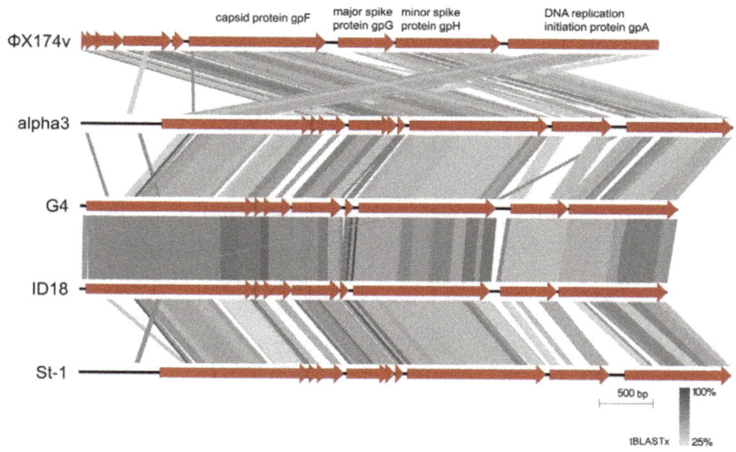

Figure 2.3 *Bullavirus* genome organization. Comparison of the genomes of representative *Escherichia* phages of the subfamily *Bullavirinae*, as made with tBLASTx and visualized with Easyfig 2.1 (Sullivan *et al.*, 2011). The major annotated proteins are indicated for phage ΦX174.

et al., 1989). The absence of an external scaffolding protein in Chp2, as well as other gokushoviruses, explains the absence of spikes on the virus particles, in contrast with the *Escherichia* microviruses (Liu *et al.*, 2000). The only sequenced member of the genus *Bdellomicrovirus*, Bdellovibrio phage phiMH2K, differs from the others in the location of ORF5 and sequence of the major capsid protein, which has large amino acid insertions between beta strands E and F (Brentlinger *et al.*, 2002).

Recently, metaviromic surveys (see Chapter 5) have greatly increased the charted diversity of the family *Microviridae*, with novel genome architectures (i.e. the gene content and organization of genes within the genome) found in oceans, freshwater lakes, the human gut, soils and sediments (Bryson *et al.*, 2015; Krupovic and Forterre, 2011; Labonté and Suttle, 2013; Quaiser *et al.*, 2015; Reavy *et al.*, 2015; Roux *et al.*, 2012). This has led to the proposal of the new subfamilies, 'Alpavirinae', 'Pichovirinae', 'Aravirinae' and 'Stokavirinae' which are currently not ratified by ICTV because no isolated representatives have been found as of yet. The proposed alpavirus genomes, which were found as prophages in *Bacteroidetes* species, are about 6 kb in length, corresponding to bullaviruses, but the proteins are more homologous to gokushoviruses, forming a separate phylogenetic clade for the major capsid protein (Krupovic and Forterre, 2011). Similar genome types have been found in a turkey gastrointestinal system (ΦCA82) as well as in a human throat swab (IME-16) (Shu *et al.*, 2015; Zsak *et al.*, 2011). The proposed pichoviruses have smaller genomes than bullaviruses and alpaviruses and present a different gene order for their core genes, i.e. VP4 – VP2 – VP1, instead of VP4 – VP1 – VP2 for all other members of the family (Roux *et al.*, 2012). The subfamilies 'Aravirinae' and 'Stokavirinae' were proposed based on the formation of new clades in the VP1 (major capsid protein) phylogenetic tree derived from sequences of genomes found in peat metaviromes (Quaiser *et al.*, 2015).

Unassigned ssDNA viruses

One ssDNA virus infecting fungi has been isolated to date, Sclerotinia sclerotiorum hypovirulence-associated DNA virus 1 (SsHADV-1) (Yu *et al.*, 2010). Its genome is 2166 nucleotides long and encodes two ORFs, a coat protein on the sense strand and a replication initiation protein on the anti-sense strand. While the proteins share conserved domains with those of geminiviruses (ssDNA plant viruses), phylogenetic analysis of the coat protein places SsHADV-1 in a separate clade, distinct from the three families of ssDNA plant viruses (*Geminiviridae, Nanoviridae, Circoviridae*).

A unique set of ssDNA viruses infecting marine diatoms are grouped together in the unassigned genus *Bacilladnavirus*. The viruses in this genus have closed circular ssDNA genomes of about 5.5 kb in length encoding for three major ORFs. Chaetoceros tenuissimus DNA virus type II (CtenDNAV type II) and Chaetoceros lorenzianus DNA virus (ClorDNAV) have reported dsDNA regions of 669 and 979 nucleotides (Kimura and Tomaru, 2015; Tomaru *et al.*, 2011). Phylogenetic analysis of the Rep protein showed the monophyletic origin of the bacilladnaviruses ClorDNAV, CsalDNAV, CtenDNAV, Csp05DNAV (Tomaru *et al.*, 2013; Toyoda *et al.*, 2012).

dsDNA viruses, no RNA stage

The order *Caudovirales*

Bacteriophages and archaeal viruses belonging to the order *Caudovirales* number over 1900 completely sequenced genomes in the NCBI database at the time of writing (Table 2.1) and 133 different ICTV-ratified genera in the three families (ICTV Taxonomy: 2015 Release, (Adriaenssens *et al.*, 2016)). Instead of trying to give a complete summary of all the genome types found within the three families of tailed phages, I have focused on one group within each family. (see Chapters 7 and 8 for additional discussion)

At the time of writing, more than 540 complete genomes classified within the family *Myoviridae* were deposited in GenBank, all describing phages with long, contractile tails. In the 2015 Taxonomy Release, ICTV recognized five subfamilies (*Eucampyvirinae, Peduovirinae, Spounavirinae, Tevenvirinae,* and *Vequintavirinae*) and a further 28 genera not assigned to a subfamily. Unsurprisingly, the genomic diversity in this family is great, with genome sizes ranging from around 30 kb for certain members of the genus *P2virus* in the subfamily *Peduovirinae* to the singleton virus (i.e. only member of its taxon) Bacillus megaterium phage G of which the genome is 497,513 bp long (NC_023719).

Arguably the most famous member of the family is *Escherichia phage T4* (Karam *et al.*, 1994; Mathews *et al.*, 1983), the type species of the genus *T4virus* in the subfamily *Tevenvirinae* (King *et al.*, 2012; Lavigne *et al.*, 2009). T4 has a circularly permuted, linear genome of 168,903 bp with 289 probable CDSs and eight tRNAs (Miller *et al.*, 2003). Comparison of an extended set of T4-like phages has led to the definition of a core set of genes (31–38) common in the genomes and mostly involved in the phage reproduction (Petrov *et al.*, 2010; Sullivan *et al.*, 2010). These core genes all belonged to one of five functional groups, 'DNA replication, repair and recombination', 'morphogenesis', 'DNA packaging', 'gene expression' or 'auxiliary metabolism' (Petrov *et al.*, 2010). An additional set of 25 genes were shared between T4-like cyanophages which have not yet received an official taxonomic assignment (Sullivan *et al.*, 2010). In a larger group of T4-related phages, a general consistency was found between single gene and whole genome phylogenies, indicating that horizontal gene transfer does not prevent inferring evolutionary relationships (Ignacio-Espinoza and Sullivan, 2012).

With over 300 genomes in the NCBI database, the family *Podoviridae* is the smallest of tailed phages, both by number of members and virion size, comprising phages with short tails. Phages belonging to the subfamily *Picovirinae* have small genomes of generally less than 20 kb, while members of the *Autographivirinae* are larger at over 35 kb and encode their own RNA polymerase (Lavigne *et al.*, 2008). A well-studied group in the latter subfamily is the genus *Phikmvvirus* with Pseudomonas phage ΦKMV as the type phage comprising a genome of 42,519 bp delineated by direct terminal repeats (Lavigne *et al.*, 2003). Although nucleotide similarity between members can be low, all share the same genome organization with ORFs on the same strand divided into an early and a late region by the location of the RNA polymerase (Adriaenssens *et al.*, 2011; Ceyssens *et al.*, 2006). Typical for many of the early genes, which are presumed to be involved in phage-adaptation to host metabolism, is their lack of sequence homology with other members of the genus, partially explained by the variation in the many different host bacteria (Adriaenssens *et al.*, 2011; Roucourt and Lavigne, 2009). Several ΦKMV-like phages (ΦKMV, ΦKF77, LKD16, LUZ19, PT2, PT5) have confirmed single-stranded nicks in the genomes, but the biological significance remains unknown (Kulakov *et al.*, 2009). Most phikmvviruses have a four gene tail fibre cassette encoding a capsid attachment protein and a lysozyme domain protein with components of the cassette also found in some siphoviruses (Ceyssens *et al.*, 2011; Karumidze *et al.*, 2012).

The *Siphoviridae* family has the most sequenced members at this time, with over 1000 genomes in GenBank. This makes it a daunting task to classify all of these genomes and so far two subfamilies and 65 genera have been recognized (Adriaenssens *et al.*, 2015; Krupovic *et al.*, 2016). A very substantial fraction of the sequenced siphovirus genomes comes from *Mycobacterium* phages isolated by the Phage Hunters, PHIRE and SEA-PHAGES programs of the University of Pittsburgh and the Howard Hughes Medical Institute (USA)(Russell and Hatfull, 2016). The statistics are currently at over 1600 sequenced phage genomes according to listings at the Actinobacteriophage Database (phagesdb.org), although not all of them have been deposited in the GenBank database (Russell and Hatfull, 2016). Interestingly, of these 1000+ sequenced *Mycobacterium* phage genomes, only 76 are myoviruses, all the others are siphoviruses. With this large amount of genome data, family-level analyses can be made.

The phages can be divided in clusters and subclusters based on nucleotide identity over 50% of the genome length (Hatfull, 2010; Hatfull *et al.*, 2008), which is now being used as the basis for formal virus taxonomy (Adriaenssens *et al.*, 2015, 2016; Krupovic *et al.*, 2016). The average genome sizes of the siphoviruses are, at this time, between 42 kb (cluster G) and 110 kb (cluster J), with a total of 20 clusters and eight singleton isolates, of which one cluster is populated by myoviruses (cluster C) (Pope *et al.*, 2015). Within clusters, GC content is generally consistent, as is genome length, while genomic diversity can still be high (232 genomes in cluster A) (Hatfull, 2012; Pope *et al.*, 2015). Many of the genomes share a common arrangement of virion structure and assembly genes which show a higher level of conservation than other regions in the genomes (Hatfull *et al.*, 2008, 2010). Throughout the *Siphoviridae* family, these genomes have a characteristic modular nature where single genes or groups of genes have been exchanged between more distantly related genomes, either relatively recently, leading to significant nucleotide sequence similarity, or in a more distant past, indicated by weak amino acid similarities (Hatfull, 2010). While no geographical patterns were found associated with the different clusters, a significant correlation between cluster and host range was established, suggesting that the host preference confers a barrier for genetic information exchange *in vivo* (Jacobs-Sera *et al.*, 2012; Pope *et al.*, 2011). While some expect to see a continuum of genomic diversity to arise after more and more virus isolates get sequenced (Hatfull, 2015), instead to date discrete groups continue to emerge, but which are still related to each other.

Membrane-containing dsDNA bacteriophages

Membrane-containing bacteriophage particles, which can be enveloped or have an internal lipid membrane, are generally sensitive to chloroform and detergents disrupting the membrane. Their genomes do not encode the lipid metabolism genes required for creation of the membrane. Rather, the lipids are derived from the host's cytoplasmic membrane, often through a process of selective enrichment (Poranen *et al.*, 2015). With standard phage isolation protocols often including a chloroform treatment step for the removal of host cellular material, there is a possible bias against the isolation of novel lipid-containing phages (Van Twest and Kropinski, 2009).

The family *Corticoviridae* has one sequenced member, Pseudoalteromonas phage PM2, which has highly supercoiled circular DNA of 10,079 bp in length (Männistö *et al.*, 1999). Combining the evidence for rolling-circle replication with the presence of a 1.2 kb region similar to plasmid pAS28 of a *Pseudoalteromonas* strain, it is possible that PM2 is maintained as a plasmid in the host system (Männistö *et al.*, 1999). Based on the CDSs annotated in PM2, other corticoviral-like elements have been found in bacterial genomes suggesting that the abundance of tailless icosahedral phages of this lineage in aquatic environments is underestimated (Krupovic and Bamford, 2007).

Acholeplasma phage L2 is currently the only recognized member of the family *Plasmaviridae*. It possesses an enveloped, quasi-spherical virion which encapsidates a superhelical, circular dsDNA genome of 11,965 bp (Maniloff *et al.*, 1994). Of the 15 CDSs which were annotated, only the putative integrase gene showed matches to the sequence database.

Although not reflected in official taxonomy, members of the family *Tectiviridae* can be divided into two groups, one group contains viruses infecting Gram-negative bacteria and the other phages of Gram-positive bacteria (Poranen *et al.*, 2015). The tectiviruses infecting Gram-negative bacteria all share a high degree of sequence similarity with the type species Enterobacteria phage PRD1 (Saren *et al.*, 2005). This icosahedral virus with inner lipid membrane has a 14,927 bp genome which has several unique features, such as the higher than random distribution of similar base groups and the low incidence of palindromic restriction sites (Bamford *et al.*, 1991). Based on comparison with the other Gram-negative tectiviruses (PR3, PR4, L17 and PR772), three protein classes with differential amino acid conservation were found. The structural proteins, certain small integral membrane proteins, and the DNA polymerase are most conserved, while the assembly proteins have a lesser degree of conservation, and the host recognition and lysis genes are the most divergent (Saren *et al.*, 2005).

The Gram-positive tectiviruses (Bam50c, GIL01, GIL16, AP50c, Wip1) all infect species of the genus *Bacillus* and are able to maintain themselves as plasmid prophages (Verheust *et al.*, 2005). Genome conservation within this group is much lower than within the PRD1 group, generally around 50% (Sozhamannan *et al.*, 2008). These phages have genomes between 14 and 15 kb in size with imperfect inverted terminal repeats (ITRs). Only a few CDSs have functional predictions, including a family B type DNA polymerase which shows the highest conservation with the PRD1-like phages, a LexA-like repressor gene, a DNA packaging protein and two endolysins (Ravantti *et al.*, 2003; Verheust *et al.*, 2003, 2005). Gene synteny is not conserved in phage Wip1, which has its DNA polymerase at the other end of the genome and on the opposite strand as the others tectiviruses (Kan *et al.*, 2013). Furthermore, interspecies exchange of endolysin domains between related *Bacillus thurengiensis* phages is a possible explanation for divergent endolysins in otherwise similar phages (Sozhamannan *et al.*, 2008).

Archaeal dsDNA viruses

While the majority of the viruses infecting moderately thermophilic bacteria and archaea are head-tailed viruses (Ackermann, 2001; Senčilo and Roine, 2014), hyperthermophilic archaeal viruses (and certain haloviruses) come in a great variety of morphologies and display a wide genomic diversity (Iranzo *et al.*, 2016; Prangishvili, 2015). This resulted in the creation of one unassigned genus, nine different families, and one order (ICTV Taxonomy: 2015 Release) which I address alphabetically in this section. Additionally, many of the archaeal viruses described in the official taxonomy have not been fully sequenced, opening the possibility that some viruses will be reassigned in the future, based on genomic data (see Chapter 8).

The family *Ampullaviridae* groups viruses with bottle-shaped morphologies. Only one isolate has been investigated in-depth so far, Acidianus bottle-shaped virus (ABV), with a linear genome of 23,814 bp and 590 bp ITRs (Peng *et al.*, 2007). A hallmark of archaeal viruses, which is described for other families as well, is the lack of homologues in sequence databases. For the 57 CDSs identified in ABSV, only three genes were functionally annotated, a glycosyl transferase, thymidylate kinase and a family B DNA polymerase (Peng *et al.*, 2007). The latter showed similarities with *polB* sequences from very diverse viruses such as the salterprovirus His1, the pleolipovirus His2, bacteriophages Φ29 and PRD1 and even human adenoviruses. This suggests that this type of protein-primed *polB* gene was acquired from a primordial gene pool which was present before the divergence of the domains of life (Peng *et al.*, 2007).

The type species of the family *Bicaudaviridae* and of the only genus in this family, *Bicaudavirus*, is *Acidianus two-tailed virus*. As the name indicates, this family groups spindle-shaped (or lemon-shaped) viruses which have two tails, one at each end (King *et al.*, 2012). Its 62,730 bp circular genome encodes 72 CDSs, with a number of them showing homology to CDSs of members the families *Fuselloviridae*, *Rudiviridae* and *Lipothrixviridae*, and even to plasmids of the thermoacidophile *Sulfolobus* (Prangishvili *et al.*, 2006b). An unusual feature in this genome is the presence of four transposase genes and the occurrence of gene duplication and duplication within genes (Prangishvili *et al.*, 2006b).

The smallest dsDNA archaeal virus genome was found in Aeropyrum pernix bacilliform virus 1 (APBV1), at 5278 bp encoding 14 CDSs. None of these CDSs show homology with previously characterized viruses (Mochizuki *et al.*, 2010). Its unusual genome and bacilliform morphology made it the first member of the family *Clavaviridae*.

The family *Fuselloviridae* groups the spindle- (or lemon-) shaped viruses without tails. The alphafusselloviruses, exemplified by Sulfolobus spindle-shaped virus 1 (SSV1) (Palm *et al.*, 1991), are a group of *Sulfolobus*-infecting viruses with collinear genomes between 14.8 and 17.6 kb in size. One of the remarkable features is that ORFs with cysteine-containing codons are clustered together on only one half of the genome, as seen in viruses SSV1, SSV2, SSV RH and SSV K1 (Palm *et al.*, 1991; Wiedenheft *et al.*, 2004). With the addition of SSV4, SSV5 and SSV7, it seems as though the genomes of all members of this group have a similar operon structure, while the distribution of ORFs with high degrees of sequence conservation is uneven along the genomes (Peng, 2008; Redder *et al.*, 2009; Wiedenheft *et al.*, 2004). The presence of a 100% identical 7.9 kb section in the genomes of SSV4 and SSV5 suggest a recent recombination event between these viruses (Redder *et al.*, 2009).

The genus *Betafusellovirus* is delineated by a difference in particle morphology, with betafuselloviruses having different tail fibre structures protruding from the virion body (Redder *et al.*, 2009). It contains the fusellovirus with the largest genome, Acidianus spindle-shaped virus 1 (ASV1) with a size of 24,186 bp. ASV1 still is closely related to the other fusellovirus genomes, but contains gene duplications and new genes (Redder *et al.*, 2009).

Two spindle-shaped viruses of *Sulfolobus tengchongensis* (STSV1 and STSV2) are similar in morphology to the fuselloviruses described above, but their 75,294 bp and 76,107 bp genomes (80% sequence identity with each other) show no similarities at either the nucleotide or amino acid levels (Erdmann *et al.*, 2014; Xiang *et al.*, 2005). Furthermore, two ORFs, a hypothetical protein and a dUTPase, show similarity with the rudiviruses SIRV1 and SIRV2 (Xiang *et al.*, 2005). This makes it difficult to classify these viruses and it is likely that they will be the first members of a new taxon.

Spherical archaeal viruses are commonly grouped in the family *Globuloviridae*. The type species Pyrobaculum spherical virus (PSV) has a linear genome of 28,337 bp with 190 bp ITRs and the genome ends are most likely covalently linked (Häring *et al.*, 2004). At the time of sequencing, its 48 CDSs had no matches in any of the sequence databases. The other sequenced spherical virus, Thermoproteus tenax spherical virus 1, has 15 CDSs with low levels of amino acid similarity with PSV, but no other database matches (Ahn *et al.*, 2006).

Of all the droplet-shaped viruses in the family *Guttaviridae*, only the betaguttavirus Aeropyrum pernix ovoid virus 1 (APOV1) has been fully sequenced. Its 13,769 bp circular genome was induced from the host genome, showing an integrase split in two parts at the borders of the proviral regions (Mochizuki *et al.*, 2011). Of the 21 CDSs, only five had database counterparts, a fusellovirus-like integrase and DnaA-like protein, and three hypothetical proteins found in hyperthermophilic crenarchaeotes (Mochizuki *et al.*, 2011). The authors noted that, based on the genome, APOV1 was not classifiable in any family, but the morphology would place it in the *Guttaviridae* family.

One order of archaeal viruses has been described so far, the order *Ligamenvirales*, comprising the families *Lipothrixviridae* and *Rudiviridae*. Lipothrixviruses are flexible filamentous viruses, enveloped by viral proteins and host-derived lipids, with different terminal structures in the different genera (King *et al.*, 2012). The betalipothrixviruses have characteristic filamentous particles with mop-like structures consisting of six tail fibres at the termini (Arnold *et al.*, 2000). The genomes of Sulfolobus islandicus filamentous virus (SIFV), the type species, and the *Acidianus* filamentous viruses AFV3, AFV6, AFV7 and AFV8, range from 36,895 bp to 40,449 bp and have long ITRs (Vestergaard *et al.*, 2008). One half of the genomes is highly conserved within this group while the other half is distinct for the separate viruses. This latter region in AFV7 is highly similar to that of deltalipothrixvirus AFV2, suggesting a recombination event between genera sharing little sequence similarity (Vestergaard *et al.*, 2008). This deltalipthrixvirus, with a genome size of 31,787 bp, does not have recognizable terminal repeats and contains an unusual region of 1 kb comprising two perfect 46 bp repeats and several imperfect ones (Häring *et al.*, 2005). AFV2 also shares similar CDSs with gammalipothrixvirus AFV1 and the rudiviruses SIRV1 and SIRV2 (Häring *et al.*, 2005). AFV1, the only sequenced member of the genus *Gammalipothrixvirus*, has a much smaller genome at 20,869 bp and 15% of its 40 CDSs have homologues in rudiviruses (Bettstetter *et al.*, 2003).

The rudiviruses Sulfolobus islandicus rod-shaped virus 1 and 2 (SIRV1, SIRV2) have collinear genomes of 32,312 bp and 35,502 bp, respectively (Peng *et al.*, 2001). They have 44

CDSs in common which are more conserved in the central part of the genome. The conservation of CDS in the central part of the genomes with greater divergence towards the genome termini is a feature the rudiviruses share with certain NCLDVs (nucleocytoplasmic large DNA viruses, discussed below). Additionally, 21 of the CDSs share homology with several NCLDVs such as poxviruses, asfarviruses, and phycodnaviruses (Peng *et al.*, 2001). The Acidianus rod-shaped virus 1 (ARV1, 24,655 bp genome), on the other hand, has very little similarity with other viruses and only shares a gene organization pattern consistent with transcription of SIRV1 and SIRV2 (Vestergaard *et al.*, 2005).

Members of the family *Sphaerolipoviridae* are icosahedral viruses containing an internal lipid membrane. The alphasphaerolipovirus Haloarcula hispanica lipid-containing virus, SH1, serves as type species with a linear genome of 30,898 bp and 309 bp ITRs (Bamford *et al.*, 2005; Porter *et al.*, 2005). Comparison with the other haloarchaeal viruses HHIV-2 and PH1 revealed a similar genome organization, but with an unequal distribution of similarity over the genome length (Jaakkola *et al.*, 2012; Porter *et al.*, 2013). Two groups of genes emerged from this comparison, a slowly evolving group coding for structural proteins (i.e. conserved capsid architectures) and a more rapidly evolving group most likely involved in virus–host interactions (Jaakkola *et al.*, 2012).

This family also comprises two thermophilic bacteriophages, Thermus flavus phage P23–77 and Thermus aquaticus phage IN93, which share 47% nucleotide sequence identity (Jalasvuori *et al.*, 2009). The IN93 circular genome is 2,567 bp longer than the 17,036 bp circular genome of P23–77, due to the presence of an integration cassette not encoded by P23–77 (Jalasvuori *et al.*, 2009). The ATPases and capsid proteins of these viruses show similarity to those of proviruses in archaeal genomes and to the other sphaerolipoviruses (Jalasvuori *et al.*, 2009).

The family *Turriviridae* was named after the elaborate turret-like structures visible on the icosahedral capsids of its members. Sulfolobus turreted icosahedral virus 1 (STIV1) has a 17,663 bp circular genome encoding 36 CDSs with little homology in databases except for two CDSs shared with *Ligamenvirales* members (Rice *et al.*, 2004). A second turrivirus, STIV2, is highly similar at 70% overall nucleotide identity, but containing two unique structural proteins (Happonen *et al.*, 2010). One of its CDSs is similar to the fusellovirus DNA-binding protein VP2, suggesting horizontal gene transfer between different archaeal viruses (Happonen *et al.*, 2010).

The unassigned genus *Salterprovirus* only has one member, the spindle-shaped Haloarcula hispanica virus His1. Its 14,464 bp linear genome has a simple genome organization with CDSs directed outward towards the termini and the DNA polymerase gene on an inverted segment (Bath and Dyall-Smith, 1998; Bath *et al.*, 2006). This DNA polymerase shows significant similarity to His2, which infects the same host, and has been classified in the new family *Pleolipoviridae* (Pietilä *et al.*, 2016).

The curious case of the newly proposed family *Pleolipoviridae*

What sets this family of pleomorphic viruses apart from other families is that it contains members with both ssDNA and dsDNA genomes. All eight of the current members infect haloarchaeal species and have a collinear genome organization with nucleotide identities only at short stretches along the genome (Pietilä *et al.*, 2016). HRPV-1, HRPV-2, HRPV-6 and HHPV-2 are ssDNA viruses, while HHPV-1 and His2 consist of dsDNA and HRPV-3 and HGPV-1 have dsDNA molecules with ssDNA stretches (Bath *et al.*, 2006; Li *et al.*, 2014a; Pietilä *et al.*, 2009, 2012; Roine *et al.*, 2010). HHPV-1 and HHPV-2 are almost perfectly collinear despite the difference in nucleic acid composition (Li *et al.*, 2014a) and a block of about 4 kb encompassing ORF4 to ORF7 is present and homologous in all eight pleolipoviruses (Pietilä *et al.*, 2016; Senčilo *et al.*, 2012). An explanation for this remarkable genomic composition has not yet been found.

Nucleocytoplasmic large DNA viruses

The NCLDVs or nucleocytoplasmic large DNA viruses collectively refer to several families of large eukaryotic DNA viruses which share a common gene set and are most likely monophyletic in origin (Iyer *et al.*, 2001, 2006) (see Chapters 10 and 11). In addition to the animal and insect virus families *Poxviridae*, *Asfarviridae* and *Iridoviridae*, this group contains several groups of protozoan viruses, including the families *Mimiviridae* and

Marseilleviridae, and the algal virus family *Phycodnaviridae*. Because of the shared origin and a conserved set of NCLDV core genes, it has been proposed to include these families in the order '*Megavirales*', but this has not yet been ratified by ICTV (Colson *et al.*, 2013).

Phycodnaviruses (members of the family *Phycodnaviridae*) were the first group of NCLDVs of microorganisms to be discovered, many of which have genomes significantly larger than that of an average bacteriophage genome. All members of this family infect unicellular algae and share a set of core genes, but the overall sequence divergence is great, leading to the creation of six different genera.

Paramecium bursaria chlorella virus 1 (PBCV-1) is the type species of the genus *Chlorovirus* and has a genome of 330,744 bp, comprising 375 CDSs and 11 tRNAs (Van Etten *et al.*, 2002). The *Chlorella* viruses NY-2A and AR158 with genomes of 368,683 bp and 344,690 bp, respectively, showed a high degree of collinearity with PBCV-1 with over 84% of the ORFs shared between the three (Fitzgerald *et al.*, 2007a). Two other chloroviruses, MT325 (314,335 bp) and FR483 (321,240 bp), had a high degree of shared ORFs with PBCV-1 (> 79%) but were modular in genome organization with respect to PBCV-1, showing little collinearity (Fitzgerald *et al.*, 2007b). Acanthocystis turfacea chlorella virus (ATCV-1), which was classified as a *Chlorovirus* member based on a 81% homologous CDSs with PBCV-1, has a smaller genome of 288,047 bp (Fitzgerald *et al.*, 2007c). All of these viruses contained a range of ORFs involved in DNA replication, recombination or repair, 6 to 11 tRNAs, and a high number of homing endonucleases (17 in ATCV-1), while no recognizable RNA polymerase could be identified.

The second genus *Coccolithovirus* is exemplified by Emiliania huxleyi virus 86 (EhV86), which has an isometric capsid of about 175 nm packaging a circular genome of 407,339 bp (Wilson *et al.*, 2005). At the time of annotation, only 66 out of 472 predicted CDSs had a functional prediction and these 66 CDSs comprised 25 NCLDV core genes. In contrast with the chloroviruses, EhV86 encodes six RNA polymerase subunits, all of which are expressed, suggesting that it replicates in the cytoplasm of its host.

Ectocarpus siliculosus virus (EsV-1), type species of the genus *Phaeovirus*, has a genome of 335,593 bp with inverted repeats of 1800 and 1560 bp at the genome ends (Delaroque *et al.*, 2001). At 231 predicted CDSs, the coding potential of this virus is low with 22% of the genome non-coding. This virus is temperate and encodes its own integrase for site-specific recombination. Two other phaeoviruses infecting *Feldmannia* species, FirrV-1 and FsV-58, have much smaller genomes (191,667 bp and 154,641 bp, respectively) and while they share gene synteny, extensive reshuffling of the genes has occurred between them and EsV-1 (Delaroque *et al.*, 2003; Schroeder *et al.*, 2009). Interesting to note is that these viruses all encode a gene related to a bacteriophage siphovirus primase/helicase, which is not similar to bacterial or eukaryotic primases (Delaroque *et al.*, 2003).

The genus *Prasinovirus* groups the viruses with the smallest genomes ranging from 184 to 196 kb, infecting marine prasinophytes. The type species is *Ostreococcus tauri virus 5* (OtV-5) has a linear genome of 186,234 bp long which is delineated by ITRs of 1695 bp and encodes 268 CDSs (Derelle *et al.*, 2008). Other prasinoviruses with similar features – such as a shared genome organization, collinearity with each other except at the genome ends, and the presence of a DNA polymerase but no RNA polymerase – were isolated from different hosts and include Micromonas pusilla virus 1 (MpV1), Ostreococcus tauri virus 1 (OtV-1), Ostreococcus lucimarinus virus (OlV1) and *Bathycoccus* viruses (BpV1, BpV2) (Moreau *et al.*, 2010; Weynberg *et al.*, 2009). Upon isolation and sequencing of a number of new *Ostreococcus lucimarinus* viruses (OlVs), two subgroups emerged which show clear gene synteny except for a central region of the genome where a 32 kb inversion has occurred (Derelle *et al.*, 2015). Phylogenetic analysis of the DNA polymerase gene clustered the OlVs based on subgroup, not geographical origin, while the *O. tauri* viruses all grouped with the same OlV subgroup, suggesting co-speciation with their host from a common ancestor (Derelle *et al.*, 2015).

The only fully sequenced and currently not officially recognized member of the genus *Prymnesiovirus* is Phaeocystis globosa virus (PgV-16T). Its large linear genome of 459,984 bp encodes 434 CDSs, 75 of which % showed homology to CDSs from Organic Lake Phycodnaviruses 1 and 2, the genomes of which were found in metaviromic data from an Antarctic lake (Santini *et al.*, 2013; Yau et al., 2011). Furthermore, 48 genes were shared between PgV and several mimiviruses, of which 18 are NCLDV core genes, indicating that the families *Phycodnaviridae* and *Mimiviridae* are more closely related to each other than to other NCLDV families. This was further substantiated by the phylogenetic analysis of the DNA polymerase gene showing that PgV-like viruses and mimiviruses cluster together and apart from other NCLDVs (Santini *et al.*, 2013).

The age of the 'megaviruses' started with the discovery and sequencing of Acanthamoeba polyphaga mimivirus (APMV), member of the family *Mimiviridae*, which has an icosahedral capsid of 400 nm in diameter and a genome of 1,181,549 bp, by far the largest at its discovery (Fig. 2.4) (Raoult *et al.*, 2004; La Scola *et al.*, 2003). Its 911 predicted coding sequences contained 42 of the NCLDV core genes but also many genes that had never been found in a viral genome before such as new DNA repair genes, glutamine metabolism genes, and lipid-manipulating enzymes (Raoult *et al.*, 2004). This virus also contains a specific early promoter motif, AAAATTGA, and experimentally validated palindromic terminators (Byrne *et al.*, 2009; Suhre *et al.*, 2005). Acanthamoeba castellanii mamavirus which shares 99% of predicted ORFs is considered to be a strain of the species *Acanthamoeba polyphaga mimivirus* and both are grouped in the genus *Mimivirus* (La Scola *et al.*, 2008). Upon isolation and sequencing of more mimiviruses, three distinct lineages have been found, the first one exemplified by APMV. This group (lineage A) is comprised of several mimivirus isolates and a group of Brazilian isolates (Samba virus, Amazonia virus, Oyster virus and Kroon virus), all of which showed >94% sequence similarity and >90% coverage with the original mimivirus with paralogous

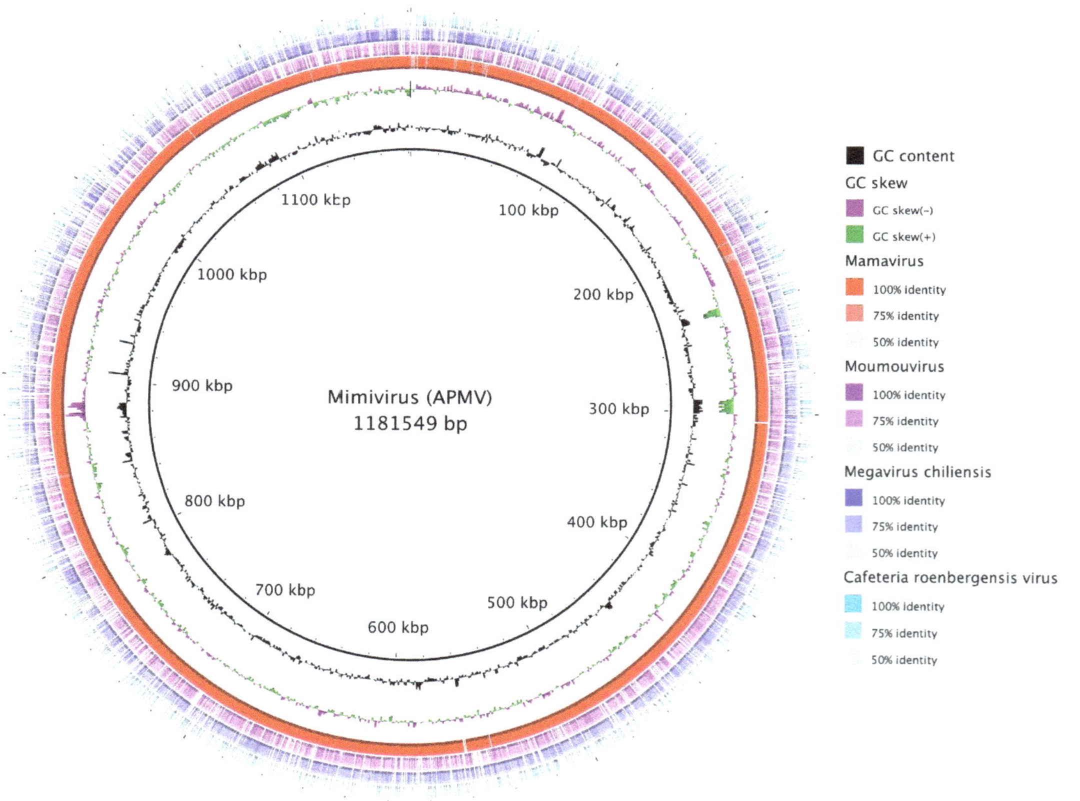

Figure 2.4 Genome representation of the family *Mimiviridae*. BLASTn-based comparison of the genome of APMV with representatives of mimivirus lineages A (Mamavirus), B (Moumouvirus), C (Megavirus chiliensis) and the Cafeteriavirus roenbergensis virus. The image was generated using BRIG software (Alikhan *et al.*, 2011).

genes predominantly distributed towards extremities of the genomes (Assis *et al.*, 2015; Campos *et al.*, 2014). Inclusion of these viruses increased the pan genome of lineage A to 1129 clusters and gave a core genome of 597–644 genes (Assis *et al.*, 2015). The type isolate of lineage B is Acanthamoeba polyphaga moumouvirus with a 1,021,348 bp linear dsDNA genome encoding 930 CDSs sharing the same early promoter and hairpin structures as mimivirus (Yoosuf *et al.*, 2012). The third lineage is exemplified by Megavirus chilensis which also infects *A. castellanii*. It has a genome size of 1,259,197 bp with 1120 putative CDSs (Arslan *et al.*, 2011). The three viruses, mimivirus, moumouvirus and megavirus, are collinear over the central 650 kb of the genome with gene shuffling and/or gene loss occurring at the termini (Arslan *et al.*, 2011; Yoosuf *et al.*, 2012).

The second genus within the family *Mimiviridae* is *Cafeteriavirus* with *Cafeteria roenbergensis virus* (CroV) as type species. CroV is much smaller than the mimiviruses with a genome of approximately 730 kb of which the central 617,453 bp were sequenced containing 544 putative CDSs (Fischer *et al.*, 2010). It contains highly repetitive regions at its termini which could serve as protective caps with similar functions as telomeres. The extent of the transcription machinery encoded on the genome suggests that transcription is host-independent and occurs in the cytoplasm. Similar to the mimiviruses, an early promoter was found in front of 31% of the genes with additionally a conserved tetramer found in front of genes expressed in the late stage of infection (Fischer *et al.*, 2010).

Members of the family *Marseilleviridae* are generally much smaller than mimiviruses, both in genome and capsid size. The type species, *Acanthamoeba castellanii marseillevirus*, has an icosahedral capsid of about 250 nm diameter and a circular dsDNA genome of 368,454 bp (Boyer *et al.*, 2009). Of the 457 predicted CDSs, 28 genes were recognized as NCLDV core genes with another 17 genes shared between mimivirus and marseillevirus alone. The genome is unusually rich in serine/threonine kinases and contains a large set of ubiquitin system proteins. The *A. castellanii* lausannevirus is slightly smaller in genome size at 346,754 bp, encoding 450 CDSs, of which 320 are shared with marseillevirus (Thomas *et al.*, 2011). Based on phylogenetic analysis of three core genes, three lineages were found in the *Marseilleviridae*. The first lineage corresponds with the genus *Marseillevirus* and comprises marseillevirus, senegalvirus, cannes 8 virus, Fontaine Saint-Charles virus and giant blood marseillevirus, the second lineage is represented by lausannevirus and the third by tunisvirus, which has a genome size of 380,011 bp and insectomimevirus (Aherfi *et al.*, 2014). The pan genome of the family *Marseilleviridae* comprises 608 protein clusters at this time and the core genome contains 233 genes (Aherfi *et al.*, 2014).

Other potential megaviruses

The pandoraviruses are the latest iteration of the ever-increasing genome size of viruses. At a staggering 2,473,870 bp, the genome of *Pandoravirus salinus* infecting *Acanthamoeba* is the largest virus genome to date (Philippe *et al.*, 2013). To accomplish its sequence determination, a combination of Illumina, 454, and PacBio sequencing was used and even then unresolved terminal repeats at the 3′ end of the contig were present leading to a total minimum genome size of 2.7 Mb. Of the 2556 predicted proteins, only 16% gave database matches, mostly to uninformative repeat-domain proteins, indicating that no relative of *P. salinus* had been previously sequenced (Philippe *et al.*, 2013). Similar to other NCLDVs, a large fraction of the identified genes encode enzymes involved in DNA processing, replication, repair, and nucleotide synthesis. Several components of the replication machinery were missing, however, suggesting that *P. salinus* replicates in the host nucleus. A second, smaller, pandoravirus, *P. dulcis*, was isolated in the same study, carrying a genome of 1,908,524 bp and showing a high degree of collinearity except for four large insertions in *P. salinus* (Philippe *et al.*, 2013). Phylogenetic clustering of the *polB* and DNA dependent RNA polymerase genes of these viruses with a third pandoravirus, *P. inopinatum* (2,246,109 bp genome), places these three viruses together on a separate branch in the NCLDV tree (Antwerpen *et al.*, 2015; Philippe *et al.*, 2013).

Recently, two amoebal viruses, infecting *A. castellani*, were resurrected from 30,000-year-old permafrost in Siberia. *Pithovirus sibericum* resembles pandoraviruses outwardly (1.5 μm by 500 nm enveloped virus with cork-like hexagonal structure at one end), but its 610,033 bp genome is more similar to the icosahedral *Marseilleviridae* and

Mimiviridae (Legendre *et al.*, 2014). Of the 467 predicted CDSs, 32.5% showed matches to the databases, with the hits equally distributed between viruses, bacteria and eukaryotes. With a maximum of 19 shared genes with marseilleviruses, *Pithovirus* is proposed as the first member of the new family 'Pithoviridae' (Abergel *et al.*, 2015). The other recovered virus, *Mollivirus sibericum*, has a spherical particle (500–600 nm diameter) and a genome of 651,523 bp (Legendre *et al.*, 2015). While it is much smaller in genome size, it was found most similar to pandoraviruses with the highest number of homologous CDSs albeit at low sequence similarity and a weak phylogenetic clustering.

Higher order genomic relationships

The rise in numbers of genomic sequences in many different families of viruses has offered the opportunity to investigate higher order evolutionary relationships between viruses of microorganisms infecting different domains of life.

Looking at the RNA viruses (found to infect bacteria, fungi, algae and protozoa, but not archaea), the diversity and abundance of eukaryote-infecting RNA viruses is far greater than that of DNA viruses, leading to the hypothesis that the emergence of a nucleus acted as a barrier for DNA virus replication (Koonin *et al.*, 2015). Koonin and colleagues identified three superfamily lineages of RNA viruses of which the picornavirus-like superfamily ancestor, including the order *Picornavirales* but also several of the dsRNA families infecting viruses of microorganisms, likely evolved at the same time or shortly after the emergence of eukaryotes and then rapidly diversified, spawning the ancestors of the other lineages (flavivirus-like superfamily and alphavirus-like superfamily). At the same time, RdRp similarity analysis points to the (+)ssRNA family *Leviviridae* as the ancestral lineage for the virion-less (+)ssRNA family *Narnaviridae* with the most likely explanation that the ancestral narnavirus evolved from an RNA bacteriophage being co-introduced into fungi along with the protomitochondrion and subsequently losing its capsid (Koonin *et al.*, 2015). Meanwhile, the dsRNA segmented reoviruses seem to be direct descendants of ancestral dsRNA cystovirus bacteriophages, retaining multiple structural genes (Koonin *et al.*, 2015). Not only evolutionary relationships have emerged from genomics. Recently, based on whole genome analysis and phylogenetics followed by transfection experiments, a (+)ssRNA virus distantly related to caliciviruses (YkV1) has been found to hijack the capsid of a totivirus relative (YnV1) to encapsidate its own RdRp and propagate (Zhang *et al.*, 2016). Metagenomic analysis has even provided evidence for a virus with a circular DNA genome encoding a rolling-circle replication initiation protein (RC-Rep) having acquired an RNA virus capsid protein (Diemer and Stedman, 2012). The discovery of totivirus sequences in the genomes of several fungi also suggests that these viruses can play a role in horizontal gene transfer in eukaryotes without actually encoding a reverse transcriptase (Taylor and Bruenn, 2009).

For the ssDNA viruses – currently found to infect bacteria, archaea, fungi and algae, but not protozoa – a clear link has been noted between these viruses and rolling-circle plasmids of bacteria and archaea (Krupovic, 2013). The polyphyletic origin of the rolling circle replication genes leads to two hypotheses. The first possibility is that the plasmids are derived from viruses which have lost their capsid. The alternative hypothesis is that viruses evolved from plasmids by the acquisition of capsid genes (Krupovic, 2013). The situation becomes more complicated when dealing with the pleolipoviruses which can have ssDNA or dsDNA genomes and it is thought that these viruses have a flexible replicative intermediate (Krupovic, 2013; Pietilä *et al.*, 2012, 2016). The discovery of the ssDNA genome of CTXΦ inserted in the genome of *Vibrio cholerae* confirms the genomic interactions between ssDNA and dsDNA organisms (Stedman, 2015).

When it comes to dsDNA viruses – infecting bacteria, archaea, algae and protozoa, but not fungi – two main structural lineages have been discovered based on a protein fold of the capsid protein (Krupovic and Bamford, 2011). The double jelly-roll fold capsid protein was detected in dsDNA virus families infecting archaea (*Turriviridae*), bacteria (*Tectiviridae, Corticoviridae*) and eukaryotes (*Asfarviridae, Iridoviridae, Mimiviridae, Phycodnaviridae, Poxviridae, Adenoviridae*) and a common ancestry dating to before the divergence in domains is the most likely explanation for this shared fold, as opposed to horizontal gene transfer which is less probable given the substantial differences in hosts belonging to different domains of life (Krupovic and Bamford, 2011; Prangishvili, 2015). The same hypothesis remains for the second lineage of viruses which contain the HK97-like capsid proteins and comprises the members of the orders *Caudovirales*, infecting bacteria and archaea, and *Herpesvirales* (infecting animals). These findings have led to the hypothesis that dsDNA viruses have arisen from three separate bacterial root lineages; two groups of bacteriophages, the tectiviruses and tailed phages gave rise to the majority of dsDNA eukaryotic viruses, while a small third lineage originated in plasmids and led to two families of small dsDNA viruses with ssDNA evolutionary intermediates (Koonin *et al.*, 2015).

The double jelly-roll lineage of the so-called PRD1–adenoviral or PRD1-like lineage has received much attention recently with the expansion of the proposed order 'Megavirales' (Forterre *et al.*, 2014). Comparative analysis of different families within this group has indicated that there is no clear correlation between genome complexity (or size) and the ability to replicate independently from the host nucleus, e.g. the pandoraviruses with the largest genomes to date do not encode a transcription apparatus, while the mimiviruses and pithoviruses do and are therefore strictly cytoplasmic (Abergel *et al.*, 2015). A question that these giant viruses posed was which model for genomic evolution they followed, 'genome degradation' or 'genome expansion'? Whole genome comparisons have shown that a third model is more likely, that of an accordion-like evolution, that is, with both gene gains and losses (Filée, 2015). Horizontal gene transfer was found to be a minor factor for evolution in certain viruses while for the phycodnaviruses it was a significant driving force suggesting that these viruses can act as shuttles for inter-kingdom gene exchange (Filée, 2015).

Take home message

While the genomic diversity of viruses of microorganisms as presented here is already high, expansion is surely expected and needed. Many of the viral taxa discussed in this chapter contain only a few members and even in the well-populated taxa, new species and genera are discovered continually. Only with a much larger number of genome sequences available will we be able to answer the question of whether the virosphere contains a continuum of genomes spanning all nucleic acid types or if evolution favours instead of the discrete groups we witness now.

The inclusion of different hosts in isolation protocols will most likely fill up a part of the unknown viral sequence space or viral sequence without database homologues, also called viral dark matter (Hatfull, 2015; Sullivan, 2015). At the time of writing, only a small subset of all bacterial phyla have been sampled and for the archaeal viruses focus has been on a very limited number of hosts (see Chapter 8). Similarly, the majority of discoveries of novel giant viruses have been made on *Acanthamoeba* species, just one genus of protozoa (see Chapter 11). Part of the expansion of the virosphere will most likely happen through metaviromics (see Chapter 5). I would like to encourage scientists to keep isolating and sequencing viruses, however.

Web resources

DNA Master; http://phagesdb.org/DNAMaster/; accessed 25 January 2017

ICTV Taxonomy: 2015 Release; EC47, London, 15 July 2015; email ratification 2016 (Master Species List #30); www.ictvonline.org/virusTaxonomy.asp; accessed 25 January 2017

Molecular Biology Tools; http://molbiol-tools.ca/; accessed 25 January 2017

ORFfinder; www.ncbi.nlm.nih.gov/orffinder/; accessed 27 February 2017

PhAnToMe; www.phantome.org/; accessed 25 January 2017

PHAST; www.phantome.org/PhageSeed/Phage.cgi?page=phast; accessed 25 January 2017

References

Abergel, C., Legendre, M., and Claverie, J.M. (2015). The rapidly expanding universe of giant viruses: Mimivirus, Pandoravirus, Pithovirus and Mollivirus. FEMS Microbiol. Rev. 39, 779–796. https://doi.org/10.1093/femsre/fuv037.

Ackermann, H.W. (2001). Frequency of morphological phage descriptions in the year 2000. Brief review. Arch. Virol. *146*, 843–857.

Adriaenssens, E.M., Ceyssens, P.J., Dunon, V., Ackermann, H.W., Van Vaerenbergh, J., Maes, M., De Proft, M., and Lavigne, R. (2011). Bacteriophages LIMElight and LIMEzero of *Pantoea agglomerans*, belonging to the 'phiKMV-like viruses'. Appl. Environ. Microbiol. 77, 3443–3450. https://doi.org/10.1128/AEM.00128-11.

Adriaenssens, E.M., Lehman, S.M., Vandersteegen, K., Vandenheuvel, D., Philippe, D.L., Cornelissen, A., Clokie, M.R., García, A.J., De Proft, M., Maes, M., *et al.* (2012). CIM(®) monolithic anion-exchange chromatography as a useful alternative to CsCl gradient purification of bacteriophage particles. Virology *434*, 265–270. https://doi.org/10.1016/j.virol.2012.09.018.

Adriaenssens, E.M., Edwards, R., Nash, J.H., Mahadevan, P., Seto, D., Ackermann, H.W., Lavigne, R., and Kropinski, A.M. (2015). Integration of genomic and proteomic analyses in the classification of the Siphoviridae family. Virology *477*, 144–154. https://doi.org/10.1016/j.virol.2014.10.016.

Adriaenssens, E.M., Krupovic, M., Knezevic, P., Ackermann, H.-W., Barylski, J., Brister, J.R., Clokie, M.R.C., Duffy, S., Dutilh, B.E., Edwards, R.A., *et al.* (2017). Taxonomy of prokaryotic viruses: 2016 update from the ICTV bacterial and archaeal viruses subcommittee. Arch. Virol. *162*, 1153–1157 https://doi.org/10.1007/s00705-016-3173-4.

Afonso, C.L., Amarasinghe, G.K., Bányai, K., Bào, Y., Basler, C.F., Bavari, S., Bejerman, N., Blasdell, K.R., Briand, F.X., Briese, T., *et al.* (2016). Taxonomy of the order *Mononegavirales*: update 2016. Arch. Virol. *161*, 2351–2360. https://doi.org/10.1007/s00705-016-2880-1.

Aherfi, S., Boughalmi, M., Pagnier, I., Fournous, G., La Scola, B., Raoult, D., and Colson, P. (2014). Complete genome sequence of Tunisvirus, a new member of the proposed family *Marseilleviridae*. Arch. Virol. *159*, 2349–2358. https://doi.org/10.1007/s00705-014-2023-5.

Ahn, D.G., Kim, S.I., Rhee, J.K., Kim, K.P., Pan, J.G., and Oh, J.W. (2006). TTSV1, a new virus-like particle isolated from the hyperthermophilic crenarchaeote *Thermoproteus tenax*. Virology *351*, 280–290.

Akopyants, N.S., Lye, L.F., Dobson, D.E., Lukeš, J., and Beverley, S.M. (2016). A narnavirus in the trypanosomatid protist plant pathogen *Phytomonas serpens*. Genome Announc. *4*, e00711–16. https://doi.org/10.1128/genomeA.00711-16.

Alikhan, N.F., Petty, N.K., Ben Zakour, N.L., and Beatson, S.A. (2011). BLAST Ring Image Generator (BRIG): simple prokaryote genome comparisons. BMC Genomics *12*, 402. https://doi.org/10.1186/1471-2164-12-402.

Altschul, S.F., Gish, W., Miller, W., Myers, E.W., and Lipman, D.J. (1990). Basic local alignment search tool. J. Mol. Biol. *215*, 403–410. https://doi.org/10.1016/S0022-2836(05)80360-2.

Altschul, S.F., Madden, T.L., Schäffer, A.A., Zhang, J., Zhang, Z., Miller, W., and Lipman, D.J. (1997). Gapped BLAST and PSI-BLAST: a new generation of protein database search programs. Nucleic Acids Res. *25*, 3389–3402.

Antwerpen, M.H., Georgi, E., Zoeller, L., Woelfel, R., Stoecker, K., and Scheid, P. (2015). Whole-genome sequencing of a *Pandoravirus* isolated from keratitis-inducing *Acanthamoeba*. Genome Announc. *3*, e00136–15. https://doi.org/10.1128/genomeA.00136-15.

Arnold, H.P., Zillig, W., Ziese, U., Holz, I., Crosby, M., Utterback, T., Weidmann, J.F., Kristjanson, J.K., Klenk, H.P., Nelson, K.E., *et al.* (2000). A novel lipothrixvirus, SIFV, of the extremely thermophilic crenarchaeon *Sulfolobus*. Virology *267*, 252–266. https://doi.org/10.1006/viro.1999.0105.

Arslan, D., Legendre, M., Seltzer, V., Abergel, C., and Claverie, J.M. (2011). Distant Mimivirus relative with a larger genome highlights the fundamental features of Megaviridae. Proc. Natl. Acad. Sci. U.S.A. *108*, 17486–17491. https://doi.org/10.1073/pnas.1110889108.

Assis, F.L., Bajrai, L., Abrahao, J.S., Kroon, E.G., Dornas, F.P., Andrade, K.R., Boratto, P.V., Pilotto, M.R., Robert, C., Benamar, S., *et al.* (2015). Pan-Genome Analysis of Brazilian Lineage A Amoebal Mimiviruses. Viruses *7*, 3483–3499. https://doi.org/10.3390/v7072782.

Atkins, J.F., Steitz, J.A., Anderson, C.W., and Model, P. (1979). Binding of mammalian ribosomes to MS2 phage RNA reveals an overlapping gene encoding a lysis function. Cell *18*, 247–256.

Attoui, H., Jaafar, F.M., Belhouchet, M., de Micco, P., de Lamballerie, X., and Brussaard, C.P.D. (2006). *Micromonas pusilla* reovirus: A new member of the family *Reoviridae* assigned to a novel proposed genus (*Mimoreovirus*). J. Gen. Virol. *87*, 1375–1383. https://doi.org/10.1099/vir.0.81584-0.

Aziz, R.K., Bartels, D., Best, A.A., DeJongh, M., Disz, T., Edwards, R.A., Formsma, K., Gerdes, S., Glass, E.M., Kubal, M., *et al.* (2008). The RAST Server: rapid annotations using subsystems technology. BMC Genomics *9*, 75. https://doi.org/10.1186/1471-2164-9-75.

Aziz, R.K., Devoid, S., Disz, T., Edwards, R.A., Henry, C.S., Olsen, G.J., Olson, R., Overbeek, R., Parrello, B., Pusch, G.D., *et al.* (2012). SEED servers: high-performance access to the SEED genomes, annotations, and metabolic models. PLOS ONE *7*, e48053. https://doi.org/10.1371/journal.pone.0048053.

Baeza, M., Bravo, N., Sanhueza, M., Flores, O., Villarreal, P., and Cifuentes, V. (2012). Molecular characterization of totiviruses in *Xanthophyllomyces* dendrorhous. Virol. J. *9*, 140. https://doi.org/10.1186/1743-422X-9-140.

Bailey, T.L., Boden, M., Buske, F.A., Frith, M., Grant, C.E., Clementi, L., Ren, J., Li, W.W., and Noble, W.S. (2009). MEME SUITE: tools for motif discovery and searching. Nucleic Acids Res. *37*, W202–208. https://doi.org/10.1093/nar/gkp416.

Baltimore, D. (1971). Expression of animal virus genomes. Bacteriol. Rev. *35*, 235–241.

Bamford, D.H., Ravantti, J.J., Rönnholm, G., Laurinavicius, S., Kukkaro, P., Dyall-Smith, M., Somerharju, P., Kalkkinen, N., and Bamford, J.K. (2005). Constituents of SH1, a novel lipid-containing virus infecting the halophilic euryarchaeon *Haloarcula hispanica*. J. Virol. *79*, 9097–9107.

Bamford, J.K., Hänninen, A.L., Pakula, T.M., Ojala, P.M., Kalkkinen, N., Frilander, M., and Bamford, D.H. (1991). Genome organization of membrane-containing bacteriophage PRD1. Virology *183*, 658–676.

Bath, C., and Dyall-Smith, M.L. (1998). His1, an archaeal virus of the *Fuselloviridae* family that infects *Haloarcula hispanica*. J. Virol. *72*, 9392–9395.

Bath, C., Cukalac, T., Porter, K., and Dyall-Smith, M.L. (2006). His1 and His2 are distantly related, spindle-shaped haloviruses belonging to the novel virus group, Salterprovirus. Virology *350*, 228–239.

Beck, E., Sommer, R., Auerswald, E.A., Kurz, C., Zink, B., Osterburg, G., Schaller, H., Sugimoto, K., Sugisaki, H., Okamoto, T., *et al.* (1978). Nucleotide sequence of bacteriophage fd DNA. Nucleic Acids Res. *5*, 4495–4503.

Beekwilder, J., Nieuwenhuizen, R., Poot, R., and van Duin, J. (1996). Secondary structure model for the first three domains of Q beta RNA. Control of A-protein synthesis. J. Mol. Biol. *256*, 8–19.

Beekwilder, M.J., Nieuwenhuizen, R., and van Duin, J. (1995). Secondary structure model for the last two domains of single-stranded RNA phage Q beta. J. Mol. Biol. *247*, 903–917.

Besemer, J., and Borodovsky, M. (2005). GeneMark: web software for gene finding in prokaryotes, eukaryotes and viruses. Nucleic Acids Res. *33*, W451–454. https://doi.org/10.1093/nar/gki487.

Bessarab, I.N., Liu, H.W., Ip, C.F., and Tai, J.H. (2000). The complete cDNA sequence of a type II *Trichomonas vaginalis* virus. Virology *267*, 350–359. https://doi.org/10.1006/viro.1999.0129.

Bessarab, I.N., Nakajima, R., Liu, H.W., and Tai, J.H. (2011). Identification and characterization of a type III *Trichomonas vaginalis* virus in the protozoan pathogen *Trichomonas vaginalis*. Arch. Virol. *156*, 285–294. https://doi.org/10.1007/s00705-010-0858-y.

Bettstetter, M., Peng, X., Garrett, R.A., and Prangishvili, D. (2003). AFV1, a novel virus infecting hyperthermophilic archaea of the genus *Acidianus*. Virology *315*, 68–79.

Boyer, M., Yutin, N., Pagnier, I., Barrassi, L., Fournous, G., Espinosa, L., Robert, C., Azza, S., Sun, S., Rossmann, M.G., *et al.* (2009). Giant Marseillevirus highlights the role of amoebae as a melting pot in emergence of chimeric microorganisms. Proc. Natl. Acad. Sci. U.S.A. *106*, 21848–21853. https://doi.org/10.1073/pnas.0911354106.

Brentlinger, K.L., Hafenstein, S., Novak, C.R., Fane, B.A., Borgon, R., McKenna, R., and Agbandje-McKenna, M. (2002). *Microviridae*, a family divided: isolation, characterization, and genome sequence of phiMH2K, a bacteriophage of the obligate intracellular parasitic bacterium *Bdellovibrio bacteriovorus*. J. Bacteriol. *184*, 1089–1094.

Bruenn, J.A. (1993). A closely related group of RNA-dependent RNA polymerases from double-stranded RNA viruses. Nucleic Acids Res. *21*, 5667–5669.

Brüssow, H., and Hendrix, R.W. (2002). Phage genomics: small is beautiful. Cell *108*, 13–16.

Bryson, S.J., Thurber, A.R., Correa, A.M., Orphan, V.J., and Vega Thurber, R. (2015). A novel sister clade to the enterobacteria microviruses (family *Microviridae*) identified in methane seep sediments. Environ. Microbiol. *17*, 3708-3721. https://doi.org/10.1111/1462-2920.12758.

Buermans, H.P., and den Dunnen, J.T. (2014). Next generation sequencing technology: Advances and applications. Biochim. Biophys. Acta *1842*, 1932–1941.

Butcher, S.J., Grimes, J.M., Makeyev, E.V., Bamford, D.H., and Stuart, D.I. (2001). A mechanism for initiating RNA-dependent RNA polymerization. Nature *410*, 235–240. https://doi.org/10.1038/35065653.

Byrne, D., Grzela, R., Lartigue, A., Audic, S., Chenivesse, S., Encinas, S., Claverie, J.M., and Abergel, C. (2009). The polyadenylation site of Mimivirus transcripts obeys a stringent 'hairpin rule'. Genome Res. *19*, 1233–1242. https://doi.org/10.1101/gr.091561.109.

Campos, J., Martínez, E., Suzarte, E., Rodríguez, B.L., Marrero, K., Silva, Y., Ledón, T., del Sol, R., and Fando, R. (2003). VGJ phi, a novel filamentous phage of *Vibrio cholerae*, integrates into the same chromosomal site as CTX phi. J. Bacteriol. *185*, 5685–5696.

Campos, R.K., Boratto, P.V., Assis, F.L., Aguiar, E.R., Silva, L.C., Albarnaz, J.D., Dornas, F.P., Trindade, G.S., Ferreira, P.P., Marques, J.T., *et al.* (2014). Samba virus: a novel mimivirus from a giant rain forest, the Brazilian Amazon. Virol. J. *11*, 95. https://doi.org/10.1186/1743-422X-11-95.

Carver, T., Berriman, M., Tivey, A., Patel, C., Böhme, U., Barrell, B.G., Parkhill, J., and Rajandream, M.A. (2008). Artemis and ACT: viewing, annotating and comparing sequences stored in a relational database. Bioinformatics *24*, 2672–2676. https://doi.org/10.1093/bioinformatics/btn529.

Ceyssens, P.J., and Lavigne, R. (2010). Bacteriophages of *Pseudomonas*. Future Microbiol. *5*, 1041–1055. https://doi.org/10.2217/fmb.10.66.

Ceyssens, P.J., Lavigne, R., Mattheus, W., Chibeu, A., Hertveldt, K., Mast, J., Robben, J., and Volckaert, G. (2006). Genomic analysis of *Pseudomonas aeruginosa* phages LKD16 and LKA1: establishment of the phiKMV subgroup within the T7 supergroup. J. Bacteriol. *188*, 6924–6931.

Ceyssens, P.J., Glonti, T., Kropinski, N.M., Lavigne, R., Chanishvili, N., Kulakov, L., Lashkhi, N., Tediashvili, M., and Merabishvili, M. (2011). Phenotypic and genotypic variations within a single bacteriophage species. Virol. J. *8*, 134. https://doi.org/10.1186/1743-422X-8-134.

Chang, B., Taniguchi, H., Miyamoto, H., and Yoshida, S.I. (1998). Filamentous bacteriophages of *Vibrio parahaemolyticus* as a possible clue to genetic transmission. J. Bacteriol. *180*, 5094–5101.

Chen, B., and Nuss, D.L. (1999). Infectious cDNA clone of hypovirus CHV1-Euro7: a comparative virology approach to investigate virus-mediated hypovirulence of the chestnut blight fungus *Cryphonectria parasitica*. J. Virol. *73*, 985–992.

Chénard, C., Chan, A.M., Vincent, W.F., and Suttle, C.A. (2015). Polar freshwater cyanophage S-EIV1 represents a new widespread evolutionary lineage of phages. ISME J. *9*, 2046–2058. https://doi.org/10.1038/ismej.2015.24.

Cheng, J., Jiang, D., Fu, Y., Li, G., Peng, Y., and Ghabrial, S.A. (2003). Molecular characterization of a dsRNA totivirus infecting the sclerotial parasite *Coniothyrium minitans*. Virus Res. *93*, 41–50.

Chevreux, B. (2005). MIRA: An automated genome and EST assembler. The Ruprecht-Karls University.

Chiba, S., Lin, Y.H., Kondo, H., Kanematsu, S., and Suzuki, N. (2013a). A novel victorivirus from a phytopathogenic fungus, *Rosellinia necatrix*, is infectious as particles and targeted by RNA silencing. J. Virol. *87*, 6727–6738. https://doi.org/10.1128/JVI.00557-13.

Chiba, S., Lin, Y.H., Kondo, H., Kanematsu, S., and Suzuki, N. (2013b). Effects of defective interfering RNA on symptom induction by, and replication of, a novel partitivirus from a phytopathogenic fungus, *Rosellinia necatrix*. J. Virol. *87*, 2330–2341. https://doi.org/10.1128/JVI.02835-12.

Chopin, M.C., Rouault, A., Ehrlich, S.D., and Gautier, M. (2002). Filamentous phage active on the gram-positive bacterium *Propionibacterium freudenreichii*. J. Bacteriol. *184*, 2030–2033.

Clasen, J.L., Hanson, C.A., Ibrahim, Y., Weihe, C., Marston, M.F., and Martiny, J.B.H. (2013). Diversity and temporal dynamics of Southern California coastal marine cyanophage isolates. Aquat. Microb. Ecol. *69*, 17–31. https://doi.org/10.3354/ame01613.

Colson, P., De Lamballerie, X., Yutin, N., Asgari, S., Bigot, Y., Bideshi, D.K., Cheng, X.W., Federici, B.A., Van Etten, J.L., Koonin, E.V., *et al.* (2013). 'Megavirales', a proposed new order for eukaryotic nucleocytoplasmic large DNA viruses. Arch. Virol. *158*, 2517–2521. https://doi.org/10.1007/s00705-013-1768-6.

Compel, P., Papp, I., Bibó, M., Fekete, C., and Hornok, L. (1999). Genetic interrelationships and genome organization of double-stranded RNA elements of *Fusarium poae*. Virus Genes *18*, 49–56.

Coutts, R.H., Covelli, L., Di Serio, F., Citir, A., Açikgöz, S., Hernández, C., Ragozzino, A., and Flores, R. (2004). Cherry chlorotic rusty spot and Amasya cherry diseases are associated with a complex pattern of mycoviral-like double-stranded RNAs. II. Characterization of a new species in the genus *Partitivirus*. J. Gen. Virol. *85*, 3399–3403. https://doi.org/10.1099/vir.0.80182-0.

Covelli, L., Coutts, R.H., Di Serio, F., Citir, A., Açikgöz, S., Hernández, C., Ragozzino, A., and Flores, R. (2004). Cherry chlorotic rusty spot and Amasya cherry diseases are associated with a complex pattern of mycoviral-like double-stranded RNAs. I. Characterization of a new species in the genus *Chrysovirus*. J. Gen. Virol. *85*, 3389–3397. https:// doi.org/10.1099/vir.0.80182-0.

D'Souza, R., and Azevedo, A. (2013). Emerging technologies for the integration and intensification of downstream bioprocesses. Pharm. Bioprocesses *1*, 423–440. https://doi.org/10.4155/pbp.13.55.

Darissa, O., Willingmann, P., Schäfer, W., and Adam, G. (2011). A novel double-stranded RNA mycovirus from *Fusarium graminearum*: nucleic acid sequence and genomic structure. Arch. Virol. *156*, 647–658. https://doi.org/10.1007/s00705-010-0904-9.

Davis, B.M., and Waldor, M.K. (2003). Filamentous phages linked to virulence of *Vibrio cholerae*. Curr. Opin. Microbiol. *6*, 35–42.

Delaroque, N., Müller, D.G., Bothe, G., Pohl, T., Knippers, R., and Boland, W. (2001). The complete DNA sequence of the *Ectocarpus siliculosus* Virus EsV-1 genome. Virology *287*, 112–132. https://doi.org/10.1006/viro.2001.1028.

Delaroque, N., Boland, W., Müller, D.G., and Knippers, R. (2003). Comparisons of two large phaeoviral genomes and evolutionary implications. J. Mol. Evol. *57*, 613–622. https://doi.org/10.1007/s00239-003-2501-y.

Delcher, A.L., Harmon, D., Kasif, S., White, O., and Salzberg, S.L. (1999). Improved microbial gene identification with GLIMMER. Nucleic Acids Res. *27*, 4636–4641.

Deng, F., and Boland, G. (2007). Natural occurrence of a partitivirus in the sapstaining fungus *Ceratocystis resinifera*. Can. J. Plant Pathol. *29*, 182–189. https://doi.org/10.1080/07060660709507455.

Deng, F., Xu, R., and Boland, G.J. (2003). Hypovirulence-Associated Double-Stranded RNA from *Sclerotinia homoeocarpa* Is Conspecific with *Ophiostoma novo-ulmi* Mitovirus 3a-Ld. Phytopathology *93*, 1407–1414. https://doi.org/10.1094/PHYTO.2003.93.11.1407.

Derelle, E., Ferraz, C., Escande, M.L., Eychenié, S., Cooke, R., Piganeau, G., Desdevises, Y., Bellec, L., Moreau, H., and Grimsley, N. (2008). Life-cycle and genome of OtV5, a large DNA virus of the pelagic marine unicellular green alga *Ostreococcus tauri*. PLOS ONE *3*, e2250. https://doi.org/10.1371/journal.pone.0002250.

Derelle, E., Monier, A., Cooke, R., Worden, A.Z., Grimsley, N.H., and Moreau, H. (2015). Diversity of Viruses Infecting the Green Microalga *Ostreococcus lucimarinus*. J. Virol. *89*, 5812–5821. https://doi.org/10.1128/JVI.00246-15.

Diemer, G.S., and Stedman, K.M. (2012). A novel virus genome discovered in an extreme environment suggests recombination between unrelated groups of RNA and DNA viruses. Biol. Direct *7*, 13. https://doi.org/10.1186/1745-6150-7-13.

Doherty, M., Coutts, R.H., Brasier, C.M., and Buck, K.W. (2006). Sequence of RNA-dependent RNA polymerase genes provides evidence for three more distinct mitoviruses in *Ophiostoma novo-ulmi* isolate Ld. Virus Genes *33*, 41–44. https://doi.org/10.1007/s11262-005-0029-5.

Ehara, M., Shimodori, S., Kojima, F., Ichinose, Y., Hirayama, T., Albert, M.J., Supawat, K., Honma, Y., Iwanaga, M., and Amako, K. (1997). Characterization of filamentous phages of *Vibrio cholerae* O139 and O1. FEMS Microbiol. Lett. *154*, 293–301.

Erdmann, S., Chen, B., Huang, X., Deng, L., Liu, C., Shah, S.A., Le Moine Bauer, S., Sobrino, C.L., Wang, H., Wei, Y., *et al.* (2014). A novel single-tailed fusiform *Sulfolobus* virus STSV2 infecting model *Sulfolobus* species. Extremophiles *18*, 51–60. https://doi.org/10.1007/s00792-013-0591-z.

Esteban, L.M., Rodriguez-Cousiño, N., and Esteban, R. (1992). T double-stranded RNA (dsRNA) sequence reveals that T and W dsRNAs form a new RNA family in *Saccharomyces cerevisiae*. Identification of 23 S RNA as the single-stranded form of T dsRNA. J. Biol. Chem. *267*, 10874–10881.

Faruque, S.M., Bin Naser, I., Fujihara, K., Diraphat, P., Chowdhury, N., Kamruzzaman, M., Qadri, F., Yamasaki, S., Ghosh, A.N., and Mekalanos, J.J. (2005). Genomic sequence and receptor for the *Vibrio cholerae* phage KSF-1phi: evolutionary divergence among filamentous vibriophages mediating lateral gene transfer. J. Bacteriol. *187*, 4095–4103.

Filée, J. (2015). Genomic comparison of closely related Giant Viruses supports an accordion-like model of evolution. Front. Microbiol. *6*, 593. https://doi.org/10.3389/fmicb.2015.00593.

Fischer, M.G., Allen, M.J., Wilson, W.H., and Suttle, C.A. (2010). Giant virus with a remarkable complement of genes infects marine zooplankton. Proc. Natl. Acad. Sci. U.S.A. *107*, 19508–19513. https://doi.org/10.1073/pnas.1007615107.

Fitzgerald, L.A., Graves, M.V., Li, X., Feldblyum, T., Nierman, W.C., and Van Etten, J.L. (2007a). Sequence and annotation of the 369-kb NY-2A and the 345-kb AR158 viruses that infect *Chlorella* NC64A. Virology *358*, 472–484.

Fitzgerald, L.A., Graves, M.V., Li, X., Feldblyum, T., Hartigan, J., and Van Etten, J.L. (2007b). Sequence and annotation of the 314-kb MT325 and the 321-kb FR483 viruses that infect *Chlorella* Pbi. Virology *358*, 459–471.

Fitzgerald, L.A., Graves, M.V., Li, X., Hartigan, J., Pfitzner, A.J., Hoffart, E., and Van Etten, J.L. (2007c). Sequence and annotation of the 288-kb ATCV-1 virus that infects an endosymbiotic chlorella strain of the heliozoon *Acanthocystis turfacea*. Virology *362*, 350–361.

Forterre, P., Krupovic, M., and Prangishvili, D. (2014). Cellular domains and viral lineages. Trends Microbiol. *22*, 554–558. https://doi.org/10.1016/j.tim.2014.07.004.

Gautheret, D., and Lambert, A. (2001). Direct RNA motif definition and identification from multiple sequence alignments using secondary structure profiles. J. Mol. Biol. *313*, 1003–1011. https://doi.org/10.1006/jmbi.2001.5102.

Ghabrial, S.A., and Suzuki, N. (2009). Viruses of plant pathogenic fungi. Annu. Rev. Phytopathol. *47*, 353–384. https://doi.org/10.1146/annurev-phyto-080508-081932.

Gibbs, M.J., Koga, R., Moriyama, H., Pfeiffer, P., and Fukuhara, T. (2000). Phylogenetic analysis of some large double-stranded RNA replicons from plants suggests they evolved from a defective single-stranded RNA virus. J. Gen. Virol. *81*, 227–233. https://doi.org/10.1099/0022-1317-81-1-227.

Gottlieb, P., Metzger, S., Romantschuk, M., Carton, J., Strassman, J., Bamford, D.H., Kalkkinen, N., and Mindich, L. (1988). Nucleotide sequence of the middle dsRNA segment of bacteriophage phi6: Placement of the genes of membrane-associated proteins. Virology *163*, 183–190. https://doi.org/10.1016/0042-6822(88)90245-0.

Gottlieb, P., Wei, H., Potgieter, C., and Toporovsky, I. (2002a). Characterization of phi12, a bacteriophage related to phi6: nucleotide sequence of the small and middle double-stranded RNA. Virology *293*, 118–124. https://doi.org/10.1006/viro.2001.1288.

Gottlieb, P., Potgieter, C., Wei, H., and Toporovsky, I. (2002b). Characterization of phi12, a bacteriophage related to phi6: nucleotide sequence of the large double-stranded RNA. Virology *295*, 266–271. https://doi.org/10.1006/viro.2002.1436.

Green, M.R., and Sambrook, J. (2012). Molecular Cloning Volume 2 (Cold Spring Harbor Laboratory Press).

Hacker, C.V., Brasier, C.M., and Buck, K.W. (2005). A double-stranded RNA from a *Phytophthora* species is related to the plant endornaviruses and contains a putative UDP glycosyltransferase gene. J. Gen. Virol. *86*, 1561–1570. https://doi.org/10.1099/vir.0.80808-0.

Happonen, L.J., Redder, P., Peng, X., Reigstad, L.J., Prangishvili, D., and Butcher, S.J. (2010). Familial relationships in hyperthermo- and acidophilic archaeal viruses. J. Virol. *84*, 4747–4754. https://doi.org/10.1128/JVI.02156-09.

Häring, M., Peng, X., Brügger, K., Rachel, R., Stetter, K.O., Garrett, R.A., and Prangishvili, D. (2004). Morphology and genome organization of the virus PSV of the hyperthermophilic archaeal genera *Pyrobaculum* and *Thermoproteus*: a novel virus family, the *Globuloviridae*. Virology *323*, 233–242. https://doi.org/10.1016/j.virol.2004.03.002.

Häring, M., Vestergaard, G., Brügger, K., Rachel, R., Garrett, R.A., and Prangishvili, D. (2005). Structure and genome organization of AFV2, a novel archaeal lipothrixvirus with unusual terminal and core structures. J. Bacteriol. *187*, 3855–3858.

Hatfull, G.F. (2010). Mycobacteriophages: genes and genomes. Annu. Rev. Microbiol. *64*, 331–356. https://doi.org/10.1146/annurev.micro.112408.134233.

Hatfull, G.F. (2012). The secret lives of mycobacteriophages. In Advances in Virus Research, Vol 82, Lobocka, M.B., and Szybalski, W.T., eds. (Academic Press, Burlington), pp. 179–288. https://doi.org/10.1016/B978-0-12-394621-8.00015-7.

Hatfull, G.F. (2015). Dark matter of the biosphere: the amazing world of bacteriophage diversity. J. Virol. *89*, 8107–8110. https://doi.org/10.1128/JVI.01340-15.

Hatfull, G.F., Cresawn, S.G., and Hendrix, R.W. (2008). Comparative genomics of the mycobacteriophages: insights into bacteriophage evolution. Res. Microbiol. *159*, 332–339. https://doi.org/10.1016/j.resmic.2008.04.008.

Hatfull, G.F., Jacobs-Sera, D., Lawrence, J.G., Pope, W.H., Russell, D.A., Ko, C.C., Weber, R.J., Patel, M.C., Germane, K.L., Edgar, R.H., *et al.* (2010). Comparative genomic analysis of 60 mycobacteriophage genomes: genome clustering, gene acquisition, and gene size. J. Mol. Biol. *397*, 119–143. https://doi.org/10.1016/j.jmb.2010.01.011.

Herrero, N., and Zabalgogeazcoa, I. (2011). Mycoviruses infecting the endophytic and entomopathogenic fungus *Tolypocladium cylindrosporum*. Virus Res. *160*, 409–413. https://doi.org/10.1016/j.virusres.2011.06.015.

Hill, D.F., and Petersen, G.B. (1982). Nucleotide sequence of bacteriophage f1 DNA. J. Virol. *44*, 32–46.

Hill, D.F., Short, N.J., Perham, R.N., and Petersen, G.B. (1991). DNA sequence of the filamentous bacteriophage Pf1. J. Mol. Biol. *218*, 349–364.

Hillman, B.I., Halpern, B.T., and Brown, M.P. (1994). A viral dsRNA element of the chestnut blight fungus with a distinct genetic organization. Virology *201*, 241–250.

Hillman, B.I., Supyani, S., Kondo, H., and Suzuki, N. (2004). A reovirus of the fungus *Cryphonectria parasitica* that is infectious as particles and related to the coltivirus genus of animal pathogens. J. Virol. *78*, 892–898.

Hong, Y., Cole, T.E., Brasier, C.M., and Buck, K.W. (1998). Evolutionary relationships among putative RNA-dependent RNA polymerases encoded by a mitochondrial virus-like RNA in the Dutch elm disease fungus, *Ophiostoma novo-ulmi*, by other viruses and virus-like RNAs and by the *Arabidopsis* mitochondrial genome. Virology *246*, 158–169.

Hong, Y., Dover, S.L., Cole, T.E., Brasier, C.M., and Buck, K.W. (1999). Multiple mitochondrial viruses in an isolate of the Dutch Elm disease fungus *Ophiostoma novo-ulmi*. Virology *258*, 118–127.

Hoogstraten, D., Qiao, X., Sun, Y., Hu, A., Onodera, S., and Mindich, L. (2000). Characterization of phi8, a bacteriophage containing three double-stranded RNA genomic segments and distantly related to phi6. Virology *272*, 218–224. https://doi.org/10.1006/viro.2000.0374.

Howitt, R.L., Beever, R.E., Pearson, M.N., and Forster, R.L. (2001). Genome characterization of *Botrytis* virus F, a flexuous rod-shaped mycovirus resembling plant 'potex-like' viruses. J. Gen. Virol. *82*, 67–78. https://doi.org/10.1099/0022-1317-82-1-67.

Howitt, R.L., Beever, R.E., Pearson, M.N., and Forster, R.L. (2006). Genome characterization of a flexuous rod-shaped mycovirus, Botrytis virus X, reveals high amino acid identity to genes from plant 'potex-like' viruses. Arch. Virol. *151*, 563–579. https://doi.org/10.1007/s00705-005-0621-y.

Huang, S., and Ghabrial, S.A. (1996). Organization and expression of the double-stranded RNA genome of *Helminthosporium victoriae* 190S virus, a totivirus infecting a plant pathogenic filamentous fungus. Proc. Natl. Acad. Sci. U.S.A. *93*, 12541–12546.

Icho, T., and Wickner, R.B. (1989). The double-stranded RNA genome of yeast virus L-A encodes its own putative RNA polymerase by fusing two open reading frames. J. Biol. Chem. *264*, 6716–6723.

Ignacio-Espinoza, J.C., and Sullivan, M.B. (2012). Phylogenomics of T4 cyanophages: lateral gene transfer in the 'core' and origins of host genes. Environ. Microbiol. *14*, 2113–2126. https://doi.org/10.1111/j.1462-2920.2012.02704.x.

Inokuchi, Y., Takahashi, R., Hirose, T., Inayama, S., Jacobson, A.B., and Hirashima, A. (1986). The complete nucleotide sequence of the group II RNA coliphage GA. J. Biochem. *99*, 1169–1180.

Inokuchi, Y., Jacobson, A.B., Hirose, T., Inayama, S., and Hirashima, A. (1988). Analysis of the complete nucleotide sequence of the group IV RNA coliphage SP. Nucleic Acids Res. *16*, 6205–6221.

Iranzo, J., Koonin, E.V., Prangishvili, D., and Krupovic, M. (2016). Bipartite network analysis of the archaeal virosphere: evolutionary connections between viruses and capsidless mobile elements. J. Virol. *90*, 11043–11055.

Iyer, L.M., Aravind, L., and Koonin, E.V. (2001). Common origin of four diverse families of large eukaryotic DNA viruses. J. Virol. *75*, 11720–11734. https://doi.org/10.1128/JVI.75.23.11720-11734.2001.

Iyer, L.M., Balaji, S., Koonin, E.V., and Aravind, L. (2006). Evolutionary genomics of nucleo-cytoplasmic large DNA viruses. Virus Res. *117*, 156–184.

Jaakkola, S.T., Penttinen, R.K., Vilén, S.T., Jalasvuori, M., Rönnholm, G., Bamford, J.K., Bamford, D.H., and Oksanen, H.M. (2012). Closely related archaeal *Haloarcula hispanica* icosahedral viruses HHIV-2 and SH1 have nonhomologous genes encoding host recognition functions. J. Virol. *86*, 4734–4742. https://doi.org/10.1128/JVI.06666-11.

Jacobs-Sera, D., Marinelli, L.J., Bowman, C., Broussard, G.W., Guerrero Bustamante, C., Boyle, M.M., Petrova, Z.O., Dedrick, R.M., Pope, W.H., Modlin, R.L., *et al.* (2012). On the nature of mycobacteriophage diversity and host preference. Virology *434*, 187–201. https://doi.org/10.1016/j.virol.2012.09.026.

Jalasvuori, M., Jaatinen, S.T., Laurinavicius, S., Ahola-Iivarinen, E., Kalkkinen, N., Bamford, D.H., and Bamford, J.K. (2009). The closest relatives of icosahedral viruses of thermophilic bacteria are among viruses and plasmids of the halophilic archaea. J. Virol. *83*, 9388–9397. https://doi.org/10.1128/JVI.00869-09.

Jiang, D., and Ghabrial, S.A. (2004). Molecular characterization of *Penicillium chrysogenum virus*: Reconsideration of the taxonomy of the genus *Chrysovirus*. J. Gen. Virol. *85*, 2111–2121. https://doi.org/10.1099/vir.0.79842-0.

Jiang, Y., Zhang, T., Luo, C., Jiang, D., Li, G., Li, Q., Hsiang, T., and Huang, J. (2015). Prevalence and diversity of mycoviruses infecting the plant pathogen *Ustilaginoidea virens*. Virus Res. *195*, 47–56. https://doi.org/10.1016/j.virusres.2014.08.022.

Kan, S., Fornelos, N., Schuch, R., and Fischetti, V.A. (2013). Identification of a ligand on the Wip1 bacteriophage highly specific for a receptor on *Bacillus anthracis*. J. Bacteriol. *195*, 4355–4364. https://doi.org/10.1128/JB.00655-13.

Kannoly, S., Shao, Y., and Wang, I.N. (2012). Rethinking the evolution of single-stranded RNA (ssRNA) bacteriophages based on genomic sequences and characterizations of two R-plasmid-dependent ssRNA phages, C-1 and Hgal1. J. Bacteriol. *194*, 5073–5079. https://doi.org/10.1128/JB.00929-12.

Karam, J.D., Drake, J.W., Kreuzer, K.N., Mosig, G., Hall, D.H., Eiserling, F.A., Black, L.W., Spicer, E.K., Kutter, E., Carlson, C., *et al.* (1994). Molecular Biology of Bacteriophage T4 (American Society for Microbiology, Washington).

Karumidze, N., Thomas, J.A., Kvatadze, N., Goderdzishvili, M., Hakala, K.W., Weintraub, S.T., Alavidze, Z., and Hardies, S.C. (2012). Characterization of lytic *Pseudomonas aeruginosa* bacteriophages via biological properties and genomic sequences. Appl. Microbiol. Biotechnol. *94*, 1609–1617. https://doi.org/10.1007/s00253-012-4119-8.

Kawasaki, T., Nagata, S., Fujiwara, A., Satsuma, H., Fujie, M., Usami, S., and Yamada, T. (2007). Genomic characterization of the filamentous integrative bacteriophages phiRSS1 and phiRSM1, which infect *Ralstonia solanacearum*. J. Bacteriol. *189*, 5792–5802. https://doi.org/10.1128/JB.00540-07.

Kazaks, A., Voronkova, T., Rumnieks, J., Dishlers, A., and Tars, K. (2011). Genome structure of caulobacter phage phiCb5. J. Virol. *85*, 4628–4631. https://doi.org/10.1128/JVI.02256-10.

Khalifa, M.E., and Pearson, M.N. (2013). Molecular characterization of three mitoviruses co-infecting a hypovirulent isolate of *Sclerotinia sclerotiorum* fungus. Virology *441*, 22–30. https://doi.org/10.1016/j.virol.2013.03.002.

Khramtsov, N.V., Woods, K.M., Nesterenko, M.V., Dykstra, C.C., and Upton, S.J. (1997). Virus-like, double-stranded RNAs in the parasitic protozoan *Cryptosporidium parvum*. Mol. Microbiol. *26*, 289–300.

Kim, J.W., Kim, S.Y., and Kim, K.M. (2003). Genome organization and expression of the *Penicillium stoloniferum* virus S. Virus Genes *27*, 249–256.

Kim, J.W., Choi, E.Y., and Lee, J.I. (2005). Genome organization and expression of the *Penicillium stoloniferum* virus F. Virus Genes *31*, 175–183. https://doi.org/10.1007/s11262-005-1793-y.

Kimura, K., and Tomaru, Y. (2015). Discovery of two novel viruses expands the diversity of single-stranded DNA and single-stranded RNA viruses infecting a cosmopolitan marine diatom. Appl. Environ. Microbiol. *81*, 1120–1131. https://doi.org/10.1128/AEM.02380-14.

King, A.M.Q., Adams, M.J., Carstens, G.B., and Lefkowitz, E.J. (2012). Virus taxonomy, classification and nomenclature of viruses. Ninth report of the International Committee on Taxonomy of Viruses. (Elsevier, London).

Kingsford, C.L., Ayanbule, K., and Salzberg, S.L. (2007). Rapid, accurate, computational discovery of Rho-independent transcription terminators illuminates their relationship to DNA uptake. Genome Biol. *8*, R22.

Klovins, J., Overbeek, G.P., van den Worm, S.H., Ackermann, H.W., and van Duin, J. (2002). Nucleotide sequence of a ssRNA phage from *Acinetobacter*: Kinship to coliphages. J. Gen. Virol. *83*, 1523–1533. https://doi.org/10.1099/0022-1317-83-6-1523.

Klumpp, J., Fouts, D.E., and Sozhamannan, S. (2012). Next generation sequencing technologies and the changing landscape of phage genomics. Bacteriophage *2*, 190–199. https://doi.org/10.4161/bact.22111.

Kondo, H., Chiba, S., Toyoda, K., and Suzuki, N. (2013a). Evidence for negative-strand RNA virus infection in fungi. Virology *435*, 201–209. https://doi.org/10.1016/j.virol.2012.10.002.

Kondo, H., Kanematsu, S., and Suzuki, N. (2013b). Viruses of the White Root Rot Fungus, *Rosellinia necatrix*. In Advances in Virus Research Vol 86, Mycoviruses (Elsevier Inc. Academic Press), pp. 177–214. https://doi.org/10.1016/B978-0-12-394315-6.00007-6.

Koonin, E.V., Dolja, V.V., and Krupovic, M. (2015). Origins and evolution of viruses of eukaryotes: The ultimate modularity. Virology 479-480, 2–25. https://doi.org/10.1016/j.virol.2015.02.039.

Kramberger, P., Honour, R.C., Herman, R.E., Smrekar, F., and Peterka, M. (2010). Purification of the *Staphylococcus aureus* bacteriophages VDX-10 on methacrylate monoliths. J. Virol. Methods *166*, 60–64. https://doi.org/10.1016/j.jviromet.2010.02.020.

Krupovic, M. (2013). Networks of evolutionary interactions underlying the polyphyletic origin of ssDNA viruses. Curr. Opin. Virol. *3*, 578–586. https://doi.org/10.1016/j.coviro.2013.06.010.

Krupovic, M., and Bamford, D.H. (2007). Putative prophages related to lytic tailless marine dsDNA phage PM2 are widespread in the genomes of aquatic bacteria. BMC Genomics *8*, 236.

Krupovic, M., and Bamford, D.H. (2011). Double-stranded DNA viruses: 20 families and only five different architectural principles for virion assembly. Curr. Opin. Virol. *1*, 118–124. https://doi.org/10.1016/j.coviro.2011.06.001.

Krupovic, M., and Forterre, P. (2011). *Microviridae* goes temperate: microvirus-related proviruses reside in the genomes of *Bacteroidetes*. PLOS ONE *6*, e19893. https://doi.org/10.1371/journal.pone.0019893.

Krupovic, M., Dutilh, B.E., Adriaenssens, E.M., Wittmann, J., Vogensen, F.K., Sullivan, M.B., Rumnieks, J., Prangishvili, D., Lavigne, R., Kropinski, A.M., *et al.* (2016). Taxonomy of prokaryotic viruses: update from the ICTV bacterial and archaeal viruses subcommittee. Arch. Virol. *161*, 1095–1099. https://doi.org/10.1007/s00705-015-2728-0.

Kuhn, J.H., Bekal, S., Caì, Y., Clawson, A.N., Domier, L.L., Herrel, M., Jahrling, P.B., Kondo, H., Lambert, K.N., Mihindukulasuriya, K.A., *et al.* (2013). *Nyamiviridae*: proposal for a new family in the order *Mononegavirales*. Arch. Virol. *158*, 2209–2226. https://doi.org/10.1007/s00705-013-1674-y.

Kulakov, L.A., Ksenzenko, V.N., Shlyapnikov, M.G., Kochetkov, V.V., Del Casale, A., Allen, C.C., Larkin, M.J., Ceyssens, P.J., and Lavigne, R. (2009). Genomes of 'phiKMV-like viruses' of *Pseudomonas aeruginosa* contain localized single-strand interruptions. Virology *391*, 1–4. https://doi.org/10.1016/j.virol.2009.06.024.

Kuo, T.T., Tan, M.S., Su, M.T., and Yang, M.K. (1991). Complete nucleotide sequence of filamentous phage Cf1c from *Xanthomonas campestris* pv. citri. Nucleic Acids Res. *19*, 2498.

Kwon, S.J., Lim, W.S., Park, S.H., Park, M.R., and Kim, K.H. (2009). Molecular characterization of a dsRNA mycovirus, *Fusarium graminearum* Virus-DK21, which is phylogenetically related to hypoviruses but has a genome organization and gene expression strategy resembling those of plant potex-like viruses. Mol. Cells *28*, 73–74. https://doi.org/10.1007/s10059-009-0112-1.

La Scola, B., Audic, S., Robert, C., Jungang, L., de Lamballerie, X., Drancourt, M., Birtles, R., Claverie, J.-M., and Raoult, D. (2003). A giant virus in amoebae. Science *299*, 2033. https://doi.org/10.1126/science.1081867.

La Scola, B., Desnues, C., Pagnier, I., Robert, C., Barrassi, L., Fournous, G., Merchat, M., Suzan-Monti, M., Forterre, P., Koonin, E., *et al.* (2008). The virophage as a unique parasite of the giant mimivirus. Nature *455*, 100–104. https://doi.org/10.1038/nature07218.

Labonté, J.M., and Suttle, C.A. (2013). Metagenomic and whole-genome analysis reveals new lineages of gokushoviruses and biogeographic separation in the sea. Front. Microbiol. *4*, 404. https://doi.org/10.3389/fmicb.2013.00404.

Labrie, S.J., Dupuis, M.È., Tremblay, D.M., Plante, P.L., Corbeil, J., and Moineau, S. (2014). A new *Microviridae* phage isolated from a failed biotechnological process driven by *Escherichia coli*. Appl. Environ. Microbiol. *80*, 6992–7000. https://doi.org/10.1128/AEM.01365-14.

Lakshman, D.K., Jian, J., and Tavantzis, S.M. (1998). A double-stranded RNA element from a hypovirulent strain of *Rhizoctonia solani* occurs in DNA form and is genetically related to the pentafunctional AROM protein of the shikimate pathway. Proc. Natl. Acad. Sci. U.S.A. *95*, 6425–6429.

Lang, A.S., Culley, A.I., and Suttle, C.A. (2004). Genome sequence and characterization of a virus (HaRNAV) related to picorna-like viruses that infects the marine toxic bloom-forming alga *Heterosigma akashiwo*. Virology *320*, 206–217. https://doi.org/10.1016/j.virol.2003.10.015.

Laslett, D., and Canback, B. (2004). ARAGORN, a program to detect tRNA genes and tmRNA genes in nucleotide sequences. Nucleic Acids Res. 32, 11–16. https://doi.org/10.1093/nar/gkh152.

Lavigne, R., Burkal'tseva, M.V., Robben, J., Sykilinda, N.N., Kurochkina, L.P., Grymonprez, B., Jonckx, B., Krylov, V.N., Mesyanzhinov, V.V., and Volckaert, G. (2003). The genome of bacteriophage phiKMV, a T7-like virus infecting *Pseudomonas aeruginosa*. Virology *312*, 49–59.

Lavigne, R., Sun, W.D., and Volckaert, G. (2004). PHIRE, a deterministic approach to reveal regulatory elements in bacteriophage genomes. Bioinformatics *20*, 629–635. https://doi.org/10.1093/bioinformatics/btg456.

Lavigne, R., Seto, D., Mahadevan, P., Ackermann, H.W., and Kropinski, A.M. (2008). Unifying classical and molecular taxonomic classification: analysis of the *Podoviridae* using BLASTP-based tools. Res. Microbiol. *159*, 406–414. https://doi.org/10.1016/j.resmic.2008.03.005.

Lavigne, R., Darius, P., Summer, E.J., Seto, D., Mahadevan, P., Nilsson, A.S., Ackermann, H.W., and Kropinski, A.M. (2009). Classification of *Myoviridae* bacteriophages using protein sequence similarity. BMC Microbiol. *9*, 224. https://doi.org/10.1186/1471-2180-9-224.

Legendre, M., Bartoli, J., Shmakova, L., Jeudy, S., Labadie, K., Adrait, A., Lescot, M., Poirot, O., Bertaux, L., Bruley, C., *et al.* (2014). Thirty-thousand-year-old distant relative of giant icosahedral DNA viruses with a pandoravirus morphology. Proc. Natl. Acad. Sci. U.S.A. *111*, 4274–4279. https://doi.org/10.1073/pnas.1320670111.

Legendre, M., Lartigue, A., Bertaux, L., Jeudy, S., Bartoli, J., Lescot, M., Alempic, J.M., Ramus, C., Bruley, C., Labadie, K., *et al.* (2015). In-depth study of *Mollivirus sibericum*, a new 30,000-y-old giant virus infecting *Acanthamoeba*. Proc. Natl. Acad. Sci. U.S.A. *112*, E5327–35. https://doi.org/10.1073/pnas.1510795112.

Li, M., Wang, R., Zhao, D., and Xiang, H. (2014a). Adaptation of the *Haloarcula hispanica* CRISPR-Cas system to a purified virus strictly requires a priming process. Nucleic Acids Res. *42*, 2483–2492. https://doi.org/10.1093/nar/gkt1154.

Li, R., Yu, C., Li, Y., Lam, T.W., Yiu, S.M., Kristiansen, K., and Wang, J. (2009). SOAP2: an improved ultrafast tool for short read alignment. Bioinformatics *25*, 1966–1967. https://doi.org/10.1093/bioinformatics/btp336.

Li, W., Zhang, T., Sun, H., Deng, Y., Zhang, A., Chen, H., and Wang, K. (2014b). Complete genome sequence of a novel endornavirus in the wheat sharp eyespot pathogen *Rhizoctonia cerealis*. Arch. Virol. *159*, 1213–1216. https://doi.org/10.1007/s00705-013-1893-2.

Linder-Basso, D., Dynek, J.N., and Hillman, B.I. (2005). Genome analysis of *Cryphonectria* hypovirus 4, the most common hypovirus species in North America. Virology *337*, 192–203.

Liu, B.L., Everson, J.S., Fane, B., Giannikopoulou, P., Vretou, E., Lambden, P.R., and Clarke, I.N. (2000). Molecular characterization of a bacteriophage (Chp2) from *Chlamydia psittaci*. J. Virol. *74*, 3464–3469.

Liu, H., Fu, Y., Jiang, D., Li, G., Xie, J., Peng, Y., Yi, X., and Ghabrial, S.A. (2009). A novel mycovirus that is related to the human pathogen hepatitis E virus and rubi-like viruses. J. Virol. *83*, 1981–1991. https://doi.org/10.1128/JVI.01897-08.

Liu, J.J., Chan, D., Xiang, Y., Williams, H., Li, X.R., Sniezko, R.A., and Sturrock, R.N. (2016). Characterization of Five Novel Mitoviruses in the White Pine Blister Rust Fungus *Cronartium ribicola*. PLOS ONE *11*, e0154267. https://doi.org/10.1371/journal.pone.0154267.

Liu, L., Xie, J., Cheng, J., Fu, Y., Li, G., Yi, X., and Jiang, D. (2014). Fungal negative-stranded RNA virus that is related to bornaviruses and nyaviruses. Proc. Natl. Acad. Sci. U.S.A. *111*, 12205–12210. https://doi.org/10.1073/pnas.1401786111.

Luiten, R.G., Putterman, D.G., Schoenmakers, J.G., Konings, R.N., and Day, L.A. (1985). Nucleotide sequence of the genome of Pf3, an IncP-1 plasmid-specific filamentous bacteriophage of *Pseudomonas aeruginosa*. J. Virol. *56*, 268–276.

Maejima, K., Himeno, M., Komatsu, K., Kakizawa, S., Yamaji, Y., Hamamoto, H., and Namba, S. (2008). Complete nucleotide sequence of a new double-stranded RNA virus from the rice blast fungus, *Magnaporthe oryzae*. Arch. Virol. *153*, 389–391. https://doi.org/10.1007/s00705-007-1101-3.

Maniloff, J., Kampo, G.J., and Dascher, C.C. (1994). Sequence analysis of a unique temperate phage: mycoplasma virus L2. Gene *141*, 1–8. https://doi.org/10.1016/0378-1119(94)90120-1.

Männistö, R.H., Kivelä, H.M., Paulin, L., Bamford, D.H., and Bamford, J.K. (1999). The complete genome sequence of PM2, the first lipid-containing bacterial virus to be isolated. Virology *262*, 355–363. https://doi.org/10.1006/viro.1999.9837.

Mathews, C.K., Kutter, E.M., Mosig, G., and Berget, P.B. (1983). Bacteriophage T4 (American Society for Microbiology, Washington).

McGraw, T., Mindich, L., and Frangione, B. (1986). Nucleotide sequence of the small double-stranded RNA segment of bacteriophage phi6: novel mechanism of natural translational control. J. Virol. *58*, 142–151.

Miller, E.S., Kutter, E.M., Mosig, G., Arisaka, F., Kunisawa, T., and Rüger, W. (2003). Bacteriophage T4 genome. Microbiol. Mol. Biol. Rev. *67*, 86–156. https://doi.org/10.1128/MMBR.67.1.86.

Mindich, L., Nemhauser, I., Gottlieb, P., Romantschuk, M., Carton, J., Frucht, S., Strassman, J., Bamford, D.H., and Kalkkinen, N. (1988). Nucleotide sequence of the large double-stranded RNA segment of bacteriophage phi6: genes specifying the viral replicase and transcriptase. J. Virol. *62*, 1180–1185.

Mochizuki, T., Yoshida, T., Tanaka, R., Forterre, P., Sako, Y., and Prangishvili, D. (2010). Diversity of viruses of the hyperthermophilic archaeal genus *Aeropyrum*, and isolation of the *Aeropyrum* pernix bacilliform virus 1, APBV1, the first representative of the family *Clavaviridae*. Virology *402*, 347–354. https://doi.org/10.1016/j.virol.2010.03.046.

Mochizuki, T., Sako, Y., and Prangishvili, D. (2011). Provirus induction in hyperthermophilic archaea: characterization of *Aeropyrum pernix* spindle-shaped virus 1 and *Aeropyrum pernix* ovoid virus 1. J. Bacteriol. *193*, 5412–5419. https://doi.org/10.1128/JB.05101-11.

Mochizuki, T., Krupovic, M., Pehau-Arnaudet, G., Sako, Y., Forterre, P., and Prangishvili, D. (2012). Archaeal virus with exceptional virion architecture and the largest single-stranded DNA genome. Proc. Natl. Acad. Sci. U.S.A. *109*, 13386–13391. https://doi.org/10.1073/pnas.1203668109.

Moreau, H., Piganeau, G., Desdevises, Y., Cooke, R., Derelle, E., and Grimsley, N. (2010). Marine prasinovirus genomes show low evolutionary divergence and acquisition of protein metabolism genes by horizontal gene transfer. J. Virol. *84*, 12555–12563. https://doi.org/10.1128/JVI.01123-10.

Morris, T.J., and Dodds, J.A. (1979). Isolation and analysis of double-stranded RNA from virus-infected plant and fungal tissue. Phytopathology *69*, 854–858. https://doi.org/10.1094/Phyto-69-854.

Nibert, M.L., Woods, K.M., Upton, S.J., and Ghabrial, S.A. (2009). *Cryspovirus*: a new genus of protozoan viruses in the family *Partitiviridae*. Arch. Virol. *154*, 1959–1965. https://doi.org/10.1007/s00705-009-0513-7.

Nogawa, M., Kageyama, T., Nakatani, A., Taguchi, G., Shimosaka, M., and Okazaki, M. (1996). Cloning and characterization of mycovirus double-stranded RNA from the plant pathogenic fungus, *Fusarium solani* f. sp. robiniae. Biosci. Biotechnol. Biochem. *60*, 784–788.

Nomura, K., Osaki, H., Iwanami, T., Matsumoto, N., and Ohtsu, Y. (2003). Cloning and characterization of a totivirus double-stranded RNA from the plant pathogenic fungus, *Helicobasidium mompa* Tanaka. Virus Genes *26*, 219–226.

Nurk, S., Bankevich, A., Antipov, D., Gurevich, A., Korobeynikov, A., Lapidus, A., Prjibelsky, A., Pyshkin, A., Sirotkin, A., Sirotkin, Y., *et al.* (2013). Assembling genomes and mini-metagenomes from highly chimeric reads. In Research in Computational Molecular Biology. RECOMB 2013. Lecture Notes in Computer Science, Deng, M., Jiang, R., Sun, F., and Zhang, X., eds. (Springer, Berlin, Heidelberg), pp. 158–170. https://doi.org/10.1007/978-3-642-37195-0_13.

Oh, C.S., and Hillman, B.I. (1995). Genome organization of a partitivirus from the filamentous ascomycete *Atkinsonella hypoxylon*. J. Gen. Virol. *76*, 1461–1470. https://doi.org/10.1099/0022-1317-76-6-1461.

Olsthoorn, R.C., Garde, G., Dayhuff, T., Atkins, J.F., and Van Duin, J. (1995). Nucleotide sequence of a single-stranded RNA phage from *Pseudomonas aeruginosa*: kinship to coliphages and conservation of regulatory RNA structures. Virology *206*, 611–625.

Ong, J.W.L., Li, H., Sivasithamparam, K., Dixon, K.W., Jones, M.G.K., and Wylie, S.J. (2016). Novel Endorna-like viruses, including three with two open reading frames, challenge the membership criteria and taxonomy of the *Endornaviridae*. Virology *499*, 203–211.

Osaki, H., Wei, C.Z., Arakawa, M., Iwanami, T., Nomura, K., Matsumoto, N., and Ohtsu, Y. (2002). Nucleotide sequences of double-stranded RNA segments from a hypovirulent strain of the white root rot fungus *Rosellinia necatrix*: possibility of the first member of the *Reoviridae* from fungus. Virus Genes *25*, 101–107.

Osaki, H., Sasaki, A., Nomiyama, K., Sekiguchi, H., Tomioka, K., and Takehara, T. (2015). Isolation and characterization of two mitoviruses and a putative alphapartitivirus from *Fusarium* spp. Virus Genes *50*, 466–473. https://doi.org/10.1007/s11262-015-1182-0.

Overbeek, R., Olson, R., Pusch, G.D., Olsen, G.J., Davis, J.J., Disz, T., Edwards, R.A., Gerdes, S., Parrello, B., Shukla, M., *et al.* (2014). The SEED and the Rapid Annotation of microbial genomes using Subsystems Technology (RAST). Nucleic Acids Res. *42*, D206–14. https://doi.org/10.1093/nar/gkt1226.

Palm, P., Schleper, C., Grampp, B., Yeats, S., McWilliam, P., Reiter, W.D., and Zillig, W. (1991). Complete nucleotide sequence of the virus SSV1 of the archaebacterium *Sulfolobus shibatae*. Virology *185*, 242–250.

Paquin, B., Laforest, M.J., Forget, L., Roewer, I., Wang, Z., Longcore, J., and Lang, B.F. (1997). The fungal mitochondrial genome project: evolution of fungal mitochondrial genomes and their gene expression. Curr. Genet. *31*, 380–395.

Park, Y., James, D., and Punja, Z.K. (2005). Co-infection by two distinct totivirus-like double-stranded RNA elements in *Chalara elegans* (*Thielaviopsis basicola*). Virus Res. *109*, 71–85.

Park, Y., Chen, X., and Punja, Z.K. (2006). Molecular and biological characterization of a mitovirus in *Chalara elegans* (*Thielaviopsis basicola*). Phytopathology *96*, 468–479. https://doi.org/10.1094/PHYTO-96-0468.

Pavesi, A. (2006). Origin and evolution of overlapping genes in the family *Microviridae*. J. Gen. Virol. *87*, 1013–1017. https://doi.org/10.1099/vir.0.81375-0.

Payet, J., and Suttle, C.A. (2013). To kill or not to kill: The balance between lytic and lysogenic viral infection is driven by trophic status. Limnol. Oceanogr. *58*, 465–474. https://doi.org/10.4319/lo.2013.58.2.

Peeters, B.P., Peters, R.M., Schoenmakers, J.G., and Konings, R.N. (1985). Nucleotide sequence and genetic organization of the genome of the N-specific filamentous bacteriophage IKe. Comparison with the genome of the F-specific filamentous phages M13, fd and f1. J. Mol. Biol. *181*, 27–39.

Peng, X. (2008). Evidence for the horizontal transfer of an integrase gene from a fusellovirus to a pRN-like plasmid within a single strain of *Sulfolobus* and the implications for plasmid survival. Microbiology *154*, 383–391. https://doi.org/10.1099/mic.0.2007/012963-0.

Peng, X., Blum, H., She, Q., Mallok, S., Brügger, K., Garrett, R.A., Zillig, W., and Prangishvili, D. (2001). Sequences and replication of genomes of the archaeal rudiviruses SIRV1 and SIRV2: relationships to the archaeal lipothrixvirus SIFV and some eukaryal viruses. Virology *291*, 226–234. https://doi.org/10.1006/viro.2001.1190.

Peng, X., Basta, T., Häring, M., Garrett, R.A., and Prangishvili, D. (2007). Genome of the *Acidianus* bottle-shaped virus and insights into the replication and packaging mechanisms. Virology *364*, 237–243.

Petrov, V.M., Ratnayaka, S., Nolan, J.M., Miller, E.S., and Karam, J.D. (2010). Genomes of the T4-related bacteriophages as windows on microbial genome evolution. Virol. J. *7*, 292. https://doi.org/10.1186/1743-422X-7-292.

Philippe, N., Legendre, M., Doutre, G., Couté, Y., Poirot, O., Lescot, M., Arslan, D., Seltzer, V., Bertaux, L., Bruley, C., *et al.* (2013). Pandoraviruses: amoeba viruses with genomes up to 2.5 Mb reaching that of parasitic eukaryotes. Science *341*, 281–286. https://doi.org/10.1126/science.1239181.

Pietilä, M.K., Atanasova, N.S., Manole, V., Liljeroos, L., Butcher, S.J., Oksanen, H.M., and Bamford, D.H. (2012). Virion architecture unifies globally distributed pleolipoviruses infecting halophilic archaea. J. Virol. *86*, 5067–5079. https://doi.org/10.1128/JVI.06915-11.

Pietilä, M.K., Roine, E., Paulin, L., Kalkkinen, N., and Bamford, D.H. (2009). An ssDNA virus infecting archaea: a new lineage of viruses with a membrane envelope. Mol. Microbiol. 72, 307–319. https://doi.org/10.1111/j.1365-2958.2009.06642.x.

Pietilä, M.K., Roine, E., Sencilo, A., Bamford, D.H., and Oksanen, H.M. (2016). *Pleolipoviridae*, a newly proposed family comprising archaeal pleomorphic viruses with single-stranded or double-stranded DNA genomes. Arch. Virol. *161*, 249–256. https://doi.org/10.1007/s00705-015-2613-x.

Polashock, J.J., and Hillman, B.I. (1994). A small mitochondrial double-stranded (ds) RNA element associated with a hypovirulent strain of the chestnut blight fungus and ancestrally related to yeast cytoplasmic T and W dsRNAs. Proc. Natl. Acad. Sci. U.S.A. *91*, 8680–8684.

Pope, W.H., Jacobs-Sera, D., Russell, D.A., Peebles, C.L., Al-Atrache, Z., Alcoser, T.A., Alexander, L.M., Alfano, M.B., Alford, S.T., Amy, N.E., *et al.* (2011). Expanding the diversity of mycobacteriophages: insights into genome architecture and evolution. PLOS ONE *6*, e16329. https://doi.org/10.1371/journal.pone.0016329.

Pope, W.H., Bowman, C.A., Russell, D.A., Jacobs-Sera, D., Asai, D.J., Cresawn, S.G., Jacobs, W.R., Hendrix, R.W., Lawrence, J.G., and Hatfull, G.F. (2015). Whole genome comparison of a large collection of mycobacteriophages reveals a continuum of phage genetic diversity. eLife *4*, e06416. https://doi.org/10.7554/eLife.06416.

Poranen, M.M., Bamford, D.H., and Oksanen, H.M. (2015). Membrane-containing bacteriophages. In eLS (John Wiley & Sons, Ltd, UK) https://doi.org/10.1002/9780470015902.a0000779.pub3.

Porter, K., Kukkaro, P., Bamford, J.K., Bath, C., Kivelä, H.M., Dyall-Smith, M.L., and Bamford, D.H. (2005). SH1: A novel, spherical halovirus isolated from an Australian hypersaline lake. Virology *335*, 22–33.

Porter, K., Tang, S.L., Chen, C.P., Chiang, P.W., Hong, M.J., and Dyall-Smith, M. (2013). PH1: an archaeovirus of *Haloarcula hispanica* related to SH1 and HHIV-2. Archaea *2013*, 456318. https://doi.org/10.1155/2013/456318.

Prangishvili, D. (2015). Archaeal viruses: living fossils of the ancient virosphere? Ann. N. Y. Acad. Sci. *1341*, 35–40. https://doi.org/10.1111/nyas.12710.

Prangishvili, D., Garrett, R.A., and Koonin, E.V. (2006a). Evolutionary genomics of archaeal viruses: unique viral genomes in the third domain of life. Virus Res. *117*, 52–67.

Prangishvili, D., Vestergaard, G., Häring, M., Aramayo, R., Basta, T., Rachel, R., and Garrett, R.A. (2006b). Structural and genomic properties of the hyperthermophilic archaeal virus ATV with an extracellular stage of the reproductive cycle. J. Mol. Biol. *359*, 1203–1216.

Preisig, O., Wingfield, B.D., and Wingfield, M.J. (1998). Coinfection of a fungal pathogen by two distinct double-stranded RNA viruses. Virology *252*, 399–406.

Preisig, O., Moleleki, N., Smit, W.A., Wingfield, B.D., and Wingfield, M.J. (2000). A novel RNA mycovirus in a hypovirulent isolate of the plant pathogen *Diaporthe ambigua*. J. Gen. Virol. *81*, 3107–3114. https://doi.org/10.1099/0022-1317-81-12-3107.

Qiao, X., Qiao, J., Onodera, S., and Mindich, L. (2000). Characterization of phi13, a bacteriophage related to phi6 and containing three dsRNA genomic segments. Virology *275*, 218–224. https://doi.org/10.1006/viro.2000.0501.

Quaiser, A., Dufresne, A., Ballaud, F., Roux, S., Zivanovic, Y., Colombet, J., Sime-Ngando, T., and Francez, A.J. (2015). Diversity and comparative genomics of *Microviridae* in Sphagnum- dominated peatlands. Front. Microbiol. *6*, 375. https://doi.org/10.3389/fmicb.2015.00375.

Raoult, D., Audic, S., Robert, C., Abergel, C., Renesto, P., Ogata, H., La Scola, B., Suzan, M., and Claverie, J.M. (2004). The 1.2-megabase genome sequence of Mimivirus. Science *306*, 1344–1350.

Ravantti, J.J., Gaidelyte, A., Bamford, D.H., and Bamford, J.K. (2003). Comparative analysis of bacterial viruses Bam35, infecting a gram-positive host, and PRD1, infecting gram-negative hosts, demonstrates a viral lineage. Virology *313*, 401–414.

Reavy, B., Swanson, M.M., Cock, P.J., Dawson, L., Freitag, T.E., Singh, B.K., Torrance, L., Mushegian, A.R., and Taliansky, M. (2015). Distinct circular single-stranded DNA viruses exist in different soil types. Appl. Environ. Microbiol. *81*, 3934–3945. https://doi.org/10.1128/AEM.03878-14.

Redder, P., Peng, X., Brügger, K., Shah, S.A., Roesch, F., Greve, B., She, Q., Schleper, C., Forterre, P., Garrett, R.A., *et al.* (2009). Four newly isolated fuselloviruses from extreme geothermal environments reveal unusual morphologies and a possible interviral recombination mechanism. Environ. Microbiol. *11*, 2849–2862. https://doi.org/10.1111/j.1462-2920.2009.02009.x.

Remmert, M., Biegert, A., Hauser, A., and Söding, J. (2012). HHblits: lightning-fast iterative protein sequence searching by HMM-HMM alignment. Nat. Methods *9*, 173–175. https://doi.org/10.1038/nmeth.1818.

Renaudin, J., Pascarel, M.C., and Bové, J.M. (1987). Spiroplasma virus 4: nucleotide sequence of the viral DNA, regulatory signals, and proposed genome organization. J. Bacteriol. *169*, 4950–4961.

Renaudin, J., Aullo, P., Vignault, J.C., and Bové, J.M. (1990). Complete nucleotide sequence of the genome of *Spiroplasma citri* virus SpV1-R8A2 B. Nucleic Acids Res. *18*, 1293.

Rice, G., Tang, L., Stedman, K., Roberto, F., Spuhler, J., Gillitzer, E., Johnson, J.E., Douglas, T., and Young, M. (2004). The structure of a thermophilic archaeal virus shows a double-stranded DNA viral capsid type that spans all domains of life. Proc. Natl. Acad. Sci. U.S.A. *101*, 7716–7720. https://doi.org/10.1073/pnas.0401773101.

Roberts, J.W., and Devoret, R. (1983). Lysogenic Induction. Cold Spring Harb. Monogr. Arch. *13*, 123–144. https://doi.org/10.1101/087969150.13.123.

Rodríguez-Cousiño, N., Solórzano, A., Fujimura, T., and Esteban, R. (1998). Yeast positive-stranded virus-like RNA replicons. 20 S and 23 S RNA terminal nucleotide sequences and 3' end secondary structures resemble those of RNA coliphages. J. Biol. Chem. *273*, 20363–20371.

Roine, E., Kukkaro, P., Paulin, L., Laurinavicius, S., Domanska, A., Somerharju, P., and Bamford, D.H. (2010). New, closely related haloarchaeal viral elements with different nucleic Acid types. J. Virol. *84*, 3682–3689. https://doi.org/10.1128/JVI.01879-09.

Rokyta, D.R., Burch, C.L., Caudle, S.B., and Wichman, H.A. (2006). Horizontal gene transfer and the evolution of microvirid coliphage genomes. J. Bacteriol. *188*, 1134–1142.

Rong, R., Rao, S., Scott, S.W., Carner, G.R., and Tainter, F.H. (2002). Complete sequence of the genome of two dsRNA viruses from *Discula destructiva*. Virus Res. *90*, 217–224.

Roucourt, B., and Lavigne, R. (2009). The role of interactions between phage and bacterial proteins within the infected cell: a diverse and puzzling interactome. Environ. Microbiol. *11*, 2789–2805. https://doi.org/10.1111/j.1462-2920.2009.02029.x.

Roux, S., Krupovic, M., Poulet, A., Debroas, D., and Enault, F. (2012). Evolution and diversity of the *Microviridae* viral family through a collection of 81 new complete genomes assembled from virome reads. PLOS ONE *7*, e40418. https://doi.org/10.1371/journal.pone.0040418.

Rumnieks, J., and Tars, K. (2012). Diversity of pili-specific bacteriophages: genome sequence of IncM plasmid-dependent RNA phage M. BMC Microbiol. *12*, 277. https://doi.org/10.1186/1471-2180-12-277.

Ruokoranta, T.M., Grahn, A.M., Ravantti, J.J., Poranen, M.M., and Bamford, D.H. (2006). Complete genome sequence of the broad host range single-stranded RNA phage PRR1 places it in the *Levivirus* genus with characteristics shared with Alloleviviruses. J. Virol. *80*, 9326–9330.

Russell, D.A., and Hatfull, G.F. (2016). PhagesDB: the actinobacteriophage database. Bioinformatics 1–3. https://doi.org/10.1093/bioinformatics/btw711.

Sanger, F., Air, G.M., Barrell, B.G., Brown, N.L., Coulson, A.R., Fiddes, C.A., Hutchison, C.A., Slocombe, P.M., and Smith, M. (1977). Nucleotide sequence of bacteriophage phi X174 DNA. Nature *265*, 687–695.

Santini, S., Jeudy, S., Bartoli, J., Poirot, O., Lescot, M., Abergel, C., Barbe, V., Wommack, K.E., Noordeloos, A.A., Brussaard, C.P., *et al.* (2013). Genome of *Phaeocystis globosa* virus PgV-16T highlights the common ancestry of the largest known DNA viruses infecting eukaryotes. Proc. Natl. Acad. Sci. U.S.A. *110*, 10800–10805. https://doi.org/10.1073/pnas.1303251110.

Saren, A.M., Ravantti, J.J., Benson, S.D., Burnett, R.M., Paulin, L., Bamford, D.H., and Bamford, J.K. (2005). A snapshot of viral evolution from genome analysis of the *Tectiviridae* family. J. Mol. Biol. *350*, 427–440.

Sasaki, A., Miyanishi, M., Ozaki, K., Onoue, M., and Yoshida, K. (2005). Molecular characterization of a partitivirus from the plant pathogenic ascomycete *Rosellinia necatrix*. Arch. Virol. *150*, 1069–1083. https://doi.org/10.1007/s00705-005-0494-0.

Schattner, P., Brooks, A.N., and Lowe, T.M. (2005). The tRNAscan-SE, snoscan and snoGPS web servers for the detection of tRNAs and snoRNAs. Nucleic Acids Res. *33*, W686–689. https://doi.org/10.1093/nar/gki366.

Schroeder, D.C., Park, Y., Yoon, H.M., Lee, Y.S., Kang, S.W., Meints, R.H., Ivey, R.G., and Choi, T.J. (2009). Genomic analysis of the smallest giant virus – *Feldmannia* sp. virus 158. Virology *384*, 223–232. https://doi.org/10.1016/j.virol.2008.10.040.

Seemann, T. (2014). Prokka: rapid prokaryotic genome annotation. Bioinformatics *30*, 2068–2069. https://doi.org/10.1093/bioinformatics/btu153.

Senčilo, A., and Roine, E. (2014). A glimpse of the genomic diversity of haloarchaeal tailed viruses. Front. Microbiol. *5*, 84. https://doi.org/10.3389/fmicb.2014.00084.

Senčilo, A., Paulin, L., Kellner, S., Helm, M., and Roine, E. (2012). Related haloarchaeal pleomorphic viruses contain different genome types. Nucleic Acids Res. *40*, 5523–5534. https://doi.org/10.1093/nar/gks215.

Sha, Y., Melcher, U., Davis, R.E., and Fletcher, J. (2000). Common elements of spiroplasma plectroviruses revealed by nucleotide sequence of SVTS2. Virus Genes *20*, 47–56.

Shapira, R., Choi, G.H., and Nuss, D.L. (1991). Virus-like genetic organization and expression strategy for a double-stranded RNA genetic element associated with biological control of chestnut blight. EMBO J. *10*, 731–739.

Shirai, Y., Takao, Y., Mizumoto, H., Tomaru, Y., Honda, D., and Nagasaki, K. (2006). Genomic and phylogenetic analysis of a single-stranded RNA virus infecting *Rhizosolenia setigera* (Stramenopiles: Bacillariophyceae). J. Mar. Biol. Assoc. UK *86*, 475–483. https://doi.org/10.1017/S0025315406013373.

Shirai, Y., Tomaru, Y., Takao, Y., Suzuki, H., Nagumo, T., and Nagasaki, K. (2008). Isolation and characterization of a single-stranded RNA virus infecting the marine planktonic diatom *Chaetoceros tenuissimus* Meunier. Appl. Environ. Microbiol. *74*, 4022–4027. https://doi.org/10.1128/AEM.00509-08.

Shu, P., Butt, A.M., Mi, Z., Wang, W., An, X., Pei, G., Zhang, Z., Huang, Y., Zhang, X., Shi, T., *et al.* (2015). Identification and genomic analysis of a novel member of *Microviridae*, IME-16, through high-throughput sequencing. Virol. Sin. *30*, 301–304. https://doi.org/10.1007/s12250-015-3596-1.

Smart, C.D., Yuan, W., Foglia, R., Nuss, D.L., Fulbright, D.W., and Hillman, B.I. (1999). *Cryphonectria hypovirus* 3, a virus species in the family hypoviridae with a single open reading frame. Virology *265*, 66–73. https://doi.org/10.1006/viro.1999.0039.

Söding, J., Biegert, A., and Lupas, A.N. (2005). The HHpred interactive server for protein homology detection and structure prediction. Nucleic Acids Res. *33*, W244–8. https://doi.org/10.1093/nar/gki408.

Sozhamannan, S., McKinstry, M., Lentz, S.M., Jalasvuori, M., McAfee, F., Smith, A., Dabbs, J., Ackermann, H.W., Bamford, J.K., Mateczun, A., *et al.* (2008). Molecular characterization of a variant of *Bacillus anthracis*-specific phage AP50 with improved bacteriolytic activity. Appl. Environ. Microbiol. *74*, 6792–6796. https://doi.org/10.1128/AEM.01124-08.

Starmer, J., Stomp, A., Vouk, M., and Bitzer, D. (2006). Predicting Shine-Dalgarno sequence locations exposes genome annotation errors. PLOS Comput. Biol. *2*, e57. https://doi.org/10.1371/journal.pcbi.0020057.

Stassen, A.P., Schoenmakers, E.F., Yu, M., Schoenmakers, J.G., and Konings, R.N. (1992). Nucleotide sequence of the genome of the filamentous bacteriophage I2-2: module evolution of the filamentous phage genome. J. Mol. Evol. *34*, 141–152.

Stedman, K.M. (2015). Deep Recombination: RNA and ssDNA Virus Genes in DNA Virus and Host Genomes. Annu. Rev. Virol. 2, 203–217. https://doi.org/10.1146/annurev-virology-100114-055127.

Storey, C.C., Lusher, M., and Richmond, S.J. (1989). Analysis of the complete nucleotide sequence of Chp1, a phage which infects avian *Chlamydia psittaci*. J. Gen.

Virol. *70*, 3381–3390. https://doi.org/10.1099/0022-1317-70-12-3381.

Strauss, E.E., Lakshman, D.K., and Tavantzis, S.M. (2000). Molecular characterization of the genome of a partitivirus from the basidiomycete *Rhizoctonia solani*. J. Gen. Virol. *81*, 549–555. https://doi.org/10.1099/0022-1317-81-2-549.

Suhre, K., Audic, S., and Claverie, J.M. (2005). Mimivirus gene promoters exhibit an unprecedented conservation among all eukaryotes. Proc. Natl. Acad. Sci. U.S.A. *102*, 14689–14693.

Sullivan, M.B. (2015). Viromes, not gene markers, for studying double-stranded DNA virus communities. J. Virol. *89*, 2459–2461. https://doi.org/10.1128/JVI.03289-14.

Sullivan, M.B., Huang, K.H., Ignacio-Espinoza, J.C., Berlin, A.M., Kelly, L., Weigele, P.R., DeFrancesco, A.S., Kern, S.E., Thompson, L.R., Young, S., *et al.* (2010). Genomic analysis of oceanic cyanobacterial myoviruses compared with T4-like myoviruses from diverse hosts and environments. Environ. Microbiol. *12*, 3035–3056. https://doi.org/10.1111/j.1462-2920.2010.02280.x.

Sullivan, M.J., Petty, N.K., and Beatson, S.A. (2011). Easyfig: a genome comparison visualizer. Bioinformatics *27*, 1009–1010. https://doi.org/10.1093/bioinformatics/btr039.

Suttle, C.A. (1993). Enumeration and isolation of viruses. In Handbook of Methods in Aquatic Microbial Ecology, Kemp, P.F., Cole, J.J., Sherr, B.F., and Sherr, E.B., eds. (CRC Press), pp. 121–134.

Suzuki, N., Supyani, S., Maruyama, K., and Hillman, B.I. (2004). Complete genome sequence of Mycoreovirus-1/Cp9B21, a member of a novel genus within the family *Reoviridae*, isolated from the chestnut blight fungus *Cryphonectria parasitica*. J. Gen. Virol. *85*, 3437–3448. https://doi.org/10.1099/vir.0.80293-0.

Tai, J.H., and Ip, C.F. (1995). The cDNA sequence of *Trichomonas vaginalis* virus-T1 double-stranded RNA. Virology *206*, 773–776.

Takao, Y., Nagasaki, K., Mise, K., Okuno, T., and Honda, D. (2005). Isolation and characterization of a novel single-stranded RNA Virus infectious to a marine fungoid protist, *Schizochytrium* sp. (*Thraustochytriaceae*, *Labyrinthulea*). Appl. Environ. Microbiol. *71*, 4516–4522.

Taylor, D.J., and Bruenn, J. (2009). The evolution of novel fungal genes from non-retroviral RNA viruses. BMC Biol. *7*, 88. https://doi.org/10.1186/1741-7007-7-88.

Taylor, D.J., Ballinger, M.J., Bowman, S.M., and Bruenn, J.A. (2013). Virus-host co-evolution under a modified nuclear genetic code. PeerJ *1*, e50. https://doi.org/10.7717/peerj.50.

Thomas, V., Bertelli, C., Collyn, F., Casson, N., Telenti, A., Goesmann, A., Croxatto, A., and Greub, G. (2011). Lausannevirus, a giant amoebal virus encoding histone doublets. Environ. Microbiol. *13*, 1454–1466. https://doi.org/10.1111/j.1462-2920.2011.02446.x.

Tomaru, Y., Takao, Y., Suzuki, H., Nagumo, T., and Nagasaki, K. (2009). Isolation and characterization of a single-stranded RNA virus infecting the bloom-forming diatom *Chaetoceros socialis*. Appl. Environ. Microbiol. *75*, 2375–2381. https://doi.org/10.1128/AEM.02580-08.

Tomaru, Y., Takao, Y., Suzuki, H., Nagumo, T., Koike, K., and Nagasaki, K. (2011). Isolation and characterization of a single-stranded DNA virus infecting *Chaetoceros lorenzianus* Grunow. Appl. Environ. Microbiol. *77*, 5285–5293. https://doi.org/10.1128/AEM.00202-11.

Tomaru, Y., Toyoda, K., Suzuki, H., Nagumo, T., Kimura, K., and Takao, Y. (2013). New single-stranded DNA virus with a unique genomic structure that infects marine diatom *Chaetoceros setoensis*. Sci. Rep. *3*, 3337. https://doi.org/10.1038/srep03337.

Toyoda, K., Kimura, K., Hata, N., Nakayama, N., Nagasaki, K., and Tomaru, Y. (2012). Isolation and characterization of a single-stranded DNA virus infecting the marine planktonic diatom *Chaetoceros* sp. (strain TG07-C28). Plankt. Benthos Res. *7*, 20–28. https://doi.org/10.3800/pbr.7.20.

Tuomivirta, T.T., and Hantula, J. (2003a). *Gremmeniella abietina* mitochondrial RNA virus S1 is phylogenetically related to the members of the genus *Mitovirus*. Arch. Virol. *148*, 2429–2436. https://doi.org/10.1007/s00705-003-0195-5.

Tuomivirta, T.T., and Hantula, J. (2003b). Two unrelated double-stranded RNA molecule patterns in *Gremmeniella abietina* type A code for putative viruses of the families *Totiviridae* and *Partitiviridae*. Arch. Virol. *148*, 2293–2305. https://doi.org/10.1007/s00705-003-0194-6.

Tuomivirta, T.T., and Hantula, J. (2005). Three unrelated viruses occur in a single isolate of *Gremmeniella abietina* var. *abietina* type A. Virus Res. *110*, 31–39.

Tuomivirta, T.T., Kaitera, J., and Hantula, J. (2009). A novel putative virus of *Gremmeniella abietina* type B (*Ascomycota*: *Helotiaceae*) has a composite genome with endornavirus affinities. J. Gen. Virol. *90*, 2299–2305. https://doi.org/10.1099/vir.0.011973-0.

Urayama, S., Kato, S., Suzuki, Y., Aoki, N., Le, M.T., Arie, T., Teraoka, T., Fukuhara, T., and Moriyama, H. (2010). Mycoviruses related to chrysovirus affect vegetative growth in the rice blast fungus *Magnaporthe oryzae*. J. Gen. Virol. *91*, 3085–3094. https://doi.org/10.1099/vir.0.025411-0.

Urayama, S., Sakoda, H., Takai, R., Katoh, Y., Minh Le, T., Fukuhara, T., Arie, T., Teraoka, T., and Moriyama, H. (2014). A dsRNA mycovirus, *Magnaporthe oryzae* chrysovirus 1-B, suppresses vegetative growth and development of the rice blast fungus. Virology *448*, 265–273. https://doi.org/10.1016/j.virol.2013.10.022.

Van, T.T., Yoshida, S., Miki, K., Kondo, A., and Kamei, K. (2014). Genomic characterization of ϕRS603, a filamentous bacteriophage that is infectious to the phytopathogen *Ralstonia solanacearum*. Microbiol. Immunol. *58*, 697–700. https://doi.org/10.1111/1348-0421.12203.

Van Duin, J., and Tsareva, N. (2006). Single-stranded RNA phages. In The Bacteriophages, Calendar, R., ed. (Oxford University Press. New York), pp. 175–196.

Van Etten, J.L., Graves, M.V., Müller, D.G., Boland, W., and Delaroque, N. (2002). *Phycodnaviridae* – large DNA algal viruses. Arch. Virol. *147*, 1479–1516. https://doi.org/10.1007/s00705-002-0822-6.

Van Twest, R., and Kropinski, A.M. (2009). Bacteriophage enrichment from water and soil. In Bacteriophages: Methods and Protocols, Clokie, M., and Kropinski,

A.M., eds. (Humana Press), pp. 15–21. https://doi.org/10.1007/978-1-60327-164-6.

van Wezenbeek, P.M., Hulsebos, T.J., and Schoenmakers, J.G. (1980). Nucleotide sequence of the filamentous bacteriophage M13 DNA genome: comparison with phage fd. Gene *11*, 129–148.

Verheust, C., Jensen, G., and Mahillon, J. (2003). pGIL01, a linear tectiviral plasmid prophage originating from *Bacillus thuringiensis* serovar israelensis. Microbiology *149*, 2083–2092. https://doi.org/10.1099/mic.0.26307-0.

Verheust, C., Fornelos, N., and Mahillon, J. (2005). GIL16, a new gram-positive tectiviral phage related to the *Bacillus thuringiensis* GIL01 and the *Bacillus cereus* pBClin15 elements. J. Bacteriol. *187*, 1966–1973.

Vestergaard, G., Häring, M., Peng, X., Rachel, R., Garrett, R.A., and Prangishvili, D. (2005). A novel rudivirus, ARV1, of the hyperthermophilic archaeal genus *Acidianus*. Virology *336*, 83–92.

Vestergaard, G., Aramayo, R., Basta, T., Häring, M., Peng, X., Brügger, K., Chen, L., Rachel, R., Boisset, N., Garrett, R.A., *et al.* (2008). Structure of the acidianus filamentous virus 3 and comparative genomics of related archaeal lipothrixviruses. J. Virol. *82*, 371–381.

Wang, Q., Kan, B., and Wang, R. (2013). Isolation and characterization of the new mosaic filamentous phage VFJphi of *Vibrio cholerae*. PLOS ONE *8*, e70934. https://doi.org/10.1371/journal.pone.0070934.

Wei, C.Z., Osaki, H., Iwanami, T., Matsumoto, N., and Ohtsu, Y. (2003). Molecular characterization of dsRNA segments 2 and and electron microscopy of a novel reovirus from a hypovirulent isolate, W370, of the plant pathogen *Rosellinia necatrix*. J. Gen. Virol. *84*, 2431–2437. https://doi.org/10.1099/vir.0.19098-0.

Wei, C.Z., Osaki, H., Iwanami, T., Matsumoto, N., and Ohtsu, Y. (2004). Complete nucleotide sequences of genome segments 1 and 3 of *Rosellinia* anti-rot virus in the family *Reoviridae*. Arch. Virol. *149*, 773–777. https://doi.org/10.1007/s00705-003-0259-6.

Weynberg, K.D., Allen, M.J., Ashelford, K., Scanlan, D.J., and Wilson, W.H. (2009). From small hosts come big viruses: the complete genome of a second *Ostreococcus tauri* virus, OtV-1. Environ. Microbiol. *11*, 2821–2839. https://doi.org/10.1111/j.1462-2920.2009.01991.x.

Wiedenheft, B., Stedman, K., Roberto, F., Willits, D., Gleske, A.K., Zoeller, L., Snyder, J., Douglas, T., and Young, M. (2004). Comparative genomic analysis of hyperthermophilic archaeal *Fuselloviridae* viruses. J. Virol. *78*, 1954–1961.

Williamson, K.E., Schnitker, J.B., Radosevich, M., Smith, D.W., and Wommack, K.E. (2008). Cultivation-based assessment of lysogeny among soil bacteria. Microb. Ecol. *56*, 437–447. https://doi.org/10.1007/s00248-008-9362-2.

Wilson, W.H., Schroeder, D.C., Allen, M.J., Holden, M.T., Parkhill, J., Barrell, B.G., Churcher, C., Hamlin, N., Mungall, K., Norbertczak, H., *et al.* (2005). Complete genome sequence and lytic phase transcription profile of a coccolithovirus. Science *309*, 1090–1092.

Wu, M.D., Zhang, L., Li, G.Q., Jiang, D.H., Hou, M.S., and Huang, H.C. (2007). Hypovirulence and double-stranded RNA in *Botrytis cinerea*. Phytopathology *97*, 1590–1599. https://doi.org/10.1094/PHYTO-97-12-1590.

Xiang, X., Chen, L., Huang, X., Luo, Y., She, Q., and Huang, L. (2005). *Sulfolobus tengchongensis* spindle-shaped virus STSV1: virus-host interactions and genomic features. J. Virol. *79*, 8677–8686.

Xie, J., Wei, D., Jiang, D., Fu, Y., Li, G., Ghabrial, S., and Peng, Y. (2006). Characterization of debilitation-associated mycovirus infecting the plant-pathogenic fungus *Sclerotina sclerotorium*. J. Gen. Virol. *87*, 241–249. https://doi.org/10.1099/vir.0.81522-0.

Xu, Z., Wu, S., Liu, L., Cheng, J., Fu, Y., Jiang, D., and Xie, J. (2015). A mitovirus related to plant mitochondrial gene confers hypovirulence on the phytopathogenic fungus *Sclerotinia sclerotiorum*. Virus Res. *197*, 127–136. https://doi.org/10.1016/j.virusres.2014.12.023.

Xue, H., Xu, Y., Boucher, Y., and Polz, M.F. (2012). High frequency of a novel filamentous phage, VCYphi, within an environmental *Vibrio cholerae* population. Appl. Environ. Microbiol. *78*, 28–33. https://doi.org/10.1128/AEM.06297-11.

Yau, S., Lauro, F.M., DeMaere, M.Z., Brown, M.V., Thomas, T., Raftery, M.J., Andrews-Pfannkoch, C., Lewis, M., Hoffman, J.M., Gibson, J.A., *et al.* (2011). Virophage control of Antarctic algal host-virus dynamics. Proc. Natl. Acad. Sci. U.S.A. *108*, 6163–6168. https://doi.org/10.1073/pnas.1018221108.

Yokoi, T., Takemoto, Y., Suzuki, M., Yamashita, S., and Hibi, T. (1999). The nucleotide sequence and genome organization of *Sclerophthora macrospora* virus B. Virology *264*, 344–349. https://doi.org/10.1006/viro.1999.0018.

Yokoi, T., Yamashita, S., and Hibi, T. (2003). The nucleotide sequence and genome organization of *Sclerophthora macrospora* virus A. Virology *311*, 394–399.

Yokoi, T., Yamashita, S., and Hibi, T. (2007). The nucleotide sequence and genome organization of *Magnaporthe oryzae* virus 1. Arch. Virol. *152*, 2265–2269. https://doi.org/10.1007/s00705-007-1045-7.

Yoosuf, N., Yutin, N., Colson, P., Shabalina, S.A., Pagnier, I., Robert, C., Azza, S., Klose, T., Wong, J., Rossmann, M.G., *et al.* (2012). Related giant viruses in distant locations and different habitats: *Acanthamoeba polyphaga moumouvirus* represents a third lineage of the *Mimiviridae* that is close to the Megavirus lineage. Genome Biol. Evol. *4*, 1324–1330. https://doi.org/10.1093/gbe/evs109.

Yu, J., Lee, K.M., Son, M., and Kim, K.H. (2011). Molecular characterization of Fusarium graminearum virus 2 isolated from *Fusarium graminearum* strain 98-8-60. Plant Pathol. J. *27*, 285–290. https://doi.org/10.5423/PPJ.2011.27.3.285.

Yu, X., Li, B., Fu, Y., Jiang, D., Ghabrial, S.A., Li, G., Peng, Y., Xie, J., Cheng, J., Huang, J., *et al.* (2010). A geminivirus-related DNA mycovirus that confers hypovirulence to a plant pathogenic fungus. Proc. Natl. Acad. Sci. U.S.A. *107*, 8387–8392. https://doi.org/10.1073/pnas.0913535107.

Zerbino, D.R., and Birney, E. (2008). Velvet: algorithms for *de novo* short read assembly using *de* Bruijn graphs. Genome Res. *18*, 821–829. https://doi.org/10.1101/gr.074492.107.

Zhang, R., Hisano, S., Tani, A., Kondo, H., Kanematsu, S., and Suzuki, N. (2016). A capsidless ssRNA virus hosted by an unrelated dsRNA virus. Nat. Microbiol. *1*, 15001. https://doi.org/10.1038/nmicrobiol.2015.1.

Zhang, T., Jiang, Y., Huang, J., and Dong, W. (2013). Complete genome sequence of a putative novel victorivirus from *Ustilaginoidea virens*. Arch. Virol. *158*, 1403–1406. https://doi.org/10.1007/s00705-013-1615-9.

Zheng, L., Zhang, M., Chen, Q., Zhu, M., and Zhou, E. (2014). A novel mycovirus closely related to viruses in the genus *Alphapartitivirus* confers hypovirulence in the phytopathogenic fungus *Rhizoctonia solani*. Virology 456-457, 220–226. https://doi.org/10.1016/j.virol.2014.03.029.

Zhong, J., Zhou, Q., Lei, X.H., Chen, D., Shang, H.H., and Zhu, H.J. (2014a). The nucleotide sequence and genome organization of two victoriviruses from the rice false smut fungus *Ustilaginoidea virens*. Virus Genes *48*, 570–573. https://doi.org/10.1007/s11262-014-1050-3.

Zhong, J., Lei, X.H., Zhu, J.Z., Song, G., Zhang, Y.D., Chen, Y., and Gao, B.D. (2014b). Detection and sequence analysis of two novel co-infecting double-strand RNA mycoviruses in *Ustilaginoidea virens*. Arch. Virol. *159*, 3063–3070. https://doi.org/10.1007/s00705-014-2144-x.

Zsak, L., Day, J.M., Oakley, B.B., and Seal, B.S. (2011). The complete genome sequence and genetic analysis of ΦCA82 a novel uncultured microphage from the turkey gastrointestinal system. Virol. J. *8*, 331. https://doi.org/10.1186/1743-422X-8-331.

Part II

Ecology of Viruses of Microorganisms

Evolutionary Ecology of the Viruses of Microorganisms

3

Brian E. Ford[1,2], Marko Baloh[1] and John J. Dennehy[1,2*]

[1]Department of Biology, Queens College, New York City, NY, USA.
[2]The Graduate Center of the City University of New York, New York City, NY, USA.

*Correspondence: john.dennehy@qc.cuny.edu

https://doi.org/10.21775/9781910190852.03

Abstract

With estimated numbers greater than 10^{31}, viruses are the most abundant organisms on the planet, and occupy all habitats: aquatic, atmospheric and terrestrial. No cellular organisms – whether animal, plant or microbe – are free from viral parasitism. Consequently, the effects and influences of viruses are pervasive, directly or indirectly affecting all organisms, populations, communities and ecosystems. Here we consider the evolutionary ecology of the viruses of microorganisms (VoMs) which, due to the abundance of their hosts, outnumber all other types of viruses. Subfields of evolutionary ecology include life history evolution, population biology, biogeography, and community ecology. Like blind men describing an elephant, each approach only describes a facet of VoM evolutionary ecology. Here we describe some of the approaches used to describe VoM evolutionary ecology in hopes that a synthesis will allow some perception of the whole.

Introduction

Whether or not one describes viruses as organisms, living entities, or instead simply as infectious agents, viruses nonetheless both evolve (evolutionary biology) and interact with their environments (ecology). In this chapter we consider the evolutionary biology and ecology of the viruses of microorganisms or VoMs, which include bacteriophages (the viruses of bacteria) (see Chapter 7), archaeal viruses (see Chapter 8), viruses of protists (see Chapters 10, 11 and 12), and various mycoviruses (viruses of fungi) (see Chapter 9). In addition, we emphasize the 'hybrid' discipline of evolutionary ecology, which covers the subdisciplines of life history evolution, population biology, biogeography, and community ecology. To our knowledge, this is the first publication to review the evolutionary ecology of VoMs as a group, although previous reviews consider these subjects as they apply to bacteriophages, archaeal or planktonic viruses (Abedon, 2008, 2009; Brüssow and Kutter, 2005; Koskella and Brockhurst, 2014; Prangishvili, 2013; Short, 2012; Weynberg *et al.*, 2017). For further discussion, especially of VoM ecology, see Chapters 4 and 6.

VoM abundance, biodiversity, and biogeography

The viruses of microorganisms may be the most numerous organisms on Earth (Angly *et al.*, 2006; Bergh *et al.*, 1989; Güemes *et al.*, 2016; Suttle, 2005). In aquatic and soil habitats, viral concentrations higher than 10^8 per millilitre or gram have been reported, even in arid desert soils (Ashelford *et al.*, 2003; Fuhrman, 1999; Güemes *et al.*, 2016; Kimura *et al.*, 2008; Suttle, 2005; Swanson *et al.*, 2009; Williamson *et al.*, 2005; Wommack and Colwell, 2000; Zablocki *et al.*, 2015) (see Chapter 4). Even the Earth's atmosphere harbours a relatively high concentration of viruses (10^7–10^8/m^3) (Prussin *et al.*, 2015; Whon *et al.*, 2012).

Given that VoMs exist in unimaginable numbers, we might expect a correspondingly high level of diversity. Studying VoM diversity, however, is complicated by the fact that, unlike cellular organisms, VoMs do not have universal ribosomal DNA sequences that facilitate the identification of discrete species and help to determine the phylogenetic relationships between these species (Rowher and Edwards, 2002). Thus, sampling VoM biodiversity had been approached in a piecemeal fashion (i.e. laboratory culturing or direct observation) until the advent of direct environmental genomic sampling (i.e. metagenomics (see Chapter 5)). Metagenomic estimates of VoM diversity entails the isolation and high-throughput sequencing of all viral nucleic acids in an environmental isolate (e.g. water, soil, tissue, faeces). To ensure that only viral genetic material is sequenced, environmental isolates are ultra-filtered and nucleases are used to digest non-capsid protected prokaryotic and eukaryotic nucleic acids (Breitbart *et al.*, 2004a; Delwart, 2007; Thurber *et al.*, 2009). Partial genetic fragments obtained from random sequence reads are computationally aligned and assembled into contigs (consensus sequences based on overlapping partial fragments). A contig spectrum is generated for a virome by counting the number of sequences that fall into each contig (Allen *et al.*, 2013). Abundant VoM types are inferred by the presence of a large number of sequences mapping to any particular contig.

By comparing the newly obtained genomic sequences with those in existing genetic sequence databases, VoM genetic richness and diversity can be estimated. Previously unknown VoM sequences (i.e. sequences not matching those in existing genetic databases) can be identified and compared among different communities (Edwards and Rohwer, 2005; Suttle, 2007). The percentage of unknown VoM sequences in recent studies has ranged from 60% to 99% (Angly *et al.*, 2006; Brum and Sullivan, 2015; Desnues *et al.*, 2008; Güemes *et al.*, 2016; Hurwitz and Sullivan, 2013; Roux *et al.*, 2016; Watkins *et al.*, 2015). As in population mark-recapture studies (Pradel, 1996), a high proportion of novel types recovered from population resampling indicates that a large number of new types remain undiscovered (Paez-Espino *et al.*, 2016). Nevertheless, recent advances in viral metagenomic techniques, both in genetic material isolation and sequencing (Brum and Sullivan, 2015; Chow *et al.*, 2014), and in the development of viral genetic sequence databases, keep increasing the number of identified VoM, particularly in the case of marine VoMs (Suttle, 2016)

Some of the first metagenomic estimates of VoM diversity were conducted on marine and human virus communities by Rohwer, Breitbart, and colleagues (Breitbart *et al.*, 2003; Breitbart *et al.*, 2002). For the marine communities, mathematical models predicted the existence of 374 to 7114 viral types in the oceans off the California coast (Breitbart *et al.*, 2002). Studies of human virome suggest that it may contain at least 1,250 distinct virus types, most of which are VoMs (Breitbart *et al.*, 2003). In retrospect, large as they may be, these numbers may be significant underestimates (Allen *et al.*, 2011; Angly *et al.*, 2006; Culley *et al.*, 2006; Güemes *et al.*, 2016; Hurwitz and Sullivan, 2013; Kristensen *et al.*, 2010; Mokili *et al.*, 2012; Rosario and Breitbart, 2011; Rosario *et al.*, 2009). For one, these estimates do not include RNA or ssDNA virus diversity, which are harder to analyse due to the difficulties in sequencing these types of nucleic acids (Güemes *et al.*, 2016). Distinct VoM types in the biosphere may number in the millions (Allen *et al.*, 2013).

Interestingly, analyses of VoM genetic material collected from geographically distinct environments (i.e. freshwater, marine, terrestrial) show that similar VoM genetic sequences can be found in widely separated ecosystems, indicating that VoMs, or at least their genes, are in constant motion through the biosphere (Breitbart *et al.*, 2004a; Breitbart *et al.*, 2004b; Breitbart and Rohwer, 2005; Danovaro *et al.*, 2016; Dutilh *et al.*, 2014; Hambly and Suttle, 2005; Kunin *et al.*, 2008; O'Keefe *et al.*, 2010; Sano *et al.*, 2004; Short and Suttle, 2005; Zhao *et al.*, 2013). Nonetheless, host habitat requirements may play a strong role in the distribution of VoMs. For example, an investigation of hot spring microbial communities found that genetically distinct viral populations were associated with each local geothermal region (Held and Whitaker, 2009). This ties into Baas Becking's idea that '*everything is everywhere, but, the environment selects*' (Baas Becking, 1934). In this context, the presence of hosts is the key aspect of the 'environment' that determines whether a virus is detectable in a specific habitat (Angly *et al.*, 2006; Paez-Espino *et al.*, 2016).

Habitat specialization and isolation can have far-reaching effects on VoM evolution because the distribution of populations across space, and the connections between them, govern population diversification. Since population genetic differentiation is usually a function of gene flow between populations, connected populations will tend to follow more similar evolutionary trajectories, whereas isolated populations will tend to diverge as they become locally adapted (Deng *et al.*, 2014; Kawecki and Ebert, 2004; Schluter and Conte, 2009). We expect that VoM distributions will be highly contingent on chance historical events and the specific details of their ecology. Overall, there appears to be a tremendous diversity of VoMs, and they tend to be broadly, but unevenly, distributed across the biosphere (Roux *et al.*, 2016; Thurber, 2009).

Population dynamics of VoMs and hosts

The number of VoM individuals present in a habitat ultimately depends on the conditions that assist or impede VoM reproduction. Since most VoMs move between hosts as particles – excepting especially fungal viruses, which for the most part appear to be transmitted either vertically from parent to offspring or instead in the course of cell-to-cell contact (Ghabrial and Suzuki, 2009; see also Chapter 9) – their macroscale movement depends on the physical properties of the medium and the forces acting upon it (e.g. bulk flow of water/air). At microscales, the laws of random diffusion by Brownian motion govern VoM movement. In any case, infection is contingent on chance collisions between VoMs and hosts, the probabilities of which may be quite low on a virus-by-virus basis. By way of illustration of the difficulties randomly diffusing virus particles can have in finding host organisms to infect, Abedon estimated that if a virus were the size of the Titanic, then a millilitre of fluid would be analogous to the volume of the Earth (Abedon, 2011). Were it not for the enormous population sizes of VoMs and their hosts, then infection events would be improbable (Dennehy, 2014).

There is a growing appreciation of how VoM morphology and capsid properties may affect mobility in different environments, thus increasing the chances of encountering a host. For example, rod-shaped viruses may diffuse faster than spherical viruses in tissues or gels (Lee *et al.*, 2013). We expect that very large virus particles, such as the *Megaviridae* of protists (see Chapter 11), would diffuse at slower rates than small, icosahedral single-stranded bacteriophages (see Chapter 7). How these factors affect infection rates is largely unknown. In addition, Barr *et al.* (2013) have documented the potential of bacteriophage particles to adhere reversibly to mucus. We expect that further investigations of the relationship between the ecological milieu of VoMs and their morphologies would be profitable.

Factors affecting rates of VoM population growth

Because of VoM random diffusion, most models assume that VoM population dynamics follow mass action principles where host infection rates are directly proportional to the concentrations of VoMs and their hosts (Dennehy, 2014). That is, the greater the concentration of VoMs and hosts in a fluid, then the greater the probability of collisions leading to infection. Growth of a population of size N can be described by a differential equation that accounts for births (b) and deaths (d) over time:

$$\frac{dN}{dT} = (b - d)\,N \tag{3.1}$$

In evolutionary biology, births minus deaths, $b - d$, over short intervals is termed the per capita growth rate, r, which often is used as an estimate of absolute fitness, a.k.a. reproductive capacity (Fisher, 1930), discussed later in Equation 3.2. Population growth in this model shows a characteristic concave-up curve, commonly termed as exponential growth (Fig. 3.1A).

Each of the major VoM life history patterns – lytic, lysogenic (or equivalent), and chronic (see Chapter 1) – will exhibit exponential growth, but at varying rates. Lytic VoMs, given sufficient environmental densities of host cells, will have growth rates greatly exceeding host growth rates. By contrast, chronic and especially lysogenic VoMs (existing as prophages or proviruses) will grow at rates more closely matching those of their hosts. The major

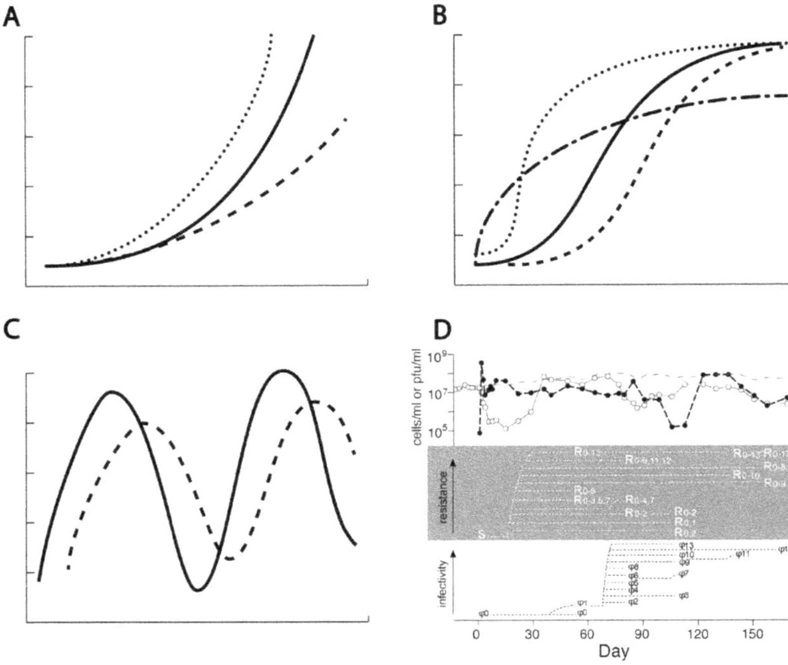

Figure 3.1 Population growth over time is depicted for three functions (A–C) and for experimental data (D). In panel A, the characteristic concave-up curve of exponential growth is shown (solid line). The shape of this function changes with increasing (dotted line) or decreasing (dashed line) growth rates. In Panel B, the incorporation of resource limitation into the population growth model tempers growth rates at higher densities, giving a distinctive S-shaped curve (solid line). Increasing (dotted line) or decreasing (dashed line) growth rate will shift the curve to the right or left, but they both approach the same maxima, the carrying capacity. Reducing resources will have the effect of reducing maximal carrying capacity (dash-dot line). Panel C shows Lotka-Volterra style predator (dashed line) and prey (solid line). Predator numbers usually correspond to prey numbers with a slight lag and a reduced magnitude. VoMs behave differently in that their numbers usually exceed that of their 'prey.' Panel D (modified from Marston *et al.*, 2012) shows the population dynamics of a bacterium, *Synechococcus* (black circles), and a virus, RIM8 (open circles), in a chemostat (top third). For reference, the dashed line is bacterial abundance in the control, virus-free chemostat. The middle and bottom panels show host and virus phenotypes found at six time points. Host-range mutants are numbered in their order of infectivity (e.g. ϕ1–ϕ12), with higher numbers indicating the ability to infect a greater number of host phenotypes. Host phenotypes are labelled by their ability to resist infection by each host-range mutant. For example, S (sensitive to RIM8) is the ancestral host, and R0–2 is resistant to ϕ0, ϕ1, and ϕ2. Dashed lines are hypothetical evolutionary histories based on the most parsimonious interpretation of the data. Reprinted with permission from Marston *et al.* (2012).

distinction between these modes of life is the degree of host exploitation effected by VoMs. Lytic viruses are parasitic, while proviruses and chronic viruses have the potential to be commensalistic or even mutualistic with their hosts. For instance, it has long been known that bacteriophage lysogens can carry genes that in many cases are likely to be beneficial to host fitness (Andersson and Banfield, 2008, Bondy-Denomy and Davidson, 2014, Feiner *et al.*, 2015). Examples include the CTXϕ phage-encoded cholera toxin (Waldor and Mekalanos, 1996) and the *bor* gene of phage λ, which provides serum resistance to *Escherichia coli* (Barondess and Beckwith, 1990; 1995). In addition, prophages often confer immunity to coinfection to their hosts, and may enable hosts to survive environmentally unfavourable periods (Brüssow *et al.*, 2004; Paul, 2008).

The exponential growth model shown in Equation 3.1 is deterministic since the only variables are initial population size and reproduction rate. In reality, VoM population dynamics depend on many other factors, including stochastic, that is, random influences. An additional complication not addressed in Equation 3.1 is that VoM per capita growth rate (r) can depend on the ratio of adsorbed VoMs to susceptible hosts found within a system (the multiplicity of infection or MOI). For many VoMs, growth rates are lower at high MOIs (i.e. negative density-dependence) because infected hosts cannot support additional virus infections while hosts that are not yet infected are not a rapidly renewable resource. That is, at MOIs $<< 1$, uninfected hosts can readily replenish themselves via replication, whereas at MOIs $>> 1$, essentially all hosts will become infected which, for lytic infections, results in a reduction of the ability of a system to replenish uninfected cells. Thus, the total number of hosts available for infection over time can be greater when MOIs are low, but ultimately more host infections over the same period means higher VoM growth rates along with higher resulting MOIs (Dennehy *et al.*, 2006).

Surprisingly, *Pseudomonas* phage $\phi 6$ shows positive density-dependence, that is, greater phage growth rates at higher MOIs under certain conditions (Dennehy *et al.*, 2006). While the mechanism for this growth pattern is not entirely clear, it seems to stem from an enhancement of infection when multiple viruses attack the same cell (Joseph *et al.*, 2009). Allee first described positive density-dependence of population growth with reference to growth limitations in small populations due to limited access to mates (Allee, 1931). Several theoretical studies suggest that positive density-dependence may be an important factor in host–parasite evolutionary ecology (Gerla *et al.*, 2013; Regoes *et al.*, 2000; Weitz and Dushoff, 2008). While positive density-dependent growth has been experimentally observed in several host–parasite systems, including bacteria–*Daphnia* (Ebert *et al.*, 2000; Little and Ebert, 2000), microsporidian–mosquito (Agnew and Koella, 1999), prion–mouse (McLean and Bostock, 2000), freshwater diatom–fungal parasite (Gerla *et al.*, 2013), Φ6–*Pseudomonas* however remains the only virus–host system where positive density-dependent growth has been observed (Dennehy *et al.*, 2006). Whether this is due to an intrinsic property of this particular phage, or simply due to multiplicity reactivation (coinfection-dependent repair of damaged virus genomes) or its genome being split into different virion particles, is unknown.

Another complication on VoM population growth is that host numbers are finite for reasons that are independent of virus infection. Limitations on environmental resources ultimately circumscribe the number of hosts that a habitat can support, which, more formally, is referred to as carrying capacity. The number of hosts for a given habitat changes over time also due to factors such as host reproduction, environmentally induced host death (including VoM-induced host lysis), and acquisition of resistance mechanisms to environmental antagonists, such as chemical and physical antagonists but also predators and viruses, which can have the effect of reducing, for example, host replication efficiency. As such, VoM population dynamics can be described by Verhulst-Pearl logistic models, where the population size over time follows a sigmoid or S-shaped function (Gause, 1934; Pearl, 1927; Pearl and Reed, 1920). This model can be formalized using the differential equation,

$$\frac{dN}{dT} = rN\left(\frac{K-N}{K}\right) \tag{3.2}$$

where N is population size and K represents the maximal possible value of N (Fig. 3.1B). Recall that r is the per capita growth rate defined as the difference between rates of births and rates of deaths, thus rN is equivalent to $(b - d)N$ in Equation 3.1. What Equation 3.2 indicates is that as N becomes larger, as ultimately bounded by K, then the rate of increase in N as a function of time (t) is expected to decrease: organismal crowding is expected to slow rates of organism population growth.

Closer inspection of growth curves derived from this model reveals an inherent asymmetry between host and VoM growth rates. This asymmetry results from the fact that hosts reproduce by binary fission while VoMs produce 10–100 progeny or more per cycle of reproduction over similar intervals. It is clear that only a few rounds of reproduction over similar time intervals will be required for VoMs to outstrip their hosts. This observation supports reports of VoM superabundance (~10–20:1) in marine (Chibani-Chennoufi *et al.*, 2004; Güemes *et al.*, 2016; Wigington *et al.*, 2016) (see also Chapter 6) and terrestrial (Ashelford *et al.*, 2003; Güemes *et al.*, 2016; Parikka *et al.*, 2017) (see also Chapter 4) habitats. Alternatively, the VoM growth advantage is dependent on host density, with a high VoM advantage predicted at higher host densities (as a result of greater similarity between VoM and host generation times) and a low or lower VoM advantage at lower host densities (where VoM generation times can be potentially much greater than those of hosts due to the substantial extensions of the virion search stages of VoM life cycles). A consequence of VoM superabundance, when it occurs in conjunction with a sufficiently large host population, can be extreme competition among VoMs for limited hosts, exerting strong evolutionary pressure on VoMs to optimize infection mechanics and host utilization.

Predator–prey dynamics

While competition among VoMs plays a role in shaping VoM life history traits, the influence of interactions with their hosts are essential, both molecularly and ecologically. In terms of ecology, a standard means of considering these virus–host interactions is in terms of predator–prey dynamics. The first explicit models of predator–prey dynamics were introduced independently by Lotka and by Volterra (Lotka, 1925; Volterra, 1926). To apply Lotka and Volterra's model to VoMs, assume that, in the absence of VoMs, host numbers increase exponentially as described in Equation 3.1. By contrast, VoM numbers will decline in the absence of prey:

$$\frac{dV}{dt} = -d_1 V \tag{3.3}$$

where V is the number of VoMs and d_1 is the VoM instantaneous death rate. That is, in the absence of hosts, and therefore of the possibility of VoM births, VoM population size will be affected by VoM death alone, with that rate of mortality described by d_1. Recall, in Equation 3.1 that d, as equivalent to d_1, is also proceeded by a minus sign. Equation 3.3 thus is simply Equation 3.1 with b set to zero (no births), d set to d_1, and N replaced with V.

If VoMs and hosts are introduced into a limited space, then the host growth rate, r_1, will be reduced by a factor dependent on VoM density,

$$\frac{dH}{dt} = (r_1 - \alpha V)\, H \left(\frac{K - H}{K}\right) \tag{3.4}$$

where α is the virus-to-host adsorption rate and H is the number of hosts. Note that here αV is literally a host mortality rate. Thus, Equation 3.4 is simply Equation 3.1 with r_1 as equivalent to r which is actually $b - d$, but with d now the virus-independent host death rate while αV is the virus-dependent host growth rate. Also note that we include, from Equation 3.2, the carrying capacity term, K, which limits maximal host growth. What α represents is the likelihood that a given virus particle will come to infect a given host cell as a function of time. On a community-wide basis, that is, taking into account that more than one virus particle (V) and more than one host cell (H) will be present, the overall rate of viral infection and therefore virus-induced host death rate is equal to αVH, while host growth rate as independent of virus action is equal to $r_1 H$, i.e. with both expressions ($r_1 H - \alpha VH$) products of multiplying Equation 3.4 out.

The VoM population will increase at a rate dependent on host density:

$$\frac{dV}{dt} = (\beta \alpha H - d_1)\, V \tag{3.5}$$

where β is the viral burst size, that is, number of virus progeny produced per infected host cell. Note, upon multiplying out Equation 3.5, that we once again see the expression, αVH. Each newly formed virus-infected host cell (as generated at rate, αVH) thus gives rise to one burst size, i.e. $\beta \alpha VH$,which in turn is the virus birth rate. Equation 3.5 thus is simply Equation 3.1 with b replaced with $\beta \alpha VH$, d replaced with d_1 and V replacing H.

Under these simplistic conditions, VoM and host populations will oscillate in a systematic manner, but showing a high degree of stability (Fig. 3.1C). That is, host populations will grow in size until brought under control by virus populations. Virus populations will grow until they significantly reduce host populations. Reduced host populations will recover once virus populations have declined given the absence of sufficient numbers of hosts. These phenomena will all happen in the indicated sequence, with each step taking a certain amount of time, and with the resulting delays giving rise to oscillations when virus and host densities are each graphed as functions of time, i.e. with Equations 3.4 and 3.5 linked together. Extensions of the basic Lotka–Volterra model have included resource limitations on host density, the time lag due to VoM infection (i.e. latent) periods, allowances for differences in host specificity (e.g. broad host ranges), and the acquisition of virus resistance by hosts (Bohannan and Lenski, 1997; Campbell, 1961; Kerr *et al.*, 2008; Korytowski and Smith, 2017; Lenski, 1988; Lenski and Levin, 1985; Levin *et al.*, 1977; Stopar and Abedon, 2008). The bottom line, however, is that it is possible, using simple models, to gain insight into virus–host community dynamics: more hosts will tend to result in more viruses. More viruses will tend to result in fewer hosts, especially assuming lytic infections.

VoM impact on host populations and communities

Darwin suggested that 'mutual relations' among organisms drives biological diversification (Darwin, 1859). It is easy to imagine that evolutionary arms races among VoMs and hosts can splinter populations into myriad resistant and counter-resistant variants. Indeed, studies have shown that diversification is a frequent occurrence in both static microcosms, i.e. artificial environments consisting of limited volumes and no mixing (Buckling and Rainey, 2002; Buckling *et al.*, 2006; Morgan *et al.*, 2009), and chemostat communities, i.e. artificial environments consisting of effectively infinite volumes with substantial mixing (Chao *et al.*, 1977; Koskella and Brockhurst, 2014; Marston *et al.*, 2012). For example, at least *four* cycles of adaptation–counteradaptation between the cyanobacterium *Synechococcus* and the VoM RIM8 were observed during 6-month chemostat experiments (Fig. 3.1D) (Marston *et al.*, 2012). In these experiments, up to 13 newly evolved RIM8 variants and 11 *Synechococcus* variants appeared in their respective coevolving populations. In this section, we consider two associated phenomena stemming from these mutual relations. The first is that of VoM–host coevolution, particularly in terms of limitations that often can be seen on that coevolution in experiments, and the second is the result that this coevolution can have on host and indeed community diversity.

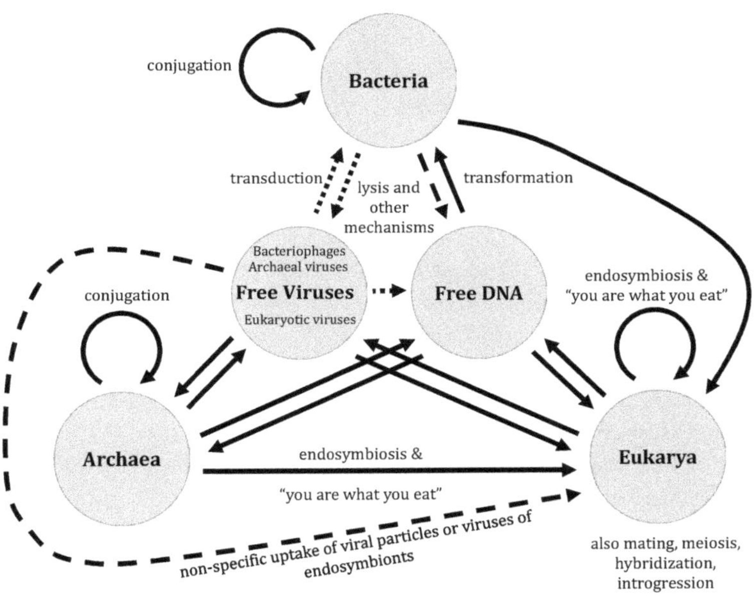

Figure 3.2 Viruses of microorganisms significantly impact the cycling of organic matter and nutrients in the ocean. The shuttling of resources among trophic levels is interrupted by viruses (virus shunt). In addition, viruses affect the biological pump between the surface and deep ocean.

VoM–host coevolution and its limitations

The general findings of predator–prey models are that (1) VoMs will limit host populations below the carrying capacity provided by available resources, (2) host populations can fluctuate in a cyclic manner and 3) coexistence can occur over long periods. Experimental support from coevolution studies for these theoretical predictions is mixed. Frequently hosts have stabilized at densities similar to controls, with VoMs persisting at relatively low densities [see Dennehy (2012), for review]. In other cases, VoMs were driven to extinction either with or without hosts first being driven to extinction. The most likely interpretation of the former result is that, although resistant hosts came to dominate the habitat, some VoMs were able to persist due to the continued presence of a small number of sensitive hosts in a spatial or numerical refuge (Chao *et al.*, 1977; Schrag and Mittler, 1996). In the latter case, given host persistence, VoMs were likely unable to counter host resistance. The main point here is that, due to the rise of resistant hosts, and resulting dampening effect that can have on overall rates of host population growth, demonstrations of repeated cycles of host–parasite coevolution in experimental microcosms have been somewhat rare.

This lack of repeated cycles given host persistence implies a lack of repeated VoM adaptation and host counter-adaptation during VoM–host co-cultivations, or at least that such dynamics typically are not dominating during VoM growth within microcosms. One possible reason for the relative lack of such arms races between co-cultured VoMs and hosts is that there is an asymmetry between VoM and host evolutionary potential. Because of their high mutation rates and large population sizes, microbial host organisms are able to search the VoM-resistance sequence space relatively rapidly, that is, acquire successfully VoM-evading adaptations. While VoMs can have even greater population sizes and rates of mutation than their hosts, the larger genomes of hosts versus VoMs may provide hosts with greater means to evolve VoM resistance. By contrast, the relatively small VoM genomes may be less well equipped to respond to challenges of host resistance (Lenski and Levin, 1985). First, the impacts of mutations may be more significant in VoMs than in their hosts; up to 40% of RNA and ssDNA virus mutations are lethal (Sanjuán *et al.*, 2004; Sanjuán, 2010). Second, with the possible exception of giant viruses, VoMs simply do not have much spare genetic material to work with. This is particularly acute when hosts can modify their resistance to VoMs via mutational loss of information (e.g. gene inactivation). In other words, if host mutations to resistance are ones of loss of function, while virus host-range mutations are ones of gains of function (even if simply change of function), then it is easier for a host to mutate to resistance than for a virus to mutate to counter that resistance. Due to their limited genomic repertoire, VoMs often must mutationally change without substantially impacting gene function, although VoMs may be more effective during VoM–host coevolution if they target, for virion adsorption, host molecules that the host is unable to do without, thereby forcing hosts as well to mutationally change also without substantially impacting gene function. In addition, various hosts have been shown to possess adaptive immune systems, i.e. CRISPR (Lundgren, 2016).

VoM genome sizes may be limited by the capacities of their capsids. Hosts, by contrast, often can easily assimilate even relatively large blocks of new genetic material, e.g. as genomic islands. This asymmetry may be exacerbated by the long history of laboratory culturing of the hosts and parasites typically used in experimental coevolution studies (e.g. *E. coli* and its phages λ, T4, T7). These laboratory-adapted strains, that is, may lack evolutionary potential for extensive coevolution because they have been selected under constrained conditions for hundreds of generations. This idea is supported by observations of repeated rounds of coevolution among hosts and parasites recently isolated from the wild (Barnet *et al.*, 1981; Buckling and Rainey, 2002; Marston *et al.*, 2012). It is believed that the wild-isolated VoMs have greater evolutionary potential, and are able to sustain more rounds of host–parasite coevolution than are laboratory-passaged strains.

How these laboratory studies relate to host–VoM evolutionary dynamics in the wild is poorly understood (Koskella and Vos, 2015). Wild populations are subject to much greater spatial, environmental, and biotic heterogeneity, significantly affecting community structure. Moreover, resistance to VoMs usually incurs fitness costs (Bohannan and Lenski, 2000; Buckling and Rainey, 2002; Chao *et al.*, 1977; Lenski, 1988; Levin *et al.*, 1977; Schrag and Mittler, 1996, Wielgoss *et al.*, 2016). Consequently, while resistant hosts may predominate in the presence of VoMs, they should be outcompeted by VoM-sensitive hosts in the absence of VoMs. This can ensure that any shifts in host-population susceptibility to VoMs may be transient. As long as some VoMs can persist during the fallow period without susceptible hosts, then their local populations may be able to recover following a precipitous drop in population size due to host resistance. VoM–host coevolution likely is an important aspect of VoM ecology, although extrapolating from laboratory studies to dynamics in the wild may require careful consideration.

VoMs as drivers of host biodiversity

A widely held notion in ecological theory is that there is an inherent tradeoff between maximizing reproduction and minimizing predation (Ludwig and Rowe, 1990; Pianka, 1976). This tension should promote the diversification of prey populations into competitive and defensive specialists (Wage *et al.*, 2014; Winter *et al.*, 2010). When predators are absent, competition specialists should dominate since they have a reproductive advantage over defensive specialists. Resulting burgeoning numbers of these competition specialists, however, will allow for the propagation of predators, freeing up resources for survival-maximizing defensive specialists. Predators under these conditions will consequently tend to decline in number, allowing competition specialists to increase again in prevalence. Such frequency-dependent selection among alternative variants is commonly referred to as 'killing the winner' (Korytowski and Smith, 2017; Maslov and Sneppen, 2017; Thingstad, 2000; Weinbauer and Rassoulzadegan, 2004). These dynamics should encourage an increase in the microbial diversity of both prey and predator species.

Given that chemostats utilized in laboratory coevolution studies are used to create minimally varying microhabitats, we might expect coevolution in the wild to result in much greater diversification given the much higher degree of environmental heterogeneity. This notion is supported by the geographic mosaic theory of coevolution, which holds that populations respond to ever-changing patterns of interactions across space and time (Thompson, 1994; 2012). Here fragmentation of the natural landscape allows local populations of one species to adapt to local populations of other species in a genotype-by-genotype-by-environment interaction. As these interactions play out across larger spatial and temporal scales, and as different subpopulations interact or fail to interact, coevolutionary changes can result in increased biodiversity. VoM hosts thus are expected to be more genetically and phenotypically diverse given the presence of VoMs versus without VoM presence in the environment.

VoM ecological adaptation

Like all organisms, VoMs are a collection of features and traits shaped by natural selection, but the story of evolution is not simply one of continual small improvements in each trait. Rather, selective forces on organismal traits often can be in direct opposition, that is, improvement in one trait can entail a reduction in performance of another. For example, capsid thickness can be correlated with virus survival, but inversely correlated with virus multiplication rate (De Paepe and Taddei, 2006; García-Villada and Drake, 2013). Natural selection, then, will favour a capsid thickness that represents the optimal tradeoff between virus survival and reproduction such that a maximum number of surviving offspring are produced over time for a given set of environmental circumstances. Since environments are often changing, natural selection is expected to be continuously recalibrating the optimal balance between specific traits.

Generation time as an optimizable trait

Nowhere is the issue of tradeoffs in the optimization of different traits more apparent than in the decision of when VoMs should release progeny from infected host cells, such as via host-cell lysis. Lysis timing for many VoMs is controlled by membrane-permeabilizing molecules, such as bacteriophage holin proteins (Saier and Reddy, 2015; Young, 2014). With phage λ, holin protein accumulates in the *E. coli* inner membrane until it reaches a critical threshold concentration, whereupon it nucleates to form holes (Young, 2014). These holes, with bacteriophages, allow lysozyme-like molecules (i.e. endolysin) to disrupt the cell wall. The cell subsequently ruptures and VoMs are released into the surrounding medium. Holin nucleation is under genetic control and phage λ holin-gene mutants expressing different lysis times have been constructed (Dennehy and Wang, 2011; Singh and Dennehy, 2014). Thus, the timing of lysis could be a trait malleable by natural selection. However, when should a VoM lyse its host?

Assuming that progeny are assembled at a linear rate (Wang *et al.*, 1996, 2006), a VoM that lyses its host early is expected to forgo exploitable host resources, i.e. host cell components which could have been converted into additional virions, and exposes itself to the risk of destruction in the external environment until another host can be acquired. While a VoM can increase progeny production by delaying lysis, the longer a VoM waits, the greater the risk that the host cell will die in a manner that also destroys accumulated virions before lysis can be effected. Moreover, the continued exploitation of host resources likely has diminishing returns over time as host resources are exhausted, that is, deviation away from linear rates of virion progeny production, which could also favour earlier rather than later lysis. Overall, natural selection likely favours genotypes that strike a balance between risk (e.g. loss of infected bacteria to predation, destruction of freely dispersing VoMs) and reward (i.e. production of additional virion progeny), and there likely are optimal lysis times which serve to maximize VoM population growth rates (Abedon, 1989; Abedon *et al.*, 2001, 2003; Wang *et al.*, 1996; Wang, 2006).

In general terms, then, the time VoMs spend infecting individual host cells can impact the evolutionary fitness of those VoMs. This can especially be the case for VoMs, versus the viruses of multicellular organisms, since virion release from infected cells and virion transmission to new hosts tends to be equivalent for VoMs but less so for non-VoM viruses (see Chapter 1). In addition, hosts of VoMs may be more vulnerable over shorter time scales than are multicellular organisms. Thus, the timing with which VoMs are released from infected cells can directly impact VoM transmission rates and thus generation time as well as the number of virions produced per infected host and also the potential for VoM survival during infection. These factors, in combination with genetically based variation in, for example, lysis timing, together can give rise to a potential for VoM infection–duration optimization. VoMs, that is, likely infect their hosts cells over time periods which are best understood in terms of combination of the specifics of VoM infection characteristics and the ecological contexts of those infections, e.g. number of hosts available, vulnerability of those hosts, etc. The exact nature of lysis time tradeoffs will depend on the exact nature of the ecological circumstances the VoM finds itself in. Making more virions may be a better strategy if extracellular risk of virion inactivation are high. By contrast, releasing virions sooner may be more important when there is strong intraspecific competition by individual virions to reach new host cells first.

A similar tradeoff may exist with respect to host attachment rates. Naively, one might expect this rate to be maximized, presumably because high attachment rates will maximize host infections. However, high attachments rates may be deleterious because they may limit the number of hosts encountered per burst, and therefore reducing productivity per host (Gallet *et al.*, 2009, Gallet *et al.*, 2011). Furthermore, mutations increasing attachment rates could have negative pleiotropic effects, that is, reducing the effectiveness of other aspects of the virus (Pepin *et al.*, 2006). Finally, high attachment rates could be disadvantageous when there are high levels of stationary phase cells or cellular debris to which too effectively adsorbing viruses might non-productively adsorb (Gallet *et al.*, 2012). On the other hand, low attachment rates would prolong search times and expose phages to increased risks of environmentally induced inactivation. How these different selective pressures balance out has been seldom studied.

Coinfection, competition and cheating

Describing the outcomes of competition between organisms is an essential feature of evolutionary ecology. Similar to other organisms, VoMs compete for the resources required to replicate additional copies of themselves. As with any competitive interaction, natural selection favours VoMs that are more efficient at acquiring and using host resources. Notably for VoMs, this competition occurs not only extracellularly, as VoMs compete for hosts, but also intracellularly as VoMs compete for host resources (Dennehy *et al.*, 2013; Nguyen *et al.*, 2014; Preisig *et al.*, 1998; Trinh, *et al.*, 2017; Turner, 2005; Yau *et al.*, 2011). In the case of coinfection or a multiple/sequential infection (superinfection), or even reactivation of a latent infection, proteins translated from a VoM's genome enter a common pool and are therefore susceptible to co-option by competing genotypes already present in the cell. That is, and in particular, some viruses can package their genomes into capsids containing proteins that they themselves did not cause to be produced. This parasitism enables the 'cheating' virus to specialize on other facets of their own reproduction, such as increasing genome replication speed and packaging efficiency. An example of this type of competition comes from coliphage P2 and its satellite P4 (Six and Klug, 1973). P4 lacks genes for head, tail and lysis proteins, but is able to package its genome in the P2 capsid (Deho and Ghisotti, 2006). As such, P4 is able to concentrate its resources on genome replication while parasitizing capsids from P2.

Viruses or virus-like elements that require other 'helper' viruses in order to reproduce include satellite viruses (Deho and Ghisotti, 2006; Krupovic *et al.*, 2016; Palukaitis, 2016; Six and Klug, 1973), viroids, virusoids (Symons, 1991), virophages (Bekliz *et al.*, 2016; Fischer and Suttle, 2011; La Scola *et al.*, 2008) (see Chapter 12), and defective-interfering (DI) particles (Horiuchi, 1983; Huang, 1973; Roux *et al.*, 1991). As a rule, these types of parasitic viruses only arise when coinfection is common – that is, especially more than one type of virus infecting individual cells. Consequently, there should be strong selection on a potentially parasitized virus to sequester a host and limit coinfection. The outcome of intense ecological competition is reflected in the numerous mechanisms of resistance to superinfection (Abedon, 2015; Dedrick *et al.*, 2017; Hyman and Abedon, 2010; Roux *et al.*, 2016; Syller and Grupa, 2016). For example, some lambdoid phages bind outer membrane protein FhuA to infect *E. coli* (Uc-Mass *et al.*, 2004). Competing phages express a protein, Cor, which blocks FhuA-mediated ferrichrome uptake, and prevents FhuA-mediated phage infection (Uc-Mass *et al.*, 2004).

Productivity versus latency

Some VoMs can enter long-term associations with host cells such as by integrating their genome into that of the host, which for bacteriophages has been described as lysogeny but more generally can be described as a form of latent infection. These VoMs, now existing as proviruses (and/or as prophages in the case of bacteriophages) only 'replicate' when the host cell replicates. As a consequence, and if uninfected host cells are otherwise fairly prevalent within an environment, then this life history strategy entails a significant reduction in reproductive output compared to the lytic strategy (Abedon *et al.*, 2001). Since natural selection typically will maximize reproduction, lysogeny is likely the product of a specific selective condition, namely where the probability of locating a new host is relatively low (Abedon *et al.*, 2001; Wang *et al.*, 1996).

The best-studied lysis–lysogeny system is that of the bacteriophage λ (see Chapter 7). When growth conditions are favourable, an important regulator of λ gene expression, *CII*, is rapidly degraded by host proteases, and the likelihood of the lytic pathway dominating is high. By contrast, when growth conditions are poor, *CII* accumulates, increasing the probability of entering the lysogenic state (Herskowitz and Hagen, 1980; Oppenheim *et al.*, 2005). Ecologically, this decision process is intuitive because the essential 'life decision' a temperate phage must make is whether to accept a low rate of replication in exchange for shelter or instead to potentially risk infecting lytically for the potential payoff of greater replication in additional hosts.

Information sources by which a lysogen can base this 'decision' include: host physiology, the presence of coinfecting viruses and small molecules from other infected hosts (Abedon, 2017). A favourable host physiological state may signal the likely presence of available hosts for progeny, whereas the opposite condition may imply that suitable hosts are in short supply. Alternatively, the presence of competing VoMs within a host implies that VoMs are common relative to hosts and uninfected hosts may not be available. Indeed, coinfection can reinforce the lysis inhibition pathway, another bacteriophage-associated lysis-delaying phenomenon, prolonging infection state (Abedon, 1990; Abedon *et al.*, 2003; Weitz *et al.*, 2008).

Interestingly, a greater penalty probably arises from not retaining infections if no new host cells are available or if all the other host cells are occupied versus choosing to stay in place (retaining infections) when permissive hosts are available. In the former, a bad decision (release without available hosts) results in eventual VoM destruction, whereas in the latter (lack of release with available hosts), the cost is a reduced reproduction rate compared to competing genotypes. This asymmetry should have interesting implications for the evolution of control mechanisms dictating the lysis–lysogeny decisions that occur at the start of infections of phages or, more generally, viruses that are able to choose between latent versus productive infection (Abedon, 2009). Related to this, exit from lysogeny is often predicated by damage to the host DNA by agents like UV light and mitomycin C (Cochran et al., 1998; Weinbauer and Suttle, 1999). As an analogy, if one's house were on fire, one would attempt escape regardless of the external conditions. Since death in the former is likely, lysis pays off regardless of the probability of survival outside the host.

Once a VoM chooses a latent-infection strategy, its fate is linked over longer periods to that of the host. If the host should die before replicating, then the VoM also dies. Not surprisingly, temperate VoMs – ones capable of displaying latent infections – often express traits that are beneficial to its host (Bondy-Denomy and Davidson, 2014; Cumby et al., 2012a). These traits may include improving host growth (Dykhuizen et al., 1978), improving host ability to colonize larger organisms (Vica Pachecho et al., 1997; Wood et al., 1996), enhancing host ability to resist infection by other VoM individuals (Soutourina et al., 2013; Vostrov et al., 1996), encoding of virulence factors (Tyler et al., 2004; Waldor and Mekalanos, 1995), protecting hosts from adverse environments including immunity (Figueroa-Bossi and Bossi, 1999; Schuch and Fischetti, 2009; Vaca-Pacheco et al., 1999; Wang et al., 2010), and improving host metabolic and photosynthetic capabilities (Rohwer and Thurber, 2009). Latent viruses also can acquire mutations abolishing their ability to exit host genomes, thus becoming permanent residents of their hosts (Casjens, 2003).

VoM genome evolution

Microorganisms span the three domains of life and their VoMs reflect this breadth in the variety of their genomes. Out of the seven virus classes within the Baltimore virus classification system, six classes are known to parasitize all major phyla of microorganisms while only one (Group 7, or non-retroviral dsDNA viruses that replicate through a ssRNA intermediate) does not contain exemplars that infect microorganisms. Despite this great diversity, not all virus classes are represented equally. Almost all of the over 100 characterized viruses infecting archaea hold their genetic information as dsDNA (Mochizuki et al., 2012; Prangishvili, 2013; Prangishvili, 2015; Snyder et al., 2015), while viruses infecting algae have genomes ranging from ssRNA to dsDNA (Van Etten et al., 1991; Short, 2012). Viruses of fungi tend to be dsRNA (Göker et al., 2011; Vainio et al., 2011, Feldman et al., 2012), while representatives of at least four types can be found among the bacteriophages (Calendar and Abedon, 2006). In this section we provide primers on VoM genomics and genome evolution. See Chapter 2 for further overview of VoM genomics.

VoM genomes

Virus nucleic acid characteristics can have a large effect on genomic attributes such as overall genome size and mutation rates. Exemplifying the upper limit of VoM genome size are the recently discovered families of amoeba viruses, *Mimiviridae*, *Marseilleviridae* and *Pandoviridae*, which have dsDNA genomes larger than 2.5 Mb, surpassing even some bacteria (La Scola et al., 2003; Monier et al., 2008; Colson et al., 2013; Philippe et al., 2013) (see Chapter 11). By contrast, some ssRNA phages have genomes as small as 3.3 Kb (Hatfull and Hendrix, 2011), a 758-fold difference. Mutation rates show tremendous variation as well, with values ranging from 1.1×10^{-3} nucleotide substitutions per site per cell infection (s/n/c) in the ssRNA phage Qβ to 9.8×10^{-8} s/n/c for the bacteriophage T2 (Sanjuán et al., 2010). Generally, VoMs that store their genetic information as RNA or ssDNA tend to have smaller genomes and higher mutation rates, while dsDNA VoMs have larger genomes and lower mutation rates (though some ssDNA genomes can still be relatively large, particularly Aeropyrum coil-shaped virus, which has a genome size of nearly 25 kb [see Chapter 8]). These differences are believed to reflect differences in polymerase fidelity and nucleic acid stability (Duffy et al., 2008).

In addition to size and stability, VoM genomes show considerable structural variation. Both circular and linear permutations are common, and do not seem to correlate with any obvious factors. In addition, ssDNA, ssRNA and dsRNA VoM genomes can be segmented (Ojosnegros et al., 2011, Iranzo and Manrubia, 2012; Moreno *et al.*, 2014). Numerous hypotheses in particular have been proposed to explain viral segmentation, including increased capsid stability (Iranzo and Manrubia, 2012; Ojosnegros *et al.*, 2011), virus inter-viral genome recombination (i.e. sex; Chao *et al.*, 1992), and copying fidelity (Pressing and Reanney, 1984) (see Chapter 9). Although a consensus is not available, observations show that the reassortment of virus segments can be an important factor in the generation of virus genetic diversity (O'Keefe *et al.*, 2010, Marshall *et al.*, 2013) and in host switching (Dennehy, 2016).

Horizontal gene transfer and genomic mosaicism

VoMs are extremely successful mediators of gene shuffling (Fig. 3.3). Regardless of reproductive strategy (lytic vs. latent vs. chronic) or type of genetic material (RNA versus DNA), many VoM genomes contain an amalgam of genetic information from a variety of sources, commonly referred to as genomic mosaicism (Casjens and Thuman-Commike, 2011, Pope *et al.*, 2015). For example, the tail genes of phages HK97 and λ show a high degree of sequence similarity, but the head genes are wholly dissimilar (Juhala *et al.*, 2000). The most parsimonious explanation for this observation is that the head genes of HK97 and λ, but not the tail genes, were acquired from two evolutionarily divergent sources.

The life history circumstances of VoMs may predispose them to genomic mosaicism since this is generated through recombination or reassortment with foreign DNA. As obligate parasites, VoMs reproduce inside cells containing separate and distinct genomes as well as coinfecting viruses, transposons, plasmids and even transformed foreign DNA (Desnues *et al.*, 2012). Given this proximity, the accidental integration of foreign DNA into VoM genomes is not surprising. While capsid size can limit the extent of integration of new genetic material, recombination and then retention of similar sized genetic fragments may be relatively common. In fact, it is frequently suggested that the modular architecture of VoM genomes specifically evolved to allow suites of genes dedicated to a common function to be traded among VoM populations, creating a vast meta-gene pool (Campbell and Botstein, 1983; Hendrix *et al.*, 2000). Some authors even go so far as to suggest that VoMs and their hosts constitute a superorganism, with access to a global gene pool (Weinbauer and Rassoulzadegan, 2004). Consistent with expectations, this genomic mixing is more pronounced in generalists – that is, VoMs capable of infecting greater diversities of cellular organisms – as they are more promiscuous in host range than specialists (Born *et al.*, 2011). A non-genetic form of mosaicism may also arise in the form of phenotypic mixing, when coinfecting VoMs can produce proteins that interact (Loverdo and Lloyd-Smith, 2013).

VoM impact on host genetics

Horizontal gene transfer and genomic mosaicism is common among hosts of VoMs as well. It has been best studied among bacteria (McDaniel, 2012, Baltrus, 2013, Boritsch *et al.*, 2016), but is known to occur in all domains (Bock, 2010; Hotopp, 2011). A majority of all sequenced bacterial genomes contain prophages (Canchaya *et al.*, 2003, Touchon *et al.*, 2016). These latent viruses comprise up to 20% of some bacterial genomes (Klumpp *et al.*, 2013) and a significant portion of the total planetary phage genes are contained within prophages (Forterre and Prangishvili, 2013). There are, in addition, numerous examples of phage-mediated gene transfer with significant implications for public health and other fields (Brüssow *et al.*, 2004; Klumpp and Fuchs, 2007; Chen and Novick, 2009; Imamovic *et al.*, 2009; Saunders *et al.*, 2001). For example, lysogenic conversion of *Corynebacterium diphtheriae* by the corynebacteriophage β renders the bacterium toxigenic as the phage encodes diphtheria toxin (Freeman, 1951). Although antibiotic resistance genes (ARGs) are thought to be widespread in phage DNA obtained from environmental isolates, and Colomer-Lluch and colleages generated resistant clones after transfection with phage DNA (Colomer-Lluch *et al.*, 2011), a recent study by Enault *et al.* suggests that ARG estimates in phage genomes are greatly overestimated (Enault *et al.*, 2016). Clearly VoMs are important components of microbial horizontal gene transfer, resulting in the gain, loss or reshuffling host genes (Andersson, 2009; Filée *et al.*, 2002; Rowher and Thurber, 2009; Krupovic *et al.*, 2011, Haaber *et al.*, 2016).

A general category of virus- and especially phage-encoded genes that appear to be new additions to

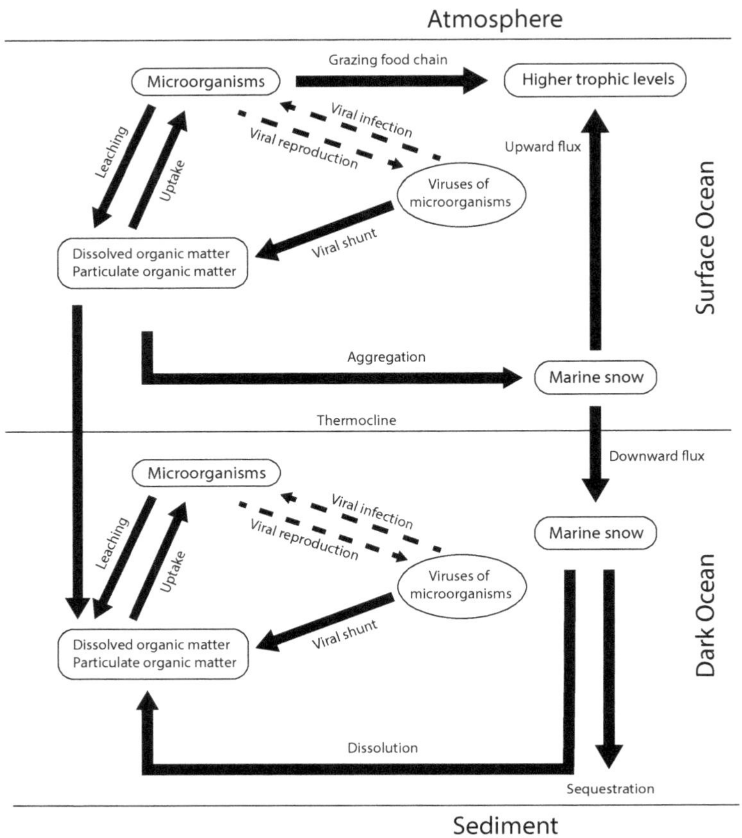

Figure 3.3 Horizontal gene transfer across domains. Phage-dominated (dotted lines) and phage-influenced (dashed lines) are emphasized. Reprinted with permission from Abedon (2009).

virus genomes and which often effect lysogenic conversion, that is, the expression of genes by prophages during lysogenic cycles, have been termed morons. Morons, intriguingly so named because they add '*more* DNA' to phage genomes, often contain their own separately regulated promoters and terminators. Although non-essential to the phage, they may allow prophages to escape deletion from host populations by providing to hosts beneficial genes such as virulence factors, restriction endonucleases, superinfection exclusion proteins and toxin–antitoxin pairs (Cumby *et al.*, 2012a,b). For example, morons encoding superoxide dismutases confer *Salmonella* resistance to reactive oxygen species produced by the mammalian immune system (Figueroa-Bossi and Bossi, 1999).

Morons, once acquired by a bacterial host, can also form pathogenicity islands (PAIs), encoding virulence factors (Hensel and Schmidt, 2008). As present in pathogenic, but not related non-pathogenic bacterial strains, these presumptive phage- or plasmid-mediated virulence factors presumably increase bacterial fitness by providing adhesins, toxins, invasins and protein secretion systems used in host infection (Hacker and Kaper, 2000). Recent theoretical modelling predicts that genes contributed by viruses can increase the evolvability of their hosts, allowing them to better explore their adaptive landscape (Williams, 2013, Zaman *et al.*, 2014). In doing so, there is no doubt that VoMs play a role in shaping microbial communities over time (Koskella and Meaden, 2013). In some cases, coevolution with VoMs can even induce a host to change its genetic code. For example, Taylor and colleagues reported several yeast strains using a modified genetic code where CUG encodes serine rather than leucine (Taylor *et al.*,

2013). Interestingly these authors also found a totivirus which possessed only one CUG codon and was able to infect these yeast strains. Presumably, this totivirus adapted to the yeast by avoiding the modified codon (Taylor *et al.*, 2013). At the other end of the spectrum, Ivanova *et al.* (2014) found a bacteriophage with a genome divided into two domains – one with a host-dissimilar codon usage and one with similar – to first infect and then take over, respectively.

VoMs as ecosystem modifiers

Microorganisms, such as autotrophic bacteria and phytoplankton, form the foundation of food webs as they convert solar energy and dissolved nutrients into particulate organic matter usable by higher organisms. Recent work indicates that more than 90% of marine biomass is composed of microorganisms (Suttle, 2007). Amazingly, VoM lysis reportedly liberates 20–55% of this biomass each day (Suttle, 1994, 2007; Wommack and Colwell, 2000, Winget *et al.*, 2011, Weitz *et al.*, 2015). The destruction of primary producers has the effect of reducing the total available resources to grazers at the next trophic level, reducing their growth, with cascading effects at all higher trophic levels. Through host cell lysis, however, VoMs also can convert organic matter from a particulate form to a dissolved form that is consumable only by osmotrophs (organisms that acquire nutrition by direct uptake of dissolved organic molecules). These osmotrophs are found at the very bottom of the food chain, with their productivity thereby contributing to the productivity of higher trophic levels (Thingstad *et al.*, 2008; Weinbauer *et al.*, 2011). This semi-closed 'viral shunt' thus may reduce the direct flow of energy between the primary producers and osmotrophs and the higher trophic levels (Bratbak *et al.*, 1992; Fuhrman, 1999; Suttle, 2007, Weitz *et al.*, 2015). Indeed, as much as 25% of the carbon fixed through photosynthesis may cycle through the viral shunt (Wilhelm and Suttle, 1999).

Due to the viral shunt, VoMs may significantly affect 'the biological pump', a process that sequesters CO_2 and other nutrients into the deep ocean because of the death and sinking of surface water microorganisms (Suttle, 2007). This effect plays a role in the global carbon cycle and any alterations of this mechanism could influence climate change via a modification of the efficiency of this sequestration process, although the exact nature of this impact is still unclear (Danovaro *et al.*, 2011). The current debate on the ultimate role of VoM-mediated lysis of marine biomass on a global scale centres on several possible outcomes (Brussaard *et al.*, 2008). First, VoMs may short-circuit the biological pump by returning nutrients back to the dissolved phase, which is hypothesized to limit their transport to the deep ocean (Poorvin *et al.*, 2004). Second, VoMs might accelerate the biological pump cycling by promoting the sinking of infected host cells to the deep ocean (Lawrence and Suttle, 2004). One way this might occur is the failure of flagella-driven motility in some infected microbes, disabling their ability to prevent sinking (Lawrence and Suttle, 2004). Finally, VoM-mediated lysis may drive the flow of nutrients to the deep sea via the aggregation and sinking of lysed cellular debris (e.g. marine snow) (Danovaro *et al.*, 2008; Mari *et al.*, 2005). These nutrients may form the basis of the food web at depths where the available sunlight is insufficient to allow photosynthesis.

We expect, due to the high degrees of organismal and environmental heterogeneity, that the activity of the viral shunt and the biological pump will exhibit considerable temporal and spatial variation. Net movement of nutrients among trophic levels and habitats thereby should show significant context dependency. A major challenge for microbial ecologists is to parse major trends in biogeochemical dynamics across widespread ecosystems, including in terms of VoM impact.

Closing remarks

Viruses of microorganisms are a spectacularly diverse, influential, and pervasive group of organisms with much of the world's genetic information potentially contained within their genomes. Our understanding of their biology, however, is not proportionate to their impact. VoMs thus may be described as the dark matter of the biosphere (Filée *et al.*, 2005) – the hidden world that directly or indirectly influences every living organism. As part of that influence, VoMs are masters of shuttling genes, manipulating genomes, and either generating or otherwise increasing biodiversity. They alter biogeochemical cycles, communities, ecosystems, and may even influence global climate and its change.

Understanding how these influences interact to produce our living world requires the application of not just ecology but also evolutionary history, i.e. understanding not just interactions between organisms and their environments but also the evolutionary underpinnings of those interactions. Further study of the evolution and ecology of VoMs can only increase our appreciation of the rich and wonderful fabric of life.

Acknowledgements

The authors are grateful for rewarding discussions with James Carpino, Gregory Lallos, Abhyudai Singh, and past and present members of the Dennehy Laboratory. Funding was provided by the National Science Foundation Faculty Early Career Award (#1148879) to JJD.

References

Abedon, S.T. (1989). Selection for bacteriophage latent period length by bacterial density: a theoretical examination. Microb. Ecol. *18*, 79–88. https://doi.org/10.1007/BF02030117.

Abedon, S.T. (1990). Selection for lysis inhibition in bacteriophage. J. Theor. Biol. *146*, 501–511.

Abedon, S.T. (2008). Bacteriophage Ecology. Advances in Molecular and Cellular Microbiology (Book 15). (Cambridge, UK: Cambridge University Press). https://doi.org/10.1017/CBO9780511541483.

Abedon, S.T. (2009). Phage evolution and ecology. Adv. Appl. Microbiol. *67*, 1–45. https://doi.org/10.1016/S0065-2164(08)01001-0.

Abedon, S.T. (2009). Bacteriophage intraspecific cooperation and defection. In Contemporary Trends in Bacteriophage Research, Adams, H.T., ed. (Hauppauge, NY: Nova Science Publishers), pp. 191-215.

Abedon, S.T. (2011). Envisaging bacteria as phage targets. Bacteriophage *1*, 228–230. https://doi.org/10.4161/bact.1.4.17281.

Abedon, S.T. (2015). Bacteriophage secondary infection. Virol. Sin. *30*, 3–10. https://doi.org/10.1007/s12250-014-3547-2.

Abedon, S.T. (2017). Commentary: Communication between viruses guides lysis-lysogeny decisions. Front. Microbiol. *8*, 983. https://doi.org/10.3389/fmicb.2017.00983.

Abedon, S.T., Herschler, T.D., and Stopar, D. (2001). Bacteriophage latent-period evolution as a response to resource availability. Appl. Environ. Microbiol. *67*, 4233–4241.

Abedon, S.T., Hyman, P., and Thomas, C. (2003). Experimental examination of bacteriophage latent-period evolution as a response to bacterial availability. Appl. Environ. Microbiol. *69*, 7499–7506.

Agnew, P., and Koella, J.C. (1999). Life history interactions with environmental conditions in a host-parasite relationship and the parasite's mode of transmission. Evol. Ecol. *13*, 67–89.

Allee, W.C. (1931). Animal Aggregations. A Study in General Sociology. (Chicago: University of Chicago Press).

Allen, H.K., Bunge, J., Foster, J.A., Bayles, D.O., and Stanton, T.B. (2013). Estimation of viral richness from shotgun metagenomes using a frequency count approach. Microbiome *1*, 5. https://doi.org/10.1186/2049-2618-1-5.

Andersson, A.F., and Banfield, J.F. (2008). Virus population dynamics and acquired virus resistance in natural microbial communities. Science *320*, 1047–1050. https://doi.org/10.1126/science.1157358.

Andersson, J.O. (2009). Gene transfer and diversification of microbial eukaryotes. Annu. Rev. Microbiol. *63*, 177–193. https://doi.org/10.1146/annurev.micro.091208.073203.

Angly, F.E., Felts, B., Breitbart, M., Salamon, P., Edwards, R.A., Carlson, C., Chan, A.M., Haynes, M., Kelley, S., Liu, H., *et al.* (2006). The marine viromes of four oceanic regions. PLOS Biol. *4*, e368.

Ashelford, K.E., Day, M.J., and Fry, J.C. (2003). Elevated abundance of bacteriophage infecting bacteria in soil. Appl. Environ. Microbiol. *69*, 285–289.

Baas Becking, L.G.M. (1934). Geobiologie of inleiding tot de milieukunde (W.P. Van Stockum & Zoon, The Netherlands).

Baltrus, D.A. (2013). Exploring the costs of horizontal gene transfer. Trends Ecol. Evol. *28*, 489–495. https://doi.org/10.1016/j.tree.2013.04.002.

Barnet, Y.M., Daft, M.J., and Stewart, W.D.P. (1981). Cyanobacteria – cyanophage interactions in continuous culture. J. Appl. Bacteriol. *51*, 541-552. https://doi.org/10.1111/j.1365-2672.1981.tb01273.x.

Barondess, J.J., and Beckwith, J. (1990). A bacterial virulence determinant encoded by lysogenic coliphage lambda. Nature *346*, 871–874. https://doi.org/10.1038/346871a0.

Barondess, J.J., and Beckwith, J. (1995). *bor* gene of phage lambda, involved in serum resistance, encodes a widely conserved outer membrane lipoprotein. J. Bacteriol. *177*, 1247–1253.

Bekliz, M., Colson, P., and La Scola, B. (2016). The Expanding Family of Virophages. Viruses *8*, E317. http://doi.org/10.3390/v8110317.

Bergh, O., Børsheim, K.Y., Bratbak, G., and Heldal, M. (1989). High abundance of viruses found in aquatic environments. Nature *340*, 467–468. https://doi.org/10.1038/340467a0.

Bläsi, U., Chang, C.Y., Zagotta, M.T., Nam, K.B., and Young, R. (1990). The lethal λ *S* gene encodes its own inhibitor. EMBO J. *9*, 981–989.

Bock, R. (2010). The give-and-take of DNA: horizontal gene transfer in plants. Trends Plant Sci. *15*, 11–22. https://doi.org/10.1016/j.tplants.2009.10.001.

Bonachela, J.A., and Levin, S.A. (2014). Evolutionary comparison between viral lysis rate and latent period. J. Theor. Biol. *345*, 32–42. https://doi.org/10.1016/j.jtbi.2013.12.006.

Bohannan, B.J.M., and Lenski, R.E. (1997). Effect of resource enrichment on a chemostat community of bacteria and bacteriophage. Ecology *78*, 2303-2315. https://doi.org/10.1890/0012-9658(1997)078[2303:EOREOA]2.0.CO;2.

Bohannan, B.J.M., and Lenski, R. (2000). Linking genetic change to community evolution: insights from studies of bacteria and bacteriophage. Ecol. Lett. *3*, 362-377. https://doi.org/10.1046/j.1461-0248.2000.00161.x.

Bondy-Denomy, J., and Davidson, A.R. (2014). When a virus is not a parasite: the beneficial effects of prophages on bacterial fitness. J. Microbiol. *52*, 235. https://doi.org/10.1007/s12275-014-4083-3.

Bondy-Denomy, J., Qian, J., Westra, E.R., Buckling, A., Guttman, D.S., Davidson, A.R., and Maxwell, K.L. (2016). Prophages mediate defense against phage infection through diverse mechanisms. ISME J. *10*, 2854–2866. https://doi.org/10.1038/ismej.2016.79.

Boritsch, E.C., Khanna, V., Pawlik, A., Honoré, N., Navas, V.H., Ma, L., Bouchier, C., Seemann, T., Supply, P., Stinear, T.P., *et al.* (2016). Key experimental evidence of chromosomal DNA transfer among selected tuberculosis-causing mycobacteria. Proc. Natl. Acad. Sci. U.S.A. *113*, 9876–9881. https://doi.org/10.1073/pnas.1604921113.

Born, Y., Fieseler, L., Marazzi, J., Lurz, R., Duffy, B., and Loessner, M.J. (2011). Novel virulent and broad-host-range *Erwinia amylovora* bacteriophages reveal a high degree of mosaicism and a relationship to *Enterobacteriaceae* phages. Appl. Environ. Microbiol. *77*, 5945–5954. https://doi.org/10.1128/AEM.03022-10.

Bratbak, G., Heldal, M., Thingstad, T.F., Riemann, B., and Haslund, O.H. (1992). Incorporation of viruses into the budget of microbial C-transfer – a first approach. Mar. Ecol. Prog. Ser. *83*, 273–280.

Breitbart, M., Felts, B., Kelley, S., Mahaffy, J.M., Nulton, J., Salamon, P., and Rohwer, F. (2004a). Diversity and population structure of a near-shore marine-sediment viral community. Proc. Biol. Sci. *271*, 565–574. https://doi.org/10.1098/rspb.2003.2628.

Breitbart, M., Hewson, I., Felts, B., Mahaffy, J.M., Nulton, J., Salamon, P., and Rohwer, F. (2003). Metagenomic analyses of an uncultured viral community from human feces. J. Bacteriol. *185*, 6220–6223. https://doi.org/10.1128/JB.185.20.6220-6223.2003.

Breitbart, M., Miyake, J.H., and Rohwer, F. (2004b). Global distribution of nearly identical phage-encoded DNA sequences. FEMS Microbiol. Lett. *236*, 249–256. https://doi.org/10.1016/j.femsle.2004.05.042.

Breitbart, M., and Rohwer, F. (2005). Here a virus, there a virus, everywhere the same virus? Trends Microbiol. *13*, 278–284. https://doi.org/10.1016/j.tim.2005.04.003.

Breitbart, M., Salamon, P., Andresen, B., Mahaffy, J.M., Segall, A.M., Mead, D., Azam, F., and Rohwer, F. (2002). Genomic analysis of uncultured marine viral communities. Proc. Natl. Acad. Sci. U.S.A. *99*, 14250–14255. https://doi.org/10.1073/pnas.202488399.

Brum, J.R., and Sullivan, M.B. (2015). Rising to the challenge: accelerated pace of discovery transforms marine virology. Nat. Rev. Microbiol. *13*, 147–159. https://doi.org/10.1038/nrmicro3404.

Brussaard, C.P., Wilhelm, S.W., Thingstad, F., Weinbauer, M.G., Bratbak, G., Heldal, M., Kimmance, S.A., Middelboe, M., Nagasaki, K., Paul, J.H., *et al.* (2008). Global-scale processes with a nanoscale drive: the role of marine viruses. ISME J. *2*, 575–578. https://doi.org/10.1038/ismej.2008.31.

Brüssow, H., Canchaya, C., and Hardt, W.D. (2004). Phages and the evolution of bacterial pathogens: from genomic rearrangements to lysogenic conversion. Microbiol. Mol. Biol. Rev. *68*, 560–602, table of contents. https://doi.org/10.1128/MMBR.68.3.560-602.2004.

Brüssow, H., and Cutter, K. 2005. Bacteriophage ecology. In Bacteriophages, Kutter, E. and Sulakvelidze, A., eds. (CRC Press, Boca Raton), pp. 129–164.

Buckling, A., and Rainey, P.B. (2002). Antagonistic coevolution between a bacterium and a bacteriophage. Proc. Biol. Sci. *269*, 931–936. https://doi.org/10.1098/rspb.2001.1945.

Buckling, A., Wei, Y., Massey, R.C., Brockhurst, M.A., and Hochberg, M.E. (2006). Antagonistic coevolution with parasites increases the cost of host deleterious mutations. Proc. Biol. Sci. *273*, 45–49. https://doi.org/10.1098/rspb.2005.3279.

Burmeister, A.R., Lenski, R.E., and Meyer, J.R. (2016). Host coevolution alters the adaptive landscape of a virus. Proc. Biol. Sci. *283*, 20161528. https://doi.org/10.1098/rspb.2016.1528.

Calendar, R., and Abedon, S.T. (2006). The Bacteriophages 2nd edn (Oxford University Press, UK).

Campbell, A. (1961). Conditions for the existence of bacteriophage. Evolution *15*, 153. https://doi.org/doi:10.2307/2406076.

Campbell, A., and Botstein, D. (1983). Evolution of the lambdoid phages. In Lambda II, Hendrix, R.W., ed. (Cold Spring Harbor: Cold Spring Harbor Laboratory Press), pp. pp. 365–380. https://doi.org/10.1002/jobm.19840240415.

Canchaya, C., Proux, C., Fournous, G., Bruttin, A., and Brüssow, H. (2003). Prophage genomics. Microbiol. Mol. Biol. Rev. 67, 238–276, https://dx.doi.org/10.1128%2FMMBR.67.2.238-276.2003.

Casjens, S.R., and Thuman-Commike, P.A. (2011). Evolution of mosaically related tailed bacteriophage genomes seen through the lens of phage P22 virion assembly. Virology *411*, 393–415. https://doi.org/10.1016/j.virol.2010.12.046.

Chantranupong, L., and Heineman, R.H. (2012). A common, non-optimal phenotypic endpoint in experimental adaptations of bacteriophage lysis time. BMC Evol. Biol. *12*, 37. https://doi.org/10.1186/1471-2148-12-37.

Chao, L., Levin, B.R., and Stewart, F.M. (1977). Complex community in a simple habitat – experimental studies with bacteria and phage. Ecology *58*, 369-378. https://doi.org/10.2307/1935611.

Chao, L., Tran, T., and Matthews, C. (1992). Muller's ratchet and the advantage of sex in the RNA virus Φ6. Evolution *46*, 289–299. https://doi.org/10.1111/j.1558-5646.1992.tb02038.x.

Chen, J., and Novick, R.P. (2009). Phage-mediated intergeneric transfer of toxin genes. Science *323*, 139–141. https://doi.org/10.1126/science.1164783.

Chibani-Chennoufi, S., Bruttin, A., Dillmann, M.L., and Brüssow, H. (2004). Phage-host interaction: an ecological perspective. J. Bacteriol. *186*, 3677–3686. https://doi.org/10.1128/JB.186.12.3677-3686.2004.

Chow, C.E., Winget, D.M., White, R.A., Hallam, S.J., and Suttle, C.A. (2015). Combining genomic sequencing methods to explore viral diversity and reveal potential virus-host interactions. Front. Microbiol. *6*, 265. https://doi.org/10.3389/fmicb.2015.00265.

Cochran, P.K., Kellogg, C.A., and Paul, J.H. (1998). Prophage induction of indigenous marine lysogenic bacteria by environmental pollutants. Mar. Sci. Fac. Pub. *64*, 125-133. https://dx.doi.org/10.3354/meps164125.

Colomer-Lluch, M., Jofre, J., and Muniesa, M. (2011). Antibiotic resistance genes in the bacteriophage DNA fraction of environmental samples. PLOS ONE *6*, e17549. https://doi.org/10.1371/journal.pone.0017549.

Colson, P., Pagnier, I., Yoosuf, N., Fournous, G., La Scola, B., and Raoult, D. (2013). 'Marseilleviridae', a new family of giant viruses infecting amoebae. Arch. Virol. *158*, 915–920. https://doi.org/10.1007/s00705-012-1537-y.

Culley, A.I., Lang, A.S., and Suttle, C.A. (2006). Metagenomic analysis of coastal RNA virus communities. Science *312*, 1795–1798. https://doi.org/10.1126/science.1127404.

Cumby, N., Davidson, A.R., and Maxwell, K.L.(2012a). The moron comes of age. Bacteriophage *2*, 225-228. https://doi.org/10.4161/bact.23146.

Cumby, N., Edwards, A.M., Davidson, A.R., and Maxwell, K.L. (2012b). The bacteriophage HK97 gp15 moron element encodes a novel superinfection exclusion protein. J. Bacteriol. *194*, 5012–5019. https://doi.org/10.1128/JB.00843-12.

Danovaro, R., Corinaldesi, C., Dell'anno, A., Fuhrman, J.A., Middelburg, J.J., Noble, R.T., and Suttle, C.A. (2011). Marine viruses and global climate change. FEMS Microbiol. Rev. *35*, 993–1034. https://doi.org/10.1111/j.1574-6976.2010.00258.x.

Danovaro, R., Dell'Anno, A., Corinaldesi, C., Magagnini, M., Noble, R., Tamburini, C., and Weinbauer, M. (2008). Major viral impact on the functioning of benthic deep-sea ecosystems. Nature *454*, 1084–1087. https://doi.org/10.1038/nature07268.

Danovaro, R., Dell'Anno, A., Corinaldesi, C., Rastelli, E., Cavicchioli, R., Krupovic, M., Noble, R.T., Nunoura, T., and Prangishvili, D. (2016). Virus-mediated archaeal hecatomb in the deep seafloor. Sci. Adv. *2*, e1600492. https://doi.org/10.1126/sciadv.1600492.

Darwin, C. (1859). On The Origin of Species by Means of Natural Selection, or the Preservation fo Favoured Races in the Struggle for Life (London: John Murray). https://doi.org/10.5962/bhl.title.68064.

Darwin, C. (1862). On the Various Contrivances by Which British and Foreign Orchids Are Fertilised by Insects (London: John Murray). https://doi.org/10.1017/CBO9780511910197.

Dedrick, R.M., Jacobs-Sera, D., Bustamante, C.A., Garlena, R.A., Mavrich, T.N., Pope, W.H., Reyes, J.C., Russell, D.A., Adair, T., Alvey, R., *et al.* (2017). Prophage-mediated defence against viral attack and viral counter-defence. Nat. Microbiol. *2*, 16251. https://doi.org/10.1038/nmicrobiol.2016.251.

De Paepe, M., and Taddei, F. (2006). Viruses' life history: towards a mechanistic basis of a trade-off between survival and reproduction among phages. PLOS Biol. *4*, e193. https://doi.org/10.1371/journal.pbio.0040273.

Deho, G., and Ghisotti, D. (2006). The satellite phage P4. In The Bacteriophages 2/e. Calendar, R.L., and Abedon, S.T., eds. (Oxford University Press, UK), pp. 391–408.

Delwart, E.L. (2007). Viral metagenomics. Rev. Med. Virol. *17*, 115–131. https://doi.org/10.1002/rmv.532.

Deng, L., Ignacio-Espinoza, J.C., Gregory, A.C., Poulos, B.T., Weitz, J.S., Hugenholtz, P., and Sullivan, M.B. (2014). Viral tagging reveals discrete populations in *Synechococcus* viral genome sequence space. Nature *513*, 242–245. https://doi.org/10.1038/nature13459.

Dennehy, J.J. (2012). What can phages tell us about host-pathogen coevolution? Int. J. Evol. Biol. *2012*, 396165. https://doi.org/10.1155/2012/396165.

Dennehy, J.J. (2014). What ecologists can tell virologists. Annu. Rev. Microbiol. *68*, 117–135. https://doi.org/10.1146/annurev-micro-091313-103436.

Dennehy, J.J., Duffy, S., O'Keefe, K.J., Edwards, S.V., and Turner, P.E. (2013). Frequent coinfection reduces RNA virus population genetic diversity. J. Hered. *104*, 704–712. https://doi.org/10.1093/jhered/est038.

Dennehy, J.J., Friedenberg, N.A., Holt, R.D., and Turner, P.E. (2006). Viral ecology and the maintenance of novel host use. Am. Nat. *167*, 429–439. https://doi.org/10.1086/499381.

Dennehy, J.J., and Wang, I.N. (2011). Factors influencing lysis time stochasticity in bacteriophage λ. BMC Microbiol. *11*, 174. https://doi.org/10.1186/1471-2180-11-174.

Taylor, D.J., Ballinger, M.J., Bowman, S.M., and Bruenn, J.A. (2013). Virus-host co-evolution under a modified nuclear genetic code. PeerJ *1*, e50. https://doi.org/10.7717/peerj.50.

Desnues, C., La Scola, B., Yutin, N., Fournous, G., Robert, C., Azza, S., Jardot, P., Monteil, S., Campocasso, A., Koonin, E.V., *et al.* (2012). Provirophages and transpovirons as the diverse mobilome of giant viruses. Proc. Natl. Acad. Sci. U.S.A. *109*, 18078–18083. https://doi.org/10.1073/pnas.1208835109.

Desnues, C., Rodriguez-Brito, B., Rayhawk, S., Kelley, S., Tran, T., Haynes, M., Liu, H., Furlan, M., Wegley, L., Chau, B., *et al.* (2008). Biodiversity and biogeography of phages in modern stromatolites and thrombolites. Nature *452*, 340–343. https://doi.org/10.1038/nature06735.

Duffy, S., Shackelton, L.A., and Holmes, E.C. (2008). Rates of evolutionary change in viruses: patterns and determinants. Nat. Rev. Genet. *9*, 267–276. https://doi.org/10.1038/nrg2323.

Dutilh, B.E., Cassman, N., McNair, K., Sanchez, S.E., Silva, G.G., Boling, L., Barr, J.J., Speth, D.R., Seguritan, V., Aziz, R.K., *et al.* (2014). A highly abundant bacteriophage discovered in the unknown sequences of human faecal metagenomes. Nat. Commun. *5*, 4498. https://doi.org/10.1038/ncomms5498.

Dykhuizen, D., Campbell, J.H., and Rolfe, B.G. (1978). The influences of a lambda prophage on the growth rate of *Escherichia coli*. Microbios *23*, 99–113.

Ebert, D., Zschokke-Rohringer, C.D., and Carius, H.J. (2000). Dose effects and density-dependent regulation of two microparasites of *Daphnia magna*. Oecologia *122*, 200–209. https://doi.org/10.1007/PL00008847.

Edwards, R.A., and Rohwer, F. (2005). Viral metagenomics. Nat. Rev. Microbiol. *3*, 504–510. https://doi.org/10.1038/nrmicro1163.

Enault, F., Briet, A., Bouteille, L., Roux, S., Sullivan, M.B., and Petit, M.A. (2017). Phages rarely encode antibiotic resistance genes: a cautionary tale for virome analyses. ISME J. *11*, 237–247. https://doi.org/10.1038/ismej.2016.90.

Feiner, R., Argov, T., Rabinovich, L., Sigal, N., Borovok, I., and Herskovits, A.A. (2015). A new perspective on lysogeny: prophages as active regulatory switches of bacteria. Nat. Rev. Microbiol. *13*, 641–650. https://doi.org/10.1038/nrmicro3527.

Feldman, T.S., Morsy, M.R., and Roossinck, M.J. (2012). Are communities of microbial symbionts more diverse than communities of macrobial hosts? Fungal. Biol. *116*, 465–477. https://doi.org/10.1016/j.funbio.2012.01.005.

Figueroa-Bossi, N., and Bossi, L. (1999). Inducible prophages contribute to *Salmonella* virulence in mice. Mol. Microbiol. *33*, 167–176.

Filée, J., Forterre, P., Sen-Lin, T., and Laurent, J. (2002). Evolution of DNA polymerase families: evidences for multiple gene exchange between cellular and viral proteins. J. Mol. Evol. *54*, 763–773. https://doi.org/10.1007/s00239-001-0078-x.

Filée, J., Tétart, F., Suttle, C.A., and Krisch, H.M. (2005). Marine T4-type bacteriophages, a ubiquitous component of the dark matter of the biosphere. Proc. Natl. Acad. Sci. U.S.A. *102*, 12471–12476. https://doi.org/10.1073/pnas.0503404102.

Fischer, M.G., and Suttle, C.A. (2011). A virophage at the origin of large DNA transposons. Science *332*, 231–234. https://doi.org/10.1126/science.1199412.

Fisher, R.A. (1930). The Genetical Theory of Natural Selection (Clarendon Press, UK).

Forterre, P., and Prangishvili, D. (2013). The major role of viruses in cellular evolution: facts and hypotheses. Curr Opin Virol *3*, 558–565. https://doi.org/10.1016/j.coviro.2013.06.013.

Freeman, V.J. (1951). Studies on the virulence of bacteriophage-infected strains of *Corynebacterium diphtheriae*. J. Bacteriol. *61*, 675–688.

Fuhrman, J.A. (1999). Marine viruses and their biogeochemical and ecological effects. Nature *399*, 541–548. https://doi.org/10.1038/21119.

Gallet, R., Lenormand, T., and Wang, I.N. (2012). Phenotypic stochasticity protects lytic bacteriophage populations from extinction during the bacterial stationary phase. Evolution *66*, 3485–3494. https://doi.org/10.1111/j.1558-5646.2012.01690.x.

Gallet, R., Kannoly, S., and Wang, I.N. (2011). Effects of bacteriophage traits on plaque formation. BMC Microbiol. *11*, 181. https://doi.org/10.1186/1471-2180-11-181.

Gallet, R., Shao, Y., and Wang, I.N. (2009). High adsorption rate is detrimental to bacteriophage fitness in a biofilm-like environment. BMC Evol. Biol. *9*, 241. https://doi.org/10.1186/1471-2148-9-241.

García-Villada, L., and Drake, J.W. (2013). Experimental selection reveals a trade-off between fecundity and lifespan in the coliphage Qß. Open Biol. 3, 130043. https://doi.org/10.1098/rsob.130043.

Gause, G.F. (1934). The Struggle for Existence (Hafner, New York).

Gerla, D.J., Gsell, A.S., Kooi, B.W., Ibelings, B.W., Van Donk, E., and Mooij, W.M. (2013). Alternative states and population crashes in a resource-susceptible-infected model for planktonic parasites and hosts. Freshwater Biol. *58*, 538-551. https://doi.org/10.1111/fwb.12010.

Göker, M., Scheuner, C., Klenk, H.P., Stielow, J.B., and Menzel, W. (2011). Codivergence of mycoviruses with their hosts. PLOS ONE *6*, e22252. https://doi.org/10.1371/journal.pone.0022252.

Gründling, A., Smith, D.L., Bläsi, U., and Young, R. (2000). Dimerization between the holin and holin inhibitor of phage lambda. J. Bacteriol. *182*, 6075–6081.

Cobián Güemes, A.G., Youle, M., Cantú, V.A., Felts, B., Nulton, J., and Rohwer, F. (2016). Viruses as winners in the game of life. Annu. Rev. Virol. 3, 197–214. https://doi.org/10.1146/annurev-virology-100114-054952.

Haaber, J., Leisner, J.J., Cohn, M.T., Catalan-Moreno, A., Nielsen, J.B., Westh, H., Penadés, J.R., and Ingmer, H. (2016). Bacterial viruses enable their host to acquire antibiotic resistance genes from neighbouring cells. Nat. Commun. *7*, 13333. https://doi.org/10.1038/ncomms13333.

Hacker, J., and Kaper, J.B. (2000). Pathogenicity islands and the evolution of microbes. Annu. Rev. Microbiol. *54*, 641–679. https://doi.org/10.1146/annurev.micro.54.1.641.

Hambly, E., and Suttle, C.A. (2005). The viriosphere, diversity, and genetic exchange within phage communities. Curr. Opin. Microbiol. *8*, 444–450. https://doi.org/10.1016/j.mib.2005.06.005.

Hatfull, G.F., and Hendrix, R.W. (2011). Bacteriophages and their genomes. Curr. Opin. Virol. *1*, 298–303. https://doi.org/10.1016/j.coviro.2011.06.009.

Heineman, R.H., and Bull, J.J. (2007). Testing optimality with experimental evolution: lysis time in a bacteriophage. Evolution *61*, 1695–1709. https://dx.doi.org/10.1111%2Fj.1558-5646.2007.00132.x.

Held, N.L., and Whitaker, R.J. (2009). Viral biogeography revealed by signatures in *Sulfolobus islandicus* genomes. Environ. Microbiol. *11*, 457–466. https://doi.org/10.1111/j.1462-2920.2008.01784.x.

Hendrix, R.W., Lawrence, J.G., Hatfull, G.F., and Casjens, S. (2000). The origins and ongoing evolution of viruses. Trends Microbiol. *8*, 504–508.

Hensel, M., and Schmidt, H., eds. (2008). Horizontal gene transfer in the evolution of pathogenesis (Cambridge: Cambridge University Press). https://doi.org/10.1017/CBO9780511541520.

Herskowitz, I., and Hagen, D. (1980). The lysis-lysogeny decision of phage lambda: explicit programming and responsiveness. Annu. Rev. Genet. *14*, 399–445. https://doi.org/10.1146/annurev.ge.14.120180.002151.

Horiuchi, K. (1983). Co-evolution of a filamentous bacteriophage and its defective interfering particles. J. Mol. Biol. *169*, 389–407.

Dunning Hotopp, J.C. (2011). Horizontal gene transfer between bacteria and animals. Trends Genet. *27*, 157–163. https://doi.org/10.1016/j.tig.2011.01.005.

Huang, A.S. (1973). Defective interfering viruses. Annu. Rev. Microbiol. *27*, 101–117. https://doi.org/10.1146/annurev.mi.27.100173.000533.

Hurwitz, B.L., and Sullivan, M.B. (2013). The Pacific Ocean virome (POV): a marine viral metagenomic dataset and associated protein clusters for quantitative viral ecology. PLOS ONE *8*, e57355. https://doi.org/10.1371/journal.pone.0057355.

Hyman, P., and Abedon, S.T. (2010). Bacteriophage host range and bacterial resistance. Adv. Appl. Microbiol. *70*, 217–248. https://doi.org/10.1016/S0065-2164(10)70007-1.

Imamovic, L., Jofre, J., Schmidt, H., Serra-Moreno, R., and Muniesa, M. (2009). Phage-mediated Shiga toxin 2 gene transfer in food and water. Appl. Environ. Microbiol. *75*, 1764–1768. https://doi.org/10.1128/AEM.02273-08.

Iranzo, J., and Manrubia, S.C. (2012). Evolutionary dynamics of genome segmentation in multipartite viruses. Proc. Biol. Sci. *279*, 3812–3819. https://doi.org/10.1098/rspb.2012.1086.

Ivanova, N.N., Schwientek, P., Tripp, H.J., Rinke, C., Pati, A., Huntemann, M., Visel, A., Woyke, T., Kyrpides, N.C., and Rubin, E.M. (2014). Stop codon reassignments in the wild. Science *344*, 909–913. https://doi.org/10.1126/science.1250691.

Joseph, S.B., Hanley, K.A., Chao, L., and Burch, C.L. (2009). Coinfection rates in Φ6 bacteriophage are enhanced by virus-induced changes in host cells. Evol. Appl. 2, 24–31. https://doi.org/10.1111/j.1752-4571.2008.00055.x.

Juhala, R.J., Ford, M.E., Duda, R.L., Youlton, A., Hatfull, G.F., and Hendrix, R.W. (2000). Genomic sequences of bacteriophages HK97 and HK022: pervasive genetic mosaicism in the lambdoid bacteriophages. J. Mol. Biol. *299*, 27–51. https://doi.org/10.1006/jmbi.2000.3729.

Kawecki, T.J., and Ebert, D. (2004), Conceptual issues in local adaptation. Ecol. Lett. *7*, 1225–1241. https://doi.org/10.1111/j.1461-0248.2004.00684.x.

Kerr, B., West, J., and Bohannan, B.J.M. (2008). Bacteriophages: models for exploring basic principles of ecology. In Bacteriophage Ecology, Abedon, S.T., ed. (Cambridge: Cambridge University Press), pp. 31-63. https://doi.org/10.1017/CBO9780511541483.005.

Kerr, P.J. (2012). Myxomatosis in Australia and Europe: a model for emerging infectious diseases. Antiviral Res. *93*, 387–415. https://doi.org/10.1016/j.antiviral.2012.01.009.

Kimura, M., Jia, Z.J., Nakayama, N., and Asakawa, S. 2008. Ecology of viruses in soils: past, present and future perspectives. Soil Sci. Plant Nutr. *54*, 1–32. https://doi.org/10.1111/j.1747-0765.2007.00197.x.

Klumpp, J., Fouts, D.E., and Sozhamannan, S. (2013). Bacteriophage functional genomics and its role in bacterial pathogen detection. Brief Funct. Genomics *12*, 354–365. https://doi.org/10.1093/bfgp/elt009.

Klumpp, J., and Fuchs, T.M. (2007). Identification of novel genes in genornic islands that contribute to *Salmonella typhimurium* replication in macrophages. Microbiology-SGM, *153*, 1207-1220. https://doi.org/10.1099/mic.0.%202006/004747-0.

Korytowski, D.A., and Smith, H. (2017). Permanence and Stability of a Kill the Winner Model in Marine Ecology. Bull. Math. Biol. *79*, 995–1004. https://doi.org/10.1007/s11538-017-0265-6.

Koskella, B., and Meaden, S. (2013). Understanding bacteriophage specificity in natural microbial communities. Viruses *5*, 806–823. https://doi.org/10.3390/v5030806.

Koskella, B., and Brockhurst, M.A. (2014). Bacteria-phage coevolution as a driver of ecological and evolutionary processes in microbial communities. FEMS Microbiol. Rev. *38*, 916–931. https://doi.org/10.1111/1574-6976.12072.

Koskella, B., and Vos, M. (2015). Adaptation in natural microbial populations. Ann. Rev. Ecol. Evol. Syst. *46*, 503-522. https://doi.org/10.1146/annurev-ecolsys-112414-054458.

Kristensen, D.M., Mushegian, A.R., Dolja, V.V., and Koonin, E.V. (2010). New dimensions of the virus world discovered through metagenomics. Trends Microbiol. *18*, 11–19. https://doi.org/10.1016/j.tim.2009.11.003.

Krupovic, M., Prangishvili, D., Hendrix, R.W., and Bamford, D.H. (2011). Genomics of bacterial and archaeal viruses: dynamics within the prokaryotic virosphere. Microbiol. Mol. Biol. Rev. *75*, 610–635. https://doi.org/10.1128/MMBR.00011-11.

Krupovic, M., Kuhn, J.H., and Fischer, M.G. (2016). A classification system for virophages and satellite viruses. Arch. Virol. *161*, 233–247. https://doi.org/10.1007/s00705-015-2622-9.

Kunin, V., He, S., Warnecke, F., Peterson, S.B., Garcia Martin, H., Haynes, M., Ivanova, N., Blackall, L.L., Breitbart, M., Rohwer, F., *et al.* (2008). A bacterial metapopulation adapts locally to phage predation despite global dispersal. Genome Res. *18*, 293–297. https://doi.org/10.1101/gr.6835308.

La Scola, B., Audic, S., Robert, C., Jungang, L., de Lamballerie, X., Drancourt, M., Birtles, R., Claverie, J.M., and Raoult, D. (2003). A giant virus in amoebae. Science, *299*, 2033-2033. https://doi.org/10.1126/science.1081867.

La Scola, B., Desnues, C., Pagnier, I., Robert, C., Barrassi, L., Fournous, G., Merchat, M., Suzan-Monti, M., Forterre, P., Koonin, E., *et al.* (2008). The virophage as a unique parasite of the giant mimivirus. Nature *455*, 100–104. https://doi.org/10.1038/nature07218.

Lawrence, J.E., and Suttle, C.A. (2004). Effect of viral infection on sinking rates of *Heterosigma akashiwo* and its implications for bloom termination. Aquat. Microb. Ecol., 1–7. https://doi.org/10.3354/ame037001.

Lee, K.L., Hubbard, L.C., Hern, S., Yildiz, I., Gratzl, M., and Steinmetz, N.F. (2013). Shape matters: the diffusion rates of TMV rods and CPMV icosahedrons in a spheroid model of extracellular matrix are distinct. Biomater. Sci. *1*. https://doi.org/10.1039/C3BM00191A.

Lenski, R.E. (1984). Coevolution of bacteria and phage – are there endless cycles of bacterial defenses and phage counterdefenses? J. Theor. Biol. *108*, 319-325.

Lenski, R.E. (1988). Dynamics of interactions between bacteria and virulent bacteriophage. Advanc. Microb. Ecol. *10*, 1–44. https://doi.org/10.1007/978-1-4684-5409-3_1.

Lenski, R.E., and Levin, B.R. (1985). Constraints on the coevolution of bacteria and virulent phage – a model, some experiments, and predictions for natural communities. Am. Nat. *125*, 585-602. https://doi.org/10.1086/284364.

Levin, B.R., Stewart, F.M., and Chao, L. (1977). Resource limited growth, competition, and predation – A model and experimental studies with bacteria and bacteriophage. Am. Nat. *111*, 3–24. https://doi.org/10.1086/283134.

Little, T.J., and Ebert, D. (2000). The cause of parasitic infection in natural populations of *Daphnia* (Crustacea:

Cladocera): the role of host genetics. Proc. Biol. Sci. *267*, 2037–2042. https://doi.org/10.1098/rspb.2000.1246.

Lotka, A.J. (1925). Elements of physical biology (Williams & Wilkins, Baltimore).

Loverdo, C., and Lloyd-Smith, J.O. (2013). Intergenerational phenotypic mixing in viral evolution. Evolution *67*, 1815–1822. https://doi.org/10.1111/evo.12048.

Ludwig, D., and Rowe, L. (1990). Life history strategies for energy gain and predator avoidance under time constraints. Am. Nat. *135*, 686-707. https://doi.org/10.1086/285069.

Lundgren, M. (2016). Exploring the ecological function of CRISPR-Cas virus defense. Commun. Integr. Biol. *9*, e1216740. https://doi.org/10.1080/19420889.2016.1216740.

Mari, X., Rassoulzadegan, F., Brussaard, C.P.D., and Wassmann, P. (2005). Dynamics of transparent exopolymeric particles (TEP) production by *Phaeocystis globosa* under N- or P-limitation: a controlling factor of the retention/export balance. Harmful Algae *4*, 895-914. https://doi.org/10.1016/j.hal.2004.12.014.

Marshall, N., Priyamvada, L., Ende, Z., Steel, J., and Lowen, A.C. (2013). Influenza virus reassortment occurs with high frequency in the absence of segment mismatch. PLOS Pathog. *9*, e1003421. https://doi.org/10.1371/journal.ppat.1003421.

Marston, M.F., Pierciey, F.J., Shepard, A., Gearin, G., Qi, J., Yandava, C., Schuster, S.C., Henn, M.R., and Martiny, J.B. (2012). Rapid diversification of coevolving marine *Synechococcus* and a virus. Proc. Natl. Acad. Sci. U.S.A. *109*, 4544–4549. https://doi.org/10.1073/pnas.1120310109.

Maslov, S., and Sneppen, K. (2017). Population cycles and species diversity in dynamic Kill-the-Winner model of microbial ecosystems. Sci. Rep. *7*, 39642. https://doi.org/10.1038/srep39642.

McDaniel, L.D. (2012). High frequency of horizontal gene transfer in the oceans. Science *337*, 911-911. https://doi.org/10.1126/science.1192243.

McLean, A.R., and Bostock, C.J. (2000). Scrapie infections initiated at varying doses: an analysis of 117 titration experiments. Phil. Trans. R. Soc. London Ser. B. Biol. Sci. *355*, 1043-1050. https://doi.org/10.1098/rstb.2000.0641.

Mochizuki, T., Krupovic, M., Pehau-Arnaudet, G., Sako, Y., Forterre, P., and Prangishvili, D. (2012). Archaeal virus with exceptional virion architecture and the largest single-stranded DNA genome. Proc. Natl. Acad. Sci. U.S.A. *109*, 13386–13391. https://doi.org/10.1073/pnas.1203668109.

Mokili, J.L., Rohwer, F., and Dutilh, B.E. (2012). Metagenomics and future perspectives in virus discovery. Curr. Opin. Virol. *2*, 63–77. https://doi.org/10.1016/j.coviro.2011.12.004.

Monier, A., Claverie, J.M., and Ogata, H. (2008). Taxonomic distribution of large DNA viruses in the sea. Genome Biol. *9*, R106. https://doi.org/10.1186/gb-2008-9-7-r106.

Moreno, E., Ojosnegros, S., García-Arriaza, J., Escarmís, C., Domingo, E., and Perales, C. (2014). Exploration of sequence space as the basis of viral RNA genome segmentation. Proc. Natl. Acad. Sci. U.S.A. *111*, 6678–6683. https://doi.org/10.1073/pnas.1323136111.

Morgan, A.D., Craig Maclean, R., and Buckling, A. (2009). Effects of antagonistic coevolution on parasite-mediated host coexistence. J. Evol. Biol. *22*, 287–292. https://doi.org/10.1111/j.1420-9101.2008.01642.x.

Nguyen, K., Simmons, K., and Tatarnikov, I. (2014). Co-inoculation of *Escherichia coli* B23 by T4 and T7 bacteriophages results in competition shown by an overall drop in phage progeny. J. Exp. Microbiol. Immunol. *18*, 156 – 161.

O'Keefe, K.J., Silander, O.K., McCreery, H., Weinreich, D.M., Wright, K.M., Chao, L., Edwards, S.V., Remold, S.K., and Turner, P.E. (2010). Geographic differences in sexual reassortment in RNA phage. Evolution *64*, 3010–3023. https://doi.org/10.1111/j.1558-5646.2010.01040.x.

Ojosnegros, S., García-Arriaza, J., Escarmís, C., Manrubia, S.C., Perales, C., Arias, A., Mateu, M.G., and Domingo, E. (2011). Viral genome segmentation can result from a trade-off between genetic content and particle stability. PLOS Genet. *7*, e1001344. https://doi.org/10.1371/journal.pgen.1001344.

Oppenheim, A.B., Kobiler, O., Stavans, J., Court, D.L., and Adhya, S. (2005). Switches in bacteriophage lambda development. Annu. Rev. Genet. *39*, 409–429. https://doi.org/10.1146/annurev.genet.39.073003.113656.

Palukaitis, P. (2016). Satellite RNAs and Satellite Viruses. Mol. Plant Microbe Interact. *29*, 181–186. https://doi.org/10.1094/MPMI-10-15-0232-FI.

Parikka, K.J., Le Romancer, M., Wauters, N., and Jacquet, S. (2017). Deciphering the virus-to-prokaryote ratio (VPR): insights into virus–host relationships in a variety of ecosystems. Biol. Rev. Camb. Philos. Soc. *92*, 1081–1100. https://doi.org/10.1111/brv.12271.

Paez-Espino, D., Eloe-Fadrosh, E.A., Pavlopoulos, G.A., Thomas, A.D., Huntemann, M., Mikhailova, N., Rubin, E., Ivanova, N.N., and Kyrpides, N.C. (2016). Uncovering Earth's virome. Nature *536*, 425–430. https://doi.org/10.1038/nature19094.

Paul, J.H. (2008). Prophages in marine bacteria: dangerous molecular time bombs or the key to survival in the seas? ISME J. *2*, 579–589. https://doi.org/10.1038/ismej.2008.35.

Pearl, R. (1927). The growth of populations. Q. Rev. Biol. *2*, 532-548.

Pearl, R., and Reed, L.J. (1920). On the Rate of Growth of the Population of the United States since 1790 and Its Mathematical Representation. Proc. Natl. Acad. Sci. U.S.A. *6*, 275–288.

Pepin, K.M., Samuel, M.A., and Wichman, H.A. (2006). Variable pleiotropic effects from mutations at the same locus hamper prediction of fitness from a fitness component. Genetics *172*, 2047–2056. https://dx.doi.org/10.1534%2Fgenetics.105.049817.

Philippe, N., Legendre, M., Doutre, G., Couté, Y., Poirot, O., Lescot, M., Arslan, D., Seltzer, V., Bertaux, L., Bruley, C., *et al.* (2013). Pandoraviruses: amoeba viruses with genomes up to 2.5 Mb reaching that of parasitic eukaryotes. Science *341*, 281–286. https://doi.org/10.1126/science.1239181.

Pianka (1976). Natural selection of optimal reproductive tactics. Am. Zool. *16*, 775-784. https://doi.org/10.1093/icb/16.4.775.

Poorvin, L., Rinta-Kanto, J.M., Hutchins, D.A., and Wilhelm, S.W. (2004). Viral release of iron and its bioavailability to marine plankton. Limnol. Oceanogr. *49*, 1734-1741. https://doi.org/10.4319/lo.2004.49.5.1734.

Pope, W.H., Bowman, C.A., Russell, D.A., Jacobs-Sera, D., Asai, D.J., Cresawn, S.G., Jacobs, W.R., Hendrix, R.W., Lawrence, J.G., and Hatfull, G.F. (2015). Whole genome comparison of a large collection of mycobacteriophages reveals a continuum of phage genetic diversity. eLife *4*, e06416. https://doi.org/10.7554/eLife.06416.

Pradel, R. (1996). Utilization of capture-mark-recapture for the study of recruitment and population growth rate. Biometrics *52*, 703-709. https://doi.org/10.2307/2532908.

Prangishvili, D. (2013). The wonderful world of archaeal viruses. Annu. Rev. Microbiol. *67*, 565–585. https://doi.org/10.1146/annurev-micro-092412-155633.

Prangishvili, D. (2015). Archaeal viruses: living fossils of the ancient virosphere? Ann. N. Y. Acad. Sci. *1341*, 35–40. https://doi.org/10.1111/nyas.12710.

Preisig, O., Wingfield, B.D., and Wingfield, M.J. (1998). Coinfection of a fungal pathogen by two distinct double-stranded RNA viruses. Virology *252*, 399–406. https://doi.org/10.1006/viro.1998.9480.

Pressing, J., and Reanney, D.C. (1984). Divided genomes and intrinsic noise. J. Mol. Evol. *20*, 135–146.

Prussin, A.J., Garcia, E.B., and Marr, L.C. (2015). Total Virus and Bacteria Concentrations in Indoor and Outdoor Air. Environ Sci. Technol. Lett. *2*, 84–88. https://doi.org/10.1021/acs.estlett.5b00050.

Rapp, F. (1969). Defective DNA animal viruses. Annu. Rev. Microbiol. *23*, 293–316. https://doi.org/10.1146/annurev.mi.23.100169.001453.

Regoes, R.R., Nowak, M.A., and Bonhoeffer, S. (2000). Evolution of virulence in a heterogeneous host population. Evolution *54*, 64–71. https://doi.org/10.1111/j.0014-3820.2000.tb00008.x.

Roux, L., Simon, A.E., and Holland, J.J. (1991). Effects of defective interfering viruses on virus replication and pathogenesis in vitro and in vivo. Adv. Virus Res. *40*, 181–211.

Rohwer, F., and Edwards, R. (2002). The Phage Proteomic Tree: a genome-based taxonomy for phage. J. Bacteriol. *184*, 4529–4535.

Rohwer, F., and Thurber, R.V. (2009). Viruses manipulate the marine environment. Nature *459*, 207–212. https://doi.org/10.1038/nature08060.

Rosario, K., and Breitbart, M. (2011). Exploring the viral world through metagenomics. Curr. Opin. Virol. *1*, 289–297. https://doi.org/10.1016/j.coviro.2011.06.004.

Rosario, K., Nilsson, C., Lim, Y.W., Ruan, Y., and Breitbart, M. (2009). Metagenomic analysis of viruses in reclaimed water. Environ. Microbiol. *11*, 2806–2820. https://doi.org/10.1111/j.1462-2920.2009.01964.x.

Roux, S., Brum, J.R., Dutilh, B.E., Sunagawa, S., Duhaime, M.B., Loy, A., Poulos, B.T., Solonenko, N., Lara, E., Poulain, J., *et al.* (2016). Ecogenomics and potential biogeochemical impacts of globally abundant ocean viruses. Nature *537*, 689–693. https://doi.org/10.1038/nature19366.

Roux, S., Enault, F., Ravet, V., Colombet, J., Bettarel, Y., Auguet, J.C., Bouvier, T., Lucas-Staat, S., Vellet, A., Prangishvili, D., *et al.* (2016). Analysis of metagenomic data reveals common features of halophilic viral communities across continents. Environ. Microbiol. *18*, 889–903. https://doi.org/10.1111/1462-2920.13084.

Saier, M.H., and Reddy, B.L. (2015). Holins in bacteria, eukaryotes, and archaea: multifunctional xenologues with potential biotechnological and biomedical applications. J. Bacteriol. *197*, 7–17. https://doi.org/10.1128/JB.02046-14.

Sanjuán, R. (2010). Mutational fitness effects in RNA and single-stranded DNA viruses: common patterns revealed by site-directed mutagenesis studies. Philos. Trans. R. Soc. Lond., B, Biol. Sci. *365*, 1975–1982. https://doi.org/10.1098/rstb.2010.0063.

Sanjuán, R., Moya, A., and Elena, S.F. (2004). The distribution of fitness effects caused by single-nucleotide substitutions in an RNA virus. Proc. Natl. Acad. Sci. U.S.A. *101*, 8396–8401. https://doi.org/10.1073/pnas.0400146101.

Sanjuán, R., Nebot, M.R., Chirico, N., Mansky, L.M., and Belshaw, R. (2010). Viral mutation rates. J. Virol. *84*, 9733–9748. https://doi.org/10.1128/JVI.00694-10.

Sano, E., Carlson, S., Wegley, L., and Rohwer, F. (2004). Movement of viruses between biomes. Appl. Environ. Microbiol. *70*, 5842–5846. https://doi.org/10.1128/AEM.70.10.5842-5846.2004.

Saunders, J.R., Allison, H., James, C.E., McCarthy, A.J., and Sharp, R. (2001). Phage-mediated transfer of virulence genes. J. Chem. Technol. Biotechnol. *76*, 662-666. https://doi.org/10.1002/jctb.437.

Schluter, D., and Conte, G.L. (2009). Genetics and ecological speciation. Proc. Natl. Acad. Sci. U.S.A. *106* (Suppl. 1), 9955–9962. https://doi.org/10.1073/pnas.0901264106.

Schrag, S.J., and Mittler, J.E. (1996). Host-parasite coexistence: The role of spatial refuges in stabilizing bacteria-phage interactions. Am. Nat. *148*, 348-377. https://doi.org/10.1086/285929.

Schuch, R., and Fischetti, V.A. (2009). The secret life of the anthrax agent *Bacillus anthracis*: bacteriophage-mediated ecological adaptations. PLOS ONE *4*, e6532. https://doi.org/10.1371/journal.pone.0006532.

Shao, Y., and Wang, I.N. (2008). Bacteriophage adsorption rate and optimal lysis time. Genetics *180*, 471–482. https://doi.org/10.1534/genetics.108.090100.

Short, C.M., and Suttle, C.A. (2005). Nearly identical bacteriophage structural gene sequences are widely distributed in both marine and freshwater environments. Appl. Environ. Microbiol. *71*, 480–486. https://dx.doi.org/10.1128%2FAEM.71.1.480-486.2005.

Short, S.M. (2012). The ecology of viruses that infect eukaryotic algae. Environ. Microbiol. *14*, 2253–2271. https://doi.org/10.1111/j.1462-2920.2012.02706.x.

Singh, A., and Dennehy, J.J. (2014). Stochastic holin expression can account for lysis time variation in the bacteriophage λ. J. R. Soc. Interface *11*, 20140140. https://doi.org/10.1098/rsif.2014.0140.

Six, E.W., and Klug, C.A. (1973). Bacteriophage P4: a satellite virus depending on a helper such as prophage P2. Virology *51*, 327–344.

Snyder, J.C., Bolduc, B., and Young, M.J. (2015). 40 Years of archaeal virology: Expanding viral diversity. Virology *479–480*, 369–378. https://doi.org/10.1016/j.virol.2015.03.031.

Soutourina, O.A., Monot, M., Boudry, P., Saujet, L., Pichon, C., Sismeiro, O., *et al.*, (2013). Genome-wide identificationof regulatory RNAs in the human pathogen *Clostridium difficile*. PLOS Genet. *9*, e1003493. https://doi.org/10.1371/journal.pgen.1003493.

Stopar, D., and Abedon, S.T. (2008). Modeling bacteriophage population growth. In Bacteriophage Ecology, Abedon, S.T., ed. (Cambridge University Press, UK). pp. 389–414.

Suttle, C.A. (1994). The significance of viruses to mortality in aquatic microbial communities. Microb. Ecol. *28*, 237–243. https://doi.org/10.1007/BF00166813.

Suttle, C.A. (2007). Marine viruses – major players in the global ecosystem. Nat. Rev. Microbiol. *5*, 801–812. https://doi.org/10.1038/nrmicro1750.

Suttle, C.A. (2016). Environmental microbiology: Viral diversity on the global stage. Nat. Microbiol. *1*, 16205. https://doi.org/10.1038/nmicrobiol.2016.205.

Swanson, M.M., Fraser, G., Daniell, T.J., Torrance, L., Gregory, P.J., and Taliansky, M. (2009). Viruses in soils: morphological diversity and abundance in the rhizosphere. Ann. Appl. Biol. *155*, 51-60. https://doi.org/10.1111/j.1744-7348.2009.00319.x.

Symons, R.H. (1991). The intriguing viroids and virusoids: what is their information content and how did they evolve? Mol. Plant Microbe. Interact. *4*, 111–121.

Thingstad, T.F. (2000). Elements of a theory for the mechanisms controlling abundance, diversity, and biogeochemical role of lytic bacterial viruses in aquatic systems. Limnol. Oceanogr. *45*, 1320-1328. https://doi.org/10.4319/lo.2000.45.6.1320.

Thingstad, T.F., Bratbak, G., and Heldal, M. (2008). Aquatic phage ecology. In Bacteriophage Ecology, Abedon, S.T., ed. (Cambridge University Press, UK), pp. 451–480.

Thompson, J.N. (1994). The Coevolutionary Process (University of Chicago Press, Chicago).

Thompson, J.N. (2012). Evolution. The role of coevolution. Science *335*, 410–411. https://doi.org/10.1126/science.1217807.

Thurber, R.V., Haynes, M., Breitbart, M., Wegley, L., and Rohwer, F. (2009). Laboratory procedures to generate viral metagenomes. Nat. Protoc. *4*, 470–483. https://doi.org/10.1038/nprot.2009.10.

Touchon, M., Bernheim, A., and Rocha, E.P. (2016). Genetic and life-history traits associated with the distribution of prophages in bacteria. ISME J. *10*, 2744–2754. https://doi.org/10.1038/ismej.2016.47.

Trinh, J.T., Székely, T., Shao, Q., Balázsi, G., and Zeng, L. (2017). Cell fate decisions emerge as phages cooperate or compete inside their host. Nat. Commun. *8*, 14341. https://doi.org/10.1038/ncomms14341.

Turner, P.E. (2005). Parasitism between co-infecting bacteriophages. Advances in Ecological Research, *37*, 309-332. https://doi.org/10.1016/S0065-2504(04)37010-8.

Tyler, J.S., Mills, M.J., and Friedman, D.I. (2004). The operator and early promoter region of the Shiga toxin type 2-encoding bacteriophage 933W and control of toxin expression. J. Bacteriol. *186*, 7670–7679. https://doi.org/10.1128/JB.186.22.7670-7679.2004.

Uc-Mass, A., Loeza, E.J., de la Garza, M., Guarneros, G., Hernández-Sánchez, J., and Kameyama, L. (2004). An orthologue of the cor gene is involved in the exclusion of temperate lambdoid phages. Evidence that Cor inactivates FhuA receptor functions. Virology *329*, 425–433. https://doi.org/10.1016/j.virol.2004.09.005.

Vaca-Pacheco, S., Paniagua-Contreras, G.L., García-González, O., and de la Garza, M. (1999). The clinically isolated FIZ15 bacteriophage causes lysogenic conversion in *Pseudomonas aeruginosa* PAO1. Curr. Microbiol. *38*, 239–243.

Våge, S., Storesund, J.E., Giske, J., and Thingstad, T.F. (2014). Optimal defense strategies in an idealized microbial food web under trade-off between competition and defense. PLOS ONE *9*, e101415. https://doi.org/10.1371/journal.pone.0101415.

Vainio, E.J., Hakanpää, J., Dai, Y.C., Hansen, E., Korhonen, K., and Hantula, J. (2011). Species of *Heterobasidion* host a diverse pool of partitiviruses with global distribution and interspecies transmission. Fungal Biol. *115*, 1234–1243. https://doi.org/10.1016/j.funbio.2011.08.008.

Van Etten, J.L., Lane, L.C., and Meints, R.H. (1991). Viruses and viruslike particles of eukaryotic algae. Microbiol. Rev. *55*, 586–620.

van Wolferen, M., Ajon, M., Driessen, A.J., and Albers, S.V. (2013). How hyperthermophiles adapt to change their lives: DNA exchange in extreme conditions. Extremophiles *17*, 545–563. https://doi.org/10.1007/s00792-013-0552-6.

Vica Pacheco, S., García González, O., and Paniagua Contreras, G.L. (1997). The lom gene of bacteriophage lambda is involved in *Escherichia coli* K12 adhesion to human buccal epithelial cells. FEMS Microbiol. Lett. *156*, 129–132.

Volterra, V. (1926). Fluctuations in the abundance of a species considered mathematically. Nature *118*, 558-560. https://doi.org/10.1038/118558a0.

Vostrov, A.A., Vostrukhina, O.A., Svarchevsky, A.N., and Rybchin, V.N. (1996). Proteins responsible for lysogenic conversion caused by coliphages N15 and phi80 are highly homologous. J. Bacteriol. *178*, 1484–1486.

Waldor, M.K., and Mekalanos, J.J. (1996). Lysogenic conversion by a filamentous phage encoding cholera toxin. Science *272*, 1910–1914.

Wang, I.-N., Dykhuizen, D.E., and Slobodkin, L.B. (1996). The evolution of phage lysis timing. Evol. Ecol. *10*, 545-558. https://doi.org/10.1007/BF01237884.

Wang, I.N., Smith, D.L., and Young, R. (2000). Holins: the protein clocks of bacteriophage infections. Annu. Rev. Microbiol. *54*, 799–825. https://doi.org/10.1146/annurev.micro.54.1.799.

Wang, X., Kim, Y., Ma, Q., Hong, S.H., Pokusaeva, K., Sturino, J.M., and Wood, T.K. (2010). Cryptic prophages help bacteria cope with adverse environments. Nat. Commun. *1*, 147. https://doi.org/10.1038/ncomms1146.

Watkins, S.C., Kuehnle, N., Ruggeri, C.A., Malki, K., Bruder, K., Elayyan, J., *et al.* (2015). Assessment of a metaviromic dataset generated from nearshore Lake Michigan. Mar. Freshwater Res. *67*, 1700-1708. https://doi.org/10.1071/MF15172.

Weinbauer, M.G., Bonilla-Findji, O., Chan, A.M., Dolan, J.R., Short, S.M., Simek, K., Wilhelm, S.W., and Suttle, C.A. (2011). *Synechococcus* growth in the ocean may depend on the lysis of heterotrophic bacteria. J. Plankton Res. 33, 1465-1476. https://doi.org/10.1093/plankt/fbr041.

Weinbauer, M.G., and Rassoulzadegan, F. (2004). Are viruses driving microbial diversification and diversity? Environ. Microbiol. *6*, 1–11.

Weinbauer, M.G., and Suttle, C.A. (1999). Lysogeny and prophage induction in coastal and offshore bacterial communities. Aquat. Microb. Ecol. 217-225. https://doi.org/10.3354/ame018217.

Winter, C., Bouvier, T., Weinbauer, M.G., and Thingstad, T.F. (2010). Trade-offs between competition and defense specialists among unicellular planktonic organisms: the 'killing the winner' hypothesis revisited. Microbiol. Mol. Biol. Rev. *74*, 42–57. https://doi.org/10.1128/MMBR.00034-09.

Weitz, J.S., and Dushoff, J. (2008). Alternative stable states in host-phage dynamics. Theor. Ecol. *1*, 13-19. https://doi.org/10.1007/s12080-007-0001-1.

Weitz, J.S., Mileyko, Y., Joh, R.I., and Voit, E.O. (2008). Collective decision making in bacterial viruses. Biophys. J. *95*, 2673–2680. https://doi.org/10.1529/biophysj.108.133694.

Weitz, J.S., Stock, C.A., Wilhelm, S.W., Bourouiba, L., Coleman, M.L., Buchan, A., Follows, M.J., Fuhrman, J.A., Jover, L.F., Lennon, J.T., *et al.* (2015). A multitrophic model to quantify the effects of marine viruses on microbial food webs and ecosystem processes. ISME J. *9*, 1352–1364. https://doi.org/10.1038/ismej.2014.220.

Weynberg, K.D., Allen, M.J., and Wilson, W.H. (2017). Marine prasinoviruses and their tiny plankton hosts: a review. Viruses *9*, E43. https://doi.org/10.3390/v9030043.

Whon, T.W., Kim, M.S., Roh, S.W., Shin, N.R., Lee, H.W., and Bae, J.W. (2012). Metagenomic characterization of airborne viral DNA diversity in the near-surface atmosphere. J. Virol. *86*, 8221–8231. https://doi.org/10.1128/JVI.00293-12.

Wielgoss, S., Bergmiller, T., Bischofberger, A.M., and Hall, A.R. (2016). Adaptation to Parasites and Costs of Parasite Resistance in Mutator and Nonmutator Bacteria. Mol. Biol. Evol. *33*, 770–782. https://doi.org/10.1093/molbev/msv270.

Wigington, C.H., Sonderegger, D., Brussaard, C.P., Buchan, A., Finke, J.F., Fuhrman, J.A., Lennon, J.T., Middelboe, M., Suttle, C.A., Stock, C., *et al.* (2016). Re-examination of the relationship between marine virus and microbial cell abundances. Nat. Microbiol. *1*, 15024. https://doi.org/10.1038/nmicrobiol.2015.24.

Wilhelm, S.W., and Suttle, C.A. (1999). Viruses and nutrient cycles in the sea – viruses play critical roles in the structure and function of aquatic food webs. Bioscience *49*, 781-788. https://doi.org/10.2307/1313569.

Williams, H.T. (2013). Phage-induced diversification improves host evolvability. BMC Evol. Biol. *13*, 17. https://doi.org/10.1186/1471-2148-13-17.

Winget, D.M., Helton, R.R., Williamson, K.E., Bench, S.R., Williamson, S.J., and Wommack, K.E. (2011). Repeating patterns of virioplankton production within an estuarine ecosystem. Proc. Natl. Acad. Sci. U.S.A. *108*, 11506–11511. https://doi.org/10.1073/pnas.1101907108.

Winter, C., Bouvier, T., Weinbauer, M.G., and Thingstad, T.F. (2010). Trade-offs between competition and defense specialists among unicellular planktonic organisms: the 'killing the winner' hypothesis revisited. Microbiol. Mol. Biol. Rev. *74*, 42–57. https://doi.org/10.1128/MMBR.00034-09.

Wommack, K.E., and Colwell, R.R. (2000). Virioplankton: viruses in aquatic ecosystems. Microbiol. Mol. Biol. Rev. *64*, 69–114.

Wood, M.W., Rosqvist, R., Mullan, P.B., Edwards, M.H., and Galyov, E.E. (1996). SopE, a secreted protein of *Salmonella dublin*, is translocated into the target eukaryotic cell via a sip-dependent mechanism and promotes bacterial entry. Mol. Microbiol. *22*, 327–338.

Yau, S., Lauro, F.M., DeMaere, M.Z., Brown, M.V., Thomas, T., Raftery, M.J., Andrews-Pfannkoch, C., Lewis, M., Hoffman, J.M., Gibson, J.A., *et al.* (2011). Virophage control of antarctic algal host-virus dynamics. Proc. Natl. Acad. Sci. U.S.A. *108*, 6163–6168. https://doi.org/10.1073/pnas.1018221108.

Young, R. (2002). Bacteriophage holins: deadly diversity. J. Mol. Microbiol. Biotechnol. *4*, 21–36.

Young, R. (2014). Phage lysis: three steps, three choices, one outcome. J. Microbiol. *52*, 243–258. https://doi.org/10.1007/s12275-014-4087-z.

Zablocki, O., Adriaenssens, E.M., and Cowan, D. (2015). Diversity and ecology of viruses in hyperarid desert soils. Appl. Environ. Microbiol. *82*, 770–777. https://doi.org/10.1128/AEM.02651-15.

Zaman, L., Meyer, J.R., Devangam, S., Bryson, D.M., Lenski, R.E., and Ofria, C. (2014). Coevolution drives the emergence of complex traits and promotes evolvability. PLOS Biol. *12*, e1002023. https://doi.org/10.1371/journal.pbio.1002023.

Zhao, Y., Temperton, B., Thrash, J.C., Schwalbach, M.S., Vergin, K.L., Landry, Z.C., Ellisman, M., Deerinck, T., Sullivan, M.B., and Giovannoni, S.J. (2013). Abundant SAR11 viruses in the ocean. Nature *494*, 357–360. https://doi.org/10.1038/nature11921

Viruses of Microorganisms in Soil Ecosystems

4

Kurt E. Williamson*

Biology Department, College of William and Mary, Williamsburg, VA, USA.

*Correspondence: kewilliamson@wm.edu

https://doi.org/10.21775/9781910190852.04

Abstract

Soils are extremely complex ecosystems that display fine-scale spatial and temporal heterogeneity. Depending on the soil type, total viral abundance in soils can range from as few as 10,000 (10^4) to over 10^9 virus particles per gram dry weight. While very limited data are available, comparative analyses suggest that soils contain a more genetically diverse array of viruses than either aquatic or sediment habitats. Viruses of Bacteria (bacteriophages) represent the most studied and best understood group of viruses of microorganisms in soils. They also are believed to be the most abundant virus type within soil viral communities and can have important impacts on host bacteria population dynamics as well as on biogeochemical processes. Bacteriophages can also impact bacterial genetic diversity through host selection, as well as host phenotypic conversion and gene transfer events. Fungi are important soil microbes and fungal viruses (mycoviruses) appear to be ubiquitous in nature as well, as most fungal lineages show evidence of viral infections. Virus-mediated lysis of fungal hosts is exceptionally rare, however, and most mycoviruses establish persistent, asymptomatic infections of their hosts. The significance of mycovirus infections therefore may lie in subtle modulations of host gene regulation but also can affect host secretion of toxins, hypovirulence, and thermotolerance. Protozoa are key players in soil microbial food webs and viruses of the protozoa, especially viruses of amoeba, have touched off a revolution in modern virology as recently discovered 'giant viruses' have been found to exceed the 0.22 μm operational size cutoff that had been historically applied to viruses. Since 2003, four novel giant virus families have been established, with two representatives isolated from soils, although the ecological impacts of these giant amoebal viruses have yet to be determined. In addition to these unknowns, almost nothing is known regarding the viruses of Archaea, cyanobacteria, algae, or diatoms with specific regard to soil habitats. These represent significant knowledge gaps and targets for future research endeavours. Soils remain an under-studied ecosystem in spite of their complexity and importance to human civilization.

Goals of this chapter

The main goal of this chapter is to provide a broad overview of our current understanding of viruses infecting microorganisms found in soil ecosystems, with specific attention to abundance, diversity and virus–host interactions. Much of what we currently understand regarding viruses of microbes in soil environments focuses on phages (i.e. viruses that infect bacterial hosts) and several excellent reviews have already been published on these topics (Abedon, 2008; Kimura *et al.*, 2008; Srinivasiah *et al.*, 2008a; Williamson, 2011) (see Chapter 7). Thus, an important goal of this chapter is to expand beyond phages as the only viruses of microbes in soil. In particular, the latter portions of this chapter summarize important gaps in our current understanding of viruses in soil with respect to non-bacterial microbes and highlights opportunities for future research.

The soil habitat

Physical structure

Soil is the layer of unconsolidated mineral and organic materials at the surface of the Earth. Despite this relatively simple definition, soils represent the most complex microbial habitat on the planet. Soils are heterogeneous, triphasic systems, consisting of solids, liquids, and gases, the proportions and specific components of which can vary considerably over both space and time. Soil solids can be made up of varying proportions of gravel (rock fragments > 2.0 mm in diameter), sand (0.05–2.0 mm), silt (0.002–0.05 mm), clay (< 0.002 mm), and organic matter – an amorphous mix of plant and animal matter in various stages of decomposition. Mineral soils, such as those found in the Sahara Desert, have very low organic matter content, while organic soils, such as those found in peat bogs, have very low mineral content. On average, only about half of the volume of any given soil is taken up by solids. The remaining volume is pore space that can be filled by liquids, gases, or a mixture of the two. Soil water can contain various ionic species and any number of soluble macromolecules including carbohydrates, proteins, and other dissolved organic matter. Soil pH, largely imparted by the solute content in soil water, can vary from 1 to 13. Soil water may completely fill soil pores or may exist as films, a few microns thick, coating the interior of a pore. Gas-filled pores may contain any mixture of oxygen, carbon dioxide, nitrogen, methane, radon, volatile organics and water vapour. Within a given soil, it is possible for water-filled, anaerobic pores and gas-filled, aerobic pores to simultaneously exist just tens of micrometres away from each other, giving rise to the microheterogeneity that is characteristic of soils. Fig. 4.1 provides a simplified view of the soil habitat, demonstrating this microheterogeneity.

Soil horizons

Soils consist of different layers, or horizons, each with distinct characteristics (Fig. 4.2). Each horizon is identified by the letters, top-to-bottom, O, A, E, B, and C, but not every soil will have all horizons represented. Some soils have an upper layer consisting entirely of organic matter, dubbed the O horizon. The A horizon is the first mineral horizon in the soil profile. Some soils will have a zone of eluviation (E horizon) where minerals and organic matter have been removed by water

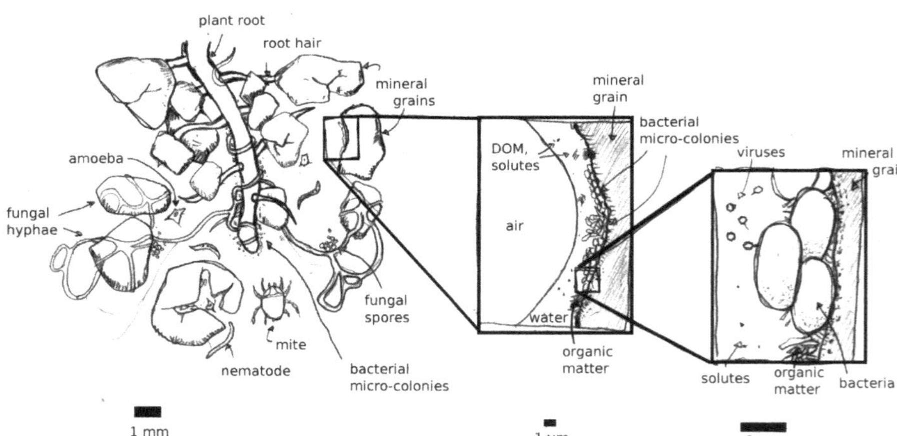

Figure 4.1 Soil aggregate. Mineral grains (sand, silt, clay) are held together by organic matter and microorganisms, including fungal mycelia, plant root exudates, and bacterial exopolymers. Increased magnification shows bacterial biofilm attached to mineral grain, organic matter, and air–water interface within the soil pore. Increased magnification shows size of typical bacteriophage (50–60-nm capsid diameter) in relation to typical bacterial cell (1–2 μm) within a water film coating the soil pore. This figure is adapted from Williamson *et al.* (2017) and used with permission.

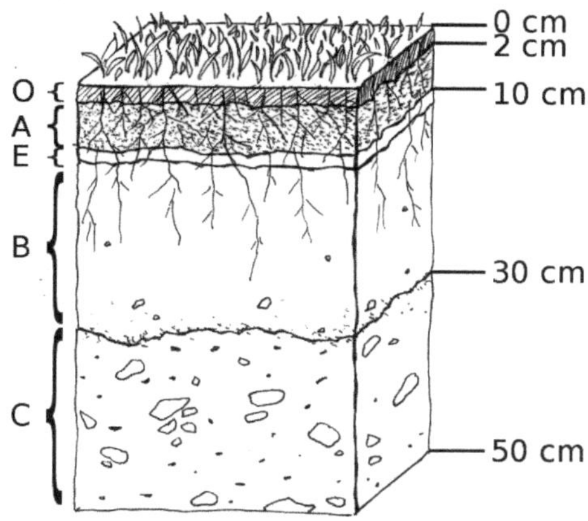

Figure 4.2 Soil horizons and their typical depth below surface within a soil profile. O horizon, 0–2 cm; A horizon, 2–10 cm; E horizon, 10–12 cm; B horizon, 12–30 cm; C horizon, 30–50 cm. R layer not shown.

infiltration. The B horizon tends to be a zone of illuviation, where nutrients and minerals transported by rainwater infiltration accumulate due to reduced permeability of the underlying material. The C horizon is made up of unconsolidated, partially weathered bedrock. The C horizon overlays the R layer, also known as the 'parent material' from which the upper horizons originated. This R layer consists of contiguous bedrock that cannot be excavated by hand. With regard to microbial ecology and viruses of soil microbes, the O and A horizons been most intensively explored, with relatively little known regarding viruses of the deep terrestrial biosphere.

Temperature

Depending on latitude, elevation, season, ground cover, and climatic factors such as precipitation and cloud-cover, soil temperatures can vary over more than two orders of magnitude. A high extreme of 93.8°C (the temperature at which DNA is melted in most laboratory polymerase chain reactions) was recorded at Furnace Creek, Death Valley National Park, USA (U.S. National Park Service), while a low extreme of −40°C was recorded at the McMurdo Dry Valleys, Antarctica (Hunt *et al.*, 2010). Even at a fixed position on Earth, surface soils in temperate zones may experience daily temperature changes of up to 4°C and annual temperature changes up to 28°C. Lower soil horizons, however, lag behind the surface horizon in terms of temperature changes and the magnitude of change decreases with depth. Given that biochemical process rates tend to double for every 10°C increase in temperature (Q_{10}), and temperature optima for enzymes tend to exhibit relatively narrow ranges, changes in soil temperature can have huge impacts on the types of organisms capable of surviving within a given soil, as well as the viruses infecting these organisms. In particular, temperature along with water availability may be an important factor in the selection of temperate as opposed to lytic replication in soils and the lower observed abundance of viruses in hot desert soils (see Chapter 1; also 'Viral abundance in soils', below).

Spatial distribution

Soils are organized into 12 different orders, depending upon the parent rock, climate, elevation, and physical, chemical, and biological factors leading to soil formation:

1. gelisols – soils with permafrost within 2 m of the surface
2. histosols – organic soils (bogs, fens)
3. spodosols – acid forest soils with subsurface accumulation of metal–humus complexes
4. andisols – soils formed in volcanic ash
5. oxisols – intensely weathered soils in tropical and subtropical environments
6. vertisols – clayey soils with high shrinking-swelling capacity
7. aridisols – calcareous soils of arid environments with subsurface horizon development
8. ultisols – strongly leached soils with a subsurface zone of clay accumulation
9. mollisols – highly fertile grassland soils
10. alfisols – moderately leached soils with a subsurface zone of clay accumulation
11. inceptisols – soils with weakly developed subsurface horizons
12. entisols – soils with little or no horizon development.

These different soil types are distributed unevenly across Earth (Fig. 4.3), and each soil may give rise to a characteristic community of microbes, presumably with characteristic cadres of viruses infecting them.

Temporal variability

In addition to the spatial heterogeneity of soils on the order of micrometres to kilometres, soils can exhibit dramatic temporal heterogeneity over varying time scales. Most soils experience some degree of drying and re-wetting over time. In the tropics, precipitation can occur on a nearly daily basis, while in the temperate zones, precipitation – which can include rain, snow, and other forms such as sleet and hail – occurs on a more variable schedule, sometimes punctuated by droughts and monsoons. Even dryland systems, which account for 1/3 of earth's continental surface (Pointing and Belnap, 2012), experience occasional precipitation or moisture inputs in the form of dew; and cold deserts may experience seasonal thawing and thin films of liquid water on ice crystals.

Temperate zones experience seasonal changes in patterns of temperature, precipitation, and vegetation cover. In particular, annual leaf drop by deciduous vegetation offers a layer of insulation against colder winter temperatures and subsidizes the maintenance of the O (organic) horizon of forest soils. Soils may also experience change due to tree fall, natural disturbances such as storms, landslides, and fires, and anthropogenic disturbances such as timber harvesting and land development. In each of these scenarios, changes in ground cover or land use will contribute to changes in soil composition and structure, leading to concomitant changes in the soil microbial community including viruses.

In summary, soils are hot-spots for interactions between several important spheres of influence: the biosphere, lithosphere, hydrosphere, atmosphere, detritosphere (dead and decaying matter), and rhizosphere (zone under the influence of plant roots) may all merge together within a given soil. The processes occurring in soils represent the net sum of the overlapping interactions between these different spheres of influence. The term 'soil' as applied to these rich ecosystems and the multiple processes occurring within them is somewhat deceptive, as 'soil' does not describe a single, well-defined object. Soil, rather, refers to a diverse grouping of individual ecosystems that share a common feature, namely in constituting the complex, porous skin of the Earth. For any scientist interested in studying soil microbes – let alone the viruses of soil microbes – it is important to grasp the complexity of the soil habitat because the

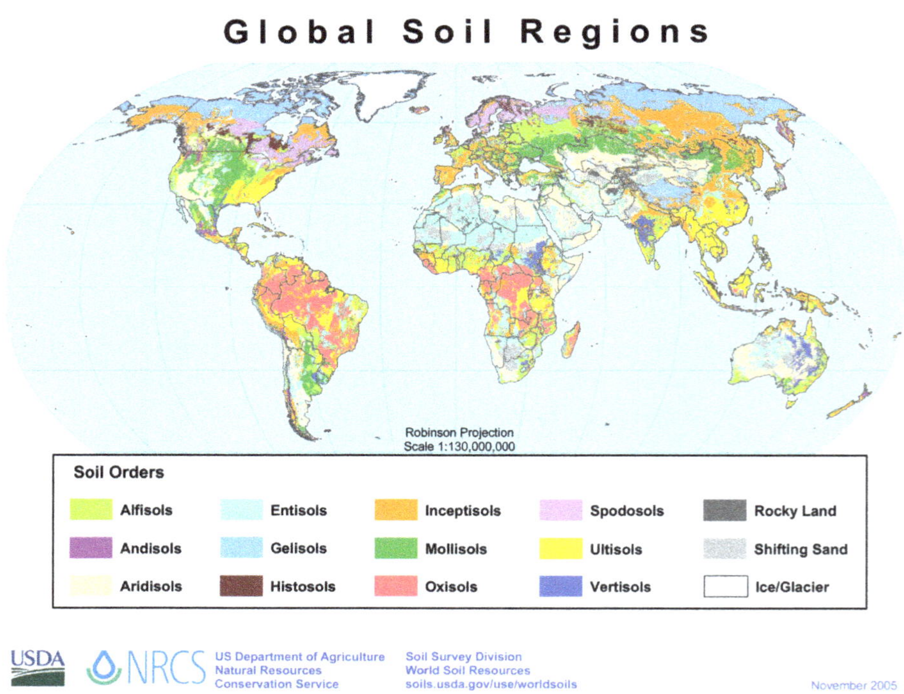

Figure 4.3 Global distribution of the 12 soil orders. Source: usda.gov

interactions between viruses and their hosts occur against this backdrop. In addition to the biological requirements that viruses must gain access to host cells, replicate, and release progeny from infected cells, the chemical and physical demands of the soil environment have provided additional selective pressures that have shaped virus–host interactions in soils over evolutionary timescales.

Viral abundance in soils

Viruses have been described as 'the most abundant biological entities on earth' (Chibani-Chennoufi *et al.*, 2004; Srinivasiah *et al.*, 2008b). In spite of its cumbersome nature, the term 'biological entity' is appropriate to describe viruses, since viruses clearly belong to the biosphere but many biologists have taken issue with viruses being considered as alive, or full-fledged 'organisms.' Regardless of terminology preferences (entities, organisms, etc.), viruses are extremely abundant in soils. Total direct count assays, performed either through transmission electron microscopy (TEM), or more frequently through epifluorescence microscopy (EFM), have revealed interesting trends in the distribution of viruses across soil types (Fig. 4.4). The prevailing view is that the majority of viruses found in soil ecosystems infect soil microbes, particularly prokaryotes. This is a fair and logical assumption, derived from several known facts, including the abundance of prokaryotes in soils (i.e. the most likely host organism for viruses in soil ecosystems (Whitman *et al.*, 1998)), the prevalence of tailed virus particles which consist almost exclusively of dsDNA phages (Williamson *et al.*, 2005) (see, however, Chapter 8), and observed linkages between viral and bacterial dynamics in soil (Marsh and Wellington, 1994; Srinivasiah *et al.*, 2008b). This assumption, however, comes with several important caveats. First, at least one report has indicated that tailed (dsDNA) phages are not prevalent in all soils (Swanson *et al.*, 2009), raising questions about the presumed dominance of dsDNA phages in soil viral communities. Second, microscopy cannot unequivocally

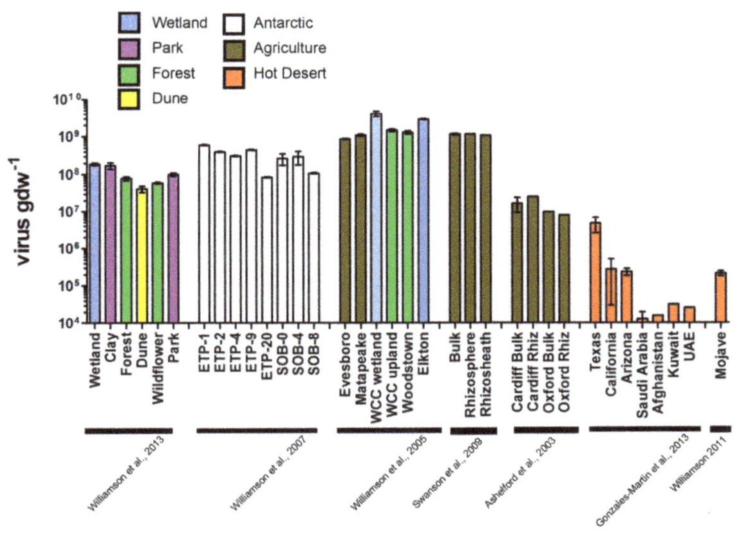

Figure 4.4 Viral abundance across soils, grouped by publication. Hot Desert soils include samples from the Mojave (California), Sonoran (Arizona), Chihuahuan (Texas), Arabian (Saudi Arabia, Kuwait, United Arab Emirates), and Registan (Afghanistan) deserts (Gonzales-Martin *et al.*, 2013). Agricultural soils include samples from England and Wales (Ashelford *et al.*, 2003), Scotland (Swanson *et al.*, 2009), and the United States (Williamson *et al.*, 2005). Forest soils include samples from Delaware (Williamson *et al.*, 2005) and Virginia (Williamson *et al.*, 2013). Wetland soils include samples from Delaware (Williamson *et al.*, 2005) and Virginia (Williamson *et al.*, 2013). Cold Desert soils include samples from the McMurdo Dry Valleys, Antarctica (Williamson *et al.*, 2007). Park (turf) soils include samples from Virginia (Williamson *et al.*, 2013). Dune soils include samples from the Outer Banks, North Carolina (Williamson *et al.*, 2013). Abbreviations: WCC, White Clay Creek; ETP, East Tom Pond (Antarctica); SOB, South Obelisk Pond (Antarctica); UAE, United Arab Emirates; Rhiz, rhizosphere; gdw, gram dry weight.

distinguish between phages and viruses that infect other host organisms. EFM reveals no structural details for morphological categorization, and the use of tail structures to specifically identify phages is problematic because not all phages have tails and phage tails may be lost during sample handling (Williamson *et al.*, 2012a). Third, EFM may not detect viruses with small RNA genomes. A comparison of virus types in marine samples indicated that RNA and DNA virus abundances were roughly equal, but that viruses with ssRNA genomes were systematically ignored by EFM-based direct count assays because the fluorescence signals emitted from their small genomes were below the limit of detection (Steward *et al.*, 2013). If the same is true for soils, then not only are viral abundances in soil significantly underestimated, but the presumption that phages comprise the majority of total viral abundance in soils would have to be re-evaluated.

Viruses of Bacteria, known as bacteriophages, or simply 'phages' (see Chapter 7) are one of the more extensively studied groups of viruses of soil microbes. The most easily recognized phages belong to the order *Caudovirales*, the tailed phages, comprised of the familes *Siphoviridae*, *Myoviridae*, and *Podoviridae*. Several additional viral families have been observed in samples prepared from soils, however, including *Microviridae*, *Leviviridae*, *Inoviridae*, and *Tectiviridae* (Swanson *et al.*, 2009; Williamson *et al.*, 2005).

Viral diversity

Only a few studies have attempted to assess the diversity of viruses in soil samples. The earliest approaches involved TEM to classify virus types based on gross particle morphology. In one study, the predominance of tailed viral morphotypes (approximately 80% of all observed morphologies) suggested that bacteriophages were the most abundant viral type within soil environments (Williamson *et al.*, 2005). A later assessment of viral diversity in a Scottish agricultural soil, however, indicated that viral assemblages of this soil were dominated by small (20 nm) spherical particles (ca. 60%), with filamentous viruses (ca. 18%) and larger (40–60 nm) spherical viruses (ca. 10%) the second and third most abundant morphotypes, respectively. Tailed viruses constituted only about 5% of the observed morphotypes (Swanson *et al.*, 2009). As mentioned above, it is possible that phage tails were broken during sample processing (Williamson *et al.*, 2012). It is also possible that some soils are not, in fact, dominated by phages but by other virus types such as those infecting plants. The small, spherical particles described in the Scottish study are consistent with many known plant viruses.

Virus metagenomics, or sequencing the total viral nucleic acids obtained from a specific environmental sample (see Chapter 5), has provided a much more powerful method than morphological TEM-based approaches to describe viral diversity and to compare diversity across samples. The construction and analysis of viral metagenomes have provided unprecedented views of the incredible genetic diversity encompassed by viruses. Viral metagenomes have been constructed, for example, from pelagic and coastal marine waters (Breitbart *et al.*, 2002, 2004; Angly *et al.*, 2006; Culley *et al.*, 2006; Williamson *et al.*, 2012), marine sediments (Breitbart *et al.*, 2004), estuarine waters (Bench *et al.*, 2007), corals (Marhaver *et al.*, 2008; Wegley *et al.*, 2007), stromatolites (Desnues *et al.*, 2008), and hot springs (Schoenfeld *et al.*, 2008), encompassing well over 1 Tbp (10^{12} base pairs) of sequence data currently available in public databases. Within this vast repository of viral metagenomic data, however, only a handful of soil viromes have been published: Joshua Tree National Park, USA; Konza Prairie LTER, USA; Manu National Park, Peru (Fierer *et al.*, 2007); rice paddy soil, Korea (Kim *et al.*, 2008); McMurdo Dry Valleys, Antarctica (Zablocki *et al.*, 2015); machair and brown earth soils, Scotland (Reavy *et al.*, 2015).

Within these soil viromes, 54.5–97.3% of all reads cannot be matched with a specific taxonomic source, a specific protein function, or both, rendering these reads unusable in most downstream analyses. This problem highlights the necessity of isolating and characterizing individual viruses from soil samples, as sequence data alone are of limited value if they cannot be matched to a known entity. Some soils are dominated by dsDNA phages (Zablocki *et al.*, 2015), while others are dominated by small, ssDNA viruses (Reavy *et al.*, 2015; Green *et al.*, 2018; Kim *et al.*, 2008). These findings challenge the dogma that all soil viromes are dominated by tailed, dsDNA phages, and opens the possibility that methodological biases might have led to erroneous conclusions regarding genetic and morphological diversity within the soil virome.

The estimated number of distinct viral genotypes (richness) in soils varies between 1000 and 1,000,000 and appears to vary according to soil type. For comparison, viral richness estimates vary from 532 to 129,000 for marine samples, and from 400 to 40,000 for freshwaters (Green *et al.*, 2015). It is interesting to note that both the lower and upper estimates of viral richness in soils are higher than the corresponding lower and upper bounds for aquatic systems. This, combined with the high percentage of unaffiliated sequences in soil metagenomes, supports the idea that the soil virome may represent the largest reservoir of genetic diversity on the planet. Given what we already know about the complex nature of soils (see 'The soil habitat' above), perhaps more coordinated efforts are warranted to obtain representative samples from a broader range of soil types. Based on the new discoveries that have arrived with each new viral metagenome, the lack of exploration of the soil virome represents a significant gap in our understanding of soil microbial ecology and of global viral diversity.

Ecological impacts of viral infection

Most phages are assumed to replicate through one of two major life cycles: lytic or lysogenic. Through virus-mediated cell lysis, phages can have significant impacts on the mortality and population dynamics of their hosts. Field experiments focusing on the population dynamics of bacteriophages in the sugar beet rhizosphere suggest that phages play important roles in the community succession patterns of soil bacteria (Ashelford *et al.*, 1999a, 1999b). Microcosm experiments using naturally occurring microbial assemblages suggest that soil bacterial communities and associated viral assemblages are highly dynamic and responsive to perturbations. Staggered fluctuations were observed between soil viruses and soil bacteria, patterns that closely follow conventional models of predator–prey dynamics (Srinivasiah *et al.*, 2008b).

Field experiments have provided additional evidence that phage-mediated mortality can be an important factor in the bacterial activity and biogeochemistry of some soils. In particular, phage-mediated mortality appears to exert top-down controls on bacterial abundance and activity in Arctic permafrost soils, limiting bacterial growth and respiration (Allen *et al.*, 2010). Attempts to replicate these experiments in temperate forest soils, however, have not been successful (Helsley *et al.*, 2013), demonstrating how the heterogeneity of soils can prevent universal application of methods and hamper progress in soil phage ecology.

Lytic replication by phages involves an obligate extracellular stage. Thus, in addition to maintaining the biological mechanisms of host entry, replication within host cells, and release from infected cells, phages must also persist in the soil environment long enough to encounter and infect a new host. Soil microbiology has a rich history of fate and transport studies determining the survival of enteric phages or animal viruses introduced into soil environments in order to track faecal pollution (Abbaszadegan *et al.*, 1993; Chu *et al.*, 2001; Ferguson *et al.*, 2003; Gerba, 1987; Goyal and Gerba, 1979; Jin and Flury, 2002; Straub *et al.*, 1993). Very few data are available regarding the survivorship and transport of naturally occurring phages in soils, however. How long can free phages indigenous to soils maintain their infectivity? And how do phage particles come into contact with their hosts in the highly structured soil environment, where both viruses and hosts are patchily distributed? These are significant, unanswered questions that establish clear directions for future research.

Lysogenic replication appears to hold a distinct advantage for viruses in soils because lysogeny supports the long-term maintenance of phages within host populations when host distribution is patchy (Stewart and Levin, 1984), when local host densities are low (Marsh and Wellington, 1994; Pantastico-Caldas *et al.*, 1992), or when host cells are nutrient-deplete (Marsh and Wellington, 1994; Williamson *et al.*, 2002). Investigations into the prevalence of lysogeny among soil bacteria isolated on laboratory media (Williamson *et al.*, 2008) and within natural assemblages of soil bacteria (Williamson *et al.*, 2007) indicate that lysogens, i.e. bacteria carrying a prophage element capable of entering a productive infection cycle, account for between 30% and 60% of the total bacterial population in a given soil. This is a significant fraction and suggests that many soil phages are temperate. This predominance of lysogeny may result in significant impacts of soil phages on the genetic diversity and evolution of hosts, because lysogenic relationships increase the probability of genetic exchange between phage and host, as well as horizontal gene transfer between hosts (Replicon *et al.*, 1995; Saye and Miller, 1989; Saye *et al.*, 1987; Wilcox and Fuhrman, 1994).

Temperate phages may also confer fitness changes upon lysogenized hosts through lysogenic conversion, a change in bacterial host phenotype resulting from expression of genes encoded by a prophage (Little, 2005). It has been speculated that host survival in a nutrient-poor environment requires an additional regulatory system provided by prophages that can suppress metabolic activity, ensuring both host and prophage survival. Without this additional control, higher host metabolic rates could result in prophage induction under conditions that would not favour the phage progeny encountering another host (Chen *et al.*, 2005; Paul, 2008). In the highly heterogeneous soil environment, where nutrient pools are often patchy and slowly available, additional metabolic control by prophage repressors would favour the long-term survival of both host and prophage. The existence of such potential fitness gains, however, has yet to be quantitatively assessed for soil lysogens, and thus represents another important realm for future exploration.

Beyond lysogenic conversion, viruses in soils may have significant impacts on host population genetics by acting as vectors for the horizontal transfer of genes between dissimilar bacterial strains through transduction. In aquatic studies, the presence of particulate matter increased transduction frequencies by providing surface area on which phages and host cells could aggregate and interact (Ripp and Miller, 1995). Soil, especially the clay and organic matter fractions, has an extremely high specific surface area [80–800 m^2/g (Hillel, 1998)), providing many potential sites for phage and host adsorption and interaction. Natural transduction, however, has never been observed in soil environments. Only a handful of experiments using cultivable phage–host systems have been performed, although all highlight the potential for transduction in soils (Deb *et al.*, 2003; Vettori *et al.*, 1999, 2000). A more recent study of the importance of phage-mediated gene transfer in soils examined viral assemblages that were obtained from the induction of soil bacteria in atrazine-treated soils. Subsequent PCR screening detected the presence of the atrazine catabolic gene, *trzN*, within the induced phage community, indicating that this important phenotypic trait could be transferred among bacterial populations through phage vectors (Ghosh *et al.*, 2008). While the frequency of transduction in soil environments is currently unknown, the possible impacts on bacterial genetic diversity and evolution are important and far-reaching.

Viruses of Archaea

The Archaea represent a domain of life phylogenetically distinct from the Bacteria (Woese, 1998). In terms of cell structure, Archaea use ether phospholipids (as opposed to ester phospholipids found in Bacteria) and their cell walls lack peptidoglycan (De Rosa *et al.*, 1986). The Archaea consist of four phyla: Euryarchaeota, Crenarchaeota, Thaumarchaeota, and Korarchaeota (Brochier-Armanet *et al.*, 2011; Woese *et al.*, 1990). The majority of known Archaea belong to the Crenarchaeaota and Euryarchaeaeota. Both thaumarchaea and korarchaea, however, appear to be widely distributed in many environments (Reigstad *et al.*, 2010), including fresh and marine waters, and soils, where they are believed to carry out the aerobic oxidation of ammonia, a key step of the nitrogen cycle (Schleper and Nicol, 2010).

Only a few members of the Thaumarchaeota have been isolated and characterized, and no korarchaea have been isolated in laboratory culture (Elkins *et al.*, 2008). As a result, archaeal viruses have so far been isolated only on crenarchaeal or euryarchaeal hosts, all originating from hot spring, hypersaline or hydrothermal environments (Pina *et al.*, 2011). There is no representation of archaeal viruses from soils among culture collections. A molecular survey of 146 soils found Crenarchaeota widely distributed across the globe, albeit at low abundances: 2–10% of all soil prokaryotes (Bates *et al.*, 2011). Based on the presence of archaeal hosts, it is highly likely that soils harbour archaeal viruses. At the time of this writing, however, nothing is known regarding the abundance, genetic diversity, or ecological roles of archaeal viruses in soil environments. See Chapter 8 for additional discussion of archaeal virus characteristics.

Viruses of Fungi

Fungal viruses, also known as mycoviruses (see Chapter 9), are likely found in most soils that support the growth of their hosts. While no published studies have sought to establish mycovirus abundance in soils, casual observances of mycovirus-like particles have been recorded (Swanson *et al.*, 2009). Based on the presence of cytoplasmic virus-like particles and dsRNA within fungal cells, mycovirus infections appear to be common in all major fungal groups. The majority of mycoviruses have dsRNA genomes, but about one-thired of all known representatives have positive-sense ssRNA genomes (Pearson *et al.*, 2009). The mycoviruses comprise four major families: *Totiviridae* (containing a monopartite genome), *Partitiviridae* (segmented genomes), *Narnaviridae*, and *Hypoviridae* (lacking capsids), with representatives in *Chrysoviridae*, *Pseudoviridae*, *Barnaviridae*, *Reoviridae*, and several taxonomically unassigned groups.

Because mycoviruses do not typically infect host cells from without, that is, existing as free viruses at any point in their life cycles, fungal viruses must both coexist with their hosts long-term and avoid elimination following death of a host from which they cannot escape. This lack of virulence among most mycoviruses precludes plaque assay methods of detection, although protoplasts generated through specific treatments in laboratory culture may be infected by virions (el-Sherbeini and Bostian, 1987; Ghabrial and Suzuki, 2009; Pearson *et al.*, 2009). While most mycovirus infections are inapparent, with no clear changes associated with host fitness, infection by certain mycoviruses can cause dramatic and noticeable changes in host phenotype, as discussed in the following subsections.

Killer phenotype

The killer phenotype or so-called 'killer viruses' encode a secreted toxin as well as an antidote molecule that confers immunity to the infected host (Breinig *et al.*, 2006; Sturley *et al.*, 1986). The mechanism of action of the secreted toxin varies by virus, but appears to fall into two main classes. Some killer viruses encode ionophoric toxins (e.g. K1, zygomycin) that disrupt normal cell membrane function by forming cation-selective channels. The ionic imbalance in the cytoplasm impairs enzyme functions and induces osmotic distress leading to cell death. Other killer viruses encode proteins that interact with nuclear factors (e.g. K28) that block the cell cycle, resulting in S-phase arrest and eventual cell death (Martinac *et al.*, 1990; Schmitt *et al.*, 1989).

The killer phenotype conferred by these viruses upon their hosts enables the infected host to kill off competitor species, resulting in clear fitness gains since the death of competitors ensures adequate resource acquisition by the host, and its parasitic mycovirus, for survival and reproduction. Because only infected hosts contain the toxin antidote molecules, killer viruses can control mating (heterokaryosis) success between infected and non-infected partners. Mating will be more successful between two infected partners, ensuring that surviving progeny carry the virus, and potentially driving evolution of the host species. This toxin–antidote system may represent an important survival strategy for the mycoviruses, potentially leading to their evolutionary success in the highly variable soil environment containing a patchy distribution of hosts.

Hypovirulence

Hypovirulence – contrasting hyp*er*virulence – as resulting from mycovirus infection can be observed in several plant-parasitic fungi, including members of the *Totiviridae*, *Narnaviridae*, *Chrysoviridae*, and *Reoviridae*, but is best characterized in the hypoviruses of chestnut blight fungus, *Cryphonectria parasitica* (Nuss, 2005). *C. parasitica* typically enters its host plant through physical wounds or breaks in the bark, infecting the underlying cambium (Anagnostakis, 2000). The fungus then secretes compounds that are toxic to the host plant, killing the exposed host plant cells. Eventually the fungal growth or canker will girdle the entire trunk, killing everything above it (Anagnostakis, 2000). Infection of *C. parasitica* by hypoviruses can severely reduce the virulence of the fungal infection, in many cases allowing the host plant to survive and retain living tissues above the fungal infection site. Infection of *C. parasitica* by hypoviruses can cause several phenotypic changes related to hypovirulence, including the loss of female fertility in sexual crosses, reduction in asexual sporulation, reduced pigment production, and reduced accumulation of oxalic acid and other metabolites (Anagnostakis,

1979; Elliston, 1985; Long, 1985). It is quite striking that such dramatic alterations in host phenotype can be induced by only 12.7 kb of viral RNA. The molecular mechanisms behind these phenotypic changes are still not fully understood, but virus-mediated alterations in G-protein signalling (Choi *et al.*, 1995; Dawe *et al.*, 2004) and reprogramming of host transcription patterns (Dawe *et al.*, 2003, 2004) have been implicated.

Because of their role in reducing the pathogenicity of fungi towards their host plants, there has been significant interest in the exploitation of hypoviruses as biological control agents of chestnut blight since the 1950s (Biraghi, 1953; Grenle and Berthelay-Sauret, 1978), well before the molecular mechanisms of hypovirulence were even dimly understood. While hypoviruses have been used with reported success in Europe (Heiniger and Rigling, 1994), the use of hypoviruses as a bio-control strategy for *C. parasitica* has not been as successful in North America (Griffin, 2000; MacDonald and Fulbright, 1991). It is not clear the extent to which viruses of other soil microbes may alter host function. But given the impacts of temperate phages and hypoviruses on their respective hosts, it is very likely that we have yet to discover other important ways in which viruses modulate their hosts, leading to differential success in the heterogeneous soil environment. See Chapter 14 for more on the use of viruses of microorganisms as microorganism control agents.

Thermotolerance

In the only known example of its kind among plants, a mycovirus has been shown to alter the fitness landscape of a host plant that has been colonized by the mycovirus' fungal host. The mycovirus, Curvularia thermal tolerance virus (CThTV), alters the phenotype of its fungal host *Curvularia protuberata* such that when the roots of the panic grass *Dichanthelium lanuginosum* are colonized by the fungus, the tripartite interaction allows both fungus and host plant to grow at high soil temperatures (Márquez *et al.*, 2007). This results in significant niche expansion, as the host plant was isolated near geothermal features in Yellowstone National Park, USA.

Laboratory experiments showed that when grass root zones were heated up to 65°C, non-symbiotic plants were not able to survive, while symbiotic plants survived. Additional experiments involving virus-infected fungi and virus-free fungal isolates revealed that the mycovirus was a crucial element in the symbiosis, as only the plants inoculated with fungal isolates carrying the mycovirus were able to survive heat stress (Márquez *et al.*, 2007). In the absence of heat stress, symbiotic plants exhibited enhanced growth rate and drought tolerance compared with non-symbiotic plants. Currently, the molecular mechanisms underlying thermotolerance and other phenotypic changes in this symbiosis have not yet been established. The existence of this tripartite relationship, however, raises important questions about how common such complex interactions might be among the mycoviruses, and indeed, among viruses of other soil microbes.

Cyanobacteria, algae, and diatoms

Algae, diatoms and cyanobacteria are common inhabitants of most soils with sufficient moisture-holding capacities to support their growth, and are widely distributed across the globe. For example, a floristic study of two Russian mountains identified 136 species of soil algae within the survey domain (Bakieva *et al.*, 2012). A similar survey in Greenland identified 81 diatom taxa in soils (van Kerckvoorde *et al.*, 2000). As also observed with the Archaea, in spite of their relatively broad global distribution, the ecology of these autotrophic soil microbes is poorly understood. Consequently, almost nothing is known regarding the viruses that infect them. See Chapter 10 for an overview of algal viruses as found in aquatic environments.

Algae and cyanobacteria are among the first microorganisms to colonize bare soil, including volcanic deposits, entisols, and inceptisols. The physical stabilization and nutrient enrichment of colonized soils enables the subsequent establishment of higher plant communities. Soil algae, especially cyanobacteria, are known to aggregate soil particles by producing extracellular polysaccharides (Lynch and Bragg, 1985) and forming water-stable aggregates that reduce the impact of wind erosion (Johansen, 1993). Soil algae and cyanobacteria – in conjunction with other soil microbes including fungi and lichens, and microphytes such as mosses – can form microbiotic crusts, which are commonly found in arid and semi-arid soils that lack vegetative cover and the soil-stabilizing properties of plant roots. Microbiotic crusts fulfill important roles in arid ecosystems by reducing wind erosion and holding soils in place (Brotherson and Rushforth, 1983; Leys and Eldridge, 1998) as well as increasing the nutrient content of underlying soils (Beymer and Klopatek, 1991; Cameron and Fuller, 1960). Through host cell lysis and mortality, viruses of soil algae and cyanobacteria have the potential to reduce this soil genesis and stabilization function.

Viruses of algae (Müller *et al.*, 1998; Reisser, 1993; Short, 2012), diatoms (Bettarel *et al.*, 2005; Bratbak *et al.*, 1996; Jacquet *et al.*, 2002), and cyanobacteria (Millard and Mann, 2006; Suttle, 2002; Wang *et al.*, 2011b) are important players in aquatic ecosystems (see Chapters 6 and 10), with significant impacts on host abundance, host population dynamics, host genetic diversity, nutrient cycles, and even cloud formation and atmospheric conditions (Arnold *et al.*, 2013). By contrast, almost nothing is known regarding the terrestrial relatives of these viruses. A handful of studies have examined the dynamics and potential impacts of cyanophages in rice paddy soils (Wang *et al.*, 2011; Lee *et al.*, 2012). Paddy systems, however, are typically flooded most of the year and it therefore is difficult to determine whether the cyanobacteria and associated phages found in these systems should be considered aquatic or terrestrial. This is largely a problem of semantics since nature cares little for such human categorizations, but the question nonetheless remains whether studies of rice paddy systems can reveal details about the ecology of terrestrial viruses.

In future studies it may be possible to generate estimates of cyanophage abundance within soil samples using a directed plaque assay approach against a battery of cyanobacterial hosts (Millard and Mann, 2006) or using a quantitative real-time PCR approach with virus-specific primers (Short and Short, 2009). Through metagenomic approaches, it also may be possible to estimate the diversity of cyanophages, algal and diatom viruses found in soil samples. These approaches, however, will only be successful if viruses of terrestrial algae, diatoms, and cyanobacteria contain similar sequence motifs as their aquatic counterparts. For example, the virus-specific primers used in qPCR were developed through multiple DNA sequence alignments obtained from several freshwater algal virus taxa (Short and Short, 2009). If the DNA sequences of terrestrial algal viruses are too divergent from the primers developed for freshwater algal viruses, then the pre-existing tools developed for aquatic viruses may not be applicable to terrestrial viruses. A related problem may exist in identifying cyanophage or algal virus sequences in soil virus metagenomes. Isolation and genome sequencing of specific terrestrial representatives of algal, diatom, and cyanobacterial viruses may be necessary before molecular approaches can be used to study the ecology of these viruses in soils. The testing of existing molecular tools (e.g. PCR primer sets), isolation of terrestrial exemplars of algal, diatom, and cyanobacterial viruses, and use of terrestrial virus models to develop specific molecular tools for analysis of soil viral communities appear to be logical first steps in addressing this information gap in the viral ecology of soils.

Viruses of amoebae and protozoa

The 2003 discovery of mimivirus (La Scola *et al.*, 2003), a 'giant' virus infecting the protozoan host, *Acanthamoeba polyphaga*, reignited important questions about what viruses are and whether or not they should be considered alive. Its discovery also raised questions about the presence and prevalence of other such giant viruses that would have been missed in any previous isolation attempts due to the standard practice of filtration through 0.22 μm. In the past decade, directed efforts using *A. polyphaga* or *A. castellanii* as hosts have resulted in the isolation of several additional giant viruses (Arslan *et al.*, 2011; Boyer *et al.*, 2009; Desnues *et al.*, 2012; Philippe *et al.*, 2013; Yoosuf *et al.*, 2012), with at least two isolates originating from soils, dubbed terravirus 1 and terravirus 2 (La Scola *et al.*, 2010). Discovery of the giant viruses has resulted in a new taxonomic grouping, the nucleocytoplasmic large DNA viruses (NCLDV), which includes the new families *Mimiviridae*, *Megaviridae*, *Marseilleviridae*, and *Pandoraviridae*, as well as the previously established virus families *Asfarviridae*, *Ascoviridae*, *Iridoviridae*, *Phycodnaviridae*, and *Poxviridae* (Koonin and Yutin, 2001). This new grouping is based on shared characteristics of both genome content and virion structure. See Chapter 11 for more detailed discussion of these viruses.

Since the deposition of mimivirus and other giant virus genome sequences into GenBank and other DNA sequence databases, environmental metagenome projects regularly turn up hits to mimivirus, suggesting that it – or its relatives – are common inhabitants of marine (Ghedin and Claverie, 2005; Monier *et al.*, 2008) and freshwater (Fancello *et al.*, 2013; López-Bueno *et al.*, 2009) ecosystems. Amoebae are common inhabitants of most soils (Anderson, 2002; Anderson and Bischoff, 1998) and have even been isolated from soils of the McMurdo Dry Valleys, Antarctica (Brown *et al.*, 1982). Since the initial isolation of Terravirus 1 and 2, 18 additional amoebal viruses have been isolated from soils (Pagnier *et al.*, 2013), allowing for comparative genomics approaches that may enable identification of specific sequences associated with terrestrial giant viruses. The identification of such sequences allows for the development of PCR primers and other molecular tools to prospect for other, related viruses in soil samples. In addition, more directed efforts to construct viral metagenomes from soils would help to establish the distribution and relative abundance of giant viruses in soil ecosystems. Currently, the abundance of amoebal viruses in soils, as well as their specific ecological impacts on host population structure and function, remain unknown.

Conclusions

Soils represent the greatest reservoir of biodiversity on the planet, with prokaryotic diversity estimated to be three orders of magnitude greater than in all other ecosystems combined (Curtis *et al.*, 2002; Kemp and Aller, 2004). Soils are critical to global biogeochemical cycles including carbon, nitrogen, and phosphorus, which support all other forms of terrestrial diversity. With regard to human civilization, soils provide ecosystem services on a global scale estimated at > $20 trillion (Boumans *et al.*, 2002), including food and fibre production, waste disposal, water purification, and production of pharmaceuticals (Daniel, 2004; Waksman and Woodruff, 1941). Most of these services are provided by microbes, and factors that influence soil microbes (e.g. viruses) will impact these ecosystem processes (Fig. 4.5). In marine ecosystems, it is well established that viruses are powerful modulators of microbial community composition, microbial activity, nutrient cycles, and host evolution (Fuhrman, 1992, 1999; Weinbauer, 2004; Wommack and Colwell, 2000). Viruses likely play similar, important roles in soil ecosystems. However, our understanding of the virus ecology of terrestrial environments lags significantly behind that of marine and aquatic environments.

This review has considered what is and is not known regarding the viruses that infect specific soil microbial groups (boxed in red, Fig. 4.5) that make up the soil microbial food web (boxed in grey, Fig. 4.5). The soil microbial food web is incredibly important in sustaining soil structure, soil fertility, and crop productivity. In marine ecosystems, viruses shunt anywhere from 6% to 25% of carbon fixed by primary production into marine microbial food webs (Kimura *et al.*, 2008; Weinbauer, 2004) (see Chapter 6). The magnitude of viral impacts on carbon flux in terrestrial ecosystems is estimated to be even greater, but very few specific impacts of viruses of soil microbes are currently known, particularly for viruses of non-bacterial microbes (Buée *et al.*, 2009; Kimura *et al.*, 2008; Srinivasiah *et al.*, 2008b).

Gaining a more holistic understanding of viruses that infect soil microbes is important in order to understand viral impacts on process rates, since viral infection of hosts can influence host community capacity to decompose soil organic matter,

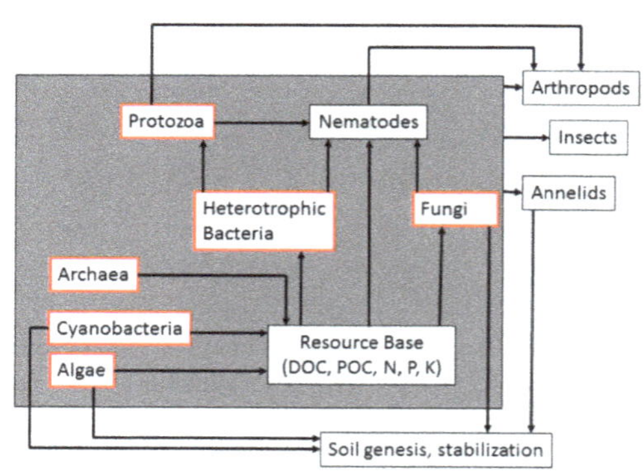

Figure. 4.5 Highly simplified terrestrial food web focused on soil organisms impacted by viral infection. Arrows point from resource/prey to main consumers; grey box includes soil microorganisms and microbially mediated processes; red boxes highlight specific soil microbes discussed within this chapter.

convert organic phosphorous into plant-available forms, and respire carbon dioxide, to name just a few specific examples. Improving our understanding of viruses that infect soil microbes will also provide critical insights to the roles viruses have played, and continue to play in shaping host fitness through selection against sensitive genotypes, lysogenic conversion, host metabolic manipulation by prophages, and horizontal gene transfer.

It has been quipped that we know more about deep space than we do about the deep ocean, ostensibly to emphasize that our home planet still contains vast realms of unexplored territory that are worthy of study. But whatever shortcomings exist in our current understanding of the global ocean, we know still less about the teeming microbial life, including viruses, literally in our own backyards. Certainly there are significant technical challenges associated with investigating this complex habitat that have impeded progress, but it may also be that the perceived commonness of soil detracts from its allure as an ecosystem worthy of study. Regardless of the reasons, soils remain the most poorly understood ecosystems on Earth. At the same time, viruses represent the largest pool of untapped genetic diversity and unexplored sequence space on the planet. In this regard, viruses of soil microbes comprise an unknown quantity within an unexplored territory: a vast new frontier, ripe with opportunities for discovery for those who dare plumb its depths.

References

Abbaszadegan, M., Huber, M.S., Gerba, C.P., and Pepper, I.L. (1993). Detection of enteroviruses in groundwater with the polymerase chain reaction. Appl. Environ. Microbiol. *59*, 1318–1324.

Abedon, S.T. (2008). Bacteriophage Ecology: Population Growth, Evolution, and Impact of Bacterial Viruses (Cambridge University Press).

Allen, B., Willner, D., Oechel, W.C., and Lipson, D. (2010). Top-down control of microbial activity and biomass in an Arctic soil ecosystem. Environ. Microbiol. *12*, 642–648. https://doi.org/10.1111/j.1462-2920.2009.02104.x.

Anagnostakis, S.L. (1979). Hypovirulence Conversion in Endothia parasitica. Phytopathology *69*, 1226–1229.

Anagnostakis, S.L. (2000). Revitalization of the Majestic Chestnut: Chestnut Blight Disease. APSnet Feature Artic. https://doi.org/10.1094/APSnetFeature-2000-1200.

Anderson, O.R. (2002). Laboratory and field-based studies of abundances, small-scale patchiness, and diversity of gymnamoebae in soils of varying porosity and organic content: evidence of microbiocoenoses. J. Eukaryot. Microbiol. *49*, 17–23.

Andrson, O.R., and Bischoff, P.J. (1998). Abundance and diversity of Gymnamoebae at varying soil sites in Northeastern U. S. A. Acta Protozool. *37*, 17–21.

Angly, F.E., Felts, B., Breitbart, M., Salamon, P., Edwards, R.A., Carlson, C., Chan, A.M., Haynes, M., Kelley, S., Liu, H., *et al.* (2006). The marine viromes of four oceanic regions. PLOS Biol. *4*, e368.

Arnold, H.E., Kerrison, P., and Steinke, M. (2013). Interacting effects of ocean acidification and warming on growth and DMS-production in the haptophyte coccolithophore *Emiliania huxleyi*. Glob. Chang. Biol. *19*, 1007–1016. https://doi.org/10.1111/gcb.12105.

Arslan, D., Legendre, M., Seltzer, V., Abergel, C., and Claverie, J.M. (2011). Distant Mimivirus relative with a larger genome highlights the fundamental features of Megaviridae. Proc. Natl. Acad. Sci. U.S.A. *108*, 17486–17491. https://doi.org/10.1073/pnas.1110889108.

Ashelford, K.E., Day, M.J., Bailey, M.J., Lilley, A.K., and Fry, J.C. (1999a). In situ population dynamics of bacterial viruses in a terrestrial environment. Appl. Environ. Microbiol. *65*, 169–174.

Ashelford, K.E., Fry, J.C., Bailey, M.J., Jeffries, A.R., and Day, M.J. (1999b). Characterization of six bacteriophages of *Serratia Liquefaciens* CP6 isolated from the sugar beet phytosphere Appl. Environ. Microbiol. *65*, 1959–1965.

Bakieva, G.R., Khaibullina, L.S., Gaisina, L.A., and Kabirov, R.R. (2012). Ecological-floristic analysis of soil algae and cyanobacteria on the Tra-Tau and Yurak-Tau Mounts, Bashkiria. Eurasian Soil Sci. *45*, 873–881.

Bates, S.T., Berg-Lyons, D., Caporaso, J.G., Walters, W.A., Knight, R., and Fierer, N. (2011). Examining the global distribution of dominant archaeal populations in soil. ISME J. *5*, 908–917. https://doi.org/10.1038/ismej.2010.171.

Bench, S.R., Hanson, T.E., Williamson, K.E., Ghosh, D., Radosovich, M., Wang, K., and Wommack, K.E. (2007). Metagenomic characterization of Chesapeake Bay virioplankton. Appl. Environ. Microbiol. *73*, 7629–7641.

Bettarel, Y., Kan, J., Wang, K., Williamson, K.E., Cooney, S., Ribblett, S., Chen, F., Wommack, K.E., and Coats, D.W. (2005). Isolation and preliminary characterisation of a small nuclear inclusion virus infecting the diatom *Chaetoceros* cf. *gracilis*. Aquat. Microb. Ecol. *40*, 103–114.

Beymer, R.J., and Klopatek, J.M. (1991). Potential contribution of carbon by microphytic crusts in pinyon-juniper woodlands. Arid Soil Res. Rehabil. *5*, 187–198.

Biraghi, A. (1953). Possible active resistance to Endothia parasitica in *Castanea crenata*. Proc. Congr. Int. Union Forest Res. Organ., Rome. p. 459.

Boumans, R., Costanza, R., Farley, J., Wilson, M.A., Portela, R., Rotmans, J., Villa, F., and Grasso, M. (2002). Modeling the dynamics of the integrated earth system and the value of global ecosystem services using the GUMBO model. Ecol. Econ. *41*, 529–560.

Boyer, M., Yutin, N., Pagnier, I., Barrassi, L., Fournous, G., Espinosa, L., Robert, C., Azza, S., Sun, S., Rossmann, M.G., *et al.* (2009). Giant Marseillevirus highlights the role of amoebae as a melting pot in emergence of chimeric microorganisms. Proc. Natl. Acad. Sci. U.S.A. *106*, 21848–21853. https://doi.org/10.1073/pnas.0911354106.

Bratbak, G., Wilson, W., and Heldal, M. (1996). Viral control of Emiliania huxleyi blooms? J. Mar. Syst. *9*, 75–81.

Breinig, F., Sendzik, T., Eisfeld, K., and Schmitt, M.J. (2006). Dissecting toxin immunity in virus-infected killer yeast uncovers an intrinsic strategy of self-protection. Proc. Natl. Acad. Sci. U.S.A. *103*, 3810–3815.

Breitbart, M., Salamon, P., Andresen, B., Mahaffy, J.M., Segall, A.M., Mead, D., Azam, F., and Rohwer, F. (2002). Genomic analysis of uncultured marine viral communities. Proc. Natl. Acad. Sci. U.S.A. *99*, 14250–14255. https://doi.org/10.1073/pnas.202488399.

Breitbart, M., Felts, B., Kelley, S., Mahaffy, J.M., Nulton, J., Salamon, P., and Rohwer, F. (2004). Diversity and population structure of a near-shore marine-sediment viral community. Proc. Biol. Sci. *271*, 565–574. https://doi.org/10.1098/rspb.2003.2628.

Brochier-Armanet, C., Forterre, P., and Gribaldo, S. (2011). Phylogeny and evolution of the Archaea: one hundred genomes later. Curr. Opin. Microbiol. *14*, 274–281. https://doi.org/10.1016/j.mib.2011.04.015.

Brotherson, J.D., and Rushforth, S.R. (1983). Influence of cryptogamic crusts on moisture relationships of soils in Navajo National Monument, Arizona. West. North Am. Nat. *43*, 73–78.

Brown, T.J., Cursons, R.T., and Keys, E.A. (1982). Amoebae from antarctic soil and water. Appl. Environ. Microbiol. *44*, 491–493.

Buée, M., Boer, W.D., Martin, F., van Overbeek, L., and Jurkevitch, E. (2009). The rhizosphere zoo: An overview of plant-associated communities of microorganisms, including phages, bacteria, archaea, and fungi, and of some of their structuring factors. Plant Soil *321*, 189–212.

Cameron, R.E., and Fuller, W.H. (1960). Nitrogen Fixation by Some Algae in Arizona Soils. Soil Sci. Soc. Am. J. *24*, 353–356.

Chen, Y., Golding, I., Sawai, S., Guo, L., and Cox, E.C. (2005). Population fitness and the regulation of *Escherichia coli* genes by bacterial viruses. PLOS Biol. *3*, e229.

Chibani-Chennoufi, S., Bruttin, A., Dillmann, M.L., and Brüssow, H. (2004). Phage-host interaction: an ecological perspective. J. Bacteriol. *186*, 3677–3686. https://doi.org/10.1128/JB.186.12.3677-3686.2004.

Choi, G.H., Chen, B., and Nuss, D.L. (1995). Virus-mediated or transgenic suppression of a G-protein alpha subunit and attenuation of fungal virulence. Proc. Natl. Acad. Sci. U.S.A. *92*, 305–309.

Chu, Y., Jin, Y., Flury, M., and Yates, M.V. (2001). Mechanisms of virus removal during transport in unsaturated porous media. Water Resour. Res. *37*, 253–263.

Culley, A.I., Lang, A.S., and Suttle, C.A. (2006). Metagenomic analysis of coastal RNA virus communities. Science *312*, 1795–1798.

Curtis, T.P., Sloan, W.T., and Scannell, J.W. (2002). Estimating prokaryotic diversity and its limits. Proc. Natl. Acad. Sci. U.S.A. *99*, 10494–10499. https://doi.org/10.1073/pnas.142680199.

Daniel, R. (2004). The soil metagenome – a rich resource for the discovery of novel natural products. Curr. Opin. Biotechnol. *15*, 199–204. https://doi.org/10.1016/j.copbio.2004.04.005.

Dawe, A.L., McMains, V.C., Panglao, M., Kasahara, S., Chen, B., and Nuss, D.L. (2003). An ordered collection of expressed sequences from *Cryphonectria parasitica* and evidence of genomic microsynteny with *Neurospora crassa* and *Magnaporthe grisea*. Microbiol. Read. Engl. *149*, 2373–2384.

Dawe, A.L., Segers, G.C., Allen, T.D., McMains, V.C., and Nuss, D.L. (2004). Microarray analysis of *Cryphonectria parasitica* Galpha- and Gbetagamma-signalling pathways reveals extensive modulation by hypovirus infection. Microbiol. *150*, 4033–4043.

Deb, C., Chakraborty, R., Ghosh, A.N., Mandal, N.C., Mukherjee, T., and Roy, P. (2003). A generalized transducing thiophage (TPC-1) of a facultative sulfur chemolithotrophic bacterium, Bosea thiooxidans CT5, of alpha-Proteobacteria, isolated from Indian soil. FEMS Microbiol. Lett. *227*, 87–92.

Desnues, C., Rodriguez-Brito, B., Rayhawk, S., Kelley, S., Tran, T., Haynes, M., Liu, H., Furlan, M., Wegley, L., Chau, B., *et al.* (2008). Biodiversity and biogeography of phages in modern stromatolites and thrombolites. Nature *452*, 340–343. https://doi.org/10.1038/nature06735.

Desnues, C., La Scola, B., Yutin, N., Fournous, G., Robert, C., Azza, S., Jardot, P., Monteil, S., Campocasso, A., Koonin, E.V., *et al.* (2012). Provirophages and transpovirons as the diverse mobilome of giant viruses. Proc. Natl. Acad. Sci. U.S.A. *109*, 18078–18083. https://doi.org/10.1073/pnas.1208835109.

Elkins, J.G., Podar, M., Graham, D.E., Makarova, K.S., Wolf, Y., Randau, L., Hedlund, B.P., Brochier-Armanet, C., Kunin, V., Anderson, I., *et al.* (2008). A korarchaeal genome reveals insights into the evolution of the Archaea. Proc. Natl. Acad. Sci. U.S.A. *105*, 8102–8107. https://doi.org/10.1073/pnas.0801980105.

Elliston, J.L. (1985). Characteristics of dsRNA-free and dsRNA-containing strains of *Endothia parasitica* in relation to hypovirulence. Phytopathology *75*, 151–158.

el-Sherbeini, M., and Bostian, K.A. (1987). Viruses in fungi: infection of yeast with the K1 and K2 killer viruses. Proc. Natl. Acad. Sci. U.S.A. *84*, 4293–4297.

Fancello, L., Trape, S., Robert, C., Boyer, M., Popgeorgiev, N., Raoult, D., and Desnues, C. (2013). Viruses in the desert: a metagenomic survey of viral communities in four perennial ponds of the Mauritanian Sahara. ISME J. *7*, 359–369. https://doi.org/10.1038/ismej.2012.101.

Ferguson, C., Husman, A.M.D., Altavilla, N., Deere, D., and Ashbolt, N. (2003). Fate and transport of surface water pathogens in watersheds. Crit. Rev. Environ. Sci. Technol. *33*, 299–361.

Fierer, N., Breitbart, M., Nulton, J., Salamon, P., Lozupone, C., Jones, R., Robeson, M., Edwards, R.A., Felts, B., Rayhawk, S., *et al.* (2007). Metagenomic and small-subunit rRNA analyses reveal the genetic diversity of bacteria, archaea, fungi, and viruses in soil. Appl. Environ. Microbiol. 73, 7059–7066.

Fuhrman, J. (1992). Bacterioplankton roles in cycling of organic matter: The microbial food web. In Primary Productivity and Biogeochemical Cycles in the Sea, Falkowski, P.G., and Woodhead, A.D., eds. (New York: Plenum Press), pp. 361–383.

Fuhrman, J.A. (1999). Marine viruses and their biogeochemical and ecological effects. Nature *399*, 541–548. https://doi.org/10.1038/21119.

Gerba, C.P. (1987). Phage as indicators of fecal pollution. In Phage Ecology, Goyal, S.M., Gerba, C.P., and Britton, G., eds. (New York, N.Y.: John Wiley and Sons), pp. 197–209.

Ghabrial, S.A., and Suzuki, N. (2009). Viruses of plant pathogenic fungi. Annu. Rev. Phytopathol. *47*, 353–384. https://doi.org/10.1146/annurev-phyto-080508-081932.

Ghedin, E., and Claverie, J.M. (2005). Mimivirus relatives in the Sargasso sea. Virol. J. *2*, 62.

Ghosh, D., Roy, K., Williamson, K.E., White, D.C., Wommack, K.E., Sublette, K.L., and Radosevich, M. (2008). Prevalence of lysogeny among soil bacteria and presence of 16S rRNA and trzN genes in viral-community DNA. Appl. Environ. Microbiol. *74*, 495–502.

Goyal, S.M., and Gerba, C.P. (1979). Comparative adsorption of human enteroviruses, simian rotavirus, and selected bacteriophages to soils. Appl. Environ. Microbiol. *38*, 241–247.

Green, J.C., Rahman, F., Saxton, M.A., and Williamson, K.E. (2015). Metagenomic assessment of viral diversity in Lake Matoaka, a temperate, eutrophic freshwater lake in southeastern Virginia, USA. Aquat. Microb. Ecol. *75*, 117–128.

Green, J.C., Rahman, F., Saxton, M.A., and Williamson, K.E. (2018). Quantifying aquatic viral community change associated with stormwater runoff in a wet retention pond using metagenomic time series data. Aquat. Microb. Ecol. *81*, 19–35.

Grenle, J., and Berthelay-Sauret, S. (1978). In Proceedings of the American Chestnut Symposium, MacDonald, W.L., Cech, F.C., Luchok, J., and Smith, C., eds. (West Virginia University Books, Morgantown, West Virginia), pp. 30–34.

Griffin, G.J. (2000). Blight control and restoration of the American chestnut. J. For. *98*, 22–27.

Heiniger, U., and Rigling, D. (1994). Biological control of chestnut blight in Europe. Annu. Rev. Phytopathol. *32*, 581–599.

Helsley, K.R., Brown, T.M., Furlong, K., and Williamson, K.E. (2013). Applications and limitations of tea extract as a virucidal agent to assess the role of phage predation in soils. Biol. Fertil. Soils *50*, 263–274.

Hillel, D. (1998). Environmental Soil Physics (San Diego: Harcourt Brace and Company).

Hunt, H.W., Fountain, A.G., Doran, P.T., and Basagic, H. (2010). A dynamic physical model for soil temperature and water in Taylor Valley, Antarctica. Antarct. Sci. *22*, 419–434.

Jacquet, S., Heldal, M., Iglesias-Rodriguez, D., Larsen, A., Wilson, W., and Bratbak, G. (2002). Flow cytometric analysis of an *Emiliana huxleyi* bloom terminated by viral infection. Aquat. Microb. Ecol. *27*, 111–124.

Jin, Y., and Flury, M. (2002). Fate and transport of viruses in porous media. Adv. Agron. *77*, 39–51.

Johansen, J. (1993). Cryptogamic crusts of semiarid and arid lands of North America. J. Phycol. *29*, 140–147.

Kemp, P.F., and Aller, J.Y. (2004). Bacterial diversity in aquatic and other environments: what 16S rDNA libraries can tell us. FEMS Microbiol. Ecol. *47*, 161–177. https://doi.org/10.1016/S0168-6496(03)00257-5.

van Kerckvoorde, A., Trappeniers, K., Nijs, I., and Beyens, L. (2000). Terrestrial soil diatom assemblages from different vegetation types in Zackenberg (Northeast Greenland). Polar Biol. *23*, 392–400.

Kim, K.H., Chang, H.W., Nam, Y.D., Roh, S.W., Kim, M.S., Sung, Y., Jeon, C.O., Oh, H.M., and Bae, J.W. (2008). Amplification of uncultured single-stranded DNA viruses from rice paddy soil. Appl. Environ. Microbiol. *74*, 5975–5985. https://doi.org/10.1128/AEM.01275-08.

Kimura, M., Jia, Z.-J., Nakayama, N., and Asakawa, S. (2008). Ecology of viruses in soils: Past, present and future perspectives. Soil Sci. Plant Nutr. *54*, 1–32.

Koonin, E.V., and Yutin, N. (2012). Nucleo-cytoplasmic Large DNA Viruses (NCLDV) of Eukaryotes. eLS. https://doi.org/10.1002/9780470015902.a0023268.

Lee, C.G., Watanabe, T., Fujita, Y., Asakawa, S., and Kimura, M. (2012). Heterotrophic growth of cyanobacteria and phage-mediated microbial loop in soil: Examination by stable isotope probing (SIP) method. Soil Sci. Plant Nutr. *58*, 161–168.

Leys, J.F., and Eldridge, D.J. (1998). Influence of cryptogamic crust disturbance to wind erosion on sand and loam rangeland soils. Earth Surf. Process. Landf. *23*, 963–974.

Long, P.E. (1985). The ecology and physiology of the fungal mycelium. Plant, Cell & Environ. *8*, 299–300. https://doi.org/10.1111/1365-3040.ep11604694.

López-Bueno, A., Tamames, J., Velázquez, D., Moya, A., Quesada, A., and Alcamí, A. (2009). High diversity of the viral community from an Antarctic lake. Science *326*, 858–861. https://doi.org/10.1126/science.1179287.

Lynch, J.M., and Bragg, E. (1985). Microorganisms and Soil Aggregate Stability. In Advances in Soil Science, Stewart, B.A. ed. (Springer, New York), pp. 133–171.

MacDonald, W.L., and Fulbright, D.W. (1991). Biological control of chestnut blight: use and limitation of transmissible hypovirulence. Plant Dis. *75*, 656–661.

Marhaver, K.L., Edwards, R.A., and Rohwer, F. (2008). Viral communities associated with healthy and bleaching corals. Environ. Microbiol. *10*, 2277–2286. https://doi.org/10.1111/j.1462-2920.2008.01652.x.

Márquez, L.M., Redman, R.S., Rodriguez, R.J., and Roossinck, M.J. (2007). A virus in a fungus in a plant: three-way symbiosis required for thermal tolerance. Science *315*, 513–515.

Marsh, P., and Wellington, E.M.H. (1994). Phage-host interactions in soil. FEMS Microbiol. Ecol. *15*, 99–107.

Martinac, B., Zhu, H., Kubalski, A., Zhou, X.L., Culbertson, M., Bussey, H., and Kung, C. (1990). Yeast K1 killer toxin forms ion channels in sensitive yeast spheroplasts and in artificial liposomes. Proc. Natl. Acad. Sci. U.S.A. *87*, 6228–6232.

Millard, A.D., and Mann, N.H. (2006). A temporal and spatial investigation of cyanophage abundance in the Gulf of Aqaba, Red Sea. J. Mar. Biol. Assoc. U. K. *86*, 507–515.

Monier, A., Larsen, J.B., Sandaa, R.A., Bratbak, G., Claverie, J.M., and Ogata, H. (2008). Marine mimivirus relatives are probably large algal viruses. Virol. J. *5*, 12. https://doi.org/10.1186/1743-422X-5-12.

Müller, D.G., Kapp, M., and Knippers, R. (1998). Viruses in marine brown algae. Adv. Virus Res. *50*, 49–67.

Nuss, D.L. (2005). Hypovirulence: mycoviruses at the fungal-plant interface. Nat. Rev. Microbiol. *3*, 632–642.

Pagnier, I., Reteno, D.G., Saadi, H., Boughalmi, M., Gaia, M., Slimani, M., Ngounga, T., Bekliz, M., Colson, P., Raoult, D., *et al.* (2013). A decade of improvements in Mimiviridae and Marseilleviridae isolation from amoeba. Intervirology *56*, 354–363. https://doi.org/10.1159/000354556.

Pantastico-Caldas, M., Duncan, K.E., Istock, C.A., and Bell, J.A. (1992). Population dynamics of bacteriophage and *Bacillus subtilis* in soil. Ecology *73*, 1888–1902.

Paul, J.H. (2008). Prophages in marine bacteria: dangerous molecular time bombs or the key to survival in the seas? ISME J. *2*, 579–589. https://doi.org/10.1038/ismej.2008.35.

Pearson, M.N., Beever, R.E., Boine, B., and Arthur, K. (2009). Mycoviruses of filamentous fungi and their relevance to plant pathology. Mol. Plant Pathol. *10*, 115–128. https://doi.org/10.1111/j.1364-3703.2008.00503.x.

Philippe, N., Legendre, M., Doutre, G., Couté, Y., Poirot, O., Lescot, M., Arslan, D., Seltzer, V., Bertaux, L., Bruley, C., *et al.* (2013). Pandoraviruses: amoeba viruses with genomes up to 2.5 Mb reaching that of parasitic eukaryotes. Science *341*, 281–286. https://doi.org/10.1126/science.1239181.

Pina, M., Bize, A., Forterre, P., and Prangishvili, D. (2011). The archeoviruses. FEMS Microbiol. Rev. *35*, 1035–1054.

Pointing, S.B., and Belnap, J. (2012). Microbial colonization and controls in dryland systems. Nat. Rev. Microbiol. *10*, 551–562. https://doi.org/10.1038/nrmicro2831.

Reavy, B., Swanson, M.M., Cock, P.J., Dawson, L., Freitag, T.E., Singh, B.K., Torrance, L., Mushegian, A.R., and Taliansky, M. (2015). Distinct circular single-stranded DNA viruses exist in different soil types. Appl. Environ. Microbiol. *81*, 3934–3945. https://doi.org/10.1128/AEM.03878-14.

Reigstad, L.J., Jorgensen, S.L., and Schleper, C. (2010). Diversity and abundance of Korarchaeota in terrestrial hot springs of Iceland and Kamchatka. ISME J. *4*, 346–356. https://doi.org/10.1038/ismej.2009.126.

Reisser, W. (1993). Viruses and virus-like particles of freshwater and marine eukaryotic algae – A review. Arch. Protistenkd *143*, 257–265.

Replicon, J., Frankfater, A., and Miller, R.V. (1995). A continuous culture model to examine factors that affect transduction among *Pseudomonas aeruginosa* strains in freshwater environments. Appl. Environ. Microbiol. *61*, 3359–3366.

Ripp, S., and Miller, R.V. (1995). Effects of suspended particulates on the frequency of transduction among *Pseudomonas aeruginosa* in freshwater environments. Appl. Env. Microbiol. *61*, 1214–1219.

De Rosa, M., Gambacorta, A., and Gliozzi, A. (1986). Structure, biosynthesis, and physicochemical properties of archaebacterial lipids. Microbiol. Rev. *50*, 70–80.

Saye, D.J., and Miller, R.V. (1989). The aquatic environment: Consideration of horizontal gene transmission in a diversified habitat. In Gene Transfer in the Environment, Levy, S.B., and Miller, R.V. eds. (McGraw-Hill, New York), pp. 223–259.

Saye, D.J., Ogunseitan, O., Sayler, G.S., and Miller, R.V. (1987). Potential for transduction of plasmids in a natural freshwater environment: effect of plasmid donor concentration and a natural microbial community on transduction in *Pseudomonas aeruginosa*. Appl. Environ. Microbiol. *53*, 987–995.

Schleper, C., and Nicol, G.W. (2010). Ammonia-oxidising archaea – physiology, ecology and evolution. Adv. Microb. Physiol. *57*, 1–41. https://doi.org/10.1016/B978-0-12-381045-8.00001-1.

Schmitt, M., Brendel, M., Schwarz, R., and Radler, F. (1989). Inhibition of DNA Synthesis in *Saccharomyces cerevisiae* by Yeast Killer Toxin KT28. J. Gen. Microbiol. *135*, 1529–1535.

Schoenfeld, T., Patterson, M., Richardson, P.M., Wommack, K.E., Young, M., and Mead, D. (2008). Assembly of viral metagenomes from Yellowstone hot springs. Appl. Environ. Microbiol. *74*, 4164–4174. https://doi.org/10.1128/AEM.02598-07.

La Scola, B., Audic, S., Robert, C., Jungang, L., de Lamballerie, X., Drancourt, M., Birtles, R., Claverie, J.M., and Raoult, D. (2003). A giant virus in amoebae. Science *299*, 2033–2033.

La Scola, B., Campocasso, A., N'Dong, R., Fournous, G., Barrassi, L., Flaudrops, C., and Raoult, D. (2010). Tentative characterization of new environmental giant viruses by MALDI-TOF mass spectrometry. Intervirology *53*, 344–353. https://doi.org/10.1159/000312919.

Short, S.M. (2012). The ecology of viruses that infect eukaryotic algae. Environ. Microbiol. *14*, 2253–2271. https://doi.org/10.1111/j.1462-2920.2012.02706.x.

Short, S.M., and Short, C.M. (2009). Quantitative PCR reveals transient and persistent algal viruses in Lake Ontario, Canada. Environ. Microbiol. *11*, 2639–2648. https://doi.org/10.1111/j.1462-2920.2009.01988.x.

Srinivasiah, S., Bhavsar, J., Thapar, K., Liles, M., Schoenfeld, T., and Wommack, K.E. (2008a). Phages across the biosphere: contrasts of viruses in soil and aquatic environments. Res. Microbiol. *159*, 349–357. https://doi.org/10.1016/j.resmic.2008.04.010.

Srinivasiah, S., Bhavsar, J., Thapar, K., Liles, M., Schoenfeld, T., and Wommack, K.E. (2008b). Phages across the biosphere: contrasts of viruses in soil and aquatic environments. Res. Microbiol. *159*, 349–357. https://doi.org/10.1016/j.resmic.2008.04.010.

Steward, G.F., Culley, A.I., Mueller, J.A., Wood-Charlson, E.M., Belcaid, M., and Poisson, G. (2013). Are we missing half of the viruses in the ocean? ISME J. *7*, 672–679. https://doi.org/10.1038/ismej.2012.121.

Stewart, F.M., and Levin, B.R. (1984). The population biology of bacterial viruses: why be temperate. Theor. Pop. Biol. *26*, 93–117.

Straub, T.M., Pepper, I.L., and Gerba, C.P. (1993). Virus survival in sewage-sludge amended desert soil. Water Sci. Technol. *27*, 421–424.

Sturley, S.L., Elliot, Q., LeVitre, J., Tipper, D.J., and Bostian, K.A. (1986). Mapping of functional domains within the *Saccharomyces cerevisiae* type 1 killer preprotoxin. EMBO J. *5*, 3381–3389.

Suttle, C.A. (2002). Cyanophages and Their Role in the Ecology of Cyanobacteria. In The Ecology of Cyanobacteria, Whitton, B.A., and Potts, M. eds. (Springer, Netherlands), pp. 563–589.

Swanson, M.M., Fraser, G., Daniell, T.J., Torrance, L., Gregory, P.J., and Taliansky, M. (2009). Viruses in soils: morphological diversity and abundance in the rhizosphere. Ann. Appl. Biol. *155*, 51–60.

Vettori, C., Stotzky, G., Yoder, M., and Gallori, E. (1999). Interaction between bacteriophage PBS1 and clay minerals and transduction of *Bacillus subtilis* by clay-phage complexes. Environ. Microbiol. *1*, 347–355.

Vettori, C., Gallori, E., and Stotzky, G. (2000). Clay minerals protect bacteriophage PBS1 of *Bacillus subtilis* against inactivation and loss of transducing ability by UV radiation. Can. J. Microbiol. *46*, 770–773.

Waksman, S.A., and Woodruff, H.B. (1941). *Actinomyces antibioticus*, a New Soil Organism Antagonistic to Pathogenic and Non-pathogenic Bacteria. J. Bacteriol. *42*, 231–249.

Wang, G., Asakawa, S., and Kimura, M. (2011a). Spatial and temporal changes of cyanophage communities in paddy field soils as revealed by the capsid assembly protein gene g20. FEMS Microbiol. Ecol. *76*, 352–359. https://doi.org/10.1111/j.1574-6941.2011.01052.x.

Wang, K., Wommack, K.E., and Chen, F. (2011b). Abundance and distribution of *Synechococcus* spp. and cyanophages in the Chesapeake Bay. Appl. Environ. Microbiol. *77*, 7459–7468. https://doi.org/10.1128/AEM.00267-11.

Wegley, L., Edwards, R., Rodriguez-Brito, B., Liu, H., and Rohwer, F. (2007). Metagenomic analysis of the microbial community associated with the coral Porites astreoides. Environ. Microbiol. *9*, 2707–2719.

Weinbauer, M.G. (2004). Ecology of prokaryotic viruses. FEMS Microbiol. Rev. *28*, 127–181. https://doi.org/10.1016/j.femsre.2003.08.001.

Whitman, W.B., Coleman, D.C., and Wiebe, W.J. (1998). Prokaryotes: the unseen majority. Proc. Natl. Acad. Sci. U.S.A. *95*, 6578–6583.

Wilcox, R.M., and Fuhrman, J.A. (1994). Bacterial viruses in coastal seawater: Lytic rather than lysogenic production. Mar Ecol Prog Ser *114*, 35–45.

Williamson, K.E. (2011). Soil phage ecology: abundance, distribution, and interactions with bacterial hosts. In Biocommunication in Soil Microorganisms, Witzany, G. ed. (Springer, Berlin, Heidelberg), pp. 113–136.

Williamson, K.E., Radosevich, M., and Wommack, K.E. (2005). Abundance and diversity of viruses in six Delaware soils. Appl. Environ. Microbiol. *71*, 3119–3125.

Williamson, K.E., Radosevich, M., Smith, D.W., and Wommack, K.E. (2007). Incidence of lysogeny within temperate and extreme soil environments. Environ. Microbiol. *9*, 2563–2574.

Williamson, K.E., Schnitker, J.B., Radosevich, M., Smith, D.W., and Wommack, K.E. (2008). Cultivation-based assessment of lysogeny among soil bacteria. Microb. Ecol. *56*, 437–447. https://doi.org/10.1007/s00248-008-9362-2.

Williamson, K.E., Helton, R.R., and Wommack, K.E. (2012a). Bias in bacteriophage morphological classification by transmission electron microscopy due to breakage or loss of tail structures. Microsc. Res. Tech. *75*, 452–457. https://doi.org/10.1002/jemt.21077.

Williamson, K.E., Fuhrmann, J.J., Wommack, K.E., and Radosevich, M. (2017). Viruses in soil ecosystems: an unknown quantity within an unexplored territory. Annu. Rev. Virol. *4*, 201–219. https://doi.org/10.1146/annurev-virology-101416-041639.

Williamson, S.J., Houchin, L.A., McDaniel, L., and Paul, J.H. (2002). Seasonal variation in lysogeny as depicted by prophage induction in Tampa Bay, Florida. Appl. Environ. Microbiol. *68*, 4307–4314.

Williamson, S.J., Allen, L.Z., Lorenzi, H.A., Fadrosh, D.W., Brami, D., Thiagarajan, M., McCrow, J.P., Tovchigrechko, A., Yooseph, S., and Venter, J.C. (2012b). Metagenomic exploration of viruses throughout the Indian Ocean. PLOS ONE *7*, e42047. https://doi.org/10.1371/journal.pone.0042047.

Woese, C. (1998). The universal ancestor. Proc. Natl. Acad. Sci. U.S.A. *95*, 6854–6859.

Woese, C.R., Kandler, O., and Wheelis, M.L. (1990). Towards a natural system of organisms: proposal for the domains Archaea, Bacteria, and Eucarya. Proc. Natl. Acad. Sci. U.S.A. *87*, 4576–4579.

Wommack, K.E., and Colwell, R.R. (2000). Virioplankton: viruses in aquatic ecosystems. Microbiol. Mol. Biol. Rev. *64*, 69–114.

Yoosuf, N., Yutin, N., Colson, P., Shabalina, S.A., Pagnier, I., Robert, C., Azza, S., Klose, T., Wong, J., Rossmann, M.G., *et al.* (2012). Related giant viruses in distant locations and different habitats: *Acanthamoeba polyphaga moumouvirus* represents a third lineage of the *Mimiviridae* that is close to the *megavirus* lineage. Genome Biol. Evol. *4*, 1324–1330. https://doi.org/10.1093/gbe/evs109.

Zablocki, O., Adriaenssens, E.M., and Cowan, D. (2015). Diversity and ecology of viruses in hyperarid desert soils. Appl. Environ. Microbiol. *82*, 770–777. https://doi.org/10.1128/AEM.02651-15.

Weather and Climate – Death Valley National Park (U.S. National Park Service)

Marine Viral Metagenomics with Emphasis on Coral Microbiomes

5

Rebecca L. Vega Thurber[1]*, Jérôme P. Payet[1,2], Lu Wang[1] and Alec O. Eastman[1]

[1]Department of Microbiology, Oregon State University, Corvallis, OR, USA.
[2]College of Earth, Ocean, and Atmospheric Science, Oregon State University, Corvallis, OR, USA.

*Correspondence: rvegathurber@gmail.com; rebecca.vega-thurber@oregonstate.edu

https://doi.org/10.21775/9781910190852.05

Abstract

Methods for analysing the viromes associated with tropical marine waters and its abundant wildlife have rapidly progressed due to advances in molecular biology methodologies, nucleic acid sequencing technology, and computational analysis. The viruses of corals, the foundational organisms of shallow tropical reefs, have been one of the most well studied symbioses using these techniques. Reef building corals are known to play host to hundreds to thousands of bacterial taxa and upwards of two dozen viral families, including many viruses of microorganisms as well those infecting host tissues. Due to the lack of single marker genes to describe viruses in these systems, however, research on these two facets of coral reef viromes has required different approaches. Most coral bacterial analysis has come from amplicon analysis while few if any studies have used this approach to study coral viruses. Instead metagenomics and meta-transcriptomics have provided the necessary tools for scientists to catalogue, characterize, and compare the genetic diversity of viruses in coral systems. This review describes the history of this field with emphasis on how methods have evolved since its inception in the early 2000s. We describe some basic findings from the past 20 years and discuss the major limitations of past and current approaches. Finally, we provide some considerations for future work in this arena of viral metagenomics and marker gene analysis.

Introduction

Coral reefs are productive and complex ecosystems (Odum and Odum, 1955) that despite only occupying <0.2% of Earth's surface (Spalding and Grenfel, 1997), contain upwards of 30% of all marine species (Fisher *et al.*, 2015). These hotspots of biodiversity and productivity are generally found in shallow, nutrient poor tropical and subtropical oceans. Coral animals make up most of the structural foundation of these reefs, by secreting calcium carbonate skeleton. Often described as the rainforest of the sea, coral reefs have thus been epicentres for both macroscopic and microscopic research in the past decades. While research on coral reefs has occurred since the time of Charles Darwin in the late 1800s, it was not until the past century that work on the microscopic components of coral reefs became a focus for biologists. The capacity of corals to thrive in nutrient poor waters is based on their mutual associations with complex and diverse microbial communities, including microalgae and other protists, bacteria, archaea, fungi and, viruses. Despite the fact that viruses are the most abundant biological entities in the oceans, the viral component, or so-called virome of the coral reef microbial communities has been understudied compared to other microorganisms, likely due to technical challenges in detection and isolation. Major technological and analytic advances, particularly in cultivation-independent detection approaches, have made it possible to investigate coral viromes and their ecological roles in coral reef health and ecosystem processes. For example, in the late 2000s several researchers started to use direct sequencing of genomic material from environments, also referred as metagenomics, to reveal and catalogue the genetic content and diversity of entire viral communities from coral reef ecosystems, with emphasis on the viruses that were found on coral hosts themselves (Willner *et al.*, 2009; Thurber *et al.*, 2009; Vega Thurber *et al.*, 2008; Marhaver *et al.*, 2008). We now know that corals and surrounding reef seawater harbour enormous viral diversity. Only a few research groups, however, have strictly focused their efforts on investigating viruses in coral reef ecosystems.

Here we aim to summarize many of the efforts on the viral side of this research, with a particular focus on how viral metagenomics, i.e. so-called viromics, has greatly accelerated the study of viral ecology and diversity in coral reef ecosystems, and how infections by these many different viruses may alter the health and functions of these special habitats. See Chapter 15 for further discussion of many of the techniques mentioned.

Single-gene surveys of viral diversity

Unlike ribosomal RNA (rRNA) genes found in cellular organisms, viruses lack of universally conserved genes that can serve as molecular markers for classification and diversity in environmental samples. This is further complicated by the mosaic nature of viral genomes, a result of numerous lateral transfer events over time (Pedulla *et al.*, 2003; Koonin *et al.*, 2006; Hendrix *et al.*, 1999). Comparative whole genome analyses, however, have revealed the presence of broadly conserved structural and functional viral genes that can serve as markers for the study of diversity of specific groups of viruses in natural environments (e.g. reviewed in Short *et al.*, 2010; Adriaenssens and Cowan, 2014) (Table 5.1). For example, single-gene based studies using polymerase chain reaction (PCR) to amplify a portion of the conserved functional DNA polymerase B (*polB*) genes (Chen and Suttle, 1995; Short and Suttle, 2003; Culley *et al.*, 2009; Short *et al.*, 2011; Gimenes *et al.*, 2012; Clasen and Suttle, 2009; Zingone *et al.*, 2006) or the major capsid protein (MCP) structural genes (Larsen *et al.*, 2008; Clerissi *et al.*, 2014; Rowe *et al.*, 2011; Rozon and Short, 2013) of viruses in the *Phycodnaviridae*, a family of nucleocytoplasmic large double stranded DNA (dsDNA) viruses (NCLDVs) that infect microalgae (Dunigan *et al.*, 2006; Van Etten *et al.*, 2002; Nagasaki and Bratbak, 2010) (see Chapter 10), have uncovered a great genetic diversity in phycodnaviruses, including many groups which are unknown from cultured isolates. Other studies have focused on structural genes encoding the *g20* portal vertex protein (Fuller *et al.*, 1998; Zhong *et al.*, 2002; Dorigo *et al.*, 2004; Wilhelm *et al.*, 2006; Sullivan *et al.*, 2008; Short and Suttle, 2005; Mühling *et al.*, 2005) or the *g23* major capsid protein (Filée *et al.*, 2005; Comeau and Krisch, 2008; Payet and Suttle, 2014; Fujii *et al.*, 2008; Wang *et al.*, 2009; Cahyani *et al.*, 2009; Bellas and Anesio, 2013; Needham *et al.*, 2017; Marston and Amrich, 2009) gene of T4-like viruses in the *Myoviridae*. These are a family of dsDNA viruses known to infect a broad range of bacterial hosts (Ackermann, 2003; Clokie *et al.*, 2011), including the unicellular cyanobacteria *Prochlorococcus* and *Synechococcus* (Sullivan *et al.*, 2003; Waterbury and Valois, 1993; Suttle and Chan, 1993; Millard and Mann, 2006; Millard *et al.*, 2009; Marston and Amrich, 2009) as well as the abundant *Candidatus Pelagibacter ubique* (Zhao *et al.*, 2013).

Studies also use degenerate primers to amplify fragments of the RNA-dependent RNA polymerase (RdRp) gene, a hallmark of positive single-strand RNA (ssRNA) viruses. Such PCR-based surveys have uncovered many novel marine viruses in the *Picornaviridae* family, whose members are known to infect protists (Culley *et al.*, 2003, 2014; Culley and Steward, 2007; Gustavsen *et al.*, 2014). There are degenerate PCR primers for capsid protein genes as well, many of which target ssDNA viruses (Labonté and Suttle, 2013a; Labonté *et al.*, 2015a). Other single gene-based studies target auxiliary metabolic genes (AMGs), such as core photosynthesis genes (*psbA*, *psbD*) (Millard *et al.*, 2006; Sullivan *et al.*, 2006; Marston and Amrich, 2009; Clasen *et al.*, 2013; Jia *et al.*, 2007; Chénard and Suttle, 2008) or phosphate assimilation genes (phoH) (Goldsmith *et al.*, 2015, 2011; Marston and Amrich, 2009). These latter endeavours have resulted in the discovery of a vast diversity of viruses infecting unicellular cyanobacteria in a wide range of aquatic environments. Indeed, to the extent that relatively common

Table 5.1 Commonly used viral marker genes and their primer sequences for the investigation of different groups of phages and eukaryotic viruses

Target viral group	Marker gene	Primer sequence	Reference
T4 phage	*g23*/major capsid protein	MZIA1bis: GAT ATT TGI GGI GTT CAG CCI ATG A	Filée *et al*. (2005)
		MZIA6: CGC GGT TGA TTT CCA GCA TGA TTT C	
T4 phage	*g20*/portal head protein	CPS1: GTA GWA TTT TCT ACA TTG AYG TTG G	Fuller *et al*. (1998)
		CPS2: GGT ARC CAG AAA TCY TCM AGC AT	
		CPS3: TGG TAY GTY GAT GGM AGA	Zhong *et al*. (2002)
		CPS4: CAT WTC WTC CCA HTC TTC	
NCLDVs	*mcp*	MCP-Fwd: GGY GGY CAR CGY ATT GA MCP-Rev: TGI ARY TGY TCR AYI AGG TA	Larsen *et al*. (2008)
NCLDVs	DNA polymerase B	VS1: GARGGIGCIACIGTIYTIGAYGC AVS2: GCIGCRTAICKYTTYTTISWRTA	Chen and Suttle (1995)
Cyanophage	*psbA*	58-VDIDGIREP-66: GTN GAY ATH GAY GGN ATH MGN GAR CC	Zeidner *et al*. (2003)
		331-MHERNAHNFP-340: GGR AAR TTR TGN GCR TTN CKY TCR TGC AT	
Megaviruses	*mutS*	MutSF1: CCW AAY RMD GCN GGM AAR WCT A.	Wilson *et al*. (2014)
		MutS8_PpVR1: CTT TGG GAT AGT CTA AAT C	
		MutS8_PgVR1: CTT CCG GAT ATT CTA AAT C	
		MutS8_CeVR1: CAT CCG GAT ATT CCA TAT C	
		MutS8_PoVR2: TTC GAT GGC AAC CAA TTG TG	
T7 phage	Sheath protein	Sheath RTF: ACA TCA GCG TTC GTT TCG G	Takashima *et al*. (2007)
		Sheath RTR: CAA TCT GGT TAG GTA GGT CG	
T7 phage	DNA polymerase	CP-DNAP-349F: CCA AAY CTY GCM CAR GT	Chen *et al*. (2009)
		CP-DNAP-533Ra: CTC GTC RTG SAC RAA SGC	
		CP-DNAP-533Rb: CTC GTC RTG DAT RAA SGC	

viral genes are known for which appropriate PCR primers can be designed, then these single-gene approaches can at least hint at an approximation of what viral diversity may exist within environments, at least as related to known viruses.

Metagenomics

Coral reef metagenomics is a fairly small but quickly growing field. Previous reviews on coral metagenomics have been written with an emphasis on methodological considerations (e.g. Wood-Charlson *et al.*, 2015). But other than a few coral-virus focused papers, most work using metagenomics in reef systems have evaluated the overlying seawaters (e.g. reviewed in Vega Thurber *et al.*, 2017) rather than corals themselves. Nevertheless, much of the methods development for coral viral metagenomics is generally applicable to all viral metagenomics. Thus, the methodological details discussed below can be broadly applied. We begin, however, with consideration of older methodologies.

Amplicon analysis of viral ecology in the sea

Parallel application of PCR-based fingerprint techniques to profile and compare the composition of viral amplicons among samples, akin to 16S/18S amplicon analysis, has provided important information into genetic structure and composition of members of viral communities across various temporal and spatial scales. When combined with multivariate analysis, these fingerprinting techniques give substantial insights into potential ecological drivers of genetic structure of viral subsets in natural assemblages. For instance, Short and

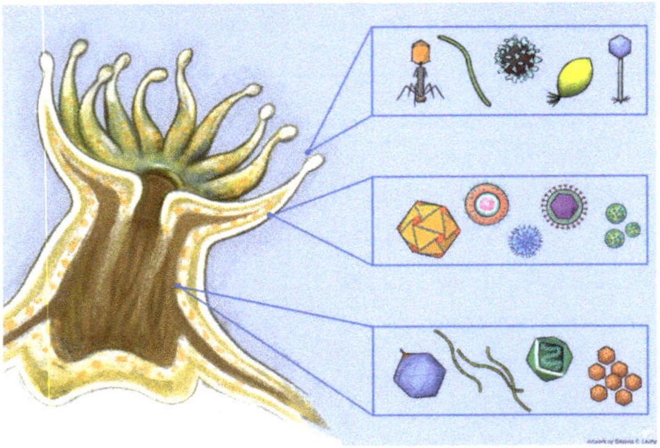

Coral Metagenomics and Virus Ecology

Figure 5.1 The coral virome based on metagenomics sequencing and TEM. Different compartments of corals contain VLPs and gene annotations similar to a variety of different viral families. The coral surface microlayer, for example, is dominated by phages. The ectoderm, by contrast, is dominated by NCLDVs, retroviruses, and atypical herpesviruses. Lastly, the gastroderm is dominated by ssRNA viruses, filamentous viruses, and megaviruses.

Suttle (2003) used PCR-targeted *polB* genes on British Columbia coastal marine waters and used DGGE (denaturing gradient gel electrophoresis) fingerprints to reveal seasonal patterns in genetic structure and composition of the phycodnaviruses present there. They reported a stable genetic structure of phycodnaviruses, despite seasonally shifting environmental conditions and host genetic structure. Frederickson *et al.* (2003) used PCR targeted *g20* genes of T4-like viruses in the Strait of Georgia marine waters, and revealed dynamic spatial patterns in genetic structure of T4-like viruses as revealed by DGGE fingerprint analysis. Likewise, Mühling *et al.* (2005) used a similar DGGE approach to separate *g20* amplicons and revealed seasonal patterns in the genetic structure of T4-like cyanophages in the Gulf of Aqaba, Red Sea.

Other PCR-based fingerprint surveys, using randomly amplified polymorphic DNA (RAPD) (Comeau *et al.*, 2004; Helton and Wommack, 2009; Winget and Wommack, 2008; Srinivasiah *et al.*, 2013; Williamson *et al.*, 2014) and endonuclease digestion of fluorescent end-labelled PCR amplicons followed by the generation of community profiles on an automated sequence analyser (T-RFLP) have confirmed dynamic and spatial variations in genetic structure of subsets in viral assemblages (Chow and Fuhrman, 2012; Pagarete *et al.*, 2013; Needham *et al.*, 2013).

Utility of shotgun metagenomics for viral diversity

Although the PCR targeted approaches discussed above are extremely powerful techniques to aid in discovery of the structure and dynamics of specific groups of viruses, unfortunately it does not sample all taxa equally, and some not at all. The large and growing number of viral genera and species also makes a PCR targeted gene approach impractical. Additionally, PCR biases including primer mismatch and chimera formation (i.e. the generation of an illegitimate sequences resulting from improper polymerase activity joining two or more different and unrelated sequences) (Suzuki and Giovannoni, 1996; Wang and Wang, 1996; Polz and Cavanaugh, 1998; Acinas *et al.*, 2005) may influence the results or skew the observed community structure. Single-gene surveys are thus not a sufficient basis to adequately investigate functional capacity and diversity of viruses in natural microbial ecosystems (Sullivan, 2015).

Limitations inherent to PCR targeted approaches prompted researchers to find new ways to analyse entire viral communities simultaneously. Metagenomics, or so-called shotgun environmental genomics, is a culture-independent approach that uses high-throughput sequencing technologies (HTS) to capture genomic information directly from environmental samples (Tringe *et al.*, 2005; Escobar-Zepeda *et al.*, 2015; Nazir, 2016). Unlike other culture-independent approaches, metagenomics produces a large amount of sequence data that allows a high-level snapshot of the overall community diversity as well as the potential metabolic capabilities or potential responses to environmental changes in complex environment settings (Figure 5.2). In some instances, draft or even complete genomes of various species are recovered, should sequencing depth be very high and/or combined with other -omic techniques (i.e. total transcriptome sequencing, also referred to as transcriptomics or RNA-seq, and mass spectrometric, i.e. mass spec protein analysis) (Sharon *et al.*, 2013). Further, new marker genes of interest, for example the ribonucleotide reductase for different viral groups, have been identified using strictly metagenomics (Sakowski *et al.*, 2014).

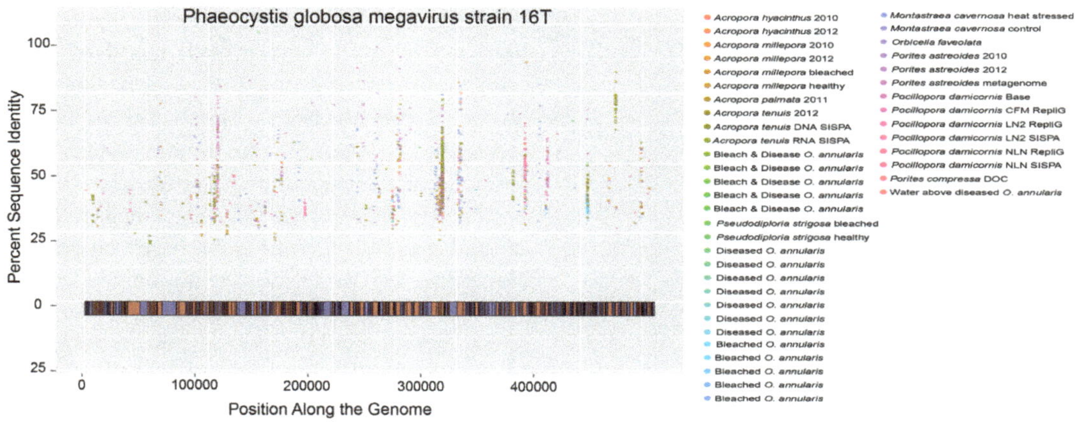

Figure 5.2 Example of metagenomic analysis methods using alignment of sequences from a variety of coral -omic datasets to a known *Phaeocystis globosa* viral genome [see Wood-Charlson *et al*. (2015) for methods].

Thus, metagenomics generates gene catalogues and, ideally, recapitulated viral genomes, allowing researchers to make inferences about the diversity and functional composition of individual viruses and viral consortia more broadly. For example, in the very first study of viral metagenomics, generated more than 14 years ago now, Breitbart *et al.* (2002) unveiled more than 7000 different phage genotypes in 200 l of coastal seawater. Since then dozens, if not hundreds, of studies have followed up on cataloguing the formidable diversity of viruses in almost all microbial ecosystems such as the oceans (Angly *et al.*, 2006; Culley *et al.*, 2006; Bench *et al.*, 2007; Hingamp *et al.*, 2013; Hurwitz and Sullivan, 2013; Kang *et al.*, 2013; Zhao *et al.*, 2013; Brum *et al.*, 2015; Roux *et al.*, 2016; Martínez Martínez *et al.*, 2014; Labonté and Suttle, 2013b), sediments (Breitbart *et al.*, 2004), soils (Paez-Espino *et al.*, 2016), deep-sea hydrothermal plumes (Anantharaman *et al.*, 2014), hot springs (Bolduc *et al.*, 2012; Gudbergsdóttir *et al.*, 2016; Bolduc *et al.*, 2015), corals (Littman *et al.*, 2011; Correa *et al.*, 2013; Soffer *et al.*, 2014, 2015; Wood-Charlson *et al.*, 2015), and polar freshwater (López-Bueno *et al.*, 2009; Aguirre de Cárcer *et al.*, 2015). For the remainder of this review we will limit our discussion to that of coral reef metagenomics in order to more comprehensively evaluate what is known about viruses of corals and coral reefs, and at the same time explain the evolution of this field, emphasizing the strengths of the approaches as well as its limitations.

Metagenomics methodologies

Metagenomic studies generally begin with experimental design and determination of the kind of questions that the analyses seek to answer. For the latter, for example, these questions include functional and taxonomic responses to environmental changes, or shifts in microbial populations in healthy versus diseased states. Afterwards, samples are collected and DNA or RNA libraries prepared. The processes of library preparation may differ depending on the type of study. For a coral viral metagenomics study, for instance, the samples are generally enriched via centrifugation or precipitation methods (Thurber *et al.*, 2009). Other pre-processing steps may include size fractionation or selective host-cell lysis. After sample libraries are prepared, nucleic acids are extracted, purified, fragmented, and then sequenced (Fig. 5.3).

When the field began, researchers generally used Sanger sequencing, which is labour-intensive and expensive compared to modern day high throughput sequencing. Currently most of the field utilizes Illumina sequencing, which is cost effective and produces limited errors, but also generates shorter read lengths of around 600bp from paired-end sequencing. After sequencing, reads undergo quality control steps in which any low-quality reads and adapters are removed. Assembly of the reads may be utilized to produce longer sequences through the production of overlapping or contiguous sequences or 'contigs'. Assembly also

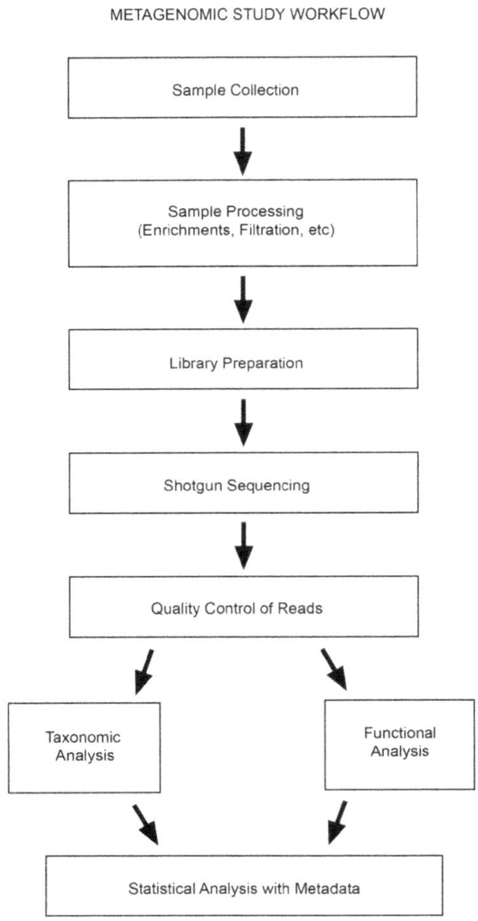

Figure 5.3 Standard pipeline for viral metagenome generation and the effects of pre-processing of coral tissue samples on metagenomics annotations. Virome generation and annotation follows a standard pipeline that in its simplest form is the isolation of a sample's nucleic acids, library generation, sequencing, library quality control measures, and then structural as well as functional analysis.

decreases the amount of computational space and power required for additional downstream analysis. Longer contigs generally provide more information for analysis of viral sequences, that often contain ORFan sequences which are sequences that have zero annotations (Wommack *et al.*, 2008), including the ability to identify more complicated units in the genome such as CRISPR elements (Oulas *et al.*, 2015). The downside to assembly is the likelihood of masking less abundant species, as well as producing erroneous chimera sequences (Sharpton, 2014).

Analysis of metagenomes is essentially separated into two parts: (1) identifying members of the community or consortia and (2) examination of potential gene functions within that community or consortia. To identify taxonomic members of the community, metagenomic reads are often 'binned' into groups of sequences that are similar to closely related taxa. This can be done based on genomic compositional data (most commonly nucleotide or k-mer composition), or based on any encoded genes within the metagenomic reads or contigs that are highly similar to other already known genes (Thomas *et al.*, 2012; Hurwitz and Sullivan, 2013; Brum *et al.*, 2015).

Functional gene annotation is the process of identifying genes of interest within the contigs and then assigning a known or hypothetical function to that gene via comparisons to protein family databases (Sharpton, 2014). The results of binning and annotation can be combined with collected metadata (parameter data collected during the sample isolation or experiment) in order to perform statistical tests to gain additional insight into the correlations between various taxa or genes and gene families and those metadata. For more detailed information on the metagenomics experimental pipeline, as well as strengths and weaknesses of the various types of annotation and binning software available, see Thomas *et al.*, 2012, Sharpton, 2014, and Oulas *et al.*, 2015.

A primer on coral biology

Corals are diverse meta-organisms such that contain not only the animal host but also a conspicuous algal partner *Symbiodinium* and a microbiome assembled of viral, bacterial, archaeal, and other eukaryotic microorganisms (Barott and Rohwer, 2012; Knowlton and Rohwer, 2003; Hernandez-Agreda *et al.*, 2017; Thompson *et al.*, 2014). Within the coral mucus layer there are approximately 2–5 × 10^6 bacteria/ml and 0.1–3 × 10^7 viruses/ml, about 10- to 100-fold more than the water column (e.g. reviewed in Vega Thurber *et al.*, 2017; Vega Thurber and Correa, 2011). Host specific differences in microbiome composition suggests that members of the microbiome are likely mutualistic (Krediet *et al.*, 2013; Hernandez-Agreda *et al.*, 2017; Oppen *et al.*, 2009; Gates and Ainsworth, 2011). Thus, many recent efforts have focused on the bacteria and their metabolic roles in coral health (Lema *et al.*, 2014; Zhang *et al.*, 2015).

The abundant bacterial communities on corals are distinct from the surrounding habitat (McKew *et al.*, 2012; Carlos *et al.*, 2013) and contain taxa that drastically differ from free living seawater microbes. In a meta-analysis of available 16S rRNA data for scleractinian or stony corals, Blackall *et al.* (2015) found that the most abundant taxa were Gammaproteobacteria, specifically *Endozoicomonas*, followed by Alphaproteobacteria. Another study found seven bacterial phylotypes broadly conserved within coral tissue across regions, depth and species (Ainsworth *et al.*, 2015). In contrast to completely species-specific associations with corals (Rohwer *et al.*, 2002; Kvennefors *et al.*, 2010; Carlos *et al.*, 2013; Chu and Vollmer, 2016), some bacteria demonstrate geographical restrictions and are only present in some corals in some locations (Littman *et al.*, 2009).

While some microbiome members may be invariant or regionally specific, other aspects of coral biology and ecology alter microbiome structure and function. For example, coral reproductive mode can determine which bacteria are acquired by coral larvae. Coral-associated bacteria can be transferred vertically from parents (Sharp *et al.*, 2012) or horizontally acquired from the environment (Sharp *et al.*, 2010; Apprill *et al.*, 2009), for example when adult resident corals release bacteria (e.g. *Altermonas* and *Roseobacter*) as a byproduct of broadcast spawning (Ceh *et al.*, 2013).

Different coral-associated bacteria are hypothesized to play varying roles in coral development, health, nutrition and survival. For example, nitrogen-fixing bacteria form species-specific associations with corals and may provide limiting fixed nitrogen to the algal partner of corals, *Symbiodinium* (Lesser *et al.*, 2004; Lema *et al.*, 2012). Bacteria also provide nitrogen directly to coral juveniles and larvae (Ceh *et al.*, 2013; Lema *et al.*, 2014), play a role in cycling sulfur (Wegley *et al.*, 2007; Raina *et al.*, 2010), and contribute to carbon fixation, metal homeostasis, organic remediation and secondary metabolism (Zhang *et al.*, 2015). Coral-associated bacteria may also defend the coral by occupying physical niche space on the coral, providing antimicrobial activities against potential pathogens (Ritchie, 2006; Rypien *et al.*, 2010; Zaneveld *et al.*, 2016; Zhang *et al.*, 2015), or alternatively directly consuming these pathogens (Krediet *et al.*, 2013; Welsh *et al.*, 2016). These interactions of the commensal or mutualistic members of the microbiome thus maintaining microbiome stability (Kvennefors *et al.*, 2010).

Metagenomic exploration of the coral virome

In 2007, with the advent of higher throughput pyrosequencing, the first work on coral host metagenomics was published (Wegley *et al.*, 2007). This paper focused mostly on the metabolic roles of coral bacteria, fungi, and archaea as determined from functional and taxonomic read assignments. Yet buried within this work was the first description of viral gene annotations associated with corals. The authors found evidence for the presence of upwards of 16 dsDNA and ssDNA viral families, dominated by tailed phages, herpes-like viruses, and NCLDVs. Within the next year, four additional metagenomics papers were published, all from the same lab, which evaluated the diversity and function of viruses associated with corals (Dinsdale *et al.*, 2008; Marhaver *et al.*, 2008; Vega Thurber *et al.*, 2008; Willner *et al.*, 2009). Together these papers quickly established that viral metagenomics could be used to uncover an unexpected diversity of viral genes within corals and other non-traditional host based systems. In fact, these four papers all used different pre-sequencing treatments, different coral species, and even different sequencing approaches, but all showed almost identical results: that sequence similarities to tailed phages and herpesviruses were the dominant annotations and common component of corals (Wegley *et al.*, 2007; Marhaver *et al.*, 2008; Vega Thurber *et al.*, 2008). In addition, these papers found that these groups of viral genes, while abundant in corals, were not abundant in the overlying seawater. Thus, the coral virome is distinct (Wegley *et al.*, 2007). Along with defining the coral virome, two of these papers showed that one group of viral sequences, those to herpes-like viral gene annotations, were more abundant in stressed corals suggesting a link between viral abundance and coral animal health (e.g. reviewed in Vega Thurber *et al.*, 2017).

Some ten years later, the community has access to cheaper, easier, and simpler approaches to study reef viral metagenomics. Nevertheless, most of these early conclusions about the dynamics of viral consortia in these systems, with some exceptions, remain the same. Importantly, at the time the authors of these earliest works used a variety of techniques that may have biased both the viral consortia present in the physical samples and the resulting viral annotations present in the resulting libraries. First, regardless of the target (virus or bacteria), all these studies used Φ29 polymerase multiple displacement amplification (MDA). MDA is a technique that uses random hexamer primers and a high-fidelity DNA polymerase to increase the amount of genomic DNA by one to two orders of magnitude (Binga *et al.*, 2008). It is known as well that MDA will alter the final sequence composition such that it preferentially amplifies circular DNA and some target sequences, resulting in a sequence library that is different than that of the true input community. While this has been debated continuously, the sequence composition that results from this form of amplification bias is not stochastic but instead remarkably reproducible (Ellegaard *et al.*, 2013). Nevertheless, the use of this approach, as it will alter the final sequence composition, should be avoided if possible. With the advent of new high-throughput sequencing technologies in the early 2010s, such as Illumina Nextera sequencing chemistry that only require nanograms of nucleic acids, avoiding MDA or whole genome amplification techniques has become easier. One major exception to this is PacBio (Pacific Biosciences) sequencing technology, which still requires large amounts of DNA.

Secondly, in the two focused metagenomic experiments, the standard caesium chloride viral purification methods at the time (e.g. reviewed in Vega Thurber *et al.*, 2009) were used but are known to introduce bias in the which kinds of viruses are selected for analysis. Lastly, in our 2009 paper chloroform was added to the coral samples in order to store and transport the material without destroying the viral particles. The addition of chloroform alters the density of the viral particles present in the samples because it can strip the virus-like particles (VLPs) of any lipid membranes. If density purification is conducted after this step, viruses that were once buoyant might become heavier and fall into a different fraction (Vega Thurber *et al.*, 2009).

Another issue that plagued the early days of research in this field was the use of low stringency expected values along with low sequence per cent identity cut offs. In other words, researchers often used less stringent bioinformatics and thus more error prone methods to predict if sequences from metagenomes were similar to any other sequence in the database. The rationale for these low stringency values stemmed from a lack of reference sequences to these new systems; restricting the analysis to high e-values specifically eliminate any signal within the system. We confirmed this suspicion using deep-sea methane seep viral metagenomics where we compared results across multiple stringency e-values (Bryson *et al.*, 2015). When we conducted our searches with low (10^{-3} or 10^{-6}) or high e-value similarity (10^{-12}) analysis, the proportion of sequences assigned to different viral types dramatically changed from annotations that represented viral families with few reference genomes (microphages) to annotations represented by viral families with large numbers of reference genomes (sipho- and myoviruses). This skew in the annotations to certain well-represented viruses in the database was somewhat removed (but not eliminated) when contiguous sequences (contigs) instead of entire reads were used for the analysis (*sensu* Wommack *et al.*, 2008).

It is thus clear from the discussion above that, although unsurprising, the use of different pre- and post-sequencing steps will alter the conclusions of a study. Using restrictively high e-values or unusually low e-values will also skew the results of each search (*sensu* Wood-Charlson *et al.*, 2015), as will using contigs versus reads or all hits versus best hits in the analysis (see, for example, Correa *et al.*, 2016). What's more, the reference library for which gene annotations are based can also alter the outcome of a virome analysis. For example, in Littman *et al.* (2011), Wood-Charlson *et al.* (2015), and Correa *et al.* (2016), the minimum e-values used were all between 10^{-5} and 10^{-7}, but different versions of BLAST (BLASTX and tBLASTX) and the target database (NCBI viral database and SEED) were used for the annotations. These differences likely resulted in slightly different conclusions about the composition and function of the viral consortia.

Metatranscriptomics, RNA viruses, and coral viral ecology

All of these initial, pre-2010 works, as well as most of the current coral viral metagenomic approaches, evaluated libraries generated only from double-stranded and single-stranded DNA. Thus, an important methodological hurdle during the evolution of the field was the isolation and sequencing of viral RNA, both genomic and transcriptional. Although the presence of a genomic signature from a metagenome is suggestive of endogenous active infection, transcripts of viral genes provide stronger evidence of a replicative and likely detrimental infections. Use of metatranscriptomics can therefore be used as a method to identify viral gene transcripts. At the same time, metatranscriptomics can provide insight into the abundance and diversity of RNA viruses. Since past microscopic evidence has inferred infection by RNA viruses of algal symbiont of corals (see Chapter 10), use of metatranscriptomics has been used to confirm the identity of these viruses.

In 2013 the first viral metatranscriptomic libraries were generated from apparently healthy as well as temperature-stressed corals. Using the Roche-454 GS FLX Titanium pyrosequencing platform, only a few assembled gene sequences (~1000 total) were identified which had any strong similarity to known viruses. Importantly, however, it was demonstrated that active infection by dsDNA viruses similar to known nucleocytoplasmic large DNA viruses (NCLDVs) was occurring in these organisms (see Chapters 10 and 11). This work also clearly showed that these NCLDV viruses were infecting the algal symbiont of corals, *Symbiodinium*, and not the coral host because the same sequences were found in previously published algal culture transcriptome data. Five sequences were also discovered that had exclusive similarity to a known +ssRNA dinoflagellate virus that infects a free-living relative of *Symbiodinium*, *Heterocapsa circularisquama* (Correa *et al.*, 2013). Although it was not known at the time if this virus was associated with *Symbiodinium* or if it had been associated with seawater or food particles ingested by the coral, this was the first -omic evidence of +ssRNA viruses in corals.

A new study has confirmed this finding, and recapitulated the genome of this virus from *Symbiodinium* transcriptomes (Levin *et al.*, 2017). The viral genome was in remarkably high abundance, confirming an active infection of some strains of *Symbiodinium*. This paper also inferred host anti-viral gene activity during infection. These *Symbiodinium* genes were found via transcript comparison against known anti-viral genes in corals, as well as genes with assigned functions of viral defence responses and viral antagonism. Additionally, transcripts were assigned to viruses using BLASTX bit scores against viral genomes and comparisons of GC content, tri-nucleotide frequency, and codon bias.

The presence of viruses in *Symbiodinium* was again confirmed by a proteomic study conducted by Weston *et al.*, (2012). In this study, the coral *Stylophora pistillata* was exposed to ambient and elevated water temperatures in conjunction with differential levels of trace metals. Symbionts were isolated at the start and end of the experiment. Samples were enriched for the endosymbiont via filtration and centrifugation to remove any coral host tissue. Proteins were extracted from the enriched samples and labelled with a tandem mass tag for downstream analysis. Though not many proteins were significantly differentially expressed, a viral replication protein, E1, was found to be 114-fold higher under the high-temperature, replete-metals treatment (Weston *et al.*, 2012). This finding contributes to the idea that thermally stressed corals can carry an increased and replicating viral load.

Identifying functional changes using metatranscriptomics

Researchers have begun to use coral metatranscriptome studies as a way to evaluate viral infection and host innate immune responses to these viral particles. One early study focused on the *Acropora millepora* polyp transcriptome via RNAseq to follow gene expression after exposure to high levels of CO_2, emulating the effects of ocean acidification on coral calcification. The authors found that high CO_2 resulted in increases in extracellular organic matrix synthesis but down-regulations in metabolism (Moya *et al.*, 2012). Another study used transcriptomics to examine the effects of the combination of changing pH and temperature on both *Symbiodinium* and *Acropora millepora*, and found decreased transcripts of genes involved in photosynthesis, respiration, and calcification (Kaniewska *et al.*, 2015). Loss of photosynthetic transcripts was also found in a microarray transcriptome study on the coral-symbiont response to pollutants (Gust *et al.*, 2014).

The holobiont response in a comparison of healthy *Orbicella faveolata* and those with signs of white plague disease found compartment-specific (Fig. 5.1) responses to the disease – with the host showing differentially expressed innate immunity, oxidative stress, and, importantly, an increase in DNA repair and retro-element transcripts (Daniels *et al.*, 2015). These researchers were able to distinguish between the host, microbiome, and symbiont by first separating reads with and without Poly (A) tails, effectively separating eukaryotic and prokaryotic transcripts. The resulting 'holotransciptome' also allowed the group to categorically assign mRNA to bacterial taxa and found different associated bacteria for healthy and diseased communities. Lastly, both the coral host and bacteria showed an increase in innate immunity factors, a presumptive response to viral infections, which may link white plague disease to viral activity as previously suggested (Soffer *et al.*, 2014).

The emerging picture of the coral virome

Since the early days of the field of metagenomics, other types of high throughput omics analysis, such as RNAseq, proteomics, and metabolomics have helped generate a synthetic view of what viruses are present in a diversity of corals, how they change under different contexts, and which members of the coral holobiont are the target(s) of these viruses (Maxwell and Frappier, 2007; Oldiges *et al.*, 2007). Here we break down what is known into the three major components of the holobiont: the microbiome, the host, and the symbiont.

Viruses and the coral microbiome

In every coral viral metagenome, the dominant viral taxonomic annotations are associated with dsDNA tailed bacteriophages within the families of *Sipho-*, *Podo-*, and *Myoviridae* of the *Caudovirales* order (Fig. 5.1) (see Chapter 7). These viruses likely have roles in the infection, mortality, and turnover of the coral-associated bacteria (Littman *et al.*, 2011). Using network analysis, it was recently shown that there are strong reproducible correlations between the presence of some phage and potential pathogens of corals (Soffer *et al.*, 2015), indicating that phage may provide this protective role to the corals as suggested by Barr *et al.* (2013). Active infection of intracellular coral bacteria by phages has also been visually confirmed in corals (Correa *et al.*, 2016), although the result of these infections remains unclear.

Viruses and the animal host

As mentioned earlier, sequences to the *Herpesvirales* order are also present in corals, with some annotations to all three of the herpesvirus families (Wood-Charlson *et al.*, 2015). Indeed, herpes-like virus particles are the dominant VLP in corals undergoing bleaching events. While there is outstanding evidence for the association of viruses which are genomically similar and physically reminiscent to *Herpesvirales* in corals, both TEM (transmission electron microscopy) and -omic evidence suggests that these viruses are distinct from all previously described herpes. The infection and replication of viral particles in the host nucleus is a defining characteristic of herpesviruses, but these particles have never been documented as present in coral cell nuclei (Correa *et al.*, 2016). Thus, these viruses found in coral cytoplasms presumably are not true herpesviruses but something distinct in their biology. For these reasons we refer to these viruses as 'atypical'. At the same time, while some phylogenetically informative marker genes, such as DNA polymerase and ribonucleotide reductase genes (Correa *et al.*, 2016; Wood Charlson *et al.*, 2015), suggest that there are herpesvirus-like annotations in these metagenomes, the bioinformatics support for this hypothesis is weak, with per cent identities to known herpesvirus genome sequences ranging in the low 30s (Wood-Charlson *et al.*, 2015). Therefore, although there remains strong -omic and microscopic support for the presence of a herpes-like virus in corals, the true genomic classification of these viruses remains an outstanding and intriguing question in the field.

Viruses and the coral algal symbiont

The metagenomic, metatranscriptomic, and TEM evidence from virus-enriched libraries also show quite clearly that a variety of viruses infect the dinoflagellate algal partner of corals, *Symbiodinium* (Fig. 5.1). The dominant evidence for this concerns the nucleocytoplasmic large DNA viruses, including members of the megaviruses, phycodnaviruses, iridoviruses and mimiviruses, (Wegley *et al.*, 2007) (see Chapters 10 and 11). Metagenomic evidence for these viruses is also found in studies on healthy, bleaching, and stressed or diseased corals (Wood-Charlson *et al.*, 2015). As mentioned above, perhaps the strongest evidence comes from metatranscriptomics that the presence of early and late gene transcripts of the viruses in heat stressed corals and *Symbiodinium* cultures (Correa *et al.*, 2013; Levin *et al.*, 2017).

Latent infections in these dinoflagellate symbionts have been induced by environmental stressors such as heat and UV radiation (Wilson *et al.*, 2001; Lawrence *et al.*, 2015). Wilson *et al.* (2001) found that heat shock induced lytic infection – production of icosahedral VLPs – in the zooxanthellae of the sea anemone, *Anemonia viridis*. Additionally, in a study of 49 *Symbiodnium* cultures, Lawrence *et al.* (2015) using flow cytometry and TEM, found that exposure to UV radiation led to an increase in production of icosahedral and filamentous VLPs. TEM from these samples found viruses resembling *Phycodnaviridae, Flexiviridae, Potyviridae* (RNA viruses that infect plants), and *Paramyxoviridae*, which are giant enveloped viruses. These latent infections can be linked to coral bleaching, as increased oceanic temperatures and other stressors combine, factors can lead to lytic infections in coral symbionts, causing a decrease in photosynthetic activity, symbiont populations, and loss of symbiosis.

Viruses and the evolution of *Symbiodinium*

Comparative genomic evidence also suggests that viruses have contributed to unique facets of dinoflagellate biology. In 2014 researchers uncovered a group of unusual non-histone like DNA condensation proteins in the dinoflagellate *Hematodinium*, a free-living relative of *Symbiodinium* (Gornik *et al.*, 2012). These novel nuclear proteins are commonly found in the NCLDV family, *Phycodnaviridae*, and thus the authors named them dinoflagellate viral nucleoproteins (DVNPs). Going further, the authors suggested that these dinoflagellate-specific nucleoproteins were horizontally transferred from algal virus to the host, leading to the unusual organization of the dinoflagellate nucleus. Given that these proteins are found in almost all dinoflagellates, including 19 homologues in *Symbiodinium minutum* (Shinzato *et al.*, 2014), these phycodnaviruses have likely infected these phytoplankton and symbionts for millions of years.

Future of coral viral metagenomics

Important questions still remain as to what are the ecological functions, biogeography, and the diversity of viruses in coral reefs. Improvements to every step of viral metagenome generation and analysis nevertheless have been significant, from the generation of libraries to their sequencing to subsequent bioinformatic analysis. Here we present some advancements that should help propel the field further and faster.

Better approaches to reducing bias

As we previously mentioned, any pre-sequencing steps are likely to add bias to a viral metagenome. Although conventional wisdom has been that viral purification methods were essential to confirm sequence occupancy within a viral genome, and not as a source of contamination, it may actually reduce the quality of the sequence library. Sequencing technology, however, has dramatically improved and new platforms give not only many more sequences but some provide dramatically longer read lengths. Both of these improve our abilities to *in silico* filter out viral sequences from host-associated libraries as opposed to those from seawater or sediments. In fact, we recently compared Illumina libraries generated from viromes that underwent different pre-extraction treatments such as filtration and centrifugation (Fig. 5.3b). Although this was not an exhaustive or replicated analysis, we found that there was no increase in the relative percentage of read annotations assigned to viruses. If anything, the number of viral reads fell when viral 'enrichment' steps were employed. Thus, we and others now conduct only whole metagenome and metatranscriptome analysis, which dramatically reduces laboratory preparatory time, pre-sequencing costs and any potential user biases (see discussion on MDA earlier in this chapter) during the processing as there are fewer steps overall.

Using CRISPR spacers to detect viruses in environmental samples

CRISPRs, or clustered regularly interspaced short palindromic repeats, are defence mechanisms found in 40% of Bacteria genomes and 90% of Archaea genomes (Snyder *et al.*, 2010). In short, CRISPR units function by incorporating pieces of invading, foreign DNA as spacers sequences, surrounded by repeat sequences. These spacer sequences can then recognize future instances of invasion from the same piece of DNA, and cleave it to prevent productive infection (Horvath and Barrangou, 2010). Thus, CRISPR spacer sequences can serve as a record of invading pieces of DNA, such as associated with viruses.

Although most CRISPR research has been focused on genome modification (Doudna and Charpentier, 2014), Snyder *et al.* (2010) has shown that microarrays consisting of spacer sequences from CRISPRs can be used as probes to detect novel viruses in the environment. Briefly, samples from acidic hot springs in Yellowstone National Park were taken for metagenomic analysis as well as viral enrichment via size selection filtration. CRISPR regions from the metagenome were identified using the web tool CRISPRFinder (Grissa *et al.*, 2007), and attached to a microarray to probe for matches in the virus-enriched samples.

One caveat to this technique is that results are not truly quantitative. This approach, however, can be used, perhaps in conjunction with metagenome analysis, to identify new viruses, as well as towards monitoring of viral population shifts in environmental samples. While this method has only been used in closed and extremely low-diversity environments, in theory it should be useful in marine environments including towards analysis of the coral holobiont. The role and significance of CRISPRs in the ocean as well as in association with corals have yet to be elucidated (Breitbart, 2012). Yet marine CRISPRs as well as this microarray virus probe technology are promising areas of future research.

Single particle sorting and viral genomics

Metagenomic and viral community studies rely on sequence availability in databases for taxonomic identification. As such, future research requires full genome sequencing of unknown viruses. Allen *et al.* (2011) described one such method, detailing a protocol for Single Virus Genomics (SVGs) in which viruses are sorted via flow cytometry onto agarose beads, stabilized in additional agar, and then lysed in preparation for MDA. One pitfall of this method, however, is contamination during initial virus sorting, which can include free genetic material as well as other viruses (Allen *et al.*, 2011).

Another study using flow cytometry and cell sorting took advantage of host physiological responses. Martínez Martínez *et al.* (2014) used lipid staining on *Emiliania huxleyi*, a marine phytoplankton, to identify and sort actively infected cells from uninfected cells. The stain binds to patchy plasma membranes caused by infections, in a phenomenon called membrane blebbing. Sorted cells then undergo MDA and sequencing to discern virus–host relationships and infection cycles. This idea can be applied to other microorganisms, as long as a physical host response is present and an appropriate tagging method is available.

Viral genomes have also been sequenced from isolated viral plaques. One such study utilized an optimized Sequence Independent Single Primer Amplification (SISPA) protocol to sequence an entire virus genome using ten picograms of starting nucleic acid material (DePew *et al.*, 2013). Changes to the protocol include increasing primer concentration for higher amplification during PCR, decreased PCR elongation times to produce fragments better optimized for library preparation, and more stringent decontamination of host DNA. When combined with high-throughput sequencing, the SISPA provides a cost-efficient and powerful tool for scientists to investigate whole genomes. This method, however, is only effective for cultivable viruses.

Culture-independent methods also include analysis of previously generated datasets including those of single amplified genomes [SAGs; e.g. reviewed in Stepanauskas (2012)]. Virus sequences can be mined from host genomic data by searching for viral marker genes, sequence anomalies such as GC content, and metagenomic fragment recruitment using existing virus sequence data. One study examined 58 single cell genomes from marine bacteria, and discovered several new viruses that infect *Thaumarchaeota*, *Verrucomicrobia*, and *Gammaproteobacteria* (Labonté *et al.*, 2015b). Lastly, with advances in bioinformatics whole new groups of viruses have been identified through extensive bioinformatic processing of previously generated metagenomes. For example, these data sorting technique has identified a new group of viral like elements called Polinton-like viruses (Yutin *et al.*, 2015).

Conclusion

To conclude, viral metagenomics continues to shed light on the diversity and functional roles of viruses in previously unexplored systems such as coral animals. As scientific, bioinformatic, and computational advances continue we will likely continue to reveal new and exciting aspects of the viral ecology associated with these basal animals. In so doing we will likely uncover many aspects of the evolution of animal viruses and their roles in the development, health, and evolution of their hosts.

Acknowledgements

This work was supported by grants from the U.S. National Science Foundation (NSF) #OCE-0960937, #OCE-1341195, and #OCE-1635913 to RVT and a grant from the Institute for Pacific Coral Reefs (IRCP) to JPP.

References

Acinas, S.G., Sarma-Rupavtarm, R., Klepac-Ceraj, V., and Polz, M.F. (2005). PCR-induced sequence artifacts and bias: insights from comparison of two 16S rRNA clone libraries constructed from the same sample. Appl. Environ. Microbiol. *71*, 8966–8969.

Ackermann, H.W. (2003). Bacteriophage observations and evolution. Res. Microbiol. *154*, 245–251.

Adriaenssens, E.M., and Cowan, D.A. (2014). Using signature genes as tools to assess environmental viral ecology and diversity. Appl. Environ. Microbiol. *80*, 4470–4480.

Aguirre de Cárcer, D., López-Bueno, A., Pearce, D.A., and Alcamí, A. (2015). Biodiversity and distribution of polar freshwater DNA viruses. Sci. Adv. *1*, e1400127. https://doi.org/10.1126/sciadv.1400127.

Ainsworth, T.D., Krause, L., Bridge, T., Torda, G., Raina, J.B., Zakrzewski, M., Gates, R.D., Padilla-Gamiño, J.L., Spalding, H.L., Smith, C., *et al.* (2015). The coral core microbiome identifies rare bacterial taxa as ubiquitous endosymbionts. ISME J. *9*, 2261–2274. https://doi.org/10.1038/ismej.2015.39.

Allen, L.Z., Ishoey, T., Novotny, M.A., McLean, J.S., Lasken, R.S., and Williamson, S.J. (2011). Single virus genomics: a new tool for virus discovery. PLOS ONE *6*, e17722. https://doi.org/10.1371/journal.pone.0017722.

Anantharaman, K., Duhaime, M.B., Breier, J.A., Wendt, K.A., Toner, B.M., and Dick, G.J. (2014). Sulfur oxidation genes in diverse deep-sea viruses. Science *344*, 757–760. https://doi.org/10.1126/science.1252229.

Angly, F.E., Felts, B., Breitbart, M., Salamon, P., Edwards, R.A., Carlson, C., Chan, A.M., Haynes, M., Kelley, S., Liu, H., *et al.* (2006). The marine viromes of four oceanic regions. PLOS Biol. *4*, e368.

Apprill, A., Marlow, H.Q., Martindale, M.Q., and Rappé, M.S. (2009). The onset of microbial associations in the coral *Pocillopora meandrina*. ISME J. 3, 685–699. https://doi.org/10.1038/ismej.2009.3.

Barott, K.L., and Rohwer, F.L. (2012). Unseen players shape benthic competition on coral reefs. Trends Microbiol. *20*, 621–628. https://doi.org/10.1016/j.tim.2012.08.004.

Barr, J.J., Auro, R., Furlan, M., Whiteson, K.L., Erb, M.L., Pogliano, J., Stotland, A., Wolkowicz, R., Cutting, A.S., Doran, K.S., *et al.* (2013). Bacteriophage adhering to mucus provide a non-host-derived immunity. Proc. Natl. Acad. Sci. U.S.A. *110*, 10771–10776. https://doi.org/10.1073/pnas.1305923110.

Bellas, C.M., and Anesio, A.M. (2013). High diversity and potential origins of T4-type bacteriophages on the surface of Arctic glaciers. Extremophiles *17*, 861–870. https://doi.org/10.1007/s00792-013-0569-x.

Bench, S.R., Hanson, T.E., Williamson, K.E., Ghosh, D., Radosovich, M., Wang, K., and Wommack, K.E. (2007). Metagenomic characterization of Chesapeake Bay virioplankton. Appl. Environ. Microbiol. 73, 7629–7641.

Binga, E.K., Lasken, R.S., and Neufeld, J.D. (2008). Something from (almost) nothing: the impact of multiple displacement amplification on microbial ecology. ISME J. *2*, 233–241. https://doi.org/10.1038/ismej.2008.10.

Blackall, L.L., Wilson, B., and van Oppen, M.J. (2015). Coral-the world's most diverse symbiotic ecosystem. Mol. Ecol. *24*, 5330–5347. https://doi.org/10.1111/mec.13400.

Bolduc, B., Shaughnessy, D.P., Wolf, Y.I., Koonin, E.V., Roberto, F.F., and Young, M. (2012). Identification of novel positive-strand RNA viruses by metagenomic analysis of archaea-dominated Yellowstone hot springs. J. Virol. *86*, 5562–5573. https://doi.org/10.1128/JVI.07196-11.

Bolduc, B., Wirth, J.F., Mazurie, A., and Young, M.J. (2015). Viral assemblage composition in Yellowstone acidic hot springs assessed by network analysis. ISME J. *9*, 2162–2177. https://doi.org/10.1038/ismej.2015.28.

Breitbart, M. (2012). Marine viruses: truth or dare. Ann. Rev. Mar. Sci. *4*, 425–448. https://doi.org/10.1146/annurev-marine-120709-142805.

Breitbart, M., Felts, B., Kelley, S., Mahaffy, J.M., Nulton, J., Salamon, P., and Rohwer, F. (2004). Diversity and population structure of a near-shore marine-sediment viral community. Proc. Biol. Sci. *271*, 565–574. https://doi.org/10.1098/rspb.2003.2628.

Breitbart, M., Salamon, P., Andresen, B., Mahaffy, J.M., Segall, A.M., Mead, D., Azam, F., and Rohwer, F. (2002). Genomic analysis of uncultured marine viral communities. Proc. Natl. Acad. Sci. U.S.A. *99*, 14250–14255. https://doi.org/10.1073/pnas.202488399.

Brum, J.R., Ignacio-Espinoza, J.C., Roux, S., Doulcier, G., Acinas, S.G., Alberti, A., Chaffron, S., Cruaud, C., de Vargas, C., Gasol, J.M., *et al.* (2015). Ocean plankton. Patterns and ecological drivers of ocean viral communities. Science *348*, 1261498. https://doi.org/10.1126/science.1261498.

Bryson, S.J., Thurber, A.R., Correa, A.M., Orphan, V.J., and Vega Thurber, R. (2015). A novel sister clade to the enterobacteria microviruses (family *Microviridae*) identified in methane seep sediments. Environ. Microbiol. *17*, 3708–3721. https://doi.org/10.1111/1462-2920.12758.

Cahyani, V.R., Murase, J., Asakawa, S., and Kimura, M. (2009). Change in T4-type bacteriophage communities during the composting process of rice straw: estimation from the major capsid gene (*g23*) sequences. Soil. Sci. Plant Nutr. 55, 468–477.

Carlos, C., Torres, T.T., and Ottoboni, L.M. (2013). Bacterial communities and species-specific associations with the mucus of Brazilian coral species. Sci. Rep. *3*, 1624. https://doi.org/10.1038/srep01624.

Ceh, J., van Keulen, M., and Bourne, D.G. (2013). Intergenerational transfer of specific bacteria in corals and possible implications for offspring fitness. Microb. Ecol. *65*, 227–231. https://doi.org/10.1007/s00248-012-0105-z.

Chen, F., and Suttle, C.A. (1995). Amplification of DNA polymerase gene fragments from viruses infecting microalgae. Appl. Environ. Microbiol. *61*, 1274–1278.

Chénard, C., and Suttle, C.A. (2008). Phylogenetic diversity of sequences of cyanophage photosynthetic gene psbA in marine and freshwaters. Appl. Environ. Microbiol. *74*, 5317–5324. https://doi.org/10.1128/AEM.02480-07.

Chow, C.E., and Fuhrman, J.A. (2012). Seasonality and monthly dynamics of marine myovirus communities. Environ. Microbiol. *14*, 2171–2183. https://doi.org/10.1111/j.1462-2920.2012.02744.x.

Chu, N.D., and Vollmer, S.V. (2016). Caribbean corals house shared and host-specific microbial symbionts over time and space. Environ Microbiol Rep *8*, 493–500. https://doi.org/10.1111/1758-2229.12412.

Clasen, J.L., Hanson, C.A., Ibrahim, Y., Weihe, C., Marston, M.F., and Martiny, J.B.H. (2013). Diversity and temporal dynamics of Southern California coastal marine cyanophage isolates. Aquat. Microb. Ecol. *69*, 17–31.

Clasen, J.L., and Suttle, C.A. (2009). Identification of freshwater *Phycodnaviridae* and their potential phytoplankton hosts, using DNA pol sequence fragments and a genetic-distance analysis. Appl. Environ. Microbiol. *75*, 991–997. https://doi.org/10.1128/AEM.02024-08.

Clerissi, C., Grimsley, N., Ogata, H., Hingamp, P., Poulain, J., and Desdevises, Y. (2014). Unveiling of the diversity of Prasinoviruses (Phycodnaviridae) in marine samples by using high-throughput sequencing analyses of PCR-amplified DNA polymerase and major capsid protein genes. Appl. Environ. Microbiol. *80*, 3150–3160. https://doi.org/10.1128/AEM.00123-14.

Clokie, M.R., Millard, A.D., Letarov, A.V., and Heaphy, S. (2011). Phages in nature. Bacteriophage *1*, 31–45. https://doi.org/10.4161/bact.1.1.14942.

Comeau, A.M., and Krisch, H.M. (2008). The capsid of the T4 phage superfamily: the evolution, diversity, and structure of some of the most prevalent proteins in the biosphere. Mol. Biol. Evol. *25*, 1321–1332. https://doi.org/10.1093/molbev/msn080.

Comeau, A.M., Short, S., and Suttle, C.A. (2004). The use of degenerate-primed random amplification of polymorphic DNA (DP-RAPD) for strain-typing and inferring the genetic similarity among closely related viruses. J. Virol. Methods *118*, 95–100. https://doi.org/10.1016/j.jviromet.2004.01.020.

Correa, A.M., Welsh, R.M., and Vega Thurber, R.L. (2013). Unique nucleocytoplasmic dsDNA and +ssRNA viruses are associated with the dinoflagellate endosymbionts of corals. ISME J. *7*, 13–27. https://doi.org/10.1038/ismej.2012.75.

Culley, A.I., Asuncion, B.F., and Steward, G.F. (2009). Detection of inteins among diverse DNA polymerase genes of uncultivated members of the *Phycodnaviridae*. ISME J. *3*, 409–418. https://doi.org/10.1038/ismej.2008.120.

Culley, A.I., Lang, A.S., and Suttle, C.A. (2003). High diversity of unknown picorna-like viruses in the sea. Nature *424*, 1054–1057. https://doi.org/10.1038/nature01886.

Culley, A.I., Lang, A.S., and Suttle, C.A. (2006). Metagenomic analysis of coastal RNA virus communities. Science *312*, 1795–1798.

Culley, A.I., Mueller, J.A., Belcaid, M., Wood-Charlson, E.M., Poisson, G., and Steward, G.F. (2014). The characterization of RNA viruses in tropical seawater using targeted PCR and metagenomics. MBio *5*, e01210–14. https://doi.org/10.1128/mBio.01210-14.

Culley, A.I., and Steward, G.F. (2007). New genera of RNA viruses in subtropical seawater, inferred from polymerase gene sequences. Appl. Environ. Microbiol. *73*, 5937–5944.

Daniels, C.A., Baumgarten, S., Yum, L.K., Michell, C.T., Bayer, T., Arif, C., Roder, C., Weil, E., and Voolstra, C.R. (2015). Metatranscriptome analysis of the reef-building coral *Orbicella faveolata* indicates holobiont response to coral disease. Front. Mar. Sci. *2*, 62. https://doi.org/10.3389/fmars.2015.00062.

Depew, J., Zhou, B., McCorrison, J.M., Wentworth, D.E., Purushe, J., Koroleva, G., and Fouts, D.E. (2013). Sequencing viral genomes from a single isolated plaque. Virol. J. *10*, 181. https://doi.org/10.1186/1743-422X-10-181.

Dinsdale, E.A., Edwards, R.A., Hall, D., Angly, F., Breitbart, M., Brulc, J.M., Furlan, M., Desnues, C., Haynes, M., Li, L., *et al.* (2008). Functional metagenomic profiling of nine biomes. Nature *452*, 629–632. https://doi.org/10.1038/nature06810.

Dorigo, U., Jacquet, S., and Humbert, J.F. (2004). Cyanophage diversity, inferred from g20 gene analyses, in the largest natural lake in France, Lake Bourget. Appl. Environ. Microbiol. *70*, 1017–1022.

Doudna, J.A., and Charpentier, E. (2014). Genome editing. The new frontier of genome engineering with CRISPR-Cas9. Science *346*, 1258096. https://doi.org/10.1126/science.1258096.

Dunigan, D.D., Fitzgerald, L.A., and Van Etten, J.L. (2006). Phycodnaviruses: a peek at genetic diversity. Virus Res. *117*, 119–132.

Ellegaard, K.M., Klasson, L., and Andersson, S.G. (2013). Testing the reproducibility of multiple displacement amplification on genomes of clonal endosymbiont populations. PLOS ONE *8*, e82319. https://doi.org/10.1371/journal.pone.0082319.

Escobar-Zepeda, A., Vera-Ponce de León, A., and Sanchez-Flores, A. (2015). The road to metagenomics: from microbiology to DNA sequencing technologies and bioinformatics. Front. Genet. *6*, 348. https://doi.org/10.3389/fgene.2015.00348.

Van Etten, J.L., Graves, M.V., Müller, D.G., Boland, W., and Delaroque, N. (2002). *Phycodnaviridae* – large DNA algal viruses. Arch. Virol. *147*, 1479–1516. https://doi.org/10.1007/s00705-002-0822-6.

Filée, J., Tétart, F., Suttle, C.A., and Krisch, H.M. (2005). Marine T4-type bacteriophages, a ubiquitous component of the dark matter of the biosphere. Proc. Natl. Acad. Sci. U.S.A. *102*, 12471–12476.

Fisher, R., O'Leary, R.A., Low-Choy, S., Mengersen, K., Knowlton, N., Brainard, R.E., and Caley, M.J. (2015). Species richness on coral reefs and the pursuit of convergent global estimates. Curr. Biol. *25*, 500–505. https://doi.org/10.1016/j.cub.2014.12.022.

Frederickson, C.M., Short, S.M., and Suttle, C.A. (2003). The physical environment affects cyanophage communities in British Columbia inlets. Microb. Ecol. *46*, 348–357. https://doi.org/10.1007/s00248-003-1010-2.

Fujii, T., Nakayama, N., Nishida, M., Sekiya, H., Kato, N., Asakawa, S., and Kimura, M. (2008). Novel capsid genes (*g23*) of T4-type bacteriophages in a Japanese paddy field. Soil. Biol. Biochem. *40*, 1049–1058.

Fuller, N.J., Wilson, W.H., Joint, I.R., and Mann, N.H. (1998). Occurrence of a sequence in marine cyanophages similar to that of T4 g20 and its application to PCR-based detection and quantification techniques. Appl. Environ. Microbiol. *64*, 2051–2060.

Gates, R.D., and Ainsworth, T.D. (2011). The nature and taxonomic composition of coral symbiomes as drivers of performance limits in scleractinian corals. J. Exp. Mar. Bio. Ecol. *408*, 94–101.

Gimenes, M.V., Zanotto, P.M., Suttle, C.A., da Cunha, H.B., and Mehnert, D.U. (2012). Phylodynamics and movement of Phycodnaviruses among aquatic environments. ISME J. *6*, 237–247. https://doi.org/10.1038/ismej.2011.93.

Goldsmith, D.B., Crosti, G., Dwivedi, B., McDaniel, L.D., Varsani, A., Suttle, C.A., Weinbauer, M.G., Sandaa, R.A., and Breitbart, M. (2011). Development of *phoH* as a novel signature gene for assessing marine phage diversity. Appl. Environ. Microbiol. *77*, 7730–7739. https://doi.org/10.1128/AEM.05531-11.

Goldsmith, D.B., Parsons, R.J., Beyene, D., Salamon, P., and Breitbart, M. (2015). Deep sequencing of the viral *phoH* gene reveals temporal variation, depth-specific composition, and persistent dominance of the same viral *phoH* genes in the Sargasso Sea. PeerJ 3, e997. https://doi.org/10.7717/peerj.997.

Gornik, S.G., Ford, K.L., Mulhern, T.D., Bacic, A., McFadden, G.I., and Waller, R.F. (2012). Loss of nucleosomal DNA condensation coincides with appearance of a novel nuclear protein in dinoflagellates. Curr. Biol. *22*, 2303–2312. https://doi.org/10.1016/j.cub.2012.10.036.

Grissa, I., Vergnaud, G., and Pourcel, C. (2007). CRISPRFinder: a web tool to identify clustered regularly interspaced short palindromic repeats. Nucleic Acids Res. *35*, W52-7.

Gudbergsdóttir, S.R., Menzel, P., Krogh, A., Young, M., and Peng, X. (2016). Novel viral genomes identified from six metagenomes reveal wide distribution of archaeal viruses and high viral diversity in terrestrial hot springs. Environ. Microbiol. *18*, 863–874. https://doi.org/10.1111/1462-2920.13079.

Gust, K.A., Najar, F.Z., Habib, T., Lotufo, G.R., Piggot, A.M., Fouke, B.W., Laird, J.G., Wilbanks, M.S., Rawat, A., Indest, K.J., *et al.* (2014). Coral-zooxanthellae meta-transcriptomics reveals integrated response to pollutant stress. BMC Genomics *15*, 591. https://doi.org/10.1186/1471-2164-15-591.

Gustavsen, J.A., Winget, D.M., Tian, X., and Suttle, C.A. (2014). High temporal and spatial diversity in marine RNA viruses implies that they have an important role in mortality and structuring plankton communities. Front. Microbiol. *5*, 703. https://doi.org/10.3389/fmicb.2014.00703.

Helton, R.R., and Wommack, K.E. (2009). Seasonal dynamics and metagenomic characterization of estuarine viriobenthos assemblages by randomly amplified polymorphic DNA PCR. Appl. Environ. Microbiol. *75*, 2259–2265. https://doi.org/10.1128/AEM.02551-08.

Hendrix, R.W., Smith, M.C., Burns, R.N., Ford, M.E., and Hatfull, G.F. (1999). Evolutionary relationships among diverse bacteriophages and prophages: all the world's a phage. Proc. Natl. Acad. Sci. U.S.A. *96*, 2192–2197.

Hernandez-Agreda, A., Gates, R.D., and Ainsworth, T.D. (2017). Defining the core microbiome in corals' microbial soup. Trends Microbiol. *25*, 125–140.

Hingamp, P., Grimsley, N., Acinas, S.G., Clerissi, C., Subirana, L., Poulain, J., Ferrera, I., Sarmento, H., Villar, E., Lima-Mendez, G., *et al.* (2013). Exploring nucleo-cytoplasmic large DNA viruses in Tara Oceans microbial metagenomes. ISME J. *7*, 1678–1695. https://doi.org/10.1038/ismej.2013.59.

Horvath, P., and Barrangou, R. (2010). CRISPR/Cas, the immune system of bacteria and archaea. Science *327*, 167–170. https://doi.org/10.1126/science.1179555.

Hurwitz, B.L., and Sullivan, M.B. (2013). The Pacific Ocean virome (POV): a marine viral metagenomic dataset and associated protein clusters for quantitative viral ecology. PLOS ONE *8*, e57355. https://doi.org/10.1371/journal.pone.0057355.

Jia, Z., Ishihara, R., Nakajima, Y., Asakawa, S., and Kimura, M. (2007). Molecular characterization of T4-type bacteriophages in a rice field. Environ. Microbiol. *9*, 1091–1096.

Kang, I., Oh, H.M., Kang, D., and Cho, J.C. (2013). Genome of a SAR116 bacteriophage shows the prevalence of this phage type in the oceans. Proc. Natl. Acad. Sci. U.S.A. *110*, 12343–12348. https://doi.org/10.1073/pnas.1219930110.

Kaniewska, P., Chan, C.K., Kline, D., Ling, E.Y., Rosic, N., Edwards, D., Hoegh-Guldberg, O., and Dove, S. (2015). Transcriptomic changes in coral holobionts provide insights into physiological challenges of future climate and ocean change. PLOS ONE *10*, e0139223. https://doi.org/10.1371/journal.pone.0139223.

Knowlton, N., and Rohwer, F. (2003). Multispecies microbial mutualisms on coral reefs: the host as a habitat. Am. Nat. *162* (Suppl. 4), S51–62.

Koonin, E.V., Senkevich, T.G., and Dolja, V.V. (2006). The ancient virus world and evolution of cells. Biol. Direct *1*, 29.

Krediet, C.J., Ritchie, K.B., Paul, V.J., and Teplitski, M. (2013). Coral-associated micro-organisms and their roles in promoting coral health and thwarting diseases. Proc. Biol. Sci. *280*, 20122328. https://doi.org/10.1098/rspb.2012.2328.

Kvennefors, E.C., Sampayo, E., Ridgway, T., Barnes, A.C., and Hoegh-Guldberg, O. (2010). Bacterial communities of two ubiquitous Great Barrier Reef corals reveals both site- and species-specificity of common bacterial associates. PLOS ONE *5*, e10401. https://doi.org/10.1371/journal.pone.0010401.

Labonté, J.M., Hallam, S.J., and Suttle, C.A. (2015a). Previously unknown evolutionary groups dominate the ssDNA gokushoviruses in oxic and anoxic waters of a coastal marine environment. Front. Microbiol. *6*, 315. https://doi.org/10.3389/fmicb.2015.00315.

Labonté, J.M., and Suttle, C.A. (2013a). Metagenomic and whole-genome analysis reveals new lineages of gokushoviruses and biogeographic separation in the sea. Front. Microbiol. *4*, 404. https://doi.org/10.3389/fmicb.2013.00404.

Labonté, J.M., and Suttle, C.A. (2013b). Previously unknown and highly divergent ssDNA viruses populate the oceans. ISME J. *7*, 2169–2177. https://doi.org/10.1038/ismej.2013.110.

Labonté, J.M., Swan, B.K., Poulos, B., Luo, H., Koren, S., Hallam, S.J., Sullivan, M.B., Woyke, T., Wommack, K.E., and Stepanauskas, R. (2015b). Single-cell genomics-based analysis of virus-host interactions in marine surface bacterioplankton. ISME J. *9*, 2386–2399. https://doi.org/10.1038/ismej.2015.48.

Larsen, J.B., Larsen, A., Bratbak, G., and Sandaa, R.A. (2008). Phylogenetic analysis of members of the *Phycodnaviridae* virus family, using amplified fragments of the major capsid protein gene. Appl. Environ. Microbiol. *74*, 3048–3057. https://doi.org/10.1128/AEM.02548-07.

Lawrence, S.A., Wilkinson, S.P., Davy, J.E., Arlidge, W.N.S., Williams, G.J., Wilson, W.H., Abey, G.S., and Davy, S.K. (2015). Influence of local environmental variables on the viral consortia associated with the coral Montipora capitata from Kaneohe Bay, Hawaii, USA. Aquat. Microb. Ecol. *74*, 251–262.

Lema, K.A., Bourne, D.G., and Willis, B.L. (2014). Onset and establishment of diazotrophs and other bacterial associates in the early life history stages of the coral Acropora millepora. Mol. Ecol. *23*, 4682–4695. https://doi.org/10.1111/mec.12899.

Lema, K.A., Willis, B.L., and Bourne, D.G. (2012). Corals form characteristic associations with symbiotic nitrogen-fixing bacteria. Appl. Environ. Microbiol. *78*, 3136–3144. https://doi.org/10.1128/AEM.07800-11.

Lesser, M.P., Mazel, C.H., Gorbunov, M.Y., and Falkowski, P.G. (2004). Discovery of symbiotic nitrogen-fixing cyanobacteria in corals. Science *305*, 997–1000. https://doi.org/10.1126/science.1099128.

Levin, R.A., Voolstra, C.R., Weynberg, K.D., and van Oppen, M.J. (2017). Evidence for a role of viruses in the thermal sensitivity of coral photosymbionts. ISME J. *11*, 808–812. https://doi.org/10.1038/ismej.2016.154.

Littman, R.A., Willis, B.L., Pfeffer, C., and Bourne, D.G. (2009). Diversities of coral-associated bacteria differ with location, but not species, for three acroporid corals on the Great Barrier Reef. FEMS Microbiol. Ecol. *68*, 152–163. https://doi.org/10.1111/j.1574-6941.2009.00666.x.

Littman, R., Willis, B.L., and Bourne, D.G. (2011). Metagenomic analysis of the coral holobiont during a natural bleaching event on the Great Barrier Reef. Environ Microbiol Rep *3*, 651–660. https://doi.org/10.1111/j.1758-2229.2010.00234.x.

López-Bueno, A., Tamames, J., Velázquez, D., Moya, A., Quesada, A., and Alcamí, A. (2009). High diversity of the viral community from an Antarctic lake. Science *326*, 858–861. https://doi.org/10.1126/science.1179287.

Marhaver, K.L., Edwards, R.A., and Rohwer, F. (2008). Viral communities associated with healthy and bleaching corals. Environ. Microbiol. *10*, 2277–2286. https://doi.org/10.1111/j.1462-2920.2008.01652.x.

Marston, M.F., and Amrich, C.G. (2009). Recombination and microdiversity in coastal marine cyanophages. Environ. Microbiol. *11*, 2893–2903. https://doi.org/10.1111/j.1462-2920.2009.02037.x.

Martínez Martínez, J., Swan, B.K., and Wilson, W.H. (2014). Marine viruses, a genetic reservoir revealed by targeted viromics. ISME J. *8*, 1079–1088. https://doi.org/10.1038/ismej.2013.214.

Maxwell, K.L., and Frappier, L. (2007). Viral proteomics. Microbiol. Mol. Biol. Rev. *71*, 398–411.

McKew, B.A., Dumbrell, A.J., Daud, S.D., Hepburn, L., Thorpe, E., Mogensen, L., and Whitby, C. (2012). Characterization of geographically distinct bacterial communities associated with coral mucus produced by *Acropora* spp. and *Porites* spp. Appl. Environ. Microbiol. *78*, 5229–5237. https://doi.org/10.1128/AEM.07764-11.

Millard, A., Clokie, M.R.J., Shub, D.A., and Mann, N.H. (2006). Genetic organization of the psbAD region in phages infecting marine *Synechococcus* strains. J. Mar. Biol. Assoc. U. K. *83*, 507–515.

Millard, A.D., and Mann, N.H. (2006). A temporal and spatial investigation of cyanophage abundance in the Gulf of Aqaba, Red Sea. J. Mar. Biol. Assoc. U. K. *86*, 507–515.

Millard, A.D., Zwirglmaier, K., Downey, M.J., Mann, N.H., and Scanlan, D.J. (2009). Comparative genomics of marine cyanomyoviruses reveals the widespread occurrence of *Synechococcus* host genes localized to a hyperplastic region: implications for mechanisms of cyanophage evolution. Environ. Microbiol. *11*, 2370–2387. https://doi.org/10.1111/j.1462-2920.2009.01966.x.

Moya, A., Huisman, L., Ball, E.E., Hayward, D.C., Grasso, L.C., Chua, C.M., Woo, H.N., Gattuso, J.P., Forêt, S., and Miller, D.J. (2012). Whole transcriptome analysis of the coral *Acropora millepora* reveals complex responses to CO_2-driven acidification during the initiation of calcification. Mol. Ecol. *21*, 2440–2454. https://doi.org/10.1111/j.1365-294X.2012.05554.x.

Mühling, M., Fuller, N.J., Millard, A., Somerfield, P.J., Marie, D., Wilson, W.H., Scanlan, D.J., Post, A.F., Joint, I., and Mann, N.H. (2005). Genetic diversity of marine *Synechococcus* and co-occurring cyanophage communities: evidence for viral control of phytoplankton. Environ. Microbiol. *7*, 499–508.

Nagasaki, K., and Bratbak, G. (2010). Isolation of viruses infecting photosynthetic and nonphotosynthetic protists. In Manual of Aquatic Viral Ecology. Weinbauer, M.G., Rowe, J.M., and Wilhelm, S.W. eds. (American Society of Limnology and Oceanography, Waco, Texas), pp. 92–101. https://doi.org/10.4319/mave.2010.978-0-9845591-0-7.92.

Nazir, A. (2016). Review on metagenomics and its applications. Imp. J. Interdiscip. Res. 2, 277–286. http://www.imperialjournals.com/index.php/IJIR/article/view/59.

Needham, D.M., Chow, C.E., Cram, J.A., Sachdeva, R., Parada, A., and Fuhrman, J.A. (2013). Short-term observations of marine bacterial and viral communities: patterns, connections and resilience. ISME J. 7, 1274–1285. https://doi.org/10.1038/ismej.2013.19.

Needham, D.M., Sachdeva, R., and Fuhrman, J.A. (2017). Ecological dynamics and co-occurrence among marine phytoplankton, bacteria and myoviruses shows microdiversity matters. ISME J. *11*, 1614–1629. https://doi.org/10.1038/ismej.2017.29.

Odum, H.T., and Odum, E.P. (1955). Trophic structure and productivity of a windward coral reef community on Eniwetok Atoll. Ecol. Monogr. *25*, 291–320.

Oldiges, M., Lütz, S., Pflug, S., Schroer, K., Stein, N., and Wiendahl, C. (2007). Metabolomics: current state and evolving methodologies and tools. Appl. Microbiol. Biotechnol. *76*, 495–511. https://doi.org/10.1007/s00253-007-1029-2.

Oppen, M.J.H., Leong, J.-A., and Gates, R.D. (2009). Coral-virus interactions: a double-edged sword? Symbiosis *47*, 1–8.

Oulas, A., Pavloudi, C., Polymenakou, P., Pavlopoulos, G.A., Papanikolaou, N., Kotoulas, G., Arvanitidis, C., and Iliopoulos, I. (2015). Metagenomics: tools and insights for analyzing next-generation sequencing data derived from biodiversity studies. Bioinform. Biol. Insights *9*, 75–88. https://doi.org/10.4137/BBI.S12462.

Paez-Espino, D., Eloe-Fadrosh, E.A., Pavlopoulos, G.A., Thomas, A.D., Huntemann, M., Mikhailova, N., Rubin, E., Ivanova, N.N., and Kyrpides, N.C. (2016). Uncovering Earth's virome. Nature *536*, 425–430.

Pagarete, A., Chow, C.E., Johannessen, T., Fuhrman, J.A., Thingstad, T.F., and Sandaa, R.A. (2013). Strong seasonality and interannual recurrence in marine myovirus communities. Appl. Environ. Microbiol. *79*, 6253–6259. https://doi.org/10.1128/AEM.01075-13.

Payet, J.P., and Suttle, C.A. (2014). Viral infection of bacteria and phytoplankton in the Arctic Ocean as viewed through the lens of fingerprint analysis. Aquat. Microb. Ecol. 72, 47–61.

Pedulla, M.L., Ford, M.E., Houtz, J.M., Karthikeyan, T., Wadsworth, C., Lewis, J.A., Jacobs-Sera, D., Falbo, J., Gross, J., Pannunzio, N.R., *et al.* (2003). Origins of highly mosaic mycobacteriophage genomes. Cell *113*, 171–182.

Polz, M.F., and Cavanaugh, C.M. (1998). Bias in template-to-product ratios in multitemplate PCR. Appl. Environ. Microbiol. *64*, 3724–3730.

Raina, J.B., Dinsdale, E.A., Willis, B.L., and Bourne, D.G. (2010). Do the organic sulfur compounds DMSP and DMS drive coral microbial associations? Trends Microbiol *18*, 101–108.

Ritchie, K.B. (2006). Regulation of microbial populations by coral surface mucus and mucus-associated bacteria. Mar. Ecol. Prog. Ser. 322, 1–14.

Rohwer, F., Seguritan, V., Azam, F., and Knowlton, N. (2002). Diversity and distribution of coral-associated bacteria. Mar. Ecol. Prog. Ser. *243*, 1–10.

Roux, S., Brum, J.R., Dutilh, B.E., Sunagawa, S., Duhaime, M.B., Loy, A., Poulos, B.T., Solonenko, N., Lara, E., Poulain, J., *et al.* (2016). Ecogenomics and potential biogeochemical impacts of globally abundant ocean viruses. Nature *537*, 689–693. https://doi.org/10.1038/nature19366.

Rowe, J.M., Fabre, M.F., Gobena, D., Wilson, W.H., and Wilhelm, S.W. (2011). Application of the major capsid protein as a marker of the phylogenetic diversity of *Emiliania huxleyi* viruses. FEMS Microbiol. Ecol. *76*, 373–380. https://doi.org/10.1111/j.1574-6941.2011.01055.x.

Rozon, R.M., and Short, S.M. (2013). Complex seasonality observed amongst diverse phytoplankton viruses in the Bay of Quinte, an embayment of Lake Ontario. Freshw. Biol. *58*, 2648–2663.

Rypien, K.L., Ward, J.R., and Azam, F. (2010). Antagonistic interactions among coral-associated bacteria. Environ. Microbiol. *12*, 28–39. https://doi.org/10.1111/j.1462-2920.2009.02027.x.

Sakowski, E.G., Munsell, E.V., Hyatt, M., Kress, W., Williamson, S.J., Nasko, D.J., Polson, S.W., and Wommack, K.E. (2014). Ribonucleotide reductases reveal novel viral diversity and predict biological and ecological features of unknown marine viruses. Proc. Natl. Acad. Sci. U.S.A. *111*, 15786–15791. https://doi.org/10.1073/pnas.1401322111.

Sharon, D., Tilgner, H., Grubert, F., and Snyder, M. (2013). A single-molecule long-read survey of the human transcriptome. Nat. Biotechnol. *31*, 1009–1014. https://doi.org/10.1038/nbt.2705.

Sharp, K.H., Distel, D., and Paul, V.J. (2012). Diversity and dynamics of bacterial communities in early life stages of the Caribbean coral *Porites astreoides*. ISME J. *6*, 790–801. https://doi.org/10.1038/ismej.2011.144.

Sharp, K.H., Ritchie, K.B., Schupp, P.J., Ritson-Williams, R., and Paul, V.J. (2010). Bacterial acquisition in juveniles of several broadcast spawning coral species. PLOS ONE *5*, e10898. https://doi.org/10.1371/journal.pone.0010898.

Sharpton, T.J. (2014). An introduction to the analysis of shotgun metagenomic data. Front. Plant Sci. *5*, 209. https://doi.org/10.3389/fpls.2014.00209.

Shinzato, C., Mungpakdee, S., Satoh, N., and Shoguchi, E. (2014). A genomic approach to coral-dinoflagellate symbiosis: studies of Acropora digitifera and Symbiodinium minutum. Front. Microbiol. *5*, 336. https://doi.org/10.3389/fmicb.2014.00336.

Short, C.M., and Suttle, C.A. (2005). Nearly identical bacteriophage structural gene sequences are widely distributed in both marine and freshwater environments. Appl. Environ. Microbiol. *71*, 480–486.

Short, S.M., Rusanova, O., and Staniewski, M.A. (2011). Novel phycodnavirus genes amplified from Canadian freshwater environments. Aquat. Microb. Ecol. *63*, 61–67.

Short, S.M., and Suttle, C.A. (2003). Temporal dynamics of natural communities of marine algal viruses and eukaryotes. Aquat. Microb. Ecol. 32, 107–119.

Short, S.W., Chen, F., and Wilhelm, S.W. (2010). The construction and analysis of marker gene libraries. In Manual of Aquatic Viral Ecology. Weinbauer, M.G., Rowe, J.M., and Wilhelm, S.W. eds. (American Society of Limnology and Oceanography, Waco, Texas), pp. 82–91. https://doi.org/10.4319/mave.2010.978-0-9845591-0-7.82.

Snyder, J.C., Bateson, M.M., Lavin, M., and Young, M.J. (2010). Use of cellular CRISPR (clusters of regularly interspaced short palindromic repeats) spacer-based microarrays for detection of viruses in environmental samples. Appl. Environ. Microbiol. *76*, 7251–7258. https://doi.org/10.1128/AEM.01109-10.

Soffer, N., Brandt, M.E., Correa, A.M., Smith, T.B., and Thurber, R.V. (2014). Potential role of viruses in white plague coral disease. ISME J. *8*, 271–283. https://doi.org/10.1038/ismej.2013.137.

Soffer, N., Zaneveld, J., and Vega Thurber, R. (2015). Phage-bacteria network analysis and its implication for the understanding of coral disease. Environ. Microbiol. *17*, 1203–1218. https://doi.org/10.1111/1462-2920.12553.

Spalding, M.D., and Grenfel, A.M. (1997). New estimates of global and regional coral reef areas. Coral Reefs, *16*, 225–230.

Srinivasiah, S., Lovett, J., Polson, S., Bhavsar, J., Ghosh, D., Roy, K., Fuhrmann, J.J., Radosevich, M., and Wommack, K.E. (2013). Direct assessment of viral diversity in soils by random PCR amplification of polymorphic DNA. Appl. Environ. Microbiol. *79*, 5450–5457. https://doi.org/10.1128/AEM.00268-13.

Stepanauskas, R. (2012). Single cell genomics: an individual look at microbes. Curr. Opin. Microbiol. *15*, 613–620. https://doi.org/10.1016/j.mib.2012.09.001.

Sullivan, M.B. (2015). Viromes, not gene markers, for studying double-stranded DNA virus communities. J. Virol. *89*, 2459–2461.

Sullivan, M.B., Coleman, M.L., Quinlivan, V., Rosenkrantz, J.E., Defrancesco, A.S., Tan, G., Fu, R., Lee, J.A., Waterbury, J.B., Bielawski, J.P., *et al.* (2008). Portal protein diversity and phage ecology. Environ. Microbiol. *10*, 2810–2823. https://doi.org/10.1111/j.1462-2920.2008.01702.x.

Sullivan, M.B., Lindell, D., Lee, J.A., Thompson, L.R., Bielawski, J.P., and Chisholm, S.W. (2006). Prevalence and evolution of core photosystem II genes in marine cyanobacterial viruses and their hosts. PLOS Biol. *4*, e234.

Sullivan, M.B., Waterbury, J.B., and Chisholm, S.W. (2003). Cyanophages infecting the oceanic cyanobacterium *Prochlorococcus*. Nature *424*, 1047–1051. https://doi.org/10.1038/nature01929.

Suttle, C.A., and Chan, A.M. (1993). Marine cyanophages infecting oceanic and coastal strains of Synechococcus: abundance, morphology, cross-infectivity and growth characteristics. Mar. Ecol. Prog. Ser. *92*, 99–109.

Suzuki, M.T., and Giovannoni, S.J. (1996). Bias caused by template annealing in the amplification of mixtures of 16S rRNA genes by PCR. Appl. Environ. Microbiol. *62*, 625–630.

Thomas, T., Gilbert, J., and Meyer, F. (2012). Metagenomics – a guide from sampling to data analysis. Microb. Inform. Exp. *2*, 3. https://doi.org/10.1186/2042-5783-2-3.

Thompson, J.R., Rivera, H.E., Closek, C.J., and Medina, M. (2014). Microbes in the coral holobiont: partners through evolution, development, and ecological interactions. Front. Cell. Infect. Microbiol. *4*, 176. https://doi.org/10.3389/fcimb.2014.00176.

Thurber, R.V., Haynes, M., Breitbart, M., Wegley, L., and Rohwer, F. (2009). Laboratory procedures to generate viral metagenomes. Nat. Protoc. *4*, 470–483. https://doi.org/10.1038/nprot.2009.10.

Tringe, S.G., von Mering, C., Kobayashi, A., Salamov, A.A., Chen, K., Chang, H.W., Podar, M., Short, J.M., Mathur, E.J., Detter, J.C., *et al.* (2005). Comparative metagenomics of microbial communities. Science *308*, 554–557.

Vega Thurber, R.L., Barott, K.L., Hall, D., Liu, H., Rodriguez-Mueller, B., Desnues, C., Edwards, R.A., Haynes, M., Angly, F.E., Wegley, L., *et al.* (2008). Metagenomic analysis indicates that stressors induce production of herpes-like viruses in the coral *Porites compressa*. Proc. Natl. Acad. Sci. U.S.A. *105*, 18413–18418. https://doi.org/10.1073/pnas.0808985105.

Vega Thurber, R., Willner-Hall, D., Rodriguez-Mueller, B., Desnues, C., Edwards, R.A., Angly, F., Dinsdale, E., Kelly, L., and Rohwer, F. (2009). Metagenomic analysis of stressed coral holobionts. Environ. Microbiol. *11*, 2148–2163. https://doi.org/10.1111/j.1462-2920.2009.01935.x.

Vega Thurber, R.L., and Correa, A.S. (2011). Viruses of reef-building scleractinian corals. J. Exp. Mar. Biol. Ecol. *408*, 102–103.

Thurber, R.V., Payet, J.P., Thurber, A.R., and Correa, A.M. (2017). Virus-host interactions and their roles in coral reef health and disease. Nat. Rev. Microbiol. *15*, 205–216. https://doi.org/10.1038/nrmicro.2016.176.

Wang, G., Hayashi, M., Saito, M., Tsuchiya, K., Asakawa, S., and Kimura, M. (2009). Survey of major capsid genes (*g23*) of T4-type bacteriophages in Japanese paddy field soils. Soil Biol. Biochem. *41*, 13–20.

Wang, G.C., and Wang, Y. (1996). The frequency of chimeric molecules as a consequence of PCR co-amplification of 16S rRNA genes from different bacterial species. Microbiology *142*, 1107–1114. https://doi.org/10.1099/13500872-142-5-1107.

Waterbury, J.B., and Valois, F.W. (1993). Resistance to co-occurring phages enables marine *Synechococcus* communities to coexist with cyanophages abundant in seawater. Appl. Environ. Microbiol. *59*, 3393–3399.

Wegley, L., Edwards, R., Rodriguez-Brito, B., Liu, H., and Rohwer, F. (2007). Metagenomic analysis of the microbial community associated with the coral *Porites astreoides*. Environ. Microbiol. *9*, 2707–2719.

Welsh, R.M., Zaneveld, J.R., Rosales, S.M., Payet, J.P., Burkepile, D.E., and Thurber, R.V. (2016). Bacterial predation in a marine host-associated microbiome. ISME J. *10*, 1540–1544. https://doi.org/10.1038/ismej.2015.219.

Weston, A.J., Dunlap, W.C., Shick, J.M., Klueter, A., Iglic, K., Vukelic, A., Starcevic, A., Ward, M., Wells, M.L., Trick, C.G., *et al.* (2012). A profile of an endosymbiont-enriched fraction of the coral *Stylophora pistillata* reveals proteins relevant to microbial-host interactions. Mol. Cell Proteomics *11*, M111.015487. https://doi.org/10.1074/mcp.M111.015487.

Wilhelm, S.W., Carberry, M.J., Eldridge, M.L., Poorvin, L., Saxton, M.A., and Doblin, M.A. (2006). Marine and freshwater cyanophages in a Laurentian Great Lake: evidence from infectivity assays and molecular analyses of g20 genes. Appl. Environ. Microbiol. 72, 4957–4963.

Williamson, K.E., Harris, J.V., Green, J.C., Rahman, F., and Chambers, R.M. (2014). Stormwater runoff drives viral community composition changes in inland freshwaters. Front. Microbiol. *5*, 105. https://doi.org/10.3389/fmicb.2014.00105.

Willner, D., Thurber, R.V., and Rohwer, F. (2009). Metagenomic signatures of 86 microbial and viral metagenomes. Environ. Microbiol. *11*, 1752–1766. https://doi.org/10.1111/j.1462-2920.2009.01901.x.

Wilson, W.H., Francis, I., Ryan, K., and Davy, S.K. (2001). Temperature induction of viruses in symbiotic dinoflagellates. Aquat. Microb. Ecol. *25*, 99–102.

Winget, D.M., and Wommack, K.E. (2008). Randomly amplified polymorphic DNA PCR as a tool for assessment of marine viral richness. Appl. Environ. Microbiol. *74*, 2612–2618. https://doi.org/10.1128/AEM.02829-07.

Wommack, K.E., Bhavsar, J., and Ravel, J. (2008). Metagenomics: read length matters. Appl. Environ. Microbiol. *74*, 1453–1463. https://doi.org/10.1128/AEM.02181-07.

Wood-Charlson, E.M., Weynberg, K.D., Suttle, C.A., Roux, S., and van Oppen, M.J. (2015). Metagenomic characterization of viral communities in corals: mining biological signal from methodological noise. Environ. Microbiol. *17*, 3440–3449. https://doi.org/10.1111/1462-2920.12803.

Yutin, N., Kapitonov, V.V., and Koonin, E.V. (2015). A new family of hybrid virophages from an animal gut metagenome. Biol. Direct *10*, 19. https://doi.org/10.1186/s13062-015-0054-9.

Zaneveld, J.R., Burkepile, D.E., Shantz, A.A., Pritchard, C.E., McMinds, R., Payet, J.P., Welsh, R., Correa, A.M., Lemoine, N.P., Rosales, S., *et al.* (2016). Overfishing and nutrient pollution interact with temperature to disrupt coral reefs down to microbial scales. Nat. Commun. *7*, 11833. https://doi.org/10.1038/ncomms11833.

Zhang, Y., Ling, J., Yang, Q., Wen, C., Yan, Q., Sun, H., Van Nostrand, J.D., Shi, Z., Zhou, J., and Dong, J. (2015). The functional gene composition and metabolic potential of coral-associated microbial communities. Sci. Rep. *5*, 16191. https://doi.org/10.1038/srep16191.

Zhao, Y., Temperton, B., Thrash, J.C., Schwalbach, M.S., Vergin, K.L., Landry, Z.C., Ellisman, M., Deerinck, T., Sullivan, M.B., and Giovannoni, S.J. (2013). Abundant SAR11 viruses in the ocean. Nature *494*, 357–360. https://doi.org/10.1038/nature11921.

Zhong, Y., Chen, F., Wilhelm, S.W., Poorvin, L., and Hodson, R.E. (2002). Phylogenetic diversity of marine cyanophage isolates and natural virus communities as revealed by sequences of viral capsid assembly protein gene g20. Appl. Environ. Microbiol. *68*, 1576–1584.

Zingone, A., Natale, F., Biffali, E., Borra, M., Forlani, G., and Sarno, D. (2006). Diversity in morphology, infectivity, molecular characteristics and induced host resistance between two viruses infecting *Micromonas pusilla*. Aquat. Microb. Ecol. *45*, 1–14.

Virus Interactions in the Aquatic World

6

Stéphan Jacquet[1], Xu Zhong[2], Peter Peduzzi[3], T. Frede Thingstad[4], Kaarle J. Parikka[5] and Markus G.Weinbauer[6,7]*

[1]INRA, UMR CARRTEL, Thonon les Bains, France.
[2]Department of Earth, Ocean and Atmospheric Sciences, University of British Columbia, Vancouver, BC, Canada.
[3]Department of Limnology and Bio-Oceanography, University of Vienna, Vienna, Austria.
[4]Department of Biology and Hjort Centre for Marine Ecosystem Dynamics, University of Bergen, Bergen, Norway.
[5]LabMCT, Belgian Department of Defence, Queen Astrid Military Hospital, Brussels, Belgium.
[6]Ocean Observatory of Villefranche-sur-Mer, Sorbonne University, Villefranche-sur-Mer, France.
[7]CNRS-INSU, Villefranche Oceanographic Laboratory, Villefranche-sur-Mer, France.

*Correspondence: wein@obs-vlfr.fr

https://doi.org/10.21775/9781910190852.06

Abstract

During the past 30 years, a vast number of articles have been published on the importance – in terms of abundance, diversity and functional roles – of viruses inhabiting the aquatic world (either marine or freshwater), which include several excellent reviews. Our knowledge, e.g. of the interactions of viruses with the living (organisms) and non-living compartments (i.e. dissolved organic matter/particles/nutrients/physical factors) in aquatic environments, has considerably increased. Furthermore, recent new ideas, concepts and technologies have shed light on host–virus interactions and the diversity of viral species and functions as discussed in this book. This chapter highlights some of these current advances in the field of aquatic viral ecology. Recently developed models and concepts will also be discussed, further emphasizing the role of defence mechanisms against viruses which appear to be important drivers in strain diversification of host species. Additionally, new environments such as anoxic and benthic sediments have received more attention and novel information has been gathered, e.g. on archaeal and RNA viruses. Interactions within entire food webs, such as with heterotrophic dinoflagellates, oysters and even cascading effects up to cyanobacteria-consuming flamingo populations represent another aspect of looking at viruses and their role in the global biosphere. The chapter also discusses knowledge gaps and proposes future research avenues.

Introduction

Viruses of microorganisms (VoMs) as found in aquatic environments consist of those infecting archaea, bacteria, or eukaryotes, the latter especially infecting algae and protists (see Chapters 7, 8, 10 and 11, respectively). This chapter focuses on findings and concepts dealing with aquatic viral ecology, virus interactions with hosts, and virus diversity. It builds on research that has been summarized in older reviews as mentioned in 'General aspects of aquatic viral ecology'. Although this review mainly deals with viruses of microorganisms, other viruses are also within the realm of viruses of aquatic environments. Rather than being comprehensive, this chapter focuses on key concepts along with some not fully considered research avenues.

In the first part we discuss the general aspects of aquatic viral ecology, focusing particularly on the role of viruses in aquatic ecology. Then we provide an updated outlook on aquatic viruses based on insights gleaned from recent findings. A glossary provides additional information on terms used in this chapter (Table 6.1).

General aspects of aquatic viral ecology

Since the discovery of high viral numbers as well as significant viral infection rates of microbes in aquatic environments (Bergh, 1989; Suttle *et al.*, 1990), it has been widely recognized that viruses are major actors for the functioning of aquatic ecosystems. Viruses have been described as 'partners' in pelagic food webs (Bratbak *et al.*, 1990), 'new players in the game' (Bratbak and Heldal, 1995), 'killers of winners' (Thingstad and Lignell, 1997; Kirchman, 2013), and 'major players' in global ecosystems (Suttle, 2007). They have also been said to 'rule the waves' (Bratbak and Heldal, 2000) as well as 'the world' (Wilson *et al.*, 2009), 'manipulate the marine environment' (Rohwer and Vega Thurber, 2009) and 'reprogram the metabolism in the sunlit and dark ocean' (Hurwitz *et al.*, 2013). Finally, viruses have been portrayed as 'the key to survival in the seas for microorganisms' (Paul, 2008) and 'catalysts of nutrients cycling' (Suttle, 2005), which 'lubricate the microbial loop' (Allen *et al.*, 2008).

Table 6.1 Glossary of terms used in this chapter

Term	Explanation
Aphotic zone	Portion of a lake or ocean where there is little or no sunlight
Archaea	One of the two domains of life forming the prokaryotes
Archaeal viruses	Viruses infecting Archaea
Attached	Microorganisms attached to particles/aggregates are operationally distinguished by filtration through 0.8–1 µm filters from their so-called free-living counterparts
Autotrophs	Here used for organisms synthesizing organic matter from carbon dioxide and releasing oxygen via photosynthesis (photoautotrophs)
Bacteria	One of the two domains of life forming the prokaryotes
Bacterial viruses	Viruses infecting Bacteria; Equivalent to bacteriophages as well as phages
Bacteriophages	See Bacterial viruses
Bloom	Originally used for phytoplankton (phytoplankton bloom), which develop in spring with the onset of water stratification and enhanced light availability, and which are then followed by a decay period
Boom–bust	Description of population growth that is followed by population decline; see similarly Bloom
Copepod	Type of small crustaceans
CRISPR	Stands for Clustered Regularly Interspaced Short Palindromic Repeats and is a type of adaptive immunity used by microorganisms against viruses
Cyanobacteria	Photoautotrophic bacteria
Diatom	Major group of unicellular microalgae
Dissolved organic matter (DOM)	Fraction of organic matter that can pass through filters with a pore-size of 0.2 – 0.45-µm; the main chemically characterizable macromolecules in the oceanic DOM pool are carbohydrates, proteins and lipids although a large fraction of DOM, usually more than 50%, is not chemically not characterizable
Diversity (of species)	Described by species richness (number of species), species evenness (abundance of individual species) and degree of species differences (taxonomic relatedness of species)
DOM	See Dissolved organic matter
Eukaryotes (Eukarya)	One of the two domains of life besides Archaea and Bacteria, comprises organisms with a nucleus in their cells
Eukaryotic viruses	Viruses infecting Eukarya
Eutrophic; Eutrophic environments	Regions where inorganic and organic nutrients are available at higher supply rates than under oligotrophic (i.e. much less productive) conditions; the nutrient supply rate influences the composition of food-web structure by favouring organisms with greater competitive abilities despite higher nutrient demand
Food web	Description of the complex trophic structure of most ecosystems, where prey species may be consumed by multiple predator species and following death all species may be consumed by multiple decomposer species

Table 6.1 Continued

Term	Explanation
Food-web efficiency	The percentage efficiency of energy transfer between trophic levels
Free-living	Contrast with Attached
Grazers	Here primarily used for protists feeding on prokaryotes as well as on other protists, although note that aquatic animals also can graze on algae and protists
Growth efficiency	Ratio between growth and uptake of substrate (i.e. nutrient) as measured in the same unit, usually in terms of carbon
Heterotrophs	Organisms that consume already assimilated organic matter (grazers and heterotrophic bacteria)
Mesopelagic	Water layer between the euphotic zone where no longer sufficient solar radiation is available to allow photosynthesis and the deep water masses. Usually the mesopelagic layer ranges between 100 m and 1000 m depth.
Microalgae	Small phytoplankton, i.e. small-in-size algae
Microorganisms	Non-taxonomic term used for microscopic organisms; here used for viruses, prokaryotes and protists
Mineraliser	Organisms which can convert organic matter to inorganic nutrients; here heterotrophic bacteria and heterotrophic protists
Oligotrophic	See Eutrophic
Organic aggregates or particles	Here used synonymously with particulate organic matter
Particulate organic matter (POM)	Fraction of organic matter retained by filters with a pore-size of 0.2 – 0.45-μm; technically, (microbial) cells belong to the POM pool
Phages	See Bacterial viruses
Photic zone	Zone within bodies of water where majority of photosynthesis occurs
Phylotype	Microorganisms identified by the sequence similarity of their small ribosomal RNA subunit
Phytoplankton	Photoautotrophic plankton
Plankton	Drifting organisms; divided into virioplankton (viruses), bacterioplankton (bacteria and archaea), phytoplankton (photoautotrophic eukaryotes), and zooplankton (heterotrophic eukaryotes)
POM	See Particulate organic matter
Primary producers	Here used for organisms assimilating organic carbon from inorganic material (photoautotrophs, although chemoauthotroph primary producers also exist within aquatic environments)
Primary production	Synthesis of organic compounds from aqueous carbon dioxide (typically by photosynthesis)
Prokaryotes	Comprised of the domains Bacteria and Archaea, i.e. organisms with no nucleus in their cells
Protists	Small, single-celled, photo- and/or heterotrophic eukaryotes
Remineralization	Breakdown or transformation of organic matter into its simplest inorganic forms
Suspended matter	Roughly equivalent to POM/aggregates/particles; usually used to describe larger material; can contain significant amounts of inorganic particles
Trophic levels	Position an organisms occupies in a food web
Upwelling	Oceanographic phenomenon that involves wind-driven motion of dense, cooler, and usually nutrient-rich water towards the ocean surface, replacing the warmer, usually nutrient-depleted surface water
Viral shunt	Release of materials from organisms by viral lysis and uptake of especially the organic component of this material by heterotrophic prokaryotes
Zooplankton	Planktonic organisms which assimilate particulate organic carbon (POM), that is, rather than DOM

A consensus thus exists as to the importance of viruses in the ecology of aquatic biomes.

More specifically, it is now well established that microorganisms play a crucial role in microbial food webs and biogeochemical cycles (Pomeroy, 1974; Azam *et al.*, 1983) (Fig. 6.1). The classical concept of a grazing food chain composed of phytoplankton (primary producers), herbivorous zooplankton (primary consumers), and carnivorous zooplankton (secondary consumers, etc.) has been extended to include additional microbial processes. A key component of this extension is production of dissolved organic matter (DOM), which is released or 'exuded' from even healthy phytoplankton, but also produced in the course of 'sloppy feeding' by zooplankton. Heterotrophic prokaryotes, that is, members of domains Bacteria and Archaea, are the main users of this DOM, and they themselves are grazed upon by flagellates and ciliates, which are consumed, in turn, by meso- and macro-zooplankton. Bacterial assimilation of DOM and subsequent grazing on bacteria by larger organisms thus can return carbon that otherwise would have been lost to food webs. This grazing action also results in an enhanced remineralization of organic matter, either directly due to the conversion of organic matter to carbon dioxide and NH_4 or indirectly due to the stimulation of bacterial activity as a consequence of sloppy feeding and release of labile organic matter from food vacuoles rich in N and P. These inorganic nutrients are then used by both primary producers and heterotrophic bacteria. These DOM-related pathways are collectively referred to as the 'microbial loop'.

Subsequent research has revealed that viral lysis is another important cause of prokaryote as well as phytoplankton mortality (Proctor and Fuhrman, 1990; Fuhrman and Noble, 1995; Noble and Fuhrman, 1997, Suttle *et al.*, 1990; Suttle and Chan, 1993). Viral lysis not only kills cells, particularly in the course of releasing new viral particles, but also converts those cells into DOM, consisting mainly of cell content, and particulate organic matter (POM), the latter consisting mainly of cell wall fragments. This conversion prevents nutrients from reaching higher trophic levels, at least in the near term. Instead, the cellular debris is consumed by heterotrophic bacteria, generating therefore a 'viral loop' (Bratbak *et al.*, 1992) or 'viral shunt' (Wilhelm and Suttle, 1999) (Fig. 6.1B), which stimulates bacterial production and respiration (Fuhrman, 1999). Viral lysis thus 'catalyses' nutrient cycling (Suttle, 2005, 2007; Wilhelm and Suttle, 1999; Weinbauer *et al.*, 2011a; Shelford *et al.*, 2012, 2014) and thereby 'lubricates' the microbial food web (Allen *et al.*, 2008).

The carbon pool found in marine DOM and POM is approximately as large as the carbon pool consisting of atmospheric CO_2 (Hedges, 2002) or in all global biota (Dittmar and Paeng, 2009). Thus, the fate of carbon bound to DOM and POM (including microorganisms) is an important factor influencing the climate. One of the most significant functions of heterotrophic bacteria in the ocean is

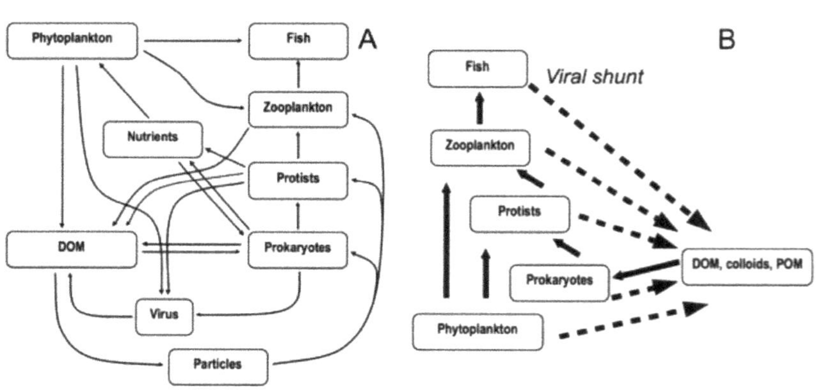

Figure 6.1 Idealized food webs. A) Simplified planktonic food web with grazing food web (phytoplankton–zooplankton), microbial loop (DOM–bacteria–protists) and viral shunt. B) details of the viral shunt and the flow of organic matter are shown (dashed lines).

the transformation of DOM and POM into biomass, on the one hand, and into CO_2 by respiration, on the other hand. The biomass is then found again in the food web (del Giorgio and Cole, 1998).

On microscales, the microbial community is operationally structured (determined experimentally by filtration) into free and particle-attached microorganisms. Attached microorganisms are embedded within nutrient-rich organic matrices (Alldredge and Cohen, 1987) and consequently bacterial abundances and activities found in association with particles can be one or two orders of magnitude higher than in surrounding waters (Simon *et al.*, 2002). Marine snow and the plume of sinking or rising particles are – presumably as a consequence – considered to be hot spots of biogeochemical transformations, e.g. generation and removal of DOM, as mediated by microorganisms (Azam, 1998; Azam and Long, 2001; Kiørboe and Jackson, 2001). Aquatic snow also impacts local prokaryotic communities e.g. by adsorbing viral particles, which can lead to diminished pressure by viral lysis of the free-living bacterioplankton (Weinbauer *et al.*, 2009; 2011b).

The fate of most of the organic carbon generated by primary production in the upper ocean, i.e. carbon fixing CO_2 via photosynthesis, is remineralization to CO_2. A small fraction of the fixed carbon, however, is not remineralized but is instead stored for millennia as 'recalcitrant' DOM (not easily biodegradable by prokaryotes). Recently, the concept of the 'microbial carbon pump' has been proposed as a conceptual framework that addresses this biogeochemical issue (Jiao *et al.*, 2010).

Likewise, metabolic balance, which is the ratio of primary production to community respiration (P/R ratio), is critical for determining whether the ocean is globally heterotrophic (in this case, gaining energy from organic carbon) or instead autotrophic (converting inorganic carbon, i.e. CO_2, to organic carbon) in terms of net trophic production. That is, whether more CO_2 is generated via cellular respiration versus fixed via processes such as photosynthesis. The P/R ratio, in other words, determines whether the ocean is globally retentive or a net source of carbon (del Giorgio and Duarte, 2002). In distinguishing these contrasting scenarios, viruses seem to play an essential role.

Virus interactions with hosts

Viruses interact with their hosts most prominently in terms of either productive or latent infections (see Chapter 1). Hosts, however, also can resist viral infection through various means. The consequences of viral infection generally involve production of new virions and modifications of the host cell, the latter being often fatal as in the case of virus-induced cell lysis. In addition, viruses can carry non-viral DNA that can become incorporated into susceptible cells (transduction).

General lifestyles

When a virus infects its host cell, it takes over its metabolism. New viral nucleic acids and capsids are assembled and packaged together into new virions. Finally, in most instances the infected cell lyses and the virions are released into the environment; these viral particles can then further re-infect other hosts. These subsequent infections of new hosts can be described epidemiologically as secondary infections (Abedon, 2015). Furthermore, the cell wall is fragmented by this lysis and cellular contents are spread into the surrounding environment. These viruses are called lytic (prokaryotes) or virulent (eukaryotes). Other viruses, called 'temperate' for viruses of prokaryotes or 'latent' for viruses of eukaryotes, can, depending upon circumstances, either follow the lytic/virulent model or become integrated into the host genome as prophages/proviruses. They then replicate together with their host genome until induced by agents (chemicals, environmental conditions, etc.) into lytic cycles or, for some viruses, chronic cycles. Chronic infection, in which viruses either bud or are otherwise extruded out of host cells without killing those cells, remain scarcely documented for aquatic viruses. Pseudolysogeny, e.g. a phenomenon when only a fraction of the infected cells produce free virions (e.g. Abedon, 2009), is also rarely found in aquatic virology literature and poorly understood in general.

Even when viruses are in a lysogenic cycle, proviruses have been shown to express transcriptional regulatory and repressor-like proteins (Paul, 2008). These proteins are hypothesized to suppress unneeded metabolic activity to help host survival in unfavourable environments. Conversely, host metabolic/physiologic status can affect (i) viral production through the modification of the latent period, the burst size and the number of infected cells, and (ii) the viral life cycle more dramatically, e.g. by leading viruses to enter the lysogenic cycle during times of host physiological stress. Changes in environmental conditions can also alter viral dynamics and replication cycles (Wommack and Colwell, 2000; Weinbauer, 2004). Recent evidence has shown the importance of host cell physiology in lifestyle decisions (Maurice *et al.*, 2011; Palesse *et al.*, 2015).

Viral infection and host resistance

Using a variety of different strategies, hosts can defend themselves against viral infection (for a review on these strategies against prokaryotic viruses, see Hyman and Abedon, 2010). Here, we present a few examples related to the aquatic environment. Viral infections can be avoided directly by production of a variety of specific enzymes such as the CRISPR-associated endonuclease, Cas. Alternatively, cells can change, reduce, or eliminate cell surface receptors, physiologically or evolutionarily, to block viral adsorption and thereby entrance, replication, and subsequent release (Labrie *et al.*, 2010; Thomas *et al.*, 2012). Host cells can also develop indirect strategies to escape viral infection by controlling their size, abundance, and growth rate to reduce encounter rates with viruses. This likely is the strategy used by SAR11, which is probably the most common bacterial group in the ocean (Morris *et al.*, 2002). It has been proposed that SAR11 is so successful because of its small cell size and the low growth rates as resistance mechanisms (so-called K-strategy) (Kirchman, 2013; Zhao *et al.*, 2013). Resistance can also take place during the change of the life cycle in algae (called metagenesis), as in the case of eukaryotic coccolithophorid phytoplankter *Emiliania huxleyi*, which can be either haploid (one chromosome) or diploid (two chromosomes). Only the latter form is susceptible to viral attack (Jacquet *et al.*, 2002; Frada *et al.*, 2008). Lysogeny, which can be interpreted as a survival strategy of viruses living at low host density or slow metabolism, can confer immunity against homologous viral infection or replication (Weinbauer and Rassoulzadegan, 2004; Fuhrman, 1999).

Tested viruses generally have been found to be able to develop solutions which allow them to fend off host anti-viral systems (Samson *et al.*, 2013). For example, advances in CRISPR-Cas studies have shown that phages can evade the CRISPR-Cas immune system by various mechanisms (Deveau *et al.*, 2008; Sapranauskas, 2011; Semenova *et al.*, 2011; Bondy-Denomy *et al.*, 2013; Seed *et al.*, 2013). Other examples of viral mechanisms of evasion of host defences are known, and evading such defences is an important means by which viruses can modify their host ranges (see e.g. Hyman and Abedon, 2010).

Despite the documented give-and-take between viral-resistance mechanisms and viral means of overcoming those mechanisms, resistance to viral infection does not appear to be a dominant process in the environment (Fuhrman, 1999; Weinbauer, 2004). This could be the result of a physiological cost related to resistance, which may cause disadvantages in the competition for resources when compared to non-resistant cells. For example, there is evidence for the concept of a cost of resistance (CoR) which is now widely accepted as an explanation for limitations on bacterial resistance to viral infections (Lennon *et al.*, 2007, 2008; Våge *et al.*, 2012, 2013). In this concept, it is assumed that there are two basic organism strategies: types not investing in resistance (low CoR) and types strongly investing into resistance (high CoR). This concept was related to the growth rate of hosts and to their life strategy. Fast-growing hosts and/or hosts possessing large cells, which have been described as 'r-selected', concur with low CoR (no investment in resistance, hence allowing for fast growth) and these hosts consequently are more susceptible to infection and lysis. K-selected hosts, i.e. slow-growing and/or small cells, have a high CoR and are less susceptible to infection (Suttle, 2007; Pagarete *et al.*, 2013). The cost of some defence systems, however, have also been considered as relatively low.

Quorum-sensing, host resistance and defence

Barrangou *et al.* (2012) suggested that CRISPR-Cas based host defence might be involved in the control of quorum-sensing signal regulatory cascades. Quorum sensing (QS) is a cell-density-dependent intercellular communication mechanism that is widespread in both prokaryotic and eukaryotic microorganisms (Kleerebezem *et al.*, 1997; Whitehead *et al.*, 2001; Nickerson *et al.*, 2006; Spraque *et al.*, 2006), which frequently occurs at high cell density (Fuqua *et al.*, 2001), and which can even be involved in cross-domain communication (e.g. Bacteria–Eukaryota; Hughes and Sperandio, 2008; Pacheco and Sperandio; Hartmann and Schikora, 2012). Cells can produce QS signal compounds (e.g. N-acyl homoserine lactones, autoinducer-2, etc.) to self-regulate their population size by auto-inducing a variety of functional genes in response to environmental stresses and adaptation (Whitehead *et al.*, 2001). Viral lysis is one of the QS-induced phenotypes and the promoting/triggering of viral lysis (prophage induction) through QS has already been detected for some bacteria (Ghosh *et al.*, 2009).

Recently, quorum-sensing has also shown to be involved in phage defence in *E. coli*, where moderately reducing the number of virus-receptor molecules on cell surfaces caused a dramatic reduction of infection rates (Høyland-Kroghsbo *et al.*, 2013). These findings support the hypothesis that virus–host interaction in cells (lysis/host defence) might be regulated by QS. It is noteworthy that the QS signal compounds have also been detected in water (e.g. wastewater; Feng *et al.*, 2013), and in waterborne bacteria such as cyanobacteria (Sharif *et al.*, 2008), macroalgae (Kim *et al.*, 2007) and *Vibrio harveyi* – *V. harveyi* quorum sensing-controlled bioluminescence is believed to be responsible for the 'milky seas' phenomenon in the ocean (Miller *et al.*, 2005; Nealson and Hastings, 2006).

Placing viral lysis and host resistance in a context of QS network regulation by hosts could help to provide answers to some pertinent questions. Firstly, the boom–bust pattern (indicative of 'Killing the Winner' KtW) for viral populations are often observed even when there are no massive blooms occurring within environments, i.e. when total bacterial abundance remains roughly stable (Chow and Fuhrman, 2012; Pagarete *et al.*, 2013; Zhong and Jacquet, 2014). This can be explained within the KtW hypothesis, where viral lysis controls the competitively fast-growing r-selected populations (Thingstad, 2000; Weinbauer, 2004; Suttle, 2007) in response to environment changes (e.g. nutrient availability), thus preventing bloom formation by the winner of the competition for nutrients. Since the maintenance of constantly elevated defences is costly, Høyland-Kroghsbo *et al.* (2013) have hypothesized that some bacteria have additionally developed mechanisms to estimate the risk of phage infection and to adjust their growth and defence strategies accordingly. Recently, three homologues of bacterial genes from the accessory gene regulator of QS systems has been detected in a *Clostridium* phage genome (phiCDHM1; Hargreaves *et al.*, 2014). This is the first time that QS genes have been reported in a phage genome and the discovery suggests that phages might influence their bacterial hosts or neighbour bacterial populations in novel ways (Hargreaves *et al.*, 2014).

Virus-mediated gene transfer, gene transfer agents, virus and non-virus genes

Viruses, along with other mobile genetic elements (e.g. plasmids, transposons and integrons), allow the transfer of genes among and between organisms in aquatic ecosystems (Frost *et al.*, 2005). The process is called 'transduction' and consists of viruses picking up DNA from one host and transferring that DNA to another host. In generalized transduction, viruses randomly acquire host DNA during DNA packaging or encapsulation, while in specialized transduction proviruses excise specific parts of host DNA, ones found adjacent to the proviral insertion site, when entering a lytic cycle. Foreign DNAs carried to and injected into recipient cells can either be digested or recombined with homologous regions of recipient chromosome resulting in exchange of genetic material. Thus, the overall effect of transduction could be to homogenize gene sequences among virus-susceptible host populations. For example, Chiura (1997) could demonstrate that phages isolated from marine bacteria could repair amino acid deficiencies even in *E. coli*. This suggests a capacity of gene transfer towards a broad range of bacterial hosts. Mounting evidence from (meta) genomic studies has shown that viruses constitute an important genetic reservoir, or gene bank, and revealed that viral particles not only contain their own genes but also a high number of host-derived genes. These include genes involved in host virulence, photosynthesis or nucleic-acid metabolism (Desnues *et al.*, 2008; Rohwer and Thurber, 2009; Clokie *et al.*, 2011). As described above, even quorum-sensing related genes were also discovered in a *Clostridium* phage genome.

Gene transfer agents (GTAs) are phage-like elements involved in horizontal gene transfer (Lang and Beatty, 2007; Lang *et al.*, 2012). They integrate genetic material into a recipient chromosome for replication and are released from host cell by lysis, similar to phages, and they are probably originally derived from phages (Lang *et al.*, 2012). Nevertheless, GTAs (i) do not carry genes that allow their own propagation (i.e. GTA encoding genes); (ii) bear insufficient DNA to encode particle-structure proteins, and (iii) carry random pieces of host genomes. Four genetically unrelated GTAs have been identified in bacteria and archaea so far (Lang *et al.*, 2012). It has been reported that genes encoding RcGTAs are found in most marine systems (Biers *et al.*, 2008). The same study also showed that these GTAs are mainly produced by marine bacteria and move genes between species of Alphaproteobacteria.

Other possible gene transfer entities include bacteriophage-like particles (BLPs) and membrane vesicles (MVs). BLPs are elements similar to GTAs (e.g. with a tailed phage morphology, packaging host DNA randomly), except for the fact that no gene transfer has yet been witnessed. In the case of PBSX-BLP of *Bacillus subtilis*, the particles kill cells that do not harbour the BLP, conferring a selective advantage for those carrying the BLP (Lang *et al.*, 2012; Wood *et al.*, 1990). Membrane vesicles, by contrast, are small (roughly the size of viruses), lipidic elements which are secreted by cells and which harbour different types of DNA, including host chromosomal, plasmid, and even viral DNA (Biller *et al.*, 2014; Gaudin *et al.*, 2014). Membrane vesicles therefore could be considered to be another indirect method for the horizontal gene transfer of host and viral DNA (Chiura *et al.*, 2011; Velimirov *et al.*, 2011). Furthermore, all the gene transfer elements cited above are difficult to distinguish from true viruses in environmental samples during enumeration using classical methods such as EFM (Forterre *et al.*, 2013).

In addition to transduction, host DNA released by viral lysis could be transferred to other organisms through uptake of environmental genetic material, i.e. natural transformation, and this constitutes another, although indirect, mechanism of the virus-mediated gene transfer (Fuhrman, 1999; Weinbauer, 2004). Uptake of exogenous cellular DNA released from viral lysis through transformation is not necessarily a host-species specific process but can also spontaneously occur between two distinct hosts, or even between prokaryotes and eukaryotes (Dunning *et al.*, 2007; Keeling and Palmer, 2008). Filée *et al.* (2008, 2009) detected bacterial-like genes in genomes of phycodnaviruses and mimivirus that infect protists. This resident bacterial DNA was acquired several times during successive acquisition events. The authors proposed that the host (or a host symbiont) of a giant virus (see Chapter 11) has provided both the necessary source of bacterial DNA (from host grazing of bacteria) and the cell compartment that enables this DNA to encounter actively replicating viruses. This bacterial DNA could have been incorporated into eukaryotic algae through transformation and thereby packaged by phycodnaviruses during replication (generalized transduction). Some studies have shown that virus-like particles in seawater and hot springs have been involved in gene transfer between the three cellular domains (Archaea, Bacteria and Eukaryota), but the nature of the particles remains unknown (Rohwer and Breitbart, 2009).

Viral infection and virus-mediated host mortality

Viral lysis is associated with death of the lysed cell and is one of the most significant ecologically functional roles of viruses. For example, it was shown that virus-mediated cell death is involved in the termination of phytoplankton blooms (Bratbak *et al.*, 1993; Tarutani *et al.*, 2000; Jacquet *et al.*, 2002). Estimations of virus-induced mortality rates, which are assumed to be associated with virus-induced cell lysis, indicate that viral lysis also controls phytoplankton biomass in non-bloom forming species (Brussaard *et al.*, 1996; Evans *et al.*, 2003; Gastrich *et al.*, 2004; Nagasaki *et al.*, 2004; Tarutani *et al.*, 2000; Baudoux *et al.*, 2006). These estimations of virus-induced cell mortality have revealed that viruses, through top-down processes, can control host population size, especially through regulation of the fastest growing taxa.

It is well established that viruses are a major cause of mortality of prokaryotes (the most abundant organisms on earth), which on average are as important as grazing by protists (Fuhrman and Noble, 1995; Weinbauer, 2004; Pradeep Ram *et al.*, 2014). For example, a recent study of 11 freshwater lakes confirms that the mechanisms of prokaryotic production regulation varied with sampling depth from grazing being more important in the euphotic zone and viral lysis being more important in the aphotic zone (Pradeep Ram *et al.*, 2013). In the euphotic zone of 19 freshwater lakes (French Massif Central), the temporal shift in viral life strategy was controlled by the availability of phosphate and chlorophyll concentrations (thus, influencing prokaryotic growth): high concentrations (stimulating growth) favoured lytic cycles and low concentrations (causing low growth) favoured lysogenic cycles (Pradeep Ram *et al.*, 2014). In natural environments without bloom episodes, viral lysis could account for approximately 20–40% of the daily loss of bacterial biomass in surface waters of oceans (Suttle, 1994), although a range of 0 up to 100% has been reported (Wilhelm and Matterson, 2008; Danovaro *et al.*, 2010).

Viral infection of phytoplankton and virus-mediated population control has been studied extensively in the past two-to-three decades (Brussaard, 2004). A recent interesting finding indicates that *Microcystis* cyanophages infection occurs in a diel cycle, which may depend on the light cycle (as previously proposed by Suttle, 2000) (Kimura *et al.*, 2012). Diel cycles of lysis have also been demonstrated in bacterioplankton (Winter *et al.*, 2004).

Population dynamics and diversity

The temporal dynamics of virus and host populations have often been studied by investigating potential correlation patterns in culture model systems (e.g. Marston and Salée, 2003; Clasen *et al.*, 2013) or in the environment (e.g. Mühling *et al.*, 2005; Sandaa *et al.*, 2008; De Corte *et al.*, 2010; 2012; Zhong *et al.*, 2013). The approach of seeking potential correlation patterns has provided some robust results in culture model systems and has been used to test ecological hypotheses. The most frequently used model system to examine the outcome of virus–host interactions has been *Synechococcus* and associated cyanomyoviruses. By monitoring the temporal dynamics between *Synechoccocus* and cyanomyoviruses, studies have shown that the abundance of the two communities are coupled, suggesting bloom-bust patterns occurring over a time scale of days to weeks (Marston and Salée, 2003; Clasen *et al.*, 2013). Using the same model system, a rapid diversification of the *Synechococcus* host (Marston *et al.*, 2013) and its co-occurring viruses (Marston and Amrich, 2009) has been demonstrated. By using qPCR to quantify cyanomyoviruses and flow cytometry to enumerate the picocyanobacteria, Sandaa and Larsen (2006) detected a coupling between the two communities in Norwegian coastal seawater across a year-long cycle, which suggests viral control over cyanobacterial abundances. In contrast, only weak (or lack of) correlations could be detected in the oligomesotrophic Lake Bourget and the oligotrophic Lake Annecy (Zhong *et al.*, 2013), except for the spring period in Lake Bourget, where the relationship between the two communities was obvious.

Other methods have included fingerprinting (e.g. T-RFLP, DGGE, RAPD) in studies between T4-like phages and their hosts (Chow and Fuhrman, 2012; Needham *et al.*, 2013; Pagarette *et al.*, 2013; Zhong *et al.*, 2014; De Corte *et al.*, 2010, 2012; Chow *et al.*, 2013). The T4-like myoviruses have been shown to exhibit seasonal dynamics and inter-annual recurrence from examination of monthly samples over two to three years (Chow and Fuhrman, 2012; Pagarette *et al.*, 2013; Chow *et al.*, 2013). Some phylotypes displayed boom–bust patterns, abundance oscillation, or persistence that could be related to shifts in either environmental parameters or host dynamics. In Norwegian coastal seawater, the dynamic pattern of total *Synechococcus* cyanobacterial communities could be related to several operational taxonomic units (OTUs) of T4-like virus, which dominated the myoviral community and whose importance differed with time (Pagarette *et al.*, 2013). Needham *et al.* (2013) examined bacteria community structure daily for 78 days via automated ribosomal intergenic spacer analysis. They found a connection between myoviruses and potential hosts (i.e. OTUs from *Actinobacterium*, *cyanobacterium*, and members of SAR11 and SAR86) over the scale of days. Furthermore, Chow *et al.* (2013) analysed the protist community using primers targeting the 18S RNA gene. They found that (i) virus–bacteria relationships were more cross-linked than protist–bacteria relationships in network association analysis (using Cytoscape), suggestive of an increased taxonomic specificity in virus–bacteria relationships, and (ii) bacteria–virus relationships showed a stronger positive correlation than bacteria–protist relationships on a monthly and seasonal scale. This in turn is suggestive of a more important viral role in controlling bacterial abundance in the ocean than protistan grazing. Winter *et al.* (2010) could demonstrate a co-variation of viral and prokaryotic community structure over a seasonal scale in the deep sea.

A few studies have provided experimental evidence that viruses can affect microbial community composition (Weinbauer and Höfle, 1998; Suttle *et al.*, 1990; Winter *et al.*, 2004; Hewson and Fuhrman, 2006; Sandaa *et al.*, 2009). For instance, increasing viral concentrations through the addition of viral concentrates induced changes in the composition of marine primary producers including cyanobacteria (Suttle *et al.*, 1990). Incubating a freshwater bacterial community with and without the viral community from the same sample, Weinbauer and Höfle (1998) detected changes in abundance and growth rate of bacterial populations as well as in the community composition. Meunier and Jacquet (2015) recently reached the same conclusion by artificially increasing or decreasing viral concentrations in lake samples. The virus-mediated host community structuring could be due to the direct lysis of specific host species. Weinbauer and Rassoulzadegan (2004) suggested that the mass release of cellular materials during cell lysis could also change the composition and bioavailability of organic materials and nutrients and thus change of the microbial community. Both quantity and composition of the material released by viral lysis affects microbial community (Suttle, 2007). Evidence has shown that cellular material released by viral lysis can indeed affect bacterial community composition (Riemann *et al.*, 2009; Holmfeldt *et al.*, 2010).

Virus–host coevolution

Using a *Synechococcus* and cyanomyoviruses model system to examine the outcome of virus–host interactions, a significant diversification in *Synechococcus* (Marston *et al.*, 2013) and its co-occurring viruses (Marston and Amrich, 2009) has been demonstrated, suggesting coevolution, that is, changes and counter changes in the genetic make up of host and viruses. Marston and Amrich (2009) revealed a significant diversity of cyanomyovriuses within a single *Synechococcus* strain, observing that viral isolates shared an average nucleotide identity of 99.3–99.8% after examining different myoviral maker genes. This diversity was believed to have been generated by point mutations, homologous recombination within a host strain, and intragenic recombination between different cyanomyovirus strains. Microdiversity among closely related viral isolates has the potential to influence the population dynamics of viruses and hosts (Marston and Amrich, 2009), and could influence overall rates of infection and viral-induced host mortality (Wilmes *et al.*, 2009). These results indicate that coevolution (in virus–hosts co-incubation) may contribute to the generation and maintenance of *Synechococcus* and viral diversity and thereby influence virus-mediated mortality of these key marine bacteria (Marston *et al.*, 2013). Thus, virus–host interactions can generate and maintain diversity within host and viral populations by exerting selective pressure on the host for viral resistance and on the virus to overcome host resistance (Andersson and Banfield, 2008; Wilmes *et al.*, 2009).

Recently, Sun *et al.* (2012) exposed a host (*Streptococcus thermophilus* DGCC7710) to a phage (phage 2972) to which the host had no resistance. After 1 week of co-culture, they examined the genomes of both viruses and host. They found newly incorporated CRISPR spacers in *S. thermophilus* DGCC7710 and recently acquired single nucleotide polymorphisms (SNPs) in phage 2972. These spacers were variable short regions in CRISPR loci, which were originally acquired from a virus and thus conferred to the host as a means of resistance to viral infection through the CRISPR-Cas immune system. Sun *et al.* (2012) suggested that the acquisition of immune elements (spacers) following phage exposure led to a genetically diverse population with multiple subdominant strain lineages. They also suggested that CRISPR-based virus-resistance can be associated with major resistance-linked viral diversity purging events that can re-direct evolutionary trajectories of both host and phage populations.

Indirect and unexpected interactions

Phage–host interactions can also have indirect as well as unexpected and even surprising effects on the food web (Miki and Jacquet, 2008). This was initially found to be the case when it was shown that viruses can potentially stimulate primary production, counteracting mortality due to grazing to the point of tipping the status of a phytoplankton population from declining numbers to growth (Staniewski and Short, 2014). It has also been shown that *Synechococcus* growth is probably dependent on regenerated nutrients, following the lysis of heterotrophic bacteria (Weinbauer *et al.*, 2011).

Grazing has been reported to stimulate viral infection of bacterioplankton (Weinbauer *et al.*, 2003, 2007; Jacquet *et al.*, 2007), although it should lower infection rates since it competes with viruses for prey (e.g. microbial cells) and grazers probably consume infected cells (Simek *et al.*, 2001). The virus-mediated effects in the upper levels of the food chain (grazing food chain; Fig. 6.1) may be variable. One example is preferential grazing by the heterotrophic dinoflagellate *Oxyrrhis marina* on virus-infected *Emiliania huxleyi* (Evans and Wilson, 2008). In this experimental study, grazing rates of *O. marina* were observed to increase in the presence of virus-infected *E. huxleyi* prey. On a large scale this would direct more carbon, nitrogen, and phosphorus into particulate forms, making it available via grazing to higher trophic levels (see Fig. 6.1). Another experimental study has provided evidence of an indirect impact of bacteriophages on bloom-forming microalgae via phage-induced lysis of algae-killing bacteria (Cai *et al.*, 2011). In this study, a lytic phage isolated from an algicidal bacterium that specifically kills the toxic dinoflagellate, *Alexandrium tamarense*, was observed to trigger algal growth by reducing the abundance of the bacterium and hence reducing the algicidal effect.

Virus-induced effects cascading up the food chain can have even more unexpected results. In a recent field study, cyanophages were shown to affect an African flamingo population in a bottom-up cascade, i.e. from the base to the top of the food web (Peduzzi *et al.*, 2014). In the alkaline-saline Lake Nakuru (Kenya), an important feeding site of the Lesser Flamingo (*Phoeniconaias minor*), cyanophage infection led to a breakdown of the main food source, the cyanobacterium *Arthrospira fusiformis*. This resulted in a dramatic reduction in flamingo abundance, indicating that viral infection at the very base of a food chain can affect the distribution of end consumers.

These different examples raise the question of how aquatic ecosystems would even function were they free of viruses. One scenario would be that primary production would slow down, resulting in much less photosynthesis and thus much less oxygen production. Hence, there is the hypothesis (see also 'General aspects of aquatic viral ecology') that viral lysis overall promotes nutrient cycling, production, and diversity in aquatic systems.

Interactions with environmental factors

In this section we focus on interactions between viruses and the physicochemical environment, including parameters of climate change, such as global warming and ocean acidification.

Interactions with dissolved organic matter

As discussed above, in planktonic food webs viral lysis products are shunted away from the grazing food chain and the microbial loop (referred to as the 'viral shunt'; Wilhelm and Suttle, 1999). Viral lysis products can be efficiently utilized by heterotrophic prokaryotes, thereby again entering the food web. The net effect is that lysis detours bacterial production away from higher trophic levels such as protists and ultimately converts organic matter into inorganic nutrients (Wilhelm and Suttle, 1999; Haaber and Middelboe, 2009; Pollard and Ducklow, 2011). Recent culture-based experiments with marine microorganisms showed that phage infection alters both host metabolism and lysate composition (Ankrah *et al.*, 2013; Sheik *et al.*, 2014). In the Northern Baltic Sea, virus-mediated cell lysates accounted for a substantial fraction of the bacterial carbon and phosphorus demand (Holmfeldt *et al.*, 2010).

The release and consumption of D-amino acids, whose origin is attributed mainly to the remains of bacterial peptidoglycan, is also anticipated to be a result of viral lysis (e.g. Azúa *et al.*, 2014). Another study also suggests the rapid use of viral lysis products after relieving phosphorus limitation (Motegi *et al.*, 2015). In an experimental study by across marine environments, the decrease of bacterial growth efficiency was explained by virus-induced conversion of bacterial biomass into dissolved organic carbon (Motegi *et al.*, 2009). More experiments with seawater cultures and the addition of monomeric and polymeric substrates revealed that bacterial communities grown under substrate-rich conditions are less subject to viral attacks, thus supporting the notion that supplies of dissolved organic matter affect the magnitude of bacteria-virus coupling in marine environments

(Motegi and Nagata, 2009; Motegi *et al.*, 2015). An increase of respiration by viral lysis of bacteria, hence a stronger conversion of DOM into CO_2, was also reported for the Australian Bremer River (Pollard and Ducklow, 2011). A previous article has already provided an overview of other significant studies on this virus-DOM coupling (Jacquet *et al.*, 2010). Nevertheless, comprehensive studies on the significance of the viral shunt in aquatic systems are strikingly scarce, particularly for flowing inland waters (Peduzzi and Luef, 2008; Peduzzi, 2015).

Interactions with particulate organic matter

The viral ecology of organic and inorganic particles in aquatic systems has been previously comprehensively reviewed (Weinbauer *et al.*, 2009). Some important conclusions from that review and new information are included in the following compilation. Quality, size and age of particles and aggregates as well as the exposure time of viruses to aggregates apparently are the most important factors regulating viral abundance (Kernegger *et al.*, 2009). Importantly, particles (mainly with organic constituents) appear to play a role as viral scavengers or reservoirs rather than viral factories (Weinbauer *et al.*, 2009; Peduzzi, 2015). Moreover, adsorption of viruses to suspended matter can stimulate the growth of the free-living prokaryotic community, e.g. by reducing viral lysis by displacement of viruses to the particulate (attached) fraction (Peduzzi and Luef, 2008; Weinbauer *et al.*, 2009, 2012). In addition, inorganic particles, such as clay and calcite, as another type of suspended matter, can also remove free-living viral particles in aquatic environments (Clasen *et al.*, 2008; Brussaard, 2004). In an Amazonian floodplain lake, flood pulses and the resulting turbidity were found to affect viral and host abundance (Barros *et al.*, 2010). Streams and rivers are often particle-rich environments, which create microscale patchiness for microorganisms. It has been estimated that up to 35% of the viruses are associated with suspended matter in water of the Danube River (Luef *et al.*, 2007; 2009). The generation of more comprehensive studies has been hampered by the complicated techniques required to reasonably enumerate particle-associated viruses (see Peduzzi, 2015). Moreover, it remains unclear to what extent viruses are truly attached to the matrix of the particles or distributed in the pore water of suspended matter and aggregates (Weinbauer *et al.*, 2009).

A specific type of suspended particle is black carbon (soot), which has both anthropogenic and non-anthropogenic origins (Weinbauer *et al.*, 2012). Significant amounts of this material, originating from the incomplete combustion of biomass and fossil fuels, enter aquatic systems via atmospheric deposition; river-runoff can transport significant amounts of this material into the ocean (and into lakes). Black carbon strongly absorbs both organic carbon and microorganisms. In experiments, addition of black carbon to seawater resulted in immediate (20–30 min) adsorption of 10–20% of the bacterial and viral abundance (Weinbauer *et al.*, 2012; Cattaneo *et al.*, 2010). A higher percentage of viruses, when compared to bacteria, were found attached to these particles, suggesting a negative effect on overall viral biomass. The experimentally observed increased bacterial production after black carbon addition could, to some extent, be due to lowered bacterial mortality stemming from reduced viral lysis (Weinbauer *et al.*, 2012). This has also been demonstrated in experiments (Malits *et al.*, 2015). Embedding bacteria into the matrix of black carbon, however, could also cause reduced infection, e.g. by 'hiding' receptors (Cattaneo *et al.*, 2010).

Viruses can also influence the formation of organic aggregates. For example, viral lysis of bacterio- and phyto-plankton may influence the formation process and stability of algal flocs (Weinbauer *et al.*, 2009). In an experimental study, the formation of fewer but larger suspended aggregates was shown to result from elevated abundances of viruses (Peduzzi and Weinbauer, 1993). The export of such diatom-derived material into the deep sea has been termed 'viral shuttle' and could be an import viral role in the ocean besides the viral shunt which rather enhances nutrient recycling (Sullivan *et al.*, 2017). A more recent experimental study reported that significant changes in the particle size distribution occurred primarily within the size range of relatively large particles (> 4 μm), likely as a result of the aggregation of smaller-sized particles originating from host microbial lysis. In a marine spring phytoplankton bloom experiment, organic micro-aggregates with attached prokaryotes were observed to form under turbulent conditions, but seemed to be reduced in the presence of viruses.

This suggests that turbulence and viruses play a significant, previously neglected role in shaping particle aggregation (Maltis and Weinbauer, 2009). Viral lysis of bacteria and subsequent particle dynamics generated large variations in the particle size distribution over a broad size range on time-scales from hours to a few days (Uitz *et al.*, 2010).

Overall, aggregates are more scavengers of viruses than sites of release of viruses into the water surrounding the particles (Weinbauer *et al.*, 2009). There are also examples, however, where aggregates serve as viral factories rather than as viral traps (Bettarel *et al.*, 2015). During aggregation, virus-trapping could predominate, while during dissolution, virus-release could become more dominant. Also, whether aggregates are already saturated with viruses (either from attachment or viral production within the particle), will influence the net outcome between trapping and release.

Finally, attached viruses can be transported with aggregates, either horizontally by floating particles or vertically by sinking particles (e.g. Weinbauer *et al.*, 2009). Viruses from sinking particles enter the sediments.

Physicochemical factors and climate change

Abiotic factors may impact viruses directly by causing viral decay, or indirectly by affecting host growth or physiology, thus affecting viral production rates. The influence of different physical factors in the environment, such as temperature, oxygen, light availability or salinity has been investigated (Weinbauer, 2004; Brussaard, 2004; Cissoko *et al.*, 2008; Bettarel *et al.*, 2011). An interesting experimental study regarding the mixing of freshwater and seawater in estuaries demonstrated that production rates of freshwater viruses sharply declined after seawater addition, followed by a rapid (within 48 h) recovery of viral populations. Conversely, marine viruses were not significantly affected by mixing with freshwater (Cissoko *et al.*, 2008). It is assumed that viruses can rapidly respond to dramatic shifts in the abundance of bacterial hosts and community composition, and that these hosts may suffer from osmotic shock (Bonilla *et al.*, 2009).

A study on the effect of oxygen on microbial and viral populations was conducted in a deep productive freshwater reservoir by Pradeep Ram *et al.* (2009). The frequency of infected bacterial cells was found to be lower in the anoxic (no oxygen present) than in the oxic zone (saturated with oxygen), caused by forcing from thermal stratification. On average, viruses were responsible for 23% loss of bacterial production in the oxygenated surface waters, but only for 9% in deep anoxic waters. In this study the viral influence on heterotrophic bacteria in the anoxic zone was remarkably low compared to other studies (Pradeep Ram *et al.*, 2009). Another study dealt with the viral community structure in the Eastern Tropical South Pacific oxygen minimum zone (Cassman *et al.*, 2012). The virus-to-microbe ratio fluctuated in the oxycline and declined in the anoxic zone to less than one. The number of viral genotypes declined from 2040 at the surface to 98 in the oxycline; only <5% of genotypes were shared between surface and anoxic core viromes. Thus, the viral community was characterized by fluctuations in abundance, taxa and diversity across the oxygen gradient, and it appeared that the reduction in oxygen coincided with a shift to unique viral genotypes in the anoxic core.

Another important physical factor influencing microbial processes is light. Some viral infections of phytoplankton in the ocean have been found to prevent a shutdown of photosynthesis by the host, e.g. as defence mechanism, in infected cells and allow it to continue and provide the energy needed for virus replication (Mann *et al.*, 2003). Other studies have demonstrated that viruses themselves are sensitive to UV-radiation (Suttle and Chen, 1992). Along a salinity gradient near the Pearl River estuary (South China Sea), viral decay rates were found to be significantly correlated with the UVA diffuse attenuation coefficient. UV-radiation increased the viral decay rates by ca. 30%, which decreased bacterial mortality, thus mitigating the inhibitory effects of UV-radiation on bacteria (Yuan *et al.*, 2011).

In flowing waters, environmental variability is especially large and clearly driven by the dynamic hydrology (Peduzzi, 2015). The high degree of spatial heterogeneity in such systems, e.g. backwaters with often lake conditions, also plays a role (Peduzzi and Luef, 2008; 2009; Jacquet *et al.*, 2010; Ma *et al.*, 2013). For example, in a river–floodplain system of the Danube and in an Amazonian floodplain lake, viruses, together with their microbial hosts, were found to be influenced in a complex manner by turbidity and hydrodynamics (Peduzzi and Luef, 2008; Barros *et al.*, 2010).

The effect of global warming on viral and host communities was experimentally studied in microcosm experiments in two contrasting Arctic marine systems (Lara *et al.*, 2013). Viral numbers, bacterial abundance and overall system production increased significantly and influenced the lytic-lysogenic decision. This showed that increases in temperature stimulate heterotrophic microbial biomass and activity, and viral lysis contributes to the increase of the DOM pool in the water column. In the review of Danovaro *et al.* (2010) on the effect of climate change on marine viruses, it was concluded that there will likely be direct and indirect consequences on viruses, including cascading effects on biogeochemical cycles, food webs, and metabolic balance of the ocean. Discussing a range of case studies, evidence was found that marine viruses interact actively with current climate change and they are considered to be a key biotic component able to influence the oceans' feedback on climate change. The limited number of data prevents incorporation of virus-related processes into current climate models on the necessary spatial and temporal scales at which climate change scenarios would respond and interact with this important compartment of the biosphere. This calls for focused research priorities and long-term data sets to enhance our knowledge about the role of aquatic viruses in the present climate change scenario (Peduzzi, 2016).

Ocean acidification is mainly due to the uptake of anthropogenic CO_2 from the atmosphere. Although the effect on microorganisms has been questioned, clear evidence has been presented that there is an effect (Liu *et al.*, 2009; Weinbauer *et al.*, 2011c). Larsen *et al.* (2008) investigated, in a mesocosm study, the effect of elevated pCO_2 on the viral community composition using flow cytometry and showed that 'high-fluorescence viruses' were stimulated while the other viral groups were not altered. Clearly the abundance of EhV (viruses infecting *Emiliania huxleyi*) decreased at elevated pCO_2 (1050 μatm). This indicates changes in the viral community composition. Evidence for moderate effects of ocean acidification of viral abundance and dynamics has been presented for the EPOCA ocean acidification experiment in Svalbard (Brussaard, 2008). This is likely due to indirect effects such as changes in host composition and activity. So far, direct effects on viruses have not been investigated.

Virivory

Consumption of viruses in the course of grazing by protists has been demonstrated in only a restricted number of studies both from marine (Suttle and Chen, 1992; Gonzales and Suttle, 1993) and fresh waters (Manage *et al.*, 2002; Bettarel *et al.*, 2005). There is no doubt that bacteria are the principal constituent in the diet of heterotrophic nanoflagellates (HNF) in most or all aquatic ecosystems (Sanders *et al.*, 1992; Sherr and Sherr, 1994; Pernthaler, 2005) but, on some occasions, viruses also can contribute to HNF nutrition or at least flagellates could be responsible for part of the decay of viral particles. Indeed, protozoa may benefit from grazing on viruses by (1) removing viruses as competitors with protozoa for bacteria and (2) obtaining additional nutrients from viruses, although the nutrient content of viruses in and of itself would be insufficient for protozoa to persist.

New and poorly studied environments or viral groups

Benthic and anoxic environments

The largest benthic ecosystem on earth, the deep-sea benthos, has revealed extremely high viral production worldwide (Danovaro *et al.*, 2008). These authors reported that viral infections can be responsible for the abatement of up to 80% of prokaryotic heterotrophic production, and they also observed a virus-induced prokaryotic mortality increasing along with depth. Beneath a depth of 1000 m, viral lysis is by far the most important mortality factor of the benthic prokaryotic heterotrophic production, indicating the important role of viral lysis in global biogeochemical cycles. This is supported by more recent evidence regarding ongoing production of viruses in deep subsurface sediments (Engelhardt *et al.*, 2013). Within a comprehensive set of globally distributed subsurface sediments (some core samples more than 300 metres below surface; mbsf), enormous numbers of viruses (up to 10^9 per ml) indicated their potential as a controlling factor for prokaryotic mortality in the marine deep biosphere. Even in the oldest sediments, microbial communities were suggested to be capable of maintaining viral populations, indicative of an ongoing viral production. Thus, viruses are likely independent indicators for microbial life in the marine deep biosphere.

The morphological diversity and community structure of benthic viruses of a deep, permanently anoxic freshwater lake revealed unique and novel putative viruses (Borrel *et al.*, 2012). Communities at the surface (–10 cm) could be discriminated from those at intermediate depths (–27 cm) or deeper (–40 cm) layers. Unusual morphotypes, new in freshwater systems, were detected, some of them resembling dsDNA viruses of hyperthermophilic and hyperhalophilic Archaea. Also, oxygen-reduced zones of the water column appear to harbour specific viral communities. For example, associated with the Eastern Tropical South Pacific oxygen minimum zone, the number of genotypes ranged from 2040 at the surface to only 98 in the oxycline (Cassman *et al.*, 2012). Within the oxygen minimum zone viromes, only 4.95% of genotypes were shared between surface and anoxic viromes. From the sampled DNA sequences within the oxygen minimum zone, more than 97.8% were novel. In this study, the viral community of the oxygen minimum zone was characterized by fluctuations in abundance, taxa and diversity across the oxygen gradient, displaying a shift to unique viral genotypes in the anoxic layer.

The atmosphere and aquatic environments

An unexplored ecosystem is the atmosphere, which is probably the 'last frontier of biological exploration on earth' (Rothschild and Mancinelli, 2001), together with the deep biosphere. Due to the few reports available, it remains unclear what fraction of the virus-like particles are microbial viruses. In a study conducted on air-samples collected from Korea, authors reported viral numbers ranging between 1.7×10^6 and $4.0 \times 10^7/m^3$ (Whon *et al.*, 2012). Genomic analysis suggested that the majority of the collected viral particles were plant and animal viruses, although sequences belonging to *Inoviridae* and *Microviridae* were also found (see Chapter 7). Another study on the Caribbean atmosphere on St. John Island (Virgin Islands, USA) enumerated 18.1 to $213/m^3$ VLPs during clear weather conditions and following an African desert dust-event, respectively (Griffin *et al.*, 2001).

Given the high abundance of microorganisms in the euphotic zone of oceans, one would expect that aerosolization transports viruses into the atmosphere with microbes that migrate with winds and rains. In fact, viruses infecting the phytoplankton species *Emiliania huxleyi* (e.g. EhVs) can spread through the air, leading to widespread algal die-offs. Sharoni *et al.* (2015) showed, indeed, that EhVs can be conveyed long distances and transmit infection to the remote locations of coccolithophore blooms. Aquatic environments can therefore be a source of viruses (and microorganisms) of the atmosphere and conversely they can be deposited from the atmosphere into aquatic environments. Hence, the investigation of the atmosphere is likely to be of major interest, in some cases, towards better understanding of host-virus dynamics and/or the distribution and traffic of some microorganisms around the globe.

Concepts and models

Lytic viruses are believed to have consequences on at least two ecologically relevant levels. One is related to viral effects on diversity and population/community structure and dynamics. The other is related to the biogeochemical consequences of diverting the flux of energy and material away from the predatory transport towards higher trophic levels. This energy and material instead is returned to non-organism aspects of environments as small particles and dissolved organic material via the viral shunt (Fig. 6.1), or instead these biogeochemical processes influence the export of organic matter into the deep sea (biological pump).

Viral control of host diversity

Several concepts have been introduced in attempts to capture essential aspects of host–virus interactions. Included in such a collection are the concepts of 'Red Queen dynamics' (Clarke *et al.*, 1994), 'Killing the winner' (KtW) (Thingstad and Lignell, 1997; Thingstad, 2000), 'Seed bank' (Breitbart and Rohwer, 2005), 'The king of the mountain' (Giovannoni *et al.*, 2013) and also 'The Cheshire cat escape strategy' (Frada *et al.*, 2008). These concepts emphasize different aspects of host–virus interactions such as the potential for rapid antagonistic coevolution, the coexistence of hosts with competition versus defence strategies, and existence of a background reservoir from which dominating viruses can be selected. Although all these aspects are likely to be present, it has proven to be remarkably difficult to obtain solid experimental evidence needed to transform many of these concepts from their status as ideas into quantitative working tools for virus ecology. This is the case even when worked out in a mathematical form such as KtW (Winter *et al.*, 2010).

The KtW concept was, in its original form, based on steady-state arguments (e.g. Thingstad, 2000) where the 'winner' (competition specialist) is the best competitor for nutrients or DOM that would 'win' in the absence of a selective loss such as from viral lysis. In the latter case, the defence (against viruses) strategist can 'win' (Fig. 6.2). The use of the KtW concept is somewhat confused as it is applied in both a dynamic version (e.g. Avrani *et al.*, 2016) and in its original steady-state (e.g. Thingstad, 2000) version. The conceptual difference is important since the 'winner' in the dynamic version is the most abundant host, killed by viruses and then replaced by another 'winner' host. In the steady-state version it is the fastest growing host, i.e. the one that would become dominating if its abundance was not controlled by lytic viruses.

Pagarete *et al.* (2013) recently used a dynamic version to explain boom–bust patterns in microbial populations and/or viruses in nature when there is no massive bloom. Here KtW was used to describe successive waves of competitively successful microbes (fast growing), which are decimated by viral infection and then replaced by defence specialists, fostering a dynamic and diverse system.

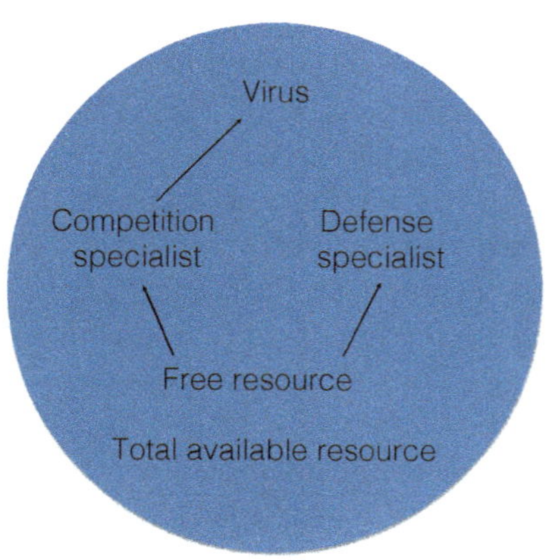

Figure 6.2 Schematic of the general structure of the 'killing-the-winner' hypothesis. Redrawn from Winter *et al.* (2010).

This fits quite well with the seed-bank hypothesis (Breitbarts and Rohwer, 2005) stating that (i) only a few members of viruses may be abundant at any given time, most of them being rather rare and forming a potential bank-population for recruitment; and (ii) viruses preying on their hosts move from the bank to the active fraction through viral production, and the previously abundant and active fraction enter back into the bank, thereby resulting in a change of the dominant types of microbial species. Recently, such dynamics and concepts were also applied for ssDNA viruses inhabiting alpine lakes (Zhong *et al.*, 2015).

The basis for both versions of KtW is that the lytic action of viruses depends largely on the probability of contact with the host (Fuhrman, 1999; 2009). In the dynamic version this makes the most abundant hosts most vulnerable. In the steady-state version, host-abundances are constrained by the steady-state requirement for their viruses: For a virus population to persist, its host population must be abundant enough to allow on average one of the viruses released per cell in a lytic event to find and successfully infect a new host before it is destroyed. Viral abundance on the other hand is constrained by the steady state requirement for the host, i.e.; hosts must divide as fast as the sum of their losses, usually assumed to be the sum of grazing and viral lysis. Since loss to viral lysis can increase with viral abundance, this links viral abundance to host activity (as in steady-state higher infection rates have to be balanced by faster host growth), rather than to host abundance. This means that the apparently plausible inference that abundant viruses belong to abundant hosts is not necessarily true. The original steady-state KtW (Thingstad and Lignell, 1997; Thingstad, 2000), has recently been refined by introducing trade-offs between competition and defence (Våge *et al.*, 2013) and strain-specificity of viruses (Thingstad *et al.*, 2014).

The introduction of a trade-off between competition and defence links the growth and loss sides for the host. As an increased defence means a reduction in the probability that a host–virus collision will lead to a successful infection, the consequence is that in steady-state conditions a higher host population is needed for defence strategists to sustain a viral population. KtW models with trade-off thus predict slow-growing defence strategists to be dominant. Experimentally reducing viral abundance in such a system would lead to positive net growth in rarer but faster-growing, otherwise virus-controlled parts of the community, while little effect would be expected in the dominant, virus-resistant part. This is seemingly consistent with experiments where previously rare host species have been observed to become dominant when viral abundance is reduced (Bouvier and delGiorgio, 2007; Cram *et al.*, 2016).

Another intriguing perspective following from this line of reasoning is that dormancy, i.e. the high proportion of low-activity cells of heterotrophic prokaryotes found in aquatic systems (e.g. Malmstrom *et al.*, 2004) could be a consequence of the need for cost-expensive defence, rather than a sign of lack of suitable nutrients. Dominance of low-activity individuals in the prokaryotic community should thus perhaps be seen more as a consequence of 'having closed the gates to keep the enemy out' rather than a result of 'there is no food out there' (Thingstad *et al.*, 2014).

Attempts to compare the KtW with experimental data have assumed that host groups would correspond to data with a resolution at 16S rDNA level, i.e. data that resolve the host community to a kind of species-level. Repeated isolation of individuals of a prokaryote 'species' from the same environment has, however, led to the concept of a core genome characterizing the 'species', but with other genomic regions being highly variable between strains. These variable regions have been suggested to play a central role in viral defence (Rodriguez-Valera *et al.*, 2009). A related discovery is how host–virus interaction matrices for isolates from natural environments can be reshuffled to reveal a nested structure (Jover *et al.*, 2013). Since arms-race dynamics are likely to lead to nested interaction matrices (Thingstad *et al.*, 2014; Haerter *et al.*, 2014), the different lines of experimental evidence converge to suggest that prokaryote population dynamics in natural systems is a blend of two fundamentally different types of dynamics: A traditional ecological dynamic, where species invade, compete and are grazed or lysed, and an evolutionary arms-race dynamics, where each host species develop successively defensive strains.

Interestingly, the incorporation of such a concept of strain diversification into a KtW-model can explains the apparent paradox suggested by Weinbauer (2004). The paradox is that in simple laboratory co-cultures of viruses and prokaryotic hosts, the cultures rapidly become dominated by resistant hosts and have a low virus-to-host ratio, while natural seem to have high virus-to-host ratios, i.e. viral abundance is higher than host abundance. Using a model which includes strain diversification, the paradox disappears. In the chemostat analysis of Thingstad *et al.* (2014), the cost of resistance creates an accumulating difference in growth rates between the original, undefended strain, and new increasingly defensive, and therefore slower growing strains. In the chemostat, an immune strain has to grow with the dilution rate; a faster growing but susceptible strain has to have a virus population size large enough to grow at a rate which exceeds the dilution rate. In an early state of the arms race with a small cost of resistance (CoR), only a few virus types are therefore needed to maintain the chemostat equilibrium. As the arms race progresses and differences in growth rates increase, the steady state in the chemostat will be characterized by more and more viruses needed to balance the increasing difference in growth rates between hosts. The natural situation described in the 'Weinbauer paradox' (Thingstad *et al.*, 2014) is thus explained by this model as the mature state of such an arms race with a high virus to host ratio (see also Fig. 6.3).

The consequences of strain diversification, for viral control of prokaryote diversity, is quite profound. The original version of the KtW concept suggested a viral top-down control of the abundance of each 'species', where defensive ability would be the major factor determining abundance of individuals in each host species. With strain diversification, abundance within a species becomes the combined function of how many strains the species can establish and the abundance of individuals within each of these established strains. Since the establishment of strains is a competitive feature, and the abundance of individuals within a strain increases with defences, a species needs a low trade-off between competitive and defensive abilities in order to become abundant (Thingstad *et al.*, 2014). Developing a better conceptual understanding of these mechanisms is central to settling the debate of whether the bacterium SAR11 has become the world's probably most abundant organism due to its competitive or due to its defensive abilities, or perhaps due to its ability to combine the two (Zhao *et al.*, 2013).

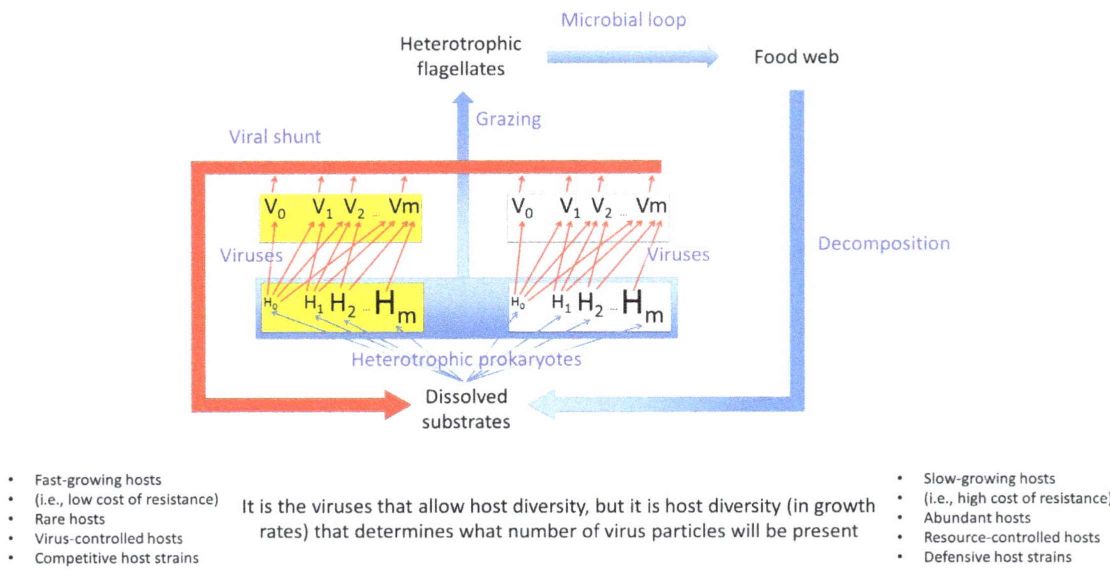

Figure 6.3 Killing-the-winner model with strain-specific viruses. V, viruses; H, hosts; n, number of viral strains; m, number of host strains. In this model (adapted from Thingstad *et al.*, 2016), prokaryote community size (blue box) is top-down controlled by grazing from heterotrophic flagellates shunting material through the microbial loop while each prokaryote species (yellow and violet) is split into strains through antagonistic host-virus arms races sending material through the viral loop. Host strains are arranged left to right from fast-growing non-defensive to slow-growing (high CoR) defensive within each species; virus strains from specialists to generalists. At steady state, the host community becomes dominated by defensive, slow growing strains. In this model, host diversity at species level becomes a function both of host competitive abilities (how many strains it can establish) and host defensive abilities (how many individuals there are per strain). Partitioning between the microbial and viral shunt depends on the difference in growth rate between strains within species and thus on CoR.

When both mechanisms, i.e. the price paid for an increase in resistance (CoR) and the ratio between the price paid and the competitive ability gained (trade-off), are incorporated into models, speculations on the consequences of the many different resistance mechanisms which have been described becomes more meaningful. For example, a modification of porin proteins, which are required for nutrient uptake, could lead to a large difference in CoR between successive host strains. In another example, a CRISPR-based defence system would require a high running and maintenance cost, that may be high at the species level; however, the price for adding new recognition sequences would be low and this could result in a low CoR for strain diversification.

Food web function

An obvious consequence of diverting energy and material into the viral shunt is that a smaller fraction of total primary production reaches higher trophic levels. This could lead to the conclusion that viral lysis in the microbial part of the food web should have a negative influence on production at higher trophic levels. However, there are two ways in which viruses could reduce this higher-level production, either by decreasing transport to higher levels (viral shunt, Fig. 6.1) or by increasing total primary production, since nutrients that are recycled could be used instead by phytoplankton. Therefore, part of the virus effect is an increase of the nutrient content in the photic zone, while another part is a reduction of the production at higher trophic levels and reduced export of organic material into the aphotic zone. How these two effects are balanced depends on factors such as where in the food chain export occurs, e.g. as sinking aggregates originating from phytoplankton debris or as faecal pellets from copepods. Another factor is how organisms are mixed between nutrient-rich and nutrient-poor water layers, such as by currents and turbulences or vertical migration of copepods. A recently published model study (Weitz *et al.*, 2015) contains such mixing effects and therefore predicts that viruses would reduce export (by reducing net import). Overall, the relative importance of the mechanisms with the viral shunt and the link to the 'viral shuttle' is poorly constrained and likely also depends on factors such as season (phytoplankton bloom or not) and system productivity (oligotrophic vs eutrophic).

Biodiversity and ecosystem functioning

Issues involving viruses are usually only poorly discussed within the framework of ecological concepts developed for plants and animals. The combination of topics discussed in 'Viral control of host diversity' and 'Food web function' could constitute the basis for a mechanistic concept combining biodiversity with ecosystem functioning, thus linking the study of viral ecology to the different traditions of general BEF (biodiversity and ecosytem functioning) research (see, for example, Duffy, 2009 and references therein). For example, an experimental study with surface seawater and with different viral abundances indicates that bacterial host diversity is maximized when bacterial productivity and disturbance (virus infection/production) are balanced (Motegi *et al.*, 2013). Bacterial diversity was related in dome-shaped patterns to the ratio of viral-to-bacterial production, with highest bacterial diversity at intermediate ratios. Thus, bacterial production and virus-induced mortality may interactively affect bacterial diversity in seawater. These results are similar to patterns often found for studies on plants and animals (intermediate disturbance hypothesis) and suggest common patterns of diversity for viruses, microorganisms, plants and animals.

Conclusions

This review, on the accumulation of information on research of viruses in aquatic systems during the last decade(s), reveals – not unexpectedly – that the development of new techniques and the adaptation of existing techniques (see Chapter 15) has been a main driving force in this progress on viral research. Major advancement has been made in research on gene transfer, host resistance, population dynamics, and coevolution of viruses and hosts. Viral biodiversity has benefited enormously from the 'Omics' area. Metagenomics (see Chapter 5) has revealed that the diversity of viruses is larger than the diversity of cells in aquatic systems. This opens a new research avenue and quest, since the identity of the vast majority of this diversity (>> 90% of the sequences) remains unknown: both viruses and their hosts remain unidentified and the functions of obtained sequences remain obscure. Pioneer data on the metabolomics of isolated virus–hosts systems are available revealing exciting insights into virus–host interactions; the extension to the environment (*in situ*) and communities (community metabolomics) is a task for the years to come.

Progress has also been made in other ways. Research during the last decades has resulted in the perception that viruses (and microorganisms) are much more tightly related and intertwined with (multi)cellular life, food webs and the inorganic environment than previously anticipated. Unexpected direct interactions such as viral predation on viruses (virophages) and indirect interactions such as cascading trophic effects through the food web have been found. Considering the vast unknown diversity of viruses, one can expect that that the understanding of the biocomplexity of such interactions has only just begun. Many environments such as organic aggregates, anoxic waters and sediments or the interactions with the environment such as physicochemical influences on viral communities *in situ* or the effect of viruses on the cycling of nutrients and organic matter *in situ* remain poorly studied.

Progress has also been made in our concepts on the role of viruses in food webs and biogeochemical cycles and this has resulted in improved models on the interactions of viruses with their hosts. While viral lysis can be detrimental at individual and population levels, viral infection seems now to be a key factor explaining how the large cellular diversity is sustained in aquatic systems and how geochemical cycles of elements such as carbon, nitrogen and phosphorous are controlled.

Acknowledgements

PP acknowledges the projects P17798 and P204604 of the Austrian Science Foundation and MGW the project ANCESTRAM of the French Science Ministry.

References

Abedon, S.T. (2009). Disambiguating bacteriophage pseudolysogeny: An historical analysis of lysogeny, pseudolysogeny and the phage carrier state. In: Contemporary Trends in Bacteriophage Research, Adams, H.T. ed. pp. 285-307.

Abedon, S.T. (2015). Bacteriophage secondary infection. Virol. Sin. *30*, 3–10. https://doi.org/10.1007/s12250-014-3547-2.

Alldredge, A.L., and Cohen, Y. (1987). Can microscale chemical patches persist in the sea? Microelectrode study of marine snow, fecal pellets. Science *235*, 689–691.

Allen, M.J., and Wilson, W.H. (2008). Aquatic virus diversity accessed through omic techniques: A route map to function. Curr. Opinion Microbiol. *1*, 226–232.

Andersson, A.F., and Banfield, J.F. (2008). Virus population dynamics and acquired virus resistance in natural microbial communities. Science *320*, 1047–1050. https://doi.org/10.1126/science.1157358.

Ankrah, N.Y.D., May, A.L., Middleton, J.L., Jones, D.R., Hadden, M.K., Gooding, J.R., LeCleir, G.R., Wilhem, S.W., Campagna, S.R., and Buchan, A. (2014). Phage infection of an environmentally relevant marine bacterium alters host metabolism and lysate composition. ISME J. *8*, 1089–1100.

Avrani, S., Schwartz, D.A., and Lindell, D. (2012). Virus-host swinging party in the oceans: Incorporating biological complexity into paradigms of antagonistic coexistence. Mob. Genet. Elements *2*, 88–95. https://doi.org/10.4161/mge.20031.

Azam, F. (1998). Microbial control of oceanic carbon flux: The plot thickens. Science, *280*, 694–696.

Azam, F., Fenchel, T., Field, J.G., Gray, J.S., Meyer-Reil, L.A., and Thingstad, F. (1983). The ecological role of water-column microbes in the sea. Mar. Ecol. Prog. Ser. *10*, 257–263.

Azam, F., and Long, R.A. (2001). Sea snow microcosms. Nature *414*, 495, 497–498. https://doi.org/10.1038/35107174.

Barrangou, R., Fremaux, C., Deveau, H., Richards, M., Boyaval, P., Moineau, S., Romero, D.A., and Horvath, P. (2007). CRISPR provides acquired resistance against viruses in prokaryotes. Science *315*, 1709–1712.

Barrangou, R., and van der Oost, J. (2012). CRISPR-Cas Systems: RNA-mediated Adaptive Immunity in Bacteria and Archaea. (Springer).

Barros, N., Farjalla, V.F., Soares, M.C., Melo, R.C., and Roland, F. (2010). Virus-bacterium coupling driven by both turbidity and hydrodynamics in an Amazonian floodplain lake. Appl. Environ. Microbiol. *76*, 7194–7201. https://doi.org/10.1128/AEM.01161-10.

Baudoux, A.-C., Noordeloos, A.A.M., Veldhuis, M.J.W., and Brussaard, C.P.D. (2006). Virally induced mortality of *Phaeocystsis globosa* during two spring blooms in temperate coastal waters. Aqua. Microb. Ecol. *44*, 207–217.

Bergh, O., Børsheim, K.Y., Bratbak, G., and Heldal, M. (1989). High abundance of viruses found in aquatic environments. Nature *340*, 467–468. https://doi.org/10.1038/340467a0.

Bettarel, Y., Bouvier, T., Bouvier, C., Carré, C., Desnues, A., Domaizon, I., Jacquet, S., Robin, A., and Sime-Ngando, T. (2011). Ecological traits of planktonic viruses and prokaryotes along a full-salinity gradient. FEMS Microbiol. Ecol. *76*, 360–372. https://doi.org/10.1111/j.1574-6941.2011.01054.x.

Bettarel, Y., Motegi, C., Weinbauer, M.G., and Mari, X. (2016). Colonization and release processes of viruses and prokaryotes on artificial marine macroaggregates. FEMS Microbiol. Lett. *363*, fnv216. https://doi.org/10.1093/femsle/fnv216.

Bettarel, Y., Sime-Ngando, T., Bouvy, M., Arfi, R., and Amblard, C. (2005). Low consumption of virus-sized particles by heterotrophic nanoflagellates in two lakes of the french massif central. Aquat. Micro. Ecol. *39*, 205–209.

Biller, S., Schubotz, F., Roggensack, S.E., Thompson, A.W., Summons, R.E., and Chisholm, S.W. (2014). Bacterial vesicles in marine ecosystems. Science *343*, 182–186.

Biers, E.J., Wang, K., Pennington, C., Belas, R., Chen, F., and Moran, M.A. (2008). Occurrence and expression of gene transfer agent genes in marine bacterioplankton. Appl. Environ. Microbiol. *74*, 2933–2939. https://doi.org/10.1128/AEM.02129-07.

Bize, A., Karlsson, E.A., Ekefjärd, K., Quax, T.E., Pina, M., Prevost, M.C., Forterre, P., Tenaillon, O., Bernander, R., and Prangishvili, D. (2009). A unique virus release mechanism in the Archaea. Proc. Natl. Acad. Sci. U.S.A. *106*, 11306–11311. https://doi.org/10.1073/pnas.0901238106.

Bondy-Denomy, J., Pawluk, A., Maxwell, K.L., and Davidson, A.R. (2013). Bacteriophage genes that inactivate the CRISPR/Cas bacterial immune system. Nature *493*, 429–432. https://doi.org/10.1038/nature11723.

Bonilla-Findji, O., Rochelle-Newall, E., Weinbauer, M.G., Pizay, M.-D., Kerros, M.-E., Gattuso, J.-P. (2009). Effect of seawater-freshwater cross-transplantations on viral dynamics and bacterial diversity and production. Aquat. Microb. Ecol. *54*, 1–11.

Borrel, G., Colombet, J., Robin, A., Lehours, A.C., Prangishvili, D., and Sime-Ngando, T. (2012). Unexpected and novel putative viruses in the sediments of a deep-dark permanently anoxic freshwater habitat. ISME J. *6*, 2119–2127. https://doi.org/10.1038/ismej.2012.49.

Bouvier, T., and del Giorgio, P.A. (2007). Key role of selective viral-induced mortality in determining marine bacterial community composition, Environ. Microbiol. *9*, 287–297.

Bratbak, G., Heldal, M., Thingstad, T.F., Riemann, B., and Haslund, O.H. (1992). Incorporation of viruses into the budget of microbial c-transfer. A first approach. Mar. Ecol. Prog. Ser. *83*, 273–280.

Bratbak, G., Egge, J.K., and Heldal, M. (1993). Viral mortality of the marine alga *Emiliania huxleyi* Haptophyceae. and termination of algal blooms. Mar. Ecol. Prog. Ser. *93*, 39–48.

Bratbak, G., Heldal, M., Norland, S., and Thingstad, T.F. (1990). Viruses as partners in spring bloom microbial trophodynamics. Appl. Environ. Microbiol. *56*, 1400–1405.

Bratbak, G., and Heldal, M. (1995). Viruses – The new players in the game, their ecological role and could they mediate genetic exchange by transduction? In Molecular Ecology of Aquatic Microbes, Joint, I. ed., (Springer-Verlag, Berlin, Heidelberg), pp. 249–264.

Bratbak, G., and Heldal, M. (2000). Viruses rule the waves ° the smallest and most abundant members of marine ecosystems. Microbiol. Today *20*, 171–173.

Breitbart, M., and Rohwer, F. (2005). Here a virus, there a virus, everywhere the same virus? Trends Microbiol. *13*, 278–284.

Brussaard, C.P.D., Kempers, R.S., Kop, A.J., Riegman, R., and Heldal, M. (1996). Virus-like particles in a summer bloom of *Emiliania huxleyi* in the North Sea. Aquat. Microb. Ecol. *10*, 105–113.

Brussaard, C.P.D. (2004). Viral control of phytoplankton populations – a review. J. Eukaryot. Microbiol. *51*, 125–138.

Brussaard, C.P.D., Noordeloos, A.A.M., Witte, H., Collenteur, M.C.J., Schulz, K., Ludwig, A., and Riebesell, U. (2013). Arctic microbial community dynamics influenced by elevated CO_2 levels. Biogeosciences *10*, 719–731.

Cai, D.-L., Tan, F.C., and Edmond, J.M. (1988). Sources and transport of particulate organic carbon in the Amazon river and estuary. Estuar. Coast. Shelf Sci. *26*, 1–14.

Cassman, N., Prieto-Davó, A., Walsh, K., Silva, G.G., Angly, F., Akhter, S., Barott, K., Busch, J., McDole, T., Haggerty, J.M., *et al.* (2012). Oxygen minimum zones harbour novel viral communities with low diversity. Environ. Microbiol. *14*, 3043–3065. https://doi.org/10.1111/j.1462-2920.2012.02891.x.

Cattaneo, R., Rouviere, C., Rassoulzadegan, F., and Weinbauer, M.G. (2010). Association of viruses and bacteria with reference black carbon particles in marine coastal waters: A scanning electron, epifluorescence and confocal laser scanning microscopy analysis. FEMS Microbiol. Ecol. *74*, 392–396.

Chiura, H.X. (1997). Generalized gene transfer by virus-like particles from marine bacteria. Aquat. Microb. Ecol. *13*, 75–83.

Chiura, H.X., Kogure, K., Hagemann, S., Ellinger, A., and Velimirov, B. (2011). Evidence for particle-induced horizontal gene transfer and serial transduction between bacteria. FEMS Microbiol. Ecol. *76*, 576–591. https://doi.org/10.1111/j.1574-6941.2011.01077.x.

Chow, C.-E.T., and Fuhrman, J.A. (2012). Seasonality and monthly dynamics of marine myovirus communities. Environ. Microbiol. *14*, 2171–2183. https://doi.org/10.1111/j.1462-2920.2012.02744.x.

Chow, C.-E.T., Kim, D.Y., Sachdeva, R., Caron, D.A., and Fuhrman, J.A. (2013). Top-down controls on bacterial community structure: microbial network analysis of bacteria, T4-like viruses and protists. ISME J. *8*, 816–829.

Cissoko, M., Desnues, A., Bouvy, M., Sime-Ngando, T., Verling, E., and Bettarel, Y. (2008). Effects of freshwater and seawater mixing on virio- and bacterioplankton in a tropical estuary. Freshwater Biol. *53*, 1154–1162.

Clarke, D.K., Duarte, E.A., Elena, S.F., Moya, A., Domingo, E., and Holland, J. (1994). The red queen reigns in the kingdom of RNA viruses. Proc. Natl. Acad. Sci. U.S.A. *91*, 4821–4824.

Clasen, J.L., Bridgen, J.S., Payet, J.P., and Suttle, C. (2008). Evidence that viral abundance across oceans and lakes is driven by different biological factors. Freshwater Biol. *53*, 1090-1100.

Clasen, J.L., Hanson, C.A., Ibrahim, Y., Weihe, C., Marston, M.F., and Martiny, J.B.H. (2013). Diversity and temporal dynamics of Southern California coastal marine cyanophage isolates. Aquat. Microb. Ecol. *69*,17–31.

Clokie, M.R., Millard, A.D., Letarov, A.V., and Heaphy, S. (2011). Phages in nature. Bacteriophage *1*, 31–45. https://doi.org/10.4161/bact.1.1.14942.

Danovaro, R., Dell'Anno, A., Corinaldesi, C., Magagnini, M., Noble, R., Tamburini, C., and Weinbauer, M. (2008). Major viral impact on the functioning of benthic deep-sea ecosystems. Nature *454*, 1084–1087. https://doi.org/10.1038/nature07268.

Cram, J.A., Parada, A.E., and Fuhrman, J.A. (2016). Dilution reveals how viral lysis and grazing shape microbial communities. Limnol. Oceanogr. *61*, 889–905. https://doi.org/10.1002/lno.10259.

Danovaro, R., Company, J.B., Corinaldesi, C., D'Onghia, G., Galil, B., Gambi, C., Gooday, A.J., Lampadariou, N., Luna, G.M., Morigi, C., *et al.* (2010). Deep-sea biodiversity in the Mediterranean Sea: the known, the unknown, and the unknowable. PLOS ONE *5*, e11832. https://doi.org/10.1371/journal.pone.0011832.

De Corte, D., Sintes, E., Winter, C., Yokokawa, T., Reinthaler, T., and Herndl, G.J. (2010). Links between viral and prokaryotic communities throughout the water column in the (sub) tropical Atlantic Ocean. ISME J. *4*, 1431-1442.

De Corte, D., Sintes, E., Yokokawa, T., Reinthaler, T., and Herndl, G.J. (2012). Links between viruses and prokaryotes throughout the water column along a North Atlantic latitudinal transect. ISME J. *6*, 1566–1577. https://doi.org/10.1038/ismej.2011.214.

del Giorgio, P.A., and Duarte, C.M. (2002). Respiration in the open ocean. Nature *420*, 379–384. https://doi.org/10.1038/nature01165.

del Giorgio, P.A., and Cole, J.J. (1998). Bacterial growth efficiency in natural aquatic systems. Ann. Rev. Ecol. Syst. *29*, 503–541.

Desnues, C., Rodriguez-Brito, B., Rayhawk, S., Kelley, S., Tran, T., Haynes, M., Liu, H., Furlan, M., Wegley, L., Chau, B., *et al.* (2008). Biodiversity and biogeography of phages in modern stromatolites and thrombolites. Nature *452*, 340–343. https://doi.org/10.1038/nature06735.

Deveau, H., Barrangou, R., Garneau, J.E., Labonté, J., Fremaux, C., Boyaval, P., Romero, D.A., Horvath, P., and Moineau, S. (2008). Phage response to CRISPR-encoded resistance in Streptococcus thermophilus. J. Bacteriol. *190*, 1390–1400.

Dittmar, T., and Paeng, J. (2009). A heat-induced molecular signature in marine dissolved organic matter. Nat. Geosci. 2, 175–179.

Duffy, J.E. (2009). Why biodiversity is important to the functioning of real-world ecosystems. Front. Ecol. Environ. 7, 437–444.

Dugdale, R.C., and Goering, J.J. (1967). Uptake of new and regenerated forms of nitrogen in primary productivity. Limnol.Oceanogr., *12*, 196–206.

Dunning Hotopp, J.C., Clark, M.E., Oliveira, D.C., Foster, J.M., Fischer, P., Muñoz Torres, M.C., Giebel, J.D., Kumar, N., Ishmael, N., Wang, S., *et al.* (2007). Widespread lateral gene transfer from intracellular bacteria to multicellular eukaryotes. Science *317*, 1753–1756.

Emerson, J.B., Thomas, B.C., Andrade, K., Allen, E.E., Heidelberg, K.B., and Banfield, J.F. (2012). Dynamic viral populations in hypersaline systems as revealed by metagenomic assembly. Appl. Environ. Microbiol. *78*, 6309–6320. https://doi.org/10.1128/AEM.01212-12.

Engelhardt, T., Sahlberg, M., Cypionka, H., and Engelen, B. (2013). Biogeography of *Rhizobium radiobacter* and distribution of associated temperate phages in deep subseafloor sediments. ISME J. *7*, 199–209. https://doi.org/10.1038/ismej.2012.92.

Evans, C., and Wilson, W.H. 2008. Preferential grazing of *Oxyrrhis marina* on virus infected *Emiliania huxleyi*. Limnol.Oceanogr. *53*, 2035–2040.

Feng, L., Wu, Z., and Yu, X. (2013). Quorum sensing in water and wastewater treatment biofilms. J. Environ. Biol. *34*, 437–444.

Filée, J. (2009). Lateral gene transfer, lineage-specific gene expansion and the evolution of Nucleo Cytoplasmic Large DNA viruses. J. Invertebr. Pathol. *101*, 169–171. https://doi.org/10.1016/j.jip.2009.03.010.

Filée, J., Pouget, N., and Chandler, M. (2008). Phylogenetic evidence for extensive lateral acquisition of cellular genes by Nucleocytoplasmic large DNA viruses. BMC Evol. Biol. *8*, 320. https://doi.org/10.1186/1471-2148-8-320.

Forterre, P., Soler, N., Krupovic, M., Marguet, E., and Ackermann, H.W. (2013). Fake virus particles generated by fluorescence microscopy. Trends Microbiol. *21*, 1–5. https://doi.org/10.1016/j.tim.2012.10.005.

Frada, M., Probert, I., Allen, M.J., Wilson, W.H., and de Vargas, C. (2008). The 'Cheshire Cat' escape strategy of the coccolithophore *Emiliania huxleyi* in response to viral infection. Proc. Natl. Acad. Sci. U.S.A. *105*, 15944–15949. https://doi.org/10.1073/pnas.0807707105.

Frost, L.S., Leplae, R., Summers, A.O., and Toussaint, A. (2005). Mobile genetic elements: the agents of open source evolution. Nat. Rev. Microbiol. *3*, 722–732.

Fuhrman, J.A., and Noble, R.T. (1995). Viruses and protists cause similar bacterial mortality in coastal seawater. Limnol. Oceanogr. *40*, 1236–1242.

Fuhrman, J.A. (1999). Marine viruses and their biogeochemical and ecological effects. Nature *399*, 541–548. https://doi.org/10.1038/21119.

Fuhrman, J.A. (2009). Microbial community structure and its functional implications. Nature *459*, 193–199. https://doi.org/10.1038/nature08058.

Fuqua, C., Parsek, M.R., and Greenberg, E.P. (2001). Regulation of gene expression by cell-to-cell communication: acylhomoserine lactone quorum sensing. Annu. Rev. Genet. *35*, 439–468.

Gastrich, M.D., Leigh-Bell, J.A., Gobler, C.J., Anderson, O.R., Wilhelm, S.W., and Bryan, M. (2004). Viruses as potential regulators of regional brown tide blooms caused by the alga, *Aureococcus anophagefferens*. Estuaries *27*, 112–119.

Gaudin, M., Krupovic, M., Marguet, E., Gauliard, E., Cvirkaite-Krupovic, V., Le Cam, E., Oberto, J., and Forterre, P. (2014). Extracellular membrane vesicles harbouring viral genomes. Environ. Microbiol. *16*, 1167–1175. https://doi.org/10.1111/1462-2920.12235.

Ghosh, D., Roy, K., Williamson, K.E., Srinivasiah, S., Wommack, K.E., and Radosevich, M. (2009). Acyl-homoserine lactones can induce virus production in lysogenic bacteria: an alternative paradigm for prophage induction. Appl. Environ. Microbiol. *75*, 7142–7152. https://doi.org/10.1128/AEM.00950-09.

Giovannoni, S.J., and Vergin, K.L. (2012). Seasonality in ocean microbial communities. Science *335*, 671–676. https://doi.org/10.1126/science.1198078.

Griffin, D., Garrison, V., Herman, J., and Shinn, E. (2001). African desert dust in the Caribbean atmosphere: microbiology and public health. Aerobiologia *17*, 203–213.

González, J.M., and Suttle, C.A. (1993). Grazing by marine nanoflagellates on viruses and viral-sized particles: ingestion and digestion. Mar. Ecol. Prog. Ser. *94*, 1–10.

Haerter, J.O., Mitarai, N., and Sneppen, K. (2014). Phage and bacteria support mutual diversity in a narrowing staircase of coexistence. ISME J. *8*, 2317–2326. https://doi.org/10.1038/ismej.2014.80.

Hartmann, A., and Schikora, A. (2012). Quorum sensing of bacteria and trans-kingdom interactions of N-acyl homoserine lactones with eukaryotes. J. Chem. Ecol. *38*, 704–713. https://doi.org/10.1007/s10886-012-0141-7.

Haaber, J., and Middelboe, M. (2009). Viral lysis of Phaeocystis pouchetii: implications for algal population dynamics and heterotrophic C, N and P cycling. ISME J. *3*, 430–441. https://doi.org/10.1038/ismej.2008.125.

Hargreaves, K.R., Kropinski, A.M., and Clokie, M.R.J. (2014). What does the talking? Quorum sensing signalling genes discovered in a bacteriophage genome. PLOS ONE *9*, e85131. 71.

Hedges, J.I. (2002). Why dissolved organics matter. In Biogeochemsitry of Marine Dissolved Organic Matter, Hansell, D.A., and Carlson, C.A. eds. (Academic Press, San Diego), pp. 1–33.

Hewson, I., and Fuhrman, J.A. (2006). Viral impacts upon marine bacterioplankton assemblage structure. J. Mar. Biol. Assoc. U.K. *86*, 577–589.

Holmfeldt, K., Titelman, J., and Riemann, L. (2010). Virus production and lysate recycling in different sub-basins of the northern Baltic Sea. Microb. Ecol. *60*, 572–580. https://doi.org/10.1007/s00248-010-9668-8.

Høyland-Kroghsbo, N.M., Maerkedahl, R.B., and Svenningsen, S.L. (2013). A quorum-sensing-induced bacteriophage defense mechanism. MBio *4*, e00362–12. https://doi.org/10.1128/mBio.00362-12.

Hughes, D.T., and Sperandio, V. (2008). Inter-kingdom signalling: communication between bacteria and their hosts. Nat. Rev. Microbiol. *6*, 111–120. https://doi.org/10.1038/nrmicro1836.

Hyman, P., and Abedon, S.T. (2010). Bacteriophage host range and bacterial resistance. Adv. Appl. Microbiol. *70*, 217–248. https://doi.org/10.1016/S0065-2164(10)70007-1.

Hurwitz, B.L., Hallam, S.J., and Sullivan, M.B. (2013). Metabolic reprogramming by viruses in the sunlit and dark ocean. Genome Biol. *14*, R123. https://doi.org/10.1186/gb-2013-14-11-r123.

Jacquet, S., Domaizon, I., Personnic, S., and Sime-Ngando, T. 2007. Do small grazers influence viral-induced bacterial mortality in Lake Bourget? Fund. Appl. Limnol. 170, *125-132.*

Jacquet, S., Heldal, M., Iglesias-Rodriguez, D., Larsen, A., Wilson, W., and Bratbak, G. (2002). Flow cytometric analysis of an *Emiliana huxleyi* bloom terminated by viral infection. Aquat. Microb. Ecol. *27,* 111-124.

Jacquet, S., Miki, T., Noble, R., Peduzzi, P., and Wilhelm, S. (2010). Viruses in aquatic ecosystems: important advancements of the last 20 years and prospects for the future in the field of microbial oceanography and limnology. Adv. Ocean Limnol. *1,* 97-141.

Jover, L.F., Cortez, M.H., and Weitz, J.S. (2013). Mechanisms of multi-strain coexistence in host-phage systems with nested infection networks. J. Theor. Biol. *332,* 65–77. https://doi.org/10.1016/j.jtbi.2013.04.011.

Jiao, N., Herndl, G.J., Hansell, D.A., Benner, R., Kattner, G., Wilhelm, S.W., Kirchman, D.L., Weinbauer, M.G., Luo, T., Chen, F., *et al.* (2010). Microbial production of recalcitrant dissolved organic matter: long-term carbon storage in the global ocean. Nat. Rev. Microbiol. *8,* 593–599. https://doi.org/10.1038/nrmicro2386.

Keeling, P.J., and Palmer, J.D. (2008). Horizontal gene transfer in eukaryotic evolution. Nat. Rev. Genet. *9,* 605–618. https://doi.org/10.1038/nrg2386.

Kimura, S., Yoshida, T., Hosoda, N., Honda, T., Kuno, S., Kamiji, R., Hashimoto, R., and Sako, Y. (2012). Diurnal infection patterns and impact of Microcystis cyanophages in a Japanese pond. Appl. Environ. Microbiol. *78,* 5805–5811. https://doi.org/10.1128/AEM.00571-12.

Kim, J.S., Kim, Y.H., Seo, Y.W., and Park, S. (2007). Quorum sensing inhibitors from the red alga, *Ahnfeltiopsis flabelliformis.* Biotechnol. Bioprocess Eng. *12,* 308-311.

Kernegger, L., Zweimüller, I., and Peduzzi, P. (2009). Effects of suspended matter quality and virus abundance on microbial parameters: experimental evidence from a large European river. Aquat. Microb. Ecol. *57,* 161–173. https://doi.org/10.3354/ame01341.

Kiørboe, T., and Jackson, G. (2001). Marine snow, organic solute plumes, and optimal chemosensory behavior of bacteria. Limnol. Oceanogr. 43,1309–1318.

Kirchman, D.L. (2013). Microbial oceanography: Killers of the winners. Nature *494,* 320–321. https://doi.org/10.1038/nature11951.

Kleerebezem, M., Quadri, L.E., Kuipers, O.P., and de Vos, W.M. (1997). Quorum sensing by peptide pheromones and two-component signal-transduction systems in Gram-positive bacteria. Mol. Microbiol. *24,* 895–904.

Labrie, S.J., Samson, J.E., and Moineau, S. (2010). Bacteriophage resistance mechanisms. Nat. Rev. Microbiol. *8,* 317–327. https://doi.org/10.1038/nrmicro2315.

Lang, A.S., and Beatty, J.T. (2007). Importance of widespread gene transfer agent genes in alpha-proteobacteria. Trends Microbiol. *15,* 54–62.

Lang, A.S., Zhaxybayeva, O., and Beatty, J.T. (2012). Gene transfer agents: phage-like elements of genetic exchange. Nat. Rev. Microbiol. *10,* 472–482. https://doi.org/10.1038/nrmicro2802.

Lara, E., Arrieta, J.M., Garcia-Zarandona, I., Boras, J.A., Duarte, C.M., Agusti, S., Wassmann, P.F., and Vaqué, D. (2013). Experimental evaluation of the warming effect on viral, bacterial and protistan communities in two contrasting Arctic systems. Aquat. Microb. Ecol. *70,* 17-32.

Larsen, J.B., Larsen, A., Thyrrhaug, R., Bratbak, G., and Sandaa, R.-A. (2008). Response of marine viral populations to a nutrient induced phytoplankton bloom at different pCO_2 levels. Biogeosciences *5,* 523-533.

Lennon, J.T., Khatana, S.A., Marston, M.F., and Martiny, J.B. (2007). Is there a cost of virus resistance in marine cyanobacteria? ISME J. *1,* 300–312.

Lennon, J.T., and Martiny, J.B.H. (2008). Rapid evolution buffers ecosystem impacts of viruses in a microbial food web. Ecol. Lett. *11,* 1178–1188. https://doi.org/10.1111/j.1461-0248.2008.01225.x.

Liu, J., Weinbauer, M.G., Maier, C., Dai, M., and Gattuso, J.-P. (2010). Effect of ocean acidification on microbial diversity and on microbe-driven biogeochemistry and ecosystem functioning. Aquat. Microb. Ecol. *61,* 291–305.

Luef, B., Aspetsberger, F., Hein, T., Huber, F., and Peduzzi, P. (2007). Impact of hydrology on free-living and particle-associated microorganisms in a river floodplain system (Danube, Austria). Fresh. Biol. *52,* 1043–1057.

Luef, B., Neu, T.R., and Peduzzi, P. (2009). Imaging and quantifying virus fluorescence signals on aquatic aggregates: a new method and its implication for aquatic microbial ecology. FEMS Microbiol. Ecol. *68,* 372–380. https://doi.org/10.1111/j.1574-6941.2009.00675.x.

Ma, L., Sun, R., Mao, G., Yu, H., and Wang, Y. (2013). Seasonal and spatial variability of virioplankton abundance in Haihe River, China. BioMed. Research International *2013,* 1–10.

Malits, A., Cattaneo, R., Sintes, E., Gasol, J.P., Herndl, J.G., and Weinbauer, M.G. (2015). Black carbon deposition reduces viral production and stimulates prokaryotic activity in surface marine waters: An experimental study with reference black carbon material. Aquat. Microb. Ecol. 75,27–42.

Malmstrom, R.R., Kiene, R.P., Cottrell, M.T., and Kirchman, D.L. (2004). Contribution of SAR11 bacteria to dissolved dimethylsulfoniopropionate and amino acid uptake in the North Atlantic ocean. Appl. Environ. Microbiol. *70,* 4129–4135. https://doi.org/10.1128/AEM.70.7.4129-4135.2004.

Manage, P.M., Kawabata, Z., and Nakanao, S. (2000). Algicidal effect of the bacterium *Alcaligenes denitrificans* on *Microcystis* spp. Aquat. Microb. Ecol. 22,111–117.

Mann, N.H., Cook, A., Millard, A., Bailey, S., and Clokie, M. (2003). Marine ecosystems: bacterial photosynthesis genes in a virus. Nature *424,* 741. https://doi.org/10.1038/424741a.

Marston, M.F., and Sallee, J.L. (2003). Genetic diversity and temporal variation in the cyanophage community infecting marine *Synechococcus* species in Rhode Island's coastal waters. Appl. Environ. Microbiol. *69,* 4639–4647.

Marston, M.F., and Amrich, C.G. (2009). Recombination and microdiversity in coastal marine cyanophages. Environ. Microbiol. *11,* 2893–2903. https://doi.org/10.1111/j.1462-2920.2009.02037.x.

Marston, M.F., Taylor, S., Sme, N., Parsons, R.J., Noyes, T.J., and Martiny, J.B. (2013). Marine cyanophages exhibit local and regional biogeography. Environ. Microbiol. *15,* 1452–1463. https://doi.org/10.1111/1462-2920.12062.

Maurice, C.F., Mouillot, D., Bettarel, Y., De Wit, R., Sarmento, H., and Bouvier, T. (2011). Disentangling the relative influence of bacterioplankton phylogeny and metabolism on lysogeny in reservoirs and lagoons. ISME J. *5*, 831–842. https://doi.org/10.1038/ismej.2010.181.

Meunier, A., and Jacquet, S. 2015. Do phages impact microbial dynamics, prokaryotic community structure and nutrient dynamics in Lake Bourget? Biol. Open *4*, 1528–1537. https://doi.org/10.1242/bio.013003.

Miki, T., and Jacquet, S. 2008. Complex interactions in the aquatic microbial world: under-explored relationships between viruses, bacteria and heterotrophic flagellates. Aquat. Microb. Ecol. 51,195–208.

Miller, S.D., Haddock, S.H., Elvidge, C.D., and Lee, T.F. (2005). Detection of a bioluminescent milky sea from space. Proc. Natl. Acad. Sci. U.S.A. *102*, 14181–14184.

Morris, R.M., Rappé, M.S., Connon, S.A., Vergin, K.L., Siebold, W.A., Carlson, C.A., and Giovannoni, S.J. (2002). SAR11 clade dominates ocean surface bacterioplankton communities. Nature *420*, 806–810.

Motegi, C., Kaiser, K., Benner, R., and Weinbauer, M.G. (2015). Effect of P-limitation on prokaryotic and viral production in surface waters of the northwestern mediterranean sea. J. Plankton Res. *37*,16–20.

Motegi, C., Nagata, T., Miki, T., Weinbauer, M.G., Legendre, L., and Rassoulzadegan, F. (2009a). Viral control of bacterial growth efficiency in marine pelagic environments. Limnol. Oceanogr. *54*,1901–1910.

Motegi, C., and Nagata, T. (2009). Addition of monomeric and polymeric organic substrates alleviates viral lytic pressure on bacterial communities in coastal seawaters. Aqua. Microb. Ecol. *57*, 343–350.

Motegi, C., Nagata, T., Miki, T., Weinbauer, M.G., Legendre, L., and Rassoulzadegan, F. (2013). Interactive effects of viral and bacterial production on marine bacterial diversity. PLOS ONE *8*, e76800. https://doi.org/10.1371/journal.pone.0076800.

Mühling, M., Fuller, N.J., Millard, A., Somerfield, P.J., Marie, D., Wilson, W.H., Scanlan, D.J., Post, A.F., Joint, I., and Mann, N.H. (2005). Genetic diversity of marine *Synechococcus* and co-occurring cyanophage communities: evidence for viral control of phytoplankton. Environ. Microbiol. *7*, 499–508.

Nagasaki, K., Tomaru, Y., Katanozaka, N., Shirai, Y., Nishida, K., Itakura, S., and Yamaguchi, M. (2004). Isolation and characterization of a novel single-stranded RNA virus infecting the bloom-forming diatom *Rhizosolenia setigera*. Appl. Environ. Microbial *70*, 704–711.

Nealson, K.H., and Hastings, J.W. (2006). Quorum sensing on a global scale: massive numbers of bioluminescent bacteria make milky seas. Appl. Environ. Microbiol. 72, 2295–2297.

Needham, D.M., Chow, C.E., Cram, J.A., Sachdeva, R., Parada, A., and Fuhrman, J.A. (2013). Short-term observations of marine bacterial and viral communities: patterns, connections and resilience. ISME J. *7*, 1274–1285. https://doi.org/10.1038/ismej.2013.19.

Nickerson, K.W., Atkin, A.L., and Hornby, J.M. (2006). Quorum sensing in dimorphic fungi: farnesol and beyond. Appl. Environ. Microbiol. *72*, 3805–3813.

Noble, R.T., and Fuhrman, J.A. (1997). Virus decay and its causes in coastal waters. Appl. Environ. Microbiol. *63*, 77–83.

Pacheco, A.R., and Sperandio, V. (2009). Inter-kingdom signaling: chemical language between bacteria and host. Curr. Opin. Microbiol. *12*, 192–198. https://doi.org/10.1016/j.mib.2009.01.006.

Palesse, S., Colombet, J., Pradeep Ram, A.S., and Sime-Ngando, T. (2014). Linking host prokaryotic physiology to viral lifestyle dynamics in a temperate freshwater lake (Lake Pavin, France). Microb. Ecol. *68*, 740–750. https://doi.org/10.1007/s00248-014-0441-2.

Pagarete, A., Chow, C.-E.T., Johannessen, T., Fuhrman, J.A., Thingstad, T.F., and Sandaa, R.A. (2013). Strong seasonality and inter-annual recurrence in marine myovirus communities. Appl. Environ. Microbiol. *79*, 6253–6259.

Paul, J.H. (2008). Prophages in marine bacteria: dangerous molecular time bombs or the key to survival in the seas? ISME J. *2*, 579–589. https://doi.org/10.1038/ismej.2008.35.

Peduzzi, P., and Weinbauer, M.G. (1993). Effect of concentrating the virus-rich 2-200 nm size fraction of seawater on the formation of algal flocs (marine snow). Limnol. Oceanogr. *38*,1562–1565.

Peduzzi, P., and Luef, B. (2008). Viruses, bacteria and suspended particles in a backwater and main channel site of the Danube (Austria). Aquat. Sci. *70*, 186–194. https://doi.org/10.1007/s00027-008-8068-3.

Peduzzi, P., Gruber, M., and Schagerl, M. 2014. The virus's tooth – cyanophages affect an African flamingo population in a bottom up cascade. ISME J. *8*, 1346–1351.

Peduzzi, P. (2015). Virus ecology of fluvial systems: a blank spot on the map? Biol. Rev. Camb. Philos. Soc. *91*, 937–949. https://doi.org/10.1111/brv.12202.

Peduzzi, P. (2016). Aquatic viruses and global climate change. In: Climate Change and Microbial Ecology: Current Research and Future Trends, Marxsen, J., ed. (Caister Academic Press, UK), pp. 73–81.

Pernthaler, J. (2005). Predation on prokaryotes in the water column and its ecological implications. Nat. Rev. Microbiol. 3, 537–546.

Pollard, P.C., and Ducklow, H.W. (2011). Ultrahigh bacterial production in a eutrophic subtropical Australian river: Does viral lysis short-circuit the microbial loop. Limnol. Oceanogr. *56*, 1115–1129.

Pomeroy, L.R. (1974). The ocean's food web, a changing paradigm. BioScience *24*,499–504.

Pradeep Ram, A.S., Sabart, M., Latour, D., and Sime-Ngando, T. (2009).Low effect of viruses on bacteria in deep anoxic water and sediment of a productive freshwater reservoir (Lake Grangent, France). Aquat. Microbial Ecol. *55*, 255–265.

Ram, A.S., Palesse, S., Colombet, J., Sabart, M., Perriere, F., and Sime-Ngando, T. (2013). Variable viral and grazer control of prokaryotic growth efficiency in temperate freshwater lakes (French Massif Central). Microb. Ecol. *66*, 906–916. https://doi.org/10.1007/s00248-013-0289-x.

Pradeep Ram, A.S., Palesse, S., Colombet, J., Thouvenot, A., and Sime-Ngando, T. (2014). The relative importance of viral lysis and nanoflagellate grazing for prokaryote mortality in temperate lakes. Freshwater Biol. *59*, 300–311.

Proctor, L.M., and Fuhrman, J.A. (1990). Viral mortality of marine bacteria and cyanobacteria. Nature *343*, 60–62.

Quax, T.E., Lucas, S., Reimann, J., Pehau-Arnaudet, G., Prevost, M.C., Forterre, P., Albers, S.V., and Prangishvili, D. (2011). Simple and elegant design of a virion egress structure in Archaea. Proc. Natl. Acad. Sci. U.S.A. *108*, 3354–3359. https://doi.org/10.1073/pnas.1018052108.

Riemann, L., and Grossart, H.P. (2008). Elevated lytic phage production as a consequence of particle colonization by a marine *Flavobacterium* (*Cellulophaga* sp.). Microb. Ecol. *56*, 505–512. https://doi.org/10.1007/s00248-008-9369-8.

Rohwer, F., and Thurber, R.V. (2009). Viruses manipulate the marine environment. Nature *459*, 207–212. https://doi.org/10.1038/nature08060.

Rodriguez-Valera, F., Martin-Cuadrado, A.B., Rodriguez-Brito, B., Pasić, L., Thingstad, T.F., Rohwer, F., and Mira, A. (2009). Explaining microbial population genomics through phage predation. Nat. Rev. Microbiol. *7*, 828–836. https://doi.org/10.1038/nrmicro2235.

Rothschild, L.J., and Mancinelli, R.L. (2001). Life in extreme environments. Nature *409*, 1092–1101. https://doi.org/10.1038/35059215.

Samson, J.E., Magadán, A.H., Sabri, M., and Moineau, S. (2013). Revenge of the phages: defeating bacterial defences. Nat. Rev. Microbiol. *11*, 675–687. https://doi.org/10.1038/nrmicro3096.

Sanders, R.W., Caron, D.A., Berninger, U.-G. (1992). Relationship between bacteria and heterotrophic nanoplankton in marine and freshwaters: An inter-ecosystem comparison Mar. Ecol. Prog. Ser. *86*, 1–14.

Sandaa, R.A., and Larsen, A. (2006). Seasonal variations in virus-host populations in Norwegian coastal waters: focusing on the cyanophage community infecting marine *Synechococcus* spp. Appl. Environ. Microbiol. *72*, 4610–4618.

Sandaa, R.A., Clokie, M., and Mann, N.H. (2008). Photosynthetic genes in viral populations with a large genomic size range from Norwegian coastal waters. FEMS Microbiol. Ecol. *63*, 2–11.

Sandaa, R.A., Gómez-Consarnau, L., Pinhassi, J., Riemann, L., Malits, A., Weinbauer, M.G., Gasol, J.M., and Thingstad, T.F. (2009). Viral control of bacterial biodiversity – evidence from a nutrient-enriched marine mesocosm experiment. Environ. Microbiol. *11*, 2585–2597. https://doi.org/10.1111/j.1462-2920.2009.01983.x.

Sapranauskas, R., Gasiunas, G., Fremaux, C., Barrangou, R., Horvath, P., and Siksnys, V. (2011). The *Streptococcus thermophilus* CRISPR/Cas system provides immunity in *Escherichia coli*. Nucleic Acids Res. *39*, 9275–9282. https://doi.org/10.1093/nar/gkr606.

Seed, K.D., Lazinski, D.W., Calderwood, S.B., and Camilli, A. (2013). A bacteriophage encodes its own CRISPR/Cas adaptive response to evade host innate immunity. Nature *494*, 489–491. https://doi.org/10.1038/nature11927.

Semenova, E., Jore, M.M., Datsenko, K.A., Semenova, A., Westra, E.R., Wanner, B., van der Oost, J., Brouns, S.J., and Severinov, K. (2011). Interference by clustered regularly interspaced short palindromic repeat (CRISPR) RNA is governed by a seed sequence. Proc. Natl. Acad. Sci. U.S.A. *108*, 10098–10103. https://doi.org/10.1073/pnas.1104144108.

Sheik, A.R., Brussaard, C.P., Lavik, G., Lam, P., Musat, N., Krupke, A., Littmann, S., Strous, M., and Kuypers, M.M. (2014). Responses of the coastal bacterial community to viral infection of the algae Phaeocystis globosa. ISME J. *8*, 212–225. https://doi.org/10.1038/ismej.2013.135.

Sharif, D.I., Gallon, J., Smith, C.J., and Dudley, E. (2008). Quorum sensing in Cyanobacteria: N-octanoyl-homoserine lactone release and response, by the epilithic colonial cyanobacterium Gloeothece PCC6909. ISME J. *2*, 1171–1182. https://doi.org/10.1038/ismej.2008.68.

Sharoni, S., Trainic, M., Schatz, D., Lehahn, Y., Flores, M.J., Bidle, K.D., Ben-Dor, S., Rudich, Y., Koren, I., and Vardi, A. (2015). Infection *of* phytoplankton by aerosolized marine viruses. Proc. Natl. Acad. Sci. U.S.A. *112*, 6643–6647. https://doi.org/10.1073/pnas.1423667112.

Shelford, E.J., Middelboe, M., Møller, E.F., and Suttle, C.A. (2012). Virus-driven nitrogen cycling enhances phytoplankton growth. Aquat. Microb. Ecol. *66*, 41–46.

Shelford, E.J., Jørgensen, N.O.G., Rasmussen, S., Suttle, C.A., and Middelboe, M. (2014). Dissecting the role of viruses in marine nutrient cycling: bacterial uptake of d- and l-amino acids released by viral lysis. Aquat. Microb. Ecol. *73*, 235243.

Sherr, E.B., Sherr, B.F., and Albright, L.J. (1987). Bacteria: link or sink. Science, *235*, 88a.

Šimek, K., Pernthaler, J., Weinbauer, M.G., Hornák, K., Dolan, J.R., Nedoma, J., Masín, M., and Amann, R. (2001). Changes in bacterial community composition, dynamics and viral mortality rates associated with enhanced flagellate grazing in a meso-eutrophic reservoir. Appl. Environ. Microbiol. *67*, 2723–2733.

Simon, M., Grossart, H.-P., Schweitzer, B., and Ploug, H. (2002). Microbial ecology of organic aggregates in aquatic ecosystems. Aquat. Microb. Ecol. *28*, 175–211.

Snyder, J.C., and Young, M.J. (2011). Advances in understanding archaea-virus interactions in controlled and natural environments. Curr. Opin. Microbiol. *14*, 497–503. https://doi.org/10.1016/j.mib.2011.07.007.

Sprague, G.F., and Winans, S.C. (2006). Eukaryotes learn how to count: quorum sensing by yeast. Genes Dev. *20*, 1045–1049.

Sullivan, M.B., Weitz, J.S., and Wilhelm, S. (2017). Viral ecology comes of age. Environ. Microbiol. Rep. *9*, 33–35. https://doi.org/10.1111/1758-2229.12504.

Sun, C.L., Barrangou, R., Thomas, B.C., Horvath, P., Fremaux, C., and Banfield, J.F. (2013). Phage mutations in response to CRISPR diversification in a bacterial population. Environ. Microbiol. *15*, 463–470. https://doi.org/10.1111/j.1462-2920.2012.02879.x.

Suttle, C.A. (1994). The significance of viruses to mortality in aquatic microbial communities. Microb. Ecol. *28*, 237–243. https://doi.org/10.1007/BF00166813.

Suttle, C.A. (2000). Ecological, evolutionary, and geochemical consequences of viral infection of cyanobacteria and eukaryotic algae. In Viral Ecology, Hurst, C.J. ed. pp. 247–296.

Suttle, C.A., Chan, A.M., and Cottrell, M.T. (1990). Infection of phytoplankton by viruses and reduction of primary productivity. Nature *347*, 467–469.

Suttle, C.A., and Chen, F. (1992). Mechanisms and rates of decay of marine viruses in seawater. Appl. Environ. Microbiol. *58*, 3721–3729.

Suttle, C.A., and Chan, A.M. (1993). Marine cyanophages infecting oceanic and coastal strains of Synechococcus: abundance, morphology, cross-infectivity and growth characteristics. Mar. Ecol. Prog. Ser. *92*, 99–109.

Suttle, C.A. (2005). Viruses in the sea. Nature *437*, 356–361.

Suttle, C.A. (2007). Marine viruses – major players in the global ecosystem. Nat. Rev. Microbiol. *5*, 801–812.

Suttle, C.A. (2013). Viruses: unlocking the greatest biodiversity on Earth. Genome *56*, 542–544. https://doi.org/10.1139/gen-2013-0152.

Tarutani, K., Nagasaki, K., and Yamaguchi, M. (2000). Viral impacts on total abundance and clonal composition of the harmful bloom-forming phytoplankton Heterosigma akashiwo. Appl. Environ. Microbiol. *66*, 4916–4920.

Thingstad, T.F., and Lignell, R. (1997). Theoretical models for the control of bacterial growth rate, abundance, diversity and carbon demand. Aquat. Microb. Ecol. *13*, 19–27.

Thingstad, T.F. (2000). Elements of a theory for the mechanisms controlling abundance, diversity, and biogeochemical role of lytic bacterial viruses in aquatic systems. Limnol. Oceanogr. *45*, 1320–1328.

Thingstad, T.F., Våge, S., Storesund, J.E., Sandaa, R.A., and Giske, J. (2014). A theoretical analysis of how strain-specific viruses can control microbial species diversity. Proc. Natl. Acad. Sci. U.S.A. *111*, 7813–7818. https://doi.org/10.1073/pnas.1400909111.

Thingstad, T.F., and Bratbak, G. (2016). Microbial oceanography: Viral strategies at sea. Nature *531*, 454–455. https://doi.org/10.1038/nature17303.

Thomas, R., Jacquet, S., Grimsley, N., and Moreau, H. (2012). Strategies and mechanisms of resistance to viruses in photosynthetic aquatic microorganisms. Adv. Limnol. Oceanogr. *3*, 1–15.

Uitz, J., Stramski, D., Baudoux, A.C., Reynolds, R.A., Wright, V.M., Dubranna, J., and Azam, F. (2009). Variations in the optical properties of a particle suspension associated with viral infection of marine bacteria. Limnol. Oceanogr. *55*: 2317–2330.

Våge, S., Storesund, J.E., and Thingstad, T.F. (2013). SAR11 viruses and defensive host strains. Nature *499*, E3–4. https://doi.org/10.1038/nature12387.

Våge, S., Storesund, J.E., and Thingstad, T.F. (2013). Adding a cost of resistance description extends the ability of virus-host model to explain observed patterns in structure and function of pelagic microbial communities. Environ. Microbiol. *15*, 1842–1852. https://doi.org/10.1111/1462-2920.12077.

Velimirov, B., and Hagemann, S. (2011). Mobilizable bacterial DNA packaged into membrane vesicles induces serial transduction. Mob. Genet. Elements. *1*, 80–81. https://doi.org/10.4161/mge.1.1.15724.

Weinbauer, M.G., and Höfle, M.G. (1998). Distribution and life strategies of two bacterial populations in a eutrophic lake Appl. Environ. Microbiol. *64*, 3776–3783.

Weinbauer, M.G., Bonilla-Findji, O., Chan, A.M., Dolan, J.R., Short, S.M., Šimek, K., Wilhelm, S.W.and Suttle, C.A. (2011a). *Synechococcus* growth in the ocean may depend on the lysis of heterotrophic bacteria. J. Plankton Res. 33, 1465–1476.

Weinbauer, M.G., Chen, F., and Wilhelm, S.W. (2011b). Virus-mediated redistribution and partitioning of carbon in the global oceans. In Microbial Carbon Pump in the Ocean, Jiao, N., Azam, F., and Sanders, S. eds. (Science/AAAS, Washington), pp. 54–56.

Weinbauer, M.G., Mari, X., and Gattuso, J. (2011c). Effect of ocean acidification on the diversity and activity of heterotrophic marine microorganisms. In Ocean Acidification, Gattuso, J.-P., and Hansson, L. eds. (Oxford Press), pp. 83–98.

Weinbauer, M.G., Cattaneo, R., Malits, A., Motegi, C., Gasol, J.M., Herndl, G.J., Mari, X., Migon, C., and Rassoulzadegan, F. (2012). Black carbon and microorganisms in aquatic systems. In Advances in Environmental Research, Daniels, J.A., ed. (Novapublishers), vol. 25, pp. 1–37.

Weinbauer, M.G., Nedoma, J., Christaki, U., and Šimek, K. (2003). Comparing the effects of resource enrichment and grazing on viral production in a meso-eutrophic reservoir. Aquat. Microb. Ecol. *31*, 137–144.

Weinbauer, M.G., Hornák, K., Jezbera, J., Nedoma, J., Dolan, J.R., and Simek, K. (2007). Synergistic and antagonistic effects of viral lysis and protistan grazing on bacterial biomass, production and diversity. Environ. Microbiol. *9*, 777–788.

Weinbauer, M.G., and Rassoulzadegan, F. (2004). Are viruses driving microbial diversification and diversity? Environ. Microbiol. *6*, 1–11.

Weinbauer, M.G. (2004). Ecology of prokaryotic viruses. FEMS Microbiol. Rev. *28*, 127–181. https://doi.org/10.1016/j.femsre.2003.08.001.

Weinbauer, M.G., Bettarel, Y., Cattaneo, R., Luef, B., Maier, C., Motegi, C., Peduzzi, P., and Mari, X. (2009). Viral ecology of organic and inorganic particles in aquatic systems: avenues for further research. Aquat. Microb. Ecol. *57*, 321–341. https://doi.org/10.3354/ame01363.

Winter, C., and Weinbauer, M.G. (2010). Randomly amplified polymorphic DNA reveals tight links between viruses and microbes in the bathypelagic zone of the Northwestern Mediterranean Sea. Appl. Environ. Microbiol. *76*, 6724–6732. https://doi.org/10.1128/AEM.00531-10.

Weitz, J.S., Stock, C.A., Wilhelm, S.W., Bourouiba, L., Coleman, M.L., Buchan, A., Follows, M.J., Fuhrman, J.A., Jover, L.F., Lennon, J.T., *et al.* (2015). A multitrophic model to quantify the effects of marine viruses on microbial food webs and ecosystem processes. ISME J. 9, 1352–1364. https://doi.org/10.1038/ismej.2014.220.

Whitehead, N.A., Barnard, A.M., Slater, H., Simpson, N.J., and Salmond, G.P. (2001). Quorum-sensing in Gram-negative bacteria. FEMS Microbiol. Rev. *25*, 365–404.

Whon, T.W., Kim, M.S., Roh, S.W., Shin, N.R., Lee, H.W., and Bae, J.W. (2012). Metagenomic characterization of airborne viral DNA diversity in the near-surface atmosphere. J. Virol. *86*, 8221–8231. https://doi.org/10.1128/JVI.00293-12.

Wilhelm, S.W., and Suttle, C.A. (1999). Viruses and nutrient cycles in the sea. Biosci. *49*, 781–788.

Wilhelm, S.W., and Matteson, A.R. (2008). Freshwater and marine virioplankton: a brief overview of commonalities and differences. Freshwater Biol. 53, 1076–89.

Wilmes, P., and Bond, P.L. (2009). Microbial community proteomics: elucidating the catalysts and metabolic mechanisms that drive the Earth's biogeochemical cycles. Curr. Opin. Microbiol. *12*, 310–317. https://doi.org/10.1016/j.mib.2009.03.004.

Wilson, W.H., Van Etten, J.L., and Allen, M.J. (2009). The Phycodnaviridae: the story of how tiny giants rule the world. Curr. Top. Microbiol. Immunol. *328*, 1–42.

Winter, C., Bouvier, T., Weinbauer, M.G., and Thingstad, T.F. (2010). Trade-offs between competition and defense specialists among unicellular planktonic organisms: the 'killing the winner' hypothesis revisited. Microbiol. Mol. Biol. Rev. *74*, 42–57. https://doi.org/10.1128/MMBR.00034-09.

Winter, C., Smit, A., Herndl, G.J., and Weinbauer, M.G. (2004). Impact of virioplankton on archaeal and bacterial community richness as assessed in seawater batch cultures. Appl. Environ. Microbiol. *70*, 804–813.

Winter, C., Herndl, G.J., and Weinbauer, M.G. (2004). Diel cycles in viral infection of bacterioplankton in the North Sea. Aqua. Microb. Ecol. *35*, 207–216.

Wommack, K.E., and Colwell, R.R. (2000). Virioplankton: viruses in aquatic ecosystems. Microbiol. Mol. Biol. Rev. *64*, 69–114.

Wood, H.E., Dawson, M.T., Devine, K.M., and McConnell, D.J. (1990). Characterization of PBSX, a defective prophage of *Bacillus subtilis*. J. Bacteriol. *172*, 2667–2674.

Zhao, Y., Temperton, B., Thrash, J.C., Schwalbach, M.S., Vergin, K.L., Landry, Z.C., Ellisman, M., Deerinck, T., Sullivan, M.B., and Giovannoni, S.J. (2013). Abundant SAR11 viruses in the ocean. Nature *494*, 357–360. https://doi.org/10.1038/nature11921.

Zhong, X., Berdjeb, L., and Jacquet, S. (2013). Temporal dynamics and structure of picocyanobacteria and cyanomyoviruses in two large and deep peri-alpine lakes. FEMS Microbiol. Ecol. *86*, 312–326. https://doi.org/10.1111/1574-6941.12166.

Zhong, X., and Jacquet, S. (2014). Contrasting diversity of phycodnavirus signature genes in two large and deep western European lakes. Environ. Microbiol. *16*, 759–773. https://doi.org/10.1111/1462-2920.12201.

Zhong, X., and Jacquet, S. (2014). Differing assemblage composition and dynamics in T4-like myophages of two neighbouring sub-alpine lakes. Freshwater Biol. *59*, 1577–1595.

Zhong, X., Ram, A.S., Colombet, J., and Jacquet, S. (2014). Variations in abundance, genome size, morphology, and functional role of the virioplankton in Lakes Annecy and Bourget over a 1-year period. Microb. Ecol. *67*, 66–82. https://doi.org/10.1007/s00248-013-0320-2.

Zhong, X., Guidoni, B., Jacas, L., and Jacquet, S. (2015). Structure and diversity of ssDNA Microviridae viruses *in* two peri-alpine lakes (Annecy and Bourget, France). Res. Microbiol. *166*, 644–654. https://doi.org/10.1016/j.resmic.2015.07.003

Part III

Diversity of Viruses of Microorganisms

Bacteriophage Diversity

Susan M. Lehman*

7

AmpliPhi Biosciences, San Diego, CA, USA.

*Correspondence: sml@ampliphibio.com

https://doi.org/10.21775/9781910190852.07

Abstract

Over 100 years of research, our understanding of bacteriophage diversity has expanded from the hypothesis of a single, highly adaptable phage species to an appreciation for the tremendous diversity that can exist in a millilitre of seawater. This progression is evident from the way that phage taxonomy has changed over time, as new tools for studying phages were developed. Basic microbiological methods have been joined by electron microscopy. Resolving the structure of DNA ushered in the genetics era that has ultimately led to rapid, inexpensive sequencing and the multi-faceted '-omics' approaches. The result has been an increasingly sophisticated understanding of phage diversity from a population level down to the dynamics of a single phage–host infection.

Introduction – an historical perspective on phage diversity

Viruses that infect bacteria (bacteriophages or phages) exist in great variety, and our understanding of this diversity is ever-changing. At the time of their discovery and early study, bacteriophages were at least appreciated to occur in both free lytic and lysogenic forms, although the biological basis and interrelatedness of lysis and lysogeny were not yet understood. Very early on, Felix d'Hérelle also described antigenic and host-range properties of phages that appeared to be characteristic of different varieties. In the 1940s, phages began to be examined by electron microscopy and their morphological diversity began to be revealed (Ackermann, 2011). By the 1950s it was common to discuss phages in terms of the bacteria they infected (e.g. coliphages, staphylococcal phages), and many individual phages, including the T-even and -odd phages of *E. coli*, had been named. After DNA was established as the mechanism of genetic inheritance (including through experiments with phages themselves), the glut of genetic experimentation with phages revealed that they are also diverse with respect to processes such as restriction-modification and genetic transfer.

The advent of moderately priced whole-genome sequencing (WGS) technologies in the mid-2000s added new complexity. In addition to allowing nucleotide-level comparisons among many phages, WGS data revealed the extent of genomic mosaicism among phages, simultaneously explaining some of the tremendous observed phage diversity and complicating inferences about shared descent. Most recently, transcriptomics and metabolomics methods have begun to reveal the tremendous diversity that exists even among genetically similar phages.

Here, phage diversity is considered from the perspectives of phage taxonomy, ecology, host interactions, microbiological methods, and -omics strategies. Most of these topics have been extensively reviewed in their own right (see Chapters 2, 3, 5, and 15; Krupovic *et al.*, 2016). The purpose of this chapter is not to repeat such reviews, but rather to examine how each of these methods has shaped our understanding of phage diversity. For detailed discussions of the biology of specific phages and phage groups, there are a number of books that provide overviews of these topics in addition to the primary literature (e.g. Catalano, 2005; Kutter and

Sulakvelidze, 2005; Calendar and Abedon, 2006; McGrath and van Sinderen, 2007).

Lessons from taxonomy

Taxonomy is the means by which we attempt to organize phage diversity. The progress of phage taxonomy reveals how the diversity of these viruses has been viewed over time.

In the 1920s, d'Hérelle suggested that the range of bacterial species able to be infected by phages, combined with observations of adaptive host-range expansion, implied that only a single species of phage existed, possessing an extreme capacity for adaptation that resulted in the generation of different phage strains (d'Hérelle, 1926). For this highly adaptive entity, he proposed the genus-species name *Protobios bacteriophagus*. His position was certainly not universal, with a number of scientists advocating that there were 'not one bacteriophage but a great many bacteriophages, as different from each other as are the bacteria' (Bruynoghe, 1924). The latter perspective won out, and the 1948 edition of Bergey's Manual listed 46 species within the genus *Phagus*.

In 1959, Adams assessed the pros and cons of a series of characteristics that might be useful for classifying phages at and above the species level, before appearing to throw up his hands in dismay at the dearth of generally applicable criteria for distinguishing true genetic relatedness from simple phenotypic similarities. For example, lytic lifestyle and virion morphology offered too few categories, host range was too easily affected by single-step mutations, chemical properties were insufficiently associated with biological activity, and both serology and mixed infection experiments had outcomes that were not always unambiguously interpretable.

In 1966, the International Committee on Taxonomy of Viruses (ICTV) was established to develop, refine, and maintain a universal system of virus taxonomy. While formal acknowledgement of taxa by the ICTV necessarily lags behind the published literature establishing their scientific bases, ICTV ratification provides a useful timeline. Prior to 1979, bacteriophage taxonomy was somewhat haphazard, with families such as 'T-even phages', 'lambda phage', and 'phiX group'. In 1978, seven families were established: *Corticoviridae*, *Cystoviridae*, *Inoviridae*, *Leviviridae*, *Microviridae*, *Plasmaviridae*, and *Tectiviridae*. Of the tailed phages, the *Myoviridae* and *Podoviridae* families were established in 1981, followed by *Siphoviridae* in 1984. In 1999, the order *Caudovirales* was established to encompass the three families of tailed phages. These classifications were heavily driven by virion morphology and genome type (DNA, RNA, single- or double-stranded). The advent of widespread phage genome sequencing launched vigorous debate about how proteomics, genome replication strategies, horizontal gene transfer, and structural gene modules could be used to improve phage classification (reviewed by Nelson, 2004). As of the 2015 virus taxonomy release, the ICTV recognizes the taxa outlined in Table 7.1 for viruses infecting bacteria. The current taxonomy is primarily driven by nucleic-acid level similarity. Morphology and sometimes genome replication strategy have also been captured in family and subfamily classifications. The basic morphology of each family is shown in Fig. 7.1.

Lessons from phage methods

Our understanding of phage diversity over time has been heavily affected by the methods used to study these viruses. Common methods of phage isolation often exclude certain types of phages. Once isolated, the methods of study again limit our observations. In itself, 'we can only see what we can see' is a trivial statement, yet it has been a crucial component of bacteriophage diversity studies.

Limitations imposed by microbiological methods

Historically, the isolation and study of phages was entirely dependent on culturing, relying heavily on the agar overlay technique in which a phage sample is mixed with actively growing bacteria plus dilute molten agar and poured over a solid agar plate. This allows visualization and isolation of single plaques, each presumably originating from a single phage, and purification by serial passage in a manner analogous to single-colony purification for bacteria.

The long-standing necessity of culturing phages in order to study them has severely restricted the study of phages that infect so-called 'unculturable' bacteria. This last issue has particularly impacted marine phages, and our understanding of their diversity has been heavily dependent on metagenomic techniques as a result (see Chapters 5 and

Table 7.1 Taxa recognized by 2015 ICTV release that encompass bacteriophages

Order	Family (morphology)	Subfamily	Number of genera[1]	Genome type	Morphology
Caudovirales (tailed)	*Myoviridae*	*Eucampyvirinae*	2	Linear dsDNA	Icosahedral capsid and contractile tail
		Peduovirinae	2		
		Spounavirinae	5		
		Tevenvirinae	7		
		Vequinvirinae	3		
		–	28		
	Podoviridae	*Autographivirinae*	4	Linear dsDNA	Icosahedral capsid and short tail
		Picovirinae	2		
		–	14		
	Siphoviridae	*Guernseyvirinae*	3	Linear dsDNA	Icosahedral capsid and flexible tail
		Tunavirinae	5		
		–	57		
Unassigned	*Corticoviridae*	–	1	Circular dsDNA	Polyhedral
	Cystoviridae	–	1	Linear multipartite dsRNA	Polyhedral
	Inoviridae	–	2	Circular ssDNA(+)	Filamentous
	Leviviridae	–	2	Linear ssRNA(+)	Polyhedral
	Microviridae	*Bullavirinae*	3	Circular ssDNA(+)	Polyhedral
		Gokushovirinae	3		
	Plasmaviridae	–	1	Circular dsDNA	Pleomorphic (enveloped)
	Sphaerolipoviridae	–	3	Linear dsDNA	Near-spherical capsid and internal lipid membrane
	Tectiviridae	–	1	Linear dsDNA	Polyhedral, surrounding lipid vesicle

[1]Not all viruses in each family and subfamily have been formally assigned to a genus and are therefore not reflected in this table.

15). The common use of chloroform during phage isolation and propagation has likely biased phage discovery in favour of tailed phages, which tend to be more chloroform resistant than filamentous phages or lipid-containing phages (Ackermann, 2004). The use of ~0.4% agar as the phage-containing phase of plate-based techniques appears to have substantially impeded discovery of phages with >200 kb genomes (Serwer *et al.*, 2007). Not all viable phages form obvious plaques and plaque formation can be affected by factors such as adsorption rates, lysis timing, burst size, and host metabolism (Gallet *et al.*, 2011), further biasing the diversity of phages that could be captured by these methods. Solutions for some of these problems do exist: chloroform is often convenient for phage manipulations but is not required; ultra-dilute agarose gels have been used to grow and isolate very large phages (Serwer *et al.*, 2007); and alternative media preparation methods might aid isolation of phages infecting previously uncultured bacteria (Tanaka *et al.*, 2014).

Diversity assessed from physical structure

The structure of a phage's genome and virion morphology have long been used as tools to assess phage similarity and diversity. Genome size and nucleic acid type provided early, gross indicators of diversity. Among sequenced phages, genome sizes range from 3.4 kb for enterobacteria phage M (NC_019707) to 497.5 kb for *Bacillus* phage G (NC_023719). Virion morphology, including capsid geometry and the presence of tails, has been

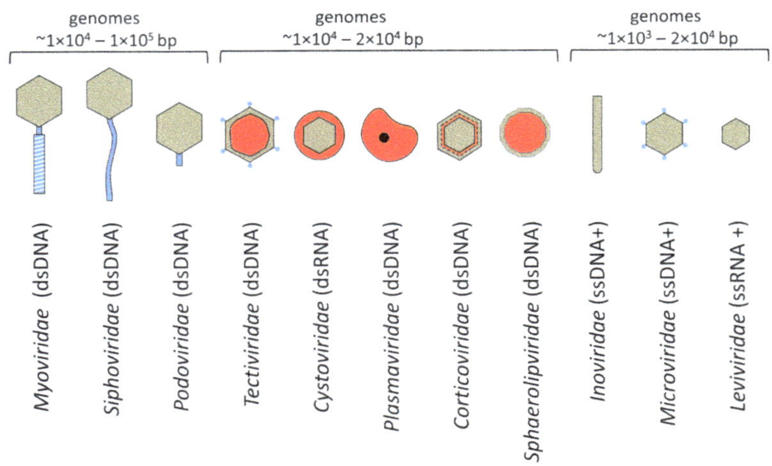

Figure 7.1 Basic morphologies of bacteriophage families. Brown represents the nucleic acid-encasing protein capsids. Blue indicates tails and other adsorption structures (tail fibres and other appendages on tailed phages are not shown). Red represents lipid structures. Capsids are depicted as isometric but some tailed phages have elongated (prolate) heads. The *Inoviridae* may be either short rigid rods or long flexible rods. *Plasmaviridae* contain a nucleoprotein granule rather than a capsid. Adapted from Hyman and Abedon (2012).

heavily used as a measure of phage diversity since the availability of electron microscopy (Ackermann *et al.*, 1978; Ackermann and DuBow, 1987; Ackermann, 2007). Even although a great diversity of phages exists within a given morphological type, the sheer number of structural proteins required to produce a virion particle means that shared virion morphology is moderately reflective of genetic similarity. For example, among the *Caudovirales* phages, the overall genome similarity supports the clustering of *Myoviridae*, *Siphoviridae*, and *Podoviridae* (Lavigne *et al.*, 2008, 2009; Holmfeldt *et al.*, 2013; Adriaenssens *et al.*, 2015) as well as shared morphological traits within those groups (Proux *et al.*, 2002; Mizrei *et al.*, 2014; Comeau *et al.*, 2012).

Susceptibility to inactivation

Phages exhibit a wide range of variability in their susceptibility to inactivation by agents such as organic solvents, detergents, temperature, pH, salinity, osmotic pressure and various wavelengths of electromagnetic radiation. These types of characteristics are illustrative of phage diversity and are likely important to the ecological factors that affect phages in their natural habitats. However, the inherent ambiguity of these traits has always made them problematic as a measure of phage diversity. Since inexpensive genome sequencing has become widely available, these traits are rarely studied as a way to characterize large groups of phages and are much more likely to receive attention when they are relevant to a specific technological application. One notable exception is a recent study that tested 83 phage isolates infecting five bacterial genera for their response to 16 different potentially inactivating conditions (Jurczak-Kurek *et al.*, 2016). Along with morphology and host infection characteristics, at least 70 of the isolates appear to be distinct phages although only seven were sequenced.

Serology

The antigenicity of phages was first documented in 1921 with the recovery of phage-neutralizing serum from rabbits that had been injected with phage lysate (Bordet and Ciuca, 1921). Phages and their hosts were quickly shown to be antigenically distinct (Otto and Winkler, 1922). In the 1950s, several tailed phages were shown to possess at least two different antigens, one associated with the phage head that was implicated in agglutination rather than neutralizing activity, and one associated with the phage tail that reacted with neutralizing antibody (Lanni and Lanni, 1953; Fodor and Adams, 1955). This understanding of neutralizing (tail-interacting) and non-neutralizing (capsid-interacting) phage antibodies has persisted, with only a small percentage of phage inactivation found to be associated with phage agglutination in the absence of antibodies specific to tail structures (Guttman *et al.*, 2005).

Serological cross-reaction was widely used to investigate phage diversity before faster and more information-rich genetic and then genomic methods reduced the appeal of time-consuming antibody production. In a variation on the classical spot test for phage activity, general groupings of serologically related phages could be determined by spotting antisera on bacterial lawns prior to spotting phages (Shirling and Speer, 1967). Combinations of phage and serum that reduced phage activity could then be investigated using kinetic inactivation studies capable of teasing out degrees of serological relatedness within each group (Adams, 1959).

Serological relatedness has consistently indicated that two phages will share additional traits. For example, serology has identified relatedness between obligately lytic and temperate *Streptococcus thermophilus* phages that share similar host ranges (Brüssow *et al.*, 1994; Brüssow and Bruttin, 1995). Early observations of serological relationships among the T-even phages were consistent with later genetic studies (e.g. Russell 1974), the serogroups of lactococcal siphoviruses correlate well with morphological and genetic traits (Jarvis 1984), and temperate *Staphylococcus aureus* siphovirus serology is highly correlated with specific molecular similarities (Rountree, 1949; Doškař *et al.*, 2000; Pantůček *et al.*, 2004; Goerke *et al.*, 2009). The reverse does not necessarily hold true, however. Among the tailed phages, it is not surprising that no cross-reactions have been observed across families (e.g. between a podovirus and a myovirus) (Guttman *et al.*, 2005; Sulakvelidze and Kutter, 2005). Mutations in a few key structural proteins, however, can eliminate antigenic similarity among closely related phages, as has been observed with certain RNA *E. coli* phages (Hirashima *et al.*, 1983).

Lessons from infection biology

Host-range diversity

Host range is perhaps the most easily explained measure of phage diversity: different phages infect different combinations of bacterial strains and species. The underlying mechanisms of host-range determination are complex, spanning adhesion, infection, intracellular replication, and lysis, but the effect is literally illustrative. When the results of pairwise phage–bacteria interactions are plotted as a matrix, the diversity of both phages and their hosts is easily visualized. A meta-analysis of 38 sets of host-range data encompassing 4365 phage–bacteria pairings revealed a nested pattern of phage susceptibility in which highly susceptible hosts are infected by both generalist and specialist phages but the rarely infected hosts are almost exclusively infected by generalist phages (Flores *et al.*, 2011). Nestedness is consistent with a gene-for-gene model of phage–host coevolution, in which phages gradually acquire the ability to infect new hosts and may lose the ability to infect others in the process, or with host range changes arising from epistatic interactions among mutations in multiple genes. These phenomena have been extensively studied through *in vitro* experimental evolution experiments (Spanakis and Horne, 1987; Duffy *et al.*, 2007; Wei *et al.*, 2010; Scanlan *et al.*, 2011), but are only beginning to be understood in nature (Koskella and Meaden, 2013). Most of these studies focused on one or a few related bacterial species, where overlapping and gradually shifting host ranges of generalist phages makes sense in the context of closely related hosts. When viewed across a larger taxonomic scale that encompasses more distantly related bacterial hosts and more geographic diversity, one might expect a more modular pattern of phage–bacteria interactions in which there are multiple non-overlapping phage–host clusters. This was, in fact, observed upon subsequent examination of a single, very large data set of cross-infection studies with marine phages and bacteria from different points around the Atlantic Ocean (Flores *et al.*, 2013; 1332 phage–bacteria pairings). On a large scale, multiple non-overlapping or minimally overlapping clusters were observed, partly correlated with geographic segregation.

Methods dependent on culturing are generally guaranteed to underestimate environmental phage diversity. They are limited to hosts that have been cultured, and to the culture conditions employed, which we know are simplified versions of the complex configurations of biotic and abiotic factors that are possible in the environment. The effects of nutrition, spatially structured environments, and other species on host metabolism are particularly likely to affect the way that phages interact with their potential hosts in nature (reviewed by Koskella *et al.*, 2013). When phages are isolated after an enrichment step, enrichments using multiple bacterial strains or species may be more likely to yield broad-host-range phages whereas single-host enrichments would tend to favour replication of strain-specific specialists over phages that infect multiple strains with slightly lower efficiency (Jensen *et al.*, 1998).

Culturing-based host-range studies can also distort phage diversity if they fail to differentiate between productive phage infection and general bactericidal effects (Ross *et al.*, 2016). Spot tests, in which a small volume of concentrated phage solution is dropped onto an agar overlay, are commonly used to rapidly screen large numbers of phage–host pairings. Most tests use minimally purified phage lysates, which may contain residual endolysins, bacterially encoded bacteriocins, or other soluble antibacterial substances that can result in cell death in the absence of phage replication and subsequent bacterial lysis. Even when highly purified phages are used, phage-mediated lysis may occur via external disruption of the cell membrane and wall, without phage replication. Such false positives can be avoided by plating or spotting serial dilutions of each phage, after which the presence of plaques at higher dilutions demonstrates that the phage is truly able to replicate on the bacterial strain in question. The absence of plaques under a particular set of culturing conditions may not be conclusive evidence against productive infection, but the combination of testing for both plaques and general bactericidal effects generates a more complete picture of phage host-range activity. Requiring plaque formation can also discriminate among closely related phages that infect a strain with different efficiencies, resulting in different apparent titres on different host strains.

For phages with overlapping host ranges, mixed infections have been used to assess phage diversity. The underlying hypothesis is that sufficiently similar phages may undergo recombination when co-infecting a single cell, whereas phages that produce no progeny during mixed infections must differ sufficiently in their intracellular reproduction as to be fundamentally incompatible (Adams *et al.*, 1959).

Broad-host-range phages capable of infecting many strains of a species and sometimes cross-species or even cross-genus infections, have been well-described among the Enterobacteriaceae, Pseudomonadaceae, and Synechococcaceae (Sullivan *et al.*, 2003; Paolozzi *et al.*, 2006; Peters *et al.*, 2015; Khawaja *et al.*, 2016; Yu *et al.*, 2016). They also tend to be tailed phages, which is perhaps not surprising as more tailed than tailless phages have been cultured. The tailed phages may also be inherently more likely to have or develop broad host ranges. Average phage genome size increases by family, in the order *Leviviridae, Microviridae, Inoviridae, Corticoviridae, Plasmaviridae, Cystoviridae, Tectiviridae, Podoviridae, Siphoviridae*, and *Myoviridae* (Hyman and Abedon, 2012), giving the *Caudovirales* larger and likely more flexible genomes with which to adapt to new hosts, even after the more complex set of tail morphogenesis genes is accounted for. The tailed phages already tend to use a wider variety of bacterial cell structures as receptors, including multi-component binding processes, as discussed in the next section. The extent to which this is the result of culturing biases will likely become better understood as other families of phages are more widely cultured and as metagenomic and bioinformatic methods allow us to infer more about the host ranges of uncultured phages.

Phage host-range diversity is intimately tied to bacterial diversity. As a basic phenotype, phage host-range diversity has long been used as a tool to assess bacterial diversity, in the form of phage typing (Lindberg and Latta, 1974; Baker and Farmer, 1982; Hickman-Brenner *et al.*, 1991; Doškař *et al.*, 2000; Williams and LeJeune, 2012). On an ecological level, phage host-range diversity influences the development of bacterial diversity (Solheim *et al.*, 2013; Koskella and Brockhurst, 2014; Betts *et al.*, 2016). This can occur directly by both predator–prey coevolution processes and transduction. For phages that do not completely degrade the bacterial host genome, it can also be indirect, by facilitating the release of bacterial DNA that may be taken up by nearby cells.

Infection, resistance, and superinfection immunity

The net host-range diversity observed among phages arises from multiple sources. The molecular mechanisms of anti-phage host defence are not the subject of this chapter and while these do contribute to an understanding of phage diversity, readers are directed elsewhere for in-depth discussions of CRISPR (Samson *et al.*, 2013; Pawluk *et al.*, 2016), restriction-modification (Tock *et al.*, 2005; Samson *et al.*, 2013), abortive infection (Chopin *et al.*, 2005; Samson *et al.*, 2013), and assembly interference (Ram *et al.*, 2012). Receptor diversity and superinfection immunity are discussed here because of their historical use for assessing phage diversity.

Initial phage–host interaction is mediated by phage recognition of bacterial cell surface receptors. Table 7.2 provides an overview of the types of bacterial structures that are known to function as phage receptors. The host ranges of known *Sphaerolipoviridae* phages are very narrow and their receptor usage is not well understood. The *Cystoviridae* are subdivided into two groups that recognize either pili or LPS (Mindich, 2006). Phages infecting Gram-positive bacteria rely on cell wall substances; the single row in Table 7.2 does not adequately reflect the molecular diversity of these receptors, which can include the incorporation of specific sugars or glycosylation in teichoic acids (Archibald, 1980; Rakhuba *et al.*, 2010). As a group, the *Caudovirales* phages appear to use the widest variety of substances as receptors, and some use more than one type. For example, T4 uses both the OmpC porin protein and LPS, and loses infection efficiency in the absence of one receptor (Yu *et al.*, 1982). Phages that recognize capsular polysaccharides do so as part of initial target recognition and then bind irreversibly to another cellular component.

Within these phage–receptor pairings, some interactions are more temperature sensitive than others, and some require divalent metal cations for efficient adsorption. In the case of flagellotropic phages, iEPS5 requires counterclockwise flagellar rotation in order to infect its *Salmonella* host (Choi *et al.*, 2013), but *Vibrio parahaemolyticus* phage OWB, which binds peritrichous flagella, exhibits superior adsorption and cellular killing when polar flagella are absent or prevented from rotating (Zhang *et al.*, 2016).

Superinfection immunity is a mechanism of infection resistance that acts downstream of both receptor binding and phage DNA uptake into the bacterial cytoplasm, and it can complicate host-range-based assessments of phage diversity. This is the phenomenon by which bacterial cells harbouring a prophage are protected from infection and lysis by certain other phages. It was observed in the 1920s and 1930s, and was a major focus of phage research in the 1950s, by which point it was understood that the mechanism of immunity was separate from phage adsorption (Adams, 1959). Lambda was the focus of most lysogeny-related research at the time, and observations led to the 'lambda immunity model', under which the superinfecting phage genome enters the cell but replication, gene expression, and productive phage infection are repressed by trans-acting repressors from the previously established prophage (Ptashne, 2004). Other mechanisms, dubbed superinfection exclusion, act earlier, blocking genome translocation (Cumby *et al.*, 2012; Bebeacua *et al.*, 2013). Observations of superinfection immunity by a bacterial lysogen against a specific phage isolate would prompt the conclusion that two phages, prophage and superinfecting phage, were largely

Table 7.2 Bacterial structures used as phage receptors

Bacterial structure	Used by[1]	Selected examples
Pili	*Leviviridae*, *Inoviridae*, *Corticoviridae*, *Tectiviridae*, *Cystoviridae*	Ff, M12, f2
LPS	*Myoviridae*, *Siphoviridae*, *Podoviridae*, *Cystoviridae*, *Microviridae*	T3, T4, T7, ΦX174
Capsule	*Myoviridae*, *Podoviridae*	K11, K29
Flagella	*Siphoviridae*, *Myoviridae*, *Podoviridae*	X, ΦAcM4
Outer membrane proteins[2]	*Myoviridae*, *Siphoviridae*, *Podoviridae*, *Tectiviridae*	T4, T5, T7, PRD1[4]
Cell wall[3]	*Caudovirales*, *Microviridae*, *Tectiviridae*	K, Φ29, γ
Plasma membrane	*Plasmaviridae*	L2

[1]Drawn from Ackermann, 2004; Calendar and Abedon, 2006; Rakhuba *et al.*, 2010. For additional mechanistic details, see also Randall and Philipson, 1980.
[2]Including structural proteins, porins, enzymes, transport proteins
[3]Including peptidoglycan and associated moieties such as teichoic acid
[4]Dependent on conjugative plasmid that encodes a transmembrane complex

homologous. Such observations are still sometimes used to infer phage relatedness (Goh *et al.*, 2005). A high degree of relatedness, however, does not guarantee that the effects of superinfection immunity will manifest. For example, at least three similar Stx phages that potentiate shiga-toxin production in *E. coli* can be harboured simultaneously (Fogg *et al.*, 2007) which, assuming sequential phage exposure, suggests a potential for establishment of additional prophages in a context where superinfection immunity would have been expected to prevent it. The genetic mechanisms that permitted each phage to occupy a different integration site in the host genome are not clear, but appear to involve reduced efficiency of repression by the cI protein.

Lifestyle and associated traits

When phages were first discovered, two general types of lysis were observed: (i) the phage acting as an external lytic agent, as when cell-free filtrate was added to a susceptible bacterial culture; and (ii) lysogenic bacteria, where a bacterial cell would spontaneously lyse without the addition of any external agent. In some early observations, lysogenic bacteria were described as taking on a 'glassy' appearance. As the particulate nature of phages was clarified, these two behaviours came to be understood as deriving from the intrinsic capabilities of a given phage and phages were divided into two groups: obligately lytic phages and temperate phages. While the fundamental distinction between these two lifestyles, whether or not the phage genome integrates into the host genome and is replicated along with it for a number of generations, still holds, a number of variations are now known to occur as well. These variations have been reviewed at length (Abedon, 2008, 2009), and include:

- classical lytic infections, where host cell destruction is contemporaneous with the release of newly replicated progeny phages a relatively short time after adsorption;
- chronic infections, during which progeny phages are released over time without abrupt destruction of the host cell;
- classical lysogeny, during which the infecting phage genome integrates into the host cell genome as is maintained during bacterial replication until a triggering event leads to phage excision, replication, and release;
- pseudolysogeny, in which the infecting phage genome is maintained in the host cell without genome integration and generally without replication;
- abortive infection, in which both the phage and cell lose viability;
- phage restriction or prophage curing, in which the infecting phage genome is destroyed and the host cell survives.

The variety of ways in which phages can replicate and kill their hosts contributes to the variety of ecological roles that phages can play. For example, chronic infections are a feature of filamentous phages. In *Pseudomonas aeruginosa* biofilms, several filamentous phages have been implicated in the complex regulation of biofilm life cycles, and even on the virulence of *P. aeruginosa* in biofilm-related human infections (Webb *et al.*, 2003, 2004; Rice *et al.*, 2009). Pseudolysogeny, as defined here, has been observed in active cells (Cenens *et al.*, 2013a,b) and has been proposed as a way of maintaining mixtures of phage-sensitive and phage-resistant bacterial cells for long-term co-existence of both phages and their host (Siringan *et al.*, 2014). It might also allow phages to survive within the cell and then begin replication once nutrient availability improves and the host cell becomes more metabolically active (Schrader *et al.*, 1997; Bryan *et al.*, 2016). This phenomenon may be important to maintaining long-term communities of phages and hosts in nutrient-poor soils or water.

Additional life cycle traits used to characterize phages include adsorption rate, latent period, and burst size. These traits have sometimes been used to differentiate phages, but their existence along a spectrum makes it difficult to distinguish among phages unless the traits in question differ quite markedly. Weinbauer and Peduzzi (1994) looked at these traits with a wider lens in an effort to expand, prior to metagenomic sequencing, the limited knowledge of phage diversity in aquatic ecosystems. After sampling multiple points within the Adriatic Sea, TEM and epifluorescence microscopy were used to determine the relative frequencies of different bacterial morphotypes (rod, cocci, spirillae), the infected proportion of each morphotype, and the quantities, sizes, and intracellular distribution patterns of those virions. While acknowledging that bacterial morphotypes will underestimate bacterial diversity, the authors were able to show that infection frequency varied with bacterial density for some bacteria, and that phage–host relationships such as burst size and virion assembly differed noticeably by morphotype, thereby painting a better picture of marine phage diversity than had been hitherto available.

Lessons from 'omics'

In his writings of 1959, Adams noted that the historical use of a type specimen to define a species had already been rejected as inadequate and had been replaced by the concept that a species is more appropriately defined as the 'total gene reservoir of an interbreeding population, which places a limit on the variability of the individuals within that population'. Acknowledging that the usual concept of an interbreeding population did not fully apply to phages, Adams advocated for both the maintenance of type specimens that will allow comparative analyses among researchers (an issue still prominent in the phage community today), and for an attempt to 'describe the limits of variability of different strains within a (phage) species.' Here, we find the tremendous potential offered by next-generation sequencing (NGS) technologies.

Metagenomics

Metagenomic sequencing, in which all genomes from an environmental sample are sequenced without being cultured or individually isolated, was a sea change in the study of phage diversity, freeing it from many of the limitations imposed by the culturing-dependent techniques described above. Table 7.3 highlights a few phage metagenomic studies describing novel uses or outcomes. Early studies had a substantial impact on our understanding of

Table 7.3 Selected metagenomic studies offering insight into phage diversity in a range of environments

Study	Environment	Highlights
Breitbart *et al.* (2002, 2004)	marine	First metagenomic analyses of uncultured marine and human faecal viral communities
Angly *et al.* (2006)	marine	184 metaviromes spanning 68 sites in four oceanic regions; prophage-like sequences were most common in Arctic; a new clade of ssDNA phages found
Desnues *et al.* (2008)	stromatolites, thrombolites	Marine stromatolite contained completely novel ssDNA phages not found in a range of water, sediment, and animal metaviromes; sequence signatures associated with marine phages were found in the freshwater thrombolite even although that environment is not believed to have been in contact with the ocean since the Jurassic period
Hurwitz and Sullivan, (2013)	marine	Introduced the Pacific Ocean Virome (POV) dataset; protein cluster analysis showed that taxonomic richness decreased in surface vs deep waters, in winter vs summer, and with distance from shore (in surface waters)
Mizuno *et al.* (2013a)	marine	Recovered complete genomes of 208 new marine phages, encompassing 21 genetically distinct groups of tailed phages (10 novel)
Dutilh *et al.* (2014)	human stool	Discovery of crAssphage, the most abundant and ubiquitous phage in human faecal microbiomes
Zablocki *et al.* (2014), Adriaenssens *et al.* (2015)	hyperarid desert soils	Cold-climate dsDNA hypolith phage communities were more diverse than in open soil; both hot- and cold-climate hypolith communities were not dominated by recognizable cyanophages, despite the abundance of cyanobacteria.
Parsons *et al.* (2012), Goldsmith *et al.* (2015)	marine	Spatiotemporal analyses of phage–host dynamics at the Bermuda Atlantic Time Series site in the Sargasso Sea. Summer peaks in viral abundance correlated with increased *Prochlorococcus* concentrations, and seasonal trends observed over three-year time frame.
Santiago-Rodriguez *et al.* (2015)	faecal culture	Metagenomic sequencing used to compare phage diversity in faecal samples and in 24-day chemostat cultures derived from those faecal samples, to assess potential utility of complete microbiome model systems.

marine phage diversity in particular (Breitbart *et al.*, 2002; Breitbart *et al.*, 2004). Despite this explosion of knowledge about the diversity of environmental phage communities, there are limitations (Bik, 2014). Early sample preparation methods selected for dsDNA phages and relied on template amplification steps that can introduce content bias and distort the relative abundance of genomes within the viral community. Few complete phage genomes could be assembled out of early viral metagenomic (metaviromic) datasets, hampering assessments of how many phage taxa were represented. Finally, not all environmental samples are created equal. Marine samples have generally been the easiest to sequence, whereas analyses of soil communities has proven more challenging due to their biological and chemical complexity (Daniel, 2005) (see Chapter 4).

As methods have improved, amplification-free metaviromes have provided insight into the relative abundance of viral community members, allowing for species richness and temporal changes in community composition to be assessed as part of overall diversity in environmental samples (Bik, 2014) as well as studies of the human microbiome (Manrique *et al.*, 2016). Longer-read sequencing, new sample preparation strategies, and improved bioinformatics methods have allowed more complete or near-complete phage genomes to be assembled within metaviromes, improving our understanding of the content and architecture of environmental phage populations (Tucker *et al.*, 2011; Mizuno *et al.*, 2013a,b; Roux *et al.*, 2012b).

As more metaviromic datasets have been generated, large percentages of genes continue to be not just genes of unknown function, but entirely novel sequences. In essence, metaviromic studies can give us a sense of just how much we do not know. Attempts to quantify this 'missing diversity' often use two methods, rarefaction curves and sequence recruitment. Rarefaction curves assess how well-sampled a community is. Curves are initially steep, as new samples uncover many new taxa. As the diversity of a sample approaches the true diversity of its source environment, the curve will become asymptotic, indicating that only the rarest taxa remain unsampled. Metaviromic studies have frequently pointed to steep rarefaction curves as an indication that only a small fraction of the total phage diversity has yet been sampled (Rohwer, 2003; Ignacio-Espinoza, 2013). Sequence recruitment refers to the proportion of sequence reads in a metavirome that show significant similarity to sequences present in non-redundant sequence databases. While read length affects this calculation and can complicate direct comparisons among studies (Wommack, 2008), proportions for published viromes range from about 28% to <1% (Roux *et al.*, 2012a), indicating a substantial quantity of novel sequence data in each new sample.

Perhaps the greatest value of metagenomic datasets is that they can be queried many times by researchers asking a variety of questions. Table 7.4 highlights a few such studies. Notable among these studies is the abundance and diversity of ssDNA, ssRNA, and dsRNA phages, highlighting the historic bias towards dsDNA phages that had been imposed by culturing-dependent methods of study.

Genomics and proteomics

Whole genome nucleotide similarity

Unlike bacterial 16S rDNA, there is no universal genetic marker that can be used to infer phylogenetic relationships among all bacteriophages. For closely related phages, nucleotide identity across an entire genome is a good measure of similarity. One of the first steps of analysing a new phage genome is often to look for highly similar genome sequences by comparing the full sequence against all publicly available phage genomes using a tool such as the BLAST algorithm from the National Center for Biotechnology Information. Whole genome nucleotide alignments can and are used to study phage relationships (e.g. Hatfull *et al.*, 2010; Pope *et al.*, 2015), although this approach has limitations. Phages with substantially different host ranges rarely share sequence similarity, and among phages that share a host genus or species, there are groups of phages with high nucleotide sequence similarity and groups with no significant similarity (Proux *et al.*, 2002; Hatfull, 2008; Holmfeldt *et al.*, 2013; Meczker *et al.*, 2014).

More broadly, summary statistics of nucleotide-level data reflect phage diversity. Phages infecting the same host species will tend to have similar GC content, with the phage genomes being slightly more AT-rich than their hosts, especially for obligately lytic and filamentous phages (Rocha and Danchin, 2002; Xia and Yuen, 2005; Mihara *et al.*,

Table 7.4 Selected re-analyses of previously published metaviromes

Study	Environment	Highlights
Tucker *et al.* (2011)	marine	Reconstructed the first two complete ssDNA phage (*Microviridae*) genomes from a marine environment; the depth and geographic distributions of ssDNA phages were more restricted than those of dsDNA phages.
Roux *et al.* (2012a)	freshwater, marine, hypersaline	31 metaviromes were analysed (2 new, 29 pre-existing); phage communities drawn from environments of the same ecological type (i.e. freshwater vs marine vs hypersaline) were more similar to each other than communities from geographically related environments of a different ecological type; small ssDNA phages comprised the vast majority of the phage communities, with tailed phages a minority
Aziz *et al.* (2015)	many	Applied multiple diversity metrics to 588 complete genomes taken from 296 metagenomes reflecting more than two dozen environments; goal was to develop methods of more comprehensively assessing ecological properties of uncultured phages.
Krishnamurthy *et al.* (2016)	many	RNA-inclusive metaviromic datasets from oceans, sewage, soil, and animal sources such as stool; identified 122 novel RNA phage phylotypes.

2016). Since phages depend on host cell processes for replication and gene expression, and can be influenced by mutational biases in their hosts, phage genomes tend to have similar codon usage as their host species (Kunisawa *et al.*, 1998; Sahu *et al.*, 2005; Sau *et al.*, 2005). Lucks *et al.* (2008) extended traditional gene-by-gene or moving-window codon usage analysis in small numbers of related phages to a 'genome landscape' concept that can be applied more systematically across taxa with different GC content and proteins with different amino acid content. Across all 74 sequenced phages of *Pseudomonas aeruginosa*, *Lactococcus lactis*, and *Escherichia coli* that were available at the time, 50 phages showed clear evidence of host-adapted codon use biases, specifically in their structural genes, and this was seen regardless of lifestyle (temperate vs obligately lytic). Tetranucleotide usage patterns in phage genomes also appear to be similar for groups of phages with common hosts (Pride *et al.*, 2003; 2006; Deschavanne *et al.*, 2010; Oakley *et al.*, 2011). The evolutionary or mechanistic significance of tetranucleotide patterns is not yet clear. Avoidance of species-specific bacterial restriction and modification systems is one possible factor (Pride *et al.*, 2003; Pristas *et al.*, 2014).

Put together, shared patterns of nucleotide sequence and compositional similarity have been used to develop methods for predicting a phage's host based solely on nucleotide-level data (Mihara *et al.*, 2016; Villarroel *et al.*, 2016). These techniques are especially valuable to culturing-independent metagenomic studies.

Proteome similarity

As the primary sequences of phage genomes diverge, similarities based on proteomics become more important. At the level of primary amino acid sequence, identity can be retained if a nucleotide change is synonymous. Similar secondary structures that are important to protein functioning can be formed from different primary amino acid sequences, and functional constraints drive protein-domain conservation in otherwise dissimilar proteins, and. In addition, related phages usually possess highly syntenic genomes, i.e. similar gene ordering. The proportion and order of proteins shared among two or more phages thus forms the basis for estimating their similarity in various ways (Table 7.5).

While there is no single protein shared among all phages, some proteins have been used to describe relationships within groups of related phages. For example, Hopkins *et al.* (2014) used the major capsid protein (MCP) to explore the diversity of *Gokushvirinae*, a subfamily of the ssDNA *Microviridae* phages. Using degenerate primers they amplified MCP genes from 12 environmental samples, ultimately identifying 316 novel MCPs that appear to arise from *Gokushvirinae* phages. MCPs from similar environments showed some phylogenetic clustering, with the greatest diversity of MCPs observed in Antarctic sediment samples. Combined with an earlier metagenomics analysis (Roux *et al.*, 2012b), the phylogeny generated from these protein sequences supports the existence of a bifurcation within the *Gokushovirinae*, separating

Table 7.5 Proteome-based methods of assessing phage diversity. Some of these methods were developed for small cellular genomes and have been subsequently used for phages.

Method	Description	Reference
CoreGenes	Identifies the proportion of proteins in annotated genomes that share a user-specified level of sequence similarity	Zafar *et al*. (2002), Turner *et al*. (2013)
GeGenees	Aligns multiple genomes to construct phylogenetic distance estimates and genetic signatures; TBLASTX mode facilitates comparison of more distantly related genomes	Ågren *et al*. (2012)
GeneOrder	Visual representations of gene synteny based on protein similarity scores.	Mazumder *et al*. (2001), Mahadevan and Seto (2010)
CLUSTAL and tree construction	Single or multiple proteins used to generate phylogenetic distance matrices	e.g. Rohwer and Edwards (2002), Smith *et al*. (2013), Grose and Casjens (2014)

eukaryote-associated isolates from environmental isolates. Among the *Caudovirales* phages, the large terminase subunit, portal protein, and major capsid protein seem to be present in all tailed phage genomes. In particular, the major capsid proteins of all tailed phages show signs of being ancient homologues (Grose and Casjens, 2014 and references therein). Yet despite our ability to identify distantly related homologues, tremendous diversity still exists. The best example of this comes from the phages of mycobacteria. More of these phages have been isolated and sequenced than any other definable phage group and more than 50,000 different proteins were identified among approximately 500 mycophages (Hatfull, 2014).

ORFans

Many open reading frames identified in phage genomes have few or no homologues among sequenced genomes and have been collectively termed ORFans. Among phages, ORFan genes tend to be shorter than non-ORFans and often have a lower GC content (Yin and Fischer, 2008). Most ORFans are now believed to be expressed proteins rather than annotation errors (Alimi *et al.*, 2000; Shmuely *et al.*, 2004), belonging either to novel protein families with unique functions or to rapidly evolving and divergent families with functions similar to known proteins (reviewed by Siew, 2003).

As discussed in the context of metaviromic sequencing, the total number of ORFans continues to grow at a rate that suggests much as-yet unsampled phage gene diversity. The percentage of ORFans relative to the number of sequenced genomes, however, is decreasing as new phages are sequenced and former ORFans find homologues that can be grouped into phage orthogous groups, or POGs (Yin and Fischer, 2008; Kristensen *et al.*, 2011).

The functions of many ORFans and newly formed POGs remains unknown. As the cost and speed of genome sequencing has decreased, the pace of ORFan discovery has increasingly outstripped the pace of functional studies. T4, P22, and the *Pseudomonas* phages offer notable highlights amid the paucity of ORFan identification. In P22, expression of a 9.3 kDa, 86 amino acid protein was found to de-repress an operon in *Salmonella* Typhimurium, triggering a transition to a pseudolysogenic state distinct from either the lytic or prophage states that have been previously confirmed for P22, in which the phage genome is maintained as an episome within the cell (Cenens *et al.*, 2013a). In T4, one former ORFan was found to inhibit topoisomerase I, improving the efficiency of mature virion accumulation prior to lysis (Mattenberger *et al.*, 2015). Another former ORFan now identified in several *E. coli* phages modulates GroES/GroEL chaperone function to favour efficient folding of structural proteins in less permissive host strains or growth conditions (Ang and Georgopoulos, 2012). Some of the most systematic ORFan characterization has been done in *Pseudomonas aeruginosa* phages, where a number of ORFans concentrated in early gene regions have been shown to target host cell processes during cellular hijacking for phage replication. Wagemans *et al.* (2014) identified six such antibacterial proteins in LUZ7 that appear to specifically target aspects of *P. aeruginosa* metabolism to restrict cell division, alter transcription and protein modification, and reduce breakdown of spermidine in the host cell, which N4-like phages require to complete productive infections. Four unknown early proteins in LUZ24 were shown to be antibacterial, one of which protects the LUZ24 genome from transcriptional silencing by a host protein that targets AT-rich DNA (Wagemans *et al.*, 2015).

Horizontal gene transfer and genome mosaicism

As early as the 1940s and 1950s, there was growing appreciation for the fact that genetic recombination could indicate relatedness among phages. Dissimilar phages will sometimes inhibit each other in mixed infections of a common host (Delbrück and Luria, 1942). When instead recombination was observed following mixed infections then this implied sufficient relatedness among those phages to permit gene exchange, such as that observed between T2 and T4 (Delbrück and Bailey, 1946; Hershey and Rotman, 1949; and others reviewed by Adams, 1959).

The generally syntenic structure of phage genomes within families and genera is believed to facilitate recombination and the generation of mosaic genomes (Casjens, 2005; Hatfull, 2008). Sudden *en bloc* changes in protein homologies or nucleotide sequence signatures such as codon and tetranucleotide usage can indicate genome mosaicism. This has frequently been observed in phages of *Mycobacterium* spp. (Pedulla *et al.*, 2003; Siranosian *et al.*, 2015), *Lactococcus* spp. (Labrie *et al.*, 2008; Murphy *et al.*, 2016), *Pseudomonas* spp. (Hertveldt *et al.*, 2005), *Burkholderia* spp. (Summer *et al.*, 2006), the *Enterobacteriaceae* (reviewed by Grose and Casjens, 2014), and more. Horizontal gene transfer and mosaicism are particularly well documented among the mycophages (Pedulla *et al.*, 2003; Hatfull, 2010; Pope *et al.*, 2015).

The *Cystoviridae* are the only phages that possess multipartite genomes. Rates of segment reassortment appear to be generally high, but geographically variable (Silander *et al.*, 2005; Díaz-Muñoz *et al.*, 2013). The geographic variability hints at ecological differences among geographically defined populations that substantially affect rates of sexual gene exchange (Díaz-Muñoz *et al.*, 2013). Since intra-segment recombination requires local sequence homology, it is, predictably, much rarer than reassortment and the limited data available does not indicate that intrasegment recombination is an important contributor to cystovirus evolution in natural populations (Onodera *et al.*, 2001; Silander *et al.*, 2005).

Transcriptomics and metabolomics

The genomics era, heralded by fast and affordable NGS technologies, has significantly advanced our understanding of phage–host interactions within virus- infected cells. The presence of early, middle, and late genes, however, was studied before genome sequencing was available at all. Generally speaking, early genes inactivate or redirect host cellular machinery to facilitate phage processes, middle genes drive phage genome replication, and late genes direct the production of progeny virions. Painstaking restriction mapping and pulse labelling experiments could overlay physical and transcription genetic maps to identify which genome regions were transcribed in clusters at various points during the lytic cycle (e.g. Pero *et al.*, 1979), and protein expression and binding assays could tease out interactions among some pairs of phage and host proteins (e.g. Van den Bossche *et al.*, 2014). Roucourt and Lavigne (2009) presented a table of 56 known interactions between phage and host proteins that alter host metabolism during infection, 36 of which are from *E. coli* tailed phages T7, T4, and λ. Clearly this does not come close to representing the full extent of phage–host interactions for these phages, let alone all of the other phage–host pairings that are available. Genomic studies have facilitated identification of ORFans involved in early infection processes, but do not themselves offer a means of studying global interactions among phages and their hosts.

Transcriptome analysis using RNA-Seq or microarrays gives a more complete picture of the diverse ways in which phages interact with their bacterial hosts. RNA-Seq in particular can characterize production of anti-sense RNAs that can regulate protein production beyond simple transcriptional timing (Chevallereau *et al.*, 2016). Do individual phages interact differently with different hosts, or during coinfection? Do different phages induce similar responses in a given host? Do the activities of viral communities differ even if the relative frequencies of their members are similar?

Giant phages such as *P. aeruginosa* phage ΦKZ are notable for encoding two sets of RNA polymerase (RNAP) subunits, one set of which is found within the mature virion. RNA-Seq analysis demonstrated that ΦKZ transcription is independent of the host transcription machinery, with the virion-associated RNAP driving early transcription including production of the second phage-encoded RNAP (Ceyssens *et al.*, 2014). Moreover, the phage RNAPs are rifampin resistant, unlike the bacterial RNAP. The giant phages are the only ones known to be transcriptionally independent of their host.

To investigate host effects on phage transcription, three phylogenetically distinct *Synechococcus* strains were grown in seawater-based medium and infected with the same phage, Syn9. Total RNA from control and infected cultures was isolated at multiple time points during infection. Phage transcription profiles were characterized by RNA-Seq and host transcription by custom microarrays. Phage gene transcription profiles were nearly identical regardless of host. Early genes made up 15% of the Syn9 genome, and are mostly unique to a small group of cyanophages. Middle genes made up 46% of the genome and include nucleic acid metabolism genes, phage-encoded tRNAs, anti-sense RNAs, and the sigma factor needed for later promoter activation. Late genes made up 39% of the genome, including structural and packaging genes. Several genes possess alternative promoters allowing, for example, a subset of the early genes to continue being transcribed at high levels well into the infection cycle. The three hosts responded noticeably differently to Syn9 infection. More than 90% of host genes were substantially down-regulated soon after infection, but of the genes that were not, most were located in hypervariable genomic islands and only two were common to all three hosts (Doron *et al.*, 2015).

Metaviromic studies explore the community composition of mixed microbial samples, but not their activity. Across multiple patients with varying states of periodontal health, RNA-Seq analysis of oral microbial communities showed that periodontal disease was associated with significantly more gene expression from myoviruses of Firmicutes but not of Proteobacteria in association (Santiago-Rodriguez *et al.*, 2015). Antirepressors, holins, and lysins were also more highly expressed in associate with periodontal disease, suggesting that periodontal health status may affect the induction of temperate phages.

The addition of metabolite analyses can improve interpretation of transcriptome data by linking transcription to biochemical outcomes or identifying biochemical changes that are not predicted by transcription profiling. For example, levels of a number of bacterial metabolites showed significant changes in PAK_P3-infected *P. aeruginosa* that appeared to contradict the observed changes in transcription of host genes for their respective pathways. This suggests that the metabolite changes were driven by phage utilization or phage-encoded auxiliary metabolic genes (Chevallereau *et al.*, 2016). De Smet *et al.* (2016) infected *P. aeruginosa* PAO1 with each of six obligately lytic phages, each from a different phage genus. Compared to the metabolite profiles of uninfected cells, approximately 25% of detected metabolites were significantly altered during phage infection. Only 2.4% of these were changed by all six phages and all of those were involved in nucleotide metabolism, presumably associated with phage genome replication. A little more than 17% of the metabolites whose levels were altered were altered by only one of the tested phages.

As yet, too few transcriptome and metabolome studies have been conducted to draw broad inferences about the correlation between genomic and transcriptomic similarities, or to explore the diversity of phage activity in different environmental situations. The potential for the existence of such correlations, however, clearly exists.

Summary – how much diversity is out there, really?

Postulating that the number of phage species is approximately an order of magnitude larger than the number of host bacterial species, it has been estimated that as many as 100 million phage species and 10 million tailed phage species may exists (Casjens, 2005). Grose and Casjens (2014) proposed a limit of 30–50 clusters of different phage types per host genus, using parameters that likely allow for multiple phage species within each cluster. Our efforts to observe and document as much diversity as possible, however, have always been constrained by the techniques available. Sequence-based methods have opened the door to understanding the immense variety of phages that had been previously been suspected to exist, but could not be confirmed or investigated with culturing-dependent options. New analytical techniques are enabling us to infer host-range relationships and ecological interactions from metagenomic data alone, and population-based genomic techniques are now being combined with single-cell techniques (Roux *et al.*, 2014). All of these developments offer us new ways to connect phage diversity to ecological processes across large and small scales. Ultimately, even this substantial phage diversity is just a fraction of the overall diversity that exists among all viruses of microorganisms.

References

Abedon, S.T. (2008). Phages, Ecology, Evolution. In: Bacteriophage Ecology: Population Growth, Evolution, and Impact of Bacterial Viruses. (Cambridge University Press), pp. 1–28.

Abedon, S.T. (2009). Disambiguating bacteriophage pseudolysogeny: An historical analysis of lysogeny, pseudolysogeny, and the phage carrier state. In Contemporary Trends in Bacteriophage Research, Adams, H.T. ed. (Nova Science Publishers, New York), pp. 385–307.

Ackermann, H.W., Audurier, A., Berthiaume, L., Jones, L.A., Mayo, J.A., and Vidaver, A.K. (1978). Guidelines for bacteriophage characterization. Adv. Virus Res. *23*, 1–24.

Ackermann, H.W., and DuBow, M.S. (1987). Viruses of Prokaryotes, Vol II, Natural Groups of Bacteriophages. (CRC Press, Boca Raton, FL)

Ackermann, H.W. (2004). Bacteriophages. In Desk Encyclopedia of Microbiology. Schaechter, M. ed. (Elsevier Academic Press), pp. 135–146.

Ackermann, H.W. (2007). 5500 Phages examined in the electron microscope. Arch. Virol. *152*, 227–243. https://doi.org/10.1007/s00705-006-0849-1.

Ackermann, H.W. (2011). The first phage electron micrographs. Bacteriophage *1*, 225–227. https://doi.org/10.4161/bact.1.4.17280.

Adams, M.H. (1959). Bacteriophages. (Interscience Publishers, New York)

Adriaenssens, E.M., Edwards, R., Nash, J.H., Mahadevan, P., Seto, D., Ackermann, H.W., Lavigne, R., and Kropinski, A.M. (2015). Integration of genomic and proteomic analyses in the classification of the *Siphoviridae* family. Virology *477*, 144–154. https://doi.org/10.1016/j.virol.2014.10.016.

Ågren, J., Sundström, A., Håfström, T., and Segerman, B. (2012). Gegenees: fragmented alignment of multiple genomes for determining phylogenomic distances and genetic signatures unique for specified target groups. PLOS ONE *7*, e39107. https://doi.org/10.1371/journal.pone.0039107.

Alimi, J.P., Poirot, O., Lopez, F., and Claverie, J.M. (2000). Reverse transcriptase-polymerase chain reaction validation of 25 "orphan" genes from *Escherichia coli* K-12 MG1655. Genome Res. *10*, 959–966. https://doi.org/10.1101/gr.10.7.959.

Ang, D., and Georgopoulos, C. (2012). An ORFan no more: the bacteriophage T4 *39.2* gene product, NwgI, modulates GroEL chaperone function. Genetics *190*, 989–1000. https://doi.org/10.1534/genetics.111.135640.

Angly, F.E., Felts, B., Breitbart, M., Salamon, P., Edwards, R.A., Carlson, C., Chan, A.M., Haynes, M., Kelley, S., Liu, H., *et al.* (2006). The marine viromes of four oceanic regions. PLOS Biol. *4*, e368. https://doi.org/10.1371/journal.pbio.0040368.

Archibald, A.R. (1980). Phage receptors in Gram-positive bacteria. In Virus Receptors Part I: Bacterial Viruses, Randall, L.L., and Philipson, L. eds. (Springer, Netherlands), pp. 5-26.

Aziz, R.K., Dwivedi, B., Akhter, S., Breitbart, M., and Edwards, R.A. (2015). Multidimensional metrics for estimating phage abundance, distribution, gene density, and sequence coverage in metagenomes. Front. Microbiol. *6*, 381. https://doi.org/10.3389/fmicb.2015.00381.

Baker, P.M., Farmer, J.J., (1982). New bacteriophage typing system for *Yersinia enterocolitica, Yersinia kristensenii, Yersinia frederiksenii*, and *Yersinia intermedia*: correlation with serotyping, biotyping, and antibiotic susceptibility. J. Clin. Microbiol. *15*, 491–502.

Bebeacua, C., Lorenzo Fajardo, J.C., Blangy, S., Spinelli, S., Bollmann, S., Neve, H., Cambillau, C., and Heller, K.J. (2013). X-ray structure of a superinfection exclusion lipoprotein from phage TP-J34 and identification of the tape measure protein as its target. Mol. Microbiol. *89*, 152–165. https://doi.org/10.1111/mmi.12267.

Betts, A., Gifford, D.R., MacLean, R.C., and King, K.C. (2016). Parasite diversity drives rapid host dynamics and evolution of resistance in a bacteria-phage system. Evolution *70*, 969–978. https://doi.org/10.1111/evo.12909.

Bik, H.M. (2014). Deciphering diversity and ecological function from marine metagenomes. Biol. Bull. *227*, 107–116. https://doi.org/10.1086/BBLv227n2p107.

Bordet, J., and Ciuca, M. (1921). Determinisme de l'autolyse microbienne transmissible. Compt. Rend. Soc. Biol. *84*, 276.

Breitbart, M., Salamon, P., Andresen, B., Mahaffy, J.M., Segall, A.M., Mead, D., Azam, F., and Rohwer, F. (2002). Genomic analysis of uncultured marine viral communities. Proc. Natl. Acad. Sci. U.S.A. *99*, 14250–14255. https://doi.org/10.1073/pnas.202488399.

Breitbart, M., Felts, B., Kelley, S., Mahaffy, J.M., Nulton, J., Salamon, P., and Rohwer, F. (2004). Diversity and population structure of a near-shore marine-sediment viral community. Proc. Biol. Sci. *271*, 565–574. https://doi.org/10.1098/rspb.2003.2628.

Brüssow, H., Fremont, M., Bruttin, A., Sidoti, J., Constable, A., and Fryder, V. (1994). Detection and classification of *Streptococcus thermophilus* bacteriophages isolated from industrial milk fermentation. Appl. Environ. Microbiol. *60*, 4537–4543.

Brüssow, H., and Bruttin, A. (1995). Characterization of a temperate *Streptococcus thermophilus* bacteriophage and its genetic relationship with lytic phages. Virology *212*, 632–640.

Bruynoghe, R. (1924). Les bacteriophages. Arch. Internat. Med. Exper. *1*, 17.

Bryan, D., El-Shibiny, A., Hobbs, Z., Porter, J., and Kutter, E.M. (2016). Bacteriophage T4 Infection of Stationary Phase *E. coli*: Life after Log from a Phage Perspective.

Front. Microbiol. *7*, 1391. https://doi.org/10.3389/fmicb.2016.01391.

Calendar, R., and Abedon, S.T. (2006). The Bacteriophages, 2nd edn, (Oxford University Press, New York)

Casjens, S.R. (2005). Comparative genomics and evolution of the tailed-bacteriophages. Curr. Opin. Microbiol. *8*, 451–458. https://dx.doi.org/10.1016/j.mib.2005.06.014.

Catalano, C.E. (2005). Viral Genome Packaging Machines: Genetics, Structure, and Mechanism. (Landes Bioscience/Eurekah, Georgetown, TX). https://doi.org/10.1007/0-387-28521-0.

Cenens, W., Mebrhatu, M.T., Makumi, A., Ceyssens, P.J., Lavigne, R., Van Houdt, R., Taddei, F., and Aertsen, A. (2013a). Expression of a novel P22 ORFan gene reveals the phage carrier state in *Salmonella* Typhimurium. PLOS Genet. *9*, e1003269. https://doi.org/10.1371/journal.pgen.1003269.

Cenens, W., Makumi, A., Mebrhatu, M.T., Lavigne, R., and Aertsen, A. (2013b). Phage-host interactions during pseudolysogeny: Lessons from the Pid/dgo interaction. Bacteriophage *3*, e25029. https://doi.org/10.4161/bact.25029.

Ceyssens, P.J., Minakhin, L., Van den Bossche, A., Yakunina, M., Klimuk, E., Blasdel, B., De Smet, J., Noben, J.P., Bläsi, U., Severinov, K., *et al.* (2014). Development of giant bacteriophage ϕKZ is independent of the host transcription apparatus. J. Virol. *88*, 10501–10510. https://doi.org/10.1128/JVI.01347-14.

Chevallereau, A., Blasdel, B.G., De Smet, J., Monot, M., Zimmermann, M., Kogadeeva, M., Sauer, U., Jorth, P., Whiteley, M., Debarbieux, L., *et al.* (2016). Next-Generation '-omics' Approaches Reveal a Massive Alteration of Host RNA Metabolism during Bacteriophage Infection of *Pseudomonas aeruginosa.* PLOS Genet. *12*, e1006134. https://doi.org/10.1371/journal.pgen.1006134.

Choi, Y., Shin, H., Lee, J.H., and Ryu, S. (2013). Identification and characterization of a novel flagellum-dependent *Salmonella*-infecting bacteriophage, iEPS5. Appl. Environ. Microbiol. *79*, 4829–4837. https://doi.org/10.1128/AEM.00706-13.

Chopin, M.C., Chopin, A., and Bidnenko, E. (2005). Phage abortive infection in lactococci: variations on a theme. Curr. Opin. Microbiol. *8*, 473–479. https://doi.org/10.1016/j.mib.2005.06.006.

Comeau, A.M., Tremblay, D., Moineau, S., Rattei, T., Kushkina, A.I., Tovkach, F.I., Krisch, H.M., and Ackermann, H.W. (2012). Phage morphology recapitulates phylogeny: The comparative genomics of a new group of myoviruses. PLOS ONE *7*, e40102. https://doi.org/10.1371/journal.pone.0040102.

Cumby, N., Edwards, A.M., Davidson, A.R., and Maxwell, K.L. (2012). The bacteriophage HK97 gp15 moron element encodes a novel superinfection exclusion protein. J. Bacteriol. *194*, 5012–5019. https://doi.org/10.1128/JB.00843-12.

Daniel, R. (2005). The metagenomics of soil. Nat. Rev. Microbiol. *3*, 470–478. https://doi.org/10.1038/nrmicro1160.

De Smet, J., Zimmermann, M., Kogadeeva, M., Ceyssens, P.J., Vermaelen, W., Blasdel, B., Bin Jang, H., Sauer, U., and Lavigne, R. (2016). High coverage metabolomics analysis reveals phage-specific alterations to *Pseudomonas aeruginosa* physiology during infection. ISME J. *10*, 1823–1835. https://doi.org/10.1038/ismej.2016.3.

Delbrück, M., and Bailey, W.T. (1946). Induced mutations in bacterial viruses. Cold Spring Harbor Symposia Quant. Biol. *11*, 33.

Delbrück, M., and Luria, S.E. (1942). Interference between bacterial viruses: I. Interference between two bacterial viruses acting upon the same host, and the mechanism of virus growth. Arch. Biochem. *1*, 111.

Deschavanne, P., DuBow, M.S., and Regeard, C. (2010). The use of genomic signature distance between bacteriophages and their hosts displays evolutionary relationships and phage growth cycle determination. Virol. J. *7*, 163. https://doi.org/10.1186/1743-422X-7-163.

Desnues, C., Rodriguez-Brito, B., Rayhawk, S., Kelley, S., Tran, T., Haynes, M., Liu, H., Furlan, M., Wegley, L., Chau, B., *et al.* (2008). Biodiversity and biogeography of phages in modern stromatolites and thrombolites. Nature *452*, 340–343. https://doi.org/10.1038/nature06735.

Díaz-Muñoz, S.L., Tenaillon, O., Goldhill, D., Brao, K., Turner, P.E., and Chao, L. (2013). Electrophoretic mobility confirms reassortment bias among geographic isolates of segmented RNA phages. BMC Evol. Biol. *13*, 206. https://doi.org/10.1186/1471-2148-13-206.

Doron, S., Fedida, A., Hernández-Prieto, M.A., Sabehi, G., Karunker, I., Stazic, D., Feingersch, R., Steglich, C., Futschik, M., Lindell, D., *et al.* (2016). Transcriptome dynamics of a broad host-range cyanophage and its hosts. ISME J. *10*, 1437–1455. https://doi.org/10.1038/ismej.2015.210.

Doškař, J., Pallová, P., Pantůček, R., Rosypal, S., Růžičková, V., Pantůčková, P., Kailerová, J., Klepárník, K., Malá, Z., and Bocek, P. (2000). Genomic relatedness of *Staphylococcus aureus* phages of the International Typing Set and detection of serogroup A, B, and F prophages in lysogenic strains. Can. J. Microbiol. *46*, 1066–1076.

Duffy, S., Burch, C.L., and Turner, P.E. (2007). Evolution of host specificity drives reproductive isolation among RNA viruses. Evolution *61*, 2614–2622. https://doi.org/10.1111/j.1558-5646.2007.00226.x.

Dutilh, B.E., Cassman, N., McNair, K., Sanchez, S.E., Silva, G.G.Z., Boling, L., Barr, J.J., Speth, D.R., Seguritan, V., Aziz, R.K., *et al.* (2014). Unknown sequences in faecal metagenomes reveal a widely distributed and highly abundant bacteriophage. Nat. Commun. *5*, 4498. https://doi.org/10.1038/ncomms5498.

Flores, C.O., Meyer, J.R., Valverde, S., Farr, L., and Weitz, J.S. (2011). Statistical structure of host-phage interactions. Proc. Natl. Acad. Sci. U.S.A. *108*, E288-E297. https://doi.org/10.1073/pnas.1101595108.

Flores, C.O., Valverde, S., and Weitz, J.S. (2013). Multi-scale structure and geographic drivers of cross-infection within marine bacteria and phages. ISME J. *7*, 520–532. https://doi.org/10.1038/ismej.2012.135.

Fodor, A.R., and Adams, M.H. (1955). Genetic control of serological specificity in bacteriophage. J. Immunol. *74*, 228–235.

Fogg, P.C., Gossage, S.M., Smith, D.L., Saunders, J.R., McCarthy, A.J., and Allison, H.E. (2007). Identification of multiple integration sites for Stx-phage Phi24B in the *Escherichia coli* genome, description of a novel integrase and evidence for a functional anti-repressor. Microbiology. *153*, 4098–4110. https://doi.org/10.1099/mic.0.2007/011205-0.

Gallet, R., Kannoly, S., and Wang, I.N. (2011). Effects of bacteriophage traits on plaque formation. BMC Microbiol. *11*, 181. https://doi.org/10.1186/1471-2180-11-181.

Goerke, C., Pantůček, R., Holtfreter, S., Schulte, B., Zink, M., Grumann, D., Bröker, B.M., Doškař, J., and Wolz, C. (2009). Diversity of prophages in dominant *Staphylococcus aureus* clonal lineages. J. Bacteriol. *191*, 3462–3468. https://doi.org/10.1128/JB.01804-08.

Goh, S., Riley, T.V., and Chang, B.J. (2005). Isolation and characterization of temperate bacteriophages of *Clostridium difficile*. Appl. Environ. Microbiol. *71*, 1079–1083. https://doi.org/10.1128/AEM.71.2.1079-1083.2005.

Goldsmith, D.B., Parsons, R.J., Beyene, D., Salamon, P., and Breitbart, M. (2015). Deep sequencing of the viral *phoH* gene reveals temporal variation, depth-specific composition, and persistent dominance of the same viral *phoH* genes in the Sargasso Sea. PeerJ 3, e997. https://doi.org/10.7717/peerj.997.

Grose, J.H., and Casjens, S.R. (2014). Understanding the enormous diversity of bacteriophages: the tailed phages that infect the bacterial family *Enterobacteriaceae*. Virology 468–470, 421–443. https://doi.org/10.1016/j.virol.2014.08.024.

Guttman, B., Raya, R., and Kutter, E. (2005). Basic phage biology, In Bacteriophages: Biology and applications, Kutter, E., and Sulakvelidze, A. eds. (CRC Press, Boca Raton, FL), pp. 29–66.

Hatfull, G.F. (2008). Bacteriophage genomics. Curr. Opin. Microbiol., *11*, 447–453. https://doi.org/10.1016/j.mib.2008.09.004.

Hatfull, G.F. (2014). Mycobacteriophages: windows into tuberculosis. PLOS Pathog. *10*, e1003953. https://doi.org/10.1371/journal.ppat.1003953.

Hatfull, G.F., Jacobs-Sera, D., Lawrence, J.G., Pope, W.H., Russell, D.A., Ko, C.C., Weber, R.J., Patel, M.C., Germane, K.L., Edgar, R.H., *et al.* (2010). Comparative genomic analysis of 60 mycobacteriophage genomes: Genome clustering, gene acquisition, and gene size. J. Mol. Biol. *397*, 119–143. https://doi.org/10.1016/j.jmb.2010.01.011.

d'Herelle, F. (1926). The Bacteriophage and its Behavior. (The Williams & Wilkins Company, Baltimore)

Hershey, A.D., and Rotman, R. (1949). Genetic recombination between host-range and plaque-type mutants of bacteriophage in single bacterial cells. Genetics *34*, 44–71.

Hertveldt, K., Lavigne, R., Pleteneva, E., Sernova, N., Kurochkina, L., Korchevskii, R., Robben, J., Mesyanzhinov, V., Krylov, V.N., and Volckaert, G. (2005). Genome comparison of *Pseudomonas aeruginosa* large phages. J. Mol. Biol. *354*, 536–545. https://doi.org/10.1016/j.jmb.2005.08.075.

Hickman-Brenner, F.W., Stubbs, A.D., and Farmer, J.J. (1991). Phage typing of *Salmonella enteritidis* in the United States. J. Clin. Microbiol. *29*, 2817–2823.

Hirashima, A., Inokuchi, Y., Harigai, H., Furuse, K., and Watanabe, I. (1983). Biochemical characterization of RNA coliphage MX1. J. Gen. Virol. *64*, 939–946. https://doi.org/10.1099/0022-1317-64-4-939.

Holmfeldt, K., Solonenko, N., Shah, M., Corrier, K., Riemann, L., Verberkmoes, N.C., and Sullivan, M.B. (2013). Twelve previously unknown phage genera are ubiquitous in global oceans. Proc. Natl. Acad. Sci. U.S.A. *110*, 12798–12803. https://doi.org/10.1073/pnas.1305956110.

Hopkins, M., Kailasan, S., Cohen, A., Roux, S., Tucker, K.P., Shevenell, A., Agbandje-McKenna, M., and Breitbart, M. (2014). Diversity of environmental single-stranded DNA phages revealed by PCR amplification of the partial major capsid protein. ISME J. *8*, 2093–2103. https://doi.org/10.1038/ismej.2014.43.

Hurwitz, B.L., and Sullivan, M.B. (2013). The Pacific Ocean Virome (POV): a marine viral metagenomic dataset and associated protein clusters for quantitative viral ecology. PLOS ONE *8*, e57355. https://doi.org/10.1371/journal.pone.0057355.

Hyman, P., and Abedon, S.T. (2012). Smaller fleas: viruses of microorganisms. Scientifica *2012*, 734023. https://doi.org/10.6064/2012/734023.

Ignacio-Espinoza, J.C., Solonenko, S.A., and Sullivan, M.B. (2013). The global virome: not as big as we thought? Curr. Opin. Virol. *3*, 566–571. https://doi.org/10.1016/j.coviro.2013.07.004.

Jarvis, A.W. (1984). Differentiation of lactic streptococcal phages into phage species by DNA-DNA homology. Appl. Environ. Microbiol. *47*, 343–349.

Jensen, E.C., Schrader, H.S., Rieland, B., Thompson, T.L., Lee, K.W., Nickerson, K.W., and Kokjohn, T.A. (1998). Prevalence of broad-host-range lytic bacteriophages of *Sphaerotilus natans*, *Escherichia coli*, and *Pseudomonas aeruginosa*. Appl. Environ. Microbiol. *64*, 575–580.

Jurczak-Kurek, A., Gąsior, T., Nejman-Faleńczyk, B., Bloch, S., Dydecka, A., Topka, G., Necel, A., Jakubowska-Deredas, M., Narajczyk, M., Richert, M., *et al.* (2016). Biodiversity of bacteriophages: morphological and biological properties of a large group of phages isolated from urban sewage. Sci. Rep. *6*, 34338. https://doi.org/10.1038/srep34338.

Koskella, B., and Meaden, S. (2013). Understanding bacteriophage specificity in natural microbial communities. Viruses *5*, 806–823. https://doi.org/10.3390/v5030806.

Koskella, B., and Brockhurst, M.A. (2014). Bacteria-phage coevolution as a driver of ecological and evolutionary processes in microbial communities. FEMS Microbiol. Rev. *38*, 916–931. https://doi.org/10.1111/1574-6976.12072.

Krishnamurthy, S.R., Janowski, A.B., Zhao, G., Barouch, D., and Wang, D. (2016). Hyperexpansion of RNA bacteriophage diversity. PLOS Biol. *14*, e1002409. https://doi.org/10.1371/journal.pbio.1002409.

Kristensen, D.M., Cai, X., and Mushegian, A. (2011). Evolutionarily conserved orthologous families in phages are relatively rare in their prokaryotic hosts. J. Bacteriol. *193*, 1806–1814. https://doi.org/10.1128/JB.01311-10.

Krupovic, M., Dutilh, B.E., Adriaenssens, E.M., Wittmann, J., Vogensen, F.K., Sullivan, M.B., Rumnieks, J., Prangishvili, D., Lavigne, R., Kropinski, A.M., *et al.* (2016). Taxonomy of prokaryotic viruses: update from the ICTV bacterial and archaeal viruses subcommittee. Arch. Virol. *161*, 1095–1099. https://doi.org/10.1007/s00705-015-2728-0.

Kunisawa, T., Kanaya, S., and Kutter, E. (1998). Comparison of synonymous codon distribution patterns of bacteriophage and host genomes. DNA Res. *5*, 319–326.

Khawaja, K.A., Rauf, M., Abbas, Z., and Rehman, S.U. (2016). A virulent phage JHP against *Pseudomonas aeruginosa* showed infectivity against multiple genera. J. Basic Microbiol. *56*, 1090–1097. https://doi.org/10.1002/jobm.201500764.

Kutter, E., and Sulakvelidze, A. (2005). Bacteriophages: Biology and Application. (CRC Press, Boca Raton, FL). https://doi.org/10.1201/9780203491751.

Labrie, S.J., Josephsen, J., Neve, H., Vogensen, F.K., and Moineau, S. (2008). Morphology, genome sequence, and structural proteome of type phage P335 from *Lactococcus lactis*. Appl. Environ. Microbiol. *74*, 4636–4644. https://doi.org/10.1128/AEM.00118-08.

Lanni, F., and Lanni, Y.T. (1953). Antigenic structure of bacteriophage. Cold Spring Harb. Symp. Quant. Biol. *18*, 159–168.

Lavigne, R., Seto, D., Mahadevan, P., Ackermann, H.W., and Kropinski, A.M. (2008). Unifying classical and molecular taxonomic classification: analysis of the *Podoviridae* using BLASTP-based tools, Res. Microbiol. *159*, 406–414. https://doi.org/10.1016/j.resmic.2008.03.005.

Lavigne, R., Darius, P., Summer, E.J., Seto, D., Mahadevan, P., Nilsson, A.S., Ackermann, H.W., and Kropinski, A.M. (2009). Classification of *Myoviridae* bacteriophages using protein sequence similarity. BMC Microbiol. *9*, 224. https://doi.org/10.1186/1471-2180-9-224.

Lindberg, R.B., and Latta, R.L. (1974). Phage typing of *Pseudomonas aeruginosa*: clinical and epidemiologic considerations. J. Infect. Dis. *130*, S33–S42. https://doi.org/10.1093/infdis/130.Supplement.S33.

Lucks, J.B., Nelson, D.R., Kudla, G.R., and Plotkin, J.B. (2008). Genome landscapes and bacteriophage codon usage. PLOS Comput. Biol. *4*, e1000001. https://doi.org/10.1371/journal.pcbi.1000001.

Mahadevan, P., King, J.F., and Seto, D. (2009). CGUG: in silico proteome and genome parsing tool for the determination of "core" and unique genes in the analysis of genomes up to ca. 1.9 Mb. BMC Res. Notes 2, 168. https://doi.org/10.1186/1756-0500-2-168.

Mahadevan, P., and Seto, D. (2010). Rapid pair-wise synteny analysis of large bacterial genomes using web-based GeneOrder4.0. BMC Res. Notes 3, 41. https://doi.org/10.1186/1756-0500-3-41.

Manrique, P., Bolduc, B., Walk, S.T., van der Oost, J., de Vos, W.M., and Young, M.J. (2016). Healthy human gut phageome. Proc. Natl. Acad. Sci. U.S.A. *113*, 10400–10405. https://doi.org/10.1073/pnas.1601060113.

Mattenberger, Y., Silva, F., and Belin, D. (2015). 55.2, a phage T4 ORFan gene, encodes an inhibitor of *Escherichia coli* topoisomerase I and increases phage fitness. PLOS ONE *10*, e0124309. https://doi.org/10.1371/journal.pone.0124309.

Mazumder, R., Kolaskar, A., and Seto, D. (2001). GeneOrder: comparing the order of genes in small genomes. Bioinformatics *17*, 162–166. https://doi.org/10.1093/bioinformatics/17.2.162.

Mc Grath, S., and van Sinderen, D. (2007). Bacteriophage: Genetics and Molecular Biology. (Caister Academic Press,UK).

Meczker, K., Dömötör, D., Vass, J., Rákhely, G., Schneider, G., and Kovács, T. (2014). The genome of the *Erwinia amylovora* phage PhiEaH1 reveals greater diversity and broadens the applicability of phages for the treatment of fire blight. FEMS Microbiol. Lett. *350*, 25–27. https://doi.org/10.1111/1574-6968.12319.

Mihara, T., Nishimura, Y., Shimizu, Y., Nishiyama, H., Yoshikawa, G., Uehara, H., Hingamp, P., Goto, S., and Ogata, H. (2016). Linking virus genomes with host taxonomy. Viruses *8*, 66. https://doi.org/10.3390/v8030066.

Mindich, L. (2006). Phages with segmented double-stranded RNA genomes. In The Bacteriophages, 2nd edn, Calendar, R., and Abedon, S.T. eds. (Oxford University Press, New York), pp. 197–209.

Mirzaei, M.K., Eriksson, H., Kasuga, K., Haggård-Ljungquist, E., and Nilsson, A.S. (2014). Genomic, proteomic, morphological, and phylogenetic analyses of vB_EcoP_SU10, a *Podoviridae* phage with C3 morphology. PLOS ONE *9*, e116294. https://doi.org/10.1371/journal.pone.0116294.

Mizuno, C.M., Rodriguez-Valera, F., Kimes, N.E., and Ghai, R. (2013a). Expanding the marine virosphere using metagenomics. PLOS Genet. *9*, e1003987. https://doi.org/10.1371/journal.pgen.1003987.

Mizuno, C.M., Rodriguez-Valera, F., Garcia-Heredia, I., Martin-Cuadrado, A.-B., and Ghai, R. (2013b). Reconstruction of novel cyanobacterial siphovirus genomes from Mediterranean metagenomic fosmids. Appl. Environ. Microbiol. *79*, 688–695. https://doi.org/10.1128/AEM.02742-12.

Murphy, J., Bottacini, F., Mahony, J., Kelleher, P., Neve, H., Zomer, A., Nauta, A., and van Sinderen, D. (2016). Comparative genomics and functional analysis of the 936 group of lactococcal *Siphoviridae* phages. Sci. Rep. *6*, 21345. https://doi.org/10.1038/srep21345.

Nelson, D. (2004). Phage taxonomy: we agree to disagree. J. Bacteriol. *186*, 7029–7031. https://doi.org/10.1128/JB.186.21.7029-7031.2004.

Oakley, B.B., Talundzic, E., Morales, C.A., Hiett, K.L., Siragusa, G.R., Volozhantsev, N.V., and Seal, B.S. (2011). Comparative genomics of four closely related *Clostridium perfringens* bacteriophages reveals variable evolution among core genes with therapeutic potential. BMC Genomics *12*, 282. https://doi.org/10.1186/1471-2164-12-282.

Onodera, S., Sun, Y., and Mindich, L. (2001). Reverse genetics and recombination in Phi8, a dsRNA bacteriophage. Virology *286*, 113–118. https://doi.org/10.1006/viro.2001.0995.

Otto, R., and Winkler, W.F. (1922). The bacteriophage. Deut. Med. Wochschr. *48*, 383.

Pantůček, R., Doškař, J., Růžičková, V., Kašpárek, P., Oráčová, E., Kvardová, V., and Rosypal, S. (2004). Identification of bacteriophage types and their carriage in *Staphylococcus aureus*. Arch. Virol. *149*, 1689–1703.

Paolozzi, L., and Ghelardini, P. (2006). The bacteriophage Mu. In The Bacteriophages, Calendar, R., and Abedon, S.T. eds (Oxford University Press, UK), pp. 469–496.

Parsons, R.J., Breitbart, M., Lomas, M.W., and Carlson, C.A. (2012). Ocean time-series reveals recurring seasonal patterns of virioplankton dynamics in the northwestern Sargasso Sea. ISME J. *6*, 273–284. https://doi.org/10.1038/ismej.2011.101.

Pawluk, A., Amrani, N., Zhang, Y., Garcia, B., Hidalgo-Reyes, Y., Lee, J., Edraki, A., Shah, M., Sontheimer, E.J., Maxwell, K.L., *et al.* (2016). Naturally Occurring Off-Switches for CRISPR-Cas9. Cell *167*, 1829–1838. https://doi.org/10.1016/j.cell.2016.11.017.

Pedulla, M.L., Ford, M.E., Houtz, J.M., Karthikeyan, T., Wadsworth, C., Lewis, J.A., Jacobs-Sera, D., Falbo, J., Gross, J., Pannunzio, N.R., *et al.* (2003). Origins of highly mosaic mycobacteriophage genomes. Cell *113*, 171–182. https://doi.org/10.1016/S0092-8674(03)00233-2.

Pero, J., Hannett, N.M., and Talkington, C. (1979). Restriction cleavage map of SP01 DNA: general location of early, middle, and late genes. J. Virol. *31*, 156–171.

Peters, D.L., Lynch, K.H., Stothard, P., and Dennis, J.J. (2015). The isolation and characterization of two *Stenotrophomonas maltophilia* bacteriophages capable of cross-taxonomic order infectivity. BMC Genomics *16*, 664. https://doi.org/10.1186/s12864-015-1848-y.

Pope, W.H., Bowman, C.A., Russell, D.A., Jacobs-Sera, D., Asai, D.J., Cresawn, S.G., Jacobs, W.R., Hendrix, R.W., Lawrence, J.G., Hatfull, G.F., *et al.* (2015). Whole genome comparison of a large collection of mycobacteriophages reveals a continuum of phage genetic diversity. eLife *4*, e06416. https://doi.org/10.7554/eLife.06416.

Pride, D.T., Meinersmann, R.J., Wassenaar, T.M., and Blaser, M.J. (2003). Evolutionary implications of microbial genome tetranucleotide frequency biases. Genome Res. *13*, 145–158. https://doi.org/10.1101/gr.335003.

Pride, D.T., Wassenaar, T.M., Ghose, C., and Blaser, M.J. (2006). Evidence of host-virus co-evolution in tetranucleotide usage patterns of bacteriophages and eukaryotic viruses. BMC Genomics. *7*, 8. https://doi.org/10.1186/1471-2164-7-8.

Pristas, P., Vandzurova, A., and Javorsky, P. (2014). Interplay between bacteriophages and restriction-modification systems in Enterococci. Nova Biotechnol. Chim. *13*, 13–20. https://doi.org/10.2478/nbec-2014-0002.

Proux, C., van Sinderen, D., Suarez, J., Garcia, P., Ladero, V., Fitzgerald, G.F., Desiere, F., and Brüssow, H. (2002). The dilemma of phage taxonomy illustrated by comparative genomics of Sfi21-like *Siphoviridae* in lactic acid bacteria. J. Bacteriol. *184*, 6026–6036. https://doi.org/10.1128/JB.184.21.6026-6036.2002.

Ptashne, M. (2004). A Genetic Switch: Phage Lambda Revisited, 3rd edn. (Cold Spring Harbor Laboratory Press, Cold Spring Harbor, NY).

Rakhuba, D.V., Kolomiets, E.I., Dey, E.S., and Novik, G.I. (2010). Bacteriophage receptors, mechanisms of phage adsorption and penetration into host cell. Pol. J. Microbiol. 59, 145–155.

Ram, G., Chen, J., Kumar, K., Ross, H.F., Ubeda, C., Damle, P.K., Lane, K.D., Penadés, J.R., Christie, G.E., and Novick, R.P. (2012). Staphylococcal pathogenicity island interference with helper phage reproduction is a paradigm of molecular parasitism. Proc. Natl. Acad. Sci. U.S.A. *109*, 16300–16305. https://doi.org/10.1073/pnas.1204615109.

Randall and Philipson. (1980). Virus Receptors Part 1: Bacterial Viruses. Receptors and Recognition Series B, Volume 7. (Chapman and Hall, UK).

Rice, S.A., Tan, C.H., Mikkelsen, P.J., Kung, V., Woo, J., Tay, M., Hauser, A., McDougald, D., Webb, J.S., and Kjelleberg, S. (2009). The biofilm life cycle and virulence of *Pseudomonas aeruginosa* are dependent on a filamentous prophage. ISME J. *3*, 271–282. https://doi.org/10.1038/ismej.2008.109.

Rocha, E.P.C., and Danchin, A. (2002). Base composition bias might result from competition for metabolic resources. Trends Genet. *18*, 291-294. https://doi.org/10.1016/S0168-9525(02)02690-2.

Rohwer, F. (2003). Global phage diversity. Cell *113*, 141. https://doi.org/10.1016/S0092-8674(03)00276-9.

Rohwer, F., and Edwards, R. (2002). The Phage Proteomic Tree: a genome-based taxonomy for phage. J. Bacteriol. *184*, 4529–4535. https://doi.org/10.1128/JB.184.16.4529-4535.2002.

Ross, A., Ward, S., and Hyman, P. (2016). More is better: selecting for broad host range bacteriophages. Front. Microbiol. *7*, 1352. https://doi.org/10.3389/fmicb.2016.01352.

Roucourt, B., and Lavigne, R. (2009). The role of interactions between phage and bacterial proteins within the infected cell: a diverse and puzzling interactome. Environ. Microbiol. *11*, 2789–2805. https://doi.org/10.1111/j.1462-2920.2009.02029.x.

Rountree, P.M. (1949). The serological differentiation of staphylococcal bacteriophages. J. Gen. Microbiol. *3*, 164–173. https://doi.org/10.1099/00221287-3-2-164.

Roux, S., Enault, F., Robin, A., Ravet, V., Personnic, S., Theil, S., Colombet, J., Sime-Ngando, T., and Debroas, D. (2012a). Assessing the diversity and specificity of two freshwater viral communities through metagenomics. PLOS ONE *7*, e33641. https://doi.org/10.1371/journal.pone.0033641.

Roux, S., Krupovic, M., Poulet, A., Debroas, D., and Enault, F. (2012b). Evolution and diversity of the *Microviridae* viral family through a collection of 81 new complete genomes assembled from virome reads. PLOS ONE *7*, e40418. https://doi.org/10.1371/journal.pone.0040418.

Roux, S., Tournayre, J., Mahul, A., Debroas, D., and Enault, F. (2014). Metavir 2: new tools for viral metagenome comparison and assembled virome analysis. BMC Bioinformatics *15*, 76. https://doi.org/10.1186/1471-2105-15-76.

Russell, R.L. (1974). Comparative genetics of the T-even bacteriophages. Genetics 78, 967–988.

Sahu, K., Gupta, S.K., Sau, S., and Ghosh, T.C. (2005). Comparative analysis of the base composition and codon usages in fourteen mycobacteriophage genomes. J. Biomol. Struct. Dyn. *23*, 63–71. https://doi.org/10.1080/07391102.2005.10507047.

Samson, J.E., Magadán, A.H., Sabri, M., and Moineau, S. (2013). Revenge of the phages: defeating bacterial defences. Nat. Rev. Microbiol. *11*, 675–687. https://doi.org/10.1038/nrmicro3096.

Santiago-Rodriguez, T.M., Naidu, M., Abeles, S.R., Boehm, T.K., Ly, M., and Pride, D.T. (2015). Transcriptome analysis of bacteriophage communities in periodontal health and disease. BMC Genomics *16*, 549. https://doi.org/10.1186/s12864-015-1781-0.

Sau, K., Sau, S., Mandal, S.C., and Ghosh, T.C. (2005). Factors influencing the synonymous codon and amino acid usage bias in AT-rich *Pseudomonas aeruginosa* phage PhiKZ. Acta Biochim. Biophys. Sin. *37*, 625–633.

Scanlan, P.D., Hall, A.R., Lopez-Pascua, L.D., and Buckling, A. (2011). Genetic basis of infectivity evolution in a bacteriophage. Mol. Ecol. *20*, 981–989. https://doi.org/10.1111/j.1365-294X.2010.04903.x.

Schrader, H.S., Schrader, J.O., Walker, J.J., Wolf, T.A., Nickerson, K.W., and Kokjohn, T.A. (1997). Bacteriophage infection and multiplication occur in *Pseudomonas aeruginosa* starved for 5 years. Can. J. Microbiol. *43*, 1157–1163.

Serwer, P., Hayes, S.J., Thomas, J.A., and Hardies, S.C. (2007). Propagating the missing bacteriophages: a large bacteriophage in a new class. Virol. J. *4*, 21. https://doi.org/10.1186/1743-422X-4-21.

Shmuely, H., Dinitz, E., Dahan, I., Eichler, J., Fischer, D., and Shaanan, B. (2004). Poorly conserved ORFs in the genome of the archaea *Halobacterium* sp. NRC-1 correspond to expressed proteins. Bioinformatics *20*, 1248–1253. https://doi.org/10.1093/bioinformatics/bth075.

Shirling, E.B., and Speer, M. (1967). Spot test method for rapid serological grouping of streptomycete bacteriophages. Appl. Microbiol.*15*, 456–457.

Siew, N., and Fischer, D. (2003). Twenty thousand ORFan microbial protein families for the biologist? Structure *11*, 7–9. https://doi.org/10.1016/S0969-2126(02)00938-3.

Silander, O.K., Weinreich, D.M., Wright, K.M., O'Keefe, K.J., Rang, C.U., Turner, P.E., and Chao, L. (2005). Widespread genetic exchange among terrestrial bacteriophages. Proc. Natl. Acad. Sci. U.S.A. *102*, 19009–19014. https://doi.org/10.1073/pnas.0503074102.

Siranosian, B., Perera, S., Williams, E., Ye, C., de Graffenried, C., and Shank, P. (2015). Tetranucleotide usage highlights genomic heterogeneity among mycobacteriophages. F1000Res *4*, 36. https://doi.org/10.12688/f1000research.6077.2.

Siringan, P., Connerton, P.L., Cummings, N.J., and Connerton, I.F. (2014). Alternative bacteriophage life cycles: the carrier state of *Campylobacter jejuni*. Open Biol. *4*, 130200. https://doi.org/10.1098/rsob.130200.

Smith, K.C., Castro-Nallar, E., Fisher, J.N.B., Breakwell, D.P., Grose, J.H., and Burnett, S.H. (2013). Phage cluster relationships identified through single gene analysis. BMC Genomics *14*, 410. https://doi.org/10.1186/1471-2164-14-410.

Solheim, H.T., Sekse, C., Urdahl, A.M., Wasteson, Y., and Nesse, L.L. (2013). Biofilm as an environment for dissemination of *stx* genes by transduction. Appl. Environ. Microbiol. *79*, 896–900. https://doi.org/10.1128/AEM.03512-12.

Spanakis, E., and Horne, M.T. (1987). Co-adaptation of *Escherichia coli* and coliphage lambda vir in continuous culture. J. Gen. Microbiol. *133*, 353–360. https://doi.org/10.1099/00221287-133-2-353.

Sulakvelidze, A., and Kutter, E. (2005). Bacteriophage therapy in humans, In Bacteriophages: Biology and Application, Kutter, E., and Sulakvelidze, A. eds. (CRC Press, Boca Raton, FL), pp 381–436.

Sullivan, M.B., Waterbury, J.B., and Chisholm, S.W. (2003). Cyanophages infecting the oceanic cyanobacterium *Prochlorococcus*. Nature *424*, 1047–1051. https://doi.org/10.1038/nature01929.

Summer, E.J., Gonzalez, C.F., Bomer, M., Carlile, T., Embry, A., Kucherka, A.M., Lee, J., Mebane, L., Morrison, W.C., Mark, L., *et al.* (2006). Divergence and mosaicism among virulent soil phages of the *Burkholderia cepacia* complex. J. Bacteriol. *188*, 255–268. https://doi.org/10.1128/JB.188.1.255-268.2006.

Tanaka, T., Kawasaki, K., Daimon, S., Kitagawa, W., Yamamoto, K., Tamaki, H., Tanaka, M., Nakatsu, C.H., and Kamagata, Y. (2014). A hidden pitfall in the preparation of agar media undermines microorganism cultivability. Appl. Environ. Microbiol. *80*, 7659–7666. https://doi.org/10.1128/AEM.02741-14.

Tock, M.R., and Dryden, D.T. (2005). The biology of restriction and anti-restriction. Curr. Opin. Microbiol. *8*, 466–472. https://doi.org/10.1016/j.mib.2005.06.003.

Tucker, K.P., Parsons, R., Symonds, E.M., and Breitbart, M. (2011). Diversity and distribution of single-stranded DNA phages in the North Atlantic Ocean. ISME J. *5*, 822–830. https://doi.org/10.1038/ismej.2010.188.

Turner, D., Reynolds, D., Seto, D., and Mahadevan, P. (2013). CoreGenes3.5: a webserver for the determination of core genes from sets of viral and small bacterial genomes. BMC Res. Notes *6*, 140. https://doi.org/10.1186/1756-0500-6-140.

Van den Bossche, A., Ceyssens, P.J., De Smet, J., Hendrix, H., Bellon, H., Leimer, N., Wagemans, J., Delattre, A.S., Cenens, W., Aertsen, A., *et al.* (2014). Systematic identification of hypothetical bacteriophage proteins targeting key protein complexes of *Pseudomonas aeruginosa*. J. Proteome Res. *13*, 4446–4456. https://doi.org/10.1021/pr500796n.

Villarroel, J., Kleinheinz, K.A., Jurtz, V.I., Zschach, H., Lund, O., Nielsen, M., and Larsen, M.V. (2016). HostPhinder: A Phage Host Prediction Tool. Viruses *8*, E116. https://doi.org/10.3390/v8050116.

Wagemans, J., Blasdel, B.G., Van den Bossche, A., Uytterhoeven, B., De Smet, J., Paeshuyse, J., Cenens, W., Aertsen, A., Uetz, P., Delattre, A.S., *et al.* (2014). Functional elucidation of antibacterial phage ORFans targeting *Pseudomonas aeruginosa*. Cell. Microbiol. *16*, 1822–1835. https://doi.org/10.1111/cmi.12330.

Wagemans, J., Delattre, A.S., Uytterhoeven, B., De Smet, J., Cenens, W., Aertsen, A., Ceyssens, P.J., and Lavigne, R. (2015). Antibacterial phage ORFans of *Pseudomonas aeruginosa* phage LUZ24 reveal a novel MvaT inhibiting protein. Front. Microbiol. *6*, 1242. https://doi.org/10.3389/fmicb.2015.01242.

Webb, J.S., Thompson, L.S., James, S., Charlton, T., Tolker-Nielsen, T., Koch, B., Givskov, M., and Kjelleberg, S. (2003). Cell death in *Pseudomonas aeruginosa* biofilm development. J. Bacteriol. *185*, 4585–4592. https://doi.org doi/10.1128/JB.185.15.4585-4592.2003.

Webb, J.S., Lau, M., and Kjelleberg, S. (2004). Bacteriophage and phenotypic variation in *Pseudomonas aeruginosa* biofilm development. J. Bacteriol. *186*, 8066–8073. https://doi.org doi/10.1128/JB.186.23.8066-8073.2004.

Wei, Y., Ocampo, P., and Levin, B.R. (2010). An experimental study of the population and evolutionary dynamics of *Vibrio cholerae* O1 and the bacteriophage JSF4. Proc. Biol. Sci. *277*, 3247–3254. https://doi.org/10.1098/rspb.2010.0651.

Weinbauer, W.G., and Peduzzi, P. (1994). Frequency, size and distribution of bacteriophages in different marine bacterial morphotypes. Mar. Ecol. Prog. Ser. *108*, 11–20.

Williams, M.L., and LeJeune, J.T. (2012). Phages and bacterial epidemiology. In Bacteriophages in Health and Disease, Hyman, P., and Abedon, S.T., eds., (AB International, UK), pp.76–85.

Wommack, K.E., Bhavsar, J., and Ravel, J. (2008). Metagenomics: read length matters. Appl. Environ. Microbiol. *74*, 1453–1463. https://doi.org/10.1128/AEM.02181-07.

Xia, X., and Yuen, K.Y. (2005). Differential selection and mutation between dsDNA and ssDNA phages shape the evolution of their genomic AT percentage. BMC Genet. *6*, 20. https://doi.org doi/10.1186/1471-2156-6-20.

Yin, Y., and Fischer, D. (2008). Identification and investigation of ORFans in the viral world. BMC Genomics *9*, 24. https://doi.org/10.1186/1471-2164-9-24.

Yu, F., and Mizushima, S. (1982). Roles of lipopolysaccharide and outer membrane protein OmpC of *Escherichia coli* K-12 in the receptor function for bacteriophage T4. J. Bacteriol. *151*, 718–722.

Yu, P., Mathieu, J., Li, M., Dai, Z., and Alvarez, P.J. (2016). Isolation of polyvalent bacteriophages by sequential multiple-host approaches. Appl. Environ. Microbiol. *82*, 808–815. https://doi.org/10.1128/AEM.02382-15.

Zablocki, O., van Zyl, L., Adriaenssens, E.M., Rubagotti, E., Tuffin, M., Cary, S.C., and Cowan, D. (2014). High-level diversity of tailed phages, eukaryote-associated viruses, and virophage-like elements in the metaviromes of Antarctic soils. Appl. Environ. Microbiol. *80*, 6888–6897. https://doi.org/10.1128/AEM.01525-14.

Zafar, N., Mazumder, R., and Seto, D. (2002). CoreGenes: a computational tool for identifying and cataloging "core" genes in a set of small genomes. BMC Bioinformatics 3, 12. https://doi.org doi/10.1186/1471-2105-3-12.

Zhang, H., Li, L., Zhao, Z., Peng, D., and Zhou, X. (2016). Polar flagella rotation in *Vibrio parahaemolyticus* confers resistance to bacteriophage infection. Sci. Rep. *6*, 26147. https://doi.org/10.1038/srep26147

Viruses of Domain Archaea

8

Stephen T. Abedon*

Department of Microbiology, The Ohio State University, Mansfield, OH, USA.

*Correspondence: abedon.1@osu.edu

https://doi.org/10.21775/9781910190852.08

Abstract

There are three cellular domains – Archaea, Bacteria, and Eukarya – each of which possesses its own distinct set of viruses. Of these viruses the least studied and, indeed, least appreciated are the archaeal viruses. This poor appreciation stems in part from the historical lateness of the recognition of domain Archaea as a distinct lineage but also the historical lateness of the isolation of the first archaeal virus, with both events occurring in the 1970s. In addition, Archaea tend to be relatively difficult to cultivate and for the most part are not responsible for diseases (with both statuses resulting in a relative lack of applied interest). Their viruses, however, seem to display a plethora of morphologies which are distinct to the Archaea. The study of archaeal viruses as a consequence of these factors is relatively new but nevertheless has become an exciting area of inquiry, with a promise of discovery of much novel biology. In this chapter I briefly consider the history of archaeal virus study, provide an overview of what hosts these viruses have been isolated against, and describe archaeal virus types. The intention overall is to provide an introduction to the idea of archaeal viruses, along with their history and characteristics, rather than, and in light of space limitations, an exhaustive overview of this otherwise still relatively new field.

Introduction

The major ecological roles of viruses are ones of exploitation, transduction, and also decomposition of the cells of their host organisms (Abedon, 2011), and it is commonly assumed that all cellular species possess associated viruses. While the latter assumption may or may not be precisely correct, there is one group of cellular organisms that was known to possess associated viruses before that group was technically understood to even exist, at least as a distinct lineage. These are the prokaryotic microorganisms known as archaea or, formally since 1990 (Woese *et al.*), as members of domain Archaea. The archaeal viruses, even given their discovery prior to the discovery of domain Archaea (Abedon and Murray, 2013), were the last of the three major groups of viruses to be discovered. Thus, while the discovery of bacterial viruses, a.k.a. bacteriophages, followed the discovery of eukaryotic viruses by multiple decades, at least as generally thought (Abedon *et al.*, 2011; Chanishvili, 2012), the discovery of archaeal viruses followed the most fully agreed upon date of discovery of bacteriophages, 1915 or 1917 (Duckworth, 1976), by more half of a century. The study of archaeal viruses, among viruses, thus is a relatively recent endeavour. Consistently, and with little exaggeration, both domain Archaea and archaeal viruses have been described as 'mysterious' (Snyder *et al.*, 2015), that is, as compared to domains Bacteria and Eukarya along with their respective viruses.

Numerous, excellent general reviews have been published on archaeal viruses – (Prangishvili *et al.*, 2006a; Stedman *et al.*, 2006; Lawrence *et al.*, 2009; Krupovic *et al.*, 2011a, 2012; Pina *et al.*, 2011; Peng *et al.*, 2012; Prangishvili, 2013; Dellas *et al.*, 2014; Pietilä *et al.*, 2014; Snyder *et al.*, 2015; Wang *et al.*, 2015a) – and I direct readers to those along with the literature there cited for additional, especially more molecular details on archaeal virus biology.

Archaeal virus DNA replication and RNA transcription along with a number of additional molecular aspects such as provirus integration and excision, for example, are extensively reviewed by Wang *et al.* (2015a). See also the supplemental material of Abedon and Murray (2013) for a somewhat comprehensive list of the archaeal virus literature through approximately 2011 and into 2012. See also Chapter 2 for consideration of archaeal virus genomics.

Brief overview of archaeal virus history (and semantics)

A surprisingly large fraction of virion structural diversity can be found among the small number of archaeal viruses, 100-plus, that so-far have been isolated and otherwise studied in much detail (Dellas *et al.*, 2014). Consistent with the small number of archaeal virus isolates – versus roughly 100 times as many bacterial viruses that have been isolated (see Chapter 7) – the history of the study of archaeal viruses is relatively brief and intimately entwined with the discovery and subsequent study of the Archaea. Archaeal viruses were officially although unknowingly first discovered by Torsvik and Dundas in 1974, with their isolation of a virus that infected the prokaryotic species, *Halobacterium salinarum*. *H. salinarum* is a simple, single-celled, salt-loving organism that only subsequently would come to be described as an archaean. Torsvik and Dundas (1980) suggest, however, that evidence of the impact of an archaeal virus may have been known since 1958, although certainly these viruses were not well considered at that time.

In 1977 the concept of archaea, then 'urkingdom... archaebacteria', was first brought to the world's attention by the late Carl Woese and colleagues (Woese and Fox, 1977). Not until 1990 was the concept of domain Archaea, however, officially unveiled (Woese *et al.*, 1990), along with the notion of domains in general (Ho *et al.*, 2013). It was only at this point, i.e. 1990, that the name 'Archaea' in fact became associated with this distinct prokaryotic lineage and therefore that the term archaeal virus could begin to rise in prominence (Abedon and Murray, 2013). Thus, in a sense, archaeal viruses as distinct viral isolates historically predate the concept of archaea as a separate lineage, while the actual descriptor, archaeal virus, came into being approximately in association with introduction of the concept of domain Archaea, and indeed with the concept of domains, period.

With the publication of this chapter and associated monograph we are only a bit more than 40 years into the archaeal virus era as generally considered, roughly 40 years post the description of archaea as a distinct lineage, and only a little more than 25 years into describing archaea as Archaea. For about as much time, the term 'archaeal virus' has been in use (Abedon and Murray, 2013). 'Archaeal virus' furthermore is now generally considered to be the proper descriptor of the viruses of domain Archaea. By contrast, the terms 'archaeal phage' (118 Google Scholar hits), 'archaeophage' (22 Google Scholar hits), 'archaeaphage' (five Google Scholar hits), 'archaephage' (five Google Scholar hits), nor 'archaeovirus' (31 Google Scholar hits) should be used as descriptors of an 'archaeal virus' (982 Google Scholar hits, all as in quotes and excluding both patents and citations). 'Archaeal virus' instead should be considered as equivalent to 'bacterial virus' (6320 Google Scholar hits) and 'eukaryotic virus' (1270 Google Scholar hits) or 'eukaryote virus' (32 Google Scholar hits) as descriptors of the viruses of the two other cellular domains. Thus, although archaeal viruses can to a degree resemble bacteriophages, the use of 'phage' or 'archaeal phage' (or equivalent) to describe these viruses generally is now frowned upon, with 'bacteriophage' (629,000 Google Scholar hits) and 'phage' therefore employed, in the present day, and in this chapter, solely as descriptors of the viruses of Bacteria (see Chapter 7) rather than also of the viruses of Archaea.

Similarly, rather than using the term 'prophage' to describe archaeal virus genomes effecting lysogenic cycles, the term provirus is preferred. Technically, one also should not employ the term 'lysogenic cycle' to describe latent infections by many temperate archaeal viruses, as well as by some bacterial viruses, since a large number of archaeal viruses in fact are released chronically rather than lytically, where 'lyso' 'genic' strictly means lysis generating (Hobbs and Abedon, 2016). Nonetheless, and despite the imprecision of the term, here 'lysogen' and its variants will be used to reflect the apparent preference in the archaeal virus literature for this term even when describing infections by temperate viruses that release their progeny chronically rather than lytically. Below, then, I present an overview of archaeal viruses, including both temperate versus non-temperate viruses as well as lytic versus chronic releasers, some of which superficially resemble bacteriophages, but many of which do not.

Hosts of archaeal viruses

The archaea typically are differentiated into kingdoms – or, less commonly, superphyla or phyla (Petitjean *et al.*, 2014) – which are known as Crenarchaeota and Euryarchaeota. There exist additional, less well-studied, potential higher taxa, but the majority of archaeal virus isolates are associated with these two major taxa, Crenarchaeota and Euryarchaeota. Alternatively, the archaea may be divided into the Euryarchaeota on the one hand and what is known as the TACK on the other, standing for thaumarchaea, aigarchaea, crenarchaea and korarchaea, the latter collectively bearing the proposed descriptor, Proteoarchaeota, with 'Proteo' referring to an ability by the constituent organisms to display a diversity of forms. Here, however, I employ the more traditional perspective of emphasizing mostly the Euryarchaeota versus the Crenarchaeota. Furthermore, generally in this chapter I will refer to the Crenarchaeota and Euryarchaeota, or crenarchaeal and euryarchaeal, without concern as to either's precise taxonomic rank, although with the understanding that these represent important higher taxa within domain Archaea.

The etymology of archaeal higher taxa names is of interest. The descriptor 'krene' comes from Greek meaning 'spring', or 'fountain', and the Crenarchaeota are associated particularly with hot springs. 'Eury' by contrast means 'broad', and appropriately the Euryarchaeota traditionally have been found to be more broadly distributed including in very salty environments, anaerobic environments as methanogens, and, like members of Crenarchaoata, also in very hot environments. It is particularly the Crenarchaeota for which substantial virion structural diversity among archaeal viruses has been observed, although historically the majority of archaeal viruses have been isolated against euryarchaeal hosts.

Archaea genera which have been shown to serve as hosts for crenarchaeal viruses include *Acidianus*, *Aeropyrum*, *Pyrobaculum*, *Stygiolobus*, *Sulfolobus*, and *Thermoproteus*, all of which are hyperthermophiles with optimal growth temperatures in excess of 80°C. These, respectively, are members of Crenarchaeota families Sulfolobaceae (order Sulfolobales), Desulfurococcaceae (order Desulfurococcales), Thermoproteaceae (order Thermoproteales), Thermococcaceae (order Thermococcales), (again) Sulfolobaceae (both *Stygiolobus* and *Sulfolobus*), and (also again) Thermoproteaceae, all of which are members of class Thermoprotei. *Acidianus*, *Stygiolobus*, and *Sulfolobus* furthermore are all acidophilic as well as thermophilic. The relatively small number of archaeal genera against which crenarchaeal viruses have been isolated likely reflects, at least in part, the relatively small number of research groups that are in the business of isolating archaeal viruses in combination with the relatively small number of archaea that have been cultured and therefore which could serve as potential hosts for virus isolation. Indeed, archaea can be difficult to culture, although thermophilic archaea, and thereby members of family Crenarchaeota, can be easier to culture than members of family Euryarchaeota (Snyder *et al.*, 2015).

Genera for which euryarchaeal viruses have been isolated include *Haloarcula*, *Halobacterium*, *Halorubrum*, *Methanobacterium*, and *Pyrococcus* which are halophilic (first three), methanogenic, and thermophilic, respectively. These genera are found respectively in classes Halobacteria (first three), Methanobacteria, and Thermococci. Roughly ten additional Euryarchaeota classes exist, however. Euryarchaeal viruses, like crenarchaeal viruses, thus have also been isolated against a relatively small subset of known host diversity. While the first archaeal viruses isolated were halophilic euryarchaeal viruses (Torsvik and Dundas, 1974), the first virus of a methanogen was not isolated until a dozen years later (Bertani and Baresi, 1986). The existence of viruses of euryarchaeal phylum Nanohaloarchaeota have also been suggested using a process of hybridization of single-cell genomes isolated from the environment against virion genomes isolated from the same environment (Martínez-García *et al.*, 2014) as well as via fosmid-generated microviromes (Garcia-Heredia *et al.*, 2012). Viruses and/or proviruses have been described as well for the archaea classified as Nanoarchaeota (Munson-McGee *et al.*, 2015), arguably also of euryarchaeal ('nano' of course meaning 'small' and these are cellular symbionts of Crenarchaeota archaeans).

Viruses have been isolated against members of the Thaumarchaeota (Brochier-Armanet *et al.*, 2008; Krupovic *et al.*, 2011b; Labonté *et al.*, 2015), with 'thaum' referring to 'miracles' or 'wonder'. These specifically are archaea that oxidize ammonia and which exist as mesophiles as well as thermophiles. As of this writing, no viruses appear to have been reported for phylum Korarchaeota ('kor' referring to 'youth'), etc. (Hedlund *et al.*, 2015). Archaea exist as well which have been described as ARMAN, standing for Archaeal Richmond Mine Acidophilic Nanoorganisms, although these have not been cultivated. They have very tiny cells with similarly small genomes and have been shown to be provirus associated (Baker *et al.*, 2010). ARMAN additionally have been seen to be infected by viruses (Comolli *et al.*, 2009).

Archaeal viruses – in comparison to both bacterial viruses and viruses of eukaryotes – in general have been isolated against only relatively few Archaea in terms of absolute numbers of genera and also in terms of the total fraction of known archaeal genetic diversity. Nonetheless, among those viruses that have been characterized, archaeal viruses appear to display greater levels of virion morphological diversity in comparison to viruses of bacteria (see Chapter 7). Indeed, only roughly 30 archaeal viruses have even been classified, but these viruses nevertheless represent 15 distinct archaeal virus families (Snyder *et al.*, 2015), compared to only ten families representing all of bacterial viruses.

Synopsis of archaeal virus characteristics

Before delving into the specifics of individual archaeal viruses, in this section we consider various general aspects of these viruses. Like bacteriophages, for example, seemingly most archaeal viruses possess dsDNA genomes while a few instead employ ssDNA. So far no RNA archaeal viruses have been isolated, although as considered subsequently that does not mean that there have not been suggestions that such viruses exist. See Chapter 2 for a more detailed summary of archaeal virus genomics. Contrasting bacterial viruses to a certain extent, as well as the viruses of eukaryotes, archaeal viruses at least at this point in their study seem in many cases as though they are somewhat less likely to be closely related to each other but rather more closely related to non-virus genetic elements (Iranzo *et al.*, 2016). This is seen particularly among those archaeal viruses which are not otherwise apparently related to bacterial or eukaryotic viruses, basically ten archaeal virus families, and suggests that especially the more unique archaeal viruses might possess non-viral origins versus having been derived evolutionarily from other virus types.

Perhaps related to this latter point, it is especially the unusualness of archaeal virus virion structure that is their most captivating feature, along with their associated morphological diversity. Indeed, despite there being roughly two orders of magnitude fewer isolates of archaeal viruses versus bacterial viruses, the archaeal viruses display what appears to be similar levels of gross structural diversity in their virions compared to the virions of bacteriophages. Why archaeal viruses should be so seemingly morphologically diverse – along with the evolutionary origins or utility of that virion-morphology diversity – likely will represent an enduring focus of the study of these viruses. Although less visually obvious, the diversity of archaeal virus infection characteristics is important as well. Perhaps not unexpectedly, we observe among archaeal viruses the three basic infection types observed with bacteriophages (see Chapter 7). These are lytic, lysogenic, and chronic.

Lytic infections may not be as common among archaeal viruses, perhaps particularly not among crenarchaeal viruses (Prangishvili, 2013; Snyder and Young, 2013), as among bacterial viruses where lytic infections are the most common type. In assessing this claim, however, it is important to keep in mind that many authors describe temperate viruses as not lytic, e.g. Gorlas and Geslin (2013), whereas in many cases temperate viruses in fact are lytic viruses, that is, rather than chronically releasing viruses (Hobbs and Abedon, 2016). See also Wang *et al.* (2015a), as well as Snyder and Young (2013), who consider what is known of viral lytic release across the three domains. Release of archaeal virions via chronic rather than lytic mechanisms furthermore is often referred to among archaeal viruses as a 'carrier state', 'stable carrier state' (Quemin and Quax, 2015) or 'stable carrier relationship' (Prangishvili, 2013), where the term 'carrier' for bacteriophages, confusingly, can have a number of additional meanings (Abedon, 2009). Indeed this term may have more than one meaning for archaeal viruses as well (Snyder and Young, 2013). See Quemin and Quax (2015) for review of mechanisms of archaeal virus adsorption and release, and Svirskaitė *et al.* (2016) as well who explore the physiology of release of a variety of euryarchaeal virions from one host, *Haloarcula hispanica*.

An additional issue that has garnered a fair amount of interest is the evolutionary relationship between archaeal viruses and viruses of the other two domains. This relationship is anything but recent, to the point that not only are nucleic acid sequence comparisons wholly unable to detect homology but so too are amino acid sequence comparisons. Instead, what is required for comparisons is the detection of homology among protein structures between as well as within domains. From a perspective simply of assigning presumptive function to archaeal virus gene products, such comparison of protein structure provides an especially useful tool in the analysis of archaeal viral biology, given the genetic diversity observed in combination with the relatively small degree that these viruses have been subject to rigorous biochemical analysis. See Dellas *et al.* (2014) for recent review of these issues, and also Krupovic *et al.* (2012).

Categories of archaeal virus types

Approximately seven virion-morphology types have been described for the euryarchaeal viruses, with at least another twelve types seen among crenarchaeal viruses. Some of these types are depicted in Fig. 8.1. Note, though, that as of this writing not all archaeal virus virion-morphology types have necessarily been formally classified into virus families. Also included among euryarchaeal viruses are families *Siphoviridae* and *Myoviridae*, familiar to bacteriophage biologists, as well as observations of what has been described as a member of family *Podoviridae* (see Chapter 7). These are tailed virions, displaying long, non-contractile tails, contractile tails, and short, non-contractile tails, respectively. Tails are the adsorption organelles of head-tailed viruses. 'Head-tailed' is the means by which these viruses tend to be referred to in the archaeal virus literature since additional archaeal viruses also possess tails but less obviously heads. Such bacteriophage-like, head-tailed viruses have not been isolated against crenarchaeal hosts, perhaps because this virion morphology is poorly suited structurally to extremely high temperatures, or instead, we can speculate, because the obligate use by head-tailed viruses of lytic productive cycles again may be less desirable for viruses replicating within hyperthermophilic niches (Snyder *et al.*, 2015). In this section I further introduce euryarchaeal and crenarchaeal virions as well as the lipid-containing archaeal viruses. There have been additional candidate archaeal viruses that have been suggested based on genomic and metagenomic data set (Roux *et al.*, 2015, 2016).

Euryarchaeal virions

Morphological types among euryarchaeal viruses are somewhat less diverse versus those found among

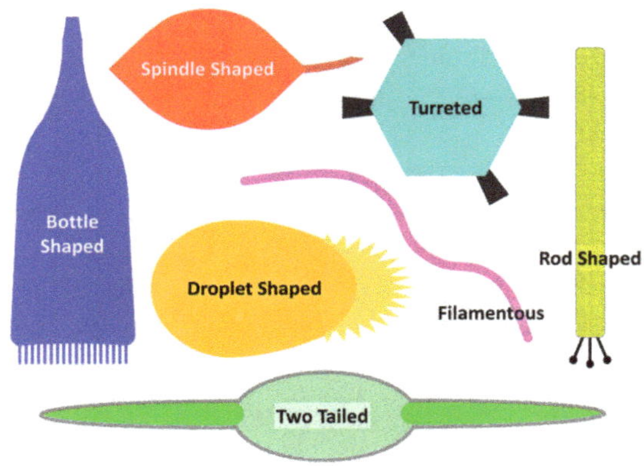

Figure 8.1 Morphologies of especially Crenarchaeota-infecting archaeal viruses. Family *Ampullaviridae* are bottle shaped and enveloped. Family *Fuselloviridae* and the otherwise unassigned genus *Salterprovirus* (a euryarchaeal virus) are both spindle shaped. Family *Lipothrixviridae* are filamentous and enveloped. Proposed family '*Turriviridae*' have icosahedral but turreted as well as internally enveloped virions (turrets shown with less elaborate detail than as actually seen at higher resolution). Family *Rudiviridae* are rod shaped and not enveloped (shown at the bottom are terminal fibres). Family *Bicaudaviridae*, that is, they are two tailed, are lemon shaped prior to growing two tails that form only once the virion has been released into the extracellular environment. Family *Guttaviridae* are droplet shaped with fibres found at one end, represented in the figure as a starburst shape. Virions are not drawn to scale. This figure is modified from that of Hyman and Abedon (2012). See Prangishvili (2013) for additional discussion of these virions.

crenarchaeal viruses. Conversely, euryarchaeal viruses include ones which, as noted, display head-tailed morphologies (i.e. virus order *Caudovirales*, consisting for archaeal viruses of families *Myoviridae* and *Siphoviridae* as well as *Podoviridae*). In addition there are virions which are icosahedral with internal membrane (family *Sphaerolipoviridae*), ones which are spindle- (lemon-) shaped (family *Fuselloviridae*), and also ones which are pleomorphic (family *Pleolipoviridae*). Two of these types are also seen among bacteriophages, particularly head-tailed but also pleomorphic, the latter as seen with family *Plasmaviridae* which infects the cell wall-less bacterium, *Acholeplasma*. Icosahedral in addition generally is a common virus morphology among the viruses of eukaryotes and also is seen, in various guises, across of number of bacteriophage virion families (see Chapter 7). Only one of these euryarchaeal virus morphological types is depicted in Fig. 8.1, however. This is the spindle-shaped virus, although this morphology is seen among Crenarchaeal viruses as well. See Table 8.1 for a summary of these virus types.

Crenarchaeal virions

Crenarchaeal viruses tend to be unusual in their virion morphologies relative to bacterial viruses or eukaryotic viruses (Fig. 8.1) (Prangishvili, 2013). In addition, and unlike euryarchaeal viruses, crenarchaeal viruses are lacking in members of virus order *Caudovirales*, that is, head-tailed virions. Crenarchaeal viruses with unusual morphologies include bottle-shaped viruses (family *Ampullaviridae*), droplet-shaped virions (family *Guttaviridae*), spindle-shaped virus (family *Fuselloviridae*), and spindle-shaped virions that grow two tails during their extracellular phase (family *Bicaudaviridae*). In addition, there are filamentous virions (family *Lipothrixviridae*), bacilliform virions (family *Clavaviridae*), rod-shaped virions (family *Rudiviridae*), and spherical virions (family *Globuloviridae*) as well as icosahedral virions (family *Turriviridae*). Prangishvili (2013) summarizes the process of isolation of many of the Crenarchaeal viruses, the characterization of as-yet uncultured Crenarchaeal viruses via a combination of electron microscopy and metagenomic analysis of enrichment cultures (see Chapter 5), as well as other aspects of these viruses such as protein function identification. See Table 8.2 for a summary of these virus types.

Lipid-containing virions

Atanasova *et al.* (2015b) review lipid-containing viruses of prokaryotic organisms; see also Roine and Bamford (2012). These viruses tend to be first identified due either to organic-solvent sensitivity (typically chloroform) or instead as a function of low virion buoyant density. The lipids themselves often are associated with the familiar envelopes which surround nucleocapsids although membranes in various viruses instead can be internal to capsid proteins. In addition there are lipids which are covalently bound to capsid proteins. Usually, although not quite always, a virus with an external envelope will exit a host cell chronically, via budding; viruses possessing internal membranes, by contrast, typically will exit hosts cells lytically. In either case, the membrane at least as a default assumption should be considered to function during virion adsorption towards virion fusion with the host plasma membrane (Dellas *et al.*, 2014), although internal membranes also can function as protein scaffolds for the assembly of capsids.

Model lipid-containing viruses among bacteriophages include phage PM2 (family *Corticoviridae*), phage PRD1 (family *Tectiviridae*), and phage ϕ6 (family *Cystoviridae*). Archaeal virus virions possessing internal membranes – family *Sphaerolipoviridae* and proposed family '*Turriviridae*' – have been shown to be 'reminiscent' of phages PM2 and PRD1. Also described as similar in terms of virion size, virion morphology, and overall virus life cycle is the *Mycoplasma* lipid-containing virus, phage L2, a member of family *Plasmaviridae*, and members of family *Pleolipoviridae* of archaea, both of which may employ their virion lipids for membrane fusion during adsorption.

Specific archaeal virus virions that have been shown to possess lipids including those which are filamentous (crenarchaeal viruses AFV1, AFV2, AFV3, AFV6, AFV7, AFV8, AFV9, DAFV, SIFV, TTV2, and TTV3; family *Lipothrixviridae*), icosahedral (euryarchaeal viruses HHIV-2, PH1, SH1, and SNJ1, i.e. family *Sphaerolipoviridae*, as well as crenarchaeal viruses STIV and STIV2 of family *Turriviridae*), pleomorphic (euryarchaeal viruses HGPV-1, HHPV-1, His2, HRPV-1, HRPV-2, HRPV-3 and HRPV-6; family *Pleolipoviridae*), spherical (crenarchaeal viruses PSV1 and TTSV1; family *Globuloviridae*), and spindle-shaped (crenarchaeal viruses ASV1, PAV1, SSV1, SSV2, SSV4,

Table 8.1 Euryarchaeota Viruses[1]

Family[2]	Genome (approximate sizes)	Examples	Virions and comments	Some known hosts (associated virus)
Myoviridae	Linear (30–80, 230 kb)	Viruses B10, BJ1 (43, kb), ϕCh1 (58.5 kb), ΦH1 (ΦH; 59 kb), ΦN (56 kb), Hh1 (37.6 kb), Hh3 (29.6 kb), HF1 (77.7 kb), HF2 (75.9 kb), HATV (and others), HJTV (and others), HRTV (and others), Hs1S, HSTV-2 (and others), Ja.1 (230 kb), ΨM1 (30.4 kb), S45, S5100	Head-tailed virions, tail is contractile	*Haloarcula sodomense* (HATV), *Halobacterium salinarum* (B10, Hs1, ΦH, S45, S5100), *Halorubrum coriense* (BJ1, HF1, HF2, Hh1, Hh3, HRTV, HSTV, Ja.1, ΦN), *Methanobacterium* (ΨsM1), *Natrialba magadii* (ϕCh1)
Pleolipoviridae	See comments	Viruses His2 (16 kb, dsDNA linear), HGPV-1 (9.7 kb), HHPV-1 (8 kb) HRPV-1 (~10 kb, ssDNA circular), HRPV-2 (~10 kb, ssDNA circular), HRPV-3 (8.7 kb), HRPV-6 (~10 kb, ssDNA circular); unless indicated otherwise then dsDNA circular	Lipid-enveloped, pleomorphic virions; dsDNA, ssDNA, linear, and circular genomes	*Acidianus*, *Halorubrum* (HRPV-1, HRPV-2, HRPV-3, HRPV-6), *Halogeometricum* (HGPV-1), *Pyrobaculum*, *Sulfolobus islandicus*, *Thermoproteus*
Podoviridae	30 kb	Virus HSTV-1 (32.2 kb)	Head-tailed virions, tail is short and non-contractile; uncommon	*Haloarcula sinaiiensis* (HSTV-1)
Sphaerolipoviridae	Linear (30 kb)	Viruses HHIV-2 (30.5 kb), PH1 (28 kb), SH1 (30.9 kb)	Icosahedral virions with internal membrane	*Haloarcula hispanica* (HHIV-2, PH1, SH1), *Halorubrum*
(*Salterprovirus*)	Linear (15 kb)	Virus His1 (14.4 kb)	Spindle-shaped virions with small tails, containing lipid	*Haloarcula hispanica* (His1)
Siphoviridae	Linear (30–80 kb)	Viruses BJ1, ΦF1 (85 kb), ΦF3 (36 kb; circular), HCTV (and others), HHTV (and others), HVTV-1 (and others), ψM1, ψM2 (26 kb), ψM100	Head-tailed virions, tail is long and non-contractile	*Haloarcula*, *Methanothermobacter*, *Methanobacterium thermoautotrophicum* Marburg (ΦF1, ΦF3, ψM2)
Unclassified	See comments	Viruses A3-VLP, HF2, HRPV1, PAV1 (18 kb), PG (50 kb), PSM1 (35 kb), SNJ1 (16.3 kb), S41, S41 Vm, S4100, S50.2, S50.2 Vm, TPV1 (21.5 kb), VTA	PAV1 and TPV1 are spindle-shaped with a short tail and circular genome; SNJ1 may be icosahedral, is spherical, and contains lipids	Known hosts include *Halobacterium* (S41, S41 Vm, S4100, S50.2, S50.2 Vm) *Haloarcula hispanica*, *Halorubrum*, *Methanobacterium* (PG, PSM1, VTA), *Methanococcus voltae* (A3-VLP), *Natrinema* (SNJ1), *Pyrococcus abyssi* (PAV1), *Thermococcus prieurii* (TPV1)

[1]For an elaboration on this table including with links, see http://www.archaealviruses.org/.
[2]Proposed taxa are in quotation marks; taxa other than at the family level are in parentheses.

Table 8.2 Crenarchaeota Viruses[1]

Family[2]	Genome (approximate sizes)	Examples	Virions and comments	Some known hosts (associated virus)
Ampullaviridae	Linear (25 kb)	Virus ABV (23.8kb)	Bottle-shaped virions with short fibres at wider end	*Acidianus convivator* (ABV)
Bicaudaviridae	Circular (65 kb)	Virus ATV (62.7 kb)	Spindle-shaped virions with two large tails which form following virion release from infected cells	*Acidianus convivator* (ATV)
Clavaviridae	Circular (5 kb)	Virus APBV1 (5.2 kb)	Rod-shaped, that is, bacillus-shaped virions	*Aeropyrum pernix* (APBV1)
Fuselloviridae	Circular (15–25 kb)	Viruses ASV1 (24.1 kb), SSV1 (15.4 kb), SSV2 (14.7 kb), SSV3 (15 kb), SSV4 (15.1 kb), SSV5 (15.3 kb), SSV6 (15.6 kb), SSV7 (17.6 kb), SSV8 (16.4 kb), SSV9 (17.3 kb), SSVk1, SSVrh	Spindle-/lemon-shaped virions with small tails; genera *Alphafusellovirus* and *Betafusellovirus*	Acidianus (ASV1), *Aeropyrum*, *Haloarcula*, *Pyrococcus*, *Stygiolobus*, *Sulfolobus* (SSV1, SSV2, SSV4, SSV5, SSV6, SSV7, SSV8, SSV9), *Sulfolobus shibatae* (a.k.a. *S. acidocaldarius*), *Thermococcus*
Globuloviridae	Linear (20–30 kb)	Viruses AOV1, PSV1 (PSV; 28.3 kb), TTSV1 (21.6 kb)	Spherical-shaped virions containing lipids	*Aeropyrum* (AOV1), *Pyrobaculum* (PSV1), *Thermoproteus tenax* (TTSV1)
Guttaviridae	Circular (15–20 kb)	Viruses APOV1 (13.7 kb), SNDV (20 kb)	Droplet-shaped virions; genera *Alphaguttavirus* and *Betaguttavirus*	*Aeropyrum pernix* (APOV1), *Sulfolobus neozealandicus* (SNDV)
Lipothrixviridae	Linear (15–40, 55 kb)	Viruses AFV1 (21 kb), AFV2 (31.7 kb), AFV3 (40.4 kb), AFV6 (39.5 kb), AFV7 (36.8 kb), AFV8 (38.1 kb), AFV9 (41.1 kb), DAFV (56 kb), SIFV (40.8 kb), TTV1 (15.9 kb), TTV2 (16 kb), TTV3 (27 kb), TTV4 (17 kb), TTV1–1	Flexible, lipid-containing linear-shaped virions with various filaments jutting from ends; order *Ligamenvirales*; genera *Alphalipothrixvirus*, *Betalipothrixvirus*, *Gammalipothrixvirus*, and *Deltalipothrixvirus*	Acidianus (AFV1, AFV2, AFV3, AFV6, AFV7, AFV8, AFV9), *Acidianus hospitalis*, *Desulfurolobus ambivalens* (DAFV), *Sulfolobus islandicus* (SIFV), *Thermoproteus tenax* (TTV1, TTV2, TTV3, TTV4, TTV1–1)
Rudiviridae	Linear (25–35 kb)	Viruses ARV1 (24.6), SIRV1 (32.3 kb), SIRV2 (35.4 kb), SRV1 (SRV; 28 kb)	Rod-shaped virions; order *Ligamenvirales*	*Acidianus* (ARV1), *Sulfolobus islandicus* (SIRV1, SIRV2), *Stygiolobus* (SRV)
Spiraviridae	ssDNA circular (25 kb)	Virus ACV (24.9 kb)	Spiral-shaped capsids	*Aeropyrum* (ACV)
'*Tristromaviridae*'	Linear (16–18 kb)	Viruses PFV1 (17.7 kb), TTV1 (15.9 kb)	Filamentous virions; see alternatively family *Lipothrixviridae* for virus TTV1	*Pyrobaculum* (PFV1), *Thermoproteus tenax* (TTV1)
'*Turriviridae*'	Circular (15–20 kb)	Viruses STIV (17.6 kb), STIV2 (16.6 kb)	Icosahedral, internal membraned virions with prominent, turret-like features on their surfaces	*Sulfolobus* (STIV, STIV2)

Table 8.2 Continued

Family[2]	Genome (approximate sizes)	Examples	Virions and comments	Some known hosts (associated virus)
Unclassified	See comments	Viruses APSV1 (38 kb), SMV1 (48.7 kb), STIV1, STIV2, STSV1 (75.2 kb), STSV2 (76 kb)	APSV1, SMV1, STSV1, and STSV2 are spindle shaped with long tails, and have circular genomes; STSV1 and STSV2 contain lipids	*Aeropyrum pernix* (APSV1), *Sulfolobus* (SMV1), *Sulfolobus solfataricus*, *Sulfolobus tengchongensis* (STSV1, STSV2)

[1]For an elaboration on this table including with links, see http://www.archaealviruses.org/.
[2]Proposed taxa are in quotation marks; taxa other than at the family level are in parentheses.

SSV5, SSV6, SSV7, SSVk1, SSVrh, and TPV1, that is, family *Fuselloviridae*; also there are crenarchaeal viruses STSV1 and STSV2 which have been suggested to be members of family *Bicaudaviridae*). All of these virions possess lipid bilayer membranes. Spindle-shaped euryarchaeal virus His1, a member of an otherwise unassigned genus, *Salterprovirus*, instead possesses lipid-modified structural proteins. Remarkably, the presence of lipids within virions can be optional among otherwise structurally related viruses, that is, not necessarily serving as a defining characteristic of viral taxa even if present in some members (Atanasova *et al.*, 2015b).

Archaeal virus taxa

In the following two sections I consider in greater detail especially the different families of archaeal viruses. These are presented in the order of Crenarchaeota viruses and then Euryarchaeota viruses. Within subsections the different types are listed predominantly in alphabetical order according to taxonomic families. The primary goal is to provide a relatively succinct overview and discussion of the different virus types as well as to indicate somewhat recent insights into the various virions types. See Fig. 8.1 for illustration of the various more unusual virion morphologies. For recent, further review and discussion of archaeal virus morphologies, see especially Dellas *et al.* (2014) and Pietilä *et al.* (2014). See Chapter 2 for an overview specifically of the genomics of these viruses.

Crenarchaeota viruses

This section considers families, or proposed families (in quotes): *Ampullaviridae, Bicaudaviridae, Clavaviridae, Fuselloviridae, Globuloviridae, Guttaviridae, Lipothrixviridae, Rudiviridae, Spiraviridae, 'Tristromaviridae'*, and *'Turriviridae'*. In addition putative RNA archaeal viruses are discussed.

Bottle- or ampulla-shaped virions (family *Ampullaviridae*)

Bottle here is as equivalent to the glass vessels within which wine is stored. Bottle-shaped virions, like many virions and also wine bottles, are radially symmetrical, with a taper at one end rather than at both ends. The latter contrasting the tapers observed at both ends as one sees with the also radially symmetrical spindle-shaped viruses (below). At the non-tapered end of bottle-shaped virions can be found roughly twenty 20 × 3 nm fibre appendages (Fig. 8.1). Only a single isolate of these viruses has been cultured, ABV1, a thermophilic virus of the crenarchaeal genus, *Acidianus* (i.e. Acidianus Bottle-Shaped Virus 1).

ABV1 was isolated from an 80°C acidic hot-spring enrichment and was one of five viral morphotypes isolated at the time. The others included both rigid and flexible filamentous or rod-shaped virions along with pleomorphic virions (Häring *et al.*, 2005b). ABV1 virions are enveloped and it is speculated that the tapered end is involved in adsorption. The broad end appears to possess a ring or disc to which the fibre appendages attach in a circle, although oddly these do not appear to be at least directly involved in virion adsorption. The length of the virion is roughly 230 nm and its width, at its widest point, approximately 75 nm. Infection by ABV1 does not appear to result in lysis and indeed virus-producing colonies of *Acidianus* could be isolated from the enrichment culture. Viral infection, however, does reduce the growth rate of the host cells.

Sequencing and discussion of ABV1's genome is provided by Peng *et al.* (2007). Related genomes in turn were found, by Gudbergsdóttir *et al.* (2016), to be quite common in acidic, high-temperature hot springs. In that study, metagenomes obtained from six hot springs were characterized for viral sequences. Among sequencing reads which were attributed to viruses, presumptive *Ampullaviridae* sequence was found ranging from approximately 1% to nearly 50% of reads, depending on the hot spring. Included in these reads were two close-to-complete genomes, which were designated ABV2 and ABV3 and for which *Sulfolobus* or *Acidianus*, both members of order Sulfolobales, were assigned as presumptive hosts based on CRISPR spacer matches. Thus, despite having been isolated only once, members of this crenarchaeal virus family may in fact be quite common.

Large spindle-shaped/pleomorphic virions with prominent tails (family *Bicaudaviridae*/superfamily '*Magnusfuselloviridae*')

Krupovic *et al.* (2014) consider the evolutionary relationships between the various spindle-shaped archaeal viruses, which although spindle-shaped nevertheless are somewhat pleomorphic in terms of width, length, and tail size. By comparing virion protein structures, they classify a number of otherwise unclassified spindle-shaped archaeal viruses among virus families *Bicaudaviridae* (this section) and *Fuselloviridae* (below). Before this analysis, Acidianus two-tailed virus (ATV) was the sole known member of family *Bicaudaviridae*. Furthermore, Krupovic *et al.* suggest the addition of the morphologically simpler single-tailed Sulfolobus tengchongensis spindle-shaped viruses, STSV1 and STSV2, to the *Bicaudaviridae* family, which they assign to a new genus, *Betabicaudavirus*. They further suggest that observed morphological similarities between the two virus families, *Bicaudaviridae* and *Fuselloviridae*, are a consequence of convergent evolution. Prangishvili (2013) indicates that members of family *Bicaudaviridae* are temperate. From a 14th-century coprolite, sequence resembling that of family *Bicaudaviridae* has been identified (Appelt *et al.*, 2014).

Hochstein *et al.* (2015, 2016) add two additional viruses to this mix. These are Acidianus tailed spindle virus (ATSV), which Hochstein *et al.* (2016) suggest should be placed in a separate family named *Fusellocaudaviridae*, and Sulfolobus monocaudavirus 1 (SMV1). Furthermore, Hochstein *et al.* (2016) propose that all five of these viruses (ATV, STSV1, STSV2, SMV1, and ATSV) could be placed in a superfamily they call *Magnusfuselloviridae*. All of these viruses possess prominent tails versus the generally although not entirely diminutive tails associated with family *Fuselloviridae*. In addition to their larger spindle-like shapes, they also possess circular dsDNA genomes, encode integrases (suggesting that they are temperate viruses), and were isolated from acidic hot springs. Furthermore, contrasting head-tailed viruses, there is no distinct head structure in the large spindle-shaped viruses but instead much more continuity between heads and tails, although tails can have tail fibres (ATV and ATSV). As based on metagenomic data, Hochstein *et al.* (2015) suggest that Yellowstone hot springs may possess at least three distinct types of large spindle-shaped viruses.

Erdmann *et al.* (2014) isolated STSV2, which is similar to the previously isolated STSV1 (Xiang *et al.*, 2005), describing a virion possessing a tail that ranges in length from 20 to 500 nm but averages 50 nm (versus 0, 133, and 68 nm as equivalent dimensions for STSV1). In addition, STSV2 virions might possess chloroform-extractable lipids. Whereas *Fuselloviridae* viruses possess genomes in the range of 15 kb, i.e. as for SSV1, the genome size of STSV2 is a comparatively massive 76 kb, and very similar to that of STSV1 at 75 kb. The genome of ATV also is similar in size to that of STSV2, at approximately 63 kb, and 71 kb is the genome size for ATSV. Indeed, these genomes are among the largest known among archaeal viruses (Hochstein *et al.*, 2015), contrasting known bacteriophage genomes ranging up to roughly an order of magnitude larger, e.g. bacteriophage G. Perhaps consistently, Erdmann *et al.* (2014) report some low but still significant sequence homology between ATV and STSV2. For STSV1, infection is found to slow the doubling time of its host from 11 h to about 30 h (Xiang *et al.*, 2005).

The virus SMV1 grows an extremely prominent tail, and sometimes two, during acellular incubation at typical growth temperatures, as may play roles in normal adsorption process (Uldahl *et al.*, 2016) (see, again equivalently, Acidianus two-tailed virus). The authors in this publication also describe a possibly sequential adsorption scheme – within an overall rapid virion-to-cell attachment process – that involves association of the 'nose' of the virion with the cell surface, subsequent association of the 'body' of the virion with the surface of the cell ('alignment stage'), and then a flattening of both body and tail against the cell surface in a manner that may result in virion–to-cell fusion. Infection with SMV1 has been found to retard host replication but not result in lysis (Uldahl *et al.*, 2016).

ATV is particularly unusual among these viruses due to its virion structure (Häring *et al.*, 2005a). Specifically, following release from its host, ATV gradually grows not one but two prominent tails at normal growth temperatures. This growth does not occur instantaneously upon exit from host but instead takes days, and can be prevented via refrigeration. As is generally the case, these tails are assumed to be involved in virus adsorption to host cells. The virions are not thought to contain lipids and infection does not appear to slow down host replication (Prangishvili *et al.*, 2006b). Recently an ATV protein capable of binding to and globally inhibiting host RNA polymerase-mediated transcription has been identified, the product of ATV ORF145. As an RNA polymerase-binding transcription inhibitor, this a first for archaeal viruses although also is odd given that archaeal viruses do not tend to encode, so far as is known, their own that is alternative RNA polymerases (Sheppard *et al.*, 2016).

Bacilliform/linear virions (family *Clavaviridae*)

A bacilliform-shaped virus, Aeropyrum pernix bacilliform virus 1 (APBV1), infects a thermophilic crenarchaean (*A. pernix*) host with optimal growth temperature of 90°C (Mochizuki *et al.*, 2010). Perhaps confusing, as other archaeal viruses also are described as rod-shaped (below), bacilliform means simply that, rod shaped, while *clava* stems from Latin meaning 'club', 'cudgel', 'staff', or 'stick'. In any case, such long-thin virions predominate hyperthermal environments (Prangishvili *et al.*, 2006a).

APVB1 virions possess ends of differing conformation, round at one end and pointed at the other. Indeed in electron micrographs these viruses look like elongated bacillus bacteria, although of course display a much smaller overall size. From the same enrichment culture that APBV1 was isolated, a second, slightly longer and slightly wider rod-shaped virus possessing additional, 45 degree-angled appendages was also found, as too were 'bean-shaped' pleomorphic particles. Indeed, more bacilliform particles along with further virion morphologies were observed as well in additional *A. pernix* cultures. APBV1 thus was isolated, by serial dilution, from a diversity of virus-like as well as bacilliform-like particles.

APBV1's virions (Mochizuki *et al.*, 2010) are rigid, 140 nm long and 20 nm wide, contain a highly glycosylated major capsid protein (along with three minor proteins), and have dsDNA, circular, as well as extremely small (5278 bp) genomes. APBV1 in fact has the smallest of known archaeal virus genomes. APBV1 furthermore was found to neither integrate into the chromosome, lyse, nor slow the growth of its host. Morphologically similar viruses may have been observed in enrichment cultures associated with a deep-sea hydrothermal vent (Geslin *et al.*, 2003a), as suggested by Mochizuki *et al.* (2010), although these other viruses may not be of equivalent size.

Note that Zillig *et al.* (1996) mention a proposed family '*Bacilloviridae*' that includes crenarchaeal viruses TTV4 (500 × 30 nm virion) and SIRV (950 × 26 nm virion) of *Thermoproteus tenax* and *Sulfolobus islandicus*, respectively. These viruses consist of 'stiff rods', no membrane, linear dsDNA genomes (17 and 33 kb), and structures found at their ends that are involved in host attachment. References are to two earlier reviews by authors also associated with Zillig (Zillig *et al.*, 1988; Reiter *et al.*, 1988). APBV1, although also a stiff rod, possesses a different virion length, different genome conformation (circular versus linear), and different genome size, plus does not appear to possess visible end structures. Thus 'bacilliform' viruses of family *Clavaviridae* and members of proposed family '*Bacilloviridae*' do not appear to be related. In fact, Ahn *et al.* (2006) list these '*Bacilloviridae*' viruses, including ARV1, SIRV1, SIRV2, and TTV4, as members of family *Rudiviridae* as discussed below.

Spindle- or lemon-shaped virions (family *Fuselloviridae*)

A spindle is a tool employed for spinning yarn and as such tends, with associated yarn, to form a radially symmetrical shape, bulging in the middle while tapered at both ends. Equivalently, the term fusiform is used to describe this shape, which to a degree can resemble as well the shape of a lemon, although pear shaped and cigar shaped virions exist as well (Redder *et al.*, 2009). At one of these ends, for smaller spindle-shaped viruses, can be found a fibre which resembles a tail. See 'Spindle Shaped' in Fig. 8.1 for visualization of this morphology. Spindle-shaped viruses may be unique to archaea, often encode integrases as consistent with temperate lifestyles (Prangishvili, 2013; Contursi *et al.*, 2014), are released from cells via budding (Quemin *et al.*, 2016), can be found in a variety of environments both extreme and not extreme, and could represent the most prevalent of archaeal virus types (Atanasova *et al.*, 2016). These viruses thus can be abundant viral constituents of environments in which archaea dominate (Hong *et al.*, 2015). As Krupovic *et al.* (2014) list, these viruses have been isolated from acid mines, anoxic freshwaters, cold Antarctic lakes, hypersaline environments, and terrestrial hot springs. In this section what is considered are spindle-shaped viruses infecting Crenarchaeota and which possess relatively small tails, the fuselloviruses (Contursi *et al.*, 2014).

Krupovic *et al.* (2014) in particular suggest that all known 'tailless', or at least small-tailed, spindle-shaped viruses should be classified into family *Fuselloviridae* – by contrast there are prominently tailed spindle-shaped viruses (above). Members of family *Fuselloviridae* include SSV1 (genus *Alphafusellovirus*); ASV1 and SSV6 (genus *Betafusellovirus*); A3-VLP, PAV1, and TPV1 (genus *Deltafusellovirus*); APSV1 (genus *Gammafusellovirus*, which in fact is only arguably tailless/diminutively tailed); and His1 (genus *Epsilonfusellovirus*, but more commonly genus *Salterprovirus*, see also below). Given the diversity of host types and environments spanned by members of proposed family *Fuselloviridae*, Krupovic *et al.* suggest further that the lineage has ancient origins.

Sulfolobus spindle-shaped virus 1 (SSV1) is considered to be one of three archaeal virus model systems (Snyder *et al.*, 2015). Stedman *et al.* (2015) provide a brief communication considering the cryo-EM virion architecture of SSV1, which possesses a spindle-shaped morphology and is released non-lytically. Stedman *et al.* (2015) found no evidence of either internal or external membranes, although Quemin *et al.* (2016) present evidence that in fact SSV1 does possess an envelope. Stedman *et al.* were, however, able to observe the hexamer-based tail. This tail juts away from one pole of the virion, although to a greater extent away from the virion's pole along the vertical axis than appears to be the case for His1 (the latter as considered in a subsequent section). This tail is assumed, again like that of His1, to be involved in virion adsorption, i.e. just as members of order *Caudovirales*, the traditionally tailed bacteriophages and archaeal viruses, adsorb largely tail first. Tail fibres found particularly at one end of fuselloviruses can be observed in electron micrographs (Redder *et al.*, 2009). Hanhijarvi *et al.* (2013, 2016) explore the biophysics of DNA ejection from His1 virion particles (Evilevitch, 2013).

The spindle shape of the overall capsid Stedman *et al.* (2015) speculate 'resembles two fused fullerene cones'. This is as equivalent to that associated with the capsid of human immunodeficiency virus (HIV), although the latter is only a single cone. They note, however, that actual virion assembly probably does *not* occur as separate cones that then fuse. In terms of actual capsid shape, the presented structure possesses nearly the exact dimensions of an American football, which can be described as a prolate spheroid, although one with six, short, closely associated 'streamers' (the tail) jutting vertically from one end (Fig. 8.1). Recently the release of SSV1 from infected bacteria was studied using dual-axis electron tomography and found to resemble the budding of eukaryotic viruses from infected cells (Quemin *et al.*, 2016).

Spherical virions (family *Globuloviridae*)

Spherical-shaped viruses have been isolated against genera *Pyrobaculum* (Pyrobaculum spherical virus, or PSV) and *Thermoproteus* (Thermoproteus tenax spherical virus 1, or TTSV1), both of which are thermophilic members of phylum Crenarchaeota. Members of this family, *Globuloviridae*, display helical nucleocapsids beneath their lipid envelopes (Prangishvili *et al.*, 2006a). PSV possesses 100-nm-diameter virions with 'a variable number of' 15 nm 'spherical protrusions' found on their surfaces, has a 28.3 kb linear dsDNA genome with nearly all genes (all but four) located on one of the two strands, does not appear to lyse its host, and also does not appear to substantially impact its host's doubling time

(Häring *et al.*, 2005). TTSV1, by contrast, possesses 70-nm-diameter virions and a 21.6 kb genome, but also shows similarity to PSV in 15 of 38 open reading frames, with only four genes located on one of the strands. Testing for TTSV1 integration into the host genome indicated that it does not (Ahn *et al.*, 2006). Confusingly, note that members of families *Sphaerolipoviridae* and '*Turriviridae*' have also been described as spherical.

Droplet-shaped virions (family *Guttaviridae*)

Droplets, like bottle and spindles, also are radially symmetrical, but rounded at one end. This is rather than flat, contrasting bottle-shaped viruses, or more distinctly tapered, contrasting spindle-shaped viruses. At the other end, droplet-shaped virions are clearly tapered and it is at the tapered end, in these viruses, that fibres can be found. These viruses have been isolated against *Aeropyrum* and *Sulfolobus* hosts.

SNDV was first described by Zillig *et al.* (1996) as 'a novel *Sulfolobus* virus from New Zealand.' It was described, in addition to having a droplet shape, as being a 180 × 80 nm virion with a closed covalently circular double-stranded DNA genome. The taxonomic name '*Guttaviridae*' also is introduced in this publication as too is the observation of fibres at the tapered end. Zillig *et al.* (1996) also, with reference to unpublished data, implicated those fibres in the adsorption process.

The first and indeed only publication of which I am aware which focuses exclusively on virus SNDV is that of Arnold *et al.* (2000), where we learn that SNDV stands for Sulfolobus neozealandicus droplet-shaped virus. Arnold *et al.* (2000) furthermore describe the virion as consisting of a combination of core and coat, with the coat possessing a 'beehive' or stacked hoops appearance. As hinted at previously, here it is confirmed that SNDV was discovered in a 'carrier state'. They describe this carrier state as unstable, however, perhaps since lysis was found to eventually occur, with lysis one form of instability of the virus–host interaction. Also, uninfected cells are readily isolated, that is, cured, with curing a second form of what can be described as a virus–host interaction instability.

Virus production is initiated early in stationary phase. Despite this delay in the initiation of virion production, the authors state that SNDV is lacking a (prophage) stage. Supporting this latter assertion, post infection but prior to virion production the growth of infected hosts is inhibited, the virus does not effectively persist within infected cells absent initiation of productive cycles (i.e. virus curing readily occurs), and integration of the virus into the host chromosome could not be demonstrated. Prangishvili (2013), however, indicates that members of family *Guttaviridae* in fact are temperate.

As of 2006, SNDV remained the only known member of family *Guttaviridae* (Lipps, 2006). Mochizuki *et al.* (2011), however, subsequently isolated and characterized a morphologically similar and therefore potentially second member of this family, Aeropyrum pernix ovoid virus 1 (APOV1), a temperate virus, and also a third morphologically similar virus, APOV2. APOV1, however, is 70 × 55 nm, so somewhat smaller than SNDV, although APOV1 also has a somewhat smaller genome, 14 kb versus 20 kb for SNDV, which is at least potentially consistent with the smaller virion size. APOV1 has been sequenced but I, as well as Mochizuki *et al.* (2011), were unable to find any indication via GenBank that SNDV had been sequenced. The Mochizuki *et al.* (2011) study is notable especially for its demonstration of the induction of proviruses which had been first identified *in silico* from the sequence of a host genome.

Rod-shaped/linear, enveloped virions (family *Lipothrixviridae*)

Virions of members of family *Lipothrixviridae* are filamentous, flexible, possess linear, dsDNA genomes and also, notably, have external envelopes. Superficially, these viruses give an impression of consisting of the tails of members of order *Caudovirales*, head-tailed viruses, without the associated heads. These viruses are not known to be temperate and are released (presumably) via some form of budding or extrusion. Members of family *Lipothrixviridae* have been isolated against *Acidianus* and *Sulfolobus* hosts. Quemin and Quax (2015) discuss the diversity of terminal structures of members of that family, structures which are thought to be involved in virion adsorption and consist of one to many filaments, so-called brushes, 'mop-like structures', and even 'claws', the latter involved in virion attachment to cell pili; see Pina *et al.* (2011) for images. Genome sizes span a broad range from as low as about 16 kb to as high as about 56 kb. DNA putatively from these viruses was identified from a 14th-century coprolite metagenome (Appelt *et al.*, 2014).

At least 14 archaeal virus isolates are considered to be members of this virus family (Dellas *et al.*, 2014). These are divided into four genera, *Alphalipothrixvirus*, *Betalipothrixvirus*, *Gammalipothrixvirus*, and *Deltalipothrixvirus*. Included in genus *Alphalipothrixvirus* is Thermoproteus tenax virus 1 (TTV1). Included in genus *Betalipothrixvirus* is Sulfolobus islandicus filamentous virus (SIFV). Included in genus *Gammalipothrixvirus* is Acidianus filamentous virus 1 and 2 (AFV1 and AFV2). Together with family *Rudiviridae* (next section), family *Lipothrixviridae* is found in order *Ligamenvirales* (Prangishvili *et al.*, 2006a; Prangishvili and Krupovic, 2012).

Gudbergsdóttir *et al.* (2016) have discovered that members of family *Lipothrixviridae* are fairly common within metagenomes gathered from terrestrial hot springs. The prevalence per hot spring analysed ranged from 2% to 82% of reads (across six metagenomes, these were 2%, 10%, 13%, 44%, 71% and 82%). The authors suggest that these viruses display substantial diversity based on a lack of successful assembly of even a single whole genome based on sequenced fragments.

The details of archaeal virus DNA replication have not been well elucidated, nor indeed have many biological aspects of individual virus isolates (Snyder *et al.*, 2015). Pina *et al.* (2014) briefly review what has been determined and in particular consider the replication of the gammalipothrixvirus, AFV1, which lives in hot springs and possesses a linear, dsDNA genome. They propose that AFV1 genome replication, which they hypothesize is largely catalysed by host DNA polymerase, both begins and ends with recombination steps, which presumably serve as means of dealing with the problems inherent to replicating ends of linear chromosomes ('end-replication problem'). They describe their proposed model as having certain similarities to the replication of coliphage T4, which also has a linear although circularly permuted genome (see Chapter 7).

Rod-shaped/linear, non-enveloped virions (family *Rudiviridae*)

Underlying the radially symmetrical, filamentous form of members of virus family *Rudiviridae* is a helical nucleocapsid and dsDNA genome of roughly 30 kb, give or take about 5 kb (with virion length proportional to genome length). As with many archaeal viruses, these rod-shaped virions also possess fibres in association with only one end. Unlike members of related family *Lipothrixviridae*, however, there is a lack of lipid and therefore no envelope associated with the virions. Like members of family *Lipothrixviridae*, however, the virions at least superficially resemble the tails of head-tailed virions, although in this case the virions are stiff rather than flexible. In addition, infections are lytic, contrasting the filamentous bacteriophages which are released non-lytically, via extrusion (see Chapter 7). As of this writing a total of five members of family *Rudiviridae* have been isolated: Acidianus rod-shaped virus 1 (ARV1), Sulfolobus islandicus rod-shaped virus 1 and 2 (SIRV1 and SIRV2), Sulfolobales Mexican rudivirus 1 (SMR1), and Stygiolobus rod-shaped virus (SRV) (Guo *et al.*, 2015). Families *Lipothrixviridae* (enveloped) and *Rudiviridae* (non-enveloped) together make up the virus order *Ligamenvirales* (Prangishvili and Krupovic, 2012). See Prangishvili *et al.* (2013) for a review of these viruses.

Sulfolobus islandicus rod-shaped virus 2 (SIRV2) is considered to be one of three archaeal virus model systems (Snyder *et al.*, 2015). Quemin *et al.* (2013) characterized SIRV2 binding to host pilus-like structures which occurs via virion-terminal fibres during adsorption, a process which they found to be quite rapid, whereas virion binding to SIRV2-preinfected cells was inhibited. Daum *et al.* (2014) consider the means by which the lytic SIRV2 and the also lytic Sulfolobus turreted icosahedral virus (STIV) exit infected cells by employing virus-associated pyramids, or VAPs. These pyramids, formed from virus proteins called PVAPs, first assemble within the infected cell plasma membrane before self-assembling into seven-fold pyramids. The pyramids then open outward, forming pores that release what until then are intracellular located virions (see Fig. 8.2 for illustration). Daum *et al.* (2014) employ whole-cell cryotomography as well as genetic and biochemical techniques to study morphological changes that occur in the course of virus-mediated production of these proteins, and via cloning were able to produce pyramids in the membranes of the non-archaeal *Escherichia coli* and *Saccharomyces cerevisiae*. Due to a lack of signal

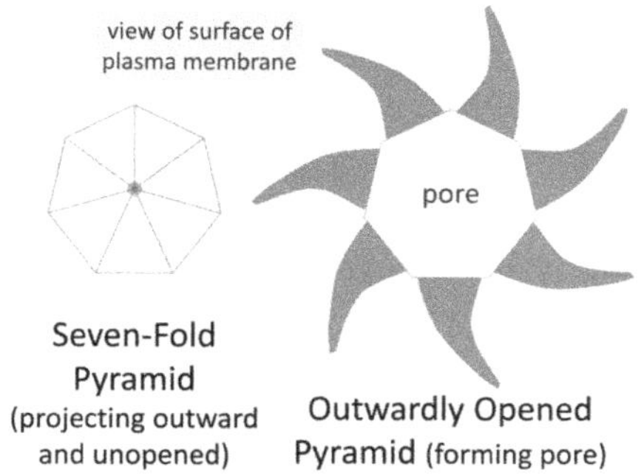

Figure 8.2 Sketch of seven-fold virus-associated pyramids (left) and then their opened form (right) that allows lytic virion egress. Images are based on those presented by Daum *et al.* (2014). These pyramids self-assemble as outward projections from the plasma membrane and then open along their seams. This mechanism of lysis is employed by viruses SIRV2 (a member of family *Rudiviridae*) and STIV, a member of proposed family *Turriviridae*.

sequence, the authors speculate that the PVAP integrate spontaneously into membranes as monomers. Opening of VAPs is thought to require a separate triggering factor since only virus-infected cells have been shown to display pyramid opening. See Quemin and Quax (2015) for further discussion of this mechanism.

SIRV2 was recently shown by DiMaio *et al.* (2015) to possess a dsDNA genome of A-form DNA rather than the more familiar B-form. *S. islandicus* is a hyperthermophilic acidophile and it is possible that the A-form DNA serves in these virions as a more stable and thereby a less heat labile DNA genome. A-form DNA is found as well within endospores. Guo *et al.* (2015) consider SIRV2 gene functions, determining that gene product 17 is a single-stranded DNA binding protein of unique structure.

Gudbergsdóttir *et al.* (2016) found rudivirus-like reads in all six of the hot spring metagenomes they characterized, but not dominating among reads. That is, they consisted of no more than about 9% of reads. One novel genome was assembled, however, which they dubbed ARV2.

Spiral- (coil-) shaped/linear virions (family *Spiraviridae*)

Isolated against *Aeropyrum pernix*, a hyperthermophilic member of the Crenarchaeota, Aeropyrum coil-shaped virus (ACV) is the only known single-stranded hyperthermophilic virus. It also, at 24.9 kb, has the distinction of having the largest ssDNA genome of any known virus. ACV also possesses, as noted, a spiral-shaped capsid surrounding its circular genome, one which forms, in the course of 40 or so turns, into an at least slightly rigid, hollow tube. Furthermore, the coils themselves consist of spirally wound nucleocapsid, and single appendages jut at 45° angles from both ends of this tube. The virions also are both naked (no envelope) and lack internal lipids, while the genome is closed circular. The virus in terms of both virion structure and nucleotide sequence has been found to be unrelated to all other known viruses (Mochizuki *et al.*, 2012).

Filamentous/linear, enveloped virions (proposed family '*Tristromaviridae*')

Filamentous/linear archaeal viruses, according to Rensen *et al.* (2016), not only all possess dsDNA genomes but represent all known dsDNA filamentous viruses, contrasting the ssDNA genome of the filamentous bacterial inoviruses. In addition, all filamentous/linear archaeal viruses infect hyperthermophilic archaea. They can differ further from filamentous bacteriophages in that in some cases for archaeal viruses their virions are enveloped; members of the bacteriophage family *Inoviridae*, by contrast, consist solely of naked virions (see Chapter 7). Four filamentous archaeal virus families include the *Clavaviridae* (as discussed above), *Lipothrixviridae* (ditto), *Rudiviridae* (ditto), and *Spiraviridae* (below), although note that terms other than filamentous are used to describe these viruses, e.g. bacilliform or rod-shaped. Families *Lipothrixviridae* (enveloped) and *Rudiviridae* (non-enveloped), together make up the virus order *Ligamenvirales* (Prangishvili and Krupovic, 2012). This section considers a yet additional filamentous archaeal virus, making up the proposed family '*Tristromaviridae*' (Rensen *et al.*, 2016).

The newly isolated Pyrobaculum filamentous virus 1 (PFV1) is a lytic, linear dsDNA crenarchaeal virus which inhabits extremely hot environments. Its genome is enclosed in two layers of proteins, an inner helical nucleocapsid as well as a surrounding helical sheath. These virions also possess a seemingly host-derived envelope (Rensen *et al.*, 2016). The organization of PFV1 virions is reported to be similar although nevertheless unrelated to that of the Ebola and Marburg animal viruses. Thermoproteus tenax virus 1 (TTV1), which was thought to have been a member of family *Lipothrixviridae*, appears to be both structurally and evolutionarily related to PFV1, and both *Pyrobaculum* and *Thermoproteus* archaeal genera are found in same archaeal order, Thermoproteales. Rensen *et al.* (2016) are sceptical of the current *Lipothrixviridae* TTV1 classification and instead propose the family '*Tristromaviridae*' for both PFV1 and TTV1.

Krupovic *et al.* (2015) discovered via 'sensitive database search methods and structural modeling' that a TTV1 nucleocapsid protein, of four nucleocapsid proteins in total, had been derived from a CRISPR-Cas system-associated nuclease (Cas4). CRISPR-Cas systems, in turn, are anti-viral defences that have been found to be encoded by a number of archaeal viruses as well as archaeal and bacterial cells. These authors argue that this is the first time that an enzyme gene has been implicated as an ancestor to a virus capsid gene. In this process of exaption – that is, a trait which evolved for one purpose is then co-opted for a different purpose – the gene was truncated and its catalytic activity lost. Thus, a catalytic protein capable of binding to DNA apparently lost its DNA cleavage function while retaining its DNA binding function.

Turreted, icosahedral virions (family '*Turriviridae*')

A virus with turret-like structures found on its capsid surface (Fig. 8.1) was isolated from an acidic Yellowstone hot spring against a *Sulfolobus* host and named Sulfolobus turreted icosahedral virus (STIV, or STIV 1). A turret is a tower that rises from the walls of especially castles, that is, a relatively small but not too small projection from what otherwise is a larger object (the castle/larger building). Alternatively, STIV turrets may be envisaged as relatively squat, cylindrical versions of the spikes familiarly associated with surfaces of many virions, here possibly playing roles in virion attachment/adsorption (Rice *et al.*, 2004). As yet, however, the role that these turrets play in the virus life cycle has not been confirmed.

STIV, as indicated in its name, is an icosahedral virion. It is distinctive, although hardly unique, in possessing an internal membrane. STIV also is a dsDNA virus with a medium-sized genome (17.7 kb) and is considered to be one of three archaeal virus model systems (Snyder *et al.*, 2015). Recently an STIV ATPase thought to be involved in DNA packaging into virions was characterized genetically by Dellas *et al.* (2015). Snyder *et al.* (2013b) discuss STIV interactions with its host that appear to be required for viral replication and virion egress. STIV additionally has been found to produce unusual, pyramidal structures during infections that are involved in host-cell lysis (Fig. 8.2) (Snyder *et al.*, 2013a), as too does archaeal virus SIRV2 (above), also of *Sulfolobus*. Fu and Johnson (2012) as well as Fulton *et al.* (2009) provide reviews of STIV biology.

A second turreted virus, dubbed STIV2, was isolated from an acidic hot spring in Iceland (Happonen *et al.*, 2010). It possesses a 16.6 kb dsDNA genome. The turrets of STIV2 possess a five-fold symmetry, as too do those of STIV (Rice *et al.*, 2004). Further, as Happonen *et al.* (2010) note (p. 4752): 'The superposition and difference imaging of the two turrets showed that there are similar barrels common to the two turrets, but STIV has in addition ear-like appendages and a cap on top of the barrel…' As a consequence of these 'ears', at lower resolution, the STIV turrets appear squatter than those of STIV2. Stedman *et al.* (2006) indicate that (p. 510) 'Similar icosahedral virus-like particles have been found in *Sulfolobus* enrichment cultures…'

Putative RNA archaeal viruses

Archaeal viruses with RNA genomes have not been isolated. Nevertheless, they at least potentially have been identified within a viral metagenome (see Chapter 5) obtained from an acidic hot spring (Bolduc *et al.*, 2012). These specifically are genes thought to encode RNA-dependent RNA polymerases which could be associated with positive-stranded RNA viruses. Analogously, ssRNA bacteriophages, members of family *Leviviridae*, also are positive stranded, meaning that their genomes serve as mRNAs (see Chapter 7). On one contig the putative polymerase was associated with putative capsid-protein genes.

A concern is that the actual host associated with these presumptive viruses has not been identified. As a consequence, at least potentially, that host therefore could be something other than a member of domain Archaea. The expectation of low stability of viruses that have merely fallen into acidic hot springs as well as the high abundance of archaeal hosts within these hot springs contends against a non-archaeal origin of these sequences, however. Stedman *et al.* (2013) nevertheless argue against an archaean hosting these putative viruses based on codon usage, although this argument is rebutted by Young *et al.* (2013). Further metagenomic exploration of the existence of RNA archaeal viruses, particularly as found within the acidic Yellowstone hot spring sample studied by Bolduc *et al.* (2012) but not in other locations, is presented by Wang *et al.* (2015b). Nevertheless, proof of the existence of RNA-genomed archaeal viruses, as reviewed here, should be assumed to be tentative.

Euryarchaeota Viruses

This section considers virus families *Myoviridae*, *Pleolipoviridae*, *Podoviridae*, *Sphaerolipoviridae*, and *Siphoviridae*. In addition is the unassigned genus *Salterprovirus*.

Myoviruses (family *Myoviridae*)

The first archaeal viruses isolated were myoviruses (Prangishvili *et al.*, 2006a). Myoviruses are members of virus family *Myoviridae*, which is found in virus order *Caudovirales*. These are head-tailed viruses in which the tail is both relatively long and contractile. This virion morphology is very familiar among bacteriophages, e.g. phage T4 (see Chapter 7), and found in some abundance among the archaeal viruses as well. Indeed, at least as of the publication of Dellas *et al.* (2014), myoviruses have been the most frequently isolated of archaeal viruses. This observation is echoed by Atanasova *et al.* (2015a), who further suggest, at least for *Halorubrum* hosts, that archaeal myoviruses tend to display somewhat broad host ranges including infection of multiple genera. These virions also, not surprisingly, display a lack of sensitivity to chloroform, which tends to be disruptive only of lipid-containing virions, which are not thought to be found among members of order *Caudovirales*. See Luk *et al.* (2014) for a list of halophilic head-tailed archaeal viruses which have been characterized.

Myoviruses, along with archaeal virus members of order *Caudovirales* more generally, have been isolated almost exclusively against members of the Euryarchaeota. Possible exceptions include a presumptive head-tailed provirus of unspecified family identified within a Thaumarchaeota host (Krupovic *et al.*, 2011b) along with a designated member of family *Myoviridae* also with a Thaumarchaeota host ('Myoviridae-like sequences' as related to those of bacteriophage T4). The latter was detected using single-celled genomics (Labonté *et al.*, 2015). One example of a Euryarchaeota-infecting myovirus is ϕCh1 (or ΦCh1) which infects *Natrialba magadii* (formerly indicated as *Natronobacterium magadii*). This virus is temperate and host-chromosome integrating. It also retains within its virions 80 to 700 nucleotide-length heterogeneous RNAs of apparent majority-host origin, along with a 55 kb dsDNA phage genome (Witte *et al.*, 1997). ϕCh1 in addition possesses an invertible host-range determinant which potentially is reminiscent of that observed in bacteriophage Mu (Klein *et al.*, 2012), where Mu is also a temperate myovirus.

Myoviruses in bacteriophages, along with members of *Caudovirales* more generally, are notoriously lytic. Consequently, readers should be cautious of suggestions that a head-tailed virus might be released via any other manner (Luk *et al.*, 2014). See Pina *et al.* (2011) for a similar argument (p. 1046): 'Determining the nature of the infection cycle and discriminating between a lytic and a carrier characteristic is not always straightforward for archeoviruses.' Alternatively, a mechanism, likely unprecedented, would have to be encoded by these relatively large, lipid-less viruses to allow release without disrupting the host plasma membrane. Larger bacterial myoviruses are also notoriously virulent, that is, not temperate, but smaller myoviruses, again as among bacteriophages, in fact can be temperate, as exemplified by the mentioned phage Mu.

Pleomorphic virions (family *Pleolipoviridae*)

These pleomorphic virions of family *Pleolipoviridae* possess lipid envelopes which are obtained from their host membranes. The surfaces of these envelopes are covered with protein spikes which, presumably, are involved in initiating the fusion between virion and target cell during adsorption. Much less conventionally for a virus, these pleomorphic virions also lack genome-associated nucleocapsid proteins. This lack of nucleocapsids can also be seen in various fungal viruses (see Chapter 9). *Acholeplasma* phage L2, which defines the lipid-enveloped pleomorphic bacteriophage family *Plasmaviridae*, on the other hand, is speculated to possess a genome that is found within a nucleoprotein condensation instead of a nucleocapsid (Maniloff and Dybvig, 2006). Release from infected cells of family *Pleolipoviridae* virions is chronic, presumably involves budding, and this ongoing virion production can impact the fitness of infected cells. These viruses have been found in association with non-methanogenic euryarchaea, i.e. halophiles, but neither with crenarchaeal nor methanogenic euryarchaea (Atanasova *et al.*, 2016).

At least three viral genera are found in family *Pleolipoviridae*. These are *Alphapleolipovirus*, *Betapleolipovirus*, and *Gammapleolipovirus*, corresponding to *Halorubrum* virus HRPV-1, *Halorubrum* virus HRPV-3, and *Haloarcula* virus His2 as type species (Krupovic *et al.*, 2016; Pietilä *et al.*, 2016). Interestingly, these viruses show substantial variation between isolates in their genome physical structure including both ssDNA and dsDNA as well both linear (dsDNA) and circular (ssDNA or dsDNA) varieties, and we can speculate that this diversity might reflect a greater flexibility in virion structure given a lack of nucleocapsid proteins. The slow-adsorbing, host growth rate slightly slowing HRPV-1 in fact is the first ssDNA archaeal virus and first circular-genomed euryarchaeal virus identified (Pietilä *et al.*, 2009). Putative pleolipoproviruses have been identified within members of genera *Haloarcula*, *Haloferax*, *Halomicrobium*, *Halopiger*, *Halorhabdus*, *Haloterrigena*, *Natrialba*, and *Natronomonas*.

Recently a pleomorphic temperate virus has been characterized that infects a member of the genus *Natrinema*, also a halophilic euryarchaean (Liu *et al.*, 2015). This virus, SNJ2, was discovered as a provirus. It also interacts with a second temperate virus, SNJ1, leading to more productive SNJ2 infection (other temperate viruses of haloarchaea are ϕCh1 and ϕH, and numerous potential additional proviruses have also been identified in haloarchaeal genomes). Unlike SNJ2, SNJ1 is not a pleomorphic virus but instead is a lipid-containing icosahedral virus. SNJ2 is unusual as well in that its circular genome possesses both nicks and ssDNA regions. Discontinuities/ssDNA regions have also been found in other pleomorphic archaeal viruses (i.e. HGPV-1 and HRPV-3) as well as various bacteriophages. Strictly, and as mentioned above, archaea possessing proviruses really should not be described as lysogenic when those viruses, as is the case for SNJ2, are released chronically from productive infections rather than lytically, although for consistency with the literature I nonetheless (reluctantly) employ the terms 'lysogen' or 'lysogenic' to describe the associated latent infections.

Podoviruses (family *Podoviridae*)

Podoviruses are also head-tailed viruses, members of order *Caudovirales*. They are less obviously tailed, however, given the short length of their non-contractile tails (contrasting siphoviruses which have much longer but also non-contractile tails). So far as is currently understood, archaeal virus members of family *Podoviridae* at best are quite rare. Especially is virus HSTV-1, which infects the host *Haloarcula sinaiiensis* (Atanasova *et al.*, 2012; Pietilä *et al.*, 2013b).

Icosahedral virions with internal membranes (family *Sphaerolipoviridae*)

Members of family *Sphaerolipoviridae* possess spiked, icosahedral, tailless virions with dsDNA genomes. These are released lytically from euryarchaeal hosts. These virions possess an internal lipid membrane that surrounds the viral genome. This membrane is acquired from the host, although lipid composition can differ from that of the host, particularly with exclusion of glycolipids from the virion. For virus SH1 of *Haloarcula hispanica*, a highly halophilic member of the Euryarchaeota, lipids are acquired during virion assembly, i.e. rather than during exit, which is as may be anticipated for a virus in which the membrane is internal rather than external to the capsid and also for which virion release is via lysis rather than budding (Svirskaitė *et al.*, 2016). Notably, these viruses appear round, thus the designation Sphaero-. Icosahedral virions with internal membranes are also seen with bacteriophages, forming family *Tectiviridae*, and members of genus *Gammasphaerolipovirus* (below) previously were classified in this family, although similarities with these bacteriophages have turned out to be superficial. For review and references, see Pawlowski *et al.*, (2014) and Wang *et al.*, (2016).

Members of family *Sphaerolipoviridae* are distinguished into three genera, *Alphasphaerolipovirus* (archaeal viruses HHIV-2, PH1, and SH1), *Betasphaerolipovirus* (archaeal virus SNJ1), and *Gammasphaerolipovirus* (bacterial viruses IN93 and P23–77), with the different genera displaying divergent genome sequences. Viruses HHIV-2, PH1, and SH1 also have been referred to using the descriptor 'halosphaerovirus' (Porter *et al.*, 2013; Luk *et al.*, 2014). *Alphaspaerolipovirus* members have linear genomes while the genomes of *Betasphaerolipovirus* and *Gammasphaerolipovirus* are circular. Virus SNJ1 exists during lysogenic cycles as a circular plasmid provirus. Hosts of genera *Alphasphaerolipovirus* and *Betasphaerolipovirus* are halophilic. The alphaspaerolipoviruses infect host *Haloarcula hispanica* along with a strain of *Halorubrum*, or even beyond this host given transfection versus normal virion adsorption. The betasphaerolipovirus instead infects members of genus *Natrinema*. A diversity of additional haloarchaeal hosts possess putative chromosome-integrated related proviruses. Members of genus *Gammasphaerolipovirus* by contrast are hosted by the hyperthermophilic bacterium, *Thermus thermophilus*. Bacterial viruses P23–72 and P23–65H serve as putative members of this latter genus.

Virus P23–77 is described as having 'turret-like protrusions' on its capsid surface (Pawlowski *et al.*, 2014), but not to be confused with those of proposed archaeal virus family '*Turriviridae*' (above). Virus SH1, meanwhile, possesses spikes that are 'horn-like' and measured adsorption rates that are quite slow. The spikes of HHIV-2, in turn, are described as 'propeller-like' (Svirskaitė *et al.*, 2016) and the virion otherwise has been structurally solved (Gil-Carton *et al.*, 2015). Also potentially horn shaped are the spikes of Haloarcula californiae icosahedral virus 1 (HCIV-1), which was recently isolated from a salt crystal, which has been proposed as an additional member of genus *Alphasphaerolipovirus*, and which possess adsorption rates described as being similar to those of HCIV-2 and SH1 (Demina *et al.*, 2016).

Siphoviruses (family *Siphoviridae*)

Siphoviruses are members of virus family *Siphoviridae*, which, like myoviruses as well as podoviruses, are members of virus order *Caudovirales*, i.e. the head-tailed viruses. Siphoviruses have tails that are relatively long, although unlike with myoviruses, these tails are not contractile. This virion morphology is very familiar among bacteriophages, e.g. phage lambda (see Chapter 7), but among the archaeal viruses are limited to euryarchaeal hosts (Pietilä *et al.*, 2016).

Siphoviruses have been isolated in association with methanogenic hosts, i.e. ϕF1, ϕF3, and ΨM1 of *Methanothermobacter thermoautotrophicus*. In addition are halophilic siphoviruses, archaeal viruses ΦN and Hh1of *Halobacterium salinarum* (Stedman *et al.*, 2006).

Spindle-shaped halophilic, euryarchaeal virions (genus *Salterprovirus*)

Spindle-shaped viruses are highly abundant within a diversity of environments in which Archaea can also be found, including hypersaline environments (Krupovic *et al.*, 2014). Nevertheless, among isolated spindle-shaped viruses, only the chronically released, linear-genomed His1 – of the otherwise unassigned virus genus *Salterprovirus* – infects halophilic euryarchaeal hosts. His1 in particular is a virus of the extreme halophilic and euryarchaeal, *Haloarcula hispanica*. Contrasting the linear genome of His1, the crenarchaeal spindle viruses of family *Fuselloviridae* instead possess circular genomes. Together these two taxa, however, represent the spindle-shaped viruses which are tailless, or at least which possess diminutive tails, contrasting family *Bicaudaviridae* (above), which are spindle-shaped viruses with prominent tails. As noted above, Krupovic *et al.* (2014) suggest, based particularly upon virion structural similarity rather than sequence similarity, that genus *Salterprovirus* instead may be classified as genus *Epsilonfusellovirus* of family *Fuselloviridae*; see also Pietilä *et al.* (2013a) for discussion of possible relationship of His1 to family *Fuselloviridae*.

Virus His1 is non-enveloped. Nevertheless, the major structural protein making up its virion to a degree appears to be lipid modified (Pietilä *et al.*, 2013a). The virion itself is flexible and the authors speculate that this flexibility derives from these lipid modifications. In addition, although virions are released chronically, the virus is able to form plaques. The His1 virion retains its infectivity upon protease digestion via bromelain, protease K, and trypsin (Hong *et al.*, 2015). It also has been found to remain infective given exposure to a wide range of NaCl concentrations, i.e. less than that of seawater (50 mM) up to 4 M, with 3.5 to 4.2 M NaCl corresponding to the optimal growth concentration for its host (Pietilä *et al.*, 2013a).

Given the extent to which spindle-shaped viruses can be found within hypersaline environments, it is odd that only one such virus, His1, has been isolated. A reportedly distant relative of His1, dubbed His2 and infecting the same host, has also been isolated (Bath *et al.*, 2006). His2, however, possesses little sequence homology with His1 and lacks gene synteny. In fact, it may not even be spindle shaped (instead existing as a pleolipovirus). Unlike His1, His2 also possesses a lipid bilayer and has a substantially different protein profile as based on SDS-PAGE analysis of purified virions (Pietilä *et al.*, 2013a). More recently, Adriaenssens *et al.* (2016) identified three contigs from a saline site that displayed amino acid-level similarity to genes found in His1.

Hong *et al.* (2015) present a cryo-electron tomography study of the His1 virion structure. They characterized in particular the somewhat diminutive His1 tail-like structure as well as six tail spikes, which they speculate are involved in virion attachment to and then alignment with host cells. Virion attachment and alignment leads to subsequent viral DNA uptake into those cells, a process that may involve transformation of the virion capsid into a genomeless tube-shaped structure. Despite variation in particle dimensions within clonal His1 populations, the tail structure remains constant. See Fig. 8.3 for a sketch of the His1 virion structure.

Another spindle-shaped 'virus-like particle' is also known, called PAV1, for Pyrococcus abyssi virus 1, and which has an 18 kb dsDNA circular genome. *Pyrococcus* is a genus of high-temperature-adapted Euryarchaeota. The virus appears to be released chronically. No sensitive secondary host was identified (Geslin *et al.*, 2003b).

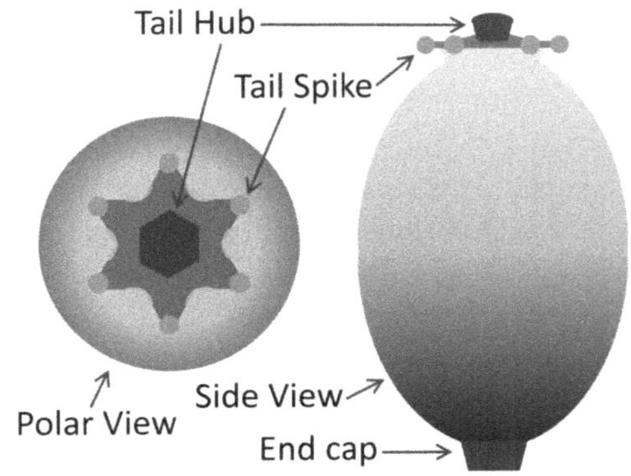

Figure 8.3 Sketch of haloarchaeal virus His1 as based on the structural analysis provided by Hong *et al.* (2015). Note the presumptive adsorption organelles shown in the centre (left-hand rendering) and at the top (right-hand rendering). The tail hub is hexagonal and attached to the tail spikes. Overall the structure is described as lemon- or spindle-shaped but is variable in width, although in association with DNA ejection from the capsid the virion takes on instead a tube shape (not shown).

Conclusion

Just as members of domain Archaea tend to be somewhat distinct from members of domain Bacteria, so too are the viruses infecting each of these domains somewhat distinct. In fact, among Archaea and particularly kingdom Crenarchaeota there appears to be at least as much morphological diversity as among the much more extensively studied bacteriophages. Although exciting, it is difficult to appreciate just why this substantial morphological diversity exists. It is perhaps telling, however, that we see this diversity and indeed archaeal viruses at all, particularly within environments in which archaea are somewhat abundant. As the latter environments often are somewhat 'extreme', and extreme in different ways (e.g. hot, acidic, and/or salty), it is perhaps that archaeal viruses have evolved divergently from both each other and from bacterial viruses at least in part to deal with substantial physical differences in environments. This thereby may have resulted in a diversity of unusual solutions, at least relative to what most bacteria face, to what can be considered to be a diversity of unusual problems. More specifically, Prangishvili (2013) suggests that it is especially differences in cell envelopes between Bacteria and Archaea, and particularly difficulty associated with viral penetration of the bacterial peptidoglycan layer of bacteria, that accounts for the morphological distinctions between the viruses infecting these two domains as well as the lower morphological diversity of bacterial viruses versus archaeal viruses (the latter, for example, may be less constrained in their morphological evolution). For now, what we know for sure is that we are still quite close to the beginning of our exploration of archaeal virus biology, with numerous answers, as well as new questions, presumably to come.

References

Abedon, S.T. (2009). Disambiguating bacteriophage pseudolysogeny: an historical analysis of lysogeny, pseudolysogeny, and the phage carrier state. In Contemporary Trends in Bacteriophage Research, Adams, H.T. ed. (Nova Science Publishers, Inc., New York), pp. 285–307.

Abedon, S.T. (2011). Communication among phages, bacteria, and soil environments. In Biocommunication of Soil Microorganisms. Soil Biology, Volume 23. Witzany, G. ed. (Springer, Berlin, Heidelberg), pp. 37-65. https://doi.org/10.1007/978-3-642-14512-4_2.

Abedon, S.T., and Murray, K.L. (2013). Archaeal viruses, not archaeal phages: An archaeological dig. Archaea, *2013*, 251245. https://doi.org/10.1155/2013/251245.

Abedon, S.T., Thomas-Abedon, C., Thomas, A., and Mazure, H. (2011). Bacteriophage prehistory: Is or is not Hankin, 1896, a phage reference? Bacteriophage *1*, 174–178. https://doi.org/10.4161/bact.1.3.16591.

Adriaenssens, E.M., van Zyl, L.J., Cowan, D.A., and Trindade, M.I. (2016). Metaviromics of Namib desert salt pans: a novel lineage of haloarchaeal salterproviruses and a rich source of ssDNA viruses. Viruses *8*, E14. https://doi.org/10.3390/v8010014.

Ahn, D.G., Kim, S.I., Rhee, J.K., Kim, K.P., Pan, J.G., and Oh, J.W. (2006). TTSV1, a new virus-like particle isolated from the hyperthermophilic crenarchaeote *Thermoproteus tenax*. Virology *351*, 280–290.

Appelt, S., Fancello, L., Le Bailly, M., Raoult, D., Drancourt, M., and Desnues, C. (2014). Viruses in a 14th-century coprolite. Appl. Environ. Microbiol. *80*, 2648–2655. https://doi.org/10.1128/AEM.03242-13.

Arnold, H.P., Ziese, U., and Zillig, W. (2000). SNDV, a novel virus of the extremely thermophilic and acidophilic archaeon *Sulfolobus*. Virology *272*, 409–416. https://doi.org/10.1006/viro.2000.0375.

Atanasova, N.S., Bamford, D.H., and Oksanen, H.M. (2016). Virus-host interplay in high salt environments. Environ. Microbiol. Rep. *8*, 431–444. https://doi.org/10.1111/1758-2229.12385.

Atanasova, N.S., Demina, T.A., Buivydas, A., Bamford, D.H., and Oksanen, H.M. (2015a). Archaeal viruses multiply: temporal screening in a solar saltern. Viruses *7*, 1902–1926. https://doi.org/10.3390/v7041902.

Atanasova, N.S., Roine, E., Oren, A., Bamford, D.H., and Oksanen, H.M. (2012). Global network of specific virus-host interactions in hypersaline environments. Environ. Microbiol. *14*, 426–440. https://doi.org/10.1111/j.1462-2920.2011.02603.x.

Atanasova, N.S., Senčilo, A., Pietilä, M.K., Roine, E., Oksanen, H.M., and Bamford, D.H. (2015b). Comparison of lipid-containing bacterial and archaeal viruses. Adv. Virus Res. *92*, 1–61. https://doi.org/10.1016/bs.aivir.2014.11.005.

Baker, B.J., Comolli, L.R., Dick, G.J., Hauser, L.J., Hyatt, D., Dill, B.D., Land, M.L., Verberkmoes, N.C., Hettich, R.L., and Banfield, J.F. (2010). Enigmatic, ultrasmall, uncultivated Archaea. Proc. Natl. Acad. Sci. U.S.A. *107*, 8806–8811. https://doi.org/10.1073/pnas.0914470107.

Bath, C., Cukalac, T., Porter, K., and Dyall-Smith, M.L. (2006). His1 and His2 are distantly related, spindle-shaped haloviruses belonging to the novel virus group, *Salterprovirus*. Virology *350*, 228–239.

Bertani, G., and Baresi, L. (1986). Looking for gene transfer mechanism in methanogenic bacteria. In Archaebacteria ’85, Kandler, O., and Zillig, W. eds. (Gustav Fischer, Stuttgart), vol. 1, p. 398.

Bolduc, B., Shaughnessy, D.P., Wolf, Y.I., Koonin, E.V., Roberto, F.F., and Young, M. (2012). Identification of novel positive-strand RNA viruses by metagenomic analysis of archaea-dominated Yellowstone hot springs. J. Virol. *86*, 5562–5573. https://doi.org/10.1128/JVI.07196-11.

Brochier-Armanet, C., Boussau, B., Gribaldo, S., and Forterre, P. (2008). Mesophilic Crenarchaeota: proposal for a third archaeal phylum, the Thaumarchaeota. Nat. Rev. Microbiol. *6*, 245–252. https://doi.org/10.1038/nrmicro1852.

Chanishvili, N. (2012). Phage therapy – history from Twort and d’Herelle through Soviet experience to current approaches. Adv. Virus Res. *83*, 3–40. https://doi.org/10.1016/B978-0-12-394438-2.00001-3.

Comolli, L.R., Baker, B.J., Downing, K.H., Siegerist, C.E., and Banfield, J.F. (2009). Three-dimensional analysis of the structure and ecology of a novel, ultra-small archaeon. ISME J. *3*, 159–167. https://doi.org/10.1038/ismej.2008.99.

Contursi, P., Fusco, S., Cannio, R., and She, Q. (2014). Molecular biology of fuselloviruses and their satellites. Extremophiles *18*, 473–489. https://doi.org/10.1007/s00792-014-0634-0.

Daum, B., Quax, T.E., Sachse, M., Mills, D.J., Reimann, J., Yildiz, Ö., Häder, S., Saveanu, C., Forterre, P., Albers, S.V., *et al.* (2014). Self-assembly of the general membrane-remodeling protein PVAP into sevenfold virus-associated pyramids. Proc. Natl. Acad. Sci. U.S.A. *111*, 3829–3834. https://doi.org/10.1073/pnas.1319245111.

Dellas, N., Snyder, J.C., Bolduc, B., and Young, M.J. (2014). Archaeal viruses: diversity, replication, and structure. Annu. Rev. Virol. *1*, 399–426. https://doi.org/10.1146/annurev-virology-031413-085357.

Dellas, N., Snyder, J.C., Dills, M., Nicolay, S.J., Kerchner, K.M., Brumfield, S.K., Lawrence, C.M., and Young, M.J. (2015). Structure-based mutagenesis of *Sulfolobus* Turreted Icosahedral Virus B204 reveals essential residues in the virion-associated DNA-packaging ATPase. J. Virol. *90*, 2729–2739. https://doi.org/10.1128/JVI.02435-15.

Demina, T.A., Pietilä, M.K., Svirskaitė, J., Ravantti, J.J., Atanasova, N.S., Bamford, D.H., and Oksanen, H.M. (2016). Archaeal *Haloarcula californiae* Icosahedral Virus 1 highlights conserved elements in icosahedral membrane-containing DNA viruses from extreme environments. MBio *7*, e00699-16. https://doi.org/10.1128/mBio.00699-16.

DiMaio, F., Yu, X., Rensen, E., Krupovic, M., Prangishvili, D., and Egelman, E.H. (2015). Virology. A virus that infects a hyperthermophile encapsidates A-form DNA.

Science *348*, 914–917. https://doi.org/10.1126/science.aaa4181.

Duckworth, D.H. (1976). 'Who discovered bacteriophage?'. Bacteriol. Rev. *40*, 793–802.

Erdmann, S., Chen, B., Huang, X., Deng, L., Liu, C., Shah, S.A., Le Moine Bauer, S., Sobrino, C.L., Wang, H., Wei, Y., *et al.* (2014). A novel single-tailed fusiform *Sulfolobus* virus STSV2 infecting model *Sulfolobus* species. Extremophiles *18*, 51–60. https://doi.org/10.1007/s00792-013-0591-z.

Evilevitch, A. (2013). Physical evolution of pressure-driven viral infection. Biophys. J. *104*, 2113–2114. https://doi.org/10.1016/j.bpj.2013.03.062.

Fu, C.Y., and Johnson, J.E. (2012). Structure and cell biology of archaeal virus STIV. Curr. Opin. Virol. *2*, 122–127. https://doi.org/10.1016/j.coviro.2012.01.007.

Fulton, J., Bothner, B., Lawrence, M., Johnson, J.E., Douglas, T., and Young, M. (2009). Genetics, biochemistry and structure of the archaeal virus STIV. Biochem. Soc. Trans. *37*, 114–117. https://doi.org/10.1042/BST0370114.

Garcia-Heredia, I., Martin-Cuadrado, A.B., Mojica, F.J., Santos, F., Mira, A., Antón, J., and Rodriguez-Valera, F. (2012). Reconstructing viral genomes from the environment using fosmid clones: the case of haloviruses. PLOS ONE *7*, e33802. https://doi.org/10.1371/journal.pone.0033802.

Geslin, C., Le Romancer, M., Gaillard, M., Erauso, G., and Prieur, D. (2003a). Observation of virus-like particles in high temperature enrichment cultures from deep-sea hydrothermal vents. Res. Microbiol. *154*, 303–307.

Geslin, C., Le Romancer, M., Erauso, G., Gaillard, M., Perrot, G., and Prieur, D. (2003b). PAV1, the first virus-like particle isolated from a hyperthermophilic euryarchaeote, '*Pyrococcus abyssi*'. J. Bacteriol. *185*, 3888–3894.

Gil-Carton, D., Jaakkola, S.T., Charro, D., Peralta, B., Castaño-Díez, D., Oksanen, H.M., Bamford, D.H., and Abrescia, N.G. (2015). Insight into the assembly of viruses with vertical single β-barrel major capsid proteins. Structure *23*, 1866–1877. https://doi.org/10.1016/j.str.2015.07.015.

Gorlas, A., and Geslin, C. (2013). A simple procedure to determine the infectivity and host range of viruses infecting anaerobic and hyperthermophilic microorganisms. Extremophiles *17*, 349–355. https://doi.org/10.1007/s00792-013-0513-0.

Gudbergsdóttir, S.R., Menzel, P., Krogh, A., Young, M., and Peng, X. (2016). Novel viral genomes identified from six metagenomes reveal wide distribution of archaeal viruses and high viral diversity in terrestrial hot springs. Environ. Microbiol. *18*, 863–874. https://doi.org/10.1111/1462-2920.13079.

Guo, Y., Kragelund, B.B., White, M.F., and Peng, X. (2015). Functional characterization of a conserved archaeal viral operon revealing single-stranded DNA binding, annealing and nuclease activities. J. Mol. Biol. *427*, 2179–2191. https://doi.org/10.1016/j.jmb.2015.03.013.

Hanhijärvi, K.J., Ziedaite, G., Hæggström, E., and Bamford, D.H. (2016). Temperature and pH dependence of DNA ejection from archaeal lemon-shaped virus His1. Eur. Biophys. J. *45*, 435–442. https://doi.org/10.1007/s00249-016-1112-7.

Hanhijärvi, K.J., Ziedaite, G., Pietilä, M.K., Hæggström, E., and Bamford, D.H. (2013). DNA ejection from an archaeal virus – a single-molecule approach. Biophys. J. *104*, 2264–2272. https://doi.org/10.1016/j.bpj.2013.03.061.

Happonen, L.J., Redder, P., Peng, X., Reigstad, L.J., Prangishvili, D., and Butcher, S.J. (2010). Familial relationships in hyperthermo- and acidophilic archaeal viruses. J. Virol. *84*, 4747–4754. https://doi.org/10.1128/JVI.02156-09.

Häring, M., Peng, X., Brügger, K., Rachel, R., Stetter, K.O., Garrett, R.A., and Prangishvili, D. (2004). Morphology and genome organization of the virus PSV of the hyperthermophilic archaeal genera *Pyrobaculum* and *Thermoproteus*: a novel virus family, the *Globuloviridae*. Virology *323*, 233–242. https://doi.org/10.1016/j.virol.2004.03.002.

Häring, M., Rachel, R., Peng, X., Garrett, R.A., and Prangishvili, D. (2005a). Viral diversity in hot springs of Pozzuoli, Italy, and characterization of a unique archaeal virus, *Acidianus* bottle-shaped virus, from a new family, the *Ampullaviridae*. J. Virol. *79*, 9904–9911.

Häring, M., Vestergaard, G., Rachel, R., Chen, L., Garrett, R.A., and Prangishvili, D. (2005b). Virology: independent virus development outside a host. Nature *436*, 1101–1102.

Hedlund, B.P., Murugapiran, S.K., Alba, T.W., Levy, A., Dodsworth, J.A., Goertz, G.B., Ivanova, N., and Woyke, T. (2015). Uncultivated thermophiles: current status and spotlight on 'Aigarchaeota'. Curr. Opin. Microbiol. *25*, 136–145. https://doi.org/10.1016/j.mib.2015.06.008.

Ho, C.C., Lau, S.K., and Woo, P.C. (2013). Romance of the three domains: how cladistics transformed the classification of cellular organisms. Protein Cell *4*, 664–676. https://doi.org/10.1007/s13238-013-3050-9.

Hobbs, Z., and Abedon, S.T. (2016). Diversity of phage infection types and associated terminology: the problem with 'Lytic or lysogenic'. FEMS Microbiol. Lett. *363*, fnw047. https://doi.org/10.1093/femsle/fnw047.

Hochstein, R., Bollschweiler, D., Engelhardt, H., Lawrence, C.M., and Young, M. (2015). Large tailed spindle viruses of archaea: a new way of doing viral business. J. Virol. *89*, 9146–9149. https://doi.org/10.1128/JVI.00612-15.

Hochstein, R.A., Amenabar, M.J., Munson-McGee, J.H., Boyd, E.S., and Young, M.J. (2016). Acidianus tailed spindle virus: A new archaeal large tailed spindle virus discovered by culture-independent methods. J. Virol. *90*, 3458–3468. https://doi.org/10.1128/JVI.03098-15.

Hong, C., Pietilä, M.K., Fu, C.J., Schmid, M.F., Bamford, D.H., and Chiu, W. (2015). Lemon-shaped halo archaeal virus His1 with uniform tail but variable capsid structure. Proc. Natl. Acad. Sci. U.S.A. *112*, 2449–2454. https://doi.org/10.1073/pnas.1425008112.

Hyman, P., and Abedon, S.T. (2012). Smaller fleas: viruses of microorganisms. Scientifica *2012*, 734023. https://doi.org/10.6064/2012/734023.

Iranzo, J., Koonin, E.V., Prangishvili, D., and Krupovic, M. (2016). Bipartite network analysis of the archaeal virosphere: Evolutionary connections between viruses and capsidless mobile elements. J. Virol. *90*, 11043–11055.

Klein, R., Rossler, N., Iro, M., Scholz, H., and Witte, A. (2012). Haloarchaeal myovirus φCh1 harbours a phase variation system for the production of protein variants with distinct cell surface adhesion specificities. Mol. Microbiol. *83*, 137–150. https://doi.org/10.1111/j.1365-2958.2011.07921.x.

Krupovic, M., Cvirkaite-Krupovic, V., Prangishvili, D., and Koonin, E.V. (2015). Evolution of an archaeal virus nucleocapsid protein from the CRISPR-associated Cas4 nuclease. Biol. Direct *10*, 65. https://doi.org/10.1186/s13062-015-0093-2.

Krupovic, M., Dutilh, B.E., Adriaenssens, E.M., Wittmann, J., Vogensen, F.K., Sullivan, M.B., Rumnieks, J., Prangishvili, D., Lavigne, R., Kropinski, A.M., *et al.* (2016). Taxonomy of prokaryotic viruses: update from the ICTV bacterial and archaeal viruses subcommittee. Arch. Virol. *161*, 1095–1099. https://doi.org/10.1007/s00705-015-2728-0.

Krupovic, M., Prangishvili, D., Hendrix, R.W., and Bamford, D.H. (2011a). Genomics of bacterial and archaeal viruses: dynamics within the prokaryotic virosphere. Microbiol. Mol. Biol. Rev. *75*, 610–635. https://doi.org/10.1128/MMBR.00011-11.

Krupovic, M., Quemin, E.R., Bamford, D.H., Forterre, P., and Prangishvili, D. (2014). Unification of the globally distributed spindle-shaped viruses of the Archaea. J. Virol. *88*, 2354–2358. https://doi.org/10.1128/JVI.02941-13.

Krupovic, M., Spang, A., Gribaldo, S., Forterre, P., and Schleper, C. (2011b). A thaumarchaeal provirus testifies for an ancient association of tailed viruses with archaea. Biochem. Soc. Trans. *39*, 82–88. https://doi.org/10.1042/BST0390082.

Krupovic, M., White, M.F., Forterre, P., and Prangishvili, D. (2012). Postcards from the edge: structural genomics of archaeal viruses. Adv. Virus Res. *82*, 33–62. https://doi.org/10.1016/B978-0-12-394621-8.00012-1.

Labonté, J.M., Swan, B.K., Poulos, B., Luo, H., Koren, S., Hallam, S.J., Sullivan, M.B., Woyke, T., Wommack, K.E., and Stepanauskas, R. (2015). Single-cell genomics-based analysis of virus-host interactions in marine surface bacterioplankton. ISME J. *9*, 2386–2399. https://doi.org/10.1038/ismej.2015.48.

Lawrence, C.M., Menon, S., Eilers, B.J., Bothner, B., Khayat, R., Douglas, T., and Young, M.J. (2009). Structural and functional studies of archaeal viruses. J. Biol. Chem. *284*, 12599–12603. https://doi.org/10.1074/jbc.R800078200.

Lipps, G. (2006). Plasmids and viruses of the thermoacidophilic crenarchaeote *Sulfolobus*. Extremophiles *10*, 17–28. https://doi.org/10.1007/s00792-005-0492-x.

Liu, Y., Wang, J., Liu, Y., Wang, Y., Zhang, Z., Oksanen, H.M., Bamford, D.H., and Chen, X. (2015). Identification and characterization of SNJ2, the first temperate pleolipovirus integrating into the genome of the SNJ1-lysogenic archaeal strain. Mol. Microbiol. *98*, 1002–1020. https://doi.org/10.1111/mmi.13204.

Luk, A.W., Williams, T.J., Erdmann, S., Papke, R.T., and Cavicchioli, R. (2014). Viruses of haloarchaea. Life *4*, 681–715. https://doi.org/10.3390/life4040681.

Maniloff, J., and Dybvig, K. (2006). Mycoplasma phages. In The Bacteriophages, Calendar, R., and Abedon, S.T. eds. (Oxford University Press, UK), pp. 636–652.

Martínez-García, M., Santos, F., Moreno-Paz, M., Parro, V., and Antón, J. (2014). Unveiling viral-host interactions within the 'microbial dark matter'. Nat. Commun. *5*, 4542. https://doi.org/10.1038/ncomms5542.

Mochizuki, T., Krupovic, M., Pehau-Arnaudet, G., Sako, Y., Forterre, P., and Prangishvili, D. (2012). Archaeal virus with exceptional virion architecture and the largest single-stranded DNA genome. Proc. Natl. Acad. Sci. U.S.A. *109*, 13386–13391. https://doi.org/10.1073/pnas.1203668109.

Mochizuki, T., Sako, Y., and Prangishvili, D. (2011). Provirus induction in hyperthermophilic archaea: characterization of *Aeropyrum pernix* spindle-shaped virus 1 and *Aeropyrum pernix* ovoid virus 1. J. Bacteriol. *193*, 5412–5419. https://doi.org/10.1128/JB.05101-11.

Mochizuki, T., Yoshida, T., Tanaka, R., Forterre, P., Sako, Y., and Prangishvili, D. (2010). Diversity of viruses of the hyperthermophilic archaeal genus *Aeropyrum*, and isolation of the *Aeropyrum pernix* bacilliform virus 1, APBV1, the first representative of the family *Clavaviridae*. Virology *402*, 347–354. https://doi.org/10.1016/j.virol.2010.03.046.

Munson-McGee, J.H., Field, E.K., Bateson, M., Rooney, C., Stepanauskas, R., and Young, M.J. (2015). *Nanoarchaeota*, their *Sulfolobales* host, and *Nanoarchaeota* virus distribution across Yellowstone National Park hot springs. Appl. Environ. Microbiol. *81*, 7860–7868. https://doi.org/10.1128/AEM.01539-15.

Pawlowski, A., Rissanen, I., Bamford, J.K., Krupovic, M., and Jalasvuori, M. (2014). Gammasphaerolipovirus, a newly proposed bacteriophage genus, unifies viruses of halophilic archaea and thermophilic bacteria within the novel family Sphaerolipoviridae. Arch. Virol. *159*, 1541–1554. https://doi.org/10.1007/s00705-013-1970-6.

Peng, X., Basta, T., Häring, M., Garrett, R.A., and Prangishvili, D. (2007). Genome of the *Acidianus* bottle-shaped virus and insights into the replication and packaging mechanisms. Virology *364*, 237–243.

Peng, X., Garrett, R.A., and She, Q. (2012). Archaeal viruses – novel, diverse and enigmatic. Sci. China Life Sci. *55*, 422–433. https://doi.org/10.1007/s11427-012-4325-8.

Petitjean, C., Deschamps, P., López-García, P., and Moreira, D. (2014). Rooting the domain archaea by phylogenomic analysis supports the foundation of the new kingdom Proteoarchaeota. Genome Biol. Evol. *7*, 191–204. https://doi.org/10.1093/gbe/evu274.

Pietilä, M.K., Atanasova, N.S., Oksanen, H.M., and Bamford, D.H. (2013a). Modified coat protein forms the flexible spindle-shaped virion of haloarchaeal virus His1. Environ. Microbiol. *15*, 1674–1686. https://doi.org/10.1111/1462-2920.12030.

Pietilä, M.K., Demina, T.A., Atanasova, N.S., Oksanen, H.M., and Bamford, D.H. (2014). Archaeal viruses and bacteriophages: comparisons and contrasts. Trends Microbiol. *22*, 334–344. https://doi.org/10.1016/j.tim.2014.02.007.

Pietilä, M.K., Laurinmäki, P., Russell, D.A., Ko, C.C., Jacobs-Sera, D., Hendrix, R.W., Bamford, D.H., and Butcher, S.J. (2013b). Structure of the archaeal head-tailed virus HSTV-1 completes the HK97 fold story. Proc.

Natl. Acad. Sci. U.S.A. *110*, 10604–10609. https://doi.org/10.1073/pnas.1303047110.

Pietilä, M.K., Roine, E., Paulin, L., Kalkkinen, N., and Bamford, D.H. (2009). An ssDNA virus infecting archaea: a new lineage of viruses with a membrane envelope. Mol. Microbiol. *72*, 307–319. https://doi.org/10.1111/j.1365-2958.2009.06642.x.

Pietilä, M.K., Roine, E., Sencilo, A., Bamford, D.H., and Oksanen, H.M. (2016). *Pleolipoviridae*, a newly proposed family comprising archaeal pleomorphic viruses with single-stranded or double-stranded DNA genomes. Arch. Virol. *161*, 249–256. https://doi.org/10.1007/s00705-015-2613-x.

Pina, M., Basta, T., Quax, T.E., Joubert, A., Baconnais, S., Cortez, D., Lambert, S., Le Cam, E., Bell, S.D., Forterre, P., *et al.* (2014). Unique genome replication mechanism of the archaeal virus AFV1. Mol. Microbiol. *92*, 1313–1325. https://doi.org/10.1111/mmi.12630.

Pina, M., Bize, A., Forterre, P., and Prangishvili, D. (2011). The archeoviruses. FEMS Microbiol. Rev. *35*, 1035–1054. https://doi.org/10.1111/j.1574-6976.2011.00280.x.

Porter, K., Tang, S.L., Chen, C.P., Chiang, P.W., Hong, M.J., and Dyall-Smith, M. (2013). PH1: an archaeovirus of *Haloarcula hispanica* related to SH1 and HHIV-2. Archaea *2013*, 456318. https://doi.org/10.1155/2013/456318.

Prangishvili, D. (2013). The wonderful world of archaeal viruses. Annu. Rev. Microbiol. *67*, 565–585. https://doi.org/10.1146/annurev-micro-092412-155633.

Prangishvili, D., Forterre, P., and Garrett, R.A. (2006a). Viruses of the Archaea: a unifying view. Nat. Rev. Microbiol. *4*, 837–848.

Prangishvili, D., Koonin, E.V., and Krupovic, M. (2013). Genomics and biology of Rudiviruses, a model for the study of virus-host interactions in Archaea. Biochem. Soc. Trans. *41*, 443–450. https://doi.org/10.1042/BST20120313.

Prangishvili, D., and Krupovic, M. (2012). A new proposed taxon for double-stranded DNA viruses, the order 'Ligamenvirales'. Arch. Virol. *157*, 791–795. https://doi.org/10.1007/s00705-012-1229-7.

Prangishvili, D., Vestergaard, G., Häring, M., Aramayo, R., Basta, T., Rachel, R., and Garrett, R.A. (2006b). Structural and genomic properties of the hyperthermophilic archaeal virus ATV with an extracellular stage of the reproductive cycle. J. Mol. Biol. *359*, 1203–1216.

Quemin, E.R., Chlanda, P., Sachse, M., Forterre, P., Prangishvili, D., and Krupovic, M. (2016). Eukaryotic-like virus budding in *Archaea*. MBio *7*, e01439–16. https://doi.org/10.1128/mBio.01439-16.

Quemin, E.R., Lucas, S., Daum, B., Quax, T.E., Kühlbrandt, W., Forterre, P., Albers, S.V., Prangishvili, D., and Krupovic, M. (2013). First insights into the entry process of hyperthermophilic archaeal viruses. J. Virol. *87*, 13379–13385. https://doi.org/10.1128/JVI.02742-13.

Quemin, E.R., and Quax, T.E. (2015). Archaeal viruses at the cell envelope: entry and egress. Front. Microbiol. *6*, 552. https://doi.org/10.3389/fmicb.2015.00552.

Redder, P., Peng, X., Brügger, K., Shah, S.A., Roesch, F., Greve, B., She, Q., Schleper, C., Forterre, P., Garrett, R.A., *et al.* (2009). Four newly isolated fuselloviruses from extreme geothermal environments reveal unusual morphologies and a possible interviral recombination mechanism. Environ. Microbiol. *11*, 2849–2862. https://doi.org/10.1111/j.1462-2920.2009.02009.x.

Reiter, W.D., Zillig, W., and Palm, P. (1988). Archaebacterial viruses. Adv. Virus Res. *34*, 143–188.

Rensen, E.I., Mochizuki, T., Quemin, E., Schouten, S., Krupovic, M., and Prangishvili, D. (2016). A virus of hyperthermophilic archaea with a unique architecture among DNA viruses. Proc. Natl. Acad. Sci. U.S.A. *113*, 2478–2483. https://doi.org/10.1073/pnas.1518929113.

Rice, G., Tang, L., Stedman, K., Roberto, F., Spuhler, J., Gillitzer, E., Johnson, J.E., Douglas, T., and Young, M. (2004). The structure of a thermophilic archaeal virus shows a double-stranded DNA viral capsid type that spans all domains of life. Proc. Natl. Acad. Sci. U.S.A. *101*, 7716–7720. https://doi.org/10.1073/pnas.0401773101.

Roine, E., and Bamford, D.H. (2012). Lipids of archaeal viruses. Archaea *2012*, 384919. https://doi.org/10.1155/2012/384919.

Roux, S., Brum, J.R., Dutilh, B.E., Sunagawa, S., Duhaime, M.B., Loy, A., Poulos, B.T., Solonenko, N., Lara, E., Poulain, J., *et al.* (2016). Ecogenomics and potential biogeochemical impacts of globally abundant ocean viruses. Nature *537*, 689–693. https://doi.org/10.1038/nature19366.

Roux, S., Hallam, S.J., Woyke, T., and Sullivan, M.B. (2015). Viral dark matter and virus-host interactions resolved from publicly available microbial genomes. eLife *4*. https://doi.org/10.7554/eLife.08490.

Sheppard, C., Blombach, F., Belsom, A., Schulz, S., Daviter, T., Smollett, K., Mahieu, E., Erdmann, S., Tinnefeld, P., Garrett, R., *et al.* (2016). Repression of RNA polymerase by the archaeo-viral regulator ORF145/RIP. Nat. Commun. *7*, 13595. https://doi.org/10.1038/ncomms13595.

Snyder, J.C., Bolduc, B., and Young, M.J. (2015). 40 Years of archaeal virology: Expanding viral diversity. Virology 479-480, 369–378. https://doi.org/10.1016/j.virol.2015.03.031.

Snyder, J.C., Brumfield, S.K., Kerchner, K.M., Quax, T.E., Prangishvili, D., and Young, M.J. (2013a). Insights into a viral lytic pathway from an archaeal virus-host system. J. Virol. *87*, 2186–2192. https://doi.org/10.1128/JVI.02956-12.

Snyder, J.C., Samson, R.Y., Brumfield, S.K., Bell, S.D., and Young, M.J. (2013b). Functional interplay between a virus and the ESCRT machinery in archaea. Proc. Natl. Acad. Sci. U.S.A. *110*, 10783–10787. https://doi.org/10.1073/pnas.1301605110.

Snyder, J.C., and Young, M.J. (2013). Lytic viruses infecting organisms from the three domains of life. Biochem. Soc. Trans. *41*, 309–313. https://doi.org/10.1042/BST20120326.

Stedman, K.M., DeYoung, M., Saha, M., Sherman, M.B., and Morais, M.C. (2015). Structural insights into the architecture of the hyperthermophilic *Fusellovirus* SSV1. Virology *474*, 105–109. https://doi.org/10.1016/j.virol.2014.10.014.

Stedman, K.M., Kosmicki, N.R., and Diemer, G.S. (2013). Codon usage frequency of RNA virus genomes from high-temperature acidic-environment metagenomes. J. Virol. *87*, 1919. https://doi.org/10.1128/JVI.02610-12.

Stedman, K.M., Prangishvili, D., and Zillig, W. (2006). Viruses of Archaea. In The Bacteriophages, Calendar, R., and Abedon, S.T. eds. (Oxford University Press, UK), pp. 499–516.

Svirskaitė, J., Oksanen, H.M., Daugelavičius, R., and Bamford, D.H. (2016). Monitoring physiological changes in haloarchaeal cell during virus release. Viruses *8*, 59. https://doi.org/10.3390/v8030059.

Torsvik, T., and Dundas, I.D. (1974). Bacteriophage of *Halobacterium salinarium*. Nature *248*, 680–681.

Torsvik, T., and Dundas, I.D. (1980). Persisting phage infection in *Halobacterium salinarium* str. 1. J. Gen. Virol. *47*, 29–36. 1980. https://doi.org/10.1099/0022-1317-47-1-29.

Uldahl, K.B., Jensen, S.B., Bhoobalan-Chitty, Y., Martínez-Álvarez, L., Papathanasiou, P., and Peng, X. (2016). Life cycle characterization of *Sulfolobus* monocaudavirus 1, an extremophilic spindle-shaped virus with extracellular tail development. J. Virol. *90*, 5693–5699. https://doi.org/10.1128/JVI.00075-16.

Wang, H., Peng, N., Shah, S.A., Huang, L., and She, Q. (2015a). Archaeal extrachromosomal genetic elements. Microbiol. Mol. Biol. Rev. *79*, 117–152. https://doi.org/10.1128/MMBR.00042-14.

Wang, H., Yu, Y., Liu, T., Pan, Y., Yan, S., and Wang, Y. (2015b). Diversity of putative archaeal RNA viruses in metagenomic datasets of a Yellowstone acidic hot spring. Springerplus *4*, 189. https://doi.org/10.1186/s40064-015-0973-z.

Wang, Y., Sima, L., Lv, J., Huang, S., Liu, Y., Wang, J., Krupovic, M., and Chen, X. (2016). Identification, characterization, and application of the replicon region of the halophilic temperate sphaerolipovirus SNJ1. J. Bacteriol. *198*, 1952–1964. https://doi.org/10.1128/JB.00131-16.

Witte, A., Baranyi, U., Klein, R., Sulzner, M., Luo, C., Wanner, G., Krüger, D.H., and Lubitz, W. (1997). Characterization of *Natronobacterium magadii* phage ϕCh1, a unique archaeal phage containing DNA and RNA. Mol. Microbiol. 23, 603–616.

Woese, C.R., and Fox, G.E. (1977). Phylogenetic structure of the prokaryotic domain: the primary kingdoms. Proc. Natl. Acad. Sci. U.S.A. *74*, 5088–5090.

Woese, C.R., Kandler, O., and Wheelis, M.L. (1990). Towards a natural system of organisms: proposal for the domains Archaea, Bacteria, and Eucarya. Proc. Natl. Acad. Sci. U.S.A. *87*, 4576–4579.

Xiang, X., Chen, L., Huang, X., Luo, Y., She, Q., and Huang, L. (2005). *Sulfolobus tengchongensis* spindle-shaped virus STSV1: virus-host interactions and genomic features. J. Virol. *79*, 8677–8686.

Young, M., Bolduc, B., Shaughnessy, D.P., Roberto, F.F., Wolf, Y.I., and Koonin, E.V. (2013). Reply to 'codon usage frequency of RNA virus genomes from high-temperature acidic-environment metagenomes'. J. Virol. *87*, 1920–1921. https://doi.org/10.1128/JVI.02883-12.

Zillig, W., Prangishvili, D., Schleper, C., Elferink, M., Holz, I., Albers, S., Janekovic, D., and Götz, D. (1996). Viruses, plasmids and other genetic elements of thermophilic and hyperthermophilic *Archaea*. FEMS Microbiol. Rev. *18*, 225–236.

Zillig, W., Reiter, W.-D., Palm, P., Gropp, F., Neumann, H., and Rettenberger, M. (1988). Viruses of archaebacteria. In The Bacteriophages, Calendar, R. ed. (Plenum Press, New York). vol. 1, pp. 517–558.

Fungal Viruses

9

Eeva J. Vainio* and Jarkko Hantula

Natural Resources Institute Finland, Helsinki, Finland.

*Correspondence: eeva.vainio@luke.fi

https://doi.org/10.21775/9781910190852.09

Abstract

Fungal viruses (mycoviruses) occur in all major fungal phyla. They are taxonomically highly diverse, and are currently classified into seven families that have mono- or multipartite dsRNA genomes and six families with ssRNA genomes. One described mycovirus species, *Sclerotinia gemycircularvirus 1*, has a genome composed of circular ssDNA and resembles plant geminiviruses. Most mycoviruses are considered to form latent infections without major phenotypic alterations in their host fungi, but both detrimental and mutualistic associations are known. As an example, a mycovirus infection causes a disease in cultivated mushrooms resulting in crop losses. On the other hand, hypoviruses of the chestnut blight fungus (*Cryphonectria parasitica*) can be used for biocontrol against their host, which is a devastating plant pathogen. There is even one mycovirus involved in a three-way symbiosis formed by a grass, its endophytic fungus, and the virus. Mycoviruses with RNA genomes are considered to be completely intracellular and to transmit laterally between fungal strains during cell-to-cell contact, and vertically via sexual or asexual spores. Most mycovirus species are regarded as host-specific, but there are also examples of horizontal virus transmission between host species. This contradicts the commonly held hypothesis of strict virus–host coevolution. The evolution of mycoviruses has likely been shaped by events of horizontal transmission between plants and fungi.

Introduction

Fungi are eukaryotes distantly related to plants and animals. Saprophytic fungi are among the most important decomposers of biological material in different ecosystems, whereas mycorrhizal fungi are symbiotic and assist trees in obtaining nutrients from the soil. In contrast, pathogenic fungi cause many diseases on other eukaryotic organisms. Some fungi, like baker's yeast, are single-celled microbes. Others are multicellular mycelial organisms and among the largest organisms on earth (Anderson, 1992). Some fungal species may even alter their way of living from yeasts to filamentous mycelia. Therefore it is not always clear whether particular fungi should be considered as micro- or macro-organisms. At the macroscopic level, fungal mycelia may form specialized structures like fruiting bodies, differentiated hyphae forming asexual spores, or resistant mycelial cords ('rhizomorphs'). The classification of the true fungi is based on genetic information as well as differences in their reproductive structures: the sexual spores of phylum Ascomycota form endogenously within specialized ascus cells, while those of basidiomycetes (phylum Basidiomycota) form exogenously. Members of phylum Chytridiomycota are aquatic fungi forming motile zoospores, while phylum Zygomycota reproduce by formation of latent zygospores. Furthermore, the fungal mycelium is very similar to that of Stramenopiles (oomycetes or water moulds), which previously were considered to be fungi, but today are known to be an unrelated group of eukaryotes. Some of their viruses are discussed in this chapter.

The first fungal virus observed was the causative agent of a disease of the cultivated mushroom, *Agaricus bisporus*. The observation was made in the farm of the La France brothers by J.W. Sinden in 1948. The nature of the causative agent as a virus was shown by M. Hollings (1962). This mycovirus genome is composed of nine dsRNA molecules, which are packed in a *ca* 35 nm particle (Wach *et al.*, 1987). The disease spreads by transfer of fungal isolates of the same vegetative compatibility group (VCG) between mushroom farms and may lead to a complete crop failure via reduced growth and cropping.

Fungal viruses can also be employed as antifungal biocontrol agents (see Chapter 14). The virocontrol of chestnut blight in Europe has been very successful (see next section), inspiring researchers throughout the world to search for possibilities to control other fungal diseases. Much of the work has involved the screening of viruses, but in several virus–host combinations considerable effort has been put to understand their biology in more detail. This has led to the discovery of many novel virus taxa that mediate hypovirulence in plant pathogenic fungal species (Chu *et al.*, 2002; Deng *et al.*, 2003; Kanematsu *et al.*, 2004; Huang and Ghabrial, 1996; Lakshman *et al.*, 1998; Preisig *et al.*, 2000; Chiba *et al.*, 2009, Yu *et al.*, 2010; Wu *et al.*, 2012;Wang *et al.*, 2015).

In addition to harmful viruses, there are also mutualistic virus–host interactions (reviewed in Roossinck, 2011). In the case of mycoviruses, the most striking one is the virus CThTV of the endophytic fungus, *Curvularia protuberata*, which participates in a three-way symbiosis allowing the grass host (*Dichanthelium lanuginosum*) to grow on geothermic soil in the Yellowstone National Park, where soil temperature may reach up to 65°C (Márquez *et al.*, 2007). Moreover, a virus of *Neonectria radicicola*, a fungal pathogen causing ginseng root rot, increases host virulence, spore production, laccase activity and pigmentation. Biochemical analyses suggest that the virus regulates fungal virulence through signal-transduction pathways involving cyclic AMP-dependent protein kinase and protein kinase C (Ahn and Lee, 2001). Increased host virulence (hypervirulence) has also been described in polymycoviruses (Kotta-Loizou and Coutts, 2017).

Despite the cases described above, most fungal viruses do not seem to cause obvious effects on their hosts. This may be due to the lack of extracellular particles, which makes mycoviruses highly dependent on the welfare of their hosts. It should, however, be noted that the benefit or harm of mycoviruses is not straightforward. Bryner and Rigling (2011) noticed that the effects of four CHV1 subtypes on the growth and sporulation of four different *C. parasitica* strains were temperature- and strain-dependent, and in some conditions host growth rate was even increased due to viral infection. Similarly, Vainio *et al.* (2010, 2012) described that the effects of *Heterobasidion* viruses on the growth of their host fungi were temperature-dependent. An even more complicated picture emerged from the work by Hyder *et al.* (2013), who compared growth rates and competitive abilities of two different viruses of *Heterobasidion* spp. They demonstrated that single virus strains mediate different effects on different *Heterobasidion* host strains and that a single virus strain may have beneficial, cryptic or detrimental effects on a single host isolate depending on environmental and ecological conditions (Table 9.1). Therefore, it may be incorrect to classify any fungal virus as solely harmful, cryptic or beneficial to its host.

This view is highlighted by the ecological and evolutionary effects of yeast (*Saccharomyces cerevisiae*) L-A virus, which is in many ways a typical fungal virus. It is, however, associated with satellite RNA-molecules, which encode a protein toxin and immunity to the toxin by the host fungus itself. This dual-activity makes infected yeast lines 'killer strains', which take over yeast populations quickly. L-A virus could therefore be considered as highly useful to its host. Immunity against L-A toxin, however, comes at a cost – it is efficient only in cells that lack RNA interference (RNAi). This likely explains why RNAi does not exist in baker's yeast and all other yeast species that contain toxin producing viruses. Drinnenberg *et al.* (2011) showed that the disappearance of RNAi has been a very recent evolutionary development among yeasts, which suggests a hypothesis that although L-A viruses are highly useful to their hosts in ecological connections, they may lead their host to a cul-de-sac in evolutionary terms. In some ways, the biology of the L-A-virus can be seen as similar

Table 9.1 Contrasting effects of mycoviruses of *Heterobasidion* species on the competitive ability of different host fungi

		Effect of virus infection on host compatibility against interacting fungi				
Virus strain	Host fungus	*Paxillus involutus* F-CY01	*Paxillus involutus* F-YF05	*Paxillus involutus* BOUX	*Meliniomyces bicolor* R-MF01	*Phlebiopsis gigantea* Rotstop
HetPV3-ec1	*H. ecrustosum* 05166	(+)[1]	+	–	0	–
	H. parviporum 7R15	(–)	0	0	–	+
	H. abietinum 00153	–	0	–	–	(–)
HetRV6-ab6	*H. abietinum* 04188	0	–	–	+	0
	H. annosum S49–5	–	(–)	0	+	(–)
	H. parviporum RK5A	+	0	0	–	–

[1]Effects on host fungi are indicated as follows: + statistically significant increase in host competiveness ($p \leq 0.05$); (+) statistically suggestive increase in host competiveness ($0.05 \leq p \leq 0.10$); 0 no effect on host competiveness; -statistically significant decrease in host competiveness ($p \leq 0.05$); (–) statistically suggestive decrease in host competiveness ($0.05 \leq p \leq 0.10$).
The data are reproduced from Hyder *et al.* (2013).

to that of temperate bacteriophages (Stewart and Levin, 1984).

Overall, fungal viruses are simple organisms with a very special relation to their hosts (reviewed in, for example, Ghabrial and Suzuki, 2009; Pearson *et al.*, 2009; Ghabrial *et al.*, 2013). Their biology is far from being fully understood, but in the light of our current knowledge, their role in ecosystems is manifold. Thus, it is clear that the close connection between mycoviruses and fungi has led to evolution of a plethora of host–virus relationships often not in accordance with the common 'human disease'-derived view of viruses as harmful parasites.

Biocontrol of chestnut blight: a pioneering case study on the biology of mycoviruses

In 1904 a diseased American chestnut (*Castanea dentata*) tree was observed in the Bronx Zoo (Merkel and Herman, 1905; Beattie and Diller, 1954). The disease was to become one of the most famous examples of plant diseases and introduced organisms, namely chestnut blight caused by an East Asian fungus, *Cryphonectria parasitica*, which had spread to North America via the plant trade from East Asia. The chestnut seedling trade continued for several decades, and introduced the fungus many times to United States until the Plant Quarantine Act of 1912 stopped it. The disease started to spread throughout the natural range of the American chestnut. By 1950 a total of 3.6 million hectares of forest was destroyed, and *Castanea dentata* had become an endangered species (Anagnostakis, 1982).

The story, however continued in Europe, where chestnut blight was first observed in 1938 in northern Italy (Pavari, 1949). It first looked like the European chestnut would follow the track of its American relative, but in 1951 an Italian researcher Antonio Biraghi noticed a group of healthy looking chestnut sprouts (Biraghi, 1953). The trees had some cankers, but they seemed to heal by themselves. The Frenchman, Jean Grente isolated in the early 1960s mycelial cultures from these cankers (Grente, 1965). Their culture morphology was slightly different from normal, and inoculation of these strains to normal cankers stopped the development of the disease. Therefore, he called them hypovirulent isolates; and later they have been used to control chestnut blight in Europe. Subsequently, the hypovirulent phenotype was shown to be associated with a fungal virus (Day, 1977). The massive use of hypovirulent isolates for decades has made viruses so frequent in, e.g. northern Italy, that the cankers are today healing via natural viral infections even without human intervention (Fig. 9.1).

European hypovirulent isolates were taken to the USA in the 1970s to be used for biocontrol against the disease killing American chestnut trees. There Peter R. Day *et al.* (1977) showed in the mid-1970's that the hypovirulent strains carry double stranded

Figure 9.1 Canker caused by *Cryphonectria parasitica* on *Castanea sativa* and ongoing inactivation of the disease by the presence of a natural hypovirus infection.

RNA (dsRNA) which did not exist in virulent isolates. This molecule was somewhat similar to some known viruses, and has later been classified as a mycovirus, in family *Hypoviridae* (see 'Mycovirus taxonomy', below).

The chestnut blight fungus, as with other fungi, forms cellular connections between hyphae. This phenomenon is called anastomosis, and takes place only between strains belonging to the same VCG. The hypoviruses of chestnut blight spread via these connections, and therefore transmission of viruses is efficient only between strains of the same VCG, although the system leaks somewhat, and transmission takes place also between isolates of different VCGs at a low frequency (Liu *et al.*, 2003). Because the chestnut blight fungus was introduced several times to North America (Beattie and Diller, 1954), there are many more VCGs in the USA than in Europe. Consequently the transmission of hypoviruses is poor, and virocontrol inefficient. In order to solve the problem, Choi and Nuss (1992) created a *C. parasitica* strain in which a cDNA copy of Cryphonectria hypovirus 1 (CHV1) was cloned and expressed. The full length cDNA has played an important tool in attempts to understand the biology of CHV1 (e.g. Suzuki *et al.*, 1999; Chen *et al.*, 2000), and hypoviruses in general.

The *C. parasitica* hypovirulence is caused by virus-induced changes in gene expression regulation. There are excellent recent reviews available (e.g. Ghabrial and Suzuki, 2009), but briefly, CHV1 affects host signal transduction pathways and induces the expression of dicer and argonaute genes. The virus, however, also encodes a papain-like protease p29 that acts as a RNA silencing suppressor (Segers *et al.*, 2007; Sun *et al.*, 2009).

Mycovirus taxonomy

The vast majority of known mycoviruses have RNA genomes although Yu *et al.* (2010) describe a ssDNA mycovirus related to plant geminiviruses, the Sclerotinia gemycircularvirus 1 (*Genomoviridae*). Moreover, a dsDNA virus (Rhizidiomyces virus) infects the hyphochytridiomycete, *Rhizidiomyces* (Dawe and Kuhn, 1983). RNA mycoviruses are currently classified according to their genomic composition into seven dsRNA virus families (*Chrysoviridae, Endornaviridae, Megabirnaviridae, Partitiviridae, Reoviridae, Totiviridae* and *Quadriviridae*), and six ssRNA virus families (*Alphaflexiviridae, Barnaviridae, Gammaflexiviridae, Hypoviridae, Mymonaviridae* and *Narnaviridae* (www.ictvonline.org; Table 9.2). In this section, we will briefly describe all currently known mycovirus families, while more detailed virus characteristics and biological features can be found in King *et al.* (2011) and Tidona and Darai (2011) and at http://ictv.global/report.

Double-stranded RNA viruses infecting fungi may have either monopartite or segmented genomes. The virus family *Totiviridae* includes viruses with monopartite linear genomes (Table 9.2) encoding a capsid protein (CP) and an RNA-dependent RNA polymerase (RdRp). The CP-coding region is located upstream of the RdRp-coding region, and translational frameshifts are often involved in the translation of the downstream RdRp. *Totiviridae* infect fungal or protozoan hosts. The family consists of five genera, three of which also include viruses infecting human parasitic protozoa (*Giardiavirus, Leishmaniavirus, Trichomonasvirus*; King *et al.*, 2012; Goodman *et al.*, 2011). Fungal totiviruses are classified in the genera *Totivirus* (type species *Saccharomyces cerevisiae virus L-A*) and *Victorivirus* (type species

Table 9.2 List of currently recognized mycovirus families[1]

Virus family	Genera infecting fungi	Type species or representative mycovirus species[2]	Nucleic acid	Number and size of genome segments	Morphology
Genomoviridae	Gemycircularvirus	Sclerotinia gemycircularvirus 1	ssDNA	1 (2.2 kb, circular)	20–22 nm virion, isometric protein capsid
Chrysoviridae	Chrysovirus	Penicillium chrysogenum virus	dsRNA	4 (2.5–3.6 kbp)	35–40 nm virion, isometric protein capsid
Endornaviridae	Alphaendornavirus	Helicobasidium mompa alphaendornavirus 1[2]	dsRNA	1 (13.9–17.6 kbp)	Vesicle-like structures, no capsid
	Betaendornavirus	Sclerotinia sclerotiorum betaendornavirus 1			
Megabirnaviridae	Megabirnavirus	Rosellinia necatrix megabirnavirus 1	dsRNA	2 (7.2–8.9 kbp)	50 nm virion, isometric protein capsid
Partitiviridae	Alphapartitivirus	Helicobasidium mompa partitivirus V70[2]	dsRNA	2 (1.4–2.4 kbp)	30–40 nm virion, isometric protein capsid
	Betapartitivirus	Atkinsonella hypoxylon virus			
	Gammapartitivirus	Penicillium stoloniferum virus S			
Reoviridae	Mycoreovirus	Mycoreovirus 1	dsRNA	*Mycoreovirus*: 11–12 (0.7–4.1 kbp)	*Mycoreovirus*: 80 nm double-shelled virion, icosahedral protein capsid
Totiviridae	Totivirus	Saccharomyces cerevisiae virus L-A	dsRNA	1 (4.6–6.7 kbp)	30–40 nm virion, isometric protein capsid
	Victorivirus	Helminthosporium victoriae virus 190S			
Quadriviridae	Quadrivirus	Rosellinia necatrix quadrivirus 1	dsRNA	4 (3.7–4.9 kbp)	45 nm virion, isometric protein capsid
Alphaflexiviridae	Botrexvirus	Botrytis virus X	ssRNA (+)	1 (7.0 kb)	720 nm flexuous rod-shaped particle
	Sclerodarnavirus	Sclerotinia sclerotiorum debilitation-associated RNA virus			
Barnaviridae	Barnavirus	Mushroom bacilliform virus	ssRNA (+)	1 (4.0 kb)	18–20 × 48–53 nm bacilliform virions
Gammaflexiviridae	Mycoflexivirus	Botrytis virus F	ssRNA (+)	1 (6.8 kb)	720 nm flexuous rod-shaped particle
Hypoviridae	Hypovirus	Cryphonectria hypovirus 1	ssRNA (+)	1 (9.1–12.7 kb)	50–80 nm pleomorphic vesicles, no true virion
Narnaviridae	Narnavirus	Saccharomyces 20S RNA narnavirus	ssRNA (+)	1 (2.3–3.6 kb)	Ribonucleoprotein complex, no true virion
	Mitovirus	Cryphonectria mitovirus 1			
Mymonaviridae	Sclerotimonavirus	Sclerotinia sclerotimonavirus	ssRNA (–)	1 (10 kb)	Filamentous helical nucleocapsid of 22 x 200–2000 nm in length
Pseudoviridae	Pseudovirus	Saccharomyces cerevisiae Ty1 virus	ssRNA	1 (5–9 kb)	60–80 nm virus-like particles
	Hemivirus	Saccharomyces cerevisiae Ty5 virus[2]			
Metaviridae	Metavirus	Saccharomyces cerevisiae Ty3 virus	ssRNA	1 (4–10 kb)	Virus-like particles, undetermined size

[1]Data are based on Howitt *et al*., 2001; Suzuki *et al*., 2004; Howitt *et al*., 2006; King *et al*., 2011; Tidona and Darai, 2011; Lin *et al*., 2012; Ghabrial *et al*., 2013; and the ICTV database (http://www.ictvonline.org/).
[2]Approved mycovirus species in a genus with a type species from a plant or insect host

Helminthosporium victoriae virus 190S). Totiviruses are mostly considered cryptic, but Helicobasidium mompa totivirus 1–17 seems to be involved in attenuation of virulence of its host fungus (Suzaki *et al.*, 2005). Currently there are twenty approved fungal totivirus species, infecting species of *Aspergillus, Beauveria, Chalara, Coniothyrium, Gremmeniella, Helicobasidium, Helminthosporium, Magnaporthe, Rosellinia, Saccharomyces, Scheffersomyces, Sphaeropsis, Tolypocladium, Tuber, Ustilago* or *Xanthophyllomyces*. There are also several novel mycovirus taxa that resemble totiviruses in their genomic organization, but are distant enough in sequence to warrant classification into separate novel virus families. Zygosaccharomyces bailii virus Z and related monosegmented viruses of proposed genus '*Unirnavirus*' with genomes sizes of ca 3 kbp resemble plant viruses of virus family *Amalgaviridae* (Martin *et al.*, 2011; Liu *et al.*, 2012; Depierreux *et al.*, 2016). Moreover, the Sclerotinia sclerotiorum non-segmented virus L (Liu *et al.*, 2012) and similar viruses of *Phlebiopsis gigantea, Fusarium graminearum* and the stramenopile *Phytophtora infestans* form a putative new family of non-segmented, possibly unencapsidated viruses with *ca* 9 kbp genomes (Hillman and Cai, 2013).

The family *Partitiviridae* includes dsRNA viruses with bisegmented genomes. Their genome segments encoding for RdRp and CP are separately encapsidated, and their relative proportions in the mycelia are virus specific and relatively stable in different hosts and temperatures (Jurvansuu *et al.*, 2014). The family includes five genera with distinct host ranges (Nibert *et al.*, 2014). The genera *Alphapartitivirus* (type species *White clover cryptic virus 1*) and *Betapartitivirus* (type species *Atkinsonella hypoxylon virus*) include viruses hosted by fungi and plants, while genus *Gammapartitivirus* (type species *Penicillium stoloniferum virus S*) comprises only mycoviruses (Fig. 9.2). The two remaining genera, *Deltapartitivirus* and *Cryspovirus* are hosted by plants or protozoa, respectively. Most partitiviruses are considered to be cryptic, but some fungal partitiviruses have deleterious effects on the reproduction, morphology, growth rate or competitive ability of their hosts (Ihrmark *et al.*, 2004; Kanematsu *et al.*, 2010; Vainio *et al.*, 2010; Magae and Sunagawa, 2010; Bhatti *et al.*, 2011; Hyder *et al.*, 2013; Chiba *et al.*, 2013; Xiao *et al.*, 2014; Vainio *et al.*, 2018). Currently there are 30 approved species of fungal partitiviruses that infect species of *Agaricus, Aspergillus, Atkinsonella, Ceratocystis, Chondrostereum, Discula, Flammulina, Fusarium, Gaeumannomyces, Gremmeniella, Helicobasidium, Heterobasidion, Ophiostoma, Penicillium, Pleurotus, Rhizoctonia, Rosellinia,* or *Sclerotinia*.

Members of the family *Megabirnaviridae* and genus *Botybirnavirus* also have bisegmented genomes, which are substantially larger than those of partitiviruses. Rosellinia necatrix megabirnavirus 1 includes four ORFs, one of which encodes an RdRp and another that encodes a major CP (Chiba *et al.*, 2009). The virus causes reduction of virulence and mycelial growth in several somatically incompatible host strains of *R. necatrix* infected by the virus via virion transfection (Chiba *et al.*, 2009). Related megabirnaviruse(s) occur in *Sclerotinia sclerotiorum* (Wang *et al.*, 2015). Botrytis porri RNA virus 1 is the first approved member of genus *Botybirnavirus* (Wu *et al.*, 2012). Based on phylogenetic analysis, members of a suggested new genus '*Phlegivirus*' (Petrzik *et al.*, 2016) infecting fungal genera *Phlebiopsis, Lentinula,* and *Thelephora* resemble megabirnaviruses, but they have monopartite genomes and no described particles (Lim *et al.*, 2011; Liu *et al.*, 2012; Magae, 2012; Petrzik *et al.*, 2016).

The family *Chrysoviridae* includes dsRNA viruses with four genomic segments (Jiang and Ghabrial, 2004; Castón *et al.*, 2013). Two of the genomic segments encode for the major CP and RdRp, and the remaining two molecules have been shown to be associated with the virions. The genus *Chrysovirus* (type species *Penicillium chrysogenum virus*) currently includes nine approved species that infect species of *Aspergillus, Cryphonecria, Fusarium, Helminthosporium, Penicillium,* and *Verticillium*. Magnaporthe oryzae chrysovirus 1 impairs vegetative growth and alters pigmentation of the rice blast fungus (Urayama *et al.*, 2010). Alternaria alternata dsRNA mycovirus (Aoki *et al.*, 2009) and Aspergillus mycovirus 341 (Hammond *et al.*, 2008) resemble members of *Chrysoviridae* in their genomic structure including four segments of 1.5–3.6 kbp, but seem to be taxonomically distinct.

The dsRNA virus family *Quadriviridae* currently includes only one approved virus species, *Rosellinia necatrix quadrivirus 1* (Lin *et al.*, 2012). This virus encompasses four dsRNA segments, one of which encodes for an RdRp, and two encode for major

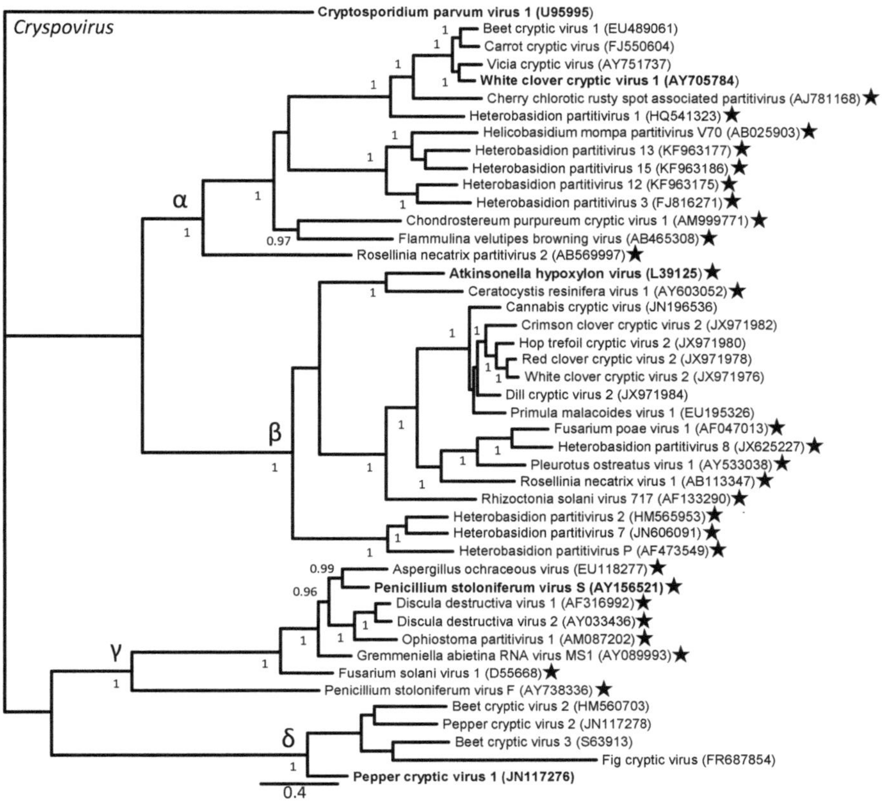

Figure 9.2 Phylogenetic analysis of ICTV-approved members of the virus family *Partitiviridae*. Fungal partitiviruses are indicated by a star symbol. The corresponding Greek letters denote the partitivirus genera: *Alphapartitivirus* (α), *Betapartitivirus* (β), *Gammapartitivirus* (γ) and *Deltapartitivirus* (δ). The scale bar shows 0.5 aa substitutions per site, and percentage posterior probabilities are indicated at branch nodes. The aa sequence alignment is based on complete RdRp sequences and was generated by MAFFT implemented in Geneious R10 (Biomatters Ltd.). Cryptosporidium parvum virus 1 was used as an outgroup. GenBank accession numbers are shown at branch ends.

capsid proteins (Lin *et al.*, 2013). The virus infection shows no obvious phenotypic effects.

The dsRNA virus family *Reoviridae* is a large virus family which includes viruses that infect fungi, plants, invertebrates, or vertebrates and comprises two subfamilies: *Sedoreovirinae* with six different genera (including for example human pathogens of genus *Rotavirus* and plant pathogens of genus *Phytoreovirus*, like the Rice dwarf virus), and subfamily *Spinareovirinae* with nine genera, including genus *Mycoreovirus* that infects fungi (type species *Mycoreovirus 1*), as well as e.g. genus *Coltivirus* (type species *Colorado tick fever virus*; King *et al.*, 2012). Three species infect fungi: Mycoreovirus 1 and 2 were found from *C. parasitica*, and Mycoreovirus 3 infects *Rosellinia necatrix* (Osaki *et al.*, 2002; Suzuki *et al.*, 2004). Mycoreoviruses are composed of 11–12 dsRNA segments encapsidated in double-shelled particles with 12 surface projections. Mycoreovirus 1 and 3 were both isolated from hypovirulent isolates of the host fungi. One of the dsRNA segments encodes an RdRp, two are involved in symptom induction, while others encode for a structural protein, a guanylyltransferase, and NTP binding protein or proteins of unknown functions (Tanaka *et al.*, 2012).

Among mycoviruses with ssRNA genomes, the family *Narnaviridae* includes viruses with non-segmented, short genomes that encode only a single protein, the RdRp. They do not form true virions, but may form ribonucleoprotein complexes including genomic RNA and the RdRp protein. Members of the genus *Mitovirus* (type species Cryphonectria mitovirus 1) are replicated inside host mitochondria and show mitochondrial codon usage, while viruses of the genus *Narnavirus* (type species *Saccharomyces 20S RNA narnavirus*) replicate in the host cytosol. Mitoviruses are very common in ascomycetous and basidiomycetous fungi (see Hillman and Cai, 2013), and narnaviruses infect yeasts and the stramenopile, *Phytophtora infestans* (Cai *et al.*, 2012). Notably, the Ophiostoma novo-ulmi mitovirus 3a is associated with hypovirulence in *Sclerotinia homoeocarpa* (Deng *et al.*, 2003). The virus genus *Ourmiavirus* encompasses trisegmented plant viruses closely related to family *Narnaviridae*. Ourmia-like viruses have recently been detected also in fungi (Marzano *et al.*, 2016a; Donaire *et al.*, 2016).

The family *Hypoviridae* includes ssRNA viruses with a genome structure resembling positive-strand RNA viruses in the plant virus family *Potyviridae* (Shapira *et al.*, 1991; Nuss, 2005). This family includes the classical hypoviruses successfully used in the biocontrol of *C. parasitica* in Europe as described above. Species of this genus differ in their capacity of causing hypovirulence; Cryphonectria hypovirus 1 (CHV1, type species of genus *Hypovirus*) and CHV2 significantly reduce host virulence, while the effects of CHV3 and CHV4 are less prominent (Linder-Basso *et al.*, 2005). Hypoviruses exist as pleomorphic vesicles containing replicative-form dsRNA and polymerase activity without forming true virions (King *et al.*, 2012). The CHV1 genome encodes the RdRp, a papain-like protease and a helicase. New hypoviruses have been recently found in *Sclerotinia sclerotiorum* (Xie *et al.*, 2011), *Valsa ceratosperma* (Yaegashi *et al.*, 2012), *Phomopsis longicolla* (Koloniuk *et al.*, 2014) and *Fusarium graminearum* (Li et al., 2015).

The virus family *Endornaviridae* currently includes two approved fungal viruses with positive-sense ssRNA genomes: Helicobasidium mompa endornavirus 1 (Osaki *et al.*, 2006) and Sclerotinia sclerotiorum endornavirus 1 (Khalifa and Pearson, 2014). They have monopartite genomes encoding for an RdRp, a putative RNA helicase, and a putative UDP glycosyltransferase. Other endornaviruses infect the stramenopile *Phytophtora* (taxon 'Douglasfir') and various plant hosts (King *et al.*, 2012; Tidona and Darai (2011). Endornaviruses do not have protein capsids, but form vesicle-like structures and replicate in the cytosol of the host fungus. The Helicobasidium mompa endornavirus 1 causes hypovirulence in its host (Ikeda *et al.*, 2003; Osaki *et al.*, 2006). Viruses with endornavirus affinities have also been found in fungal genera *Gremmeniella* (Tuomivirta *et al.*, 2009), *Tuber* (Stielov *et al.*, 2011), *Rhizoctonia* (Li *et al.*, 2014), *Botrytis, Alternaria,* and *Chalara* (see Hao *et al.*, 2017).

The family *Barnaviridae* includes only one mycovirus species: the disease-causing mushroom bacilliform virus infecting *Agaricus bisporus* (Revill *et al.*, 1994). The ssRNA genome including an RdRp and CP somewhat resembles plant luteoviruses and carmoviruses.

Two families of the ssRNA virus order *Tymovirales* include fungal viruses. The family *Alphaflexiviridae* includes two mycovirus genera, genus *Botrexvirus* includes Botrytis virus X (BVX; Howitt *et al.*, 2006) and genus *Sclerodarnavirus* includes Sclerotinia sclerotiorum debilitation-associated RNA virus (Xie *et al.*, 2006). BVX has a genome that contains genes for RdRp and CP, and forms flexuous rod-shaped particles. Moreover, the family *Gammaflexiviridae* includes the genus *Mycoflexivirus*, type species *Botrytis virus F* (BVF; Howitt *et al.*, 2001). BVF was isolated from the same *Botrytis cinerea* isolate as BVX and also has a flexuous rod-shaped particle. Members of *Tymovirales* seem to be more or less latent in their natural hosts but may cause considerable phenotypic alterations when transfected into originally virus-free isolates (Boine, 2012). The Oyster mushroom spherical virus somewhat resembles BVF (Yu *et al.*, 2003).

The virus family *Mymonaviridae* includes one approved species, *Sclerotinia sclerotimonavirus*, which is the first negative-sense ssRNA virus described among fungi (Liu *et al.*, 2014). Yet unclassified negative-sense ssRNA viruses have been described also in *Fusarium poae* (Osaki *et al.*, 2016) as well as *Macrophomina phaseolina* and *Rhizoctonia solani* (Marzano et al., 2016b).

Besides the recognized mycovirus families described above, there are some interesting new virus taxa that await formal classification. One of these novel taxonomical groups includes the Curvularia thermal tolerance virus that is the only strictly mutualistic mycovirus described so far (Márquez *et al.*, 2007). Related viruses have been recently described for example in *Fusarium graminearum* (Yu *et al.*, 2009), species of *Heterobasidion* (Vainio *et al.*, 2012), and in *Rhizoctonia solani* (Zheng *et al.*, 2013). Most of these viruses encompass two genome segments of 1.7–2.4 kbp, one of which encodes an RdRp. Moreover, family name 'Fusariviridae' has been proposed to accommodate positive-sense ssRNA viruses found in *Fusarium graminearum* (Kwon *et al.*, 2007) and *Rosellinia necatrix* (Zhang *et al.*, 2014). The proposed virus family 'Alternaviridae' includes quadripartite dsRNA viruses (Kozlakidis *et al.*, 2013), whereas family name 'Polymycoviridae' has been suggested for a group of viruses consisting of four major dsRNAs and variable number of additional genome segments (Kotta-Loizou and Coutts, 2017).

Viral integration into fungal genomes

Many fungal species have retrotransposable elements incorporated in their genome. These elements, also known as LTR-retrotransposons of the Ty1-*copia* or Ty3-*gypsy* families, replicate by virus-like intermediate particles (VLPs) but are not considered infective and do not spread horizontally between hosts (King *et al.*, 2012). Currently these retrotransposons are classified by the ICTV as members of *Pseudoviridae* and *Metaviridae.* The family *Pseudoviridae* includes Saccharomyces cerevisiae Ty-viruses 1, 2, 4 and 5, and the family *Metaviridae* includes e.g. Saccharomyces cerevisiae Ty3 virus, Fusarium oxysporum skippy virus, and the Schizosaccharomyces pombe viruses Tf1 and Tf2. Recently, members of *Pseudoviridae* and *Metaviridae* have been proposed to be abolished from ICTV databases.

In addition to retrotransposable elements, some non-retroviral RNA virus genes have been found integrated into fungal genomes. Thus totivirus polymerase and/or capsid gene homologues occur in the nuclear genomes of e.g. *Candida* and *Penicillium* spp. (Taylor and Bruenn, 2009; Koonin, 2010; Liu *et al.*, 2010). Surveys on sequence databases have also revealed geminivirus-related sequences in the genomes of e.g. *Laccaria* and *Tuber* spp. (Liu *et al.*, 2011), and negative sense ssRNA viruses (mononegaviruses) in chromosomes of *Erysiphe pisi* and *Slerotinia homoeocarpa* (Kondo *et al.*, 2013). It was hypothesized that integrated viral polymerase genes might be involved in host antiviral immunity via RNA interference.

Mycovirus ecology and dispersal

Most mycoviruses propagate within the cytoplasm or mitochondria of their host fungus and are considered to lack extracellular particles. Sclerotinia gemycircularvirus 1, however, is infectious as a free particle (Yu *et al.*; 2013), and transmittable by an insect vector (Liu *et al.*, 2016). Gemycircularviruses have also been detected in urban river sediments (Kraberger *et al.*, 2013) and faecal matter of various animals (Sikorski *et al.*, 2013). Unlike these novel findings on DNA viruses, mycoviruses with RNA genomes are considered to be completely intracellular. They transmit laterally between fungal strains during anastomosis contacts, and vertically during hyphal growth, and within sexual or asexual spores. Asexual conidial spores are often infected by mycoviruses, although the transmission rate of viruses into conidia may vary considerably even among isolates of the same host species. Besides *C. parasitica,* conidia of the ascomycete species *Ophiostoma ulmi* (Rogers *et al.*, 1986) and *Aspergillus nidulans* (Coenen *et al.*, 1997), as well as the basidiomycetous *Heterobasidion* spp. (Ihrmark *et al.*, 2002), have been reported to contain dsRNA elements. Sexual spores may also be virus carriers, although less commonly than asexual ones. For example, ascospores of *Magnaporthe grisea* and basidiospores of *Heterobasidion* species harbour viral dsRNA (Chun and Lee, 1997; Ihrmark *et al.*, 2004; Vainio *et al.*, 2015b), while ascospores of *C. parasitica* and basidiospores of *Helicobasidium mompa* are devoid of viruses (Carbone *et al.*, 2004; Ikeda *et al.*, 2004). As with conidia, the virus transmission frequency to sexual spores varies greatly between fruiting bodies of the same host species.

Dispersal patterns of mycoviruses in natural conditions have often been studied by artificially introducing viruses into host populations. Hoegger *et al.* (2003) describe successful dissemination of CHV1 between somatically compatible host strains under field conditions in Switzerland. Despite limitations caused by host somatic incompatibility, introduced *Cryphonectria* hypoviruses have also been shown to persist in American chestnut trees for over a decade (Griffin *et al.*, 2004). Moreover, an introduced alphapartitivirus of the conifer root rot pathogen, *Heterobasidion parviporum,* was transmitted via mycelial contact into indigenous

Heterobasidion strains in field conditions (Vainio *et al.*, 2013). Overall, the natural virus infection frequency seems to be higher among *H. parviporum* strains collected from forest stands with a long history of root disease (Vainio *et al.*, 2015b).

Ikeda *et al.* (2005) described the accumulation of dsRNA elements in natural strains of the violet root rot fungus, *Helicobasidium mompa*, during the course of five years. Interestingly, novel dsRNA virus infections from unknown sources accumulated during time in isolates of the root rot pathogen, *Rosellinia necatrix*, both in the field and in laboratory conditions (Yaegashi *et al.*, 2013; Yaegashi and Kanematsu, 2015). The authors hypothesize that mycoviruses might transmit by micro-arthropods or nematodes feeding on the host fungus. Drenkhan *et al.* (2013; 2016) recently showed that mycoviruses are able to persist within their fungal host during the passage through the digestive system of the large pine weevil (*Hylobius abietis*), but the ecological significance of these observations remains unknown.

Overall, there is an obvious need for large-scale sequence-based investigations on natural mycovirus communities. In some cases, host epidemiology can be inferred based on mycovirus population structure (Thapa *et al.*, 2016; Schoebel *et al.*, 2017). The metagenomics approach (see Chapter 5) has only been used in a limited number of studies so far (Coetzee *et al.*, 2010; Al Rwahnih *et al.*, 2010; Feldman *et al.*, 2012; Marzano *et al.*, 2016a), but the use of next generation sequencing techniques has already revealed many new aspects on mycovirus populations. It has, for example, broadened the view of virus host ranges and revealed previously unknown mycovirus taxa (Nerva *et al.*, 2016; Schoebel *et al.*, 2014; Vainio *et al.*, 2015a; Marzano *et al.*, 2016b; Donaire and Ayllón, 2016).

Evolution of fungal viruses

The origin of mycoviruses may be ancient or alternatively they may have originated as plant or other eukaryote viruses that have changed hosts sometime during their evolution. This question was considered by Ghabrial (1998), who concluded that totiviruses probably existed prior to the divergence of fungi and protozoa, and partitiviruses would have been derived from a totivirus ancestor. On the other hand, hypoviruses and barnaviruses appear to have common ancestry with plant positive-strand RNA viruses. Similarly, the La France isometric virus might be of recent origin since it differs from its host in codon usage. Notably, alphapartitiviruses infecting fungi and plants are closely related, suggesting horizontal virus transmission between these host groups (Li *et al.*, 2009; Roossinck, 2010; Fig. 9.2). Recently it was shown that fungal partitiviruses are capable of replicating in plant protoplast cells (Nerva *et al.*, 2017).

The transmission of many mycoviruses is restricted even among strains of single host species. This was already mentioned in connection with hypoviruses of *C. parasitica* (Anagnostakis and Day, 1979; Liu and Milgroom, 1996; Milgroom and Cortesi, 2004), and is also true among viruses of *Ophiostoma ulmi* (Rogers *et al.*, 1986) and *Aspergillus nidulans* (Coenen *et al.*, 1997). In *Gremmeniella abietina*, mitoviruses seem to have crossed species borders only in exceptional conditions (Botella *et al.*, 2012); and closely related sympatric biotypes of the fungus are inhabited by polyphyletic viruses (Hantula and Tuomivirta, 2003; Botella *et al.*, 2012). This intimate linkage between the virus and host may have led to loss of virulence and even development of commensalism for fungal viruses.

Still it may be a too narrow a view to consider viruses as restricted to coevolution with their hosts, as there are also well documented cases of interspecies transmission of mycoviruses. Mycoviruses infecting *Heterobasidion* spp. seem to cross species borders rather easily wherever different species of this genus occur sympatrically or when paired in laboratory conditions (Ihrmark, 2002; Vainio *et al.*, 2010, 2011a,b, 2012). This has led to a situation where the taxonomy of the viruses does not follow the taxonomy of the host fungi, which is in contrast with the scenario resulting from strict coevolution (Vainio *et al.*, 2011b, 2013b). Similarly, viruses of *Sclerotinia* spp. can be transmitted between species in the laboratory (Melzer *et al.*, 2010), and phylogenetic analysis supports the occurrence of interspecies virus transfer also among species of *Ophiostoma* (Brasier *et al.*, 2004) and *Rhizoctonia* (Charlton *et al.*, 2008). Moreover, conspecific mitoviruses occur in *Ophiostoma novo-ulmi* and *Sclerotinia homoeocarpa* (Deng *et al.*, 2003). Notably, highly similar viral sequences have also been detected in distantly related fungal species occurring in the same habitat (Feldman *et al.*, 2012; Vainio *et al.*, 2017). Lateral transfer of viruses sometimes results in viral coinfections, either by distantly related mycoviruses (Lakshman *et al.*, 1998; Ghabrial *et al.*, 2002; Tuomivirta and Hantula, 2005; Vainio *et al.*, 2011b, 2012), but also by congeneric viruses (Hong *et al.*, 1999; Osaki *et al.*, 2004; Peever *et al.*, 1997; Preisig *et al.*, 1998, Park *et al.*, 2005). This may lead to recombination, and therefore affect the phylogeny and evolution of mycoviruses (Liu *et al.*, 2012).

Whether restricted to their hosts or not, the importance of viruses as an evolutionary force in fungal evolution was demonstrated by Brasier and others (2001), who showed that *Ophiostoma novo-ulmi*, which spread to Europe and North-America in late 20th century acquired quickly VCG genes from its relative *O. ulmi* in order to restrict virus spread within the population. Interestingly, the viruses also had spread from *O. ulmi* to the newcomer.

Intricate systems have been developed to artificially broaden the host range of mycoviruses. These include the transformation system mentioned above for *C. parasitica* hypoviruses (involving plasmid-mediated introduction of hypoviral cDNA into the host genome; Choi and Nuss, 1992), as well as a related protocol involving transfection of synthetic hypovirus transcripts into *C. parasitica* spheroplasts by electroporation (Chen *et al.*, 1994). Moreover, Hillman *et al.* (2004) described a virion transfection protocol where they introduced purified reovirus particles into protoplasts of *C. parasitica* by the use of polyethylene glycol. Protoplast fusion has been used for species of *Aspergillus* (van Diepeningen *et al.*, 1998), but this includes cytoplasmic mixing between the two host strains involved. Recently, Ikeda *et al.* (2013) described chemically mediated anastomoses between incompatible, conspecific strains of *Rosellinia necatrix* by the use of $ZnCl_2$. On the other hand, virus removal methods ('curing') have been designed in order to be able to study the effects of viruses in isogenic hosts. Virus removal can be achieved by chemical compounds like cycloheximide (Elias and Cotty, 1996) or temperature treatment (Vainio *et al.*, 2010, 2012). Moreover, isolation of single hyphal tips or conidial spores often results in virus-free cultures, as sections of growing mycelia are often free of viruses (Sasaki *et al.*, 2007; Yaegashi *et al.*, 2010). Experimental broadening of the host range of mycoviruses has been shown to result in phenotypic alterations in novel host species (Chen *et al.*, 1996; Kanematsu *et al.*, 2010; Chiba *et al.*, 2013).

Conclusion and future prospects

Mycoviruses have been a relatively unexplored field of the biological realm, but research activity on this topic has recently increased. This has led to novel insights into biological diversity and ecological significance of these viruses. The simple view of fungal viruses as dsRNA elements restricted to mycelia has during the last two decades evolved to a diverse community of viruses with dsRNA, ssRNA and even DNA genomes, sometimes capable of forming infective extracellular particles. The metagenomics approach has recently been applied to this field, and can be expected to further change our understanding of the diversity and evolution of mycoviruses. Also detailed studies on the ecology and population biology of these viruses are under way, and will change our understanding on their dispersal within the host fungus population. Finally, these viruses have usually been classified as cryptic, beneficial or detrimental to their hosts, but recent developments might suggest that this view is too narrow. The association of viruses and their hosts may turn out to be considerably more complicated and additionally affected by external factors. As more a detailed picture emerges, it would not be surprising if fungal viruses would have a considerable impact on the ecology of many ecosystems in a way analogous to viruses of pelagic microbes, the role of which was only fully realized in 1980s, and completely changed our understanding on the carbon cycle in oceans (Suttle *et al.*, 1990).

References

Ahn, I.P., and Lee, Y.H. (2001). A viral double-stranded RNA up regulates the fungal virulence of *Nectria radicicola*. Mol. Plant Microbe Interact. *14*, 496–507. https://doi.org/10.1094/MPMI.2001.14.4.496.

Al Rwahnih, M., Daubert, S., Urbez-Torres, J.R., Cordero, F., and Rowhani, A. (2011). Deep sequencing evidence from single grapevine plants reveals a virome dominated by mycoviruses. Arch. Virol. *156*, 397–403. https://doi.org/10.1007/s00705-010-0869-8.

Anagnostakis, S.L. (1982). Biological control of chestnut blight. Science *215*, 466–471.

Anderson, S.M.B.J. (1992). The fungus *Armillaria bulbosa* is among the largest and oldest living organisms. Nature *356*, 428–431.

Aoki, N., Moriyama, H., Kodama, M., Arie, T., Teraoka, T., and Fukuhara, T. (2009). A novel mycovirus associated with four double-stranded RNAs affects host fungal growth in *Alternaria alternata*. Virus Res. *140*, 179–187. https://doi.org/10.1016/j.virusres.2008.12.003.

Beattie, R.K., and Diller, J.D. (1954). Fifty years of chestnut blight in America. J. Forest. *52*, 323–329.

Bhatti, M.F., Jamal, A., Petrou, M.A., Cairns, T.C., Bignell, E.M., and Coutts, R.H. (2011). The effects of dsRNA mycoviruses on growth and murine virulence of *Aspergillus fumigatus*. Fungal Genet. Biol. *48*, 1071–1075. https://doi.org/10.1016/j.fgb.2011.07.008.

Biraghi, A. (1953). Possible active resistance to *Endothia parasitica* in *Castanea sativa*. In Reports to 11th Congress of the International Union of Forest Research Organizations. (International Union of Forest Research Organizations, Rome), p. 643.

Boine, B. (2012). A study of the interaction between the plant pathogenic fungus *Botrytis cinerea* and the filamentous ssRNA mycoviruses *Botrytis* virus X and *Botrytis* virus F (Doctoral dissertation, The University of Auckland, New Zealand).

Botella, L., Tuomivirta, T.T., Vervuurt, S., Diez, J.J., and Hantula, J. (2012). Occurrence of two different species of mitoviruses in the European race of *Gremmeniella abietina* var. *abietina*, both hosted by the genetically unique Spanish population. Fungal Biol. *116*, 872–882. https://doi.org/10.1016/j.funbio.2012.05.004.

Brasier, C.M. (2001). Rapid evolution of introduced plant pathogens via interspecific hybridization. BioScience *51*, 123–133.

Brasier, C., Buck, K., Paoletti, M., Crawford, L., and Kirk, S. (2004). Molecular analysis of evolutionary changes in populations of *Ophiostoma novo-ulmi*. Invest. Agrar. Sist. Recur. For. *13*, 93–103.

Bruenn, J.A. (1993). A closely related group of RNA-dependent RNA polymerases from double-stranded RNA viruses. Nucleic Acids Res. *21*, 5667–5669.

Bryner, S.F., and Rigling, D. (2011). Temperature-dependent genotype-by genotype interaction between a pathogenic fungus and its hyperparasitic virus. Am. Nat. *177*, 65–74.

Cai, G., and Hillman, B.I. (2013). *Phytophthora* viruses. Adv. Virus Res. *86*, 327–350. https://doi.org/10.1016/B978-0-12-394315-6.00012-X.

Cai, G., Myers, K., Fry, W.E., and Hillman, B.I. (2012). A member of the virus family *Narnaviridae* from the plant pathogenic oomycete *Phytophthora infestans*. Arch. Virol. *157*, 165–169. https://doi.org/10.1007/s00705-011-1126-5.

Carbone, I., Liu, Y.C., Hillman, B.I., and Milgroom, M.G. (2004). Recombination and migration of *Cryphonectria* hypovirus 1 as inferred from gene genealogies and the coalescent. Genetics *166*, 1611–1629.

Castón, J.R., Luque, D., Gómez-Blanco, J., and Ghabrial, S.A. (2013). Chrysovirus structure: repeated helical core as evidence of gene duplication. Adv. Virus Res. *86*, 87–108. https://doi.org/10.1016/B978-0-12-394315-6.00004-0.

Charlton, N.D., Carbone, I., Tavantzis, S.M., and Cubeta, M.A. (2008). Phylogenetic relatedness of the M2 double-stranded RNA in *Rhizoctonia* fungi. Mycologia *100*, 555–564.

Chen, B., Choi, G.H., and Nuss, D.L. (1994). Attenuation of fungal virulence by synthetic infectious hypovirus transcripts. Science *264*, 1762–1764.

Chen, B., Chen, C.H., Bowman, B.H., and Nuss, D.L. (1996). Phenotypic changes associated with wild-type and mutant hypovirus RNA transfection of plant pathogenic fungi phylogenetically related to *Cryphonectria parasitica*. Phytopathology *86*, 301–310.

Chen, B., Geletka, L.M., and Nuss, D.L. (2000). Using chimeric hypoviruses to fine-tune the interaction between a pathogenic fungus and its plant host. J. Virol. *74*, 7562–7567.

Chiba, S., Salaipeth, L., Lin, Y.H., Sasaki, A., Kanematsu, S., and Suzuki, N. (2009). A novel bipartite double-stranded RNA Mycovirus from the white root rot Fungus *Rosellinia necatrix*: molecular and biological characterization, taxonomic considerations, and potential for biological control. J. Virol. *83*, 12801–12812. https://doi.org/10.1128/JVI.01830-09.

Chiba, S., Lin, Y.H., Kondo, H., Kanematsu, S., and Suzuki, N. (2013). Effects of defective interfering RNA on symptom induction by, and replication of, a novel partitivirus from a phytopathogenic fungus, *Rosellinia necatrix*. J. Virol. *87*, 2330–2341. https://doi.org/10.1128/JVI.02835-12.

Choi, G.H., and Nuss, D.L. (1992). Hypovirulence of chestnut blight fungus conferred by an infectious viral cDNA. Science *257*, 800–803.

Chu, Y.M., Jeon, J.J., Yea, S.J., Kim, Y.H., Yun, S.H., Lee, Y.W., and Kim, K.H. (2002). Double-stranded RNA mycovirus from *Fusarium graminearum*. Appl. Environ. Microbiol. *68*, 2529–2534.

Chun, S.J., and Lee, Y.H. (1997). Inheritance of dsRNAs in the rice blast fungus, *Magnaporthe grisea*. FEMS Microbiol. Lett. *148*, 159–162.

Coenen, A., Kevei, F., and Hoekstra, R.F. (1997). Factors affecting the spread of double-stranded RNA viruses in *Aspergillus nidulans*. Genet. Res. *69*, 1–10.

Coetzee, B., Freeborough, M.J., Maree, H.J., Celton, J.M., Rees, D.J., and Burger, J.T. (2010). Deep sequencing analysis of viruses infecting grapevines: Virome of a vineyard. Virology *400*, 157–163. https://doi.org/10.1016/j.virol.2010.01.023.

Dawe, V.H., and Kuhn, C.W. (1983). Isolation and characterization of a double-stranded DNA mycovirus infecting the aquatic fungus, *Rhizidiomyces*. Virology *130*, 21–28.

Dawe, A.L., and Nuss, D.L. (2013). Hypovirus molecular biology: from Koch's postulates to host self-recognition genes that restrict virus transmission. Adv. Virus Res. *86*, 109–147.

Day, P.R., Dodds, J.A., Elliston, J.E., Jaynes, R.A., and Anagnostakis, S.L. (1977). Double-stranded RNA in *Endothia parasitica*. Phytopathology *67*, 1393–1396.

Deng, F., Xu, R., and Boland, G.J. (2003). Hypovirulence-Associated Double-Stranded RNA from *Sclerotinia homoeocarpa* Is Conspecific with *Ophiostoma novo-ulmi* Mitovirus 3a-Ld. Phytopathology *93*, 1407–1414. https://doi.org/10.1094/PHYTO.2003.93.11.1407.

Depierreux, D., Vong, M., and Nibert, M.L. (2016). Nucleotide sequence of Zygosaccharomyces bailii virus Z: Evidence for +1 programmed ribosomal frameshifting and for assignment to family *Amalgaviridae*. Virus Res. *217*, 115–124. https://doi.org/10.1016/j.virusres.2016.02.008.

Donaire, L., and Ayllón, M.A. (2016). Deep sequencing of mycovirus-derived small RNAs from *Botrytis* species. Mol. Plant Pathol. *18*, 1127–1137. https://doi.org/10.1111/mpp.12466.

Donaire, L., Rozas, J., and Ayllón, M.A. (2016). Molecular characterization of Botrytis ourmia-like virus, a mycovirus close to the plant pathogenic genus *Ourmiavirus*. Virology *489*, 158–164. https://doi.org/10.1016/j.virol.2015.11.027.

Drenkhan, T., Sibul, I., Kasanen, R., and Vainio, E.J. (2013). Viruses of *Heterobasidion parviporum* persist within their fungal host during passage through the alimentary tract of *Hylobius abietis*. Forest Pathol. *43*, 317–323.

Drenkhan, T., Kasanen, R., and Vainio, E.J. (2016). *Phlebiopsis gigantea* and associated viruses survive passing through the digestive tract of *Hylobius abietis*. Biocontrol Sci. Techn. *26*, 320–330.

Drinnenberg, I.A., Fink, G.R., and Bartel, D.P. (2011). Compatibility with killer explains the rise of RNAi-deficient fungi. Science *333*, 1592. https://doi.org/10.1126/science.1209575.

Elias, K.S., and Cotty, P.J. (1996). Incidence and stability of infection by double-stranded RNA genetic elements in *Aspergillus* section flavi and effects on aflatoxigenicity. Can. J. Bot. *74*, 716–725.

Feldman, T.S., Morsy, M.R., and Roossinck, M.J. (2012). Are communities of microbial symbionts more diverse than communities of macrobial hosts? Fungal Biol. *116*, 465–477. https://doi.org/10.1016/j.funbio.2012.01.005.

Ghabrial, S.A. (1998). Origin, adaptation and evolutionary pathways of fungal viruses. Virus Genes *16*, 119–131.

Ghabrial, S.A., Soldevila, A.I., and Havens, W.M. (2002). Molecular genetics of the viruses infecting the plant pathogenic fungus *Helminthosporium victoriae*. In dsRNA Genetic Elements: Concepts and Applications in Agriculture, Forestry and Medicine, Tavantzis, S.M., ed. (CRC Press, Boca Raton), pp. 213–236.

Ghabrial, S.A., and Suzuki, N. (2009). Viruses of plant pathogenic fungi. Annu. Rev. Phytopathol. *47*, 353–384. https://doi.org/10.1146/annurev-phyto-080508-081932.

Goodman, R.P., Ghabrial, S.A., Fichorova, R.N., and Nibert, M.L. (2011). *Trichomonasvirus*: a new genus of protozoan viruses in the family *Totiviridae*. Arch. Virol. *156*, 171–179. https://doi.org/10.1007/s00705-010-0832-8.

Grente, J. (1965). Les formes Hypovirulentes d'Endothia *parasitica* et les espoirs de lutte contre le chancre du châtaignier. Académie d'Agriculture de France, Extrait du Proces-verbal de la Séance. *51*,1033–1037.

Griffin, G.J., Robbins, N., Hogan, E.P., and Farias-Santopietro, G. (2004). Nucleotide sequence identification of *Cryphonectria* hypovirus 1 infecting *Cryphonectria parasitica* on grafted American chestnut trees 12–18 years after inoculation with a hypovirulent strain mixture. Forest Pathol. *34*, 33–46.

Hammond, T.M., Andrewski, M.D., Roossinck, M.J., and Keller, N.P. (2008). *Aspergillus* mycoviruses are targets and suppressors of RNA silencing. Eukaryotic Cell *7*, 350–357.

Hantula, J., and Tuomivirta, T.T. (2003). The species complex of *Gremmeniella abietina* – intertype hybridization, viruses and gene flow in northen Europe. Can. J. Bot *81*, 1213–1215.

Hao, F., Zhou, Z., Wu, M., and Li, G. (2017). Molecular characterization of a novel endornavirus from the phytopathogenic fungus *Botrytis cinerea*. Arch. Virol. *162*, 313–316. https://doi.org/10.1007/s00705-016-3106-2.

Hillman, B.I., Supyani, S., Kondo, H., and Suzuki, N. (2004). A reovirus of the fungus *Cryphonectria parasitica* that is infectious as particles and related to the coltivirus genus of animal pathogens. J. Virol. *78*, 892–898.

Hillman, B.I., and Cai, G. (2013). The family narnaviridae: simplest of RNA viruses. Adv. Virus Res. *86*, 149–176. https://doi.org/10.1016/B978-0-12-394315-6.00006-4.

Hoegger, P.J., Heiniger, U., Holdenrieder, O., and Rigling, D. (2003). Differential transfer and dissemination of hypovirus and nuclear and mitochondrial genomes of a hypovirus-infected *Cryphonectria parasitica* strain after introduction into a natural population. Appl. Environ. Microbiol. *69*, 3767–3771.

Hollings, M. (1962). Viruses associated with die-back disease of cultivated mushrooms. Nature *196*, 962-965.

Hong, Y., Dover, S.L., Cole, T.E., Brasier, C.M., and Buck, K.W. (1999). Multiple mitochondrial viruses in an isolate of the Dutch Elm disease fungus *Ophiostoma novo-ulmi*. Virology *258*, 118–127.

Howitt, R.L., Beever, R.E., Pearson, M.N., and Forster, R.L. (2001). Genome characterization of *Botrytis* virus F, a flexuous rod-shaped mycovirus resembling plant 'potex-like' viruses. J. Gen. Virol. *82*, 67–78.

Howitt, R.L., Beever, R.E., Pearson, M.N., and Forster, R.L. (2006). Genome characterization of a flexuous rod-shaped mycovirus, *Botrytis* virus X, reveals high amino acid identity to genes from plant 'potex-like' viruses. Arch. Virol. *151*, 563–579. https://doi.org/10.1007/s00705-005-0621-y.

Huang, S., and Ghabrial, S.A. (1996). Organization and expression of the double-stranded RNA genome of *Helminthosporium victoriae* 190S virus, a totivirus infecting a plant pathogenic filamentous fungus. Proc. Natl. Acad. Sci. U.S.A. *93*, 12541–12546.

Hyder, R., Pennanen, T., Hamberg, L., Vainio, E.J., Piri, T., and Hantula, J. (2013). Two viruses of *Heterobasidion* confer beneficial, cryptic or detrimental effects to their hosts in different situations. Fungal Ecol. *6*, 387–396.

Ihrmark, K., Johannesson, H., Stenström, E., and Stenlid, J. (2002). Transmission of double-stranded RNA in *Heterobasidion annosum*. Fungal Genet. Biol. *36*, 147–154. https://doi.org/10.1016/S1087-1845(02)00011-7.

Ihrmark, K., Stenström, E., and Stenlid, J. (2004). Double-stranded RNA transmission through basidiospores of *Heterobasidion annosum*. Mycol. Res. *108*, 149–153.

Ikeda, K.I., Nakamura, H., and Matsumoto, N. (2003). Hypovirulent strain of the violet root rot fungus *Helicobasidium mompa*. J. Gen. Plant Pathol. *69*, 385–390.

Ikeda, K.-I., Nakamura, H., Arakawa, M., and Matsumoto, N. (2004). Diversity and vertical transmission of double-stranded RNA elements in root rot pathogens of trees, *Helicobasidium mompa* and *Rosellinia necatrix*. Mycol. Res.*108*, 626–634.

Ikeda, K.-I., Nakamura, H., Arakawa, M., Koiwa, T., and Matsumoto, N. (2005). Dynamics of double-stranded RNA segments in a *Helicobasidium mompa* clone from a tulip tree plantation. FEMS Microbiol. Ecol. *51*, 293–301.

Ikeda, K., Inoue, K., Kida, C., Uwamori, T., Sasaki, A., Kanematsu, S., and Park, P. (2013). Potentiation of mycovirus transmission by zinc compounds via attenuation of heterogenic incompatibility in *Rosellinia necatrix*. Appl. Environ. Microbiol. *79*, 3684–3691.

Jiang, D., and Ghabrial, S.A. (2004). Molecular characterization of *Penicillium chrysogenum* virus: reconsideration of the taxonomy of the genus *Chrysovirus*. J. Gen. Virol. *85*, 2111–2121.

Jurvansuu, J., Kashif, M., Vaario, L., Vainio, E., and Hantula, J. (2014). Partitiviruses of a fungal forest pathogen have species-specific quantities of genome segments and transcripts. Virology 462-463, 25–33. https://doi.org/10.1016/j.virol.2014.05.021.

Kanematsu, S., Arakawa, M., Oikawa, Y., Onoue, M., Osaki, H., Nakamura, H., Ikeda, K., Kuga-Uetake, Y., Nitta, H., Sasaki, A., Suzaki, K., Yoshida, K., and Matsumoto, N. (2004). A reovirus causes hypovirulence of *Rosellinia necatrix*. Phytopathology *94*, 561–568.

Kanematsu, S., Sasaki, A., Onoue, M., Oikawa, Y., and Ito, T. (2010). Extending the fungal host range of a partitivirus and a mycoreovirus from *Rosellinia necatrix* by inoculation of protoplasts with virus particles. Phytopathology *100*, 922–930. https://doi.org/10.1094/PHYTO-100-9-0922.

Khalifa, M.E., and Pearson, M.N. (2014). Molecular characterisation of an endornavirus infecting the phytopathogen *Sclerotinia sclerotiorum*. Virus Res. *189*, 303–309. https://doi.org/10.1016/j.virusres.2014.06.010.

King, A.M., Adams, M.J., Lefkowitz, E.J., and Carstens, E.B., eds. (2011). Virus Taxonomy: IXth Report of the International Committee on Taxonomy of Viruses, vol. 9.

Koloniuk, I., El-Habbak, M.H., Petrzik, K., and Ghabrial, S.A. (2014). Complete genome sequence of a novel hypovirus infecting *Phomopsis longicolla*. Arch. Virol. *159*, 1861–1863. https://doi.org/10.1007/s00705-014-1992-8.

Kondo, H., Chiba, S., Toyoda, K., and Suzuki, N. (2013). Evidence for negative-strand RNA virus infection in fungi. Virology *435*, 201–209. https://doi.org/10.1016/j.virol.2012.10.002.

Koonin, E.V. (2010). Taming of the shrewd: novel eukaryotic genes from RNA viruses. BMC Biol. *8*, 2. https://doi.org/10.1186/1741-7007-8-2.

Kotta-Loizou, I., and Coutts, R.H. (2017). Studies on the virome of the entomopathogenic fungus *Beauveria bassiana* reveal novel dsRNA elements and mild hypervirulence. PLOS Pathog. *13*, e1006183. https://doi.org/10.1371/journal.ppat.1006183.

Kozlakidis, Z., Herrero, N., Ozkan, S., Kanhayuwa, L., Jamal, A., Bhatti, M.F., and Coutts, R.H. (2013). Sequence determination of a quadripartite dsRNA virus isolated from *Aspergillus foetidus*. Arch. Virol. *158*, 267–272. https://doi.org/10.1007/s00705-012-1362-3.

Kraberger, S., Stainton, D., Dayaram, A., Zawar-Reza, P., Gomez, C., Harding, J.S., and Varsani, A. (2013). Discovery of *Sclerotinia sclerotiorum* hypovirulence-associated virus-*1* in urban river sediments of Heathcote and Styx rivers in Christchurch City, New Zealand. Genome Announc. *1*, e00559-13. https://doi.org/10.1128/genomeA.00559-13.

Kwon, S.J., Lim, W.S., Park, S.H., Park, M.R., and Kim, K.H. (2007). Molecular characterization of a dsRNA mycovirus, Fusarium graminearum virus-DK21, which is phylogenetically related to hypoviruses but has a genome organization and gene expression strategy resembling those of plant potex-like viruses. Mol. Cells *23*, 304–315.

Lakshman, D.K., Jian, J., and Tavantzis, S.M. (1998). A double-stranded RNA element from a hypovirulent strain of *Rhizoctonia solani* occurs in DNA form and is genetically related to the pentafunctional AROM protein of the shikimate pathway. Proc. Natl. Acad. Sci. U.S.A. *95*, 6425–6429.

Li, L., Tian, Q., Du, Z., Duns, G.J., and Chen, J. (2009). A novel double-stranded RNA virus detected in *Primula malacoides* is a plant-isolated partitivirus closely related to partitivirus infecting fungal species. Arch. Virol. *154*, 565–572. https://doi.org/10.1007/s00705-009-0342-8.

Li, W., Zhang, T., Sun, H., Deng, Y., Zhang, A., Chen, H., and Wang, K. (2014). Complete genome sequence of a novel endornavirus in the wheat sharp eyespot pathogen *Rhizoctonia cerealis*. Arch. Virol. *159*, 1213–1216. https://doi.org/10.1007/s00705-013-1893-2.

Li, P., Zhang, H., Chen, X., Qiu, D., and Guo, L. (2015). Molecular characterization of a novel hypovirus from the plant pathogenic fungus *Fusarium graminearum*. Virology *481*, 151–160. https://doi.org/10.1016/j.virol.2015.02.047.

Lin, Y.H., Chiba, S., Tani, A., Kondo, H., Sasaki, A., Kanematsu, S., and Suzuki, N. (2012). A novel quadripartite dsRNA virus isolated from a phytopathogenic filamentous fungus, *Rosellinia necatrix*. Virology *426*, 42–50. https://doi.org/10.1016/j.virol.2012.01.013.

Lin, Y.H., Hisano, S., Yaegashi, H., Kanematsu, S., and Suzuki, N. (2013). A second quadrivirus strain from the phytopathogenic filamentous fungus *Rosellinia necatrix*. Arch. Virol. *158*, 1093–1098. https://doi.org/10.1007/s00705-012-1580-8.

Linder-Basso, D., Dynek, J.N., and Hillman, B.I. (2005). Genome analysis of *Cryphonectria* hypovirus 4, the most common hypovirus species in North America. Virology *337*, 192–203.

Lim, J.M., Jamal, A., Phoon, X., Korhonen, K., and Coutts, R.H. (2011). Incidence of *Phlebiopsis gigantea* large virus-1 in a collection of *Phlebiopsis gigantea* isolates. Arch. Virol. *156*, 2091–2094. https://doi.org/10.1007/s00705-011-1086-9.

Liu, Y.-C., and Milgroom, M.G., (1996). Correlation between hypovirus transmission and the number of vegetative incompatibility (vic) genes different among isolates from a natural population of *Cryphonectria parasitica*. Phytopathology *86*, 79–86.

Liu, Y.C., Linder-Basso, D., Hillman, B.I., Kaneko, S., and Milgroom, M.G. (2003). Evidence for interspecies transmission of viruses in natural populations of filamentous fungi in the genus *Cryphonectria*. Mol. Ecol. *12*, 1619–1628.

Liu, H., Fu, Y., Jiang, D., Li, G., Xie, J., Cheng, J., Peng, Y., Ghabrial, S.A., and Yi, X. (2010). Widespread horizontal gene transfer from double-stranded RNA viruses to eukaryotic nuclear genomes. J. Virol. *84*, 11876–11887. https://doi.org/10.1128/JVI.00955-10.

Liu, H., Fu, Y., Li, B., Yu, X., Xie, J., Cheng, J., Ghabrial, S.A., Li, G., Yi, X., and Jiang, D. (2011). Widespread horizontal gene transfer from circular single-stranded DNA viruses to eukaryotic genomes. BMC Evol. Biol. *11*, 276. https://doi.org/10.1186/1471-2148-11-276.

Liu, H., Fu, Y., Xie, J., Cheng, J., Ghabrial, S.A., Li, G., Peng, Y., Yi, X., and Jiang, D. (2012). Evolutionary genomics of mycovirus-related dsRNA viruses reveals cross-family horizontal gene transfer and evolution of diverse viral lineages. BMC Evol. Biol. *12*, 91. https://doi.org/10.1186/1471-2148-12-91.

Magae, Y., and Sunagawa, M. (2010). Characterization of a mycovirus associated with the brown discoloration of edible mushroom, *Flammulina velutipes*. Virol. J. *7*, 342. https://doi.org/10.1186/1743-422X-7-342.

Magae, Y. (2012). Molecular characterization of a novel mycovirus in the cultivated mushroom, *Lentinula edodes*. Virol. J. *9*, 60. https://doi.org/10.1186/1743-422X-9-60.

Márquez, L.M., Redman, R.S., Rodriguez, R.J., and Roossinck, M.J. (2007). A virus in a fungus in a plant: three-way symbiosis required for thermal tolerance. Science *315*, 513–515.

Martin, R.R., Zhou, J., and Tzanetakis, I.E. (2011). Blueberry latent virus: an amalgam of the *Partitiviridae* and *Totiviridae*. Virus Res. *155*, 175–180. https://doi.org/10.1016/j.virusres.2010.09.020.

Marzano, S.L., and Domier, L.L. (2016a). Novel mycoviruses discovered from metatranscriptomics survey of soybean phyllosphere phytobiomes. Virus Res. *213*, 332–342.

Marzano, S.Y., Nelson, B.D., Ajayi-Oyetunde, O., Bradley, C.A., Hughes, T.J., Hartman, G.L., Eastburn, D.M., and Domier, L.L. (2016b). Identification of diverse mycoviruses through metatranscriptomics characterization of the viromes of five major fungal plant pathogens. J. Virol. *90*, 6846–6863. https://doi.org/10.1128/JVI.00357-16.

Melzer, M.S., Ikeda, S.S., and Boland, G.J. (2002). Interspecific transmission of double-stranded RNA and hypovirulence from *Sclerotinia sclerotiorum* to *S. minor*. Phytopathology *92*, 780–784. https://doi.org/10.1094/PHYTO.2002.92.7.780.

Merkel, H.W. (1905). A deadly fungus on the American chestnut. N.Y. Zoological Society, 10th Annual Report, pp. 97–103.

Milgroom, M.G., and Cortesi, P. (2004). Biological control of chestnut blight with hypovirulence: a critical analysis. Annu. Rev. Phytopathol. *42*, 311–338. https://doi.org/10.1146/annurev.phyto.42.040803.140325.

Nakatsukasa-Akune, M., Yamashita, K., Shimoda, Y., Uchiumi, T., Abe, M., Aoki, T., Kamizawa, A., Ayabe, S., Higashi, S., and Suzuki, A. (2005). Suppression of root nodule formation by artificial expression of the TrEnodDR1 (coat protein of White clover cryptic virus 1) gene in *Lotus japonicus*. Mol. Plant Microbe Interact. *18*, 1069–1080. https://doi.org/10.1094/MPMI-18-1069.

Nerva, L., Ciuffo, M., Vallino, M., Margaria, P., Varese, G.C., Gnavi, G., and Turina, M. (2016). Multiple approaches for the detection and characterization of viral and plasmid symbionts from a collection of marine fungi. Virus Res. *219*, 22–38. https://doi.org/10.1016/j.virusres.2015.10.028.

Nerva, L., Varese, G.C., Falk, B.W., and Turina, M. (2017). Mycoviruses of an endophytic fungus can replicate in plant cells: evolutionary implications. Sci. Rep. *7*, 1908. https://doi.org/10.1038/s41598-017-02017-3.

Nuss, D.L. (2005). Hypovirulence: mycoviruses at the fungal-plant interface. Nat. Rev. Microbiol. *3*, 632–642.

Osaki, H., Wei, C.Z., Arakawa, M., Iwanami, T., Nomura, K., Matsumoto, N., and Ohtsu, Y. (2002). Nucleotide sequences of double-stranded RNA segments from a hypovirulent strain of the white root rot fungus *Rosellinia necatrix*: possibility of the first member of the *Reoviridae* from fungus. Virus Genes *25*, 101–107.

Osaki, H., Nomura, K., Matsumoto, N., and Ohtsu, Y. (2004). Characterization of double-stranded RNA elements in the violet root rot fungus *Helicobasidium mompa*. Mycol. Res. *108*, 635–640.

Osaki, H., Nakamura, H., Sasaki, A., Matsumoto, N., and Yoshida, K. (2006). An endornavirus from a hypovirulent strain of the violet root rot fungus, *Helicobasidium mompa*. Virus Res. *118*, 143–149.

Osaki, H., Sasaki, A., Nomiyama, K., and Tomioka, K. (2016). Multiple virus infection in a single strain of *Fusarium poae* shown by deep sequencing. Virus Genes *52*, 835–847. https://doi.org/10.1007/s11262-016-1379-x.

Park, Y., James, D., and Punja, Z.K. (2005). Co-infection by two distinct totivirus-like double-stranded RNA elements in *Chalara elegans* (*Thielaviopsis basicola*). Virus Res. *109*, 71–85.

Pavari, A. (1949). Chestnut blight in Europe. Unsasylva 3, 8–13.

Peever, T.L., Liu, Y.-C., and Milgroom, M.G. (1997). Diversity of hypoviruses and other double-stranded RNAs in *Cryphonectria parasitica* in North America. Phytopathology *87*, 1026–1033.

Petrzik, K., Sarkisova, T., Starý, J., Koloniuk, I., Hrabáková, L., and Kubešová, O. (2016). Molecular characterization of a new monopartite dsRNA mycovirus from mycorrhizal *Thelephora terrestris* (Ehrh.) and its detection in soil oribatid mites (Acari: Oribatida). Virology *489*, 12–19. https://doi.org/10.1016/j.virol.2015.11.009.

Preisig, O., Wingfield, B.D., and Wingfield, M.J. (1998). Coinfection of a fungal pathogen by two distinct double-stranded RNA viruses. Virology *252*, 399–406.

Preisig, O., Moleleki, N., Smit, W.A., Wingfield, B.D., and Wingfield, M.J., 2000. A novel RNA mycovirus in a hypovirulent isolate of the plant pathogen *Diaporthe ambigua*. J. Gen. Virol. *81*, 3107–3114.

Revill, P.A., Davidson, A.D., and Wright, P.J. (1994). The nucleotide sequence and genome organization of mushroom bacilliform virus: a single-stranded RNA virus of *Agaricus bisporus* (Lange) Imbach. Virology *202*, 904–911.

Rogers, H.J., Buck, K.W., and Brasier, C.M. (1986). Transmission of double-stranded RNA and a disease factor in *Ophiostoma ulmi*. Plant Pathol. 35, 227–287.

Roossinck, M.J. (2010). Lifestyles of plant viruses. Philos. Trans. R. Soc. Lond., B, Biol. Sci. *365*, 1899–1905. https://doi.org/10.1098/rstb.2010.0057.

Roossinck, M.J. (2011). The good viruses: viral mutualistic symbioses. Nat. Rev. Microbiol. *9*, 99–108. https://doi.org/10.1038/nrmicro2491.

Sasaki, A., Kanematsu, S., Onoue, M., Oikawa, Y., Nakamura, H., and Yoshida, K. (2007). Artificial infection of *Rosellinia necatrix* with purified viral particles of a member of the genus *Mycoreovirus* reveals its uneven distribution in single colonies. Phytopathology *97*, 278–286. https://doi.org/10.1094/PHYTO-97-3-0278.

Schoebel, C.N., Zoller, S., and Rigling, D. (2014). Detection and genetic characterisation of a novel mycovirus in *Hymenoscyphus fraxineus*, the causal agent of ash dieback. Infect. Genet. Evol. *28*, 78–86. https://doi.org/10.1016/j.meegid.2014.09.001.

Schoebel, C.N., Botella, L., Lygis, V., and Rigling, D. (2017). Population genetic analysis of a parasitic mycovirus to infer the invasion history of its fungal host. Mol. Ecol. *26*, 2482–2497. https://doi.org/10.1111/mec.14048.

Segers, G.C., Zhang, X., Deng, F., Sun, Q., and Nuss, D.L. (2007). Evidence that RNA silencing functions as an antiviral defense mechanism in fungi. Proc. Natl. Acad. Sci. U.S.A. *104*, 12902–12906.

Shapira, R., Choi, G.H., and Nuss, D.L. (1991). Virus-like genetic organization and expression strategy for a double-stranded RNA genetic element associated with biological control of chestnut blight. EMBO J. *10*, 731–739.

Sikorski, A., Massaro, M., Kraberger, S., Young, L.M., Smalley, D., Martin, D.P., and Varsani, A. (2013). Novel myco-like DNA viruses discovered in the faecal matter of various animals. Virus Res. *177*, 209–216. https://doi.org/10.1016/j.virusres.2013.08.008.

Stewart, F.M., and Levin, B.R. (1984). The population biology of bacterial viruses: why be temperate. Theor. Popul. Biol. *26*, 93–117.

Stielow, B., Klenk, H.P., and Menzel, W. (2011). Complete genome sequence of the first endornavirus from the ascocarp of the ectomycorrhizal fungus *Tuber aestivum* Vittad. Arch. Virol. *156*, 343–345. https://doi.org/10.1007/s00705-010-0875-x.

Sun, Q., Choi, G.H., and Nuss, D.L. (2009). A single Argonaute gene is required for induction of RNA silencing antiviral defense and promotes viral RNA recombination. Proc. Natl. Acad. Sci. U.S.A. *106*, 17927–17932. https://doi.org/10.1073/pnas.0907552106.

Suttle, A., Chan, M., and Cottrell, M.T. (1990). Infection of phytoplankton by viruses and reduction of primary productivity. Nature *347*, 467–469.

Suzaki, K., Ikeda, K.I., Sasaki, A., Kanematsu, S., Matsumoto, N., and Yoshida, K. (2005). Horizontal transmission and host-virulence attenuation of totivirus in violet root rot fungus *Helicobasidium mompa*. J. Gen. Plant Pathol. *71*, 161–168.

Suzuki, N., Chen, B., and Nuss, D.L. (1999). Mapping of a hypovirus p29 protease symptom determinant domain with sequence similarity to potyvirus HC-Pro protease. J. Virol. 73, 9478–9484.

Suzuki, N., Supyani, S., Maruyama, K., and Hillman, B.I. (2004). Complete genome sequence of Mycoreovirus-1/Cp9B21, a member of a novel genus within the family *Reoviridae*, isolated from the chestnut blight fungus *Cryphonectria parasitica*. J. Gen Virol. *85*, 3437–3448.

Tanaka, T., Eusebio-Cope, A., Sun, L., and Suzuki, N. (2012). Mycoreovirus genome alterations: similarities to and differences from rearrangements reported for other reoviruses. Front. Microbiol. 3, 186. https://doi.org/10.3389/fmicb.2012.00186.

Taylor, D.J., and Bruenn, J. (2009). The evolution of novel fungal genes from non-retroviral RNA viruses. BMC Biol. 7, 88. https://doi.org/10.1186/1741-7007-7-88.

Thapa, V., Turner, G.G., Hafenstein, S., Overton, B.E., Vanderwolf, K.J., and Roossinck, M.J. (2016). Using a novel partitivirus in *Pseudogymnoascus destructans* to understand the epidemiology of white-nose syndrome. PLOS Pathog. *12*, e1006076. https://doi.org/10.1371/journal.ppat.1006076.

Tidona, C., and Darai, G. (2011). The Springer Index of Viruses. (Springer)

Tuomivirta, T.T., and Hantula, J. (2005). Three unrelated viruses occur in a single isolate of *Gremmeniella abietina* var. *abietina* type A. Virus Res. *110*, 31–39.

Tuomivirta, T.T., Kaitera, J., and Hantula, J. (2009). A novel putative virus of *Gremmeniella abietina* type B (*Ascomycota*: *Helotiaceae*) has a composite genome with endornavirus affinities. J. Gen. Virol. *90*, 2299–2305.

Urayama, S., Kato, S., Suzuki, Y., Aoki, N., Le, M.T., Arie, T., Teraoka, T., Fukuhara, T., and Moriyama, H. (2010). Mycoviruses related to chrysovirus affect vegetative growth in the rice blast fungus *Magnaporthe oryzae*. J. Gen. Virol. *91*, 3085–3094.

Vainio, E.J., Korhonen, K., Tuomivirta, T.T., and Hantula, J. (2010). A novel putative partitivirus of the saprotrophic fungus *Heterobasidion ecrustosum* infects pathogenic species of the *Heterobasidion annosum* complex. Fungal Biol. *114*, 955–965.

Vainio, E.J., Keriö, S., and Hantula, J. (2011a). Description of a new putative virus infecting the conifer pathogenic fungus *Heterobasidion parviporum* with resemblance to *Heterobasidion* annosum P-type partitivirus. Arch. Virol. *156*, 79–86. https://doi.org/10.1007/s00705-010-0823-9.

Vainio, E.J., Hakanpää, J., Dai, Y.C., Hansen, E., Korhonen, K., and Hantula, J. (2011b). Species of *Heterobasidion* host a diverse pool of partitiviruses with global distribution and interspecies transmission. Fungal Biol. *115*, 1234–1243. https://doi.org/10.1016/j.funbio.2011.08.008.

Vainio, E.J., Hyder, R., Aday, G., Hansen, E., Piri, T., Doğmuş-Lehtijärvi, T., Lehtijärvi, A., Korhonen, K., and Hantula, J. (2012). Population structure of a novel putative mycovirus infecting the conifer root-rot fungus *Heterobasidion annosum* sensu lato. Virology *422*, 366–376. https://doi.org/10.1016/j.virol.2011.10.032.

Vainio, E.J., Piri, T., and Hantula, J. (2013a). Virus community dynamics in the conifer pathogenic fungus *Heterobasidion parviporum* following an artificial introduction of a partitivirus. Microb. Ecol. *65*, 28–38. https://doi.org/10.1007/s00248-012-0118-7.

Vainio, E.J., Capretti, P., Motta, E., and Hantula, J. (2013b). Molecular characterization of HetRV8-ir1, a partitivirus of the invasive conifer pathogenic fungus Heterobasidion *irregulare*. Arch. Virol. *158*, 1613–1615. https://doi.org/10.1007/s00705-013-1643-5.

Vainio, E.J., Jurvansuu, J., Streng, J., Rajamäki, M., Hantula, J., and Valkonen, J.P.T. (2015a). Diagnosis and discovery of fungal viruses by deep sequencing of small RNAs. J. Gen. Virol. *96*, 714–725.

Vainio, E.J., Müller, M.M., Korhonen, K., Piri, T., and Hantula, J. (2015b). Viruses accumulate in aging infection centers of a fungal forest pathogen. ISME J. *9*, 497–507. https://doi.org/10.1038/ismej.2014.145.

Vainio, E.J., Pennanen, T., Rajala, T., and Hantula, J. (2017). Occurrence of similar mycoviruses in pathogenic, saprotrophic and mycorrhizal fungi inhabiting the same forest stand. FEMS Microbiol. Ecol. *93*. https://doi.org/10.1093/femsec/fix003.

Vainio, E.J., Jurvansuu, J., Hyder, R., Kashif, M., Piri, T., Tuomivirta, T., Poimala, A., Xu, P., Mäkelä, S., *et al.* (2018). The partitivirus Het PV13-an1 mediates severe growth debilitation and major alterations in the gene expression of a fungal forest pathogen. J. Virol. *92*, e/1744-17.

van Diepeningen, A.D., Debets, A.J., and Hoekstra, R.F. (1998). Intra- and interspecies virus transfer in *Aspergilli* via protoplast fusion. Fungal Genet. Biol. *25*, 171–180.

Wach, M.P., Sriskantha, A., and Romaine, C.P. (1987). Double-stranded RNAs associated with La France disease of the commercial mushroom. Phytopathology *77*, 1321–1325.

Wang, S., Kondo, H., Liu, L., Guo, L., and Qiu, D. (2013). A novel virus in the family *Hypoviridae* from the plant pathogenic fungus *Fusarium graminearum*. Virus Res. *174*, 69–77. https://doi.org/10.1016/j.virusres.2013.03.002.

Wang, M., Wang, Y., Sun, X., Cheng, J., Fu, Y., Liu, H., Jiang, D., Ghabrial, S.A., and Xie, J. (2015). Characterization of a novel megabirnavirus from *Sclerotinia sclerotiorum* reveals horizontal gene transfer from single-stranded RNA virus to double-stranded rna virus. J. Virol. *89*, 8567–8579. https://doi.org/10.1128/JVI.00243-15.

Wu, M., Jin, F., Zhang, J., Yang, L., Jiang, D., and Li, G. (2012). Characterization of a novel bipartite double-stranded RNA mycovirus conferring hypovirulence in the phytopathogenic fungus *Botrytis porri*. J. Virol. *86*, 6605–6619. https://doi.org/10.1128/JVI.00292-12.

Xie, J., Wei, D., Jiang, D., Fu, Y., Li, G., Ghabrial, S., and Peng, Y. (2006). Characterization of debilitation-associated mycovirus infecting the plant-pathogenic fungus *Sclerotinia sclerotiorum*. J. Gen. Virol. *87*, 241–249.

Xie, J., Xiao, X., Fu, Y., Liu, H., Cheng, J., Ghabrial, S.A., Li, G., and Jiang, D. (2011). A novel mycovirus closely related to hypoviruses that infects the plant pathogenic fungus *Sclerotinia sclerotiorum*. Virology *418*, 49–56. https://doi.org/10.1016/j.virol.2011.07.008.

Yaegashi, H., Sawahata, T., Ito, T., and Kanematsu, S. (2011). A novel colony-print immunoassay reveals differential patterns of distribution and horizontal transmission of four unrelated mycoviruses in *Rosellinia necatrix*. Virology *409*, 280–289. https://doi.org/10.1016/j.virol.2010.10.014.

Yaegashi, H., Kanematsu, S., and Ito, T. (2012). Molecular characterization of a new hypovirus infecting a phytopathogenic fungus, *Valsa ceratosperma*. Virus Res. *165*, 143–150. https://doi.org/10.1016/j.virusres.2012.02.008.

Yaegashi, H., Nakamura, H., Sawahata, T., Sasaki, A., Iwanami, Y., Ito, T., and Kanematsu, S. (2013). Appearance of mycovirus-like double-stranded RNAs in the white root rot fungus, *Rosellinia necatrix*, in an apple orchard. FEMS Microbiol. Ecol. *83*, 49–62. https://doi.org/10.1111/j.1574-6941.2012.01454.x.

Yaegashi, H., and Kanematsu, S. (2015). Natural infection of the soil-borne fungus *Rosellinia necatrix* with novel mycoviruses under greenhouse conditions. Virus Res. *219*, 83–91.

Yu, H.J., Lim, D., and Lee, H.S. (2003). Characterization of a novel single-stranded RNA mycovirus in *Pleurotus ostreatus*. Virology *314*, 9–15.

Yu, J., Kwon, S.J., Lee, K.M., Son, M., and Kim, K.H. (2009). Complete nucleotide sequence of double-stranded RNA viruses from *Fusarium graminearum* strain DK3. Arch. Virol. *154*, 1855–1858. https://doi.org/10.1007/s00705-009-0507-5.

Yu, X., Li, B., Fu, Y., Jiang, D., Ghabrial, S.A., Li, G., Peng, Y., Xie, J., Cheng, J., Huang, J., *et al.* (2010). A geminivirus-related DNA mycovirus that confers hypovirulence to a plant pathogenic fungus. Proc. Natl. Acad. Sci. U.S.A. *107*, 8387–8392. https://doi.org/10.1073/pnas.0913535107.

Yu, X., Li, B., Fu, Y., Xie, J., Cheng, J., Ghabrial, S.A., Li, G., Yi, X., and Jiang, D. (2013). Extracellular transmission of a DNA mycovirus and its use as a natural fungicide. Proc. Natl. Acad. Sci. U.S.A. *110*, 1452–1457. https://doi.org/10.1073/pnas.1213755110.

Zhang, R., Liu, S., Chiba, S., Kondo, H., Kanematsu, S., and Suzuki, N. (2014). A novel single-stranded RNA virus isolated from a phytopathogenic filamentous fungus, *Rosellinia necatrix*, with similarity to hypo-like viruses. Front. Microbiol. *5*, 360. https://doi.org/10.3389/fmicb.2014.00360.

Zheng, L., Liu, H., Zhang, M., Cao, X., and Zhou, E. (2013). The complete genomic sequence of a novel mycovirus from *Rhizoctonia solani* AG-1 IA strain B275. Arch. Virol. *158*, 1609–1612. https://doi.org/10.1007/s00705-013-1637-3

Diversity of Viruses Infecting Eukaryotic Algae

10

Steven M. Short[1,2*], Michael A. Staniewski[1], Yuri V. Chaban[2], Andrew M. Long[1], and Donglin Wang[1]

[1]Department of Ecology and Evolutionary Biology, University of Toronto, Toronto, ON, Canada.
[2]Department of Biology, University of Toronto Mississauga, Mississauga, ON, Canada.

*Correspondence: steven.short@utoronto.ca

https://doi.org/10.21775/9781910190852.10

Abstract

Algae are photosynthetic organisms that drive aquatic ecosystems, e.g. fuelling food webs or forming harmful blooms. The discovery of viruses that infect eukaryotic algae has raised many questions about their influence on aquatic primary production and their role in algal ecology and evolution. Although the full extent of algal virus diversity is still being discovered, this review summarizes current knowledge of this topic. Where possible, formal taxonomic classifications are referenced from the International Committee on Taxonomy of Viruses (ICTV); since the pace of virus discovery has far surpassed the rate of formal classification, however, numerous unclassified viruses are discussed along with their classified relatives. In total, we recognized 61 distinct algal virus taxa with highly variable morphologies that include dsDNA, ssDNA, dsRNA, and ssRNA genomes ranging from approximately 4.4 to 560 kb, with virion sizes from approximately 20 to 210 nm in diameter. These viruses infect a broad range of algae and, although there are a few exceptions, they are generally lytic and highly species or strain specific. Dedicated research efforts have led to the appreciation of algal viruses as diverse, dynamic, and ecologically important members of the biosphere, and future investigations will continue to reveal the full extent of their diversity and impact.

Introduction

Algae are the engines of aquatic ecosystems that convert inorganic matter into organic matter and fuel aquatic food webs. Their growth influences the productivity of aquatic environments, and also environmental quality by either supporting or impeding beneficial human use, e.g. by sustaining fisheries, or forming nuisance blooms. A few decades ago, scientists discovered that viruses infecting eukaryotic algae are present in nearly all aquatic environments, and that most of these pathogens are both species-specific and cause the lysis, or death, of their hosts. As an aside, throughout this review we will use the term 'algae' specifically to refer to eukaryotic algae; in some discussions of aquatic primary production cyanobacteria are also included under the umbrella term algae, such as in the first two sentences of this paragraph, but here we are focusing only on eukaryotes. Discovery of ubiquitous algal viruses in aquatic environments raised many questions about the influence of viruses on aquatic primary production and their role in algal ecology and evolution, and spurred the development of algal virus ecology as a research field unto itself. Work on this topic has taught us much about the importance of algal viruses, and over the years numerous reviews have summarized this work (Brussaard, 2004; Brussaard and Martínez Martínez, 2008; Dunigan *et al.*, 2006; Kang *et al.*, 2005; Nagasaki, 2008; Short, 2012; Suttle, 2000; Van Etten *et al.*, 2002; Van Etten and Meints, 1999; Wilson *et al.*, 2009; Yamada *et al.*,

2006). Yet we still lack a deep understanding of algal virus diversity and the range of algae that are susceptible to viral infections.

At this stage of algal virus research, more than 40 years after the first virus was isolated from an infected eukaryotic alga (Mayer and Taylor, 1979), observer effects are rife so it is challenging to discern if any observations reflect natural patterns of diversity. For example, viral metagenomes (see Chapter 5) and PCR-based diversity surveys (see Chapter 15) of large dsDNA algal viruses have been dominated by a single group within the *Phycodnaviridae*, the Prasinoviruses (Clasen and Suttle, 2009; Hingamp *et al.*, 2013; Short and Short, 2008; Zhong and Jacquet, 2014). Moreover, almost all isolated viruses that infect diatoms, an algal group with an enormous role in global biogeochemistry, have been isolated from species of *Chaetoceros*, and the only viruses of macrophytic algae infect the reproductive spores of certain brown algae within the order Ectocarpales. These observations could be interpreted to suggest that these types of algae are particularly permissive hosts for viral parasites, or that most other algae are resistant to viral infections. In reality, it is more likely that surveys of virus diversity are far from complete, and ongoing isolation and genomic efforts will reveal even greater diversity than is currently known; patterns of diversity inferred from current literature in other words likely do not capture fully the patterns of nature.

Given the caveats that it is not clear what should be considered a virus 'species', and most algal viruses are not yet formally classified by the International Committee on Taxonomy of Viruses (ICTV), we have compiled an up-to-date list of all algal virus 'species' reported in the literature (Table 10.1). In this list we have included viruses that can be considered distinct because of hosts they infect, or because of other viral characteristics, or both. In some cases viruses that might represent distinct strains or species are grouped into a single line in the table. For example, Table 10.1 lists MpV as a single virus type infecting the marine prasinophyte *Micromonas pusilla*, but seven MpV viruses were actually described that fell into four groups based on molecular weights of major proteins (Cottrell and Suttle, 1991). A similar situation exists for the chloroviruses such as those that infect *Chlorella* NC64a (*Chlorella variabilis*); these cases will be discussed in more detail in the sections that follow.

In total, 61 viruses are listed in Table 10.1 infecting algal species belonging to six phyla and ten classes. Because virus systematics is a work in progress and discovery is ongoing, this number however merely represents the minimum number of unique viruses isolated from eukaryotic algae. It is true that these viruses have been isolated from diverse algal hosts, but the number of isolates and the diversity of their hosts pales in comparison to the estimated 64,500 species belonging to 14 phyla and 51 classes of eukaryotic algae (Guiry, 2012). It is reasonable to think that most, if not all algal species are hosts to one or more types of viruses, but of course even that notion is purely speculative.

Despite the obvious gaps in our knowledge of algal virus diversity, ongoing isolation efforts over the past few decades have continuously revealed novel algal viruses. Indeed, the rate of algal virus discovery, inferred from literature reports, has remained relatively constant since the 1990s and can be expected to continue as long as there is interest in how these parasites influence or possibly control aquatic food webs (Fig. 10.1). It is clear that there is much to be learned about the diversity of these globally important algal parasites, and we are far from knowing the extent of algal virus diversity.

No matter how incomplete our current knowledge, since their first discovery an impressive volume of information about algal viruses has been obtained and the field has expanded far beyond the early days of research when the only known algal viruses were dsDNA viruses assigned to the order *Phycodnaviridae*. Currently, dsDNA, ssDNA, dsRNA, and ssRNA viruses have been isolated from algae, and some isolates remain uncharacterized with respect to their genome type, or the nature of some genomes is speculative (Fig. 10.2). Thus, while much needs to be learned, enough is known about algal viruses to warrant the following review. In line with the classical Baltimore virus classification scheme (Baltimore, 1971) (see also Chapter 2), we have organized the sections of this review by genome type. Since the majority of known algal viruses are dsDNA viruses, that section is split into subsections describing each recognized genera of dsDNA algal viruses followed by a brief note identifying the many dsDNA viruses not formally classified by the ICTV; many of these unclassified viruses are closely related to or are even widely accepted by researchers as species of recognized

Table 10.1 Reports describing isolations and/or preliminary characterizations of algal viruses

Virus	Host(s)	Host phylum/class	References
AaV (BtV)	*Aureococcus anophagefferens*	Ochrophyta/Pelagophyceae	Milligan and Cosper (1994)
AglaRNAV	*Asterionellopsis glacialis*	Bacillariophyta/Bacillariophyceae	Tomaru *et al.* (2012)
ATCV	*Chlorella heliozoae* (SAG 3.83)	Chlorophyta/Trebouxiophyceae	Bubeck and Pfitzner (2005)
BpV	*Bathycoccus* sp.	Chlorophyta/Prasinophyceae	Bellec *et al.* (2009)
CbV	*Chrysochromulina brevifilum, C. strobilus*	Haptophyta/Coccolithophyceae	Suttle and Chan (1995)
CdebDNAV	*Chaetoceros debilis*	Bacillariophyta/Mediophyceae	Tomaru *et al.* (2008b)
CeV	*Chrysochromulina ericina*	Haptophyta/Coccolithophyceae	Sandaa *et al.* (2001)
ClorDNAV	*Chaetoceros lorenzianus*	Bacillariophyta/Mediophyceae	Tomaru *et al.* (2011b)
CpV	*Chrysochromulina parva*	Haptophyta/Coccolithophyceae	Mirza *et al.* (2015)
CsetDNAV	*Chaetoceros setoensis*	Bacillariophyta/Mediophyceae	Tomaru *et al.* (2013b)
CsfrRNAV	*Chaetoceros socialis* f. *radians*	Bacillariophyta/Mediophyceae	Tomaru *et al.* (2009b)
CsalDNAV (CsNIV)	*Chaetoceros salsugineum*	Bacillariophyta/Mediophyceae	Nagasaki *et al.* (2005c)
Csp03RNAV	*Chaetoceros* sp.	Bacillariophyta/Mediophyceae	Tomaru *et al.* (2013a)
Csp05DNAV	*Chaetoceros* sp.	Bacillariophyta/Mediophyceae	Toyoda *et al.* (2012)
Csp07DNAV	*Chaetoceros* sp.	Bacillariophyta/Mediophyceae	Kimura and Tomaru (2013)
CspNIV	*Chaetoceros* cf. *gracilis*	Bacillariophyta/Mediophyceae	Bettarel *et al.* (2005)
CtenDNAV	*Chaetoceros tenuissimus*	Bacillariophyta/Mediophyceae	Tomaru *et al.* (2011a)
CtenDNAV type II	*Chaetoceros tenuissimus*	Bacillariophyta/Mediophyceae	Kimura and Tomarua (2015)
CtenRNAV	*Chaetoceros tenuissimus*	Bacillariophyta/Mediophyceae	Shirai *et al.* (2008)
CtenRNAV type II	*Chaetoceros* sp., *C. tenuissimus, C. setoensis, C. socialis* f. *radians*	Bacillariophyta/Mediophyceae	Kimura and Tomarua (2015)
CVM	*Micractinium conductrix* (*Chlorella* Pbi)	Chlorophyta/Trebouxiophyceae	Reisser *et al.* (1986), Reisser *et al.* (1988)
CwNIV	*Chaetoceros* cf. *wighamii*	Bacillariophyta/Mediophyceae	Eissler *et al.* (2009)
EfasV	*Ectocarpus fasciculatus*	Ochrophyta/Phaeophyceae	Müller *et al.* (1996), Kapp *et al.* (1997)
EhV	*Emiliania huxleyi*	Haptophyta/Coccolithophyceae	Castberg *et al.* (2002)
EsV	*Ectocarpus siliculosus, Kuckuckia kylinii*	Ochrophyta/Phaeophyceae	Müller *et al.* (1992), Müller (1996), Lanka *et al.* (1993)
FirrV	*Feldmannia irregularis*	Ochrophyta/Phaeophyceae	Kapp *et al.* (1997)
FlexV	*Feldmannia simplex*	Ochrophyta/Phaeophyceae	Friess-klebl *et al.* (1994)
FsV	*Feldmannia* sp.	Ochrophyta/Phaeophyceae	Henry and Meints (1992)
HaNIV	*Heterosigma akashiwo*	Ochrophyta/Raphidophyceae	Lawrence *et al.* (2001)
HaRNAV	*Heterosigma akashiwo*	Ochrophyta/Raphidophyceae	Nagasaki and Yamaguchi (1997)
HaV	*Heterosigma akashiwo*	Ochrophyta/Raphidophyceae	Tai *et al.* (2003)
HcDNAV	*Heterocapsa circularisquama*	Miozoa/Dinophyceae	Tarutani *et al.* (2001)
HcRNAV	*Heterocapsa circularisquama*	Miozoa/Dinophyceae	Tomaru *et al.* (2004a)
HcV	*Heterocapsa circularisquama*	Miozoa/Dinophyceae	Tarutani *et al.* (2001)
HeV RF02	*Haptolina ericina, Prymnesium kappa*	Haptophyta/Coccolithophyceae	Johannessen *et al.* (2015)

Table 10.1 Continued

Virus	Host(s)	Host phylum/class	References
HincV	*Hincksia hinckiae*	Ochrophyta/Phaeophyceae	Kapp *et al*. (1997)
HpygDNAV	*Heterocapsa pygmaea*	Miozoa/Dinophyceae	Kim *et al*. (2012)
HVCV	*Chlorella*-like endosymbiont of *Hydra viridis*	Chlorophyta/Trebouxiophyceae	Van Etten *et al*. (1981)
MclaV	*Myriotrichia clavaeformis*	Ochrophyta/Phaeophyceae	Kapp *et al*. (1997)
MiV	*Micromonas* sp.	Chlorophyta/Prasinophyceae	Bellec *et al*. (2009)
MpRV	*Micromonas pusilla*	Chlorophyta/Prasinophyceae	Brussaard *et al*. (2004a)
MpV	*Micromonas pusilla*	Chlorophyta/Prasinophyceae	Mayer and Taylor (1979), Cottrell and Suttle (1991)
MpVN1	*Micromonas pusilla*	Chlorophyta/Prasinophyceae	Zingone *et al*. (2006)
MpVN2	*Micromonas pusilla*	Chlorophyta/Prasinophyceae	Zingone *et al*. (2006)
Ols1	*Heterosigma akashiwo*	Ochrophyta/Raphidophyceae	Lawrence *et al*. (2006)
OtV	*Ostreococcus* sp.	Chlorophyta/Prasinophyceae	Bellec *et al*. (2009)
PBCV	*Chlorella variabilis* (*Chlorella* NC64A)	Chlorophyta/Trebouxiophyceae	Van Etten *et al*. (1982), Van Etten *et al*. (1983)
PgV Group I	*Phaeocystis globosa*	Haptophyta/Coccolithophyceae	Brussaard *et al*. (2004b), Baudoux and Brussaard (2005)
PgV Group II	*Phaeocystis globosa*	Haptophyta/Coccolithophyceae	Brussaard *et al*. (2004b), Baudoux and Brussaard (2005)
PgV-102P	*Phaeocystis globosa*	Haptophyta/Coccolithophyceae	Wilson *et al*. (2006)
PkV RF01	*Prymnesium kappa*, *Haptolina ericina*	Haptophyta/Coccolithophyceae	Johannessen *et al*. (2015)
PkV RF02	*Prymnesium kappa*	Haptophyta/Coccolithophyceae	Johannessen *et al*. (2015)
PlitV	*Pilayella littoralis*	Ochrophyta/Phaeophyceae	Maier *et al*. (1998)
PoV	*Pyramimonas orientalis*	Chlorophyta/Prasinophyceae	Sandaa *et al*. (2001)
PpV	*Phaeocystis pouchetii*	Haptophyta/Coccolithophyceae	Jacobsen *et al*. (1996)
RsetRNAV (RsRNAV)	*Rhizosolenia setigera*	Bacillariophyta/ Coscinodiscophyceae	Nagasaki *et al*. (2004)
ScosV	*Skeletonema costatum*	Bacillariophyta/Mediophyceae	Kim *et al*. (2015a)
SpalV	*Stephanopyxis palmeriana*	Bacillariophyta/ Coscinodiscophyceae	Kim *et al*. (2015b)
TampV	*Teleaulax amphioxeia*	Cryptophyta/Cryptophyceae	Nagasaki *et al*. (2009)
TnitDNAV	*Thalassionema nitzschioides*	Bacillariophyta/Bacillariophyceae	Tomaru *et al*. (2012)
TsV-N1	*Tetraselmis striata*	Chlorophyta/ Chlorodendrophyceae	Pagarete *et al*. (2015)

Algae were classified at the level of phylum and class according to information from AlgaeBase (http://algaebase.org/, accessed 26 May 2016).

genera, so these are discussed in corresponding sections on classified taxa. On the other hand, there are many unclassified dsDNA algal viruses that were presumed to be members of the *Phycodnaviridae*, but based on more recent evidence should not be included in this family. To respect the history of algal virus discovery, many of these cases will be discussed in sections on classified viruses, and a separate section on unclassified viruses will note exceptions and questions that remain in algal virus systematics.

It is probable that this accounting of known algal viruses is incomplete since it seems likely that novel viruses have been isolated but have not yet been

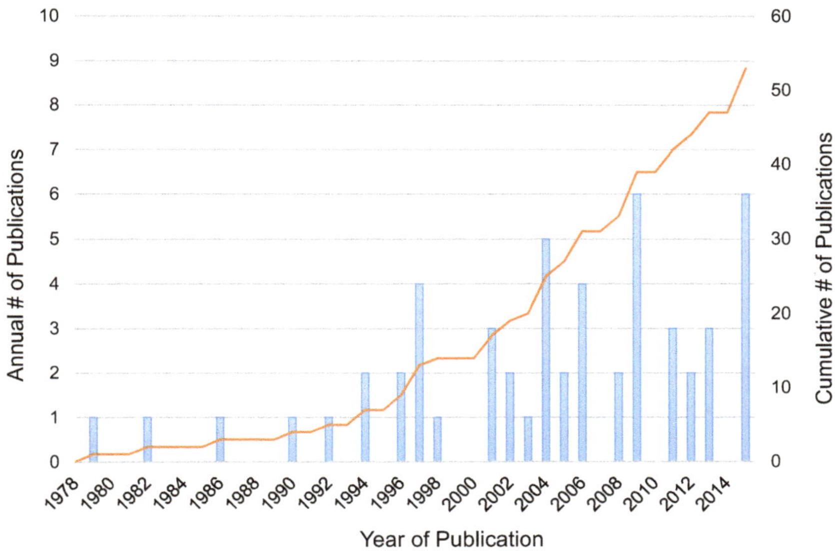

Figure 10.1 Annual and cumulative number of publications describing novel algal viruses. The blue bars show the individual number of publications each year, while the orange line represents the cumulative number of publications, as of year-end 2015. It should be noted that although most research articles focus on a single virus isolate, the number of viruses isolated does not directly correspond to the number of publications shown in this figure since some articles reported the isolation of several different viruses; the 61 virus isolates listed in Table 10.1 were reported in 51 unique publications. The publications that were used to generate this figure were the initial reports shown for each virus listed in Table 10.1.

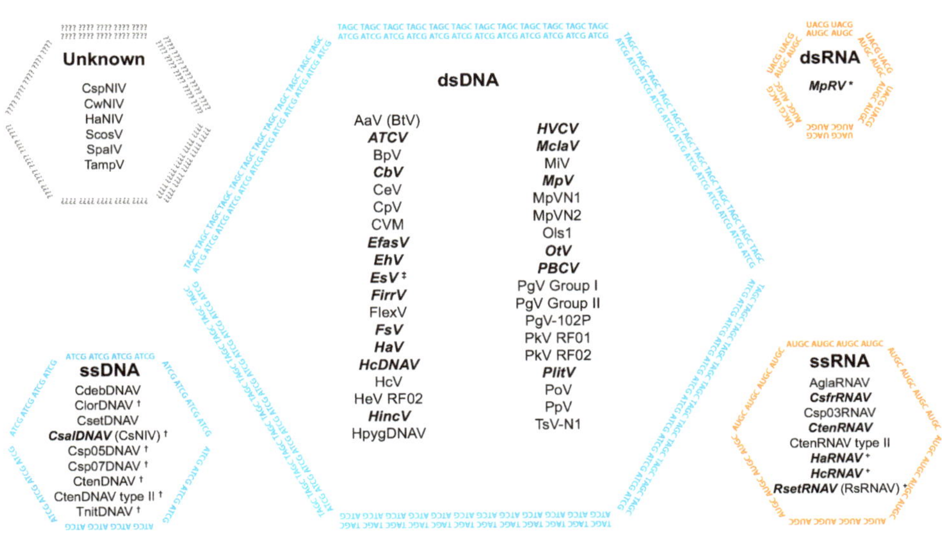

Figure 10.2 Isolated viruses infecting eukaryotic algae categorized by genome classification. Symbols that follow certain virus names highlight specific features of the viral genomes according to the following scheme: (*) segmented genome; (+) positive-sense RNA; (†) circular genome with partially double-stranded regions; (‡) circular genome with partially single-stranded regions. Virus names are simplified for brevity (e.g. MpV represents MpV-SPI and its closely related strains), and viruses that have been formally classified by the ICTV are highlighted with bold italic fonts.

described in the literature. For example, in their article describing RNA viruses in tropical waters, Culley and co-authors (2014) included sequences from two novel viruses (ScloRNAV01 and 02) that infect pennate diatoms that, to our knowledge, have not been described elsewhere. Similarly, the genomic sequence of a virus that infects the Chlorophyte alga *Dunaliella viridis* has been determined and is available in GenBank (accession # NC_023841), but no reports of this virus's isolation have been published. On that note, since molecular and genomic approaches have also been applied to the study of algal virus diversity, this review would not be complete without also drawing attention to these efforts that have provided major insights into the world of algal viruses.

Diversity inferred from environmental nucleic acids

PCR studies

Not long after viruses were observed to be abundant in marine environments (Bergh *et al.*, 1989), researchers turned to cultivation-free molecular studies of environmental DNA (eDNA) to examine algal virus diversity (see Chapters 5 and 15). The first attempt to uncover the hidden diversity of algal viruses in natural communities was based on PCR amplification and sequence analysis of fragments of virus DNA polymerase genes (abbreviated as *polB*, or PolB when referring to the protein) from dsDNA algal viruses (Chen *et al.*, 1996); at the time all known algal viruses were dsDNA viruses, classified as *Phycodnaviridae*. Moreover, development of universal, algal-virus-specific PCR primers (Chen and Suttle, 1995) demonstrated that virus DNA polymerase (i.e. *polB*) sequences were a robust taxonomic marker for phycodnaviruses and related dsDNA viruses (Chen and Suttle, 1996). Because several unknown phycodnaviruses were detected in Gulf of Mexico eDNA, this early research also highlighted the need for intensified work on algal virus diversity (Chen *et al.*, 1996).

Subsequent virus diversity studies based on PCR amplification of algal virus (i.e. phycodnavirus) *polB* genes demonstrated that remarkably similar virus genes could be detected in different oceans (Short and Suttle, 2002), some virus genes were constantly present in some environments (Short and Suttle, 2003), and freshwater environments are also inhabited by diverse algal viruses (Clasen and Suttle, 2009; Short *et al.*, 2011; Short and Short, 2008; Zhong and Jacquet, 2014). These and other studies of phycodnavirus diversity based on PCR amplification with *polB* primers have led to the surprisingly consistent observation that *polB* libraries of natural water samples are dominated by sequences related to cultivated members of the genus *Prasinovirus* (e.g. Clasen and Suttle, 2009; Culley *et al.*, 2009; Park *et al.*, 2011; Short and Short, 2008; Zhong and Jacquet, 2014). Although the most parsimonious explanation for the preponderance of prasinovirus-like sequences is biased PCR methods, there is evidence from amplification-free metagenomic studies that the bias actually reflects the natural dominance of these types of viruses (Clasen and Suttle, 2009; Short, 2012). Whatever the case, it is apparent that algal-virus-specific *polB* primers are not truly universal because some cultivated algal viruses such as *Emiliania huxleyi* viruses cannot be amplified with this primer set.

Knowledge that the algal-virus-specific *polB* primers could not target viruses of the coccolithophore *Emiliania huxleyi* (EhVs) motivated development of PCR primers targeting their major capsid protein (MCP) (Schroeder *et al.*, 2002). These primers were subsequently used to demonstrate that diverse populations of EhVs can be detected in the water column, but only a few genotypes dominate following termination of *E. huxleyi* blooms (Martínez Martínez *et al.*, 2007). Moreover, some EhV genotypes (strains) seem cosmopolitan despite evidence for distinct EhV communities associated with specific localities (Rowe *et al.*, 2011). Universal MCP primers that would target more than just coccolithoviruses were also developed with the aim to examine more broadly the diversity of viruses missed by the algal-virus-specific *polB* primers (Larsen *et al.*, 2008).

Generally, amplification of eDNA with the algal virus MCP primers reveals greater diversity than can be detected with *polB* primers alone, and they are a complementary tool for examining algal virus diversity (Short, 2012). For example, studies of South Korean coastal waters (Park *et al.*, 2011) and an embayment of Lake Ontario, Canada (Rozon and Short, 2013) using both MCP and *polB* primers demonstrated that the MCP primers readily amplified genes encoded by uncultured algal viruses that are not most closely related to known prasinoviruses. Instead, the MCP primers seem particularly effective for the amplification of genes from a clade of algal viruses that infect microalgae such as *Phaeocystis globosa*, *Chrysochromulina ericina*, or *Pyramimonas orientalis*. Currently, these unclassified algal viruses are considered members of the '*Megaviridae*' group (Arslan *et al.*, 2011; Santini *et al.*, 2013) that are also known as 'extended *Mimiviridae*' (Fischer, 2016) (see Chapter 11). The *Megaviridae* are, like their relatives in the *Phycodnaviridae*, part of a diverse group of dsDNA viruses termed nucleo-cytoplasmic large DNA viruses (NCLDVs) (Iyer *et al.*, 2001; Iyer *et al.*, 2006). The name '*Megavirales*' has been proposed as a virus order encompassing the NCLDVs (Colson *et al.*, 2013), but as of this writing the names *Megaviridae* and *Megavirales* have not yet been approved by the ICTV. Although their use is still debated among researchers studying algal viruses and their giant amoebae-infecting relatives (Fischer, 2016), these names are used widely in the literature. Hence, because of their wide use and the fact that these terms seem to have a legitimate taxonomic basis, *Megavirales* and NCLDVs, and *Megaviridae* and extended *Mimiviridae* will be used interchangeably here. Most recently, Moniruzzaman *et al.* (2016) developed an MCP primer set that specifically targets algae-infecting *Megaviridae* and monitored the diversity of this group over the course of a bloom of the brown tide causing alga *Aureococcus anophagefferens*. Hundreds of unique amplicons were observed that clustered into 145 OTUs (operational taxonomic units) at 97% nucleotide identity highlighting the huge but previously unobserved diversity of megaviruses. Community analysis of the sequence data highlighted the role of these viruses in the bloom as the viral community composition apparently changed from the beginning to the end of the event (Moniruzzaman *et al.*, 2016).

PCR amplification of algal virus *polB* and MCP gene fragments has culminated in deep-amplicon sequencing projects facilitated by 'next generation' sequencing technologies. Newly designed primers that specifically target close relatives of *Prasinovirus* (see below) were used in studies that uncovered immense diversity; using a very conservative cut-off of 26% nucleotide similarity to define OTUs, 17 OTUs with hundreds of different *polB* genotypes were observed in surface and deep chlorophyll maximum samples from three locations in the Indian Ocean (Clerissi *et al.*, 2014). Beyond the remarkable diversity of prasinoviruses that were observed, deep-amplicon sequencing corroborated earlier results that showed that some phycodnaviruses are globally distributed; different OTUs amplified from Indian Ocean eDNA samples were most closely related to other environmental sequences from either the Southern Ocean; coastal British Columbia, Canada; coastal Hawai'i, USA; and even freshwater lakes in Ontario, Canada. A subsequent study by the same researchers expanded these results and attempted to link prasinovirus and host diversity, and determined that the unknown virus sequences that did not cluster with cultivated prasinoviruses of *Bathycoccus*, *Ostreococcus*, and *Micromonas* (order Mamiellales) probably infect prasinophyte species within the closely related algal order, Dolichomastigales (Clerissi *et al.*, 2015). Clearly, ongoing isolation efforts are essential to fully appreciate the host range of uncultivated prasinoviruses, which appear to be the most diverse and abundant group of phycodnaviruses.

Most recently, a gene involved in DNA mismatch repair, *MutS*, has been identified as a diagnostic marker for the *Megaviridae*. Two new homologues of *MutS* (*MutS7* and *MutS8*) were identified in the genomes of mimivirus and the algae-infecting megaviruses, and are also abundant in the Global Ocean Sampling (GOS) expedition metagenomic sequence data (Ogata *et al.*, 2011). Following this discovery, Wilson and colleagues (2014) developed a set of PCR primers that specifically target *MutS8*, a *MutS* clade that discriminates algae-infecting megaviruses. Amplification of eDNA samples from coastal Maine, USA, revealed wide-spread occurrence of *MutS8* homologues demonstrating the efficacy of this primer set for the detection of viruses most closely related to algae-infecting viruses of the *Megaviridae* (Wilson *et al.*, 2014). These new primers are a welcome addition to the molecular toolkit for detecting large dsDNA algal viruses, and they will undoubtedly help resolve taxonomic questions about the megaviruses and the extent of their natural diversity.

At the other end of the viral–genome spectrum, research on the diversity of uncultivated RNA viruses in the sea began with the development of PCR primers targeting the RNA-dependent-RNA-polymerase (RdRp) genes of viruses belonging to the order *Picornavirales* (Culley *et al.*, 2003). Amplification of cDNA synthesized from environmental RNA samples from coastal British Columbia, Canada, revealed surprising diversity of picorna-like viruses of unidentified eukaryote hosts, as well as numerous sequences closely related to the RdRp sequence of HaRNAV (Culley *et al.*, 2003). HaRNAV was the first ssRNA virus isolated that infects a microalga, the economically important, bloom forming raphidophyte *Heterosigma akashiwo*. Subsequent refinement of this cultivation-free approach for assessing marine RNA virus diversity extended these results by demonstrating the existence of a diverse clade of picorna-like viruses that likely infect marine protists, of which many are probably microalgae (Culley and Steward, 2007).

Metagenomic studies

Cully and colleagues (2014) also conducted a whole-genome shotgun sequencing (metagenomic) study of marine RNA virus communities. This approach was fundamentally different than the PCR-based methods mentioned above since it did not rely on amplification of targeted genes, yet the results confirmed the existence of a distinct marine picornavirus-like clade with cultivated viruses that infect microalgae (Culley *et al.*, 2006) (]see Chapter 5). More recent work in this area coupled RdRp-PCR and metagenomics via 'next-generation' sequencing, and suggested that marine RNA virus communities are dominated by positive-sense single-stranded RNA viruses. Once again, many of the RdRp sequences and genes from several assembled virus genomes were most closely related to cultivated algal viruses that infect several species of diatoms and the raphidophyte *Heterosigma akashiwo* (Culley *et al.*, 2014).

The earliest metagenomic studies of marine RNA viruses used Sanger-based shotgun sequencing and were facilitated by the relatively small size of RNA virus genomes. Although more technically challenging, the earliest metagenomic study of DNA viruses was also based on shotgun sequencing clone libraries of viral DNA. This examination of DNA viruses in the ocean suggested that bacteriophages dominated marine virus assemblages, yet putative algal virus genes were also observed even although the filtration steps during eDNA sample prep could have excluded larger viruses of eukaryotes (Breitbart *et al.*, 2002). Following the pioneering work of Breitbart and colleagues, other high-throughput sequencing projects targeting viruses have allowed investigators to examine virus diversity across a range of aquatic environments. All together, these efforts have led to the consistent conclusion that the extent of global viral diversity is largely unknown, and most (i.e. >60%) of the sequences obtained in such studies are not homologous to known sequences (Breitbart *et al.*, 2007; Martínez Martínez *et al.*, 2014; Mokili *et al.*, 2012). Moreover, most of the identifiable virus sequences discovered through metagenomic studies appear to be derived from bacteriophages, but NCLDV sequences and, in particular, sequences related to phycodnaviruses and members of the 'extended mimivirus' or *Megaviridae* clade with algal hosts are frequently observed in metagenomes (e.g. Angly *et al.*, 2006; Bench *et al.*, 2007; Breitbart *et al.*, 2004; Breitbart *et al.*, 2002; Breitbart *et al.*, 2007; Chow *et al.*, 2015; Edwards and Rohwer, 2005; Kristensen *et al.*, 2010; Mokili *et al.*, 2012; Rosario *et al.*, 2009; Steward and Preston, 2011; Williamson *et al.*, 2008; Yoshida-Takashima *et al.*, 2012). Again, this is in spite of the fact that most eDNA preparations for virus-specific metagenomic studies include filtration steps that should lead to bias against large dsDNA viruses.

Algal virus sequences have also been observed in metagenomic studies that did not specifically target viruses such as the high-profile, whole-genome shotgun sequencing of eDNA from the Sargasso Sea (Venter *et al.*, 2004) and the Atlantic and Pacific oceans during the GOS expedition (Rusch *et al.*, 2007). The virus sequences originally observed in these studies presumably originated from bacteriophages, and algal virus genes were not specifically noted. Furthermore analysis of the Sargasso Sea metagenome, however, revealed that unknown megaviruses with close homology to mimivirus are abundant in the Sargasso (Ghedin and Claverie, 2005). Similarly, an independent data mining project revealed that small but relevant fractions of GOS viral protein sequences were identified as eukaryote viruses (5% of all sequences) or mimiviruses (0.7%) suggesting that numerous algal virus sequences were captured in the GOS eDNA samples (Williamson *et al.*, 2008). Using DNA polymerase sequences (*polB*) as a reference, 'phylogenetic mapping' was also used to examine the distribution of large DNA viruses in the GOS sequence data (Monier *et al.*, 2008a). Although most (78%) of the putative viral *polB* sequences (811 total) were of phage origin, 14% of the sequences were mimivirus-like, and a large fraction of these (84%) clustered with the 'extended mimiviruses' known to infect the haptophyte and chlorophyte algae *Chrysochromulina ericina*, *Phaeocystis pouchetii*, and *Pyramimonas orientalis*. Additionally, numerous phycodnavirus *polB* fragments were also observed in the GOS sequences, and most clustered with putative prasinoviruses and followed the general trend of increased abundances in temperate versus tropical waters (Monier *et al.*, 2008a,b). Expanding upon and complementing this work, Kristensen *et al.* (2010) queried the Sargasso Sea and GOS metagenomes using three NCLDV core genes that have no known homologues outside this group of viruses and found that *Phycodnaviridae* and *Mimiviridae* dominated representing roughly 70 and 20%, respectively, of the NCLDV sequences in these metagenomes.

In general, sequences from viruses belonging to the *Phycodnaviridae* tend to dominate the eukaryotic virus sequences obtained from metagenomic studies of a wide range of aquatic habitats including the marine locations noted above, an Antarctic Lake (López-Bueno *et al.*, 2009), a temperate eutrophic lake in the continental USA (Green *et al.*, 2015), and even perennial ponds in the central Saharan desert (Fancello *et al.*, 2013). Analysis of 17 metagenomes collected during the Tara Oceans expedition from the Atlantic and Indian Oceans, and the Mediterranean, Red, and Arabian Seas also demonstrated the dominance of phycodnaviruses, and more specifically prasinoviruses, among NCLDVs (Hingamp *et al.*, 2013). Based on 16 NCLDV marker genes, 52% of putative NCLDV sequences were identified as *Phycodnaviridae* (85% of the phycodnaviruses were presumed prasinoviruses), 36% were '*Megaviridae*' and the remainder were related to unclassified NCLDVs, *Iridoviridae*, or *Asfarviridae*; it is noteworthy that the *polB* sequence of a virus infecting the dinoflagellate *Heterocapsa circularisquama* is most closely related to the *Asfarviridae* (Ogata *et al.*, 2009), suggesting that some of the NCLDVs detected in these metagenomes besides phycodnaviruses and megaviruses may actually infect algae. The results of this incredible survey of viruses across the world's oceans largely corroborated earlier conclusions from metagenomic studies and highlighted the fact that most of the dsDNA viruses of eukaryotes in the oceans infect algae, and frequently most of the algal viruses infect chlorophytes, or green algae, that are closely related to the family *Prasinophyceae*. Hence, the predominance of prasinophyte sequences in some PCR-based studies cannot be easily dismissed as due to PCR biases alone.

Many other fascinating observations of algal viruses have stemmed from metagenomic studies of viruses. For example, viral metagenomes associated with corals have revealed a remarkable diversity of sequences related to a broad range of eukaryote viruses including phycodnaviruses such as *Chlorovirus* and *Coccolithovirus* (Littman *et al.*, 2011; Marhaver *et al.*, 2008; Vega Thurber *et al.*, 2008; Wegley *et al.*, 2007). These observations permit speculation that all components of the coral holobiont, the assemblage composed of the coral animal and its symbiotic algae, protists, fungi, and bacteria, could be infected by viruses (see Chapter 5). Motivated by speculation that the phenomenon of coral bleaching may be caused, in part, by viral infections of symbiotic algae, Vega Thurber and coworkers coupled a study of DNA virus metagenomes of the coral *Montastraea cavernosa* with sequences from EST (expressed sequence tag) libraries of the coral's algal symbiont, the dinoflagellate *Symbiodinium* (Correa *et al.*, 2013). This study provided compelling evidence that NCLDVs, particularly phycodnaviruses and megaviruses, and ssRNA viruses closely related to the dinoflagellate-infecting virus HcRNAV were associated with the coral algal symbiont, and therefore that viruses could be a major source of mortality for coral symbionts and play a significant role in coral health.

Metagenomics studies have also enabled assembly of nearly complete algal virus and virophage genomes from eDNA. Metagenome sequences from the 0.1–0.8 μm size fraction of Organic Lake, Antarctica, were dominated by putative phycodnavirus genes (>60% of reads), and three large scaffolds representing different virus genomes were identified as phycodnaviruses based on conserved *polB* sequences; at the time these virus genomes were classified as phycodnaviruses (Yau *et al.*, 2011), but with current understanding of NCLDVs and the extended-*Mimiviridae*, it is now apparent that these genomes are more closely related to the algae-infecting megaviruses and are likely not part of the family *Phycodnaviridae*. Another surprising observation from this metagenomic project was the assembly of a complete genome of a virophage (Yau *et al.*, 2011) that could parasitize algal viruses when infecting the same host cell (Claverie and Abergel, 2009; Fischer and Suttle, 2011; La Scola *et al.*, 2008) (see Chapter 12). Virophage sequences have now been detected in metagenomic data sets collected from all over the world (Zhou *et al.*, 2013), and seven presumably complete virophage genomes (Zhou *et al.*, 2015) and four nearly complete algal virus genomes were assembled from the metagenomes of Yellowstone Lake, USA (Zhang *et al.*, 2015). Three of these algal virus genomes were nested among viruses of green algae and were most closely related to prasinoviruses. The fourth algal virus genome assembly was shorter and missing several core genes, but based on its MCP sequence, it appears to be a close relative of the algal megaviruses *Phaeocystis globosa* virus PgV-16T and OLPV1 and 2, the Organic Lake 'phycodnaviruses' mentioned above (Zhang *et al.*, 2015). Considering the relative dearth of isolated viruses, the ability to study nearly complete algal virus genomes assembled from environmental metagenomic sequences is remarkable and bodes well for rapid growth in knowledge of algal viruses.

Likewise, by coupling other technologies to metagenomics some researchers have opened up new avenues of investigation in algal virus research. For example, over the course of an induced algal bloom, next-generation sequencing was coupled with DNA separation methods (CsCl gradient and pulsed field gel electrophoresis) to specifically monitor changes in coccolithovirus (i.e. EhV) protein diversity (Pagarete *et al.*, 2014). The results confirmed earlier PCR studies (see above), demonstrating EhV genotype diversity decreases during a bloom, but the question of how high levels of diversity can be maintained in natural EhV communities still remains. Martínez Martínez *et al.* (2014) combined fluorescence activated cell sorting (FACS) and metagenomics and were able to separate the virus community into distinct size classes to enrich metagenomes for specific kinds of viruses, such as giant algal viruses. This methodology minimized cellular contamination, yet many reads from the virus metagenomes were still homologous to bacterial genes, strengthening the argument that the high proportion of cellular genes in metagenomes is an artefact due to the relatively low representation of reference viral genomes in databases. Nevertheless, libraries from the largest viruses based on flow cytometry were dominated by mimiviruses and phycodnaviruses, as expected, but genotypic richness (52–163 genotypes) was surprisingly much lower than observed in previous metagenomic studies, suggesting that this method could facilitate the assembly of individual virus genomes (Martínez Martínez *et al.*, 2014). Refinement of metagenomic flow cytometry coupled methods will undoubtedly permit exploration of algal virus diversity and ecology with unprecedented resolution.

Clearly, investigations of algal virus diversity in environmental nucleic acid samples have greatly expanded our knowledge of these globally important parasites. These studies would be much less enlightening, however, without sequence information from cultivated viruses. Indeed, the basis for identifying any gene obtained from an environmental sample is comparison to sequences from cultivated organisms. Thus, viruses that can be maintained and studied in the laboratory provide context for identifying gene sequences obtained through gene-specific PCR methods or metagenomics. Hence, the remainder of this discussion of algal virus diversity will focus on cultivated viruses – they are the basis for algal virus taxonomy and provide the most detailed glimpse into the biology of these fascinating microbial parasites.

Diversity of cultivated algal viruses

For the sake of simplicity, the following discussion of isolated viruses is organized around a hierarchical scheme based on the Baltimore classification system followed by formal taxonomy. The formal classification schemes discussed below are based on the most recent published report on the taxonomy of viruses (King *et al.*, 2012), as well as current (as of October 2016) and historical taxonomy releases from the ICTV (http://www.ictvonline.org/virusTaxonomy.asp). As illustrated throughout this review of algal virus diversity, however, the pace of virus discovery in general has far surpassed the rate of formal classification. The diversity of all then-known viruses noted in the first report of the ICTV (ca. 1971) included only two families of viruses with 43 genera and 290 species. This pales in comparison to a recent report which includes seven orders, 111 families, 609 genera, and 3701 species (http://www.ictvonline.org/taxonomyReleases.asp; website accessed July 19, 2016). Given this pace of virus discovery, it is not surprising that most of the algal viruses isolated to date have not been formally classified. There are many cases, however, where the presumed classification of a virus, based on genetic relationships and host ranges, seems relatively straightforward. For these cases, the unclassified viruses will be discussed along with their formally classified close relatives.

On the other hand, the taxonomy of many algal virus isolates is not clear-cut, and notions about algal virus taxonomy have matured and changed with the discovery of new viruses. For example, before the turn of the millennium algal viruses were generally considered to be dsDNA algal viruses and were tentatively considered phycodnaviruses. Following the discovery of mimivirus (La Scola *et al.*, 2003) and analysis of its genome (Raoult *et al.*, 2004) (see Chapter 11), and the more recent discovery of many other *Acanthamoeba*-infecting viruses, scientific perspectives on taxonomy of the larger group of dsDNA viruses (NCLDVs) that encompasses phycodnaviruses and mimiviruses have, however, rapidly changed. These discoveries (reveiwed in Abergel *et al.*, 2015) have led to the establishment of new families in the NCLDVs such as the *Marseilleviridae*, and to the notion that many dsDNA algal viruses are actually members of the 'extended *Mimiviridae*' (Fischer, 2016), and are not as closely related to the *Phycodnaviridae* as originally assumed. Hence, within the following section on dsDNA viruses we have included a section describing unclassified isolates. Furthermore, because many viruses have now been isolated but not characterized with respect to their genomic composition, we have concluded our discussion of cultivated algal viruses by considering those viruses whose genome types remain unknown. This is then followed with a section on ssDNA algal viruses and then RNA algal viruses.

Double-stranded DNA (dsDNA) viruses

Phycodnaviridae

The family *Phycodnaviridae* is the first formal classification adopted for any group of algal viruses. The name *Phycodnaviridae* was proposed in 1990 and was established as a family of polyhedral dsDNA viruses that infect *Chlorella*-like green algae with the virus PBCV-1 as a type species within the genus *Phycodnavirus*. This nomenclature stood until 1998, when the ICTV accepted the proposal to rename the genus *Phycodnavirus* to *Chlorovirus*, and added the new genera *Prasinovirus*, *Prymnesiovirus*, and *Phaeovirus* to the *Phycodnaviridae* (Pringle, 1998). In addition to these *four* genera, the family *Phycodnaviridae* also includes the genera *Coccolithovirus* and *Raphidovirus* that were both added in 2004. Broadly, phycodnaviruses are defined as large (ca. 100 to 200 μm diameter) icosahedral dsDNA viruses that infect algae, and that have genomes ranging from 160 to 560 kb (Dunigan *et al.*, 2006; Wilson *et al.*, 2009). Along with the recognized species of these genera, isolated viruses presumed to be members of these genera are discussed in more detail below.

Chlorovirus

As noted above, chloroviruses were the first algal viruses classified by the ICTV and they remain the only formally recognized freshwater viruses in the *Phycodnaviridae.* Viruses within this genus infect unicellular eukaryotic *Chlorella*-like algae that exist symbiotically with the protozoan *Paramecium bursaria*, the coelenterate *Hydra viridis*, or the heliozoon *Acanthocystis turfacea* (Bubeck and Pfitzner, 2005; Meints *et al.*, 1981; Van Etten *et al.*, 1982). In their symbiotic state these algae, or zoochlorellae, appear to be resistant to viral infection due to their enclosure in host vacuoles that exclude viruses (Dunigan *et al.*, 2006). The *Hydra viridis* endosymbiont has not been cultured independently from its host, so its viruses have only been isolated from cells released from *Hydra* (Kang *et al.*, 2005). Three strains of zoochlorellae that can be cultivated separately from their symbiont, however, have been used to isolate viruses; these particular *Chlorella* strains, NC64A, SAG 3.83, and Pbi, are now classified as *Chlorella variabilis*, *Chlorella heliozoae*, and *Micractinium conductrix*, respectively (Jeanniard *et al.*, 2013). *Chlorella* NC64A and *Chlorella* Pbi are both endosymbionts of *Paramecium bursaria* and the viruses that infect them are generally referred to as NC64A viruses and Pbi viruses, respectively (Kang *et al.*, 2005). For the sake of consistency in the names of viruses we list in Table 10.1 and Fig. 10.2, we have used CVM in place of Pbi viruses because CVM-1 was the first Pbi virus isolated. The *Acanthocystis turfacea* symbiont *Chlorella* SAG 3.83 is host to the most recently discovered *Chlorovirus* species, referred to as ATCV (Bubeck and Pfitzner, 2005). The list of recognized species within the genus *Chlorovirus* has expanded and contracted a number of times over the years, but in the 2015 taxonomy release the ICTV recognizes 19 species of *Chlorovirus* including the type species *Paramecium bursaria Chlorella virus 1* (PBCV-1) and 16 other NC64A viruses, as well *Hydra viridis Chlorella virus 1* (HVCV-1) and *Acanthocystis turfacea Chlorella virus 1* (ATCV-1). Although *Chlorella* Pbi viruses are distinct from NC64A viruses and are generally recognized as another species of *Chlorovirus* (e.g. Kang *et al.*, 2005; Van Etten and Meints, 1999; Yamada *et al.*, 2006), they are not yet formally classified by the ICTV.

The first observation of viruses infecting zoochlorellae stemmed from attempts to cultivate the *Chlorella*-like symbiont of *Hydra viridis* (Meints *et al.*, 1981). A subsequent study demonstrated that similar dsDNA viruses could be isolated from *Chlorella*-like symbionts from other sources of *Hydra* and from the protozoan *Paramecium* as well (Van Etten *et al.*, 1982). Interestingly, the source of these viruses, named HVCV-1 for the Hydra viridis-Chlorella virus and PBCV-1 for the Paramecium bursaria-Chlorella virus, remained unknown throughout these studies as only cells that were freshly liberated from the symbiont showed any signs of infection. Fortunately, unlike the *Hydra* symbiont, the *Paramecium* zoochlorellae (*Chlorella* strain NC64A) could be maintained in culture and used to screen water samples for lytic activity. Shortly thereafter, *Chlorovirus* host specificity was tested by inoculating NC64A and ten different free-living *Chlorella* species with PBCV-1 and lysis was observed for NC64A only (Van Etten *et al.*, 1985a). A separate study also determined that NC64A viruses could be readily isolated from freshwater bodies across the USA, demonstrating the widespread distribution of these viruses (Van Etten *et al.*, 1985b). The next discovery of viruses of *Chlorella*-like algae came from studies of German freshwaters using *Chlorella* Pbi, an ex-symbiont from a European strain of *Paramecium*, as the host (Reisser *et al.*, 1988). Reisser and colleagues (1991) found that the Pbi-infecting viruses (e.g. CVM) could not infect NC64A and vice versa, despite the fact that *Paramecium* could establish a symbiotic relationship with North American or European *Chlorella* strains. Hence, it was apparent that the determinants of virus specificity were not the same as the recognition factors for symbiosis (Reisser *et al.*, 1991). Most recently, viruses of the *Chlorella*-like symbionts of *Acanthocystis turfacea* were also isolated from a German freshwater pond. This research demonstrated that these viruses, called ATCV, were closely related to, but nonetheless distinct from PBCV-1, and that the host range of the chloroviruses were limited to the hosts used for isolation; Pbi viruses only infected *Chlorella* Pbi, NC64A viruses only infected *Chlorella* NC64A, and ATCV only infected *Chlorella* SAG 3.83 ex-symbionts of *A. turfacea* (Bubeck and Pfitzner, 2005). A number of studies have demonstrated that all of these viruses of symbiotic *Chlorella*-like algae can be readily isolated from freshwaters all around the world (Dunigan *et al.*, 2006; Kang *et al.*, 2005; Van Etten and Meints, 1999). Unexpectedly, since it is not known if zoochlorellae can exist outside of their hosts, chloroviruses have been observed at high abundance (e.g. $> 10^5$ plaque-forming units/ml) in natural waters despite the fact that green paramecia are relatively rare. It is possible therefore that chloroviruses have other hosts besides the strains used for their isolation (Kang *et al.*, 2005).

While the natural history of chloroviruses is poorly understood, they are arguably the best-studied algal viruses and many details of their structure and biology have been examined. The type species of *Chlorovirus*, PBCV-1, is the best-studied *Chlorovirus* and is the source of most information about *Chlorovirus* biology. As summarized by Van Etten and Meints (1999), PBCV-1 attaches to the surface of the NC64A cell wall, digests the wall at the point of attachment, injects its DNA into the cell, and leaves behind an empty capsid. Indirect evidence suggests that the DNA, and possibly some viral proteins, move quickly to the nucleus where early transcription can be detected within 5 to 10 minutes post infection (p.i.). Viral DNA replication begins after an hour. At two hours p.i. the host cell ultrastructure is altered and capsids begin to assemble at virus assembly centres localized throughout the cytoplasm. At 5 hours p.i. completed virions, presumably filled with genomic material, can be observed throughout the cytoplasm and most particles are released from the cell by 8 hours p.i.; typical burst sizes (number of viruses produced per infected host cell) range from 200 to 350 plaque-forming units/cell, but the number of particles released is presumably much higher since only 25–50% of particles are infectious. This bacteriophage-like mechanism of infection and lytic growth is also shared by the other known *Chlorella* viruses (Yamada *et al.*, 2006).

Genome sequencing of chloroviruses began with portions of the PBCV-1 genome nearly 20 years ago (Li *et al.*, 1997), and a decade later genome sequences were obtained for at least one representative of each of the species of chlorovirus, NC64A, SAG 3.83, and Pbi, that infect each cultivated chlorella ex-symbiont (Fitzgerald *et al.*, 2007a,b,c). These and related studies have revealed the fascinating features and coding potential of chlorovirus genomes, which contain roughly 330 to 380 protein coding sequences (Jeanniard *et al.*, 2013). As a few examples, these algal viruses encode DNA methyltransferases and DNA site specific endonucleases (i.e. type II restriction endonuclease systems), and are the only viruses known to encode components required to glycosylate their own proteins and synthesize polysaccharides. They also encode several tRNAs. Some *Chlorella* virus-encoded proteins are among the smallest in their families including DNA ligases and topoisomerases, and a K^+ channel protein, Kcv, that was the first virus encoded K^+ channel discovered. Interestingly, some of these viruses encode a protein (vSET) that methylates cellular histones and is packaged in virions, suggesting a mechanism for viral control of host transcription. These and other remarkable features of chloroviruses have been summarized in several excellent reviews (Dunigan *et al.*, 2006; Kang *et al.*, 2005; Yamada *et al.*, 2006).

Most recently, complete genome sequences for an additional 35 chloroviruses were obtained and an in-depth comparison was conducted for isolates from five continents with each subgenera, or species (i.e. NC64A, SAG 3.83, or Pbi infecting viruses), represented by at least 11 unique strains from at least two continents (Jeanniard *et al.*, 2013). Genomes ranged from 287 to 348 kb, with a range of sizes within each subgenera, but GC content was on average highest for the SAG viruses (49%) and lowest for NC64A (40%) with the Pbi viruses between. Furthermore, genome collinearity as well as a phylogeny based on concatenated sequences of 32 core protein families demonstrated that chlorovirus genomes clustered cleanly into the three known species: Pbi viruses, NC64A viruses (PBCV-1 and relatives), and SAG viruses (ATCV-1 and relatives), but the relationship among chlorovirus species was not echoed among the hosts, demonstrating that host-virus coevolution did not drive chlorovirus divergence. A single Pbi virus isolate, NE-JV-1, was distinct from the other three species with only 73.7% amino acid identity to the other Pbi viruses compared with averages of 92.6%, 95.0%, and 97.4% identity for the other NC64A, SAG and Pbi viruses, respectively. Hence, NE-JV-1 may represent a previously unknown subgroup of chloroviruses, one that serves to highlight our limited knowledge of the full extent of their diversity. With respect to the CDSs (coding sequences) themselves, their lack of homology to host genes suggests that gene capture, or horizontal gene transfer (HGT) of host genes was a rare event in the evolutionary history of chloroviruses. Fascinatingly, more than half of the gene families in chloroviruses originated after the chlorovirus species diverged and many of these have no known homologues in public databases, or even among the other chlorovirus species, leaving questions about the genomic origins of chlorovirus CDSs unanswered (Jeanniard *et al.*, 2013).

Research on the chloroviruses, undoubtedly the best studied algal viruses, has revealed the incredible complexity and coding capacity of the phycodnaviruses. Despite a wealth of information about the biochemistry, metabolic capabilities, genomics, and life histories of chloroviruses, it appears that many novel chloroviruses await discovery, and we cannot yet fully comprehend their evolutionary history. An even more fundamental aspect of chloroviruses, the identity of their hosts in natural waters, is not yet well established; it seems likely that there are other hosts for these viruses that have been isolated using permissive ex-symbiotic zoochlorellae hosts. Hence a deeper understanding of their role in aquatic ecosystems likely awaits researchers studying this fascinating group of algal viruses.

Coccolithovirus

The genus *Coccolithovirus* consists of a group of viruses that infect *Emiliania huxleyi*, a unicellular haptophyte alga (algal division Haptophyta, also referred to as Prymnesiophyta). Haptophyte algae such as *E. huxleyi* produce calcium carbonate coccoliths and can form vast coastal and mid-oceanic blooms (Holligan *et al.*, 1993). *E. huxleyi* has a global distribution from the tropic to subarctic waters, and plays a key role in CO_2 cycling and dimethyl sulphide production, thus making it a globally important species in marine primary productivity, sediment formation, and climate (e.g. Charlson *et al.*, 1987; Franklin *et al.*, 2010; Westbroek *et al.*, 1993). The first indication that viruses could play a role in the bloom dynamics and ecology of *E. huxleyi* came from observations of abundant virus-like particles following the demise of blooms in mesocosms manipulated with nutrients (Bratbak *et al.*, 1993).

Coccolithoviruses were isolated independently by different research teams and were found in numerous samples from the English Channel and Norwegian coastal waters (Castberg *et al.*, 2002; Schroeder *et al.*, 2002; Wilson *et al.*, 2002). Wilson *et al.* (2002) initially reported the isolation of two strains of a virus (EhV-84 and EhV-86) that could infect *Emiliania huxleyi* strain CCMP1516. A subsequent study of 10 different EhV strains found that they had similar host ranges, infecting at least three strains of *E. huxleyi* (CCMP374, CCMP1516, and strain L), but some *E. huxleyi* strains (CCMP 370, 373, and 379) were not susceptible to infection (Schroeder *et al.*, 2002). Similarly, Castberg *et al.* (2002) tested the lytic activity of three EhV isolates (EhV-99B1, EhV-2KB1, and EhV-2KB2) and found that they all infected the three strains of *E. huxleyi* tested, but other algae including *Phaeocystis pouchetii*, *Pyramimonas orientalis*, *Chrysochromulina ericina*, and *Micromonas pusilla* were not susceptible. Phylogenetic analysis of *E. huxleyi* virus *polB* genes demonstrated a close affiliation with other phycodnaviruses (Castberg *et al.*, 2002; Schroeder *et al.*, 2002). It was suggested, however, that the *E. huxleyi* viruses represent a novel *Phycodnaviridae* genus that was not closely related to *Prymnesiovirus*, a genus of viruses which also infect haptophyte algae (described below). Nevertheless, phylogenetic analysis of several NCLDV core genes of EhV-86 corroborated its placement in the *Phycodnaviridae* (Allen *et al.*, 2006b).

E. huxleyi viruses were classified by the ICTV in 2004 as a new *Phycodnaviridae* genus, *Coccolithovirus* with *Emiliania huxleyi virus 86* (EhV-86) designated as the type species. Although 12 other tentative species were listed in the proposal, EhV-86 is the only recognized *Coccolithovirus*. The EhV-86 virion is between 170 and 175 nm in diameter, and encapsidates a circular dsDNA genome of 407,339 bp (Wilson *et al.*, 2005). Within this large genome are 25 of the 40 to 50 conserved genes within the NCLDVs, but EhV-86 also encodes six RNA polymerase subunits, all of which are expressed, a unique feature within the *Phycodnaviridae*. This, in combination with a novel promoter, could indicate that EhV-86 encodes its own transcription machinery, allowing transcription to occur within the host cytoplasm. Additionally, at least four genes within EhV-86 have been linked to sphingolipid biosynthesis (Wilson *et al.*, 2005). Sphingolipid biosynthesis produces ceramide, a compound that suppresses cell growth and signals apoptosis. This, in combination with the fact that EhV-68 also encodes eight proteases, which are also linked to ceramide-induced apoptosis, suggests that the virus induces host cell apoptosis for progeny dissemination (Frada *et al.*, 2008; Wilson *et al.*, 2005).

Nearly complete genome sequences have been determined for several other virus strains including EhV-18, EhV-84, EhV-88, EhV-145, EhV-156, EhV-163, EhV-164, EhV-201, EhV-202, EhV-203, EhV-207, and EhV-208. This work has revealed not only high levels of homology, but also considerable CDS diversity among these isolates and a fascinating complement of genes such as those encoding tRNAs, methyltransferases, glycosylases, and a sphingolipid biosynthesis pathway, for example (Allen *et al.*, 2006a; Nissimov *et al.*, 2011a,b, 2012a,b; Pagarete *et al.*, 2013). Over the last decade *Coccolithovirus* has emerged as one of the most thoroughly studied algal virus model systems; coupled with available genomic information, studies of these viruses in the wild and in manipulated mesocosm experiments will undoubtedly help drive advances in algal virus research.

Phaeovirus

As of the 2015 release of viral taxonomy from ICTV, genus *Phaeovirus* contains nine species that all infect reproductive cells of certain macrophytic brown algae (class Phaeophyceae): *Ectocarpus fasciculatus virus a* (EfasV-a), *Ectocarpus siliculosus virus 1* (EsV-1), *Ectocarpus siliculosus virus a* (EsV-a), *Feldmannia irregularis virus a* (FirrV-a), *Feldmannia species virus* (FsV), *Feldmannia species virus a* (FsV-a), *Hincksia hinckiae virus a* (HincV-a), *Myriotrichia clavaeformis virus a* (MclaV-a), and *Pilayella littoralis virus 1* (PlitV-1). EsV-1 has been designated as the type species, and at least one other virus closely related to members of the genus *Phaeovirus* has been isolated, but is not classified by the ICTV.

The unclassified, putative *Phaeovirus*, Feldmannia simplex virus 1 (FlexV-1), infects the filamentous brown algae *F. simplex*. It has an icosahedral capsid with a diameter of ca. 120 to 150 nm and a 220 kb dsDNA genome (Friess-klebl *et al.*, 1994). Similarly, all phaeoviruses have icosahedral capsids ranging from 120 to 180 nm in diameter and dsDNA genomes that range in size from ca. 150 to 340 kb (Kapp *et al.*, 1997; Maier *et al.*, 1998). Unique among members of family *Phycodnaviridae*, phaeoviruses infect macroalgae, specifically brown filamentous algae belonging to the order Ectocarpales. Generally, the unicellular gametes or swimming spores of these filamentous brown algae lack cell walls and are susceptible to viral infection while the mature, multicellular brown algae are resistant to new infections. While initial infection occurs in the unicellular gamete or spore, virus multiplication occurs in the gametangia or sporangia tissues of mature algae (Kapp *et al.*, 1997). Additionally, the only reports of latent infections in the family *Phycodnaviridae* are of phaeoviruses, where evidence for a latent lifestyle stems from observations of Mendelian segregation of viral genomes following meiosis (Müller, 1991). Like most algal viruses, host specificity in phaeoviruses is generally rigid, with most studies reporting viruses infecting only a single host species. EsV-1, however, can infect and successfully replicate in the brown algae *Kuckuckia kylinii* (Müller, 1992) as well as in its main host, *E. siliculosus*. EsV-1 can also infect *F. simplex* (based on observation of abnormal morphological symptoms), although TEM observations did not reveal any intact, recognizable virions (Müller and Parodi, 1993). EfasV can similarly infect *E. siliculosus* (based on both observed pathologic symptoms and PCR detection of a viral gene fragment), although the virions produced were apparently not released and, moreover, were not infectious; attempted infection experiments using the infected thalli as virus donors were unsuccessful (Müller *et al.*, 1996).

The complete genome sequences of EsV-1 (Delaroque *et al.*, 2001) and FsV-158 (Schroeder *et al.*, 2009) and a partial genome of FirrV-1 (Delaroque *et al.*, 2003) have been determined. Interestingly the viruses infecting *Feldmannia* sp. (FsV) exist in two forms that have genomes of either 158 or 178 kb. The genome size of the smaller FsV, FsV-158 (the entire genome is actually 154,641 bp), makes it one of the smallest genomes in all of *Phycodnaviridae*, and, moreover, it shares only ten core genes with other NCLDVs, far less than the typical NCLDV complement of 31 core genes. FsV-158 has the fewest CDSs of any phycodnavirus (150), but almost 90% of these have orthologues in FirrV-1 and EsV-1. On the other hand, as is the case for most large dsDNA algal viruses, only a small proportion of these CDSs (25%) have significant homologies to genes in sequence databases that have known cellular functions including DNA replication, recombination, repair, and modification; integration and transposition; transcription; nucleotide metabolism; protein and lipid synthesis, modification, and degradation; and signalling (for example, Schroeder *et al.*, 2009).

Phylogenetic relationships of the core genes shared among these sequenced phaeoviruses have confirmed the recent evolutionary history of the phaeoviruses relative to other NCLDVs. Furthermore, the phylogenetic relationships between phaeoviruses and other phycodnavirus genera reflect the evolutionary relationships of their host algae (Schroeder *et al.*, 2009), suggesting an evolutionary history of phycodnaviruses that is at least as old as the lineages of algae they infect; the separation of Chlorophyte and Phaeophyte algae is estimated at ca. 1.5 billion years ago (Yoon *et al.*, 2004), so it is reasonable to assume that the common ancestor of some phycodnaviruses, in particular *Chlorovirus* and *Phaeovirus*, is at least this old. Like other types of algal viruses, the full extent of *Phaeovirus* diversity and the breadth of hosts they infect cannot yet be fully appreciated. Nevertheless, it is easy to imagine that this ancient group of viruses has had an important role in the speciation and microevolution of at least some types of brown algae.

Prasinovirus

According to the ICTV, the genus *Prasinovirus* currently includes two species: the type species *Micromonas pusilla virus SP1* (MpV-SP1) and *Ostreococcus tauri virus OtV5*. In 1998, *Prasinovirus* was named as the second genus in the *Phycodnaviridae* with MpV-SPI as the sole species (Pringle, 1998), and OtV5 was added as a second species in 2011. Although there are only two species formally classified as *Prasinovirus*, it is clear that there are many other viruses infecting prasinophyte algae that are similar in morphology and genome size to MpV-SP1 and OtV5, and they share closely related homologues of the NCLDV core gene, *polB*. All of these putative prasinoviruses infect algae within the order Mamiellales including the genera *Micromonas*, *Ostreococcus*, and *Bathycoccus*, so it seems apparent that these closely related viruses should all be classified as prasinoviruses and, hence, they will be discussed in the following paragraphs. To clarify the term Prasinophyceae, this formerly recognized class of chlorophyte algae (i.e. green algae) has been shown to be polyphyletic and classification schemes within the green algae are the subject of ongoing debate (Brodie and Lewis, 2007). According to the Taxonomy Browser on the National Center for Biotechnology Information (NCBI), U.S. National Library of Medicine website (www.ncbi.nlm.nih.gov/taxonomy/; accessed 20 July 2016), the phylum Chlorophyta includes the unranked group 'prasinophytes' within which the class Mamiellophyceae and the order Mamiellales reside. All of the hosts of the prasinoviruses are species within the order Mamiellales.

As noted in the introduction, the first reports of the isolation of a eukaryotic algal virus described viruses that infect the marine prasinophyte alga *Micromonas pusilla* (Mayer and Taylor, 1979; Waters and Chan, 1982). A decade later, Cottrell and Suttle (1991) isolated seven genetically distinct strains of *M. pusilla* viruses from coastal waters of British Columbia, California, New York, and Texas, as well as the central Gulf of Mexico. A few years later, viruses that were morphologically similar to MpV-SP1 were isolated from the North Atlantic, and were highly specific for *M. pusilla*; 11 other diverse microalgal species were challenged with these viruses, but no infection was observed (Sahlsten, 1998). Similar viruses were also isolated from the Gulf of Naples in the Mediterranean Sea (Zingone *et al.*, 1999) suggesting that MpVs, like their hosts (Slapeta *et al.*, 2006), are ubiquitous in the world's oceans. For this study of MpVs in the Mediterranean, five different strains of the host *M. pusilla* were used to titre MpVs revealing intraspecific variability in the host's susceptibility to MpV (Zingone *et al.*, 1999). Eventually, two novel *M. pusilla* viruses, MpVN1 and MpVN2, were isolated by Zingone *et al.* (2006) from the Gulf of Naples. Surprisingly, these viruses were morphologically and genetically distinct, one featuring a tail-like structure, and they had different patterns of infectivity against the 11 strains of *M. pusilla* isolates that were screened. Based on amplification with the conserved algal virus *polB* PCR primers, MpVN2 seems related to known prasinoviruses, but the primers were unsuccessful on MpVN1, suggesting that it might not be closely related (Zingone *et al.*, 2006). Clearly, MpVs are widely distributed in nature, and the extent of their diversity is not yet known.

Following the isolation of these *M. pusilla* viruses, the next *Prasinovirus* isolate reported in the literature was OtV5, the second formally recognized species of *Prasinovirus* that infects the Mamiellophyceae *Ostreococcus tauri*. Like the MpVs, OtV5 appeared to be highly host specific; the virus did not lyse any other algal species tested including other strains of *Ostreococcus tauri* (Derelle *et al.*, 2008). A more recent study by Clerissi *et al.* (2012) demonstrated that most but not all isolates of OtV are species or strain specific. Although most (34) of the 40 virus strains examined were *O. tauri*-specific, the others were able to lyse other species of *Ostreococcus* such as *O. lucimarinus* or *O. mediterraneus* and one viral isolate was able to lyse all three *Ostreococcus* species. None of the viruses, however, could infect the strains of the other Mamiellophyceae, *Micromonas* and *Bathycoccus*, that were tested (Clerissi *et al.*, 2012). Nevertheless, novel prasinoviruses have been isolated from a strain of *Micromonas pusilla* (strain RCC1109) other than the one originally used to isolate MpVs (strain UTEX 991, or CCMP 491).

In general, prasinoviruses appear to be widespread in nature. For example, hundreds of *Ostreococcus* viruses that can be distinguished based on sequence information or host specificity have been obtained from marine locations around the world including the South and North Pacific Ocean,

North Atlantic Ocean, Mediterranean Sea, coastal Argentina, and the English Channel (Bellec *et al.*, 2010; Derelle *et al.*, 2008; Manrique *et al.*, 2012; Weynberg *et al.*, 2009). That said, it is likely that some prasinoviruses are cosmopolitan while others are geographically constrained. A recent study of the global distribution of *Ostreococcus* viruses provided the observation that *O. tauri* viruses were restricted to coastal lagoons whereas *O. lucimarinus* viruses were detected in distant ocean samples from across the planet. Interestingly, there was no evidence for biogeography among *O. lucimarinus* viruses, i.e. no relationship was found between the phylogenetic and geographic distances of individual virus strains (Bellec *et al.*, 2010).

Besides *Micromonas* and *Ostreococcus*, another prasinophyte genus is a host to viruses presumed to be prasinoviruses. Samples collected from the Mediterranean Sea over a 2-year period harboured diverse viruses that could infect several mamiellophycean algae including *Micromonas*, *Ostreococcus*, and *Bathycoccus* (strain RCC1105) (Bellec *et al.*, 2009). Based on their partial *polB* sequences, all of these viruses were clustered among the formally recognized prasinoviruses MpV-SP1 and OtV5, suggesting that they are all bona fide prasinoviruses. The novel *Micromonas* viruses that grouped as their own distinct clade separate from MpV were named MiV as they were genetically distinct from other MpVs. Likewise, the *Bathycoccus* viruses, named BpV, also formed a distinct clade suggesting that they should be recognized as a new species of *Prasinovirus*. Recent discoveries of diverse, novel prasinoviruses such as the ones noted here highlight a recurring theme in algal virus research, that the extent of algal virus diversity in general, and prasinoviruses specifically, is largely unknown.

Prymnesiovirus

Describing the diversity of *Prymnesiovirus* is challenging because many different viruses infecting prymnesiophyte algae (also interchangeably referred to as haptophytes) have been isolated, but so far only one, CbV-PW1, has been formally classified. Because of their dsDNA genomes and similar morphology, most viruses of prymnesiophytes were originally assumed to be phycodnaviruses, but it is now apparent that while they are all NCLDVs, they do not all share a recent common ancestor and only some should be classified within the *Phycodnaviridae*. It is also noteworthy that the *Coccolithovirus* EhV (described above) also infects a haptophyte algal species and is classified as a phycodnavirus, but it is more closely related to other phycodnaviruses which do not infect haptophytes. Although it seems certain that many prymnesiophyte viruses are not species of *Prymnesiovirus* and await formal classification, it would be an oversight to not mention their discovery within this section. Hence, to recognize the history of and progress in this area of algal virus research, the unclassified algal viruses of prymnesiophytes will be briefly described here, and their taxonomic status will also be noted below in the section 'Unclassified dsDNA algal viruses'.

The type species of *Prymnesiovirus*, CbV-PW1, is an icosahedral virion ca. 170 nm in diameter that infects the marine haptophyte *Chrysochromulina brevifilum*, a member of the algal class Prymnesiophyceae. Two strains of CbV (PW1 and PW3) were isolated from numerous seawater samples collected from coastal Texas, USA. Interestingly, lytic viruses for *C. brevifilum* were detected at all sampling locations, but were only present at certain times of the year (Suttle and Chan, 1995), thus providing some of the earliest evidence for seasonality of algal viruses. The next haptophyte virus discovered was PpV01, a virus with phycodnavirus morphology from Norwegian fjords that infected two strains of *Phaeocystis pouchettii* but did not infect strains of *P. globosa* or *P. antarctica* (Jacobsen *et al.*, 1996). Soon after, two viral isolates were described that infected the haptophyte *Chrysochromulina ericina* (CeV-01B and PoV-01B). Surprisingly, one of these infected the prasinophyte algae *Pyramimonas orientalis* as well and was thus named PoV (Sandaa *et al.*, 2001). PoV and CeV were tested against a range of other algae and were not infectious, but PoV was one of the first algal viruses reported to infect more than a single species of algae. Both of these viruses were morphologically similar to phycodnaviruses and were thought to belong with the *Phycodnaviridae*, but it was notable that they had much larger genomes (> 500 kb) compared to typical phycodnaviruses (Sandaa *et al.*, 2001).

Following the discovery of CeV and PoV, 24 strains of viruses infecting the haptophyte *Phaeocystis globosa* were isolated from Dutch coastal waters and a subset were genetically characterized (Brussaard *et al.*, 2004). Based on *polB* sequences, seven

P. globosa viruses (PgV) clustered most closely with CbV-PW1 suggesting they belonged in the *Phycodnaviridae*. Also, it was noted in that study that conserved phycodnavirus PCR primers for *polB*, an important phycodnavirus marker gene (see 'PCR studies' above), did not amplify genes for some types of prymnesiophyte-infecting viruses such as PpV (Brussaard *et al.*, 2004), suggesting that they were taxonomically distinct from prymnesiophyte-infecting phycodnaviruses. Further studies on twelve other Dutch isolates of PgV demonstrated that there were actually two groups of morphologically distinct viruses that infected twelve *P. globosa* strains with variable host ranges, but they did not infect any of the other 27 algae tested. The so-called Group I PgVs had larger genomes and capsid diameters (466 kb and 150 nm, respectively), whereas Group II PgVs had 177 kb genomes packaged in 100 nm virions (Baudoux and Brussaard, 2005). To further complicate the natural history of *Phaeocystis globosa* viruses, another virus was isolated from the UK (PgV-102P) that, based on *polB* sequences, did not cluster with the PgVs characterized by Brussaard *et al.* (2004), but was instead most closely related to CbV-PW1 (Wilson *et al.*, 2006).

Most recently, three new haptophyte viruses, HeV RF02, PkV RF01, and PkV RF02, that infect *Haptolina ericina* and *Prymnesium kappa* (formerly called *Chrysochromulina ericina* and *C. kappa*, respectively) were isolated from Norwegian waters (Johannessen *et al.*, 2015). Two of these viruses, HeV RF02 and PkV RF01, could infect both of these haptophyte species, but they did not infect other haptophytes. Nonetheless, these viruses represent one of the few cases of an algal virus that can infect multiple species of hosts, in this case belonging to two different genera (Johannessen *et al.*, 2015). Around the same time, the first freshwater haptophyte virus was isolated from Lake Ontario, Canada (Mirza *et al.*, 2015). This virus, CpV-BQ1, that infects the freshwater alga *Chrysochromulina parva*, has typical phycodnavirus morphology and its *polB* sequence clusters with CbV-PW1 suggesting that it should be classified in the *Phycodnaviridae*. On the other hand, the fact that CpV's major capsid protein sequences and large genome size (approximately 485 kb) are more similar to the Group 1 PgVs and the other haptophyte viruses (CeV, PpV, HeV, and PkV) and are not most closely related to CbV-PW1 blurs its taxonomic affiliation (Mirza *et al.*, 2015).

Traditionally, viruses like CeV and PoV that had large dsDNA genomes and infected haptophyte algae were tentatively classified as phycodnaviruses (Sandaa *et al.*, 2001). Others, such as PpV were not classified at the time of their discovery (Jacobsen *et al.*, 1996), mainly because the genus *Prymnesiovirus* had not been established by the ICTV until 1998. As noted below in the section 'Unclassified dsDNA algal viruses', the discovery of other types of NCLDVs and the advent of viral genomics has helped refine the taxonomic relationships of many dsDNA algal viruses.

For example, with respect to its *polB* sequence, PpV is undoubtedly an NCLDV but it does not cluster within the *Phycodnaviridae*. Rather, it clusters among other types of large dsDNA viruses (Larsen *et al.*, 2008; Monier *et al.*, 2008b) and should be considered a member of the so-called 'extended *Mimiviridae*' (Fischer, 2016). Likewise, ongoing research has demonstrated that all of the haptophyte viruses mentioned in this section are related and can all be considered NCLDVs, or 'megaviruses', but only some are closely related to the *Prymnesiovirus* type species CbV-PW1 and can be considered members of the *Phycodnaviridae*. Although some PgVs, namely Group II isolates, are closely related to CbV (Brussaard *et al.*, 2004; Santini *et al.*, 2013), the genome sequence of the PgV Group I isolate 16T demonstrated that this isolate shared features found only in mimivirus genomes. Moreover, based on its *polB* sequence, PgV-16T clustered with the other prymnesiophyte-infecting viruses that are most closely related to mimiviruses, namely CeV and PoV (Santini *et al.*, 2013).

To summarize, the genus *Prymnesiovirus* includes only one recognized species, CbV-PW1, but based on *polB* phylogenies several other prymnesiophyte-infecting viruses such as PgV Group II, and possibly CpV-BQ1 appear to be closely related and could be classified within this genus. On the other hand, several prymnesiophyte-infecting viruses with large genomes, greater than 400 kb such as CeV, HeV, PgV Group I, PkV, PoV, and PpV, are more closely related to *Mimivirus* and should not be considered as members of the *Phycodnaviridae*.

Raphidovirus

The genus *Raphidovirus* currently consists of only one species, HaV (Heterosigma akashiwo virus 01), which infects *Heterosigma akashiwo*, a widespread microalga that is found in coastal waters of the temperate and subarctic Atlantic and Pacific Oceans. Importantly, *H. akashiwo* has been associated with recurring red tides in Japan, aquaculture losses in British Columbia, Washington, Chile, and New Zealand (Hallegraeff and Hara, 2004; Tomaru *et al.*, 2008), and its lytic viruses have been implicated in bloom termination (Nagasaki *et al.*, 1994a; Nagasaki *et al.*, 1994b). A virus infecting *H. akashiwo* was initially isolated using a single host strain, but its host range was tested by screening lysates against *four* other strains of *H. akashiwo* and a range of other species belonging to several different classes of algae. Only two strains of *H. akashiwo* were sensitive to the virus that was described as a 202 nm diameter naked icosahedral virion with a dsDNA genome (Nagasaki and Yamaguchi, 1997). Shortly after, several different strains of morphologically similar viruses of *Heterosigma* were isolated and the strain HaV01 was characterized in greater detail revealing that the virus has a latent period of 30 to 33 hours with mature particles appearing within the host cytoplasm 24 hours post infection, and approximately 770 infectious particles released per lysed cell (Nagasaki *et al.*, 1999). Based on pulsed field gel electrophoresis, the dsDNA genome of HaV01 was approximately 294 kb, and sequence analysis of the virus's *polB* gene revealed its close taxonomic affiliation with other dsDNA algal viruses within the *Phycodnaviridae* (Nagasaki *et al.*, 2005). The genome is now partially sequenced and annotated, and many of its genes are highly similar to other phycodnavirus or, interestingly, mimivirus genes (Tomaru *et al.*, 2008). Although Heterosigma akashiwo virus 01 is recognized by the ICTV as a member of the *Phycodnaviridae*, it is worth noting that another dsDNA virus of *H. akashiwo* has been isolated from British Columbia, Canada, coastal waters. The Canadian isolate, named OIs1, has only been partially characterized and may not exist in pure culture, but since it has a smaller genome and particle size than HaV (Lawrence *et al.*, 2006), it is not likely that OIs1 is related to *Raphidovirus*.

Ecological studies of HaV demonstrated that Japanese waters contained numerous strains of both *H. akashiwo* and its co-occurring lytic viruses, and complex spectra of host-strain susceptibility and virus-strain infectivity has been observed during several studies. Thus, HaV infectivity is beyond species specific, but is actually strain specific and moreover, depending on the host–virus pair, not all infections lead to complete lysis of the host culture (Nagasaki and Yamaguchi, 1998). Together, complicated patterns of lytic activity with some host strains displaying some degree of resistance to infection as well as complex virus–host succession patterns observed during the course of *H. akashiwo* blooms demonstrate that this lytic virus plays an intimate but incredibly complex role in the ecology of its harmful bloom-forming algal host (Tarutani *et al.*, 2000; Tomaru *et al.*, 2008; Tomaru *et al.*, 2004). The fact that there are multiple types of viruses infecting *H. akashiwo*, such as *Marnavirus* (an ssRNA virus described below), means that this virus–host consortia is even more complex and fascinating and could serve as an excellent model system to address questions related to competition among viruses.

Dinodnavirus, a dsDNA algal virus not assigned to a family

The proposed genus *Dinodnavirus* has one species, *Heterocapsa circularisquama DNA virus 01* (HcDNAV01). HcDNAV01 is one of two viruses isolated that infect the dinoflagellate *H. circularisquama*, and is thus far the only dsDNA virus isolated that infects any dinoflagellate species (Nagasaki, 2008). Genomic data (Correa *et al.*, 2016; Correa *et al.*, 2013) and evidence from electron microscopy, however, suggest that other dsDNA viruses related to NCDLVs are associated with endosymbiotic dinoflagellates (*Symbiodinium*) of scleractinian corals (Correa *et al.*, 2016; Lawrence *et al.*, 2014; Wilson *et al.*, 2001). Originally classified within *Phycodnaviridae*, HcDNAV01 has several characteristics in common with other NCDLVs such as cytoplasmic replication in a viroplasm (an organized virus replication inclusion body seen in infected cells) and large (180 to 210 nm dia.) icosahedral virions (Tarutani *et al.*, 2001) encapsidating a large (approximately 364 kb) dsDNA genome (Nagasaki *et al.*, 2005). Analysis of HcDNAV01's *polB* sequence suggests that it is most closely related to African swine fever virus (ASFV), the only member of the *Asfarviridae*, a distinct family of NCLDVs closely related to *Phycodnaviridae* (Ogata *et al.*, 2009). It was upon this realization that *Dinodnavirus* was proposed as a genus, but whether it is close enough to ASFV to warrant its classification as a genus in *Asfarviridae* is not yet known and will require further characterization.

So far, HcDNAV01 is known to infect only *H. circularisquama*; no lysis was observed when HcDNAV was used to challenge other types of algae including nine other species of Dinophyceae, four of Bacillariophyceae, two of Prymnesiophyceae, four of Raphidophyceae, and one each of Chlorophyceae, Cryptophyceae, Euglenophyceae, and Eustigmatophyceae (Tarutani *et al.*, 2001). HcDNAV01 however can infect multiple strains of *H. circularisquama* and successfully infects its hosts at a wide range of temperatures (15–30°C), has a latent period (time from initial infection until production of infectious virions) of 40–56 h, and a burst size of 1.8×10^3 to 2.4×10^3 infectious particles/cell (Nagasaki *et al.*, 2003). The ecological importance of HcDNAV01 has not been fully explored, but it has a widespread distribution along the western coast of Japan (Tomaru and Nagasaki, 2004) and often co-occurs with HcRNAV, suggesting an interesting ecological relationship between these two very different types of *H. circularisquama*-infecting viruses (Tomaru *et al.*, 2009).

Unclassified dsDNA algal viruses

As noted previously, many dsDNA viruses remain unclassified (Fig. 10.2). Whereas the taxonomic affiliation seems clear for some unclassified algal viruses, the evolutionary history and therefore the classification scheme is less clear or completely unknown for many others. For example, the dsDNA viruses Ols1 that infects the raphidophyte *Heterosigma akashiwo* (Lawrence *et al.*, 2006; Lawrence *et al.*, 2002) and MpVN1 that infects the prasinophyte *Micromonas pusilla* (Zingone *et al.*, 2006) were not characterized in detail beyond reporting their genome as dsDNA, so it is not yet known if these viruses are phycodnaviruses, or even members of the NCLDVs, or *Megavirales* (Colson *et al.*, 2013). For other viruses such as the Tetraselmis striata virus (TsV-N1), more detailed characterization has not provided a clear taxonomic affiliation. TsV is a small (60 nm dia.) virus with a 31 kb genome that is distinct from other dsDNA algae viruses allowing the conclusion that it is not a member of the *Phycodnaviridae*, but because this virus is genetically unique it cannot be placed within extant virus genera or families (Pagarete *et al.*, 2015).

As well, the viruses CeV, HeV, PgV Group I, PkV, PoV, and PpV that infect prymnesiophyte, or haptophyte algae should not be considered prymnesioviruses (see 'Pyrmnesiovirus' above) because they are more closely related to mimiviruses than the *Prymnesiovirus* type species CbV (see Chapter 11). Genomic characterization of the Aureococcus anophagefferens virus AaV provided clear evidence that it was member of the '*Megaviridae*' or 'extended *Mimiviridae*' (Moniruzzaman *et al.*, 2014) and is another example of a giant dsDNA algal virus within the NCLDVs but outside of the *Phycodnaviridae*. Similarly, the dinoflagellate-infecting virus HcV (*Dinodnavirus*, described above) has characteristics of NCLDVs, but does not appear to be a close relative of either the *Mimiviridae* or *Phycodnaviridae*. Instead, based on its *polB* sequence, HcV appears most closely related to *Asfarviridae*. It is notable that a recent analysis of DNA mismatch repair genes (*MutS*) in several dsDNA algal viruses including CeV, HcDNAV, PoV, and PpV (Ogata *et al.*, 2011) provided additional evidence supporting the hypothesis that these algal viruses are more closely related to mimiviruses than to phycodnaviruses. On the other hand, based on several diagnostic characteristics, the most recently discovered dsDNA dinoflagellate virus which infects *Heterocapsa pygmaea*, HpygDNAV, was considered a putative phycodnavirus (Kim *et al.*, 2012). Because its genome awaits detailed characterization, however, this suggested classification is tentative.

The taxonomic affiliation of some unclassified dsDNA algal viruses seems clear and places them within the *Phycodnaviridae*, yet they await formal classification by the ICTV. For the most part, host taxonomy, viral morphology, and sequences of the NCLDV core gene *polB* have provided evidence that these viruses are close relatives of species within established genera and have permitted their tentative classification. Most of these viruses were already discussed above in the sections describing their closest relatives within established genera. For example, as noted above in the section describing *Prasinovirus*, the prasinophyte-infecting viruses BpV, MiV, and MpVN2 that infect *Bathycoccus* and *Micromonas*, respectively, are most closely related to the *Prasinovirus* type species MpV-SP1 and should be classified as such. Likewise, the virus FlexV of the macrophytic brown algae *Feldmannia* is a putative

Phaeovirus, strains of *Phaeocystis globosa* virus in group II and possibly the freshwater haptophyte virus CpV are putative *Prymnesiovirus* species, and the *Chlorella* Pbi-infecting virus CVM-1 and its close relatives are most certainly species of *Chlorovirus*. Algal virus discovery over the last couple of decades has far outpaced the rate of virus classification, so algal virus researchers have assigned tentative classifications based on the known characteristics of the viruses they study. In many cases these informal classification schemes are well justified and it seems likely that they will eventually be adopted by the ICTV.

It is the hallmark of scientific progress that we can now look back at the early years of algal virus discovery and realize how naïve we were and how far the study of these fascinating parasites has come. It is now apparent that there are many more types of viruses that infect algae besides the dsDNA viruses of the *Phycodnaviridae* and, moreover, not even all dsDNA algal viruses are phycodnaviruses. Only through ongoing isolation and characterization efforts will the taxonomy of dsDNA algal viruses be well resolved. It is becoming more apparent with every new isolate that the 'tree' of algal viruses is actually a 'shrub' with many new viruses filling in the gaps between known taxa creating an almost continuous spectrum of algal viruses with shared characteristics.

Single-stranded DNA (ssDNA) viruses

Bacilladnavirus and similar viruses

Currently, all isolated algal viruses with ssDNA genomes infect various species of diatoms. The only formally classified ssDNA algal virus is the *Chaetoceros salsugineum nuclear inclusion virus* (CsalDNAV), isolated by Nagasaki *et al.* (2005c), and it is therefore the type species of the genus *Bacilladnavirus*. At least eight other genetically similar diatom viruses have been isolated, however, and proposed as additional species of *Bacilladnavirus*, including: CdebDNAV (Tomaru *et al.*, 2008b), ClorDNAV (Tomaru *et al.*, 2011b), CtenDNAV (Tomaru *et al.*, 2011a), TnitDNAV (Tomaru *et al.*, 2012), Csp05DNAV (Toyoda *et al.*, 2012), Csp07DNAV (Kimura and Tomaru, 2013), CsetDNAV (Tomaru *et al.*, 2013b), and CtenDNAV type II (Kimura and Tomarua, 2015). All of these viruses were isolated from Japanese coastal waters and, except for TnitDNAV which infects *Thalassiosira nitzschioides*, infect various *Chaetoceros* species. Only a single strain was identified for each virus with two exceptions: 118 CdebDNAV and eight CtenDNAV type I virus clones were isolated and although some of these had different infection ranges against the host strains tested, only CdebDNAV18 and CtenDNAV06 type I were thoroughly characterized. For all of these diatom viruses, infectivity is highly host specific and these viruses infect only a single diatom species, or only certain strains of an individual species. The virions of these ssDNA diatom viruses are icosahedral, 32-38 nm in diameter, and lack envelopes or tails. The particles assemble in the host nucleus where, interestingly, rod-shaped particles are often also observed (e.g. Kimura and Tomarua, 2015; Tomaru *et al.*, 2011a,b). Whether these rod-shaped particles represent a precursor involved in mature virion assembly and/or a co-infecting virus remains unclear.

Viral genomes in this group are composed of both a single molecule of covalently closed, circular ssDNA nucleic acid ca. 5500 to 6000 bp in length and a smaller (600–1000 bp) linear segment of ssDNA; CdebDNAV is an exception and instead harbours a ssDNA genome ca. 7 kb long with undetermined structure. Interestingly, in each virus the linear segment is complimentary to a region on the closed circle and forms a partially dsDNA region; the genome of CsetDNAV is another exception, however, as it includes eight rather than one complimentary ssDNA segments ranging from 67 to 145 bp in length. The functional significance of the dsDNA viral region remains unclear, although the region may contain an ORF (open reading frame) coding for an unknown protein and has also been suggested to serve as a primer for strand elongation during viral replication. Where determined, viral genomes have at least three ORFs coding for various viral proteins with unknown function (VP1), or are recognizable as a replication protein (VP3), or possibly the viral structural protein (VP2). Burst sizes and latent periods for these viruses vary, ranging from 29.1 infectious units/cell to 2.2×10^4 infectious units/cell, and from less than 12 to 96 h, respectively (Kimura and Tomarua, 2015). Importantly, since the genetic relationships inferred between these viruses vary depending on the viral ORFs used for analysis, the phylogeny among putative bacilladnaviruses is unclear and would benefit from further investigation (Kimura and Tomaru, 2013). Nonetheless, it is clear that all of these viruses are closely related to each other and only distantly related to other known viruses, such as avian circoviruses, and therefore they likely represent distinct taxa within an evolutionarily conserved group of diatom viruses.

RNA viruses

Historically, research on algal viruses has focused on dsDNA viruses, the earliest isolated algal viruses, but recent work (as described in the section 'Diversity inferred from environmental nucleic acids' above) has demonstrated that RNA viruses are also abundant in marine environments and their impacts on plankton mortality warrant further investigation (Chow and Suttle, 2015, Steward *et al.*, 2013). Currently, MpRV (Brussaard *et al.*, 2004a) remains the only dsRNA algal virus formally recognized by the ICTV, whereas several algal viruses with positive-strand (+)ssRNA genomes have been isolated, including viruses of raphidophyte algae such as HaRNAV (Tai *et al.*, 2003), the dinoflagellate virus HcRNAV (Tomaru *et al.*, 2004a), and various diatom viruses that infect species of *Rhizosolenia* (Nagasaki *et al.*, 2004), *Chaetoceros* (Kimura and Tomarua, 2015; Shirai *et al.*, 2008; Tomaru *et al.*, 2009b, 2013a) and *Asterionellopsis* (Tomaru *et al.*, 2012). Notably, no negative-sense ssRNA viruses infecting eukaryotic algae have been isolated to date.

Double-stranded RNA (dsRNA) viruses

Mimoreovirus

Viruses infecting the green alga *Micromonas pusilla* provided the first evidence for infective algal viruses in seawater (Mayer and Taylor, 1979), and some of the *M. pusilla* viruses represent the only known dsRNA viruses that infect eukaryotic algae (Brussaard *et al.*, 2004a). *M. pusilla* is a globally important primary producer throughout the world's oceans, and is a member of the class Prasinophyceae, a primitive chlorophyte that may be ancestral to all other green algae and land plant classes (Martínez Martínez *et al.*, 2015); isolations of viruses infecting this primitive algal host therefore could be important for understanding virus–host interactions over long evolutionary time scales (Brussaard *et al.*, 2004a).

Although most viruses isolated from *M. pusilla* are members of the family *Phycodnaviridae*, dsRNA viruses that also infect *M. pusilla* were isolated from coastal waters in Bergen, Norway (Attoui *et al.*, 2006; Brussaard *et al.*, 2004a). These viruses (MpRV) have small, ca. 90 to 95 nm diameter, icosahedral virions with thick outer layers surrounding smaller electron-dense inner cores. Viral infectivity is retained following treatment with chloroform suggesting that the virions lack a lipid component. MpRV encodes a 22,563 bp genome composed of 11 linear dsRNA segments ranging in size from 741 bp to 5,792 bp (Attoui *et al.*, 2006; Brussaard *et al.*, 2004a). Genome segments each encode a single ORF except segment 5, which may encode two related protein products. Hence, the MpRV genome encodes at least 11 protein products. The segmented dsRNA genome of MpRV is characteristic of *Reoviridae*, a family of viruses that infect a huge range of eukaryote hosts and that have 18,500 to 29,210 bp genomes composed of single copies of 9 to 12 linear dsRNA segments. *Reoviridae* capsids are non-enveloped, 60 to 100 nm in diameter, composed of one to concentric protein layers surrounding the genome, and exhibit icosahedral symmetry, all of which are characteristics of MpRV. Additionally, all genomic segments and the putative RNA-dependent RNA-polymerase (RdRp) of MpRV contain, respectively, conserved terminal sequences and the signature motifs of family RdRps, further supporting the notion that MpRV is a member of *Reoviridae* (Attoui *et al.*, 2006; Brussaard *et al.*, 2004a).

Among other factors, the number of genomic segments distinguishes *Reoviridae* genera, yet MpRV possesses features unusual among other *Reoviridae* members including unique termini sequences, an algal host, and an additional, constitutively expressed outer protein coat. These features led to the classification of MpRV as belonging to a genus of *Reoviridae* designated *Mimoreovirus* (Micromonas pusilla reovirus, MpRV). Phylogenetic analysis of MpRV RNA polymerase sequences demonstrated that it does not cluster with any known reoviruses (≤ 21% amino acid identity). Instead, the MpRV branch bisects the *Reoviridae*

RdRp phylogenetic tree into the two recognized *Reoviridae* subfamilies, the *Sedoreovirnae* within which the *Mimoreovirus* resides, and the *Spinareovirinae.* Based on its RdRp phylogeny and the older evolutionary age of *M. pusilla* compared with other *Reoviridae* hosts, MpRV may represent a third, possibly ancestral, branch of reoviruses (Attoui *et al.*, 2006).

Currently, MpRV is the first and only isolated dsRNA algal virus, thereby demonstrating the need for ongoing isolation efforts to more fully appreciate the diversity of algal viruses. Additionally, the co-occurrence of MpRV in a water sample with larger, 100 to 140 nm diameter dsDNA viruses that also infect *M. pusilla* is noteworthy (Brussaard *et al.*, 2004a). Although mechanisms allowing the co-existence of these two viruses infecting the same host remain unclear, the unique latent periods, burst sizes, infectivity of virions produced (the proportion of infectious particles to intact virions), and strategies to prevent superinfection (i.e. simultaneous infection of a single host cell by both MpRV and the dsDNA virus) are likely important factors. Ultimately, future work should investigate the presumed stable co-existence of different viruses infecting the same host to further understanding of the ecology of viruses infecting phytoplankton (Brussaard *et al.*, 2004a).

Single-stranded RNA (ssRNA) viruses

Marnavirus

Isolated from the Strait of Georgia, British Columbia, Canada, HaRNAV is a picornavirus (order: *Picornavirales*; family: *Marnaviridae*; genus: *Marnavirus*) that lyses the toxic bloom-forming alga *Heterosigma akashiwo* and is also the first ssRNA virus isolated that infects phytoplankton (Tai *et al.*, 2003). The host range of HaRNAV was tested using host strains obtained from distant locations such as coastal North America, Japan, and the Northwest Atlantic and demonstrated that the virus has a distinct host range compared with other *H. akashiwo* viruses. The fact that HaRNAV can infect host strains from both North American and Japanese waters suggest that this virus may be widespread in nature, at least within the Pacific Ocean (Tai *et al.*, 2003). Although the original classification of HaRNAV was to an unassigned family, the *Marnaviridae*, it was recognized as an ssRNA picorna-like virus. Subsequent work on picorna-like viruses led to the eventual creation of the order *Picornavirales* in 2005 and the inclusion of *Marnaviridae* as a family with the order (Le Gall *et al.*, 2008).

Infectivity of HaRNAV is host strain specific, and virions are icosahedral, approximately 25 nm in diameter, lack envelopes or tails, and undergo lytic release. Cytopathic effects such as endoplasmic reticulum swelling begin approximately 48 h after infection, and the latent period and burst size of HaRNAV are estimated to be 29 h and 2.1×10^4 viral particles/cell, respectively (Lawrence *et al.*, 2006). HaRNAV houses an approximately 8.6 kb positive sense ssRNA genome containing a single ORF that is polyadenylated at the 3′ terminus (Lang *et al.*, 2004). Both the 5′ and 3′ untranslated regions (UTRs) share the same 123 bp sequence and, interestingly, this pair of repeats may facilitate viral RNA replication or polyprotein translation. The HaRNAV polyprotein, in order from 5′ to 3′, is predicted to have the following protein domains: helicase, protease, RNA-dependent RNA polymerase, and structural viral proteins 2, 3, and 1. While sequencing of the genome revealed the relation of HaRNAV to viruses from the picorna-like superfamily (now recognized as the order *Picornavirales*), HaRNAV did not clearly belong to any defined families at the time of isolation. As such, the family *Marnaviridae* was established within which HaRNAV is currently the only member (Le Gall *et al.*, 2008).

Bacillarnavirus

Bacillarnavirus is a genus within the *Picornavirales* that has not yet been assigned to a family. Currently, three virus species are formally recognized within this genus including *Rhizosolenia setigera RNA virus 01* (RsetRNAV01), *Chaetoceros tenuissimus RNA virus 01* (CtenRNAV01), and *Chaetoceros socialis f. radians RNA virus 01* (CsfrRNAV01), but based on phylogenetic relationships of the replicase (i.e. ORF1 or RdRp) or structural protein (ORF2), it appears that the genus should also include the diatom viruses Csp03RNAV that infects *Chaetoceros* sp. (SS08-C03), AglaRNAV that infects *Asterionella glacialis,* and CtenRNAV type II that infects *Chaetoceros tenuissimus* (Kimura and Tomarua, 2015; Tomaru *et al.*, 2013a).

All of these RNA viruses were isolated from water bodies near western Japan and all infect *Chaetoceros* species except RsetRNAV and AglaRNAV, which infect *Rhizosolenia setigera* and *Asterionellopsis glacialis*, respectively. Interestingly, all viruses infect centric (radially symmetrical) diatoms except AglaRNAV, which is the first virus isolated that infects a pennate (bilaterally symmetric) diatom (Tomaru *et al.*, 2012). Virions are non-enveloped, non-tailed, icosahedral, approximately 22 to 36 nm in diameter, and assemble in the cytoplasm. Only a single viral strain was identified for each virus except RsetRNAV, where nine strains (RsetRNAV01–09) were obtained from the original water sample; only strains RsetRNAV01 and RsetRNAV06, however, were further characterized (Nagasaki *et al.*, 2004). The infectivity of most *Bacillarnavirus* members seems to be highly host strain specific; for instance, unique infectivity spectra were observed for each of the nine RsetRNAV strains isolated when they were tested against 25 strains of *R. setigera* (Nagasaki *et al.*, 2004). One exception is CtenRNAV type II which, unlike CtenRNAV type I (CtenRNAV), lyses four other *Chaetoceros* species in addition to *C. tenuissimus*. Virus latent periods are less than 48 h for CsfrRNAV and Csp03RNAV, less than 24 h for CtenRNAV type I, 24 to 48 h for CtenRNAV type II, and 48 h for RsetRNAV. The burst sizes for RsetRNAV, CtenRNAV type I, CtenRNAV type II, CsfrRNAV, and Csp03RNAV are 3.1×10^3, 1.0×10^4, 287, and 66 infectious units/cell, and unknown, respectively. Both the latent period and burst size for AglaRNAV have yet to be determined (Kimura and Tomarua, 2015). *Bacillarnavirus* members have linear, positive ssRNA genomes ranging from 8.4 to 9.6 kb in length with 3′ poly(A) tails and two ORFs, the first encodes the replicase – which includes the viral RNA helicase and RNA-dependent RNA polymerase – while the second encodes a structural polyprotein (Kimura and Tomarua, 2015; Nagasaki, 2008; Shirai *et al.*, 2006, 2008).

From an ecological perspective, it is interesting to note that at least four distinct virus species, CtenRNAV types I and II as well as CtenDNAV types I and II (see 'ss DNA viruses' above), have been isolated that share the same diatom host, *C. tenuissimus*. Importantly, like *Heterosigma akashiwo* and its viruses, *C. tenuissimus* and its viruses could serve as an important model system for studies of algal host–virus relationships in order to enhance understanding of competition between viruses and what impact this might have on their host dynamics (Kimura and Tomarua, 2015). Furthermore, given that several algae such as *Micromonas pusilla*, *Heterosigma akashiwo*, and *Chaetoceros tenuissimus*, for example, are known to be infected by multiple types of viruses, it is easy to speculate that this is a widespread phenomenon in algal ecology, and algal virus research concerning such intra–virus interactions will remain probably in a 'discovery phase' for many years to come.

Dinornavirus

HcRNAV, a virus that infects the dinoflagellate *Heterocapsa circularisquama* and was originally isolated from various sites in coastal western Japan, is the first ssRNA dinoflagellate virus isolated and is the only species of the newly established genus and family *Dinornavirus* and *Alvernaviridae*. Moreover, the family name *Alvernaviridae* was chosen to recognize the fact that *Dinornavirus* was the first virus isolated to infect an Alveolate, a broad group of protozoans that includes ciliates, apicomplexa, and dinoflagellates. HcRNAV is a tailless icosahedron particle approximately 30 nm in diameter that lacks an external membrane and replicates in the host cytoplasm by forming crystalline arrays or unordered aggregations. The latent period and burst size of HcRNAV are estimated to be 24 to 48 h and 3.4×10^3 to 2.1×10^4 infectious units/cell. The virus harbours a +ssRNA genome that is approximately 4.4 kb in size, lacks a 3′ poly(A) tail (instead having a 3′ stem–loop structure), and contains two ORFs: ORF-1 encodes a polyprotein containing protease and RNA-dependent RNA polymerase domains, while ORF-2 encodes a capsid protein gene within which four variable regions are present that determine strain-specific infectivity (Nagasaki *et al.*, 2005a).

More than 100 HcRNAV virus strains were isolated using four *H. circularisquama* host strains (HU9433-P, HA92–1, HCLG-1, or HY9423) and, after testing their infection spectra against 56 different host strains, were ultimately classified into two virus types, CY and UA, depending on whether they were isolated using either of the HCLG-1 and HY9423, or HU9433-P and HA92-1 host strain pairs, respectively. These viruses were also used to challenge 32 other algal species belonging to the families Bacillariophyceae, Chlorophyceae, Dinophyceae, Euglenophyceae, Eustigmatophyceae, and Raphidophyceae, but no infection was observed for any alga besides *H. circularisquama* (Tomaru *et al.*, 2004a). Strains HcRNAV34 (type CY) and HcRNAV109 (type UA) were further characterized demonstrating that these two viruses were 97.0% similar with respect to their nucleotide sequences, and also demonstrated their complimentary, non-overlapping host strain-specificity (Nagasaki *et al.*, 2005a). Interestingly, the co-existence of many distinct virus and host strains in a particular geographic region suggests that HcRNAV can influence the population dynamics of *H. circularisquama* in terms of both biomass and strain composition (Tomaru *et al.*, 2007). How these two distinct virus types evolved, were selected for, and can stably co-exist with each other as well as with HcV, a dsDNA virus also infecting *H. circularisquama* (see 'dsDNA viruses' above) remains unclear, however, and warrants further study (Tomaru *et al.*, 2007).

Genome classification unknown

Several eukaryotic algal viruses with unknown or unverified genome types, none of which have been classified by the ICTV, have been isolated and partially characterized. These include the nuclear inclusion viruses CspNIV (Bettarel *et al.*, 2005), CwNIV (Eissler *et al.*, 2009) and HaNIV (Lawrence *et al.*, 2001), as well as the diatom viruses ScosV (Kim *et al.*, 2015a) and SpalV (Kim *et al.*, 2015b), and the virus TampV (Nagasaki *et al.*, 2009) which infects a Cryptophyte alga.

The nuclear inclusion viruses CspNIV, CwNIV, and HaNIV are all characterized by small virus particle sizes (approximately 20–30 nm in diameter), viral replication within the nucleoplasm of their host cells, and formation of crystalline or paracrystalline arrays of virus-like particles during infection. Both CspNIV and CwNIV were isolated from the Chesapeake Bay, USA and lyse the diatoms *Chaetoceros* cf. *gracilis* and *Chaetoceros* cf. *Wighamii*, respectively (Bettarel *et al.*, 2005; Eissler *et al.*, 2009). Infectivity of the viruses appears to be specific to these species, although strain-specificity is unknown. HaNIV was isolated from coastal British Columbia, Canada and lyses the Raphidophyte *Heterosigma akashiwo* (Lawrence *et al.*, 2001). Infectivity is host-specific, lysing some strains of *H. akashiwo* but not others. All three viruses are icosahedral, although in the case of CwNIV-infected cells, virus-like particles are also observed in a rod-like structural arrangement. It has been suggested that CwNIV virus maturation involves the formation of such rod-like structures (observed in cross-section as paracrystalline arrays) which fragment to produce free virus particles (as no paracrystalline structures have been observed in lysates), although it is also possible that two different viruses may instead co-infect the same cell. The latent periods for CspNIV and CwNIV are estimated to be less than 24 and 8 h, respectively. The burst size for CwNIV is estimated to be approximately 2.64×10^4 viruses/cell.

ScosV and SpalV are diatom viruses, isolated from Jaran Bay, Korea, which lyse *Skeletonema costatum* and *Stephanopyxis palmeriana*, respectively (Kim *et al.*, 2015a; 2015b). Infectivity appears to be strain-specific for ScosV and species-specific for SpalV (though strain-specificity has not been tested). While ScosV is icosahedral in shape (hexagonal in outline) and approximately 40 to 50 nm in diameter, SpalV is round-shaped and only 20 to 30 nm in diameter. Both viruses propagate in the cytoplasm of their host, form no crystalline arrays, and have no visible outer membrane or tail-like structure. The latent periods for ScosV and SpalV are estimated to be less than 48 and 80 h, respectively, while their burst sizes are estimated to be approximately 90 to 250 and 92 infectious units/cell, respectively.

TampV is a Cryptophyte virus, isolated from the Yasushiro Sea in Japan, which lyses *Teleaulax amphioxeia* (Nagasaki *et al.*, 2009). It is approximately 203 nm in diameter, icosahedral in shape, and its infectivity appears to be strain-specific. The virus propagates in the cytoplasm of its host, forms no crystalline arrays, and has no visible outer membrane or tail-like structure. The latent period for the virus was estimated to be less than 24 h, while its burst size was estimated at approximately 430 to 530 infectious units/cell (though it has also been estimated that there may be upwards of 760 virus particles within an infected cell). It is thought that TampV harbours a dsDNA genome, although this has not yet been confirmed.

Summary

Clearly, our knowledge of algal viruses has expanded rapidly over the last few decades. We have come a long way since the discovery and isolation of the dsDNA viruses that infect *Chlorella*-like algae and that led to the establishment of the prominent algal virus family, the *Phycodnaviridae*. Currently, dsDNA, ssDNA, dsRNA, and ssRNA viruses that infect a diverse array of algae have been isolated. Although there are a few notable exceptions, the vast majority of these algal pathogens are lytic viruses that are highly species or strain specific. We must be cautious, however, when drawing such broad conclusions, as strain-specificity tests cannot be exhaustive, and environmental conditions that promote latent infections may not be adequately mimicked under laboratory conditions. Nonetheless, due to the dedicated research efforts of the many individuals cited herein, we have come to appreciate that algal viruses are a diverse, dynamic, and ecologically important part of the biosphere. Whether only a few, many, or all species of algae are susceptible to one or more viral infections is impossible to know at this stage of research. Whatever the case, it can be assumed that many viruses await discovery and characterization, and it is undoubtable that the world of algal virus research will bring many more surprises. As active, major participants in the Earth's biogeochemical cycles, knowledge of the major aquatic primary producers (i.e. algae) is critical for our understanding of the living planet we inhabit. By extension, and based on the knowledge that diverse algae are infected by diverse viruses, knowing and understanding the viruses that infect algae is a vitally important scientific endeavour that reaches beyond the aquatic sciences.

References

Abergel, C., Legendre, M., and Claverie, J.M. (2015). The rapidly expanding universe of giant viruses: mimivirus, pandoravirus, pithovirus and mollivirus. FEMS Microbiol. Rev. *39,* 779–796. https://doi.org/10.1093/femsre/fuv037.

Allen, M.J., Schroeder, D.C., Donkin, A., Crawfurd, K.J., and Wilson, W.H. (2006a). Genome comparison of two coccolithoviruses. Virol. J. *3,* 15.

Allen, M.J., Schroeder, D.C., Holden, M.T., and Wilson, W.H. (2006b). Evolutionary history of the *Coccolithoviridae*. Mol. Biol. Evol. *23,* 86–92.

Angly, F.E., Felts, B., Breitbart, M., Salamon, P., Edwards, R.A., Carlson, C., Chan, A.M., Haynes, M., Kelley, S., Liu, H., *et al.* (2006). The marine viromes of four oceanic regions. PLOS Biol. *4,* e368.

Arslan, D., Legendre, M., Seltzer, V., Abergel, C., and Claverie, J.M. (2011). Distant Mimivirus relative with a larger genome highlights the fundamental features of Megaviridae. Proc. Natl. Acad. Sci. U.S.A. *108,* 17486–17491. https://doi.org/10.1073/pnas.1110889108.

Attoui, H., Jaafar, F.M., Belhouchet, M., de Micco, P., de Lamballerie, X., and Brussaard, C.P.D. (2006). *Micromonas pusilla* reovirus: a new member of the family *Reoviridae* assigned to a novel proposed genus (*Mimoreovirus*). J. Gen. Virol. *87,* 1375–1383.

Baltimore, D. (1971). Expression of animal virus genomes. Bacteriol. Rev. *35,* 235–241.

Baudoux, A.C., and Brussaard, C.P. (2005). Characterization of different viruses infecting the marine harmful algal bloom species Phaeocystis globosa. Virology *341,* 80–90.

Bellec, L., Grimsley, N., and Desdevises, Y. (2010). Isolation of prasinoviruses of the green unicellular algae *Ostreococcus* spp. on a worldwide geographical scale. Appl. Environ. Microbiol. *76,* 96–101. https://doi.org/10.1128/AEM.01799-09.

Bellec, L., Grimsley, N., Moreau, H., and Desdevises, Y. (2009). Phylogenetic analysis of new Prasinoviruses (*Phycodnaviridae*) that infect the green unicellular algae *Ostreococcus*, *Bathycoccus* and *Micromonas*. Environ Microbiol. Rep. *1,* 114–123. https://doi.org/10.1111/j.1758-2229.2009.00015.x.

Bench, S.R., Hanson, T.E., Williamson, K.E., Ghosh, D., Radosovich, M., Wang, K., and Wommack, K.E. (2007). Metagenomic characterization of Chesapeake Bay virioplankton. Appl. Environ. Microbiol. *73,* 7629–7641.

Bergh, O., Børsheim, K.Y., Bratbak, G., and Heldal, M. (1989). High abundance of viruses found in aquatic environments. Nature *340,* 467–468. https://doi.org/10.1038/340467a0.

Bettarel, Y., Kan, J., Wang, K., Williamson, K.E., Cooney, S., Ribblett, S., Chen, F., Wommack, K.E., and Coats, D.W. (2005). Isolation and preliminary characterisation of a small nuclear inclusion virus infecting the diatom *Chaetoceros* cf. *gracilis*. Aquat. Microb. Ecol. *40,* 103–114.

Bratbak, G., Egge, J.K., and Heldal, M. (1993). Viral mortality of the marine alga *Emiliania huxleyi* (Haptophyceae) and termination of algal blooms. Mar. Ecol. Prog. Ser. *93,* 39–48.

Breitbart, M., Felts, B., Kelley, S., Mahaffy, J.M., Nulton, J., Salamon, P., and Rohwer, F. (2004). Diversity and population structure of a near-shore marine-sediment viral community. Proc. Biol. Sci. *271,* 565–574. https://doi.org/10.1098/rspb.2003.2628.

Breitbart, M., Salamon, P., Andresen, B., Mahaffy, J.M., Segall, A.M., Mead, D., Azam, F., and Rohwer, F. (2002). Genomic analysis of uncultured marine viral communities. Proc. Natl. Acad. Sci. U.S.A. *99,* 14250–14255. https://doi.org/10.1073/pnas.202488399.

Breitbart, M., Thompson, L.R., Suttle, C.A., and Sullivan, M.B. (2007). Exploring the vast diversity of marine viruses. Oceanography *20,* 135–139.

Brodie, J., and Lewis, J., eds. (2007). Unravelling the algae: the past, present, and future of algal systematics (CRC Press, Boca Raton).

Brussaard, C.P. (2004). Viral control of phytoplankton populations – a review. J. Eukaryot. Microbiol. *51,* 125–138.

Brussaard, C.P.D., and Martínez Martínez, J. (2008). Algal bloom viruses. Plant Viruses *2,* 1–13.

Brussaard, C.P., Noordeloos, A.A., Sandaa, R.A., Heldal, M., and Bratbak, G. (2004a). Discovery of a dsRNA virus infecting the marine photosynthetic protist *Micromonas pusilla*. Virology *319,* 280–291. https://doi.org/10.1016/j.virol.2003.10.033.

Brussaard, C.P., Short, S.M., Frederickson, C.M., and Suttle, C.A. (2004b). Isolation and phylogenetic analysis of novel viruses infecting the phytoplankton *Phaeocystis globosa* (Prymnesiophyceae). Appl. Environ. Microbiol. *70,* 3700–3705. https://doi.org/10.1128/AEM.70.6.3700-3705.2004.

Bubeck, J.A., and Pfitzner, A.J.P. (2005). Isolation and characterization of a new type of chlorovirus that infects an endosymbiotic *Chlorella* strain of the heliozoon *Acanthocystis turfacea*. J. Gen. Virol. *86,* 2871–2877.

Castberg, T., Thyrhaug, R., Larsen, A., Sandaa, R.A., Heldal, M., Van Etten, J.L., and Bratbak, G. (2002). Isolation and characterization of a virus that infects *Emiliania huxleyi* (Haptophyta). J. Phycol. *38,* 767–774.

Charlson, R.J., Lovelock, J.E., Andreae, M.O., and Warren, S.G. (1987). Oceanic phytoplankton, atmospheric sulfur, cloud albedo and climate. Nature *326,* 655–661.

Chen, F., and Suttle, C.A. (1995). Amplification of DNA polymerase gene fragments from viruses infecting microalgae. Appl. Environ. Microbiol. *61,* 1274–1278.

Chen, F., and Suttle, C.A. (1996). Evolutionary relationships among large double-stranded DNA viruses that infect microalgae and other organisms as inferred from DNA polymerase genes. Virology *219,* 170–178.

Chen, F., Suttle, C.A., and Short, S.M. (1996). Genetic diversity in marine algal virus communities as revealed by sequence analysis of DNA polymerase genes. Appl. Environ. Microbiol. *62,* 2869–2874.

Chow, C.E., and Suttle, C.A. (2015). Biogeography of viruses in the sea. Annu. Rev. Virol. *2,* 41–66. https://doi.org/10.1146/annurev-virology-031413-085540.

Chow, C.E., Winget, D.M., White, R.A., Hallam, S.J., and Suttle, C.A. (2015). Combining genomic sequencing methods to explore viral diversity and reveal potential virus-host interactions. Front. Microbiol. *6,* 265. https://doi.org/10.3389/fmicb.2015.00265.

Clasen, J.L., and Suttle, C.A. (2009). Identification of freshwater Phycodnaviridae and their potential phytoplankton hosts, using DNA pol sequence fragments and a genetic-distance analysis. Appl. Environ. Microbiol. *75,* 991–997. https://doi.org/10.1128/AEM.02024-08.

Claverie, J.M., and Abergel, C. (2009). Mimivirus and its virophage. Annu. Rev. Genet. *43,* 49–66. https://doi.org/10.1146/annurev-genet-102108-134255.

Clerissi, C., Desdevises, Y., and Grimsley, N. (2012). Prasinoviruses of the marine green alga Ostreococcus tauri are mainly species specific. J. Virol. *86,* 4611–4619. https://doi.org/10.1128/JVI.07221-11.

Clerissi, C., Desdevises, Y., Romac, S., Audic, S., de Vargas, C., Acinas, S.G., Casotti, R., Poulain, J., Wincker, P., Hingamp, P., *et al.* (2015). Deep sequencing of amplified *Prasinovirus* and host green algal genes from an Indian Ocean transect reveals interacting trophic dependencies and new genotypes. Environ Microbiol Rep *7,* 979–989. https://doi.org/10.1111/1758-2229.12345.

Clerissi, C., Grimsley, N., Ogata, H., Hingamp, P., Poulain, J., and Desdevises, Y. (2014). Unveiling of the diversity of Prasinoviruses (*Phycodnaviridae*) in marine samples by using high-throughput sequencing analyses of PCR-amplified DNA polymerase and major capsid protein genes. Appl. Environ. Microbiol. *80,* 3150–3160. https://doi.org/10.1128/AEM.00123-14.

Colson, P., De Lamballerie, X., Yutin, N., Asgari, S., Bigot, Y., Bideshi, D.K., Cheng, X.W., Federici, B.A., Van Etten, J.L., Koonin, E.V., *et al.* (2013). 'Megavirales', a proposed new order for eukaryotic nucleocytoplasmic large DNA viruses. Arch. Virol. *158,* 2517–2521. https://doi.org/10.1007/s00705-013-1768-6.

Correa, A.M., Ainsworth, T.D., Rosales, S.M., Thurber, A.R., Butler, C.R., and Vega Thurber, R.L. (2016). Viral outbreak in corals associated with an in situ bleaching event: atypical herpes-like viruses and a new megavirus infecting *Symbiodinium*. Front. Microbiol. *7,* 127. https://doi.org/10.3389/fmicb.2016.00127.

Correa, A.M., Welsh, R.M., and Vega Thurber, R.L. (2013). Unique nucleocytoplasmic dsDNA and +ssRNA viruses are associated with the dinoflagellate endosymbionts of corals. ISME J. *7,* 13–27. https://doi.org/10.1038/ismej.2012.75.

Cottrell, M.T., and Suttle, C.A. (1991). Wide-spread occurrence and clonal variation in viruses which cause lysis of a cosmopolitan, eukaryotic marine phytoplankter, *Micromonas pusilla*. Mar. Ecol. Prog. Ser. *78,* 1–9.

Culley, A.I., Asuncion, B.F., and Steward, G.F. (2009). Detection of inteins among diverse DNA polymerase genes of uncultivated members of the *Phycodnaviridae*. ISME J. *3,* 409–418. https://doi.org/10.1038/ismej.2008.120.

Culley, A.I., Lang, A.S., and Suttle, C.A. (2003). High diversity of unknown picorna-like viruses in the sea. Nature *424,* 1054–1057. https://doi.org/10.1038/nature01886.

Culley, A.I., Lang, A.S., and Suttle, C.A. (2006). Metagenomic analysis of coastal RNA virus communities. Science *312,* 1795–1798.

Culley, A.I., Mueller, J.A., Belcaid, M., Wood-Charlson, E.M., Poisson, G., and Steward, G.F. (2014). The characterization of RNA viruses in tropical seawater using targeted PCR and metagenomics. MBio *5,* e01210–14. https://doi.org/10.1128/mBio.01210-14.

Culley, A.I., and Steward, G.F. (2007). New genera of RNA viruses in subtropical seawater, inferred from polymerase gene sequences. Appl. Environ. Microbiol. *73,* 5937–5944.

Delaroque, N., Boland, W., Müller, D.G., and Knippers, R. (2003). Comparisons of two large phaeoviral genomes and evolutionary implications. J. Mol. Evol. *57,* 613–622. https://doi.org/10.1007/s00239-003-2501-y.

Delaroque, N., Müller, D.G., Bothe, G., Pohl, T., Knippers, R., and Boland, W. (2001). The complete DNA sequence of the *Ectocarpus siliculosus* virus EsV-1 genome. Virology *287,* 112–132. https://doi.org/10.1006/viro.2001.1028.

Derelle, E., Ferraz, C., Escande, M.L., Eychenié, S., Cooke, R., Piganeau, G., Desdevises, Y., Bellec, L., Moreau, H., and Grimsley, N. (2008). Life-cycle and genome of OtV5, a large DNA virus of the pelagic marine unicellular green alga *Ostreococcus tauri*. PLOS ONE 3, e2250. https://doi.org/10.1371/journal.pone.0002250.

Dunigan, D.D., Fitzgerald, L.A., and Van Etten, J.L. (2006). Phycodnaviruses: a peek at genetic diversity. Virus Res. *117,* 119–132.

Edwards, R.A., and Rohwer, F. (2005). Viral metagenomics. Nat. Rev. Microbiol. 3, 504–510.

Eissler, Y., Wang, K., Chen, F., Eric Wommack, K., and Wayne Coats, D. (2009). Ultrastructural characterization of the lytic cycle of an intranuclear virus infecting the diatom *Chaetoceros* Cf. *Wighamii* (Bacillariophyceae) from Chesapeake Bay, USA(1). J. Phycol. *45,* 787–797. https://doi.org/10.1111/j.1529-8817.2009.00705.x.

Fancello, L., Trape, S., Robert, C., Boyer, M., Popgeorgiev, N., Raoult, D., and Desnues, C. (2013). Viruses in the desert: a metagenomic survey of viral communities in four perennial ponds of the Mauritanian Sahara. ISME J. *7,* 359–369. https://doi.org/10.1038/ismej.2012.101.

Fischer, M.G. (2016). Giant viruses come of age. Curr. Opin. Microbiol. *31,* 50–57.

Fischer, M.G., and Suttle, C.A. (2011). A virophage at the origin of large DNA transposons. Science *332,* 231–234. https://doi.org/10.1126/science.1199412.

Fitzgerald, L.A., Graves, M.V., Li, X., Feldblyum, T., Hartigan, J., and Van Etten, J.L. (2007a). Sequence and annotation of the 314-kb MT325 and the 321-kb FR483 viruses that infect *Chlorella* Pbi. Virology *358,* 459–471.

Fitzgerald, L.A., Graves, M.V., Li, X., Feldblyum, T., Nierman, W.C., and Van Etten, J.L. (2007b). Sequence and annotation of the 369-kb NY-2A and the 345-kb AR158 viruses that infect *Chlorella* NC64A. Virology *358,* 472–484.

Fitzgerald, L.A., Graves, M.V., Li, X., Hartigan, J., Pfitzner, A.J., Hoffart, E., and Van Etten, J.L. (2007c). Sequence and annotation of the 288-kb ATCV-1 virus that infects an endosymbiotic chlorella strain of the heliozoon *Acanthocystis turfacea.* Virology *362,* 350–361.

Frada, M., Probert, I., Allen, M.J., Wilson, W.H., and de Vargas, C. (2008). The 'Cheshire Cat' escape strategy of the coccolithophore *Emiliania huxleyi* in response to viral infection. Proc. Natl. Acad. Sci. U.S.A. *105,* 15944–15949. https://doi.org/10.1073/pnas.0807707105.

Franklin, D.J., Steinke, M., Young, J., Probert, I., and Malin, G. (2010). Dimethylsulphoniopropionate (DMSP), DMSP-lyase activity (DLA) and dimethylsulphide (DMS) in 10 species of coccolithophore. Mar. Ecol. Prog. Ser. *410,* 13–23.

Friess-Klebl, A.K., Knippers, R., and Muller, D.G. (1994). Isolation and characterization of a DNA virus infecting *Feldmannia simplex* (Phaeophyceae). J. Phycol. *30,* 653–658.

Ghedin, E., and Claverie, J.M. (2005). Mimivirus relatives in the Sargasso sea. Virol. J. *2,* 62.

Green, J.C., Rahman, F., Saxton, M.A., and Williamson, K.E. (2015). Metagenomic assessment of viral diversity in Lake Matoaka, a temperate, eutrophic freshwater lake in Southeastern Virginia, USA. Aquat. Microb. Ecol. *75,* 117–128.

Guiry, M.D. (2012). How many species of algae are there? J. Phycol. *48,* 1057–1063. https://doi.org/10.1111/j.1529-8817.2012.01222.x.

Hallegraeff, G.M., and Hara, Y. (2004). Taxonomy of harmful marine raphidophytes. In Manual on Harmful Marine Microalgae, Hallegraeff, G.M., Anderson, D.M., and Cembella, A.D., eds. (UNESCO Publishing, Paris), p. 793.

Henry, E.C., and Meints, R.H. (1992). A persistent virus-infection in *Feldmannia* (Phaeophyceae). J. Phycol. *28,* 517–26.

Hingamp, P., Grimsley, N., Acinas, S.G., Clerissi, C., Subirana, L., Poulain, J., Ferrera, I., Sarmento, H., Villar, E., Lima-Mendez, G., *et al.* (2013). Exploring nucleo-cytoplasmic large DNA viruses in Tara Oceans microbial metagenomes. ISME J. *7,* 1678–1695. https://doi.org/10.1038/ismej.2013.59.

Holligan, P.M., Fernandez, E., Aiken, J., Balch, W.M., Boyd, P., Burkill, P.H., Finch, M., Groom, S.B., Malin, G., Muller, K., *et al.* (1993). A biogeochemical study of the coccolithophore, *Emiliania huxleyi,* in the North Atlantic. Global Biogeochem. Cycles *7,* 879–900.

Iyer, L.M., Aravind, L., and Koonin, E.V. (2001). Common origin of four diverse families of large eukaryotic DNA viruses. J. Virol. *75,* 11720–11734. https://doi.org/10.1128/JVI.75.23.11720-11734.2001.

Iyer, L.M., Balaji, S., Koonin, E.V., and Aravind, L. (2006). Evolutionary genomics of nucleo-cytoplasmic large DNA viruses. Virus Res. *117,* 156–184.

Jacobsen, A., Bratbak, G., and Heldal, M. (1996). Isolation and characterization of a virus infecting *Phaeocystis pouchetii* (Prymnesiophyceae). J. Phycol. *32,* 923–927.

Jeanniard, A., Dunigan, D.D., Gurnon, J.R., Agarkova, I.V., Kang, M., Vitek, J., Duncan, G., McClung, O.W., Larsen, M., Claverie, J.M., *et al.* (2013). Towards defining the chloroviruses: a genomic journey through a genus of large DNA viruses. BMC Genomics *14,* 158. https://doi.org/10.1186/1471-2164-14-158.

Johannessen, T.V., Bratbak, G., Larsen, A., Ogata, H., Egge, E.S., Edvardsen, B., Eikrem, W., and Sandaa, R.A. (2015). Characterisation of three novel giant viruses reveals huge diversity among viruses infecting Prymnesiales (Haptophyta). Virology *476,* 180–188. https://doi.org/10.1016/j.virol.2014.12.014.

Kang, M., Dunigan, D.D., and van Etten, J.L. (2005). *Chlorovirus*: a genus of Phycodnaviridae that infects certain chlorella-like green algae. Mol. Plant Pathol. *6,* 213–224. https://doi.org/10.1111/j.1364-3703.2005.00281.x.

Kapp, M., Knippers, R., and Muller, D.G. (1997). New members of a group of DNA viruses infecting brown algae. Phycol. Res. *45,* 85–90.

Kim, J., Kim, C.H., Takano, Y., Jang, I.K., Kim, S.W., and Choi, T.J. (2012). Isolation and physiological characterization of a new algicidal virus infecting the harmful dinoflagellate *Heterocapsa pygmaea.* Plant. Pathol. J. *28,* 433–438.

Kim, J., Kim, C.H., Youn, S.H., and Choi, T.J. (2015a). Isolation and physiological characterization of a novel algicidal virus infecting the marine diatom *Skeletonema costatum.* Plant Pathol. J. *31,* 186–191. https://doi.org/10.5423/PPJ.NT.03.2015.0029.

Kim, J., Yoon, S.H., and Choi, T.J. (2015b). Isolation and physiological characterization of a novel virus infecting *Stephanopyxis palmeriana* (Bacillariophyta). Algae *30,* 81–87.

Kimura, K., and Tomaru, Y. (2013). Isolation and characterization of a single-stranded DNA virus infecting the marine diatom *Chaetoceros* sp. strain SS628-11 isolated from western Japan. PLOS ONE *8,* e82013. https://doi.org/10.1371/journal.pone.0082013.

Kimura, K., and Tomaru, Y. (2015). Discovery of two novel viruses expands the diversity of single-stranded DNA and single-stranded RNA viruses infecting a cosmopolitan marine diatom. Appl. Environ. Microbiol. *81,* 1120–1131. https://doi.org/10.1128/AEM.02380-14.

King, A.M.Q., Adams, M.J., Carstens, E.B., and Lefkowitz, E.J., eds. (2012). Virus Taxonomy: Classification and Nomenclature of Viruses: Ninth Report of the International Committee on Taxonomy of Viruses (Elsevier Academic Press, San Diego).

Kristensen, D.M., Mushegian, A.R., Dolja, V.V., and Koonin, E.V. (2010). New dimensions of the virus world discovered through metagenomics. Trends Microbiol. *18,* 11–19. https://doi.org/10.1016/j.tim.2009.11.003.

La Scola, B., Audic, S., Robert, C., Jungang, L., de Lamballerie, X., Drancourt, M., Birtles, R., Claverie, J.M., and Raoult, D. (2003). A giant virus in amoebae. Science *299,* 2033–2033.

La Scola, B., Desnues, C., Pagnier, I., Robert, C., Barrassi, L., Fournous, G., Merchat, M., Suzan-Monti, M., Forterre, P., Koonin, E., *et al.* (2008). The virophage as a unique parasite of the giant mimivirus. Nature *455,* 100–104. https://doi.org/10.1038/nature07218.

Lang, A.S., Culley, A.I., and Suttle, C.A. (2004). Genome sequence and characterization of a virus (HaRNAV) related to picorna-like viruses that infects the marine toxic bloom-forming alga *Heterosigma akashiwo*. Virology *320,* 206–217. https://doi.org/10.1016/j.virol.2003.10.015.

Lanka, S.T.J., Klein, M., Ramsperger, U., Muller, D.G., and Knippers, R. (1993). Genome structure of a virus infecting the marine brown alga *Ectocarpus siliculosus*. Virology *193,* 802–11.

Larsen, J.B., Larsen, A., Bratbak, G., and Sandaa, R.A. (2008). Phylogenetic analysis of members of the *Phycodnaviridae* virus family, using amplified fragments of the major capsid protein gene. Appl. Environ. Microbiol. *74,* 3048–3057. https://doi.org/10.1128/AEM.02548-07.

Lawrence, J.E., Brussaard, C.P., and Suttle, C.A. (2006). Virus-specific responses of *Heterosigma akashiwo* to infection. Appl. Environ. Microbiol. *72,* 7829–7834.

Lawrence, J.E., Chan, A.M., and Suttle, C.A. (2001). A novel virus (HaNIV) causes lysis of the toxic bloom-forming alga *Heterosigma akashiwo* (Raphidophyceae). J. Phycol. *37,* 216–222.

Lawrence, J.E., Chan, A.M., and Suttle, C.A. (2002). Viruses causing lysis of the toxic bloom-forming alga *Heterosigma akashiwo* (Raphidophyceae) are widespread in coastal sediments of British Columbia, Canada. Limnol. Oceanogr. *47,* 545–550.

Lawrence, S.A., Wilson, W.H., Davy, J.E., and Davy, S.K. (2014). Latent virus-like infections are present in a diverse range of *Symbiodinium* spp. (Dinophyta). J. Phycol. *50,* 984–997. https://doi.org/10.1111/jpy.12242.

Le Gall, O., Christian, P., Fauquet, C.M., King, A.M., Knowles, N.J., Nakashima, N., Stanway, G., and Gorbalenya, A.E. (2008). Picornavirales, a proposed order of positive-sense single-stranded RNA viruses with a pseudo-T = 3 virion architecture. Arch. Virol. *153,* 715–727. https://doi.org/10.1007/s00705-008-0041-x.

Li, Y., Lu, Z., Sun, L., Ropp, S., Kutish, G.F., Rock, D.L., and Van Etten, J.L. (1997). Analysis of 74 kb of DNA located at the right end of the 330-kb chlorella virus PBCV-1 genome. Virology *237,* 360–377.

Littman, R., Willis, B.L., and Bourne, D.G. (2011). Metagenomic analysis of the coral holobiont during a natural bleaching event on the Great Barrier Reef. Environ. Microbiol. Rep. *3,* 651–660. https://doi.org/10.1111/j.1758-2229.2010.00234.x.

López-Bueno, A., Tamames, J., Velázquez, D., Moya, A., Quesada, A., and Alcamí, A. (2009). High diversity of the viral community from an Antarctic lake. Science *326,* 858–861. https://doi.org/10.1126/science.1179287.

Maier, I., Wolf, S., Delaroque, N., Muller, D.G., and Kawai, H. (1998). A DNA virus infecting the marine brown alga *Pilayella littoralis* (Ectocarpales, Phaeophyceae) in culture. Eur. J. Phycol. *33,* 213–220.

Manrique, J.M., Calvo, A.Y., and Jones, L.R. (2012). Phylogenetic analysis of *Ostreococcus* virus sequences from the Patagonian Coast. Virus Genes *45,* 316–326. https://doi.org/10.1007/s11262-012-0762-5.

Marhaver, K.L., Edwards, R.A., and Rohwer, F. (2008). Viral communities associated with healthy and bleaching corals. Environ. Microbiol. *10,* 2277–2286. https://doi.org/10.1111/j.1462-2920.2008.01652.x.

Martínez Martínez, J., Boere, A., Gilg, L., van Lent, J.W.M., Witte, H.J., van Bleijswijk, J.D.L., and Brussaard, C.P.D. (2015). New lipid envelope-containing dsDNA virus isolates infecting *Micromonas pusilla* reveal a separate phylogenetic group. Aquat. Microb. Ecol. *74,* 17–28.

Martínez Martínez, J., Schroeder, D.C., Larsen, A., Bratbak, G., and Wilson, W.H. (2007). Molecular dynamics of *Emiliania huxleyi* and cooccurring viruses during two separate mesocosm studies. Appl. Environ. Microbiol. *73,* 554–562.

Martínez Martínez, J., Swan, B.K., and Wilson, W.H. (2014). Marine viruses, a genetic reservoir revealed by targeted viromics. ISME J. *8,* 1079–1088. https://doi.org/10.1038/ismej.2013.214.

Mayer, J.A., and Taylor, F.J.R. (1979). A virus which lyses the marine nanoflagellate *Micromonas pusilla*. Nature *281,* 299–301.

Meints, R.H., Van Etten, J.L., Kuczmarski, D., Lee, K., and Ang, B. (1981). Viral infection of the symbiotic chlorella-like alga present in *Hydra viridis*. Virology *113,* 698–703.

Milligan, K.L.D., and Cosper, E.M. (1994). Isolation of virus capable of lysing the brown tide microalga, *Aureococcus anophagefferens*. Science *266,* 805–7.

Mirza, S.F., Staniewski, M.A., Short, C.M., Long, A.M., Chaban, Y.V., and Short, S.M. (2015). Isolation and characterization of a virus infecting the freshwater algae *Chrysochromulina parva*. Virology *486,* 105–115. https://doi.org/10.1016/j.virol.2015.09.005.

Mokili, J.L., Rohwer, F., and Dutilh, B.E. (2012). Metagenomics and future perspectives in virus discovery. Curr. Opin. Virol. *2,* 63–77. https://doi.org/10.1016/j.coviro.2011.12.004.

Monier, A., Claverie, J.M., and Ogata, H. (2008a). Taxonomic distribution of large DNA viruses in the sea. Genome Biol. *9,* R106. https://doi.org/10.1186/gb-2008-9-7-r106.

Monier, A., Larsen, J.B., Sandaa, R.A., Bratbak, G., Claverie, J.M., and Ogata, H. (2008b). Marine mimivirus relatives are probably large algal viruses. Virol. J. *5,* 12. https://doi.org/10.1186/1743-422X-5-12.

Moniruzzaman, M., Gann, E.R., LeCleir, G.R., Kang, Y., Gobler, C.J., and Wilhelm, S.W. (2016). Diversity and dynamics of algal Megaviridae members during a harmful brown tide caused by the pelagophyte, *Aureococcus anophagefferens*. FEMS Microbiol. Ecol. *92,* fiw058. https://doi.org/10.1093/femsec/fiw058.

Moniruzzaman, M., LeCleir, G.R., Brown, C.M., Gobler, C.J., Bidle, K.D., Wilson, W.H., and Wilhelm, S.W. (2014). Genome of brown tide virus (AaV), the little giant of the Megaviridae, elucidates NCLDV genome expansion and host-virus coevolution. Virology 466-467, 60–70. https://doi.org/10.1016/j.virol.2014.06.031.

Müller, D.G. (1991). Mendelian segregation of a virus genome during host meiosis in the marine brown alga *Ectocarpus siliculosus*. J. Plant. Physiol. *137,* 739–743.

Müller, D.G. (1992). Intergeneric transmission of a marine plant DNA virus. Naturwissenschaften *79,* 37–39.

Müller, D.G., and Parodi, E. (1993). Transfer of a marine DNA virus from *Ectocarpus* to *Feldmannia* (Ectocarpales, Phaeophyceae): aberrant symptoms and restitution of the host. Protoplasma *175,* 121–125.

Müller, D.G., Sengco, M., Wolf, S., Bräutigam, M., Schmid, C.E., Kapp, M., and Knippers, R. (1996). Comparison of two DNA viruses infecting the marine brown algae *Ectocarpus siliculosus* and *E. fasciculatus*. J. Gen. Virol. *77,* 2329–2333. https://doi.org/10.1099/0022-1317-77-9-2329.

Nagasaki, K. (2008). Dinoflagellates, diatoms, and their viruses. J. Microbiol. *46,* 235–243. https://doi.org/10.1007/s12275-008-0098-y.

Nagasaki, K., Ando, M., Imai, I., Itakura, S., and Ishida, Y. (1994a). Virus-like particles in *Heterosigma akashiwo* (Raphidophyceae) – a possible red tide disintegration mechanism. Mar. Biol. *119,* 307–312.

Nagasaki, K., Ando, M., Itakura, S., Imai, I., and Ishida, Y. (1994b). Viral mortality in the final stage of *Heterosigma akashiwo* (Raphidophyceae) red tide. J. Plankton. Res. *16,* 1595–1599.

Nagasaki, K., Kim, J.J., Tomaru, Y., Takao, Y., and Nagai, S. (2009). Isolation and characterization of a novel virus infecting *Teleaulax amphioxeia* (Cryptophyceae). Plankton Benthos Res. *4,* 122–124.

Nagasaki, K., Shirai, Y., Takao, Y., Mizumoto, H., Nishida, K., and Tomaru, Y. (2005a). Comparison of genome sequences of single-stranded RNA viruses infecting the bivalve-killing dinoflagellate *Heterocapsa circularisquama*. Appl. Environ. Microbiol. *71,* 8888–8894.

Nagasaki, K., Shirai, Y., Tomaru, Y., Nishida, K., and Pietrokovski, S. (2005b). Algal viruses with distinct intraspecies host specificities include identical intein elements. Appl. Environ. Microbiol. *71,* 3599–3607.

Nagasaki, K., Tarutani, K., and Yamaguchi, M. (1999). Growth characteristics of *Heterosigma akashiwo* virus and its possible use as a microbiological agent for red tide control. Appl. Environ. Microbiol. *65,* 898–902.

Nagasaki, K., Tomaru, Y., Katanozaka, N., Shirai, Y., Nishida, K., Itakura, S., and Yamaguchi, M. (2004). Isolation and characterization of a novel single-stranded RNA virus infecting the bloom-forming diatom *Rhizosolenia setigera*. Appl. Environ. Microbiol. *70,* 704–711.

Nagasaki, K., Tomaru, Y., Takao, Y., Nishida, K., Shirai, Y., Suzuki, H., and Nagumo, T. (2005c). Previously unknown virus infects marine diatom. Appl. Environ. Microbiol. *71,* 3528–3535.

Nagasaki, K., Tomaru, Y., Tarutani, K., Katanozaka, N., Yamanaka, S., Tanabe, H., and Yamaguchi, M. (2003). Growth characteristics and intraspecies host specificity of a large virus infecting the dinoflagellate *Heterocapsa circularisquama*. Appl. Environ. Microbiol. *69,* 2580–2586.

Nagasaki, K., and Yamaguchi, M. (1997). Isolation of a virus infectious to the harmful bloom causing microalga *Heterosigma akashiwo* (Raphidophyceae). Aquat. Microb. Ecol. *13,* 135–140.

Nagasaki, K., and Yamaguchi, M. (1998). Intra-species host specificity of HaV (*Heterosigma akashiwo* virus) clones. Aquat. Microb. Ecol. *14,* 109–112.

Nissimov, J.I., Worthy, C.A., Rooks, P., Napier, J.A., Kimmance, S.A., Henn, M.R., Ogata, H., and Allen, M.J. (2011a). Draft genome sequence of the coccolithovirus EhV-84. Stand. Genomic Sci. *5,* 1–11. https://doi.org/10.4056/sigs.1884581.

Nissimov, J.I., Worthy, C.A., Rooks, P., Napier, J.A., Kimmance, S.A., Henn, M.R., Ogata, H., and Allen, M.J. (2011b). Draft genome sequence of the Coccolithovirus *Emiliania huxleyi* virus 203. J. Virol. *85,* 13468–13469. https://doi.org/10.1128/JVI.06440-11.

Nissimov, J.I., Worthy, C.A., Rooks, P., Napier, J.A., Kimmance, S.A., Henn, M.R., Ogata, H., and Allen, M.J. (2012a). Draft genome sequence of four coccolithoviruses: *Emiliania huxleyi* virus EhV-88, EhV-201, EhV-207, and EhV-208. J. Virol. *86,* 2896–2897. https://doi.org/10.1128/JVI.07046-11.

Nissimov, J.I., Worthy, C.A., Rooks, P., Napier, J.A., Kimmance, S.A., Henn, M.R., Ogata, H., and Allen, M.J. (2012b). Draft genome sequence of the coccolithovirus *Emiliania huxleyi* virus 202. J. Virol. *86,* 2380–2381. https://doi.org/10.1128/JVI.06863-11.

Ogata, H., Ray, J., Toyoda, K., Sandaa, R.A., Nagasaki, K., Bratbak, G., and Claverie, J.M. (2011). Two new subfamilies of DNA mismatch repair proteins (MutS) specifically abundant in the marine environment. ISME J. *5,* 1143–1151. https://doi.org/10.1038/ismej.2010.210.

Ogata, H., Toyoda, K., Tomaru, Y., Nakayama, N., Shirai, Y., Claverie, J.M., and Nagasaki, K. (2009). Remarkable sequence similarity between the dinoflagellate-infecting marine girus and the terrestrial pathogen African swine fever virus. Virol. J. *6,* 178. https://doi.org/10.1186/1743-422X-6-178.

Pagarete, A., Grébert, T., Stepanova, O., Sandaa, R.A., and Bratbak, G. (2015). Tsv-N1: a novel DNA algal virus that infects *Tetraselmis striata*. Viruses 7, 3937–3953. https://doi.org/10.3390/v7072806.

Pagarete, A., Kusonmano, K., Petersen, K., Kimmance, S.A., Martínez Martínez, J., Wilson, W.H., Hehemann, J.H., Allen, M.J., and Sandaa, R.A. (2014). Dip in the gene pool: metagenomic survey of natural coccolithovirus communities. Virology 466-467, 129–137. https://doi.org/10.1016/j.virol.2014.05.020.

Pagarete, A., Lanzén, A., Puntervoll, P., Sandaa, R.A., Larsen, A., Larsen, J.B., Allen, M.J., and Bratbak, G. (2013). Genomic sequence and analysis of EhV-99B1, a new coccolithovirus from the Norwegian fjords. Intervirology *56*, 60–66. https://doi.org/10.1159/000341611.

Park, Y., Lee, K., Lee, Y.S., Kim, S.W., and Choi, T.J. (2011). Detection of diverse marine algal viruses in the South Sea regions of Korea by PCR amplification of the DNA polymerase and major capsid protein genes. Virus Res. *159*, 43–50. https://doi.org/10.1016/j.virusres.2011.04.024.

Pringle, C.R. (1998). Virus taxonomy – San Diego 1998. Arch. Virol. *143*, 1449–1459.

Raoult, D., Audic, S., Robert, C., Abergel, C., Renesto, P., Ogata, H., La Scola, B., Suzan, M., and Claverie, J.M. (2004). The 1.2-megabase genome sequence of Mimivirus. Science *306*, 1344–1350.

Reisser, W., Becker, B., and Klein, T. (1986). Studies on ultrastructure and host range of a *Chlorella* attacking virus. Protoplasma *135*, 162–5.

Reisser, W., Burbank, D.E., Meints, R.H., Becker, B., and Vanetten, J.L. (1991). Viruses distinguish symbiotic *Chlorella* spp. of *Paramecium bursaria*. Endocytobiosis Cell Res. *7*, 245–251.

Reisser, W., Burbank, D.E., Meints, S.M., Meints, R.H., Becker, B., and Van Etten, J.L. (1988). A comparison of viruses infecting two different *Chlorella*-like green algae. Virology *167*, 143–149.

Rosario, K., Nilsson, C., Lim, Y.W., Ruan, Y., and Breitbart, M. (2009). Metagenomic analysis of viruses in reclaimed water. Environ. Microbiol. *11*, 2806–2820. https://doi.org/10.1111/j.1462-2920.2009.01964.x.

Rowe, J.M., Fabre, M.F., Gobena, D., Wilson, W.H., and Wilhelm, S.W. (2011). Application of the major capsid protein as a marker of the phylogenetic diversity of *Emiliania huxleyi* viruses. FEMS Microbiol. Ecol. *76*, 373–380. https://doi.org/10.1111/j.1574-6941.2011.01055.x.

Rozon, R.M., and Short, S.M. (2013). Complex seasonality observed amongst diverse phytoplankton viruses in the Bay of Quinte, an embayment of Lake Ontario. Freshwater Biol. *58*, 2648–2663.

Rusch, D.B., Halpern, A.L., Sutton, G., Heidelberg, K.B., Williamson, S., Yooseph, S., Wu, D., Eisen, J.A., Hoffman, J.M., Remington, K., *et al.* (2007). The Sorcerer II Global Ocean Sampling expedition: northwest Atlantic through eastern tropical Pacific. PLOS Biol. *5*, e77.

Sahlsten, E. (1998). Seasonal abundance in Skagerrak-Kattegat coastal waters and host specificity of viruses infecting the marine photosynthetic flagellate *Micromonas pusilla*. Aquat. Microb. Ecol. *16*, 103–108.

Sandaa, R.A., Heldal, M., Castberg, T., Thyrhaug, R., and Bratbak, G. (2001). Isolation and characterization of two viruses with large genome size infecting *Chrysochromulina ericina* (Prymnesiophyceae) and *Pyramimonas orientalis* (Prasinophyceae). Virology *290*, 272–280.

Santini, S., Jeudy, S., Bartoli, J., Poirot, O., Lescot, M., Abergel, C., Barbe, V., Wommack, K.E., Noordeloos, A.A., Brussaard, C.P., *et al.* (2013). Genome of *Phaeocystis globosa* virus PgV-16T highlights the common ancestry of the largest known DNA viruses infecting eukaryotes. Proc. Natl. Acad. Sci. U.S.A. *110*, 10800–10805. https://doi.org/10.1073/pnas.1303251110.

Schroeder, D.C., Oke, J., Malin, G., and Wilson, W.H. (2002). Coccolithovirus (*Phycodnaviridae*): characterisation of a new large dsDNA algal virus that infects *Emiliania huxleyi*. Arch. Virol. *147*, 1685–1698.

Schroeder, D.C., Park, Y., Yoon, H.M., Lee, Y.S., Kang, S.W., Meints, R.H., Ivey, R.G., and Choi, T.J. (2009). Genomic analysis of the smallest giant virus – *Feldmannia* sp. virus 158. Virology *384*, 223–232. https://doi.org/10.1016/j.virol.2008.10.040.

Shirai, Y., Takao, Y., Mizumoto, H., Tomaru, Y., Honda, D., and Magasaki, K. (2006). Genomic and phylogenetic analysis of a single-stranded RNA virus infecting *Rhizosolenia setigera* (Stramenopiles: Bacillariophyceae). J. Mar. Biol. Assoc. U.K. *86*, 475–483.

Shirai, Y., Tomaru, Y., Takao, Y., Suzuki, H., Nagumo, T., and Nagasaki, K. (2008). Isolation and characterization of a single-stranded RNA virus infecting the marine planktonic diatom *Chaetoceros tenuissimus* Meunier. Appl. Environ. Microbiol. *74*, 4022–4027. https://doi.org/10.1128/AEM.00509-08.

Short, S.M. (2012). The ecology of viruses that infect eukaryotic algae. Environ. Microbiol. *14*, 2253–2271. https://doi.org/10.1111/j.1462-2920.2012.02706.x.

Short, S.M., Rusanova, O., and Staniewski, M.A. (2011). Novel phycodnavirus genes amplified from Canadian freshwater environments. Aquat. Microb. Ecol. *63*, 61–67.

Short, S.M., and Short, C.M. (2008). Diversity of algal viruses in various North American freshwater environments. Aquat. Microb. Ecol. *51*, 13–21.

Short, S.M., and Suttle, C.A. (2002). Sequence analysis of marine virus communities reveals that groups of related algal viruses are widely distributed in nature. Appl. Environ. Microbiol. *68*, 1290–1296.

Short, S.M., and Suttle, C.A. (2003). Temporal dynamics of natural communities of marine algal viruses and eukaryotes. Aquat. Microb. Ecol. *32*, 107–119.

Slapeta, J., López-García, P., and Moreira, D. (2006). Global dispersal and ancient cryptic species in the smallest marine eukaryotes. Mol. Biol. Evol. *23*, 23–29.

Steward, G.F., Culley, A.I., Mueller, J.A., Wood-Charlson, E.M., Belcaid, M., and Poisson, G. (2013). Are we missing half of the viruses in the ocean? ISME J. *7*, 672–679. https://doi.org/10.1038/ismej.2012.121.

Steward, G.F., and Preston, C.M. (2011). Analysis of a viral metagenomic library from 200 m depth in Monterey Bay, California constructed by direct shotgun cloning. Virol. J. *8*, 287. https://doi.org/10.1186/1743-422X-8-287.

Suttle, C.A. (2000). Ecological, evolutionary, and geochemical consequences of viral infection of cyanobacteria and eukaryotic algae. In Viral Ecology, Hurst, C.J. ed. (Academic Press, San Diego), pp. 247–296.

Suttle, C.A., and Chan, A.M. (1995). Viruses infecting the marine Prymnesiophyte *Chrysochromulina* spp.: Isolation, preliminary characterization and natural abundance. Mar. Ecol. Prog. Ser. *118*, 275–282.

Tai, V., Lawrence, J.E., Lang, A.S., Chan, A.M., Culley, A.I., and Suttle, C.A. (2003). Characterization of HaRNAV, a single-stranded RNA virus causing lysis of *Heterosigma akashiwo* (Raphidophyceae). J. Phycol. *39*, 343–352.

Tarutani, K., Nagasaki, K., Itakura, S., and Yamaguchi, M. (2001). Isolation of a virus infecting the novel shellfish-killing dinoflagellate *Heterocapsa circularisquama*. Aquat. Microb. Ecol. *23*, 103–111.

Tarutani, K., Nagasaki, K., and Yamaguchi, M. (2000). Viral impacts on total abundance and clonal composition of the harmful bloom-forming phytoplankton *Heterosigma akashiwo*. Appl. Environ. Microbiol. *66,* 4916–4920.

Tomaru, Y., Hata, N., Masuda, T., Tsuji, M., Igata, K., Masuda, Y., Yamatogi, T., Sakaguchi, M., and Nagasaki, K. (2007). Ecological dynamics of the bivalve-killing dinoflagellate *Heterocapsa circularisquama* and its infectious viruses in different locations of western Japan. Environ. Microbiol. *9,* 1376–1383.

Tomaru, Y., Katanozaka, N., Nishida, K., Shirai, Y., Tarutani, K., Yamaguchi, M., and Nagasaki, K. (2004a). Isolation and characterization of two distinct types of HcRNAV, a single-stranded RNA virus infecting the bivalve-killing microalga *Heterocapsa circularisquama*. Aquat. Microb. Ecol. *34,* 207–218.

Tomaru, Y., Mizumoto, H., Takao, Y., and Nagasaki, K. (2009a). Co-occurrence of DNA and RNA viruses infecting the bloom-forming dinoflagellate *Heterocapsa circularisquama*, on the Japan Coast. Plankton Benthos Res. *4,* 129–134.

Tomaru, Y., and Nagasaki, K. (2004). Widespread occurrence of viruses lytic to the bivalve-killing dinoflagellate *Heterocapsa circularisquama* in the Western coast of Japan. Plankton Biol. Ecol. *51,* 1–6.

Tomaru, Y., Shirai, Y., and Nagasaki, K. (2008a). Ecology, physiology and genetics of a phycodnavirus infecting the noxious bloom-forming raphidophyte *Heterosigma akashiwo*. Fish. Sci. *74,* 701–711.

Tomaru, Y., Shirai, Y., Suzuki, H., Nagumo, T., and Nagasaki, K. (2008b). Isolation and characterization of a new single-stranded DNA virus infecting the cosmopolitan marine diatom Chaetoceros debilis. Aquat. Microb. Ecol. *50,* 103–112.

Tomaru, Y., Shirai, Y., Toyoda, K., and Nagasaki, K. (2011a). Isolation and characterisation of a single-stranded DNA virus infecting the marine planktonic diatom *Chaetoceros tenuissimus*. Aquat. Microb. Ecol. *64,* 175–184.

Tomaru, Y., Takao, Y., Suzuki, H., Nagumo, T., Koike, K., and Nagasaki, K. (2011b). Isolation and characterization of a single-stranded DNA virus infecting *Chaetoceros lorenzianus* Grunow. Appl. Environ. Microbiol. *77,* 5285–5293. https://doi.org/10.1128/AEM.00202-11.

Tomaru, Y., Takao, Y., Suzuki, H., Nagumo, T., and Nagasaki, K. (2009b). Isolation and characterization of a single-stranded RNA virus infecting the bloom-forming diatom *Chaetoceros socialis*. Appl. Environ. Microbiol. *75,* 2375–2381. https://doi.org/10.1128/AEM.02580-08.

Tomaru, Y., Tarutani, K., Yamaguchi, M., and Nagasaki, K. (2004b). Quantitative and qualitative impacts of viral infection on a *Heterosigma akashiwo* (Raphidophyceae) bloom in Hiroshima Bay, Japan. Aquat. Microb. Ecol. *34,* 227–238.

Tomaru, Y., Toyoda, K., Kimura, K., Hata, N., Yoshida, M., and Nagasaki, K. (2012). First evidence for the existence of pennate diatom viruses. ISME J. *6,* 1445–1448. https://doi.org/10.1038/ismej.2011.207.

Tomaru, Y., Toyoda, K., Kimura, K., Takao, Y., Sakurada, K., Nakayama, N., and Nagasaki, K. (2013a). Isolation and characterization of a single-stranded RNA virus that infects the marine planktonic diatom *Chaetoceros* sp (SS08-C03). Phycol. Res. *61,* 27–36.

Tomaru, Y., Toyoda, K., Suzuki, H., Nagumo, T., Kimura, K., and Takao, Y. (2013b). New single-stranded DNA virus with a unique genomic structure that infects marine diatom *Chaetoceros setoensis*. Sci. Rep. *3,* 3337. https://doi.org/10.1038/srep03337.

Toyoda, K., Kimura, K., Hata, N., Nakayama, N., Nagasaki, N., and Tomaru, Y. (2012). Isolation and characterization of a single-stranded DNA virus infecting the marine planktonic diatom *Chaetoceros* sp. (Strain Tg07-C28). Plankton Benthos Res. *7,* 20–28.

Van Etten, J.L., Burbank, D.E., Schuster, A.M., and Meints, R.H. (1985a). Lytic viruses infecting a *Chlorella*-like alga. Virology *140,* 135–143.

Van Etten, J.L., Burbank, D.E., Xia, Y., and Meints, R.H. (1983). Growth cycle of a virus, PBCV-1, that infects *Chlorella*-like algae. Virology *126,* 117–25.

Van Etten, J.L., Graves, M.V., Müller, D.G., Boland, W., and Delaroque, N. (2002). Phycodnaviridae – large DNA algal viruses. Arch. Virol. *147,* 1479–1516. https://doi.org/10.1007/s00705-002-0822-6.

Van Etten, J.L., and Meints, R.H. (1999). Giant viruses infecting algae. Annu. Rev. Microbiol. *53,* 447–494. https://doi.org/10.1146/annurev.micro.53.1.447.

Van Etten, J.L., Meints, R.H., Burbank, D.E., Kuczmarski, D., Cuppels, D.A., and Lane, L.C. (1981). Isolation and characterization of a virus from the intracellular green-alga symbiotic with *Hydra Viridis*. Virology *113,* 704–11.

Van Etten, J.L., Meints, R.H., Kuczmarski, D., Burbank, D.E., and Lee, K. (1982). Viruses of symbiotic *Chlorella*-like algae isolated from *Paramecium bursaria* and *Hydra viridis*. Proc. Natl. Acad. Sci. U.S.A. *79,* 3867–3871.

Van Etten, J.L., Van Etten, C.H., Johnson, J.K., and Burbank, D.E. (1985b). A survey for viruses from fresh water that infect a eucaryotic chlorella-like green alga. Appl. Environ. Microbiol. *49,* 1326–1328.

Venter, J.C., Remington, K., Heidelberg, J.F., Halpern, A.L., Rusch, D., Eisen, J.A., Wu, D., Paulsen, I., Nelson, K.E., Nelson, W., *et al.* (2004). Environmental genome shotgun sequencing of the Sargasso Sea. Science *304,* 66–74. https://doi.org/10.1126/science.1093857.

Vega Thurber, R.L., Barott, K.L., Hall, D., Liu, H., Rodriguez-Mueller, B., Desnues, C., Edwards, R.A., Haynes, M., Angly, F.E., Wegley, L., *et al.* (2008). Metagenomic analysis indicates that stressors induce production of herpes-like viruses in the coral Porites compressa. Proc. Natl. Acad. Sci. U.S.A. *105,* 18413–18418. https://doi.org/10.1073/pnas.0808985105.

Waters, R.E., and Chan, A.T. (1982). *Micromonas pusilla* virus: the virus growth cycle and associated physiological events within the host-cells; host range mutation. J. Gen. Virol. *63,* 199–206.

Wegley, L., Edwards, R., Rodriguez-Brito, B., Liu, H., and Rohwer, F. (2007). Metagenomic analysis of the microbial community associated with the coral *Porites astreoides*. Environ. Microbiol. *9,* 2707–2719.

Westbroek, P., Brown, C.W., Vanbleijswijk, J., Brownlee, C., Brummer, G.J., Conte, M., Egge, J., Fernandez, E., Jordan, R., Knappertsbusch, M., *et al.* (1993). A model system approach to biological climate forcing: the example of *Emiliania huxleyi*. Glob. Planet Change *8,* 27–46.

Weynberg, K.D., Allen, M.J., Ashelford, K., Scanlan, D.J., and Wilson, W.H. (2009). From small hosts come big viruses: the complete genome of a second *Ostreococcus tauri* virus, OtV-1. Environ. Microbiol. *11*, 2821–2839. https://doi.org/10.1111/j.1462-2920.2009.01991.x.

Williamson, S.J., Rusch, D.B., Yooseph, S., Halpern, A.L., Heidelberg, K.B., Glass, J.I., Andrews-Pfannkoch, C., Fadrosh, D., Miller, C.S., Sutton, G., *et al.* (2008). The Sorcerer II Global Ocean Sampling Expedition: metagenomic characterization of viruses within aquatic microbial samples. PLOS ONE *3*, e1456. https://doi.org/10.1371/journal.pone.0001456.

Wilson, W.H., Francis, I., Ryan, K., and Davy, S.K. (2001). Temperature induction of viruses in symbiotic dinoflagellates. Aquat. Microb. Ecol. *25*, 99–102.

Wilson, W.H., Gilg, I.C., Duarte, A., and Ogata, H. (2014). Development of DNA mismatch repair gene, MutS, as a diagnostic marker for detection and phylogenetic analysis of algal Megaviruses. Virology 466–467, 123–128. https://doi.org/10.1016/j.virol.2014.07.001.

Wilson, W.H., Schroeder, D.C., Allen, M.J., Holden, M.T., Parkhill, J., Barrell, B.G., Churcher, C., Hamlin, N., Mungall, K., Norbertczak, H., *et al.* (2005). Complete genome sequence and lytic phase transcription profile of a *Coccolithovirus*. Science *309*, 1090–1092.

Wilson, W.H., Schroeder, D.C., Ho, J., and Canty, M. (2006). Phylogenetic analysis of PgV-102P, a new virus from the English Channel that infects *Phaeocystis globosa*. J. Mar. Biol. Assoc. U. K. *86*, 485–490.

Wilson, W.H., Tarran, G.A., Schroeder, D., Cox, M., Oke, J., and Malin, G. (2002). Isolation of viruses responsible for the demise of an *Emiliania huxleyi* bloom in the English Channel. J. Mar. Biol. Assoc. U. K. *82*, 369–377.

Wilson, W.H., Van Etten, J.L., and Allen, M.J. (2009). The *Phycodnaviridae*: the story of how tiny giants rule the world. In Lesser known large dsDNA viruses, pp. 1–42.

Yamada, T., Onimatsu, H., and Van Etten, J.L. (2006). Chlorella viruses. Adv. Virus Res. *66*, 293–336.

Yau, S., Lauro, F.M., DeMaere, M.Z., Brown, M.V., Thomas, T., Raftery, M.J., Andrews-Pfannkoch, C., Lewis, M., Hoffman, J.M., Gibson, J.A., *et al.* (2011). Virophage control of antarctic algal host-virus dynamics. Proc. Natl. Acad. Sci. U.S.A. *108*, 6163–6168. https://doi.org/10.1073/pnas.1018221108.

Yoon, H.S., Hackett, J.D., Ciniglia, C., Pinto, G., and Bhattacharya, D. (2004). A molecular timeline for the origin of photosynthetic eukaryotes. Mol. Biol. Evol. *21*, 809–818. https://doi.org/10.1093/molbev/msh075.

Yoshida-Takashima, Y., Yoshida, M., Ogata, H., Nagasaki, K., Hiroishi, S., and Yoshida, T. (2012). Cyanophage infection in the bloom-forming cyanobacteria *Microcystis aeruginosa* in surface freshwater. Microbes Environ. *27*, 350–355.

Zhang, W., Zhou, J., Liu, T., Yu, Y., Pan, Y., Yan, S., and Wang, Y. (2015). Four novel algal virus genomes discovered from Yellowstone Lake metagenomes. Sci. Rep. *5*, 15131. https://doi.org/10.1038/srep15131.

Zhong, X., and Jacquet, S. (2014). Contrasting diversity of phycodnavirus signature genes in two large and deep western European lakes. Environ. Microbiol. *16*, 759–773. https://doi.org/10.1111/1462-2920.12201.

Zhou, J., Sun, D., Childers, A., McDermott, T.R., Wang, Y., and Liles, M.R. (2015). Three novel virophage genomes discovered from Yellowstone Lake metagenomes. J. Virol. *89*, 1278–1285. https://doi.org/10.1128/JVI.03039-14.

Zhou, J., Zhang, W., Yan, S., Xiao, J., Zhang, Y., Li, B., Pan, Y., and Wang, Y. (2013). Diversity of virophages in metagenomic data sets. J. Virol. *87*, 4225–4236. https://doi.org/10.1128/JVI.03398-12.

Zingone, A., Natale, F., Biffali, E., Borra, M., Forlani, G., and Sarno, D. (2006). Diversity in morphology, infectivity, molecular characteristics and induced host resistance between two viruses infecting *Micromonas pusilla*. Aquat. Microb. Ecol. *45*, 1–14.

Zingone, A., Sarno, D., and Forlani, G. (1999). Seasonal dynamics in the abundance of *Micromonas pusilla* (Prasinophyceae) and its viruses in the Gulf of Naples (Mediterranean Sea). J. Plankton Res. *21*, 2143–2159.

Protozoal Giant Viruses

11

Dorine G.I. Reteno[1†], Leena H. Bajrai[1,2†], Sarah Aherfi[1], Philippe Colson[1] and Bernard La Scola[1*]

[1]L' Institute de recherche pour le développement (IRD), Microbes Evolution Phylogeny and Infections (MEPHI), Institut Hospitalo-Universitaire Méditerranée Infection, Aix Marseille Université, Marseille, France.
[2]Department of Biochemistry, Faculty of Science, King Abdulaziz University, Jeddah, Saudi Arabia.

[†]These authors contributed equally to the work.

*Correspondence: bernard.la-scola@univ-amu.fr

https://doi.org/10.21775/9781910190852.11

Abstract

Giant viruses that infect protists have been discovered during the past decade. They are named giant viruses due to the large size of their viral particle (over 100 nm in diameter but up to 1.5 μm) and to the extraordinary size of their genome, ranging from 150 kilobase pairs (kbp) to nearly 2.8 megabase pairs (Mbp). Their genomes, generally in the form of double-stranded DNA, may be linear or circular. They have been grouped into the nucleocytoplasmic large DNA virus (NCLDV) superfamily. The NCLDVs are divided into six families. Among these six viral families, three infect protozoa: the *Mimiviridae* and *Marseilleviridae* families infect amoebae and algae, and the newly discovered *Pandoravirus salinus* and *P. dulcis*, infect amoebae.

Introduction

Viruses are obligate parasites of Eukarya, Archaea and Bacteria (Raoult and Forterre, 2008). Formerly, viruses were considered to be small particles with small genomes containing a relatively limited number of protein-encoding genes and the ability to traverse through 0.2-μm-pore filters. Giant viruses that infect protists living in aqueous environments, however, were discovered relatively recently (Raoult *et al.*, 2004). These viruses do not pass through filters classically used to stop bacteria. Moreover, they have double-stranded DNA genomes larger than 150 kilobase pairs (kbp) and ranging up to over 2.5 megabase pairs (Mbp), which encode hundreds of proteins (Colson *et al.*, 2012; Van Etten *et al.*, 2010; Yutin and Koonin, 2009). These protist-associated giant viruses have been linked to the nucleocytoplasmic large DNA viruses superfamily (NCLDVs) (Iyer *et al.*, 2001, 2006; Yutin and Koonin, 2009).

The NCLDVs are a group of giant viruses infecting many organisms, including both animals and protists. They are characterized in part by the extraordinary size of their genomes and in part by high levels of genetic diversity. Based on phylogenetic and phyletic analyses, NCLDVs nevertheless compose a monophyletic group, and they were proposed to be a fourth domain of life (Boyer *et al.*, 2010). NCLDVs are divided into six families: the *Poxviridae*, whose members infect insects and some vertebrates; the *Asfarviridae*, one of which, the African swine fever virus, infects pigs; the *Asco-Iridoviridae*, which encompass viruses infecting some insects, fish and amphibians; the *Phycodnaviridae*, whose members infect algae; and finally *Mimiviridae* and *Marseilleviridae*, which encompass giant viruses of amoebae as well as algae (Boyer *et al.*, 2010; Iyer *et al.*, 2006; Yutin and Koonin, 2009). To these six families were added the recently discovered *Pandoravirus salinus* and *P. dulcis*, which constitute a full-fledged branch of the NCLDVs phylogenetic tree, although their exact phylogenetic position needs to be clarified (Philippe *et al.*, 2013). NCLDVs were recently proposed to be reclassified in a new viral order named 'Megavirales' (Colson *et al.*, 2012, 2013a).

The first mimivirus was discovered and sequenced in 2003 (La Scola *et al.*, 2003; Raoult *et al.*, 2004). It was isolated from water of a cooling tower in England by co-culture with *Acanthamoeba polyphaga* and was named Acanthamoeba polyphaga mimivirus. The discovery of mimivirus by both its particle size (approximately 750 nm) and its genetic repertoire (1.2-Mbp genome containing 911 protein-coding genes) seemingly negate the dogma that viruses are completely dependent on their host for macromolecular synthesis (La Scola *et al.*, 2003; Raoult *et al.*, 2004). With the presence of numerous unique genes that were not found in other NCLDVs, mimiviruses were classified into a new NCLDV family, named *Mimiviridae*. From 2003 to 2008, Acanthamoeba polyphaga mimivirus remained the only member of the *Mimiviridae*. Since then, several other giant viruses related to mimivirus have been discovered, and the genetic diversity observed within family *Mimiviridae* allows us to divide these viruses into two groups. The first group includes three lineages: lineage A with Acanthamoeba polyphaga mimivirus; lineage B with Acanthamoeba polyphaga moumouvirus; and lineage C with Megavirus chilensis. The second group includes Cafeteria roenbergensis virus (CroV) (Arslan *et al.*, 2011; Boughalmi *et al.*, 2013b; Colson *et al.*, 2011a, 2012; La Scola *et al.*, 2010; Pagnier *et al.*, 2013).

The first member of family *Marseilleviridae* was isolated in 2007 from a sample of water collected in Paris, France. Currently, this family comprises three lineages, with marseillevirus, lausannevirus and tunisvirus as the leading members of lineages A, B and C, respectively (Table 11.1) (Boughalmi *et al.*, 2013a; Boyer *et al.*, 2009; Colson *et al.*, 2013c; Thomas *et al.*, 2011).

The vast majority of giant viruses have been isolated from various types of environmental samples (such as freshwater, cooling towers, seawater, and soil), but recent data show the presence of these viruses in human samples (stool, blood, bronchoalveolar fluid) (Boughalmi *et al.*, 2013a,b; Colson *et al.*, 2013c; Lagier *et al.*, 2012; Pagnier *et al.*, 2013; Saadi *et al.*, 2013a,b). These findings have fuelled the debate about the pathogenicity of mimiviruses and marseilleviruses. Because of the content and the size of its genome, the discovery of mimivirus has generated considerable interest in giant viruses within the scientific community.

This review focuses on the *Megavirales* members that parasitize protozoa, chiefly the mimiviruses, including Cafeteria roenbergensis virus, the marseilleviruses and the more recently described pandoraviruses, pithoviruses, faustoviruses, mollivirus, and kaumoebavirus.

Genetic classification of *Mimiviridae* and *Marseilleviridae* families

The *Mimiviridae* and *Marseilleviridae* families have been associated with the NCLDV superfamily (Colson *et al.*, 2013c; Raoult *et al.*, 2004). This superfamily was first described in 2001 (Iyer *et al.*, 2001). It includes four other families outside *Mimiviridae* and *Marseilleviridae*: the *Phycodnaviridae*, the *Poxviridae*, the *Asfarviridae* and the *Asco-Iridoviridae* (Fig. 11.1 and Table 11.1). The NCLDVs are a group of viruses infecting animals and diverse unicellular eukaryotes. Based on phylogenetic and phyletic analysis, these viruses compose a monophyletic group; i.e. they likely have a common ancestor and share a set of genes encoding proteins that may allow relative independence from the cell for their replication, expression and morphogenesis. *Megavirales* members have in their gene repertoire nine core genes that define them as a monophyletic group, including viral hallmark genes, and 22 genes that are found in at least three of the four viral families initially defined by Iyer *et al.* (Iyer *et al.*, 2001, 2006; Yutin and Koonin, 2009). The nine core genes encode a VV D5-type ATPase, a DNA polymerase, a VV A32 virion-packaging ATPase, a VV A18 helicase, a capsid protein, a thiol oxidoreductase, a VV DR6/D11L-like helicase, an S/T protein kinase and a VLTF2 transcription factor. In addition, *Megavirales* genes are classified into clusters of orthologous groups of proteins (COGs), which allows the annotation of newly sequenced viral genomes and optimizes the analyses of their evolution. These clusters were named NCVOGs (nucleo-cytoplasmic virus orthologous groups) (Yutin *et al.*, 2009).

In the 45 NCLDV genomes analysed, 1445 NCVOGs have been described. 177 NCVOGs are present in at least two *Megavirales* families, while others are family-specific proteins. In addition, only

Table 11.1 The essential characteristics of main giant viruses' representatives

Virus name	Host	Particle diameter (nm)	Genome size	Genome conformation	G+C (%)	Number of predicted protein coding genes	References
Paramecium bursaria chlorella virus 1 (PBCV-1)	*Chlorella* NC64A	190	313–370 kb	Closed linear dsDNA	40	802	Dunigan *et al.* (2006)
Emiliania huxleyi virus 86 (EhV-86)	*Emiliania huxleyi*	160–200	407–415 kb	Circular/ circular dsDNA	40	472	Dunigan *et al.* (2006), Wilson *et al.* (2009)
Ectocarpus siliculosus virus 1 (EsV-1)	*Ectocarpus siliculosus*	130–200	160–340 kb	Open linear ssDNA	52	231	Dunigan *et al.* (2006)
Micromonas sp. RCC1109 virus (MpV1)	*Micromonas* sp.	115–200	184,09 kb	Linear dsDNA	40	244	ICTV, Dunigan *et al.* (2006)
Chrysochromulina brevifilum virus PW1 (CbV-PW1)	*Chrysochromulina brevifilum*	120–160	485–510 kb	Linear dsDNA	n.a.	n.a.	Dunigan *et al.* (2006)
Heterosigma akashiwo virus 01 (HaV01)	*Heterosigma akashiwo*	202	294 kb	Linear DNA	n.a.	n.a.	Dunigan *et al.* (2006)
Canarypox virus	Birds	200	365 kb	Linear dsDNA	30	328	ICTV, Tulman *et al.* (2004)
Anomala cuprea entomopoxvirus	*Anomala cuprea*	220–265	246 kb	Linear dsDNA	20	263	ICTV; Mitsuhashi *et al.* (2014)
African swine fever virus	Pigs	172–191	170–182 kb	Linear dsDNA	39	151	ICTV, Carrascosa *et al.* (1984)
Diadromus pulchellus ascovirus 4a	*Diadromus pulchellus*	130	119 kb	Circular dsDNA	50	119	ICTV, Bigot *et al.* (1997)
Heliothis virescens ascovirus 3e	*Heliothis virescens*	130	186 kb	Linear dsDNA	45	180	ICTV, Asgari *et al.* (2006)
Trichoplusia ni ascovirus 2c	*Trichoplusia ni*	100–150	174 kb	Circular dsDNA	35	164	ICTV, Wang *et al.* (2006)
Spodoptera frugiperda ascovirus 1a	*Spodoptera frugiperda*	100–150	157 kb	Circular dsDNA	49	123	ICTV, Bideshi *et al.* (2006)
Aedes taeniorhynchus iridescent virus	*Aedes taeniorhynchus*	180	191 kb	Linear dsDNA	47	126	ICTV, Delhon *et al.* (2006)
Invertebrate iridescent virus 6	Insects	120–130	212 kb	Circular dsDNA	28	468	ICTV, Jenkins *et al.* (2011)
Lymphocystis disease virus – isolate china	Fish	50–200	186 kb	Linear dsDNA	27	239	ICTV, Yan *et al.* (2011)
Lymphocystis disease virus 1	Fish	50–200	103 kb	Linear dsDNA	29	110	ICTV, Williams, (1998)
Infectious spleen and kidney necrosis virus	*Siniperca chuatsi*	150–250	111 kb	Linear dsDNA	54	125	ICTV, He *et al.* (2001)

Table 11.1 Continued

Virus name	Host	Particle diameter (nm)	Genome size	Genome conformation	G+C (%)	Number of predicted protein coding genes	References
Ambystoma tigrinum virus	*Ambystoma tigrinum*	150	106 kb	Linear dsDNA	54	95	ICTV, Collins *et al.* (2004), Jancovich *et al.* (1997)
Acanthamoeba polyphaga mimivirus	*Acanthamoeba polyphaga*	400	1.2 Mb	Linear dsDNA	28	979	Raoult *et al.* (2004)
Acanthamoeba castellanii mamavirus	*Acanthamoeba castellanii*	400	1,191 kb	Linear dsDNA	28	1023	La Scola *et al.* (2008)
Acanthamoeba polyphaga moumouvirus	*Acanthamoeba polyphaga*	420	1,021 kb	Linear dsDNA	25	930	La Scola *et al.* (2010), Yoosuf *et al.* (2012)
Megavirus chilensis	*Acanthamoeba polyphaga* *Acanthamoeba castellanii* *Acanthamoeba griffini*	520	1,259 kb	Linear dsDNA	25	1120	Legendre *et al.* (2012)
Megavirus LBA	*Acanthamoeba polyphaga*	544 ± 10	1,230 kb	Linear dsDNA	25	1.176	Saadi *et al.* (2013a)
Cafeteria roenbergensis virus (Crov)	*Cafeteria roenbergensis*	300	730 kb	Linear dsDNA	23	544	Fischer *et al.* (2011); Colson *et al.* (2011a)
Marseillevirus	*Acanthamoeba polyphaga* *Acanthamoeba castellanii*	250	368 kb	Circular dsDNA	45	428	Boyer *et al.* (2009)
Lausannevirus	*Acanthamoeba polyphaga*	190-220	347 kb	Linear / circular dsDNA	43	444	Thomas *et al.* (2011)
Cannes 8 virus	*Acanthamoeba polyphaga*	190	374 bp	Circular dsDNA	45	484	La Scola *et al.* (2010), Aherfi *et al.* (2013)
Pandoravirus dulcis	*Acanthamoeba castellanii*	500	1.908 Mb	Linear dsDNA	64	1487	ICTV, Philippe *et al.* (2013)
Pandoravirus salinus	*Acanthamoeba castellanii*	500	2.47 Mb	Linear dsDNA	62	2541	ICTV, Philippe *et al.* (2013)

five NCVOGs, including proteins from all 45 analysed viruses, have been found. These proteins are the major capsid protein (orthologues of vaccinia virus D13 protein), a primase-helicase (VV D5), a family B DNA polymerase (VV E9), a packaging ATPase (VV A32) and a transcription factor (VV A2). In addition to their monophyletic origin, these giant viruses were suggested to compose a new branch of life aside Bacteria, Archaea, and Eukarya (Boyer *et al.*, 2010). In fact, unconventional characteristics of giant viruses, especially mimiviruses and marseilleviruses – regarding the structure of the viral particle (icosahedral particle with 150 to 500 nm in diameter), the genome size (between 103 and 1259 kb and harbouring 95 to 1120 genes) and the composition of the genome (viral messenger

Fig. 1

a.

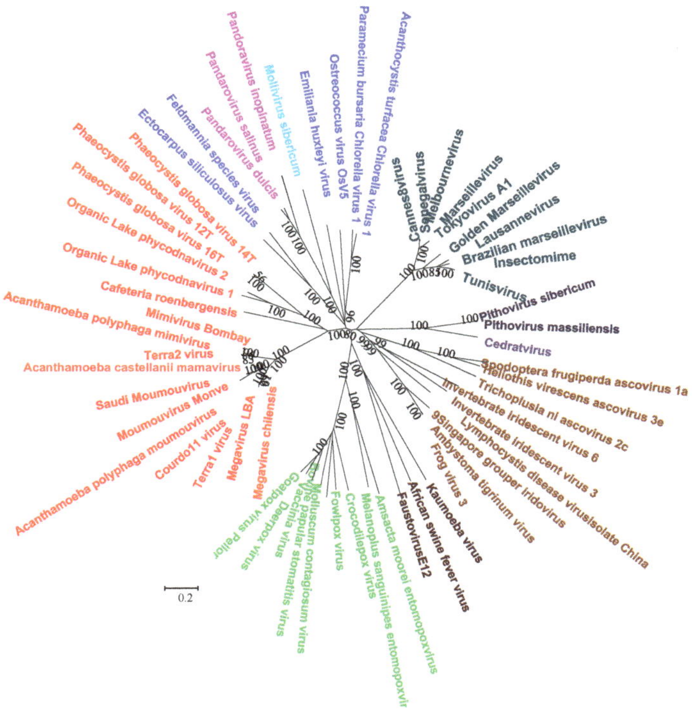

Figure 11.1 Phylogenetic reconstruction of the proposed order *Megavirales*, including the families *Mimiviridae* and *Marseilleviridae*. The evolutionary history was inferred based on the family B DNA polymerase gene using the neighbour joining method based on the JTT matrix-based model with MEGAv5. The percentage of trees in which the associated taxa clustered together is shown next to the branches. The analysis involved 61 amino acid sequences and a total of 3532 positions in the final dataset.

RNAs; proteins involved in translation) as well as major biological features (virus reproduction within cytoplasmic factories) – indicate that the current definition of virus is inappropriate and led to the proposal to reclassify the NCLDV families into the new order *Megavirales* (Colson *et al.*, 2012, 2013a). In addition, the NCLDV superfamily has no recognized taxonomic status according to the International Committee on Taxonomy of Viruses (ICTV). In the current ICTV classification, none of the NCLDV families are assigned to a viral order.

Mimiviridae family

Acanthamoeba polyphaga mimivirus

It was in 1992 that mimivirus (for 'microbe-mimicking virus') was isolated from amoebal cultures for the first time. These amoebae came from a sample of cooling-tower water during an outbreak of pneumonia in England. The virus was mistakenly identified as a bacterium because it could be stained Gram-positive. It was only in 2003 that the viral nature of this lytic microorganism for *Acanthamoeba polyphaga* was revealed by electron microscopy and named APMV (Acanthamoeba polyphaga mimivirus) (La Scola *et al.*, 2003; Raoult *et al.*, 2007).

Phylogenetic analysis classified APMV within the group of nucleocytoplasmic large DNA viruses. APMV is closely related to *Phycodnaviridae*, *Iridoviridae* and *Poxviridae*, all members of NCLDV. (Colson *et al.*, 2012, 2013a; Iyer *et al.*, 2006). With the presence of numerous unique genes not found in other NCLDVs, including eight that putatively encode proteins involved in translation such as four aminoacyl-tRNA synthetases (for Arg, Cys, Met and Tyr); five that putatively encode proteins involved in DNA repair; and three that putatively encode chaperone proteins, mimivirus is classified into the new NCLDV family *Mimiviridae* (Koonin and Yutin, 2010; Raoult *et al.*, 2004).

The *Mimiviridae* family can be divided into two groups. The first group corresponds to mimiviruses that infect Acanthamoebae and includes three phylogenetic lineages: lineage A with Acanthamoeba polyphaga mimivirus and Acanthamoeba polyphaga mamavirus; lineage B with moumouvirus; and lineage C with Megavirus chilensis (Arslan *et al.*, 2011; Colson *et al.*, 2012; Desnues *et al.*, 2012a). The second group includes Cafeteria roenbergensis virus (CroV) (Fischer *et al.*, 2011) and additional smaller mimiviruses that were recently reclassified into the family *Mimiviridae* and include Organic Lake phycodnaviruses, infect marine or freshwater eukaryotic algae, and Phaeocystis globosa viruses (Yutin *et al.*, 2013).

Until 2013 and the discovery of pandoraviruses, mimiviruses were the largest known viruses and among those the most extensively studied. They are ubiquitous in environments such as freshwater, seawater and soil (Boughalmi *et al.*, 2013b; Ghedin and Claverie, 2005; La Scola *et al.*, 2010; Monier *et al.*, 2008a,b; Pagnier *et al.*, 2013; Raoult *et al.*, 2007).

Mimivirus particles observed by transmission electron microscopy show a size of approximately 650 nm with fibrils, an icosahedral capsid of 450 nm (Fig. 11.2) and the presence of a gap approximately 300–500 Å between the genome sac and the outer capsid. The presence of external fibrils is specific to mimivirus (Suzan-Monti *et al.*, 2006; Van Etten *et al.*, 2002).The cryo-EM reconstruction shows that the nucleocapsid sac keeps the virus in a rigid and consistent morphology. Lengthy internal fibrils have been observed by atomic force microscopy (AFM) after application of mechanical force to break the outer layers of mimivirus capsid. The fibrils, which have a diameter of about 0.9 to 1 nm, display a 7 nm periodicity of dark and light bands. It has been suggested that the nucleocapsid could be supported by these internal fibres (Klose *et al.*, 2010; Xiao *et al.*, 2009). The unique feature of mimiviruses is a star-shaped area that occupies a fivefold vertex of each icosahedral virion. This star-shaped apparatus has a starfish shape and was named a 'stargate' by Zauberman (Zauberman *et al.*, 2008). It confers a 5-fold symmetry on the mimivirus particle. The nucleocapsid has a concave depression facing the 'starfish'-associated vertex, which suggests a specialized organization that might be necessary for host infection through the arms of the starfish. These

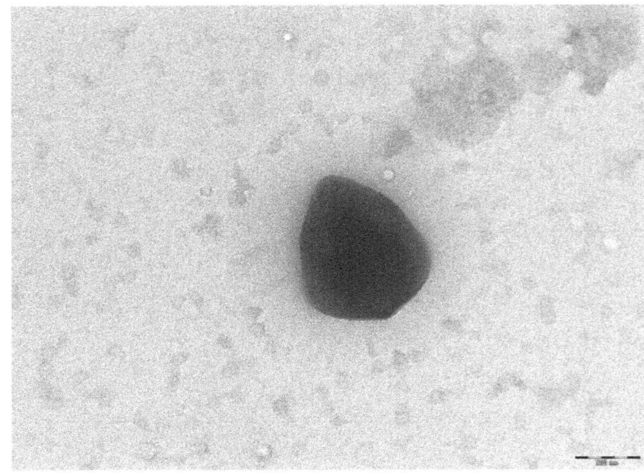

Figure 11.2 Transmission electron microscopy of mimivirus showing the morphology of a typical mimivirus particle. Scale bar indicates 300 nm.

arms are located around the five faces of the unique viral vertex (Klose *et al.*, 2010; Kuznetsov *et al.*, 2010; Xiao *et al.*, 2009; Zauberman *et al.*, 2008).

From the outside to the inside, the capsid is composed of several membrane layers. *Mimiviridae* viruses are composed of an outer layer of dense fibrils 120 to 140 nm long. These surface fibrils are resistant to proteases, suggesting they are protected by peptidoglycan. That could explain why mimivirus can be positive for Gram stain. Staple fibre components with peptidoglycan components may act as a lure to attract amoebas and thus facilitate the attachment to the host cell (Claverie *et al.*, 2006; Xiao *et al.*, 2009). Similar to other NCLDVs, mimivirus has an internal lipid membrane surrounding the central core. This central core is dense when viewed using electron microscopy and contains the 1.2 Mb linear double-stranded (ds)DNA viral genome. Similar to other NCLDVs, mimivirus has hexameric capsomers consisting of double jelly-roll structures. Jelly-roll structures are structural motifs with a role in the assembly of viral capsids. It consists of two opposing β-sheets, each consisting of four anti-parallel β-strands creating a wedge-shaped structure. Mimivirus major capsid protein (MCP) is similar to that of PBCV-1 (Klose *et al.*, 2010). Indeed, it has *four* genes that are homologous to the double jelly-roll capsomer-containing PBCV1 Vp54, including L425 and R441, and to the MCPs of other large dsDNA viruses (Klose *et al.*, 2010; Mutsafi *et al.*, 2013; Xiao *et al.*, 2009).

Mimiviruses have a membrane sac surrounding the genomic material. This genetic material is linear double-stranded DNA of approximately 1.2 Mbp (Klose *et al.*, 2010; Raoult *et al.*, 2004). The mimivirus genome might adopt a circular topology, as some other NCLDV genomes do, resulting in the annealing of inverted repeats found towards the genome's extremities. The nucleotide composition of the mimivirus genome shows an A+T content of 72%.

A total of 90.5% of the mimivirus genome is predicted to encode proteins. Thus, 1262 putative open reading frames (ORFs) with lengths ≥100 amino acids have been described. Among the 911 proteins initially annotated, 298 were found to have a defined function. Of the predicted mimivirus ORFs, 194 matched significantly with 108 COGs (Raoult *et al.*, 2004; Suzan-Monti *et al.*, 2006). The mimivirus genome contains homologues of all NCLDV class I core genes, six homologues of the eight class II core genes, 11 homologues of the 14 class III core genes and 16 homologues of the 30 class IV core genes (Iyer *et al.*, 2001; Raoult *et al.*, 2004). In addition, several genes never identified in a viral genome have been found in the mimivirus genome. These genes can be classified into *four* functional categories, namely protein translation, DNA repair enzymes, chaperones and new enzymatic pathways (Raoult *et al.*, 2004). Additionally, mimivirus is the only virus and one of the few microorganism to harbour genes encoding a set of three topoisomerases, namely type IA, type IB and type II (Raoult *et al.*, 2004).

It has been suggested that mimivirus infects its host by triggering a phagocytic-like process, in which fibrils play an essential role (Kuznetsov *et al.*, 2010). Shortly after the exposure of amoeba to viruses, we can observe virions attached to cell surface of amoeba. Several virions can then be attached to the same cell. On the other hand, a stargate has not been observed on these particles, possibly because it was turned towards the host's surface. This is consistent with the hypothesis that the stargate has a role in cell entry.

Between 1.5 and 2 hours post infection (PI), we can observe virus activity, and most cells have already formed small membranous sacs called transport vesicles. These transport vesicles accumulate in the cell over a period of 1 to 2 hours. At 4 hours PI, much of the cell is filled with the vesicles. The transport vesicles are significantly different from virus particles at any of the latter several stages of mimivirus development and the precise origin of these vesicles is not entirely clear. They have not been observed in uninfected cells, and they do not seem to come from a rearrangement of preexisting vesicles or organelles. Also at 4 hours PI, there is an eclipse phase where viral particles have completely disappeared from cytoplasm. During the eclipse phase there is genome replication, transcription and translation of mRNA into protein by the cell machinery. At this stage, it is impossible to observe and to isolate a virus particle. (Kuznetsov *et al.*, 2010, 2013; Mutsafi *et al.*, 2013; Suzan-Monti *et al.*, 2006).

From 5 hours to 8 hours post-infection, virus factories (VF) appear in the cytoplasm. The accumulation of transport vesicles precedes the appearance of a recognizable virus factory. VFs are distinct bodies that form in the cell cytoplasm and are responsible for the synthesis of the icosahedral capsids and filling the capsids with viral DNA along with associated proteins and most likely RNA, although they vary in size and appearance during infection. Initially they are relatively small (5 to 10 μm in diameter) and dense but grow with time. When the cell is filled with viruses, VF become disorganized and diffuse, filling almost an entire cell (Fig. 11.3). During their most productive phase, they are approximately 10 to 20 μm in diameter. At the periphery of the cytoplasmic VF, we can observe many membranes and partially finished icosahedral shaped particles at 10 hours post infection (PI) using transmission electron microscopy

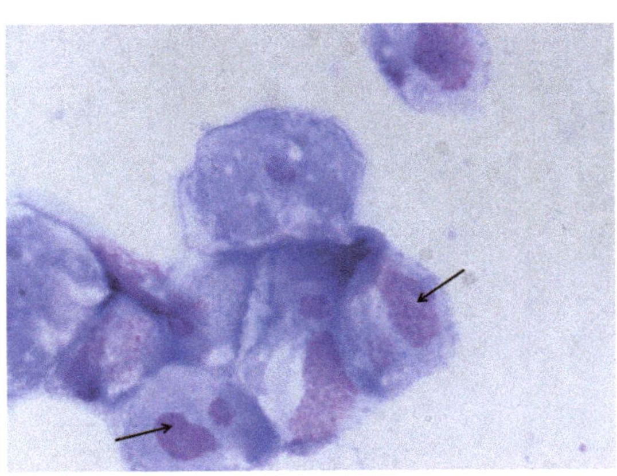

Figure 11.3 Hemacolor staining of *A. polyphaga* infected with marseillevirus visualized by optical microscopy at x1000 magnification. The arrows show the virus factories.

(TEM) and electron tomography (ET) (Fig. 11.4). These particles are covered with inner membrane and outer capsid layers A thin gap separates the two layers. The outer layer seems to give the icosahedral shape to the viral particle; it presumably is the mimivirus capsid protein. Furthermore, the inner membrane layer is connected with membranes that accumulate at the periphery of the virus factory. These connected membranes have open ends. Thus, it has been hypothesized that these open membrane intermediates are used for viral membrane assembly (Mutsafi *et al.*, 2013; Suárez *et al.*, 2013). At 8 hours to 12 hours PI, we can observe the presence of an icosahedral virus capsid completely assembled at the periphery of viral factory. At 24 hours PI, amoebal lysis occurs (Mutsafi *et al.*, 2010, 2013; Suzan-Monti *et al.*, 2006; Zauberman *et al.*, 2008).

Like any other virus, mimiviruses are obligate intracellular parasites that require the metabolism and functions of the host cell to multiply. Currently they can infect some amoebae, such as *A. polyphaga*, *A. castellanii* and *A. mauritaniensis*, but according to recent studies, mimiviruses are able to infect other organisms, such as the phagocytic protist *Cafeteria roenbergensis*, a widespread marine heterotrophic flagellate (Colson *et al.*, 2011a; La Scola *et al.*, 2008; Suzan-Monti *et al.*, 2006). Mimiviruses are also able to infect macrophages via phagocytosis and replicate within the macrophages (Silva *et al.*, 2014). In fact, Ghigo *et al.* (2008) show that mimivirus particles can be internalized by macrophages but not by non-professional phagocytic cells. They also demonstrated that mimiviruses infect macrophages via phagocytosis; this was the first evidence that mimiviruses are internalized by macrophages by a mechanism normally used by bacteria and parasites.

Until 2008, APMV remained the lone member of the family *Mimiviridae*. In 2008, a new strain of APMV was isolated by inoculating *A. polyphaga* with water from a cooling tower in Paris. This strain was named A. polyphaga mamavirus (Colson *et al.*, 2011b; La Scola *et al.*, 2008). At the same time, a small viral particle of 50 μm in diameter was observed in cytoplasm and in virus factories of infected cells (Fig. 11.5). This small viral particle was named Sputnik and is considered a virophage (Claverie and Abergel, 2009; La Scola *et al.*, 2008) (see Chapter 12). Since then, several other mimiviruses have been isolated from different environments (Boughalmi *et al.*, 2013b; Colson *et al.*, 2011a,b; Ghedin and Claverie, 2005; La Scola *et al.*, 2008, 2010; Lagier *et al.*, 2012). Pagnier *et al.* (2013) recently described the collection of giant viruses they have isolated from various environmental samples such as soil samples, freshwater, seawater and hypersaline soil and hypersaline water. This collection encompasses a wide range of mimiviruses and marseilleviruses. Thus, their team currently has 43 strains of mimiviruses (14 of lineage A, 6 of lineage B, and 23 of lineage C) and 17 strains of marseilleviruses. They have accumulated the largest collection of giant viruses in the world (Boughalmi *et al.*, 2013a,b; Ngounga *et al.*, 2013; Pagnier *et al.*, 2013; Saadi *et al.*, 2013a,b).

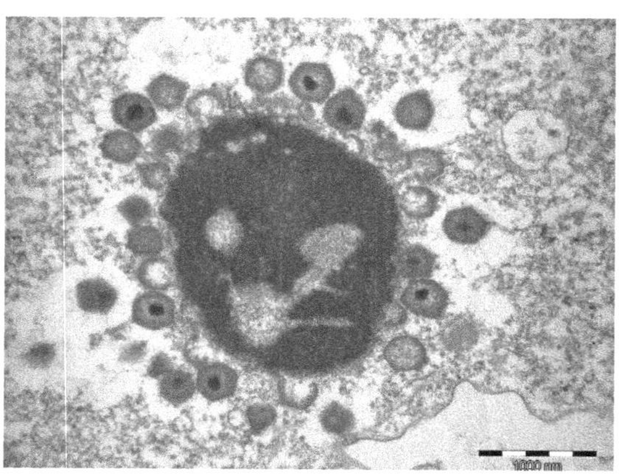

Figure 11.4 Transmission electron microscopy images showing mimivirus viral factory.

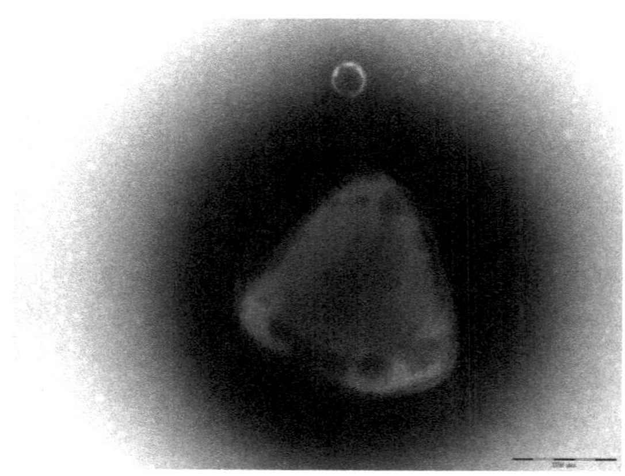

Figure 11.5 Transmission electron microscopy images showing mamavirus and its virophage. Scale bar indicates 200 nm.

Acanthamoeba castellanii mamavirus

Acanthamoeba castellanii mamavirus is a new isolate of Acanthamoeba polyphaga mimivirus. It was first isolated by inoculating *A. polyphaga* with water from a cooling tower in Paris. This virus was named Acanthamoeba castellanii mamavirus because, after its isolation, all the subsequent experiments were carried out in *Acanthamoeba castellanii* and because it seemed to be larger than mimivirus after its observation under transmission electron microscopy (Fig. 11.5) (Colson *et al.*, 2011b; La Scola *et al.*, 2008). The main morphological and culture characteristics of mamavirus are the same as mimivirus. The particle of mamavirus is an icosahedral capsid of approximately 500 nm with dense fibrils of 120 to 140 nm. Similar to mimivirus, mamavirus has one icosahedral five-fold vertex with a starfish-shaped structure, presumably a stargate.

The 1,191,693 bp mamavirus genome includes 1,023 ORFs identified as putative protein coding genes. Its genome is highly AT-rich with an A+T content of approximately 72%. Mamavirus ORFs are 98.3% similar to mimivirus ORFs. Orthologs to mimivirus open reading frames (ORFs) were detected for 99% of the predicted mamavirus genes, with amino acid identity ranging from 75% to 100%. Therefore, mamavirus is closely related to mimivirus and could be considered a second strain of mimivirus. In the mamavirus genome, the genes are nearly equally distributed between the two DNA strands, with 497 genes on the direct strand and 526 on the reverse strand. The average size of intergenic regions is 133 nt, with a predicted protein-coding density of 0.86 genes/kb. Comparison of the predicted proteins of mamavirus with the predicted protein sequences of mimivirus using an all-against-all BLASTP search yielded 833 bidirectional best hits (BBHs), displaying an average of 98.3% amino acid identity and 98.8% nucleotide identity. Additionally, the genomes of mamavirus and mimivirus are highly similar but present differences in their terminal regions. Notably, the mamavirus genome has an approximately 13,000 bp fragment at the 5′-terminal fragment with disrupted duplicated genes, which is absent in the mimivirus genome. These differences, however small, may reveal pathways of mimivirus genome evolution. In addition, comparison of the mamavirus and mimivirus genomes showed 29 mimivirus ORFs and 46 mamavirus ORFs that were partially or completely absent in the counterpart genome.

Acanthamoeba polyphaga moumouvirus

Acanthamoeba polyphaga moumouvirus is a new member of the *Mimiviridae* family that infects the amoeba *A. polyphaga.* Moumouvirus composes a distinct phylogenetic branch in the *Mimiviridae.* It belongs to *Mimiviridae* lineage B (Desnues *et al.*, 2012b), phylogenetically related to Megavirus chilensis but with the smallest genome among the three lineages of mimivirus. It was isolated in February, 2008, from water from an industrial cooling tower located in southeast France, by coculture with the amoeba, *Acanthamoeba polyphaga* (La Scola *et al.*, 2010; Suzan-Monti *et al.*, 2006; Yutin *et al.*, 2013). Like all other *Mimiviridae,* moumouvirus has an icosahedral capsid (Fig. 11.6). This capsid is approximately 420 nm in diameter and is also covered by a dense layer of fibrils that are approximately 100 nm thick. Moumouvirus particles exhibit a distinctive starfish-like vertex, as do other *Mimiviridae.* Viral factories were also observed within the *A. polyphaga* cytoplasm during the replication cycle of the moumouvirus, and morphology of these moumouvirus factories is similar to that observed for Acanthamoeba polyphaga mimivirus, Acanthamoeba castellanii mamavirus and Megavirus chilensis.

The moumouvirus genome is a linear dsDNA molecule that is 1,021,348 bp in length and contains 930 ORFs identified as putative protein-coding genes. These are uniformly distributed on both DNA strands, encoding proteins with a mean length of 290 amino acids. The average size of intergenic regions is 130 nucleotides, and the coding density is 0.91 genes/kb. Similar to other mimiviruses, moumouvirus contains eight group I self-splicing introns and three inteins. The moumouvirus genome is >100 kb shorter than the mimivirus genome and >200 kb shorter than the Megavirus chiliensis genome. Of the 930 moumouvirus ORFs, 879 have homologues in protein sequence databases, and the most similar homologue is from Megavirus chiliensis in a majority of cases. The median amino acid identity between moumouvirus and Megavirus chiliensis orthologous protein pairs is 62%. An analogous comparison between

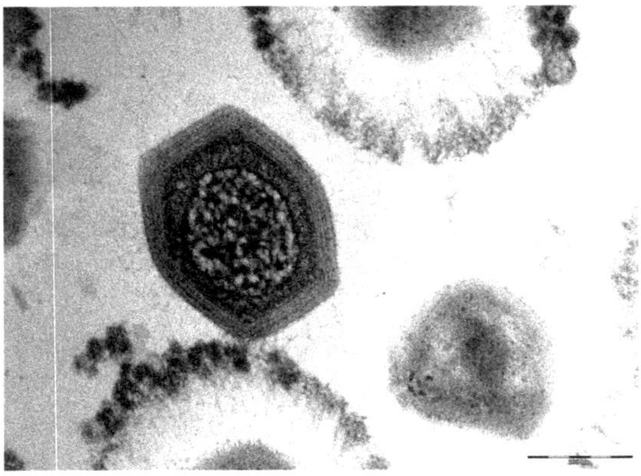

Figure 11.6 Transmission electron microscopy of a moumouvirus particle. Scale bar indicates 200 nm.

moumouvirus and mimivirus identified 548 putative orthologues, with a median 52% identity, which indicates that moumouvirus is more similar to Megavirus chiliensis than to mimivirus.

Saudi moumouvirus

Saudi moumouvirus (SDMV) is a new giant virus belonging to mimivirus lineage B, isolated from a sewage sample taken from the King Abdulaziz University hospital in Jeddah, Saudi Arabia (Bajrai *et al.*, 2016a). This virus was detected using an amoeba-associated virus co-culture (*A. polyphaga*) as previously described (La Scola *et al.*, 2001). It could also be grown in *A. griffini*, but not in *Acanthamoeba* sp. E4, *V. vermiformis* and *D. discoideum.* Observation by electron microscopy showed that this strain has a capsid of 500 nm which at least 80 nm larger than other members of mimivirus lineage B (Pagnier *et al.*, 2013). The SDMV genome consists of a DNA molecule with around 1,030,056 bp, which is larger than all previously described moumouvirus-like genomes (Yoosuf *et al.*, 2012; Pagnier *et al.*, 2013). Even when partially sequenced, this genome was predicted to encode 953 ORFs (considering putative ORFs, ORFans and Pseudo-ORFs), which is more than the gene content of Moumouvirus type species (Yoosuf *et al.*, 2012). In addition, it has two tRNA genes, encoding tRNAs corresponding to histidine and cysteine. Interestingly, SDMV contains 48 sequences with no similarity to any sequence available in the NCBI nr database, which is the non-redundant set of reference sequences database in NCBI. The transcription of the ORFan 850 (ORF_850), one of these 48 sequences, is confirmed by the qPCR which suggests other ORFans are also transcribed.

Mimivirus Bombay

In India, a new *Mimiviridae* family member was isolated from sewage water in Bombay by infecting *A. castellanii* using a method described previously (Chatterjee *et al.*, 2016; Raoult *et al.*, 2004). The virus named mimivirus Bombay (MVB) is 435 nm in diameter with a 1,182,200-bp genome possessing a 28% G+C content, and identified as a new member of lineage A of the *Mimiviridae* family according to the phylogeny of the DNA polymerase sequence. The MVB genome contains the following features: six tRNAs, nine transposons, and six mimiviral CRISPR-like elements that are associated with virophage infection immunity (see Chapter 12).

Megavirus chilensis

Megavirus chilensis is currently the largest mimivirus. This virus belongs to lineage C of the family *Mimiviridae*, which was initially found to encompass Courdo7 virus and Terra1 virus (Colson *et al.*, 2012; La Scola *et al.*, 2010). Megavirus chilensis was isolated from marine coastal water in Chile by culturing on a panel of *Acanthamoeba* species.

Megavirus chilensis has the same morphology and structure as other mimiviruses. The Megavirus chilensis capsid has an icosahedral form of 520 nm in diameter and is covered with a layer of fibrils of approximately 75 nm in length, corresponding to a total particle diameter of approximately 680 nm. The megavirus particles show a special vertex corresponding to the stargate previously described for mimivirus. Megavirus chilensis can infect *Acanthamoeba griffini*, *Acanthamoeba polyphaga*, and *Acanthamoeba castellanii.* Unlike the mimivirus cycle, during which lysis can occur 12 hours post-infection, for Megavirus chilensis new viral particles are observed at 17 hours post-infection (Arslan *et al.*, 2011). Megavirus chilensis replicates from large circular intracytoplasmic virion factories, as previously described for other *Mimiviridae* viruses. Similarly to Mimivirus, Megavirus chilensis induces a rounding of the infected *A. castellanii* cells. However, this phenomenon occurs more than 6 hours post-infection. Unlike Mimivirus infection, where rounded cells remain adherent, Megavirus chilensis infection causes most of them to detach from the culture plate.

Megavirus chilensis is a linear double-stranded DNA with a size of 1,259,197 bp. The genome of Megavirus chilensis is highly AT-rich, with an A+T content of 74.76%. A total of 1120 putative protein-coding sequences (CDSs) and three tRNAs have been annotated (Legendre *et al.*, 2012). This genome has a high coding density of 90.14%. Megavirus chilensis and mimivirus share 594 orthologous proteins. In total, 826 of 1120 (77%) putative protein-coding sequences have homologues in Mimivirus, while 793 of the Mimivirus CDSs have homologues in Megavirus chilensis (81%). Megavirus chilensis contains 258 proteins with no homologue in Mimivirus, and only 34 of them have a predicted function. Among the 186 Mimivirus genes without homologues in Megavirus chilensis, 149 (80%) are not similar to any other ORF. Point mutations appear to be the leading cause of the difference between the Megavirus chilensis and Mimivirus genomes. Megavirus chilensis has all of the genomic features of Mimivirus, in particular its genes encoding key elements of translation. In addition to homologues of the four aminoacyl-tRNA synthetases (aaRS) encoded by Mimivirus, Megavirus chilensis has three additional small aaRS, bringing the total of known virus-encoded aaRS to seven: IleRS, TrpRS, AsnRS, ArgRS, CysRS, MetRS, and TyrRS. These data suggest that large DNA viruses came from an ancestral cellular genome by reductive evolution (Arslan *et al.*, 2011; Legendre *et al.*, 2012).

Cafeteria roenbergensis virus (CroV)

Cafeteria roenbergensis virus is a giant lytic virus that infects *Cafeteria roenbergensis*, a marine heterotrophic flagellate and phagotrophic protist that is widespread in marine environments and is found in habitats such as surface waters, deep-sea sediments, and hydrothermal vents. *Cafeteria roenbergensis* belongs to the Chromalveolata phylum and grazes on bacteria and viruses. Cafeteria roenbergensis virus is a nucleocytoplasmic large DNA virus (Fischer *et al.*, 2011). Phylogenetic analysis indicates that it is a unique member of the second group of the *Mimiviridae* family. Indeed, Colson *et al.* (2011a) have demonstrated that in the phylogenetic tree of the NCLDVs based on a concatenated alignment of *four* universal NCVOGs (primase-helicase, DNA polymerase, packaging ATPase, and A2L-like transcription factor), CroV is related to the *Mimiviridae* family and is deeply located in the *Mimiviridae* branch.

This large DNA virus was isolated in Texas, USA, from coastal waters in 1990. The main morphological and structural characteristics of CroV are the same those of other mimiviruses (Figs. 11.7 and 11.8). Like all giant viruses, CroV's genome is a linear dsDNA molecule. It is 730 kb long and surrounded by a icosahedral capsid of 300 nm in diameter (Fischer *et al.*, 2011). The genome of this giant virus is AT-rich, with 77% A+T. A total of 544 putative protein-coding sequences (CDSs) have been identified in the 618-kb sequenceable central region of the CroV genome, amounting to a coding density of 90.1%. These CDSs have a mean length of 341 amino acids (range 141–3337). CroV encodes an isoleucyl-tRNA synthetase and putative homologues of eukaryotic translation initiation factors. It also encodes 22 tRNA genes and two putative tRNA-modifying enzymes. In addition, multiple DNA repair proteins are identified in CroV. These include a presumably complete base excision repair pathway, an NAD-dependent DNA ligase, the DNA mismatch repair protein MutS, XPG endonuclease, a homologue of the alkylated DNA repair protein AlkB, and two DNA photolyases. The CroV genome also encodes several proteins that regulate gene expression, including eight DNA-dependent RNA polymerase II subunits, at least six transcription factors, a trifunctional mRNA capping enzyme, a polyA polymerase, and multiple helicases. Other proteins are annotated as sugar-modifying enzymes. Because this giant virus has many predicted genes implicated in DNA replication, transcription, translation, protein modification, and carbohydrate metabolism, it seems it might have a highly autonomous propagation strategy during infection.

Virophages

The first virophage was discovered in 2008 at the same time as mamavirus and was named Sputnik. La Scola and his team observed icosahedral small viral particles of 50 nm in size in virus factories and in the cytoplasm of cells infected by mamavirus (Fig. 11.5). These particles did not multiply when inoculated into *A. castellanii* but grew in *A. castellanii* co-infected with mimivirus or mamavirus. Sputnik

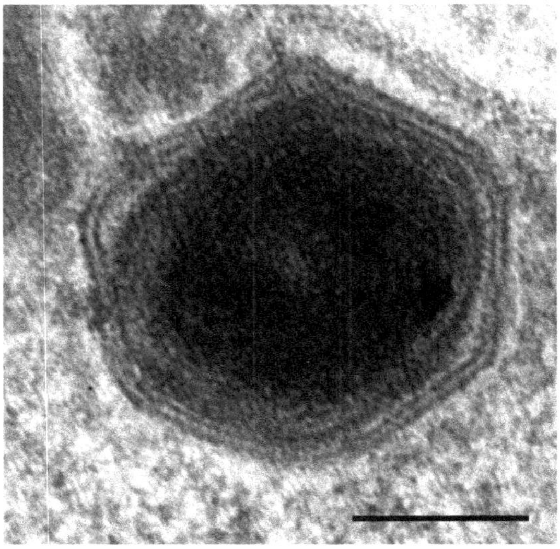

Figure 11.7 Transmission electron microscopy images of a Cafeteria roenbergensis virus (CroV) particle. Acknowledgements to Matthias Fischer and Ulrike Mersdorf.

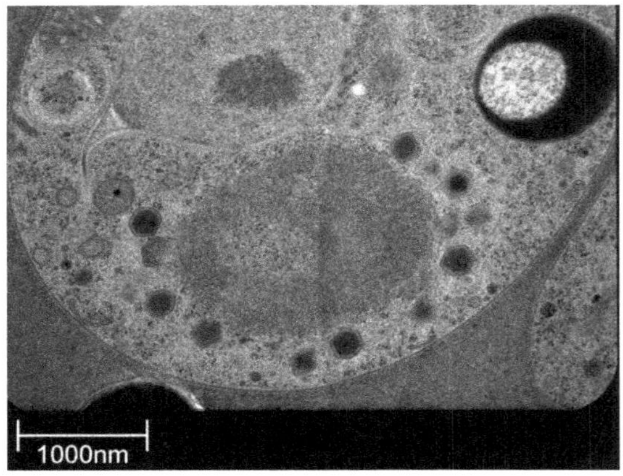

Figure 11.8 Transmission electron microscopy images of a CroV viral factory. Acknowledgements to Matthias Fischer and Ulrike Mersdorf.

coinfection was associated with formation of abnormal mamavirus or mimivirus virions, characterized by partial thickening of the capsid (Gaia *et al.*, 2013; La Scola *et al.*, 2008). The virophage possesses an 18-kbp, double-stranded, circular, highly A+T-rich genome. It encodes 21 ORFs with sizes that range between 88 and 779 amino acids. The entering process of the virophage into host cells is not yet well defined. It seems that Sputnik attaches at the virus fibrils. Thus, according to that hypothesis, Mimivirus and its virophage would infect the host cell at the same time. After the eclipse phase, 6 to 8 hours PI, we can observe a virus factory containing Mimivirus and Sputnik viral particles. At 24 hours PI, more than half of the co-infected cells are lysed and new viral particles are released (Boughalmi *et al.*, 2013b; Desnues *et al.*, 2012a; Desnues and Raoult, 2010, 2012; Sun *et al.*, 2010). Apart from Sputnik, five other virophages have been described: mavirus the CroV virophage (Fischer, 2011; Fischer and Suttle, 2011), Sputnik 2 (Desnues *et al.*, 2012), Sputnik 3 (Gaia *et al.*, 2013), Zamilon (Gaia *et al.*, 2014), and a recent isolated named Zamilon 2 (Bekliz *et al.*, 2015). Further discussion can be found in Chapter 12.

Mimivirus lineage A specifically displays a viral defence system, named mimivirus virophage resistance element (MIMIVIRE), and which has been identified as a nucleic-acid-based immunity against Zamilon virophage infection (Levasseur *et al.*, 2016a). It contains putative *Cas*-like genes and four copies of 15-nucleotides Zamilon sequence. Its essential role as a mimivirus defence against virophage as confirmed by two methods: (1) silencing of MIMIVIRE genes and (2) functional characterization of MIMIVIRE proteins. The enzymatic activities that may apply in the foreign nucleic acid cleavage include a helicase-nuclease and an endonuclease. After examining multiple strains of mimiviruses, the relation between the presence of repeated Zamilon sequences in group A mimivirus ($p<0.001$) and Zamilon resistance is linked to the immunity against Zamilon virophage.

Marseilleviridae family

Marseillevirus

The first marseillevirus was isolated in 2007 by *A. polyphaga* co-culture of a cooling-tower water sample collected in Paris (Boyer *et al.*, 2009). By transmission electron microscopy and cryo-electron microscopy 3D reconstruction using images of purified virus, it was shown that the marseillevirus capsid is icosahedral with an approximately 250 nm diameter covered by fibrils 12 nm in length with globular ends (Boyer *et al.*, 2009; Colson *et al.*, 2013c).

The genome of Marseillevirus is a circular dsDNA molecule of 368,453 bp with a G+C content of 44.73%. A total of 457 ORFs are predicted to encode proteins, and these ORFs are nearly equally distributed on both strands, 233 and 224 ORFs being detected on the negative and positive strands, respectively. Thus, coding sequences of Marseillevirus represent 89% of the genome, and approximately 1.2 genes are present per kilobase. Among the 41 *Megavirales* core genes, 28 have been identified in Marseillevirus. In addition, phylogenetic analyses of conserved proteins reveal that Marseillevirus was the founder *Marseille viridae* of a new NCLDV family. Marseillevirus is unusually rich in serine/threonine protein kinases, a large set of kinases for a virus, and also contains proteins not yet found in *Megavirales*, including three histone-like proteins encoded by ORF166, ORF413 and ORF414. Notably, the genome repertoire of marseillevirus contains genes derived from several distinct sources, in large part via horizontal gene transfer (HGT). Indeed, some predicted marseillevirus proteins show sequence similarity to homologues from viruses, bacteria, eukaryotes, and archaea. *Acanthamoeba* protein homologues were identified concurrently with 17 genes shared with mimivirus/mamavirus but not with other *Megavirales* members (Boyer *et al.*, 2009; Colson *et al.*, 2013c).

The replication cycle of Marseillevirus begins between 30 minutes and 1 hour after attachment/post-infection (PI), by the entry of the virus into the amoeba. Shortly after host cell entry, we can observe a virus factory (VF) in the amoeba's cytoplasm (Fig. 11.3) with a diffuse aspect observed close to the amoeba nucleus, where both capsid assembly and viral DNA encapsidation seem to occur simultaneously, resulting in formation of immature and mature viral particles. Therefore, the viral replication cycle of Marseillevirus is quicker than that of Mimivirus (Aherfi *et al.*, 2013; Colson *et al.*, 2013c; Thomas *et al.*, 2011). After the discovery of the first marseillevirus isolated in 2007 by co-culture with amoebae from a sample of cooling-tower water in Paris, other marseilleviruses have been isolated from various and diverse samples. All members of *Marseilleviridae* share similar morphological features, such as the size of their capsids, ranging from 190 to 250 nm (Fig. 11.9); the thickness of capsid, at approximately 10 nm; 12-nm-long fibrils on the viral surface; and a membrane surrounding the nucleocapsid, which is separated from the capsid shell by a gap of 52 nm. For all members of this family, viral factories can be observed during the replication cycle, but these viral factories have different aspects than those observed for mimiviruses.

Senegalvirus

Senegalvirus was isolated from stools of a healthy young Senegalese man. This virus was the first giant virus of amoebae isolated from a human sample (Colson *et al.*, 2013c; Lagier *et al.*, 2012). Since then, other close relatives of Marseillevirus have been detected in human samples, including blood, pharyngeal samples, and a lymph node (Aherfi *et al.*, 2016a,b; Popgeorgiev *et al.*, 2013a,b,c). Similar to other marseilleviruses, the size of its capsid varies from 190 to 250 nm. This capsid has an icosahedral form and a thickness of approximately 10 nm. In amoebae infected by senegalvirus, viral factories can be observed during the replication cycle, as for all members of the family *Marseilleviridae*. The genome of senegalvirus is ~372,690 base pairs (bp) in length, making it ~4200 bp larger than the Marseillevirus genome, and is predicted to encode 479 proteins, among which 351 and 253 are included in pairs of bidirectional best hits with Marseillevirus and lausannevirus proteins, respectively (Colson *et al.*, 2013c; Lagier *et al.*, 2012). Senegalvirus and marseillevirus orthologous proteins share 97% amino acid identity, while the identity is only 59% between Senegalvirus and Lausannevirus orthologs. Additionally, phylogeny based on the

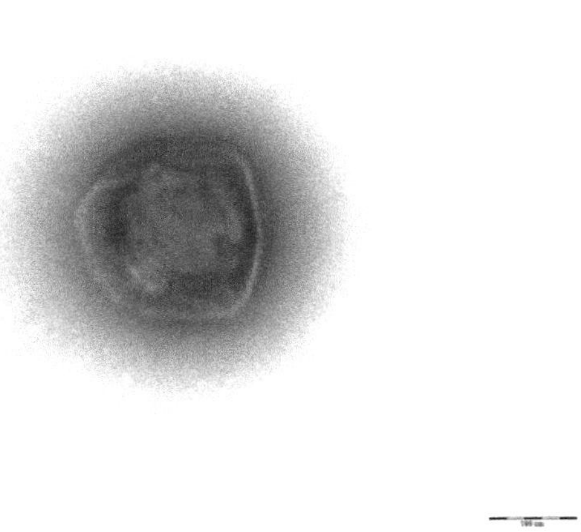

Figure 11.9 Transmission electron microscopy showing the morphology of a typical marseillevirus particle. Scale bar indicates 100 nm.

family B DNA polymerase showed that Senegalvirus clusters with Marseillevirus within the family *Marseilleviridae*.

Cannes 8 virus

Cannes 8 virus is a marseillevirus isolated by amoebal co-culture of a freshwater sample collected in Cannes in south eastern France. The major morphological and structural characteristics of Cannes 8 virus are identical to those of other marseilleviruses (Fig. 11.10A). Its viral particle is an icosahedral capsid measuring less than 250 nm. Its infectious cycle is similar to other marseilleviruses. The Cannes 8 virus genome is a circular double-stranded DNA molecule of 374,041 base pairs (Aherfi *et al.*, 2013). The G+C content is ~45%. This genome contains 483 putative genes that cover 92% of the genome. No tRNA genes have been detected. Cannes 8 virus has 84% and 70% of its gene repertoire in common with Marseillevirus and Lausannevirus, respectively. In addition, the Cannes 8 virus genome has a high level of collinearity with the Marseillevirus genome, whereas its orthologues have a lower collinearity (30%) with the Lausannevirus genome.

Lausannevirus

Lausannevirus is a large virus discovered by *Acanthamoeba castellanii* co-culture of an environmental sample collected in 2005 from the Seine River in France. This virus infects various *Acanthamoeba* spp. but is unable to infect other amoebae (such as *Hartmannella vermiformis* or *Dictyostelium discoideum*). Lausannevirus and Marseillevirus share 89% gene similarity and therefore belong to the same family as confirmed by core gene phylogeny. In fact, all NCLDV core genes found in Marseillevirus have highly similar orthologs in Lausannevirus. Lausannevirus viral particles are icosahedral with an average diameter of 190 to 220 nm. Lausannevirus particles have no fibrils and are mainly distributed within cytoplasmic vacuoles.

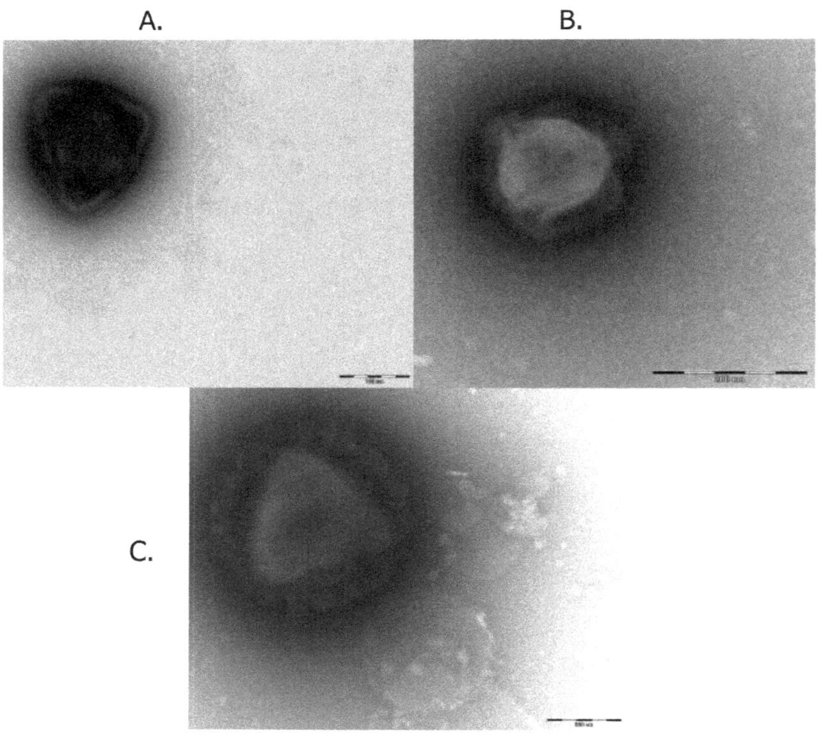

Figure 11.10 Transmission electron microscopy images of three protozoal viruses. (A) Cannes 8 virus. Scale bar indicates 100 nm. (B) Tunisvirus. Scale bar indicates 200 nm. (C). Megavirus LBA111. Scale bar indicates 200 nm.

Viral factory-containing maturating virus particles were observed in infected cells, as described for other NCLDVs. Further out at the periphery of this granular area, most viral particles are mature.

Lausannevirus has a genome of 346,754 bp with a G+C content of 43%. This genome can take two possible conformations: linear with terminal repeats or circularized. A total of 450 genes are predicted, having an average length of 716 bp, covering 93% of the genome, and being almost equally distributed on both strands (47% and 52% on the positive and negative strands, respectively).

The developmental cycle of Lausannevirus was studied by Thomas *et al.* (2011). Thirty minutes after infection, a few viral particles are observed in the cytoplasm of the infected cell. Two hours post-infection, viral particles are no longer observed within amoebae. This corresponds to the eclipse phase described above in *Mimiviridae* viruses. After 4 hours, large vesicles filled with viruses are present within the infected cell. After 8 hours, all amoebae are filled with viral particles, and 16 hours post infection, the cells are lysed and clusters of released viruses are detected (Thomas *et al.*, 2011).

Tunisvirus

Tunisvirus was recovered from water collected from a decorative fountain in Tunis, Tunisia (Boughalmi *et al.*, 2013b). The results of the PCR and sequencing analyses indicated 88% similarity to lausannevirus and thus confirmed that this virus belongs to the family *Marseilleviridae*. Transmission electron microscopy shows that the viral particle is icosahedral with an ~226 nm diameter and the same main morphological and structural characteristics of marseilleviruses (Fig. 11.10 B). The genome of Tunisvirus is circular dsDNA with a length of 380,011 bp and a G+C content of 43%, as seen in other marseilleviruses (Aherfi *et al.*, 2014a). A total of 484 protein-encoding genes are predicted, among which 320 and 358 have significant homology to a Marseillevirus and a Lausannevirus protein, respectively. In addition, substantial levels of collinearity exist between the Tunisvirus genome and Marseillevirus and Lausannevirus genomes. A total of 30% of the Tunisvirus gene content returns matches in organisms other than Marseilleviruses, including 11% in other members of the proposed order *Megavirales*, 11% in eukaryotes (2% in amoebae) and 7% in bacteria, which suggests extensive lateral gene transfer that has built up a mosaic gene content. A total of 21 genuine ORFans have been identified in the Tunisvirus genome.

Insectomime virus

Insectomime virus was isolated by amoebal co-culture of samples obtained from the internal organs and digestive tract of *Eristalis tenax* (Diptera, Syrphidae) larvae (Boughalmi *et al.*, 2013a). This insect has a complex metamorphosis process. Eggs are typically laid on the surface of water, and the larvae have three stages of development. *E. tenax* is an important vector for multiple bacterial species, such as *Mycobacterium* spp., and the possibility that the virus was ingested by the insect larva might explain its presence in only the internal organs of the animal and not on the surface. Some insect species can be colonized by amoebae which could be the host for Insectomime virus. Indeed in one study, Otta and his team showed that amoeba *Acanthamoeba* spp. may be found in *Aedes aegypti* mosquito larvae, demonstrating the possible interaction between protozoa and mosquitoes in their natural habitat. (Otta *et al.*, 2012). Regarding the virus' presence in the insect, it has been hypothesized that the virus was directly ingested, or it could have entered the insect through the ingestion of an infected amoeba.

Insectomime virus is the first marseillevirus isolated from an insect. The PCR results and sequencing analysis indicated that insectomime virus belongs to *Marseilleviridae* family. Transmission electron microscopy revealed a viral particle diameter of 225 nm and an icosahedral structure typical of giant viruses. The insectomime virus genome is a dsDNA molecule of 386,000 bp, with a G+C content of 43%. A total of 477 proteins are predicted. Protein sizes vary from 46 to 1643 amino acids. Among these proteins, 435 have as best hit either a Marseillevirus or a Lausannevirus protein. Tunisvirus seems to be the most closely related to Insectomime virus, as the two viruses share 446 orthologs. Otherwise, 329 Insectomime virus proteins are annotated as hypothetical proteins. The best BLASTP hits for eight Insectomime virus proteins are in *A. castellanii*, Acanthamoeba polyphaga moumouvirus, amoeba-associated mimiviruses of lineage B, *Cercomonas* spp., two opisthokonts (*Capsaspora owczarzaki* and *Hydra magnipapillata*), *Cellulophaga algicola*, Azospirillum phage Cd, and *Acyrthosiphon pisum*, an insect from the family *Aphididae*. The infectious cycle of Insectomime virus is similar to those of viruses belonging to *Marseilleviridae*, with an eclipse phase at 2 hours post-infection. Formation of virus factories in the cytoplasm of the infected cell a few hours after infection and the presence of mature or immature viral particles have also been reported. The availability of the Insectomime virus and Tunisvirus genomes led to the proposal that the family *Marseilleviridae* may encompass three phylogenetic lineages, A, B and C, with Marseillevirus, Lausannevirus and Tunisvirus as leading members (Aherfi *et al.*, 2014b).

Brazilian marseillevirus

Brazilian marseillevirus was isolated in 2014 from a sewage sample collected in Brazil (Lopes Assis *et al.*, 2016). The analysis of its gene content and phylogenetic reconstructions showed that Brazilian marseillevirus is divergent from other marseilleviruses previously described. Indeed, its genome of 362,000 bp in length is predicted to encode 491 proteins. Brazilian marseillevirus shares 311 (63%), 365 (74%) and 364 (74%) best reciprocal hits with Marseillevirus, Lausannevirus and Tunisvirus, the representatives of the lineages A, B, and C, respectively. A phylogeny based on core genes showed that a fourth clade composed of Brazilian marseillevirus is clearly delineated besides the three other lineages. Brazilian marseillevirus is thus the founding member of a fourth lineage among the family *Marseilleviridae.* Congruent to these findings, analysis of the size of the pan genome of family *Marseilleviridae* increases greatly when adding the genome of Brazilian marseillevirus, especially due to the high number of ORFans found in association with this virus (29). Moreover, unlike other previously described marseilleviruses, Brazilian marseillevirus was not able to replicate in *Acanthamoeba polyphaga*, and its multiplication was restricted to *Acanthamoeba castellanii* strains.

Golden marseillevirus

Golden marseillevirus was isolated from golden mussels, *Limnoperna fortunei*, in south of Brazil, by co-culture on *Acanthamoeba polyphaga* (Dos Santos *et al.*, 2016). The replicative cycle showed the appearance of viral factories in the cytoplasm of *Acanthamoeba castellanii*, 3 hours post infection. It harbours a genome of 360,610 bp in length with a GC% of 43.1%, encoding for 483 predicted proteins, the half consisting in uncharacterized proteins. A total of 43% of the golden marseillevirus gene content is shared with the representatives of the four lineages of the family *Marseilleviridae*, namely Marseillevirus, Lausannevirus, Tunisvirus, and Brazilian marseillevirus. Phylogeny and comparative genomic analyses showed that golden marseillevirus is clearly different from other previously described marseilleviruses. It was proposed to classify this new isolate as the founding member of a fifth lineage, named lineage E.

Tokyovirus

Tokyovirus is a member of the family *Marseilleviridae,* isolated from a muddy freshwater sample collected in a river in Tokyo (Takemura, 2016a). Virions encompass a genome of 372,000 bp in length, predicted to encode 487 proteins. Interestingly, the genome of Tokyovirus is predicted to encode two tRNA genes (Leu and His-tRNAs), one of which is a pseudogene. It is the first demonstration that a marseillevirus genome could encode tRNA genes. Genomic architecture comparisons and phylogenetic analysis based on three core genes, demonstrated that Tokyovirus shares similarities with members from the clade A, while still being divergent, but not with those of clades B, C and D. Interestingly, a genomic region of 50,000 bp in length is inverted and positioned 300,000 bp downstream in three marseillevirus genomes from the lineage A. These findings suggest that it may have evolved separately from the other marseilleviruses (Takemura, 2016b).

Pandoravirus

Pandoravirus salinus and Pandoravirus dulcis

Pandoravirus salinus and Pandoravirus dulcis are recently described amoebal viruses (Philippe *et al.*, 2013). Pandoravirus salinus was isolated from the superficial marine sediment layer on the coast of central Chile, and Pandoravirus dulcis was isolated from mud collected at the bottom of a shallow freshwater pond in Australia. They are currently the largest dsDNA viruses isolated, having particle sizes of 1 μm in length. In addition, the genome size of Pandoravirus salinus is estimated to reach 2.77

Mb, and this DNA molecule is predicted to encode 2,556 putative proteins. The genome of Pandoravirus dulcis is 1.91 Mb long and is predicted to encode 1502 proteins. Pandoravirus genomes have a high G+C content (61.7% and 63.7% for Pandoravirus dulcis and Pandoravirus salinus, respectively), which differs substantially from the G+C contents of mimiviruses and marseilleviruses.

A total of 93% of the proteins identified in the Pandoravirus salinus particles have a homologue in the Pandoravirus dulcis genome. Pandoravirus salinus predicted genes have a mean size of 258 amino acids and represent 80% of the genome size. Only 16% of these genes return significant hits in GenBank, among which 54% are proteins containing ankyrin, membrane occupation and recognition nexus (MORN) and F-box motifs. When excluding these latter proteins, best matches are proteins from eukaryotes, bacteria, and viruses in 101, 43 and 42 cases, respectively. Similar features are described for the Pandoravirus dulcis gene content. The Pandoravirus salinus genome has 14 of the 31 core genes shared by *Megavirales* members and three of the nine class I core genes (found in all viral genomes). Pandoravirus salinus possesses 54 proteins involved in DNA processing, which is typical of amoebal viruses. In addition, two amino acyl-tRNA ligases (TyrRS and TrpRS) are found in mimiviruses but are more similar to those from *Acanthamoeba* spp., and a elF-4E translation initiation factor, a SUA5-like tRNA modification enzyme, and three tRNA genes (Pro-tRNA, Met-tRNA and Trp-tRNA) are detected. Yutin and Koonin reanalysed the genomes of pandoraviruses and identified 17/47 conserved genes they had defined previously as most likely present in the genome of the Megavirales common ancestor (Yutin and Koonin, 2009). Phylogeny reconstruction for these 17 genes showed that pandoraviruses were grouped with phycodnaviruses in six cases, marseilleviruses in one case and eukaryotes in ten cases. Overall, phylogenetic analyses of the NCVOGs showed that pandoraviruses are highly derived phycodnaviruses, with the greatest affinity to coccolithoviruses (Philippe *et al.*, 2013; Yutin and Koonin, 2013).

The replication cycle of pandoraviruses begins by the internalization of viral particles. A channel through which the particle proteins and DNA content can be delivered is then formed by fusion of the internal lipid membrane with the vacuole. This fusion process leads to an eclipse phase similar to that of mimivirus. From two to four hours after infection, the host nucleus undergoes loss of its spherical form, and eight to 10 hours after the start of infection, the cells become rounded and lose their adherence. Later, new particles appear at the periphery of the region formerly occupied by the nucleus. Ten to 15 hours after infection, infected cells are lysed and new viral particles are released.

Pandoravirus inopinatum

The first pandoravirus genome (Pandoravirus inopinatum) was described in 2015, after being mistakenly thought to be a novel type of parasite, an endocytobiont, when it was isolated from *Acanthamoeba* isolated from a patient's eye inflamed with keratitis in 2008 (Antwerpen *et al.*, 2015; Scheid *et al.*, 2008). Because of the morphology resemblance of two previous *Pandoravirus* species, a complete whole genome of a new pandoravirus isolate was reinvestigated and sequenced (Antwerpen *et al.*, 2015). Its nucleotide sequence is 2,243,109 bp with 60.66% G+C content. The nucleotide sequence identity of this strain is 89% and 85% to Pandoravirus dulcis and Pandoravirus salinus, respectively, by search on BLASTn against non-redundant database. Moreover, there are 1389 homologous proteins with other species of *Pandoravirus;* and this isolate clusters with the published members of *Pandoravirus* on the protein level of B-family DNA polymerase and DNA-directed RNA polymerase II.

Pithoviruses

Pithovirus sibericum

Reflecting the diversity of giant viruses, pithovirus members were isolated from environmental samples, thawed permafrost and other soils, by co-culturing into *Acanthamoeba castellanii.* The fact that two viruses have different prehistorically permafrost and infect same host cell with many genomes details is concerned further studies.

Pandoravirus and Pithovirus sibericum were described in 2013 and 2014, respectively, both growing in *A. castellanii* (Legendre *et al.*, 2014; Philippe *et al.*, 2013). The first virus was isolated from a stream in Melbourne, Australia, and also from the coast of Chile, while the second virus was isolated from Siberian dirt that had been frozen for more than 30,000 years (Legendre *et al.*, 2014; Philippe *et al.*, 2013). The virions of Pithovirus sibericum resemble pandoraviruses morphologically but are larger (1.5 μm in length) and have an envelope of hexagonal structure about 60 nm thick; the capsid is missing in pandoraviruses. Pandoravirus and Pithovirus sibericum both have the apical pore, a small pore in the outer wall. Although the virion size of Pithovirus sibericum is slightly larger than pandoraviruses, it has a short dsDNA genome (610 kbp) and 36% of G+C- content shared among those of mimiviruses and marseilleviruses, and pandoraviruses (Aherfi *et al.*, 2016). The genome has ORFans encoding two thirds of the 467 predicted proteins, of which 159 were detected inside the virion by proteomics. Moreover, the greatest number of best BLAST viral matches was with marseilleviruses (19), then mimiviruses (15), and iridoviruses (10) as determined through the phylogenetic and phyletic analyses (Legendre *et al.*, 2014; Sharma *et al.*, 2015a). The pithovirus replicative cycle is similar to that of pandoraviruses except that the cycle duration lasts 10–20 hours and the appearance of mature virions occurs in 6–8 hours and release may be by exocytosis (Birtles *et al.*, 1997).

Pithovirus massiliensis

In 2016, Pithovirus massiliensis, a pithovirus-like virus with a 'cork' structure at one end of the 1.4 ± 0.2 μm capsid, was isolated, described and compared with Pithovirus sibericum genome (Levasseur *et al.*, 2016b). This strain was grown in *A. castellanii* strain Neff (ATCC number 30010) using the same method as Reteno *et al.* (2015). The genome set was sequenced and displayed a unique mosaic sequence due to horizontal gene transfer (HGT) of *Megavirales* core genomes. Moreover, this genome conserved the same mosaicism as Pithovirus sibericum, which is an indication of high degree of conservation between pithovirus strains. Based on the studying of all orthologous genes, which countered 93.5% of Pithovirus sibericum, the 683 kbp scaffold of Pithovirus massiliensis genome was identified and revealed the compression features between the modern and fossil strains. First, the synteny of seven blocks of genes between two strains were conserved, indicating substantial stability for both genomes. Second, The Average Nucleotide Identity (ANI) values indicated that the stains were similar with ~83.6% of nucleotide identity. Lately, Legendre and his colleagues, in the study about Pithovirus sibericum in 2015, mentioned a number of ORFans that were found and conserved in Pithovirus massiliensis.

Cedratvirus

Cedratvirus is a new virus, ovoid-shaped and double-corked isolated by *A. castellanii* co-culture from an Algerian environmental sample (Andreani *et al.*, 2016). The genomic analysis of Cedratvirus revealed that the 589,068 bp genome is similar to Pithovirus sibericum and Pithovirus massiliensis, sharing 104 and 113 genes, respectively, and includes 445 unique genes. Moreover, the hypothetical proteins of Cedratvirus, with unknown function, are represented in more than 50% of its gene content. Compared to *Pithovirus* members, the intergenic regions of Cedratvirus have been shown to not have the same large palindromic repeats like those in Pithovirus sibericum. Ultimately, the conservation of gene patterns among giant viruses demonstrates the relation of Cedratvirus to pithoviruses.

Faustovirus

The use of different and unique amoebae other than *Acanthamoeba* spp. as hosts leads to isolation of a variety of giant virus strains. For instance, faustovirus strain E12 was the first giant virus isolated from amoebae named *Vermamoeba vermiformis* (Bradbury, 2014; Reteno *et al.*, 2015). The faustovirus E12 genome is 466,265 bp long consisting of circular dsDNA containing 36% of G+C content with 451 genes predicted. About 13% of these genes have homologues in some *Megavirales* representatives, primarily in asfarviruses (a genera of family *Asfarviridae*; infects insects and swine; has double stranded genome), then in phycodnaviruses, mimiviruses, marseilleviruses, and ascoviruses (a genera of *Ascoviridae* family, infects lepidopteran larvae, and has circular dsDNA genome) while more than two-thirds have no homologues in the GenBank non-redundant protein sequence database (Aherfi *et al.*, 2016c; Reteno *et al.*, 2015; Federici *et al.*, 2009). It has other best-hit sequences from bacteria and eukaryotes, and rarely, sequences from archaea and phages. Comparative genomics and phylogeny analysis show that faustoviruses are closely related to asfarviruses, but their genome size is three times larger than those of asfarviruses. Many genes that are shared with asfarviruses encode proteins such as a ribosomal protein acetyltransferase, a bacteriophage tail fibre protein, and two polyproteins (Reteno *et al.*, 2015). Moreover, using proteomics procedures, one-third of the faustovirus E12 predicted proteins were detected inside its virions. The faustovirus replicative cycle takes 18–20 hours and it shows a viral factory that is close to the host cell nucleus with deformation of nuclear morphology.

Mollivirus

In addition to Pithovirus sibericum, inoculating *A. castellanii* with 30,000-year-old Siberian permafrost sample resulted in isolation of another new giant virus named Mollivirus sibericum (Legendre *et al.*, 2015). The size and genome length of Mollivirus sibericum virion are 500–600 nm in diameter and 651,523 bp, respectively, and the virus is covered with 2–4 layers of fibres (Legendre *et al.*, 2015). The virion of Mollivirus sibericum has a spherical shape covered with 2–4 layers of fibres. The linear dsDNA genome of Mollivirus sibericum harbours ~10 kbp (in length) inverted terminal repeats with 523 genes and three tRNAs predicted (most of the genes are ORFans). The G+C content is 60%, similar to that of pandoraviruses. The major proportion (91%) of the viral best matches to Mollivirus sibericum proteins correspond to the pandoraviruses. Among proteins with homologues, 18% and 3% were most similar to proteins from viruses and prokaryotes, respectively, and 14% were most similar to eukaryotic proteins including 4% from *Acanthamoeba*. Moreover, among proteins with a functional annotation, the main group encompassed proteins containing ankyrin repeats, then proteins involved in DNA processing and nucleotide biosynthesis. A total of 230 proteins were identified in virions including 60% from Mollivirus sibericum and 40% from *Acanthamoeba* (Legendre *et al.*, 2015). Remarkably, Mollivirus sibericum is not like other giant viruses of amoebas because the viral factory is not cytoplasmic but instead perinuclear, and the replication cycle is not lytic for the amoebal host when the new virions are released after 6 hours PI (Legendre *et al.*, 2015). The synthesis of the envelope and inner contents of the virions occurs simultaneously as for pandoraviruses and Pithovirus sibericum virions (Aherfi *et al.*, 2016c).

Kaumoebavirus

Another new giant virus, Kaumoebavirus, was obtained by co-culture in *V. vermiformis* of a sewage water sample from the southern area of Jeddah (Saudi Arabia) (Bajrai *et al.*, 2016b). Kaumoebavirus has a size of ~250 nm with a large icosahedral capsid, DNA genome length of 350,731 bp, and morphological resemblance to faustoviruses (Reteno *et al.*, 2015). Faustovirus and kaumoebavirus virions have an icosahedral capsid without fibres. Like other virions of giant viruses, the internal lipid membrane surrounding the core fuses with the vacuole membrane to release the dsDNA genome during infection (Aherfi *et al.*, 2016c). In addition, kaumoebavirus genome has circular dsDNA (350,731 bp), 43.7% of GC content, and 465 identified proteins with homology to virus (59%), bacteria (15%), other eukaryotes (25%), and Archaea proteins (2%). Regarding viral protein homology, Kaumoebavirus proteins mostly matched to faustoviruses (43%), and asfarviruses 23%, and lesser percentage to mimiviruses, phycodnaviruses, poxviruses, marseilleviruses and baculoviruses. In contrast to other giant viruses, kaumoebavirus has a 28 hour replicative cycle with a distinct separation between the viral factory and nucleus. Also, clustering of the virions of kaumoebaviruses inside its viral factory is another unusual feature of this virus.

Pathogenicity of mimiviruses and marseilleviruses

Amoeba-associated microorganisms cause pneumonia in different patient populations. *Acanthamoeba polyphaga mimivirus*, the most studied giant virus, and marseillevirus are amoeba-associated microorganisms. Thus, it has been suggested that mimiviruses and marseilleviruses have potential pathologic roles in humans, but this hypothesis remains controversial. Currently, several works support this hypothesis. Many studies have detected mimivirus antibodies in patients with pneumonia, and DNA of mimivirus and marseillevirus have been detected, respectively, in bronchoalveolar lavage (BAL) and blood of patients with pneumonia

(Colson *et al.*, 2013b; Vincent *et al.*, 2010). In 20–50% of pneumonia cases, the causative agent is unknown (Marrie *et al.*, 1989). Thus, in 2005, with the goal of identifying new causative agents of both community- and hospital-acquired pneumonia, La Scola *et al.* (2005) have searched for antibodies against several amoeba-associated microorganisms in serum samples from 376 Canadian patients with community-acquired pneumonia and from 511 healthy control subjects. They found 12 (2.3%) who exhibited a substantial titer of antibodies to mimivirus in the controls and 36 (9.66%) among community-acquired pneumonia patients. In an earlier study, the same team found mimivirus DNA in bronchoalveolar lavage specimens from patients in the ICU of the Ste. Marguerite Hospital, Marseille, France (Marrie *et al.*, 2001). In 2006, the same team screened amoeba-associated microorganisms and conventional microorganisms in 210 pneumonia patients in intensive care units using culture, polymerase chain reaction, and serological assays. In that study, Acanthamoeba polyphaga Mimivirus was the fourth most common etiologic agent. It has been detected in eight patients among 15 patients in whom the infectious agents were identified (Berger *et al.*, 2006). The same year, other studies showed that Mimivirus can infect macrophages through phagocytosis and cause pneumonia in mouse models (Ghigo *et al.*, 2008; Khan *et al.*, 2007).

To investigate the pathogenicity of Acanthamoeba polyphaga Mimivirus in the occurrence of pneumonia infections in mammalian hosts, Khan inoculated adult BALB/c and C57BL/6 mice with Acanthamoeba polyphaga Mimivirus via an intracardiac route. The mice developed histopathological features of pneumonia, characterized by the presence of thickened alveolar walls with cellular infiltrates comprising mononuclear leucocytes, macrophages and lymphocytes; diffuse alveolar damages with the formation of a hyaline membrane; and erythrocytes in the alveolar lumen (Khan *et al.*, 2007). A second case of infection with mimivirus has been observed in one laboratory technician who worked with mimiviruses. This technician presented an unexplained respiratory infection. By two-dimensional gel electrophoresis and Western blotting analyses on his two serum samples collected before and during pneumonia, the technician had seroconverted for total antibodies against 23 mimivirus proteins, including four unique to mimivirus (Raoult *et al.*, 2006). Another case associating a giant virus of amoebae with a human infection led to the discovery of lentillevirus, a mimivirus that was recovered from the contact lens storage case liquid of a 17-year-old myopic woman presenting with keratitis. Also isolated from this liquid was a new virophage of giant viruses, which was named Sputnik 2 (Cohen *et al.*, 2011).

The third piece of evidence of human exposure to giant viruses is the detection of antibodies to marseillevirus in an eleven-month-old boy suffering from adenitis. Marseillevirus DNA was also found in his blood using PCR and in his lymph node using FISH and immunohistochemistry. The authors of that case study speculated that adenitis during early childhood may be caused by a marseillevirus-like virus (Popgeorgiev *et al.*, 2013d). Recently, mimiviruses have been isolated from other human samples (stool and BAL of a patient with pneumonia) (Saadi *et al.*, 2013a,b).

By cultivating on *Acanthamoeba polyphaga* from several pulmonary samples from Tunisian patients having pneumonia, Saadi *et al.* (2013a) isolated a giant virus in human samples from a 72-year-old woman. This giant virus was named LBA11. Polymerase chain reaction and sequencing analysis showed that LBA111 is a mimivirus group C virus. It is the largest virus ever isolated from a human. Transmission electron microscopy revealed a mimivirus-like virion with a size of 554 ± 10 nm and a dense layer of fibrils (Fig. 11.10C). The genome of LBA111 is composed of 1,230,522 base pairs. A total of 1178 protein-encoding genes are predicted, with 930 included in pairs of reciprocal best BLAST hits with Megavirus chilensis genes. The mean length of these aligned pairs is 318 amino acids, and their mean identity is 98%. Thus, LBA111 is closely related to Megavirus chilensis.

Another giant virus was also isolated in a different type of human sample by the same team: shan virus. This virus was isolated from the stool of a Tunisian patient with pneumonia. Transmission electron microscopy indicates that shan virus has an icosahedral shape typical of *Mimiviridae* and a mimivirus-like virion with a size of 640 ± 10 nm with a dense layer of fibrils, a starfish-shaped structure, and the presence of viral factories detected in amoeba-infected cells (Saadi *et al.*, 2013b).

More recent work has shown that environmental exposure to mimivirus represents a risk factor in triggering autoimmunity to collagens and promotes arthritis. The authors have investigated the localization of collagens in Acanthamoeba polyphaga mimivirus particles and the response to collagen of mice immunized with mimivirus particles. By a protein biotinylation process, they showed that exposure to mimivirus collagens generates the production of anti-collagen antibodies in DBA/1 mice immunized intra-dermally with mimivirus protein extracts. This antibody response also targeted the mouse collagen type II, as observed in collagen-induced arthritis (Shah *et al.*, 2014). Taken together, these facts indicate that amoeba-infecting viruses are emerging pathogens in human disease.

Conclusion

The mimivirus, the first known giant virus of amoebae, was discovered by T.J. Rowbotham in a cooling tower in Bradford, UK, and subsequently characterized by the team of La Scola. Since then, several other mimiviruses and additional protozoal viruses, such as marseilleviruses and pandoraviruses, were discovered thanks to the development of new strategies that facilitate the isolation of these protozoal giant viruses. In fact, some of these giant viruses have been detected and isolated by innovative procedures such as a high-throughput agar method that allows for the detection of the cytopathic effect of the virus on agar plates, in addition to the standard protocols of co-culture with protozoa and specific real-time PCR. We believe that, with such innovative methods, other families of protozoal viruses will likely be discovered.

References

Aherfi, S., Pagnier, I., Fournous, G., Raoult, D., La Scola, B., and Colson, P. (2013). Complete genome sequence of Cannes 8 virus, a new member of the proposed family 'Marseilleviridae'. Virus Genes *47*, 550–555. https://doi.org/10.1007/s11262-013-0965-4.

Aherfi, S., Boughalmi, M., Pagnier, I., Fournous, G., La Scola, B., Raoult, D., and Colson, P. (2014a). Complete genome sequence of Tunisvirus, a new member of the proposed family *Marseilleviridae*. Arch. Virol. *159*, 2349–2358. https://doi.org/10.1007/s00705-014-2023-5.

Aherfi, S., La Scola, B., Pagnier, I., Raoult, D., and Colson, P. (2014b). The expanding family *Marseilleviridae*. Virology 466-467, 27–37. https://doi.org/10.1016/j.virol.2014.07.014.

Aherfi, S., Colson, P., and Raoult, D. (2016a). Marseillevirus in the pharynx of a patient with neurologic disorders. Emerg. Infect. Dis. *22*, 2008–2010. https://doi.org/10.3201/eid2211.160189.

Aherfi, S., Colson, P., Audoly, G., Nappez, C., Xerri, L., Valensi, A., Million, M., Lepidi, H., Costello, R., and Raoult, D. (2016b). Marseillevirus in lymphoma: A giant in the lymph node. Lancet Infect. Dis. *16*, 225–234.

Arslan, D., Legendre, M., Seltzer, V., Abergel, C., and Claverie, J.M. (2011). Distant Mimivirus relative with a larger genome highlights the fundamental features of Megaviridae. Proc. Natl. Acad. Sci. U.S.A. *108*, 17486–17491. https://doi.org/10.1073/pnas.1110889108.

Aherfi, S., Colson, P., La Scola, B., and Raoult, D. (2016c). Giant viruses of amoebas: An update. Front. Microbiol. *7*, 349. https://doi.org/10.3389/fmicb.2016.00349.

Andreani, J., Aherfi, S., Bou Khalil, J.Y., Di Pinto, F., Bitam, I., Raoult, D., Colson, P., and La Scola, B. (2016). Cedratvirus, a double-cork structured giant virus, is a distant relative of pithoviruses. Viruses *8*, E300.

Antwerpen, M.H., Georgi, E., Zoeller, L., Woelfel, R., Stoecker, K., and Scheid, P. (2015). Whole-genome sequencing of a pandoravirus isolated from keratitis-inducing acanthamoeba. Genome Announc. 3, e00136-15. https://doi.org/10.1128/genomeA.00136-15.

Asgari, S. (2006). Replication of Heliothis virescens ascovirus in insect cell lines. Arch. Virol. *151*, 1689–99. https://doi.org/10.1007/s00705-006-0762-7.

Bajrai, L.H., de Assis, F.L., Azhar, E.I., Jardot, P., Robert, C., Abrahão, J., Raoult, D., and La Scola, B. (2016a). Saudi Moumouvirus, the first group B mimivirus isolated from Asia. Front. Microbiol. *7*, 2029. https://doi.org/10.3389/fmicb.2016.02029.

Bajrai, L.H., Benamar, S., Azhar, E.I., Robert, C., Levasseur, A., Raoult, D., and La Scola, B. (2016b). Kaumoebavirus, a new virus that clusters with Faustoviruses and *Asfarviridae*. Viruses *8*, E278.

Bekliz, M., Verneau, J., Benamar, S., Raoult, D., La Scola, B., and Colson, P. (2015). A new Zamilon-like virophage partial genome assembled from a bioreactor metagenome. Front. Microbiol. *6*, 1308. https://doi.org/10.3389/fmicb.2015.01308.

Berger, P., Papazian, L., Drancourt, M., La Scola, B., Auffray, J.P., and Raoult, D. (2006). Ameba-associated microorganisms and diagnosis of nosocomial pneumonia. Emerg. Infect. Dis. *12*, 248–255.

Bideshi, D.K., Demattei, M.V., Rouleux-Bonnin, F., Stasiak, K., Tan, Y., Bigot, S., Bigot, Y., and Federici, B.A. (2006). Genomic sequence of *Spodoptera frugiperda Ascovirus 1a*, an enveloped, double-stranded DNA insect virus that manipulates apoptosis for viral reproduction. J. Virol. *80*, 11791–805. https://doi.org/10.1128/JVI.01639-06.

Bigot, Y., Rabouille, A., Doury, G., Sizaret, P.Y., Delbost, F., Hamelin, M.H., and Periquet, G. (1997). Biological and molecular features of the relationships between Diadromus pulchellus ascovirus, a parasitoid hymenopteran wasp (*Diadromus pulchellus*) and its lepidopteran host, *Acrolepiopsis assectella*. J. Gen. Virol. 78, 1149–63. https://doi.org/10.1099/0022-1317-78-5-1149.

Birtles, R.J., Rowbotham, T.J., Storey, C., Marrie, T.J., and Raoult, D. (1997). Chlamydia-like obligate parasite of free-living amoebae. Lancet 349, 925–926.

Boughalmi, M., Pagnier, I., Aherfi, S., Colson, P., Raoult, D., and La Scola, B. (2013a). First isolation of a Marseillevirus in the Diptera Syrphidae *Eristalis tenax*. Intervirology 56, 386–394. https://doi.org/10.1159/000354560.

Boughalmi, M., Saadi, H., Pagnier, I., Colson, P., Fournous, G., Raoult, D., and La Scola, B. (2013b). High-throughput isolation of giant viruses of the *Mimiviridae* and *Marseilleviridae* families in the Tunisian environment. Environ. Microbiol. *15*, 2000–2007. https://doi.org/10.1111/1462-2920.12068.

Boyer, M., Madoui, M.A., Gimenez, G., La Scola, B., and Raoult, D. (2010). Phylogenetic and phyletic studies of informational genes in genomes highlight existence of a 4 domain of life including giant viruses. PLOS ONE *5*, e15530. https://doi.org/10.1371/journal.pone.0015530.

Boyer, M., Yutin, N., Pagnier, I., Barrassi, L., Fournous, G., Espinosa, L., Robert, C., Azza, S., Sun, S., Rossmann, M.G., *et al.* (2009). Giant Marseillevirus highlights the role of amoebae as a melting pot in emergence of chimeric microorganisms. Proc. Natl. Acad. Sci. U.S.A. *106*, 21848–21853. https://doi.org/10.1073/pnas.0911354106.

Bradbury, R.S. (2014). Free-living amoebae recovered from human stool samples in *Strongyloides* agar culture. J. Clin. Microbiol. 52, 699–700. https://doi.org/10.1128/JCM.02738-13.

Carrascosa, J.L., Carazo, J.M., Carrascosa, A.L., García, N., Santisteban, A., and Viñuela, E. (1984). General morphology and capsid fine structure of African swine fever virus particles. Virology *15*, 160–72.

Claverie, J.M., and Abergel, C. (2009). Mimivirus and its virophage. Annu. Rev. Genet. *43*, 49–66. https://doi.org/10.1146/annurev-genet-102108-134255.

Claverie, J.M., Ogata, H., Audic, S., Abergel, C., Suhre, K., and Fournier, P.E. (2006). Mimivirus and the emerging concept of 'giant' virus. Virus Res. *117*, 133–144.

Chatterjee, A., Ali, F., Bange, D., and Kondabagil, K. (2016). Isolation and complete genome sequencing of *Mimivirus bombay*, a giant virus in sewage of Mumbai, India. Genom. Data *9*, 1–3. https://doi.org/10.1016/j.gdata.2016.05.013.

Cohen, G., Hoffart, L., La Scola, B., Raoult, D., and Drancourt, M. (2011). Ameba-associated keratitis, France. Emerg. Infect. Dis. *17*, 1306–1308. https://doi.org/10.3201/eid1707.100826.

Collins, J.P., Brunner, J.L., Jancovich, J.K., and Schock, D.M. (2004). A model host-pathogen system for studying infectious disease dynamics in amphibians: tiger salamanders (*Ambystoma tigrinum*) and *Ambystoma tigrinum* virus. Herpetol. J., *14*, 195–200.

Colson, P., de Lamballerie, X., Fournous, G., and Raoult, D. (2012). Reclassification of giant viruses composing a fourth domain of life in the new order *Megavirales*. Intervirology 55, 321–332. https://doi.org/10.1159/000336562.

Colson, P., De Lamballerie, X., Yutin, N., Asgari, S., Bigot, Y., Bideshi, D.K., Cheng, X.W., Federici, B.A., Van Etten, J.L., Koonin, E.V., *et al.* (2013a). 'Megavirales', a proposed new order for eukaryotic nucleocytoplasmic large DNA viruses. Arch. Virol. *158*, 2517–2521. https://doi.org/10.1007/s00705-013-1768-6.

Colson, P., Gimenez, G., Boyer, M., Fournous, G., and Raoult, D. (2011a). The giant Cafeteria roenbergensis virus that infects a widespread marine phagocytic protist is a new member of the fourth domain of Life. PLOS ONE *6*, e18935. https://doi.org/10.1371/journal.pone.0018935.

Colson, P., La Scola, B., and Raoult, D. (2013b). Giant viruses of amoebae as potential human pathogens. Intervirology *56*, 376–385. https://doi.org/10.1159/000354558.

Colson, P., Pagnier, I., Yoosuf, N., Fournous, G., La Scola, B., and Raoult, D. (2013c). 'Marseilleviridae', a new family of giant viruses infecting amoebae. Arch. Virol. *158*, 915–920. https://doi.org/10.1007/s00705-012-1537-y.

Colson, P., Yutin, N., Shabalina, S.A., Robert, C., Fournous, G., La Scola, B., Raoult, D., and Koonin, E.V. (2011b). Viruses with more than 1,000 genes: Mamavirus, a new *Acanthamoeba polyphaga* mimivirus strain, and reannotation of Mimivirus genes. Genome Biol. Evol. 3, 737–742. https://doi.org/10.1093/gbe/evr048.

Desnues, C., Boyer, M., and Raoult, D. (2012a). Sputnik, a virophage infecting the viral domain of life. Adv. Virus Res. *82*, 63–89. https://doi.org/10.1016/B978-0-12-394621-8.00013-3.

Desnues, C., La Scola, B., Yutin, N., Fournous, G., Robert, C., Azza, S., Jardot, P., Monteil, S., Campocasso, A., Koonin, E.V., *et al.* (2012b). Provirophages and transpovirons as the diverse mobilome of giant viruses. Proc. Natl. Acad. Sci. U.S.A. *109*, 18078–18083. https://doi.org/10.1073/pnas.1208835109.

Desnues, C., and Raoult, D. (2010). Inside the lifestyle of the virophage. Intervirology 53, 293–303. https://doi.org/10.1159/000312914.

Desnues, C., and Raoult, D. (2012). Virophages question the existence of satellites. Nat. Rev. Microbiol. *10*, 234. https://doi.org/10.1038/nrmicro2676-c3.

Dornas, F.P., Assis, F.L., Aherfi, S., Arantes, T., Abrahão, J.S., Colson, P., and La Scola, B. (2016). A Brazilian Marseillevirus Is the Founding Member of a Lineage in Family *Marseilleviridae*. Viruses *8*, 76. https://doi.org/10.3390/v8030076.

Dornas, F.P., Khalil, J.Y., Pagnier, I., Raoult, D., Abrahão, J., and La Scola, B. (2015). Isolation of new Brazilian giant viruses from environmental samples using a panel of protozoa. Front. Microbiol. *6*, 1086. https://doi.org/10.3389/fmicb.2015.01086.

Dos Santos, R.N., Campos, F.S., Medeiros de Albuquerque, N.R., Finoketti, F., Côrrea, R.A., Cano-Ortiz, L., Assis, F.L., Arantes, T.S., Roehe, P.M., and Franco, A.C. (2016). A new marseillevirus isolated in Southern Brazil from *Limnoperna fortunei*. Sci. Rep. *6*, 35237. https://doi.org/10.1038/srep35237.

Dunigan, D.D., Fitzgerald, L.A., Van Etten, J.L. (2006). Phycodnaviruses: a peek at genetic diversity. Papers in Plant Pathology *92*. https://digitalcommons.unl.edu/plantpathpapers/92/.

Federici, B.A., Bideshi, D.K., Tan, Y., Spears, T., and Bigot, Y. (2009). Ascoviruses: superb manipulators of apoptosis for viral replication and transmission. Curr. Top. Microbiol. Immunol. *328*, 171–196.

Fischer, M.G. (2011). Sputnik and Mavirus: more than just satellite viruses. Nat. Rev. Microbiol. *10*, 78. https://doi.org/10.1038/nrmicro2676-c1.

Fischer, M.G., Allen, M.J., Wilson, W.H., and Suttle, C.A. (2010). Giant virus with a remarkable complement of genes infects marine zooplankton. Proc. Natl. Acad. Sci. U.S.A. *107*, 19508–19513. https://doi.org/10.1073/pnas.1007615107.

Fischer, M.G., and Suttle, C.A. (2011). A virophage at the origin of large DNA transposons. Science *332*, 231–234. https://doi.org/10.1126/science.1199412.

Gaia, M., Pagnier, I., Campocasso, A., Fournous, G., Raoult, D., and La Scola, B. (2013). Broad spectrum of mimiviridae virophage allows its isolation using a mimivirus reporter. PLOS ONE *8*, e61912. https://doi.org/10.1371/journal.pone.0061912.

Gaia, M., Benamar, S., Boughalmi, M., Pagnier, I., Croce, O., Colson, P., Raoult, D., and La Scola, B. (2014). Zamilon, a novel virophage with Mimiviridae host specificity. PLOS ONE *9*, e94923. https://doi.org/10.1371/journal.pone.0094923.

Ghedin, E., and Claverie, J.M. (2005). Mimivirus relatives in the Sargasso sea. Virol. J. *2*, 62.

Ghigo, E., Kartenbeck, J., Lien, P., Pelkmans, L., Capo, C., Mege, J.L., and Raoult, D. (2008). Ameobal pathogen mimivirus infects macrophages through phagocytosis. PLOS Pathog. *4*, e1000087. https://doi.org/10.1371/journal.ppat.1000087.

He, J.G., Deng, M., Weng, S.P., Li, Z., Zhou, S.Y., Long, Q.X., Wang, X.Z., and Chan, S.M. (2001). Complete genome analysis of the mandarin fish infectious spleen and kidney necrosis iridovirus. Virology *291*, 126–39. https://doi.org/10.1006/viro.2001.1208.

Iyer, L.M., Aravind, L., and Koonin, E.V. (2001). Common origin of four diverse families of large eukaryotic DNA viruses. J. Virol. *75*, 11720–11734. https://doi.org/10.1128/JVI.75.23.11720-11734.2001.

Iyer, L.M., Balaji, S., Koonin, E.V., and Aravind, L. (2006). Evolutionary genomics of nucleo-cytoplasmic large DNA viruses. Virus Res. *117*, 156–184.

Jancovich, J.K., Davidsson, E., Morado, J.F., Jacobs, B.L., and Collins, J.P. (1997). Isolation of a lethal virus from the endangered tiger salamander *Ambystoma tigrinum stebbinsi*. Dis. of Aquat. Organ. *31*, 161–7. http://www.int-res.com/articles/dao/31/d031p161.pdf.

Jenkins, D.A., Hunter, W.B., and Goenaga, R. (2011). Effects of invertebrate iridescent virus 6 in *Phyllophaga vandinei* and its potential as a biocontrol delivery system. J. Insect Sci. *11*, 44. https://doi.org/10.1673/031.011.0144.

Khan, M., La Scola, B., Lepidi, H., and Raoult, D. (2007). Pneumonia in mice inoculated experimentally with *Acanthamoeba polyphaga mimivirus*. Microb. Pathog. *42*, 56–61.

Klose, T., Kuznetsov, Y.G., Xiao, C., Sun, S., McPherson, A., and Rossmann, M.G. (2010). The three-dimensional structure of Mimivirus. Intervirology *53*, 268–273. https://doi.org/10.1159/000312911.

Koonin, E.V., and Yutin, N. (2010). Origin and evolution of eukaryotic large nucleo-cytoplasmic DNA viruses. Intervirology *53*, 284–292. https://doi.org/10.1159/000312913.

Kuznetsov, Y.G., Klose, T., Rossmann, M., and McPherson, A. (2013). Morphogenesis of mimivirus and its viral factories: an atomic force microscopy study of infected cells. J. Virol. *87*, 11200–11213. https://doi.org/10.1128/JVI.01372-13.

Kuznetsov, Y.G., Xiao, C., Sun, S., Raoult, D., Rossmann, M., and McPherson, A. (2010). Atomic force microscopy investigation of the giant mimivirus. Virology *404*, 127–137.

La Scola, B., Audic, S., Robert, C., Jungang, L., de Lamballerie, X., Drancourt, M., Birtles, R., Claverie, J.M., and Raoult, D. (2003). A giant virus in amoebae. Science *299*, 2033.

La Scola, B., Campocasso, A., N'Dong, R., Fournous, G., Barrassi, L., Flaudrops, C., and Raoult, D. (2010). Tentative characterization of new environmental giant viruses by MALDI-TOF mass spectrometry. Intervirology *53*, 344–353. https://doi.org/10.1159/000312919.

La Scola, B., Desnues, C., Pagnier, I., Robert, C., Barrassi, L., Fournous, G., Merchat, M., Suzan-Monti, M., Forterre, P., Koonin, E., *et al.* (2008). The virophage as a unique parasite of the giant mimivirus. Nature *455*, 100–104. https://doi.org/10.1038/nature07218.

La Scola, B., Marrie, T.J., Auffray, J.P., and Raoult, D. (2005). Mimivirus in pneumonia patients. Emerg. Infect. Dis. *11*, 449–452.

Lagier, J.C., Armougom, F., Million, M., Hugon, P., Pagnier, I., Robert, C., Bittar, F., Fournous, G., Gimenez, G., Maraninchi, M., *et al.* (2012). Microbial culturomics: paradigm shift in the human gut microbiome study. Clin. Microbiol. Infect. *18*, 1185–1193. https://doi.org/10.1111/1469-0691.12023.

Legendre, M., Lartigue, A., Bertaux, L., Jeudy, S., Bartoli, J., Lescot, M., Alempic, J.M., Ramus, C., Bruley, C., Labadie, K., *et al.* (2015). In-depth study of *Mollivirus sibericum*, a new 30,000-y-old giant virus infecting Acanthamoeba. Proc. Natl. Acad. Sci. U.S.A. *112*, E5327–35. https://doi.org/10.1073/pnas.1510795112.

Legendre, M., Bartoli, J., Shmakova, L., Jeudy, S., Labadie, K., Adrait, A., Lescot, M., Poirot, O., Bertaux, L., Bruley, C., *et al.* (2014). Thirty-thousand-year-old distant relative of giant icosahedral DNA viruses with a pandoravirus morphology. Proc. Natl. Acad. Sci. U.S.A. *111*, 4274–4279. https://doi.org/10.1073/pnas.1320670111.

Legendre, M., Arslan, D., Abergel, C., and Claverie, J.M. (2012). Genomics of Megavirus and the elusive fourth domain of Life. Commun. Integr. Biol. *5*, 102–106.

Levasseur, A., Bekliz, M., Chabrière, E., Pontarotti, P., La Scola, B., and Raoult, D. (2016a). MIMIVIRE is a defence system in mimivirus that confers resistance to virophage. Nature *531*, 249–252. https://doi.org/10.1038/nature17146.

Levasseur, A., Andreani, J., Delerce, J., Bou Khalil, J., Robert, C., La Scola, B., and Raoult, D. (2016b). Comparison of a modern and fossil *Pithovirus* reveals its genetic conservation and evolution. Genome Biol. Evol. *8*, 2333–2339. https://doi.org/10.1093/gbe/evw153.

Marrie, T.J., Durant, H., and Yates, L. (1989). Community-acquired pneumonia requiring hospitalization: 5-year prospective study. Rev. Infect. Dis. *11*, 586–599.

Marrie, T.J., Raoult, D., La Scola, B., Birtles, R.J., and de Carolis, E. (2001). *Legionella*-like and other amoebal pathogens as agents of community-acquired pneumonia. Emerging Infect. Dis. 7, 1026–1029.

Mitsuhashi, W., Miyamoto, K., and Wada, S. (2014). The complete genome sequence of the *Alphaentomopoxvirus Anomala cuprea entomopoxvirus*, including its terminal hairpin loop sequences, suggests a potentially unique mode of apoptosis inhibition and mode of DNA replication. Virology *452–3*, 95–116. https://doi.org/10.1016/j.virol.2013.12.036.

Monier, A., Claverie, J.M., and Ogata, H. (2008a). Taxonomic distribution of large DNA viruses in the sea. Genome Biol. *9*, R106. https://doi.org/10.1186/gb-2008-9-7-r106.

Monier, A., Larsen, J.B., Sandaa, R.A., Bratbak, G., Claverie, J.M., and Ogata, H. (2008b). Marine mimivirus relatives are probably large algal viruses. Virol. J. *5*, 12. https://doi.org/10.1186/1743-422X-5-12.

Mutsafi, Y., Shimoni, E., Shimon, A., and Minsky, A. (2013). Membrane assembly during the infection cycle of the giant Mimivirus. PLOS Pathog. *9*, e1003367. https://doi.org/10.1371/journal.ppat.1003367.

Mutsafi, Y., Zauberman, N., Sabanay, I., and Minsky, A. (2010). Vaccinia-like cytoplasmic replication of the giant Mimivirus. Proc. Natl. Acad. Sci. U.S.A. *107*, 5978–5982. https://doi.org/10.1073/pnas.0912737107.

Ngounga, T., Pagnier, I., Reteno, D.G., Raoult, D., La Scola, B., and Colson, P. (2013). Real-time PCR systems targeting giant viruses of amoebae and their virophages. Intervirology *56*, 413–423. https://doi.org/10.1159/000354563.

Otta, D.A., Rott, M.B., Carlesso, A.M., and da Silva, O.S. (2012). Prevalence of *Acanthamoeba* spp. (Sarcomastigophora: Acanthamoebidae) in wild populations of *Aedes aegypti* (Diptera: Culicidae). Parasitol. Res. *111*, 2017–2022. https://doi.org/10.1007/s00436-012-3050-3.

Pagnier, I., Reteno, D.G., Saadi, H., Boughalmi, M., Gaia, M., Slimani, M., Ngounga, T., Bekliz, M., Colson, P., Raoult, D., *et al.* (2013). A decade of improvements in Mimiviridae and Marseilleviridae isolation from amoeba. Intervirology *56*, 354–363. https://doi.org/10.1159/000354556.

Philippe, N., Legendre, M., Doutre, G., Couté, Y., Poirot, O., Lescot, M., Arslan, D., Seltzer, V., Bertaux, L., Bruley, C., *et al.* (2013). Pandoraviruses: amoeba viruses with genomes up to 2.5 Mb reaching that of parasitic eukaryotes. Science *341*, 281–286. https://doi.org/10.1126/science.1239181.

Popgeorgiev, N., Boyer, M., Fancello, L., Monteil, S., Robert, C., Rivet, R., Nappez, C., Azza, S., Chiaroni, J., Raoult, D., *et al.* (2013a). Marseillevirus-like virus recovered from blood donated by asymptomatic humans. J. Infect. Dis. *208*, 1042–1050. https://doi.org/10.1093/infdis/jit292.

Popgeorgiev, N., Colson, P., Thuret, I., Chiarioni, J., Gallian, P., Raoult, D., and Desnues, C. (2013b). Marseillevirus prevalence in multitransfused patients suggests blood transmission. J. Clin. Virol. *58*, 722–725. https://doi.org/10.1016/j.jcv.2013.10.001.

Popgeorgiev, N., Michel, G., Lepidi, H., Raoult, D., and Desnues, C. (2013c). Marseillevirus adenitis in an 11-month-old child. J. Clin. Microbiol. *51*, 4102–4105. https://doi.org/10.1128/JCM.01918-13.

Raoult, D., Audic, S., Robert, C., Abergel, C., Renesto, P., Ogata, H., La Scola, B., Suzan, M., and Claverie, J.M. (2004). The 1.2-megabase genome sequence of Mimivirus. Science *306*, 1344–1350.

Raoult, D., and Forterre, P. (2008). Redefining viruses: lessons from Mimivirus. Nat. Rev. Microbiol. *6*, 315–319. https://doi.org/10.1038/nrmicro1858.

Raoult, D., La Scola, B., and Birtles, R. (2007). The discovery and characterization of Mimivirus, the largest known virus and putative pneumonia agent. Clin. Infect. Dis. *45*, 95–102.

Raoult, D., Renesto, P., and Brouqui, P. (2006). Laboratory infection of a technician by mimivirus. Ann. Intern. Med. *144*, 702–703.

Reteno, D.G., Benamar, S., Khalil, J.B., Andreani, J., Armstrong, N., Klose, T., Rossmann, M., Colson, P., Raoult, D., and La Scola, B. (2015). Faustovirus, an asfarvirus-related new lineage of giant viruses infecting amoebae. J. Virol. *89*, 6585–6594.

Saadi, H., Pagnier, I., Colson, P., Cherif, J.K., Beji, M., Boughalmi, M., Azza, S., Armstrong, N., Robert, C., Fournous, G., *et al.* (2013a). First isolation of Mimivirus in a patient with pneumonia. Clin. Infect. Dis. *57*, e127–34. https://doi.org/10.1093/cid/cit354.

Saadi, H., Reteno, D.G., Colson, P., Aherfi, S., Minodier, P., Pagnier, I., Raoult, D., and La Scola, B. (2013b). Shan virus: a new mimivirus isolated from the stool of a Tunisian patient with pneumonia. Intervirology *56*, 424–429. https://doi.org/10.1159/000354564.

Scheid, P., Zöller, L., Pressmar, S., Richard, G., and Michel, R. (2008). An extraordinary endocytobiont in *Acanthamoeba* sp. isolated from a patient with keratitis. Parasitol. Res. *102*, 945–950. https://doi.org/10.1007/s00436-007-0858-3.

Shah, N., Hülsmeier, A.J., Hochhold, N., Neidhart, M., Gay, S., and Hennet, T. (2014). Exposure to mimivirus collagen promotes arthritis. J. Virol. *88*, 838–845. https://doi.org/10.1128/JVI.03141-13.

Silva, L.C., Almeida, G.M., Oliveira, D.B., Dornas, F.P., Campos, R.K., La Scola, B., Ferreira, P.C., Kroon, E.G., and Abrahão, J.S. (2014). A resourceful giant: APMV is able to interfere with the human type I interferon system. Microbes Infect. *16*, 187–195. https://doi.org/10.1016/j.micinf.2013.11.011.

Suárez, C., Welsch, S., Chlanda, P., Hagen, W., Hoppe, S., Kolovou, A., Pagnier, I., Raoult, D., and Krijnse Locker, J. (2013). Open membranes are the precursors for assembly of large DNA viruses. Cell. Microbiol. *15*, 1883–1895. https://doi.org/10.1111/cmi.12156.

Sun, S., La Scola, B., Bowman, V.D., Ryan, C.M., Whitelegge, J.P., Raoult, D., and Rossmann, M.G. (2010). Structural studies of the Sputnik virophage. J. Virol. *84*, 894–897. https://doi.org/10.1128/JVI.01957-09.

Suzan-Monti, M., La Scola, B., and Raoult, D. (2006). Genomic and evolutionary aspects of Mimivirus. Virus Res. *117*, 145–155.

Takemura, M. (2016a). Draft genome sequence of *Tokyovirus*, a member of the family *Marseilleviridae* isolated from the Arakawa River of Tokyo, Japan.

Genome Announc. *4* pii: e00429-16. https://doi.org/10.1128/genomeA.00429-16.

Takemura, M. (2016b). Morphological and taxonomic properties of *Tokyovirus*, the first *Marseilleviridae* member isolated from Japan. Microbes Environ. *31*, 442–448. https://doi.org/10.1264/jsme2.ME16107.

Thomas, V., Bertelli, C., Collyn, F., Casson, N., Telenti, A., Goesmann, A., Croxatto, A., and Greub, G. (2011). Lausannevirus, a giant amoebal virus encoding histone doublets. Environ. Microbiol. *13*, 1454–1466. https://doi.org/10.1111/j.1462-2920.2011.02446.x.

Tulman, E.R., Afonso, C.L., Lu, Z., Zsak, L., Kutish, G.F., and Rock, D.L. (2004). The genome of canarypox virus. J Virol. *78*, 353–66. https://doi.org/10.1128/JVI.78.1.353-366.2004.

Van Etten, J.L., Graves, M.V., Müller, D.G., Boland, W., and Delaroque, N. (2002). *Phycodnaviridae* – large DNA algal viruses. Arch. Virol. *147*, 1479–1516. https://doi.org/10.1007/s00705-002-0822-6.

Van Etten, J.L., Gurnon, J.R., Yanai-Balser, G.M., Dunigan, D.D., and Graves, M.V. (2010). Chlorella viruses encode most, if not all, of the machinery to glycosylate their glycoproteins independent of the endoplasmic reticulum and Golgi. Biochim. Biophys. Acta. *1800*, 152–159.

Vincent, A., La Scola, B., and Papazian, L. (2010). Advances in Mimivirus pathogenicity. Intervirology *53*, 304–309. https://doi.org/10.1159/000312915.

Wang, L., Xue, J., Seaborn, C.P., Arif, B.M., and Cheng, X.W. (2006). Sequence and organization of the *Trichoplusia ni* ascovirus 2c (*Ascoviridae*) genome. Virology *10*, 167–77. https://ac.els-cdn.com/S0042682206004326/1-s2.0-S0042682206004326-main.pdf?_tid=cbb93e36-a1df-4c7e-bd61-78612b03b3e7&acdnat=1525032451_104f42d5f02312b63c855dd45e8150b5.

Williams, T. (1998). Invertebrate iridescent viruses. In: The Insect Viruses, Miller, L., and Ball, A., eds. (Plenum, New York, NY), pp. 31–68.

Xiao, C., Kuznetsov, Y.G., Sun, S., Hafenstein, S.L., Kostyuchenko, V.A., Chipman, P.R., Suzan-Monti, M., Raoult, D., McPherson, A., and Rossmann, M.G. (2009). Structural studies of the giant mimivirus. PLOS Biol. *7*, e92. https://doi.org/10.1371/journal.pbio.1000092.

Yan, X.Y., Wu, Z.H., Jian, J.C., Lu, Y.S., and Sun, X.Q. (2011). Analysis of the genetic diversity of the lymphocystis virus and its evolutionary relationship with its hosts. Virus Genes. *43*, 358–66. https://doi.org/10.1007/s11262-011-0646-0.

Yoosuf, N., Yutin, N., Colson, P., Shabalina, S.A., Pagnier, I., Robert, C., Azza, S., Klose, T., Wong, J., Rossmann, M.G., *et al.* (2012). Related giant viruses in distant locations and different habitats: *Acanthamoeba polyphaga* moumouvirus represents a third lineage of the *Mimiviridae* that is close to the megavirus lineage. Genome Biol. Evol. *4*, 1324–1330. https://doi.org/10.1093/gbe/evs109.

Yutin, N., Colson, P., Raoult, D., and Koonin, E.V. (2013). Mimiviridae: clusters of orthologous genes, reconstruction of gene repertoire evolution and proposed expansion of the giant virus family. Virol. J. *10*, 106. https://doi.org/10.1186/1743-422X-10-106.

Yutin, N., and Koonin, E.V. (2009). Evolution of DNA ligases of nucleo-cytoplasmic large DNA viruses of eukaryotes: a case of hidden complexity. Biol. Direct *4*, 51. https://doi.org/10.1186/1745-6150-4-51.

Yutin, N., and Koonin, E.V. (2013). Pandoraviruses are highly derived phycodnaviruses. Biol. Direct *8*, 25. https://doi.org/10.1186/1745-6150-8-25.

Yutin, N., Wolf, Y.I., Raoult, D., and Koonin, E.V. (2009). Eukaryotic large nucleo-cytoplasmic DNA viruses: clusters of orthologous genes and reconstruction of viral genome evolution. Virol. J. *6*, 223. https://doi.org/10.1186/1743-422X-6-223.

Zauberman, N., Mutsafi, Y., Halevy, D.B., Shimoni, E., Klein, E., Xiao, C., Sun, S., and Minsky, A. (2008). Distinct DNA exit and packaging portals in the virus *Acanthamoeba polyphaga mimivirus*. PLOS Biol. *6*, e114. https://doi.org/10.1371/journal.pbio.0060114

The Virophage Family *Lavidaviridae* 12

Matthias G. Fischer[1]*

[1]Department of Biomolecular Mechanisms, Max Planck Institute for Medical Research, Heidelberg, Germany.

*Correspondence: mfischer@mpimf-heidelberg.mpg.de

https://doi.org/10.21775/9781910190852.12

Abstract

Double-stranded (ds) DNA viruses of the family *Lavidaviridae*, commonly known as virophages, are a fascinating group of eukaryotic viruses that depend on a coinfecting giant dsDNA virus of the *Mimiviridae* for their propagation. Instead of replicating in the nucleus, virophages multiply in the cytoplasmic virion factory of a coinfecting giant virus inside a phototrophic or heterotrophic protistal host cell. Virophages are parasites of giant viruses and can inhibit their replication, which may lead to increased survival rates of the infected host cell population. The genomes of virophages are 17–33 kilobase pairs (kbp) long and encode 16–34 proteins. Genetic signatures of virophages can be found in metagenomic datasets from various saltwater and freshwater environments around the planet. Most virophages share a set of conserved genes that code for a major and a minor capsid protein, a cysteine protease, a genome-packaging ATPase, and a superfamily 3 helicase, although the genomes are otherwise diverse and variable. Lavidaviruses share genes with other mobile genetic elements, suggesting that horizontal gene transfer and recombination have been major forces in shaping these viral genomes. Integrases are occasionally found in virophage genomes and enable these DNA viruses to persist as provirophages in the chromosomes of their viral and cellular hosts. As we watch the genetic diversity of this new viral family unfold through metagenomics, additional isolates are still lacking and critical questions regarding their infection cycle, host range, and ecology remain to be answered.

Introduction

Virophages are a recently discovered class of double-stranded (ds) DNA viruses that have evolved a dependency on complex dsDNA viruses of eukaryotes, so-called giant viruses. The discovery of giant viruses was therefore a prerequisite for the isolation and characterization of virophages (see Fig. 12.1), as shall be reviewed here briefly (see also Chapter 11).

In 1992, following a pneumonia outbreak in Bradford, England, an intra-amoebal parasite was isolated by the team of T.J. Rowbotham and given the name 'Bradford coccus'. This microorganism was initially assumed to be a bacterium because of its size and positive Gram-stain reaction; however, attempts to amplify and analyse its 16S ribosomal DNA sequences failed. A decade later in the laboratory of Didier Raoult in Marseille, electron microscopy revealed that Bradford coccus was in fact a giant virus with a fibre-studded, 0.75 µm-diameter particle (La Scola *et al.*, 2003). The serendipitous discovery of Acanthamoeba polyphaga mimivirus (family *Mimiviridae*, genus *Mimivirus*, species *Acanthamoeba polyphaga mimivirus*) and the analysis of its 1.2 million base pair dsDNA genome have profoundly changed our view of the viral world (Raoult *et al.*, 2004). This virus exhibited a particle size and genome length that exceeded those of the smallest cellular organisms, blurring the boundary between viruses and cells. The finding that some mimivirus genes were homologous to genes that were previously known to occur only in cellular genomes, in particular four aminoacyl-tRNA synthetases and four translation factors, spurred evolutionary scenarios in which giant viruses were hypothesized to

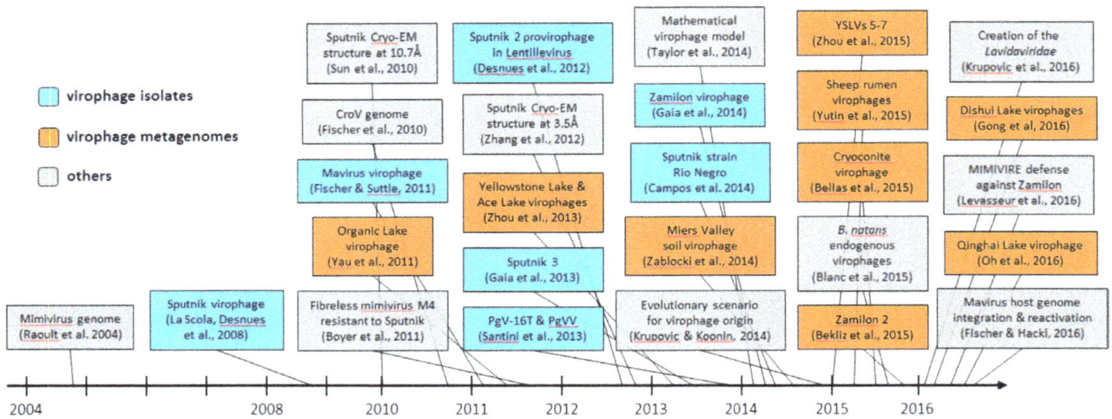

Figure 12.1 Historical time line of virophage-related publications.

be descendants of a cellular ancestor (Boyer *et al.*, 2010; Nasir *et al.*, 2012; Raoult *et al.*, 2004). Such hypotheses, however, have been widely criticized and the origin of giant viruses remains a matter of ongoing debate (Williams *et al.*, 2011; Yutin *et al.*, 2014). Akin to poxviruses (Broyles, 2003), mimivirus and related giant viruses replicate solely in the cytoplasm of their unicellular eukaryotic host (Mutsafi *et al.*, 2010), which is made possible by hundreds of virus-encoded proteins that provide nuclear functions to the virion factory (VF), such as DNA replication and transcription. For instance, the mimivirus-encoded transcription apparatus consists of at least eight DNA-dependent RNA polymerase subunits, a trifunctional mRNA capping enzyme, a polyadenylate polymerase, and several transcription factors.

A targeted search for giant viruses was launched following the characterization of mimivirus and resulted in the isolation of dozens of new viral strains (La Scola *et al.*, 2010). One of the first strains to be isolated was Acanthamoeba castellanii mamavirus, a close relative of mimivirus that was recovered from a cooling tower in Paris. Electron microscopy analysis of mamavirus-infected amoebae revealed the presence of a second, smaller icosahedral virus that was named Sputnik (La Scola *et al.*, 2008). Sputnik was unable to infect *Acanthamoeba* cells on its own, and replicated only when the cells were coinfected with mamavirus. In coinfected cells, Sputnik colocalized to the mamavirus VF, providing the first evidence that Sputnik uses giant virus-encoded enzymes for its propagation. The presence of Sputnik also elicited a morphological phenotype in mamavirus, with visible deformations such as partial capsid thickening in many of the newly synthesized virions. The yield of mamavirus progeny from coinfected cells was reduced by ≈70% (La Scola *et al.*, 2008). Sputnik thus acted as a parasite of mimiviruses and the term 'virophage' was coined to reflect the relationship as a 'virus of a virus'. However, the parasitic interaction appears to be restricted to the intracellular phase of the virus life cycle, when the biochemical complexity of giant viruses unfolds in the form of the VF organelle. In the years following the discovery of Sputnik, additional virophages such as mavirus (Fischer and Suttle, 2011) and Zamilon (Gaia *et al.*, 2014) were isolated, and more than a dozen putative virophage genomes were found in metagenomic datasets (Gong *et al.*, 2016; Oh *et al.*, 2016; Yau *et al.*, 2011; Yutin *et al.*, 2015a; Zhou *et al.*, 2013, 2015) (see Table 12.1). These eukaryotic dsDNA viruses are geographically widespread, genetically diverse, and appear to be commonly associated with giant virus-infected protist populations. The interested reader is referred to previously published review articles on this topic (Bekliz *et al.*, 2016; Claverie and Abergel, 2009b; Desnues and Raoult, 2010; Gaia *et al.*, 2013a).

Virion structure

Only few virophage representatives (Sputnik, mavirus, and Zamilon) have been isolated in laboratory culture and are thus amenable to structural studies.

The virophage particles that have been examined so far are 50–75 nm in diameter and possess icosahedral symmetry (Fig. 12.2). A 3.5 Å resolution cryo-electron microscopy (cryo-EM) reconstruction of the 75 nm wide Sputnik particle suggested the lack of a membrane component (Zhang *et al.*, 2012), contradicting earlier reports that assumed a lipid component in the Sputnik particle (Desnues *et al.*, 2012a; Sun *et al.*, 2010). The Sputnik capsid structure is composed of the major capsid protein (MCP) encoded by the V20 gene, and the minor capsid protein (mCP) encoded by the V18/V19 gene. The MCP contains a double jelly-roll fold and forms trimeric capsomers (hexons) with pseudo-hexagonal symmetry that build the 20 faces of the icosahedral particle (Fig. 12.2C), whereas the mCP is a single jelly-roll protein that forms pentameric capsomers (pentons) occupying the 12 vertices. The mature capsid consists of 260 hexons and 12 pentons that are arranged in a lattice with a triangulation (T) number of 27 ($h=3$; $k=3$) (Sun *et al.*, 2010; Zhang *et al.*, 2012).

The capsid proteins of different virophages show low levels of conservation on the amino acid level (e.g. ~40% for the MCPs of Sputnik and mavirus) (Krupovic *et al.*, 2016). The mCP in particular is highly diverse, which hinders sequence similarity-based identification of novel virophages. Despite the high sequence divergence, all MCPs and mCPs are assumed to adopt the double and single jelly-roll folds, respectively, and preliminary data shows that the three-dimensional structure of the mavirus MCP is very similar to that of Sputnik (D. Born, L. Reuter, U. Mersdorf, M. Mueller, M.G. Fischer, A. Meinhart and J. Reinstein, under review). However, since most of the sequence space occupied by virophages remains unknown, the existence of different capsid architectures cannot be excluded. For instance, the virophage genomes assembled from a sheep rumen metagenome apparently lack a penton gene (Yutin *et al.*, 2015a). In addition to the main capsid components (MCP and mCP), other virophage-encoded proteins may be present in the mature virion, such as the mavirus MV13

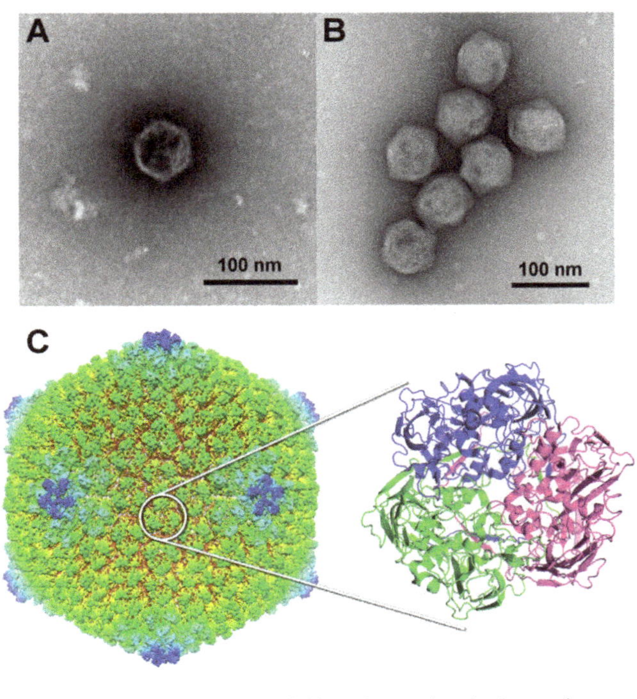

Figure 12.2 The structure of lavidavirus capsids. A) Negative stain electron micrograph of a Sputnik particle (courtesy of M. Gaia and B. La Scola, Univ. Aix-Marseille, France). B) Negative stain EM image of CsCl gradient-purified mavirus particles (U. Mersdorf and M. Fischer, Max Planck Institute for Medical Research, Germany). C) Cryo-EM reconstruction of the Sputnik virion at 3.5 Å resolution with a magnified major capsid protein trimer (PDB entry 3J26). Modified from (Zhang *et al.*, 2012).

protein, which could have lipase activity (Fischer *et al.*, 2014).

Virophage genomics

All virophages described so far possess dsDNA genomes ranging in size from 17 to more than 30 kbp. Fig. 12.3 depicts the genome organization of virophages, and their properties are listed in Table 12.1. Cultured lavidaviruses (Sputnik, mavirus, Zamilon) have circular genomes while those that were assembled from metagenomic sequences are either circular, linear, or have unknown genome topologies (Gong *et al.*, 2016; Oh *et al.*, 2016; Yau *et al.*, 2011; Yutin *et al.*, 2015a; Zhou *et al.*, 2013, 2015). The G+C content of virophage genomes tends to fall between 27% and 37%, with the exception of YSLV 5 (51.1% G+C). Similarly low G+C contents are found in the virophage-associated giant viruses, e.g. mimivirus (28% G+C), CroV (23%), PgV-16T (32% G+C), and OLPV-1 (30% G+C) (Fischer *et al.*, 2010; Raoult *et al.*, 2004; Santini *et al.*, 2013; Yau *et al.*, 2011). In contrast, the G+C contents of the nuclear genomes in the eukaryotic hosts are typically higher, e.g. 59% for *Acanthamoeba polyphaga* (GenBank Acc. #CDFK00000000) and 65–70% for *Cafeteria roenbergensis* (T. Hackl and M. Fischer, unpublished data). This may indicate that virophage and giant virus genomes are subject to similar evolutionary forces that favour high A+T contents.

Virophage genomes contain from 16 to 34 protein-coding sequences, which are for the most part spaced in a non-overlapping manner, similar to other dsDNA viruses and bacterial genomes (Fig. 12.3). The Organic Lake virophage (OLV) has five annotated ORFs that are nested within larger predicted coding sequences (CDSs) (Yau *et al.*,

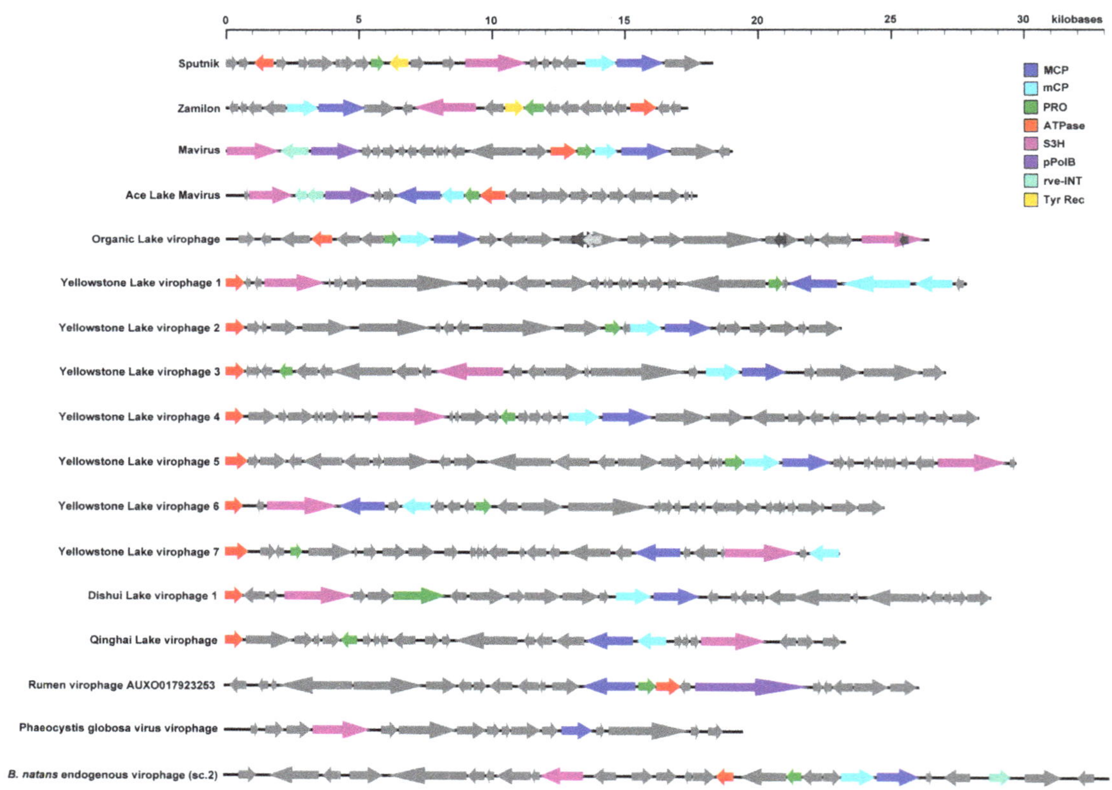

Figure 12.3 Comparative genomics of virophages. Complete and near-complete virophage genomes are drawn to scale and their open reading frames are indicated by arrows. Selected genes are shown in colour. MCP: major capsid protein, mCP: minor capsid protein, PRO: cysteine protease, ATPase: FtsK-HerA-family genome packaging ATPase, Prim/HEL: primase/superfamily 3 helicase, pPolB: protein-primed DNA polymerase B, rve-INT: retrovirus-family integrase, Tyr Rec: tyrosine recombinase-family integrase.

2011). There is no indication that they encode proteins, however, since one of those ORFs (OLV26) has no significant similarity to sequences in public databases, and the other four ORFs (OLV14-16, 21) contain highly repetitive amino acid patterns, which may give false-positive homologues in BLAST searches.

Conserved genes

Nearly all virophage genomes encode a structural module that consists of a major capsid protein (MCP), a minor capsid or penton protein (mCP), a cysteine protease (PRO), and an FtsK-HerA-like ATPase (coloured in Fig. 12.3). These proteins are involved in virion formation. The highly conserved MCP–mCP gene pair is found in all virophages including the virophage-like element PgVV and also in polinton-like viruses (PLVs), which encode poorly conserved versions of these capsid genes (Yutin *et al.*, 2015b). These two genes are syntenic in most virophages (Fig. 12.3), indicative of their importance. The MCP adopts a double jelly-roll fold, whereas the mCP gene encodes the single jelly-roll penton protein of virophage capsids. Yellowstone Lake virophage (YSLV) 1 encodes two versions of the mCP, coding sequences (CDSs) 26 and 27, of which CDS 26 contains a 390 aa long insertion that is missing in other virophage mCPs. Because mCP pentamers are located at the vertices of the icosahedral capsid, they are most likely to interact with host components (e.g. membrane proteins), which may explain their high sequence variability due to adaptation to new hosts. The virophage-encoded PRO bears similarity to maturation proteases from adenoviruses and PRD1, and this protease is most likely required for virion morphogenesis (D. Born, L. Reuter, U. Mersdorf, M. Mueller, M.G. Fischer, A. Meinhart and J. Reinstein, under review). The cryo-EM reconstruction of the Sputnik virion revealed that the assembled MCP was 87 aa shorter than predicted from the DNA sequence (Zhang *et al.*, 2012). Therefore, the C-terminus must be cleaved off at a diglycine motif that is conserved across all virophages except for PgVV and the rumen virophages (RVPs). PRO is thus responsible for C-terminal processing of the MCP in at least Sputnik and mavirus. The last conserved virophage gene is a predicted genome-translocating ATPase that belongs to the FtsK-HerA superfamily. These NTPases usually form hexamers at a unique vertex of the viral capsid and pump the genome through the central pore of this hexamer, a process that is powered by ATP hydrolysis (Burroughs *et al.*, 2007). The mavirus ATPase differs from those of all other virophages by a unique 54 aa insertion between Walker A and Walker B motifs.

Most virophage genomes encode a superfamily 3 helicase (S3H) domain (Fig. 12.3), but the genetic relationships among these genes are complicated. Virophage S3H genes are apparently polyphyletic, and the versions found in Sputnik, Zamilon, OLV, YSLV 1, YSLV 6, QLV, and PgVV are coupled to an amino-terminal domain with predicted primase-polymerase activity (TVpol, related to bacterial DNA polymerase I enzymes) (Iyer *et al.*, 2008; Yutin *et al.*, 2013). The S3H may thus have adopted different functions in different virophages, e.g. as a helicase in mavirus (which encodes a separate DNA polymerase), and as the primary DNA replication enzyme in Sputnik. Short zinc ribbon domains are also frequently found in virophages, where they can be fused with GIY-YIG endonucleases (e.g. mavirus, OLV), or occur separately (e.g. Sputnik) (Yutin *et al.*, 2013).

Non-conserved genes

Many of the genes found in virophage genomes exhibit diverse evolutionary affiliations, including eukaryotes, bacteria, different viral families, and diverse mobile genetic elements (Fischer and Suttle, 2011; Krupovic and Koonin, 2015; La Scola *et al.*, 2008; Yau *et al.*, 2011; Yutin *et al.*, 2013). Examples are the bacteriophage lambda-type tyrosine recombinase encoded by Sputnik or the retroviral integrase encoded by mavirus. Although virophages are genetically distinct from their associated host viruses, they encode one to several genes with clear affiliations to giant DNA viruses, and these genes often contain repetitive motifs. The genomes of Sputnik, Zamilon, and predicted algae-infecting virophages contain collagen-like repeat proteins that are also prominent in genomes of *Mimiviridae* members (Gaia *et al.*, 2014; La Scola *et al.*, 2008; Suhre, 2005; Yau *et al.*, 2011). Sputnik ORF V12 codes for a low-complexity protein that is similar to mimivirus ORF R546; and Sputnik ORFs V4, V14, V16, and V17 display similarities to proteins encoded mainly by moumouviruses. The most

Table 12.1 Properties of partial and complete virophage genomes

Virophage	Length (bp)	% GC	Topology	ORFs	Cultured or sequenced	Associated giant virus(es)	Host organism	GenBank accession no.	Reference
Sputnik	18,342	27.04	Circular	21	In culture	Mimi-, Moumou-, Megaviruses	*Acanthamoeba* sp.	EU606015.1	La Scola *et al*. (2008)
Sputnik 2	18,338	27.04	Circular	21	In culture	Mimi-, Moumou-, Megaviruses	*Acanthamoeba* sp.	JN603369.1	Desnues *et al*. (2012b)
Sputnik 3	18,338	27.04	Circular	21	In culture	Mimi-, Moumou-, Megaviruses	*Acanthamoeba* sp.	JN603370.1	Gaia *et al*. (2013b)
Sputnik strain Rio Negro	1788 (partial)	33.45	Unknown	Unknown	In culture	Samba virus (SMBV)	*Acanthamoeba* sp.	KJ183141.1	Campos *et al*. (2014)
Zamilon	17,276	29.67	Circular	20	In culture		*Acanthamoeba* sp.	HG531932.1	Gaia *et al*. (2014)
Zamilon 2	6717 (partial)	32	Unknown	15	Sequence only	Unknown	Unknown	N/A	Bekliz *et al*. (2015)
Mavirus (exogenous)	19,063	30.26	Circular	20	In culture	Cafeteria roenbergensis virus (CroV)	*Cafeteria roenbergensis*	HQ712116.1	Fischer and Suttle (2011)
Mavirus (endogenous)	20,190	30.67	Integrated	20	In culture	Cafeteria roenbergensis virus (CroV)	*Cafeteria roenbergensis*	KU052222.1	Fischer and Hackl (2016)
Ace Lake Mavirus (ALM)	17,767	26.73	Unknown	22	Sequence only	Unknown	Unknown	KC556923.1	Zhou *et al*. (2013)
Organic Lake virophage (OLV)	26,421	36.51	Unknown	26	Sequence only	Unknown	Unknown	HQ704801.1	Yau *et al*. (2011)
Yellowstone Lake virophage (YSLV) 1	27,851	33.37	Unknown	28	Sequence only	Unknown	Unknown	KC556924.1	Zhou *et al*. (2013)

YSLV2	23,184	33.16	Unknown	21	Sequence only	Unknown	Unknown	KC556925.1	Zhou *et al*. (2013)
YSLV3	27,050	35.86	Unknown	23	Sequence only	Unknown	Unknown	KC556926.1	Zhou *et al*. (2013)
YSLV4	28,306	37.07	Unknown	34	Sequence only	Unknown	Unknown	KC556922.1	Zhou *et al*. (2013)
YSLV5	29,767	51.10	Unknown	32	Sequence only	Unknown	Unknown	KM502589.1	Zhou *et al*. (2015)
YSLV6	24,837	26.84	Unknown	29	Sequence only	Unknown	Unknown	KM502590.1	Zhou *et al*. (2015)
YSLV7	23,193	27.33	Unknown	26	Sequence only	Unknown	Unknown	KM502591.1	Zhou *et al*. (2015)
Dishui Lake virophage (DSLV)	28,788	43.20	Circular	28	Sequence only	Unknown	Unknown	KT894027.1	Gong *et al*. (2016)
Qinghai Lake virophage (QLV)	23,379	33.17	Circular	25	Sequence only	Unknown	Unknown	KJ854379.1	Oh *et al*. (2016)
Rumen virophages (RVP)	varies, longest: 26,209	varies, longest: 34.64	Linear with TIRs	varies, longest: 22	Sequence only	Unknown	Unknown	longest AUXO017923253	Yutin *et al*. (2015a)
Cryoconite virophage (CryV)	12,572 (partial)	32.80	unknown	15	Sequence only	Unknown	Unknown	LDNN01000025.1	Bellas *et al*. (2015)
Miers Valley soil virophage (MVSV)	1035 (partial)	34.69	unknown	unknown	sequence only	unknown	unknown	CZR14427.1	Zablocki *et al*. (2014)
Phaeocystis globosa virus virophage (PgVV)	19,527	35.78	linear or integrated	16	in culture	Phaeocystis globosa virus 16T (PgV-16T)	*Phaeocystis globosa*	NC_021333	Santini *et al*. (2013)
Bigelowiella natans virophages (BnVP)	varies, up to 33.3 kb	varies, ≈36.4	integrated	varies, up to 27	sequence only	unknown	*Bigelowiella natans*	Assembly GCA_000320545	Blanc *et al*. (2015)

frequent repetitive element in the CroV genome is called FNIP or IP22 (Fischer *et al.*, 2010; O'Day *et al.*, 2006) and occupies more than 10% of the 692 kbp genome (Hackl and Fischer, in preparation), and the same repeat is also found in mavirus ORF MV20. These observations suggest that DNA or protein repeats may be involved in virophage–giant virus interactions. Repetitive proteins are also found in the genomes of uncultivated virophages, e.g. the Kelch, ankyrin, and adhesin domain-containing proteins in the endogenous *Bigelowiella natans* virophages (Blanc *et al.*, 2015).

Diversity and taxonomy of virophages

After the isolation and characterization of Sputnik and mavirus, it became apparent that these viruses were the first representatives of a previously unknown group of dsDNA viruses. Classification of virophages was initially complicated by their requirement for coinfecting giant dsDNA viruses for their propagation. This property, which virophages share with satellite viruses, led to the initial classification of Sputnik as a satellite virus, placed within the obscure category of 'subviral agents' in the Ninth Report of the International Committee on Taxonomy of Viruses (ICTV) (King *et al.*, 2011). The distinction between satellite viruses and virophages has caused some confusion and debate (Desnues and Raoult, 2012; Fischer, 2011; Krupovic and Cvirkaite-Krupovic, 2011) and shall be discussed here briefly.

Virophages and satellite viruses

Virophages as well as satellite viruses can exert a negative effect on their associated giant viruses, or helper viruses, respectively. This effect varies significantly depending on the virus and multiple infection parameters. Satellite viruses also share some other aspects with virophages, such as subcellular localization or overlap with their helper viruses regarding polyadenylation sites (Krupovic and Cvirkaite-Krupovic, 2011). On the other hand, virophages are genetically distinct from satellite viruses. While the latter have short ssRNA genomes (satellite viruses infecting plants and arthropods) or short ssDNA genomes (adeno-associated viruses) (King *et al.*, 2011), virophages possess considerably longer dsDNA genomes encoding genes that are conserved in the PRD1–adenovirus lineage. Unlike satellite viruses and owing to the greater complexity of virophage particles, virophages encode multiple morphogenesis proteins (two capsid proteins, maturation protease, DNA-packaging ATPase) and DNA replication proteins. As discussed in more detail below, virophages probably depend on the giant virus-provided transcription apparatus in a way that is similar to how other small dsDNA viruses depend on the cellular transcription system.

Thus, while several phenomena are shared between virophages and satellite viruses (dependency on another virus, variable hypovirulence), their larger particle size and coding potential, genetic and infection properties, and relatedness to viruses of the PRD1–adenovirus lineage clearly set virophages apart from those much simpler and genetically distinct groups of satellite viruses. Virophages can be viewed as eukaryotic dsDNA viruses that became evolutionarily adapted to giant dsDNA viruses, and that use the viral transcription enzymes in the cytoplasm instead of accessing the host nucleus for cellular transcription.

The *Lavidaviridae*

The family *Lavidaviridae* was created to reflect the unique properties of virophages, which distinguish them from other groups of viruses including satellite viruses (Krupovic *et al.*, 2016). 'Lavida-' is an acronym for 'large virus dependent or associated' and highlights the most striking feature of these viruses, i.e. their co-dependency on giant dsDNA viruses. This family currently contains two genera, *Sputnikvirus* and *Mavirus*, which were created based on shared subsets of genes that are specific to each genus, in addition to phylogenetic analysis of genes that are conserved in all virophages.

The topologies of the resulting trees vary somewhat depending on the chosen marker gene (Fig. 12.4). The most highly conserved gene and thus arguably the best phylogenetic marker is the major capsid protein (Fig. 12.4A), but the ATPase is also conserved enough to be useful for phylogenetic studies (Fig. 12.4B). Both marker genes resolve the two genera with high branch support, and additional well-supported clusters can be observed, as discussed below. Given the rapid accumulation of virophage genomes from metagenomes, which

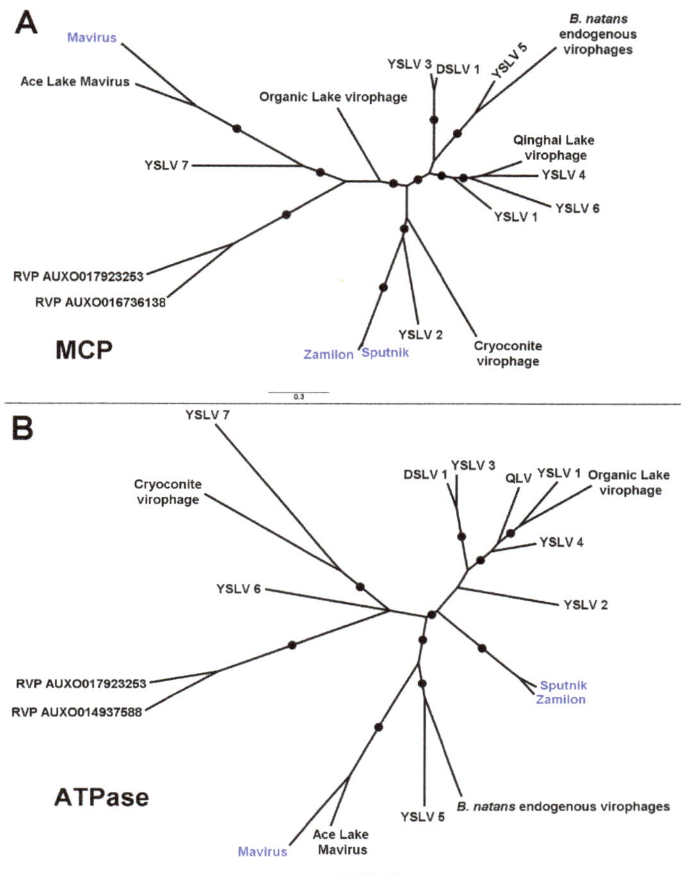

Figure 12.4 Phylogenetic relationships among virophages. Major capsid protein (A) and FtsK-HerA-family ATPase (B) sequences were aligned with PROMALS3D (Pei *et al.*, 2008), and the manually edited alignment was used to create an unrooted Bayesian phylogenetic tree using MrBayes v3.1.2 (Ronquist and Huelsenbeck, 2003) with 1 million generations and a burn-in of 1000. Branches with posterior probabilities less than 0.5 were collapsed; those with posterior probabilities higher than 0.90 are marked by black dots. Cultured virophages are printed in blue.

leads to a better resolution of the phylogenetic reconstructions, additional *Lavidaviridae* genera will likely be added in the near future.

The genus *Sputnikvirus*

Sputnik, the first virophage to be discovered, was isolated from water of a cooling tower near Paris, France (La Scola *et al.*, 2008). The same sample also contained a new mimivirus strain named mamavirus with a 1,191,693 bp dsDNA genome. Based on transmission electron microscopy, the Sputnik particles were initially reported to be 50 nm wide, in contrast to cryo-EM studies that were conducted later and reported a capsid diameter of 74 nm (Sun *et al.*, 2010). After Sputnik virions were purified from mamavirus by filtration through 0.2 μm pore-size filters and inoculated on amoebae, no Sputnik multiplication was observed. This virus instead only replicated when the amoebae were coinfected with mamavirus. Inspection of coinfected amoebal cells by electron microscopy further revealed that Sputnik and mamavirus particles emerged from the same cytoplasmic virion factory (Fig. 12.5). Sputnik multiplied faster than mamavirus and progeny Sputnik virions appeared at distinct locations of the virion factory. In addition, it was noted that Sputnik had a negative effect on mamavirus production, as the yield of the latter was reduced by ~70% and abnormal giant virus capsids were observed in the presence of Sputnik (La Scola *et al.*, 2008). This

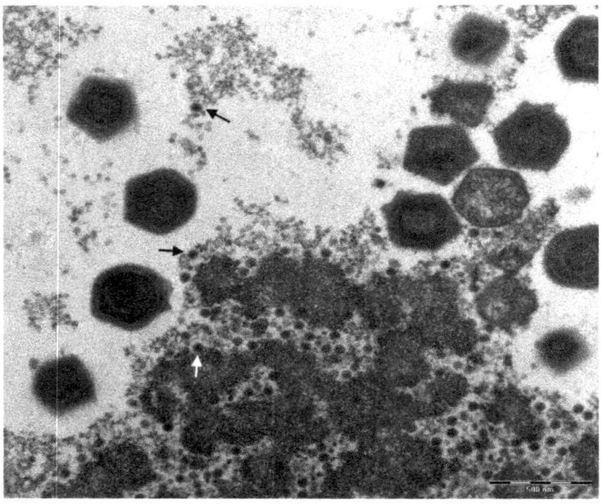

Figure 12.5 Sputnik particles (arrows) produced in *Acanthamoeba polyphaga* cytoplasmic mimivirus factories. This thin-section transmission electron micrograph is courtesy of M. Gaia and B. La Scola, Univ. Aix-Marseille, France.

led to the classification of Sputnik as a parasite of mamavirus and the introduction of the term 'virophage', as a virus infecting another virus.

The Sputnik genome consists of a circular dsDNA molecule 18,343 bp in length with an A+T content of 73%, and encodes 21 predicted protein-coding sequences (CDSs). Two other Sputnik strains have been isolated and genetically characterized. Sputnik 2 was found associated with lentillevirus, a mimivirus strain isolated from lens fluid (Desnues *et al.*, 2012b; La Scola *et al.*, 2010) and Sputnik 3 was recovered via a reporter system from a soil sample near Marseille, France during a screen for new virophages (Gaia *et al.*, 2013b). The Sputnik 2 and 3 genomes are four base pairs shorter than Sputnik 1, which results from single base pair deletions at positions 877, 7949, 7958, and 12936. Initially, the Sputnik 1 genome displayed another additional base at position 14047. This, however, was found to be a sequencing artifact that resulted in a frame shift artificially separating ORFs 18 and 19 (Gaia *et al.*, 2013b; Zhang *et al.*, 2012). The corrected Sputnik 1 sequence encodes the minor capsid protein V18/19 in a single ORF, which is not only in agreement with the gene topology in Sputnik 2 and 3, but also enables full-length sequence alignments with homologous mCP genes in other virophages. The genome of Sputnik 3 further differs from Sputnik 1 and 2 by two T to A transversions at positions 8986 and 8991 (located in the intergenic region between V12 and V13) and two A to G transitions at positions 16098 and 17666, resulting in a A464T change in the MCP V20 and a D371N change in V21, respectively. Due to their nearly identical genomes, Sputniks 1, 2 and 3 are closely related strains.

Another Sputnik isolate is the Rio Negro virophage (RNV) that was found in the Rio Negro River of the Brazilian Amazon (Campos *et al.*, 2014). RNV is associated with a mimivirus variant called Samba virus. The RNV particles were reported to be only 35 nm in diameter, even though the partial major capsid gene sequence was 100% identical to the Sputnik MCP gene, which makes it likely that the true capsid size of RNV is comparable to that of Sputnik.

The Zamilon virophage was isolated from Tunisian soil together with its host virus, the Mont1 mimivirus (Boughalmi *et al.*, 2013; Gaia *et al.*, 2014). Zamilon (Arabic for 'neighbour') is a relative of the Sputnik virophages, and its circular 17,276-bp-long genome shares 76% nucleotide identity with Sputnik. The Zamilon genome encodes 20 predicted proteins, 17 of which have homologues in Sputnik. In contrast to Sputnik, Zamilon does not replicate with all members of the *Mimivirus* genus. Zamilon replication was only detected during coinfection with viruses of lineage B (moumouviruses) and lineage C (megaviruses), but not with lineage A members (mimiviruses) (Gaia *et al.*, 2014). In addition, Zamilon differs from Sputnik by not inhibiting the replication of its coinfecting giant virus.

The genus *Mavirus*

Mavirus

The second described virophage was mavirus (meaning, maverick-related virus; mavericks are large transposons found in eukaryotic genomes that are also called polintons from the presence of DNA polymerase and integrase genes; see 'Evolutionary connections of virophages' below), a parasite of Cafeteria roenbergensis virus (CroV). The exact geographic origin of mavirus is not clear. CroV was isolated from coastal waters of the Gulf of Mexico near Aransas Pass, Texas, USA, in 1989 (Garza and Suttle, 1995). The flagellate host strain, on which CroV was isolated, originated from the Pacific coast near Yaquina Bay, Oregon, USA

(Gonzalez and Suttle, 1993). After mavirus was discovered in CroV-infected flagellate cultures (Fischer and Suttle, 2011), it was assumed that mavirus had been co-isolated with CroV from the same Gulf of Mexico water sample and that it had been propagated alongside CroV until the virophage was identified in 2009. Recent findings of endogenous mavirus genomes integrated in flagellate chromosomal DNA (Fischer and Hackl, 2016), however, have opened up an alternative explanation, according to which some cells of the Oregon host strain could have harboured a mavirus provirophage that was subsequently reactivated upon contact with CroV.

The mavirus genome is a circular 19,063-bp-long molecule that is able to linearize for integration. The terminal inverted repeats (TIRs) of exogenous mavirus genomes are ≈50 bp long (Fischer and Suttle, 2011), whereas the TIRs of the endogenous mavirus genome are 615/616 bp long, resulting in a total length of 20,190 bp for the provirophage genome (Fischer and Hackl, 2016). Mavirus encodes 20 predicted proteins and has a GC content of 30%, which is similar to the properties of Sputnik and Zamilon.

The particles of mavirus are icosahedral in shape and have a diameter of 60–65 nm in thin-section transmission electron micrographs, and 70–75 nm in negative stain electron micrographs (Fig. 12.2B). The burst size of mavirus is roughly ten times higher than that of CroV, the latter as measured during a mavirus-free CroV infection of *C. roenbergensis*.

Ace Lake mavirus

Although known only from metagenomic sequences, the genome of Ace Lake Mavirus (ALM) clearly falls within the genus *Mavirus*. The ALM genome was assembled from short metagenomic reads that originated from the Antarctic Ace Lake (Zhou *et al*., 2013). The genome sequence is not complete and may contain sequence errors due to erroneous assembly and low sequence coverage. Despite these potential problems, the ALM genome provided a first glance at sequence diversity within the mavirus subgroup of virophages. ALM and mavirus share 13 of their 22 and 20 predicted genes, respectively. The gene order is also conserved; however, an inversion affecting seven genes occurred in one of the genomes. In addition to the four highly conserved morphogenesis genes, mavirus and ALM encode homologues for the predicted pPolB, rve-INT, S3H, as well as six genes of unknown function.

ALM ORF14 probably starts at the second ATG start codon, as suggested by alignments with its homologue mavirus ORF12 and the presence of a putative promoter motif directly upstream of the second ATG site, whereas no such promoter motif is present in the immediate upstream region of the first start codon. Like mavirus, ALM encodes an rve-INT. The ALM homologue is annotated in two adjacent reading frames and at present it is not known whether this gene is split into two separate ORFs, or whether an artificial frameshift was introduced during assembly of the metagenomic reads.

Additional virophages from metagenomes

Virophages appear to be distributed world-wide and several groups of virophage-like genomes have been discovered in metagenomes in recent years (Figs. 12.1, 12.3 and 12.6). This was made possible by seeding BLAST search with genes from fully sequenced genomes of the few virophage isolates that are currently available. Interestingly, whereas Sputnik originated from a freshwater cooling tower, Zamilon from soil, and mavirus from an oceanic environment, most uncultured virophages were found in lakes, suggesting that lake ecosystems provide conditions that allow certain virophage populations to thrive.

The Organic Lake virophage

The first virophage genome assembled from a metagenome was reported in 2011 (Yau *et al*., 2011). A genomic study of the microbial community in East Antarctica's hypersaline Organic Lake revealed the presence of a virophage-like genome. The Organic Lake Virophage (OLV) is probably associated with NCLDVs of the extended *Mimiviridae* family that infects a photosynthetic host[2]. Its genome was assembled into a 26,421 bp long sequence, encoding 24 ORFs. The phylogenetic position of OLV depends on the gene used (Fig.

2 The same publication reports the partial assembly of two giant virus genomes, named Organic Lake phycodnavirus (OLPV) 1 and 2 (Yau *et al*., 2011). Despite the name, the OLPVs are not part of the *Phycodnaviridae*, but phylogenetically more closely related to the *Mimiviridae*.

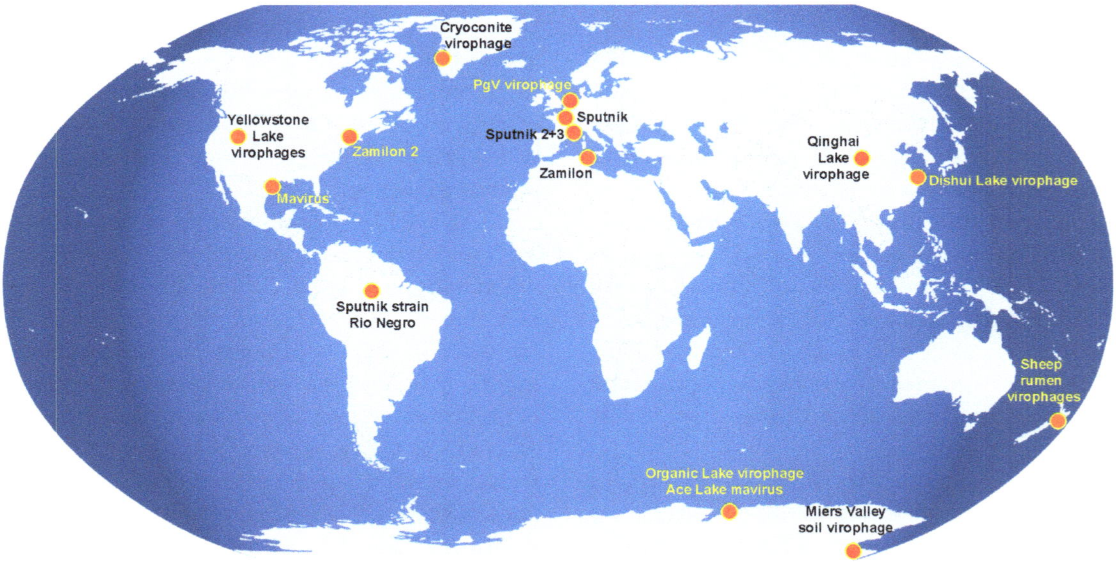

Figure 12.6 Geographic origin of published virophage genomes.

12.4), but based on MCP phylogeny, there are currently no known close relatives of OLV.

The Yellowstone Lake virophages and Ace Lake mavirus

In 2013, Zhou *et al.* reported the assembly of four virophage genomes (YSLV1–4) from a metagenome of Yellowstone Lake water samples (Zhou *et al.*, 2013). These findings were expanded by the same group of researchers with the publication of three additional Yellowstone Lake virophage sequences (YSLV5–7) 2 years later (Zhou *et al.*, 2015). The YSLV genomes are 23–30 kb long and appear to belong to different subgroups, as shown by phylogenetic reconstruction (Fig. 12.4). In particular, YSLV5 and YSLV7 do not cluster with other YSLVs. In their 2013 publication, the authors also described the genome of Ace Lake mavirus (ALM), assembled from the saline meromictic Ace Lake in the Vestfold Hills of Antarctica (Zhou *et al.*, 2013). Interestingly, this lake is in close geographic proximity to Organic Lake, where OLV was detected (Yau *et al.*, 2011).

The Dishui Lake and Qinghai Lake virophages

In 2016, virophage genomes were reconstructed from metagenomic data of two Asian freshwater lakes, one in Shanghai and one in Tibet. The Dishui Lake virophages (DSLVs) occur in an artificial freshwater lake in Shanghai, China, and are closely related to YSLV 3 (Fig. 12.4) (Gong *et al.*, 2016). They are presumed to replicate in combination with algae-infecting giant viruses. Another population of virophages was found in a surface water metagenome of Lake Qinghai, in the mountains of Tibet (Oh *et al.*, 2016). The Qinghai Lake virophages (QLVs) are most closely related to YSLVs 1 + 4 and could infect phototrophic protists.

Virophages from a sheep rumen metagenome

A novel lineage of virophages was identified in the metagenome of sheep rumen (Yutin *et al.*, 2015a). The longest of these was 26,209 bp long and encoded 22 ORFs. The rumen virophage (RVP) genomes appear to be a hybrid between bona fide virophages (with genes for MCP, ATPase, PRO) and polintons (mavericks, encoding a protein-primed DNA polymerase B). Such a scenario has also been proposed for the origin of mavirus (Yutin *et al.*, 2013); however, RVPs and mavirus are phylogenetically distinct based on pPolB and MCP analyses and RVPs branch separately from all other virophages (Yutin *et al.*, 2015a) (Fig. 12.4). So far, no minor capsid protein has been found in RVP genomes, suggesting that their capsid architecture differs from that of Sputnik and mavirus.

Other virophage sightings

A metagenomic study of viromes from cryoconite holes in Greenland yielded a partial virophage genome (Bellas *et al.*, 2015). Cryoconite ('ice dust') holes are cylindrical depressions at the surface of glaciers which are filled with dark coloured sediments. The cryoconite virophage (CryV) partial genome is 12,572 bp long and includes the four morphogenesis genes (MCP, mCP, PRO, ATPase). Virophage genome fragments were also identified at the opposite pole, from surface soil of the dry Miers Valley in Eastern Antarctica (Zablocki *et al.*, 2014). Gene fragments of this Miers Valley soil virophage (MVSV) suggested that viruses of the Sputnik clade exist in Antarctica, as well as mimiviruses and phycodnaviruses, whose genomic signatures were also found in the same study. Antarctica therefore harbours a diverse virophage community in its aquatic and terrestrial environments, as evidenced by OLV, ALM, and MVSV. Not surprisingly, genetic signatures of giant viruses are also present in Antarctica (Kerepesi and Grolmusz, 2017; Yau *et al.*, 2011). The endogenous virophage-like elements found in the nuclear genome of *Bigelowiella natans* (Blanc *et al.*, 2015) are discussed later in this chapter. A recent study identified 25 uncultivated virophage populations using time-series metagenomes from two North-American freshwater lakes (Roux *et al.*, 2017). This report will help refine the taxonomy of virophages and confirms that most of their genetic diversity remains to be described.

The virophage-like element PgVV and the polinton-like viruses

The large DNA virus PgV-16T belongs to the extended *Mimiviridae* family and infects the bloom-forming microalga *Phaeocystis globosa* (Santini *et al.*, 2013). During assembly of the 459,984-bp-long, linear dsDNA PgV-16T genome, an additional contig was found. This 19,527 bp long linear dsDNA molecule with 1 kb terminal inverted repeats contained 16 predicted CDSs that are all located on the same strand. Based on the similarity of three of these CDSs to genes in OLV and mavirus, this genetic element was termed PgV-associated virophage (PgVV). However, no 50–80 nm sized capsids were found and the authors proposed that PgVV replicates either as a linear plasmid or as a provirophage that is integrated in the PgV-16T genome (Santini *et al.*, 2013). Although the PgVV genome was initially reported to lack any recognizable capsid genes, ORF PgVV_00012 likely encodes a distant version of a double-jelly roll major capsid protein and ORF PgVV_00010 could encode a minor capsid protein (Krupovic *et al.*, 2014). At the time of discovery, PgVV was most closely related to mavirus and OLV. Based on the identification of polinton-like viruses (PLVs) in metagenomic datasets using the PgVV-predicted MCP as bait, however, PgVV is considered to be a PLV rather than a virophage (Yutin *et al.*, 2015b). The distinction is justified by the lack of a cysteine protease that is conserved in virophages, as well as by the distinct versions of MCP, mCP, and ATPase genes found in PgVV and PLVs. (Krupovic *et al.*, 2014; Yutin *et al.*, 2015b).

Evolutionary connections of virophages

Virophages may have evolved from an ancestral virus of the PRD1–adenovirus lineage by multiple recombination and gene replacement events (Krupovic and Koonin, 2015). Genome analysis of the first virophage, Sputnik, suggested mixed origins for its genes which exhibited links to viruses infecting archaea, bacteria, and eukaryotes (La Scola *et al.*, 2008). Surprisingly, seven of the 20 genes from the second described virophage, mavirus, had their closest homologues in one particular group of genetic elements, the maverick/polinton elements (MPEs) (Fischer and Suttle, 2011). The MPEs were originally described as a class of DNA transposons widespread in eukaryotic genomes with highly variable copy numbers (Jordan *et al.*, 2004; Kapitonov and Jurka, 2006; Pritham *et al.*, 2007). They are 15–20 kb large and contain conserved rve-INT integrase and pPolB DNA polymerase genes in addition to other virus-affiliated genes such as an FtsK/HerA-like genome packaging ATPase and an adenovirus-like protease. Recently, it was shown that MPEs also code for putative mCP and MCP genes (Krupovic *et al.*, 2014), which resulted in a shift of perception, and capsid-encoding MPEs are now regarded as endogenous viruses, called polintoviruses (Krupovic and Koonin, 2015), rather than as transposons. Roughly one-third of the genome of the human parasite *Trichomonas vaginalis* consists of MPEs, indicating that at least in some eukaryotic lineages these elements are able to spread intragenomically.

MPEs and virophages of the genus *Mavirus* share seven homologous proteins: S3H, rve-INT, pPolB, FtsK/HerA-type ATPase, PRO, mCP and MCP. In addition, virophages and MPEs have similar genome length (~20 kb), overlapping host range (mainly protists, although MPEs are also found in animals), and the mavirus genome exhibits terminal inverted repeats similar to those found in MPEs. The genes shared by mavirus and MPEs have thus evolved from a common ancestor; however, the nature of this ancestor is unclear. One hypothesis states that ancestral mavirus-like virophages integrated into a eukaryotic host genome, where they conferred resistance to the host against infection by giant viruses and were positively selected for in the host cell population (Fischer and Suttle, 2011). An alternative hypothesis, which is mainly based on tree topologies in phylogenetic reconstructions of conserved MPE/virophage proteins, suggests that virophages evolved from ancestral MPEs through multiple gene transfers and rearrangements and that they subsequently became dependent on giant DNA viruses (Krupovic and Koonin, 2015; Yutin *et al.*, 2013).

Recent data mining approaches in metagenomic sequences have revealed several additional groups of viral elements with ties to virophages and MPEs. The RVP virophages found in a sheep rumen metagenome share with MPEs the pPolB, ATPase, PRO and MCP genes, but lack the rve-INT and mCP. In phylogenetic trees based on multiple sequence alignments of MCP, the RVP virophages cluster as a monophyletic group, distinct from Sputnik-, mavirus-, and OLV-like virophages (Yutin *et al.*, 2015a) (Fig. 12.4). Another group of related viruses discovered in metagenomes are the polinton-like viruses (PLVs) (Yutin *et al.*, 2015b), which share a highly diverged version of the double-jelly roll MCP with PgVV. The PLVs are 18–28 kb in length, encode ATPase, mCP, and MCP (but no PRO) and some of them contain long TIRs. Notably, PLVs were also found in the genomes of photosynthetic protists, which suggests – together with the endogenous *B. natans* virophages – that virophage integration into eukaryotic genomes may be a common phenomenon (Blanc *et al.*, 2015; Fischer, 2015; Yutin *et al.*, 2015b). For a more detailed overview of the evolutionary connections between virophages, MPEs, PLVs, and a class of mimivirus-associated parasitic linear DNA elements called transpovirons, see Koonin and Krupovic (2017) and Krupovic and Koonin (2016).

The virophage infection cycle

Owing to the paucity of culture-based virophage systems, many details of their replication cycle remain unknown. Studies on the *Acanthamoeba*–mimivirus–Sputnik and *Cafeteria*–CroV–mavirus systems, however, have given us a vague idea about the intracellular events that occur during virophage infection (Fig. 12.7). The most striking feature about these viruses is that, despite their respectable size and coding potential, they cannot reproduce inside a host cell without a suitable coinfecting giant virus. Hence, the replication cycle of virophages is tightly linked to that of their host viruses.

Virion entry

Virions enter the host cell either by endocytosis, or as a composite with their host virus by phagocytosis. An example for the former entry mode is mavirus, which attaches to the host cell surface via an unknown receptor (Fig. 12.8B). The membrane at the attachment site then starts to invaginate and the virus is endocytosed (Fig. 12.8C). Inside infected host cells, mavirus particles can be observed within coated (Fig. 12.8A and D) as well as uncoated vesicles (Fig. 12.8E) by electron microscopy, which implies that virions are internalized by clathrin-mediated endocytosis (Fischer and Suttle, 2011).

In contrast, Sputnik has never been observed to enter the amoeba independently of a coinfecting virus of the *Mimiviridae* family. Sputnik is able to adhere to the 1250 Å long, heavily glycosylated surface fibres that cover the mimivirus capsid (Piacente *et al.*, 2012; Xiao *et al.*, 2009). This interaction may be mediated by the ~100 Å long mushroom-like surface fibres that decorate the Sputnik virion (Sun *et al.*, 2010). The virophage–giant virus composite is then phagocytosed by the amoeba and Sputnik particles entangled within mimivirus surface fibres are occasionally observed by electron microscopy (Desnues and Raoult, 2010). In particular, the mimivirus R135 gene product, a glycosylated fibre protein related to GMC-type oxidoreductases that elicits an antigenic response (Klose *et al.*, 2015;

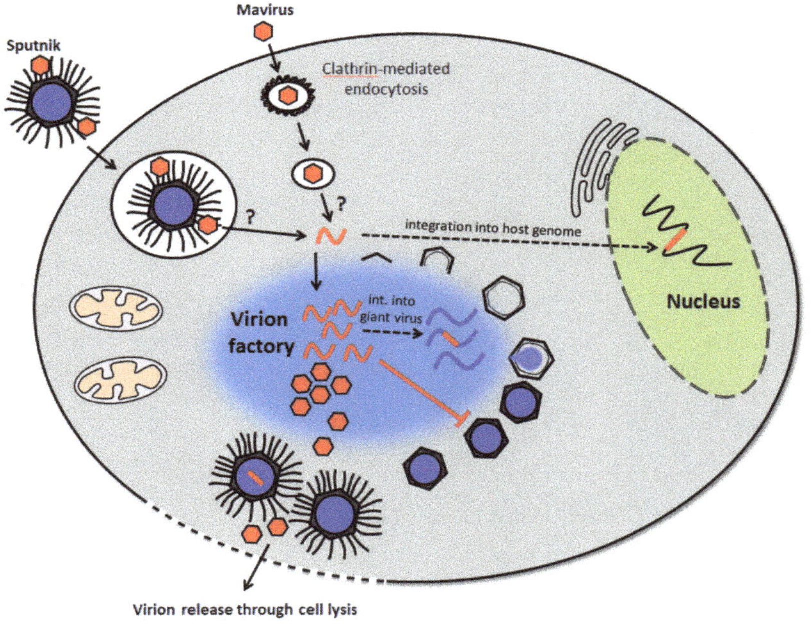

Figure 12.7 Simplified diagram depicting the infection cycle of lavidaviruses. Virophages enter the cell either via phagocytosis as a composite with a giant virus (Sputnik), or independently via endocytosis (mavirus). The mechanism by which they escape the vesicle and target the cytoplasmic virion factory of their giant viruses is unknown. Virophages occasionally integrate their genomes into the nuclear host genome or 'int.' into giant virus genomes. The virophage genome is replicated in the virion factory during the late phase of giant virus infection; virophage proteins are synthesized and progeny virions are assembled. During any of these processes, giant virus replication may be impaired. Finally, virophage and giant virus particles are released upon cell lysis.

Pelletier *et al.*, 2009), appears to have a high affinity for Sputnik particles, as this protein was identified by mass spectrometry in a purified Sputnik preparation (La Scola *et al.*, 2008). Additional support for this hypothesis stems from the analysis of a mimivirus deletion mutant, which lacks the surface fibres due to the deletion of 155 genes. This bald form of mimivirus, termed M4, no longer supports Sputnik replication (Boyer *et al.*, 2011). The R135 gene is among the genes that are deleted in mimivirus M4.

The composite entry model of Sputnik and mimivirus via phagocytosis creates the need for a mechanism that would allow Sputnik particles to escape from the phagosome. Uncoating of mimivirus particles in the phagosome is initiated by opening of the stargate portal, a unique structural feature of mimiviruses (see Chapter 11). The open stargate exposes the inner viral membrane, which subsequently fuses with the phagosomal membrane, thereby creating a gateway to the cytoplasm for the viral core (Zauberman *et al.*, 2008). This uncoating mechanism, however, grants cytoplasmic access only to molecules that are located inside the mimivirus membrane. Any other particles inside the phagosome, such as Sputnik capsids, would have to use an alternative escape route. For this reason, a mimivirus-independent entry pathway of Sputnik cannot be excluded, and the exact mechanism of viral genome delivery from the particle into the host cytoplasm remains to be determined. Cryo-EM studies of empty Sputnik virions suggested that low pH conditions (pH ~5.5) can trigger the dissociation of some of the penton structures, which would create a portal through which the viral DNA could exit the virion (Zhang *et al.*, 2012). Capsid dismantling of Sputnik may thus resemble that of adenoviruses, which are known to lose penton bases in acidified endosomes during cell entry (Greber *et al.*, 1993). The uncoating process of mavirus, by contrast, is not known.

As suggested by mathematical modelling, the two different entry modes may have consequences for long-term population dynamics of viruses and hosts (Taylor *et al.*, 2014).

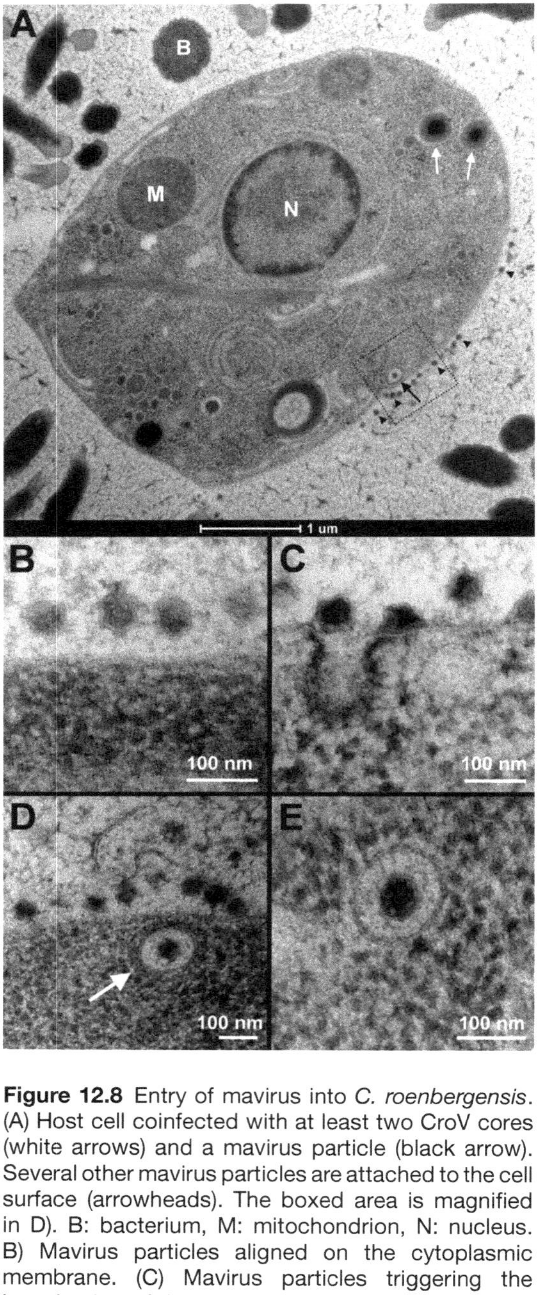

Figure 12.8 Entry of mavirus into *C. roenbergensis*. (A) Host cell coinfected with at least two CroV cores (white arrows) and a mavirus particle (black arrow). Several other mavirus particles are attached to the cell surface (arrowheads). The boxed area is magnified in D). B: bacterium, M: mitochondrion, N: nucleus. B) Mavirus particles aligned on the cytoplasmic membrane. (C) Mavirus particles triggering the invagination of the cytoplasmic membrane, creating a coated pit. (D) Mavirus particle inside a coated vesicle. Other virions are visible outside the cell. (E) Mavirus particle inside an uncoated vesicle in the host cytoplasm. All images by M. Fischer, C. Suttle, and the Bioimaging Facility at the University of British Columbia, Canada. Images C) – E) modified from Fischer and Suttle, 2011.

Virus factories

Following entry, the virophage genome is targeted to the developing virion factory of the coinfecting giant virus. After an eclipse phase, which lasts 2–4 h for Sputnik, newly synthesized virophage particles are visible within the virion factory (Figs. 12.5 and 12.9). Sputnik particle production usually begins at one pole of the mimivirus factory (La Scola *et al.*, 2008). The phenotype of mavirus-infected CroV factories varies, depending on parameters such as cell line, multiplicity of infection (MOI), and whether a coinfection with external mavirus particles or a reactivation of endogenous mavirus takes place. Occasionally, factories producing both CroV and mavirus particles can be observed (Fig. 12.9A). More frequently, however, and especially during coinfection with high MOIs of mavirus, the CroV factories will develop to occupy a large part of the cytoplasm, but no CroV particle production occurs. Instead, nested electron-dense subfactories synthesizing mavirus virions can be seen (Fig. 12.9B).

Coinfection dependence

Sputnik, Zamilon, mavirus, and presumably other virophages to be isolated in the future cannot replicate in their host cells without coinfection of their respective host viruses. At first glance, the gene content of virophages does not offer any obvious explanation for such a dependency. The conserved morphogenetic gene module consists of MCP, mCP, PRO and ATPase, and virophages are thus presumably able to synthesize their own capsids. Virophage genomes also code for DNA replication proteins, which should enable them to copy their genetic information. Therefore, virophages appear to be self-sufficient for DNA replication and particle assembly.

Several observations, in contrast, suggest that transcription is the process for which virophages depend on their coinfecting host viruses. First, the computational analysis of the mavirus genome revealed that all 20 genes contain a conserved sequence motif in their immediate 5′ upstream region (Fischer and Suttle, 2011). This motif consists of a 'TCTA' core which is flanked by AT-rich sequences and is located approximately 14 nt upstream of the start codon. A highly similar conserved sequence motif had been found earlier at the same upstream position of those CroV genes that are expressed late during infection (Fischer *et al.*,

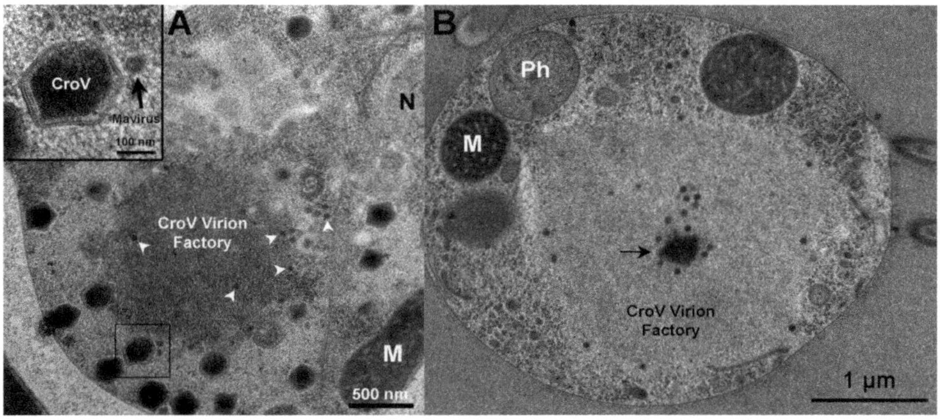

Figure 12.9 Variable CroV phenotypes during coinfection with mavirus. A) *C. roenbergensis* cell displaying a virion factory that produces both CroV and mavirus particles. Modified from (Fischer and Suttle, 2011). The inset shows a magnification of the boxed area. Mavirus particles are marked by white arrowheads (large image) or a black arrow (inset). B) *C. roenbergensis* cell featuring a large virion factory that produces no CroV virions, but contains a nested factory (arrow) that produces mavirus particles. Thin-section EM image by U. Mersdorf and M. Fischer, Max Planck Institute for Medical Research, Germany. M: mitochondrion, N: nucleus, Ph: phagosome.

2010). Second, quantitative real-time PCR analysis of cDNA generated from RNA samples isolated at different time points post infection revealed that mavirus genes become active at the beginning of late phase of CroV infection (≈4 h p.i.), whereas no mavirus gene activity was detected before the onset of late phase (K. Fenzl and M. Fischer, unpublished data). For Sputnik, a less obvious promoter motif was identified, which also resembled a putative late promoter motif of mimivirus (Legendre *et al.*, 2010). Third, Sputnik and mimivirus share similar transcription termination signals. In mimivirus, the location of polyadenylation sites appear to be defined by palindromic sequences, leading to the proposition of a 'hairpin rule' for mimivirus polyadenylation (Byrne *et al.*, 2009). In the Sputnik genome, 16 such palindromic sequences were identified, 14 of which were located in intergenic regions, which makes them candidates for polyadenylation signals (Claverie and Abergel, 2009a). In summary, these findings strongly suggest that virophages use the giant virus-encoded transcription machinery, and in particular the late phase-specific transcription factors to express their genes.

DNA replication

There are currently no data to indicate that the DNA replication proteins encoded by giant viruses may play a role in replicating virophage genomes. Members of the *Mavirus* genus encode a predicted protein-primed B-family DNA polymerase, whereas a distinct version of bacterial DNA polymerase I (TVPol) with a predicted primase activity is found in the Sputnik and OLV clades (Iyer *et al.*, 2008; Yutin *et al.*, 2013). Protein-primed family B DNA polymerases are encoded by various parasitic elements such as adenoviruses, phi29-like viruses, tectiviruses, some archaeal viruses, and mavericks/polintons (Redrejo-Rodríguez and Salas, 2014). These polymerases use a terminal protein that is covalently attached to the 5′ ends of the genome as a primer. Due to low sequence conservation, however, a terminal protein for mavirus has not been identified yet. In addition, protein-primed replication requires a linear DNA template and the mavirus genome has circular topology, thus the exact mechanism of DNA replication in viruses of the Mavirus genus remains to be explored.

The protist–virus–virophage triangle

Host range

Lavidaviruses require two distinct hosts for successful replication, a susceptible host cell and a permissive host virus in the form of the cytoplasmic virion factory. Each of the two hosts must fulfil certain requirements for virophage replication to occur. The host cell must be susceptible to the virophage, i.e. the virophage must be able to enter the cell. In addition, the virophage relies on the host cell for energy, metabolites, protein translation, and other essential systems. As a second host component, the virophage requires the presence of a permissive coinfecting giant virus, which provides the components (presumably transcription proteins) that are necessary for the virophage to complete its replication cycle. Currently, the only known cellular hosts for virophages are phagotrophic protists. With only two such virophage–virus–host systems in culture (*Acanthamoeba* sp. and *Cafeteria* sp.), however, the information on virophage host cell range is extremely low. Most of the environmental virophage genomes that have been assembled from metagenomic sequences, or host-integrated virophage genomes (Blanc *et al.*, 2015; Gong *et al.*, 2016; Oh *et al.*, 2016; Yau *et al.*, 2011; Zhou *et al.*, 2013) are assumed to be associated with giant viruses of the extended *Mimiviridae* family (Fischer, 2016), which mostly infect phototrophic protists. We can predict that it will only be a matter of time until the first algal virophages are isolated and grown in laboratory cultures.

Cellular host range of virophages will mainly be determined by the ability of the virophage to enter the cell. Members of the genus *Sputnikvirus* are proposed to enter the amoebal host by attachment to the external fibres of mimiviruses and via subsequent uptake of the virus-virophage composite by phagocytosis. Since this entry mode presumably does not depend on virophage–host cell contact, cellular host range should be determined solely by the giant virus host requirements, and the host cell might not be able to become resistant to the virophage while remaining susceptible to the giant virus. In contrast, members of the genus *Mavirus* enter their flagellate host cells by receptor-mediated endocytosis, a process that requires specific virus–host interactions at the cell surface. Here, it is conceivable that a cellular host becomes resistant to the virophage by mutations in the cell surface receptor, while remaining susceptible to the giant virus. Evolutionarily, this would probably result in a selective disadvantage for the host cell, since mavirus can strongly inhibit CroV replication and increase the survival of CroV-infected flagellate populations (Fischer and Hackl, 2016).

The type of entry of virophages may reflect the host specificity of their associated giant viruses. Mimiviruses are assumed to have a broad host range and can replicate in various amoebae (see Chapter 11). Although mimiviruses replicate well in *Acanthamoeba* sp., their main natural host is unknown. Hence, mimivirus-associated virophages may be maladapted to *Acanthamoeba*, which might explain why they are not capable of independent cell entry. Mavirus, on the other hand, is able to trigger specific uptake by *C. roenbergensis* cells via clathrin-mediated endocytosis. Its viral host CroV infects members of the genus *Cafeteria*, as well as closely related genera such as *Picophagus* (M. Fischer, unpublished data). Unlike mimiviruses, CroV appears to enter host cells not by phagocytosis, but by docking to the cytoplasmic membrane and releasing only the viral core into the cytoplasm (U. Mersdorf and M. Fischer, unpublished results). This suggests that marine heterotrophic nanoflagellates such as *Cafeteria* are the natural hosts for CroV and mavirus.

In general, virophage replication depends on the availability of cytoplasmic transcription. Therefore, the giant virus specificity of lavidaviruses is likely determined by the ability of the giant virus-encoded transcription machinery to synthesize virophage transcripts. Based on the high similarity of late gene promoters as well as transcription termination motifs between giant viruses and their virophages, it is predicted that this interaction largely depends on the recognition of virophage gene promoters by a giant virus-encoded transcription factor. The *Mimiviridae* is not the only family of large dsDNA viruses that encode their own transcription system, and it is possible that virophage-like parasites also exist for other members of the proposed viral order *Megavirales*[3], such as pithoviruses and poxviruses.

Virophages are unlikely to replicate in human cells or cause disease (Desnues *et al.*, 2012b; Parola *et al.*, 2012). The case of Sputnik 2, however, shows that virophages may be part of the human virome.

3 The order *Megavirales* was proposed to replace the informal grouping of 'Nucleocytoplasmic large DNA viruses', which includes eukaryotic large dsDNA viruses that are of possible monophyletic origin, with a proper taxon (Colson *et al.*, 2013; Iyer *et al.*, 2001, 2006). However, at the time of writing, the ICTV has not approved this new order, nor does the proposal appear to be scheduled for discussion any time soon.

Sputnik 2 was isolated together with lentillevirus from the contact lens of a keratitis patient, whose symptoms were probably caused by *Acanthamoeba* cells that also happened to harbour the giant virus and the virophage (Desnues *et al.*, 2012b).

Virophage–giant virus interactions

Virophages that encode integrases are able to physically link their genomes to those of other organisms. The tyrosine recombinase found in Sputnik virophages may catalyse the occasional integration into mimivirus genomes. Analysis of paired-end Illumina reads of lentillevirus DNA revealed that Sputnik 2 may also exist as a provirophage in the genome of its host virus (Desnues *et al.*, 2012b). Similarly, the PgVV genome likely integrates at multiple sites within its associated Phaeocystis globosa virus 16T (Santini *et al.*, 2013). Persistence as a provirophage in a giant virus genome has clear advantages for the smaller virus genome, as it will automatically coinfect and colocalize with its supporting giant virus.

Even although virophages are known to inhibit the production of giant virus particles during a coinfection, their parasitic effect on the host virus appears to vary considerably. Whereas mavirus is a potent inhibitor of CroV and coinfected cells often produce no CroV virions at all, the negative impact of Sputnik on mimivirus is less pronounced and viral factories producing both types of capsids are frequently observed. The Zamilon virophage on the other hand has no discernible negative impact on the replication of a coinfecting mimivirus. The reason for this variation is unknown, but the degree of virus-against-virus pathogenicity most certainly affects the cellular and viral host dynamics, and the resulting ecological consequences will differ for different virophages. As is to be expected for a host–parasite system, the host can become resistant to the parasite, or acquire means to control the parasite. Mimivirus thus has been shown to become resistant to Sputnik through large-scale deletions at the genome termini, resulting in a 'bald' capsid phenotype where external fibres are no longer present (Boyer *et al.*, 2011). This may prevent Sputnik from entering the amoeba via a composite mechanism, although it cannot be excluded that other genes essential for virophage replication are located in the deleted regions.

A general defence mechanism against virophages was recently proposed in mimivirus (Levasseur *et al.*, 2016). Of the three mimivirus lineages (A: mimiviruses, B: moumouviruses, C: megaviruses), Zamilon can only replicate with lineage B and C strains, but not with lineage A mimiviruses. Genome analysis of lineage A viruses revealed an open reading frame (R349) with four 15-nt long repeats of a Zamilon DNA sequence. This so-called 'mimivirus virophage resistance element' (MIMIVIRE) was proposed to represent a CRISPR-Cas-like adaptive immune system against virophages. Silencing of the R349 gene allowed Zamilon to replicate in lineage mimiviruses. Two additional genes located close to the R349 gene, the helicase R350 and the nuclease R354, were also proposed to be part of this defence system, because silencing any of these two genes increased Zamilon replication in lineage A mimiviruses. Criticism has been voiced, however, over the alleged similarity to CRISPR-Cas systems, owing, amongst other things, to the lack of regularly spaced or flanking repeats that are typical for CRISPR-Cas (Claverie and Abergel, 2016; Mohanraju *et al.*, 2016).

Instead, Claverie and Abergel proposed a protein–protein interaction scenario, according to which the repeated Zamilon sequences in the R349 protein would compete for binding partners with the Zamilon ORF 4 protein, in which the repeated sequences are present (Claverie and Abergel, 2016). The R350 and R354 proteins would not be directly involved in virophage restriction, but be required for general replication or transcription processes and their silencing would thus indirectly benefit the replication of Zamilon. Clearly, the mechanism of the observed resistance to Zamilon in lineage A mimiviruses remains to be elucidated. Regardless of the specific workings of such a defence system, it becomes evident that host–parasite arms races are not restricted to cellular hosts and are also being waged between giant viruses and their viral parasites.

Virophage–host cell interactions

From the beginning, virophages were hypothesized to integrate into host genomes. This was based on the presence of integrases in Sputnik and mavirus, and the genetic relatedness between mavirus and the endogenous eukaryotic maverick/polinton elements (Fischer and Suttle, 2011; La Scola *et al.*, 2008). Mavirus/ALM and Sputnik/Zamilon encode two different types of integrases. Whereas members of the genus *Sputnikvirus* have a bacteriophage lambda-like tyrosine recombinase (Gaia *et al.*, 2014; La Scola *et al.*, 2008), the mavirus-like virophages encode a retrovirus RVE-type integrase (Fischer and Suttle, 2011; Zhou *et al.*, 2013). The benefits of genome integration for virophages are apparent: this process physically ties the virophage to one of its two host components and increases its chances to encounter conditions favourable for replication. At the same time, integrated viral DNA is maintained and replicated by the host cell, whereas free virions are exposed to various environmental parameters that may destroy infectivity, such as ultraviolet irradiation, pH, salinity, etc.

A targeted computational search for virophages in published eukaryotic genomes revealed a group of closely related viral elements in the nuclear genome of the chlorarachniophyte alga *Bigelowiella natans* (Blanc *et al.*, 2015). The authors of this study identified 38 virophage-like elements with high similarity to each other (average 91% nucleotide similarity), ranging in size from 0.1 kb to 33.3 kb. Based on phylogenetic reconstruction of MCP and ATPase, the *B. natans* virophages are most closely related to YSLV 5 (Fig. 12.4). Interestingly, many of these virophage-like genes were found to be transcribed, including conserved genes such as MCP, mCP, and ATPase. The presence of integrated genomic fragments from putative viruses of the NCLDV clade suggests that *B. natans* is host to both giant DNA viruses as well as virophages (Blanc *et al.*, 2015). It remains unclear, however, how old these endogenous virophage-like elements are and how frequently they integrate into host cell genomes.

An experimental study in the *Cafeteria*–CroV–mavirus system was able to shed first light on these questions by showing that mavirus integrates very efficiently into the nuclear genome of *C. roenbergensis* during a coinfection with CroV (Fischer and Hackl, 2016). The authors sequenced and assembled the nuclear genome of a host strain before and after coinfection, which allowed them to determine the exact integration sites of mavirus. A least 11 distinct mavirus integrations were found within a single host genome and the integration sites displayed no obvious consensus motif, suggesting that integration is not sequence-specific. Host and virus population data from this study showed that nearly every CroV-infected host cell is destined to lyse. The high percentage of *de novo* provirophages can be explained by the ability of mavirus to integrate into the host genome in the absence of CroV (M. Fischer, unpublished data). Unlike the *B. natans* virophages, the endogenous mavirus genes were found to be transcriptionally silent. Infection of a mavirus-bearing cell line with CroV, however, resulted in gene expression, DNA replication, and particle formation of the endogenous mavirus genomes.

The presence of the endogenous virophage did not benefit the CroV-infected host cell directly, nor was CroV replication inhibited as a result of mavirus reactivation. During subsequent rounds of infection, however, the reactivated mavirus particles inhibited CroV replication during regular coinfections, which led to a significantly reduced host cell mortality rate, depending on the MOI of mavirus and CroV (Fischer and Hackl, 2016). Thus, endogenous virophages can in principle protect host populations from giant virus infection, as proposed earlier (Fischer and Suttle, 2011). A crucial condition to ensure host survival, however, is that the spread of the lytic virus can be stopped before all cells are infected. The initially infected provirophage-bearing cells would be sacrificed in an altruistic manner to release virophage particles and eventually protect the population. Not all virophages may act in a way similar to mavirus though, and the sporadic distribution of integrases in virophage genomes, as well as the moderate effect of Zamilon on mimiviruses, suggests that not all virophages are beneficial to their cellular hosts. This should not, however, detract from the possible ecological importance of virophages on their viral and cellular hosts, which was proposed soon after the discovery of virophages (Fischer and Suttle, 2011; Yau *et al.*, 2011).

Several studies have explored host–virus–virophage dynamics *in silico* using mathematical models. Yau *et al.* (2011) presented a Lotka–Volterra simulation with OLV as a predator of a lytic giant virus which in turn preys on an alga. In this model, the presence of the virophage reduced the recovery time of the host population after giant virus lysis. Algal blooms were predicted to occur more frequently, with an overall increase in secondary production due to the action of the virophage. Wodarz explored the evolutionary dynamics of a virophage-containing tripartite system by modelling cells infected with one, both, or none of the two viruses (Wodarz, 2013). This model suggested that the virophage evolves towards higher levels of pathogenicity on the giant virus, which may lead to instability and eventual extinction of giant virus and virophage. Thus, a refined model is needed to allow for persistence of the virophage, which is apparently the case for naturally occurring virophages and their hosts. Taylor *et al.* (2014) constructed a model in which they paid particular attention to different entry modes of virophages, i.e. entry either as a composite with the giant virus (Sputnik) or independently of the giant virus (mavirus). The authors found that both entry modes allow for stable coexistence of such a tripartite system, albeit with slight differences. Also, entry mode did not influence the beneficial effect of virophages on host populations as long as the virophage was pathogenic for the giant virus. Clearly more experimental data are needed to validate these theoretical predictions and explore the ecological role of virophages.

Conclusion

Viruses are often perceived as the ultimate parasites, encoding just the bare essentials for successfully infecting and manipulating a host cell. The discovery of giant DNA viruses with genome lengths upwards of 1 million base pairs contradicted this paradigm. Increasing complexity allows these viruses to become independent from the host cell in many biochemical pathways, such as DNA replication and repair, transcription, or glycosylation. On the other hand, building enzymatically rich replication compartments also bears the risk of becoming a target for other parasites. Virophages of the family *Lavidaviridae* have evolved to utilize the giant virus-encoded transcription machinery for their gene expression, instead of locating to the nucleus to use the DNA-dependent RNA polymerase of the host. Until recently, virophages have escaped human notice because they are primarily associated with unicellular eukaryotic hosts and cause no disease in plants or animals. Nevertheless, these viruses are probably very ancient and can be found all over the globe in diverse marine and freshwater environments, where they replicate within various heterotrophic and presumably also phototrophic protists, and are associated with different types of giant DNA viruses. Many aspects of their infection biology, ecology and evolution remain to be studied, but from the very beginning of their discovery, these fascinating 'viruses of viruses' have captured the interest of scientists and the general public alike.

References

Bekliz, M., Verneau, J., Benamar, S., Raoult, D., La Scola, B., and Colson, P. (2015). A new Zamilon-like virophage partial genome assembled from a bioreactor metagenome. Front. Microbiol. *6*, 1308. https://doi.org/10.3389/fmicb.2015.01308.

Bekliz, M., Colson, P., and La Scola, B. (2016). The expanding family of virophages. Viruses *8*, E317.

Bellas, C.M., Anesio, A.M., and Barker, G. (2015). Analysis of virus genomes from glacial environments reveals novel virus groups with unusual host interactions. Front. Microbiol. *6*, 656. https://doi.org/10.3389/fmicb.2015.00656.

Blanc, G., Gallot-Lavallée, L., and Maumus, F. (2015). Provirophages in the Bigelowiella genome bear testimony to past encounters with giant viruses. Proc. Natl. Acad. Sci. U.S.A. *112*, E5318–26. https://doi.org/10.1073/pnas.1506469112.

Boughalmi, M., Saadi, H., Pagnier, I., Colson, P., Fournous, G., Raoult, D., and La Scola, B. (2013). High-throughput isolation of giant viruses of the *Mimiviridae* and *Marseilleviridae* families in the Tunisian environment. Environ. Microbiol. *15*, 2000–2007. https://doi.org/10.1111/1462-2920.12068.

Boyer, M., Madoui, M.A., Gimenez, G., La Scola, B., and Raoult, D. (2010). Phylogenetic and phyletic studies of informational genes in genomes highlight existence of a 4 domain of life including giant viruses. PLOS ONE *5*, e15530. https://doi.org/10.1371/journal.pone.0015530.

Boyer, M., Azza, S., Barrassi, L., Klose, T., Campocasso, A., Pagnier, I., Fournous, G., Borg, A., Robert, C., Zhang, X., *et al.* (2011). Mimivirus shows dramatic genome reduction after intraamoebal culture. Proc. Natl. Acad. Sci. U.S.A. *108*, 10296–10301. https://doi.org/10.1073/pnas.1101118108.

Broyles, S.S. (2003). Vaccinia virus transcription. J. Gen. Virol. *84*, 2293–2303.

Burroughs, A.M., Iyer, L.M., and Aravind, L. (2007). Comparative genomics and evolutionary trajectories of viral ATP dependent DNA-packaging systems. Genome Dyn 3, 48–65. https://doi.org/10.1159/000107603.

Byrne, D., Grzela, R., Lartigue, A., Audic, S., Chenivesse, S., Encinas, S., Claverie, J.M., and Abergel, C. (2009). The polyadenylation site of Mimivirus transcripts obeys a stringent 'hairpin rule'. Genome Res. *19*, 1233–1242. https://doi.org/10.1101/gr.091561.109.

Campos, R.K., Boratto, P.V., Assis, F.L., Aguiar, E.R., Silva, L.C., Albarnaz, J.D., Dornas, F.P., Trindade, G.S., Ferreira, P.P., Marques, J.T., *et al.* (2014). Samba virus: a novel mimivirus from a giant rain forest, the Brazilian Amazon. Virol. J. *11*, 95. https://doi.org/10.1186/1743-422X-11-95.

Claverie, J.M., and Abergel, C. (2009a). Mimivirus and its virophage. Annu. Rev. Genet. *43*, 49–66. https://doi.org/10.1146/annurev-genet-102108-134255.

Claverie, J.M., and Abergel, C. (2016). CRISPR-Cas-like system in giant viruses: why MIMIVIRE is not likely to be an adaptive immune system. Virol. Sin. *31*, 193–196. https://doi.org/10.1007/s12250-016-3801-x.

Claverie, J.M., and Abergel, C. (2009b). Mimivirus and its virophage. Annu. Rev. Genet. *43*, 49–66. https://doi.org/10.1146/annurev-genet-102108-134255.

Colson, P., De Lamballerie, X., Yutin, N., Asgari, S., Bigot, Y., Bideshi, D.K., Cheng, X.W., Federici, B.A., Van Etten, J.L., Koonin, E.V., *et al.* (2013). 'Megavirales', a proposed new order for eukaryotic nucleocytoplasmic large DNA viruses. Arch. Virol. *158*, 2517–2521. https://doi.org/10.1007/s00705-013-1768-6.

Desnues, C., and Raoult, D. (2010). Inside the lifestyle of the virophage. Intervirology *53*, 293–303. https://doi.org/10.1159/000312914.

Desnues, C., and Raoult, D. (2012). Virophages question the existence of satellites. Nat. Rev. Microbiol. *10*, 234. https://doi.org/10.1038/nrmicro2676-c3.

Desnues, C., Boyer, M., and Raoult, D. (2012a). Sputnik, a virophage infecting the viral domain of life. Adv. Virus Res. *82*, 63–89. https://doi.org/10.1016/B978-0-12-394621-8.00013-3.

Desnues, C., La Scola, B., Yutin, N., Fournous, G., Robert, C., Azza, S., Jardot, P., Monteil, S., Campocasso, A., Koonin, E.V., *et al.* (2012b). Provirophages and transpovirons as the diverse mobilome of giant viruses. Proc. Natl. Acad. Sci. U.S.A. *109*, 18078–18083. https://doi.org/10.1073/pnas.1208835109.

Fischer, M.G. (2011). Sputnik and Mavirus: more than just satellite viruses. Nat. Rev. Microbiol. *10*, 78. https://doi.org/10.1038/nrmicro2676-c1.

Fischer, M.G. (2015). Virophages go nuclear in the marine alga *Bigelowiella natans*. Proc. Natl. Acad. Sci. U.S.A. *112*, 11750–11751. https://doi.org/10.1073/pnas.1515142112.

Fischer, M.G. (2016). Giant viruses come of age. Curr. Opin. Microbiol. *31*, 50–57.

Fischer, M.G., and Hackl, T. (2016). Host genome integration and giant virus-induced reactivation of the virophage mavirus. Nature *540*, 288–291. https://doi.org/10.1038/nature20593.

Fischer, M.G., and Suttle, C.A. (2011). A virophage at the origin of large DNA transposons. Science *332*, 231–234. https://doi.org/10.1126/science.1199412.

Fischer, M.G., Allen, M.J., Wilson, W.H., and Suttle, C.A. (2010). Giant virus with a remarkable complement of genes infects marine zooplankton. Proc. Natl. Acad. Sci. U.S.A. *107*, 19508–19513. https://doi.org/10.1073/pnas.1007615107.

Fischer, M.G., Kelly, I., Foster, L.J., and Suttle, C.A. (2014). The virion of *Cafeteria roenbergensis* virus (CroV) contains a complex suite of proteins for transcription and DNA repair. Virology 466-467, 82–94. https://doi.org/10.1016/j.virol.2014.05.029.

Gaia, M., Colson, P., Desnues, C., and La Scola, B. (2013a). Virophage Concept, The. In eLS, (Chichester, UK: John Wiley & Sons, Ltd), pp. 1–12.

Gaia, M., Pagnier, I., Campocasso, A., Fournous, G., Raoult, D., and La Scola, B. (2013b). Broad spectrum of mimiviridae virophage allows its isolation using a mimivirus reporter. PLOS ONE *8*, e61912. https://doi.org/10.1371/journal.pone.0061912.

Gaia, M., Benamar, S., Boughalmi, M., Pagnier, I., Croce, O., Colson, P., Raoult, D., and La Scola, B. (2014). Zamilon, a novel virophage with Mimiviridae host specificity. PLOS ONE *9*, e94923. https://doi.org/10.1371/journal.pone.0094923.

Garza, D.R., and Suttle, C.A. (1995). Large double-stranded DNA viruses which cause the lysis of a marine heterotrophic nanoflagellate (*Bodo* sp) occur in natural marine viral communities. Aquat. Microb. Ecol. *9*, 203–210.

Gong, C., Zhang, W., Zhou, X., Wang, H., Sun, G., Xiao, J., Pan, Y., Yan, S., and Wang, Y. (2016). Novel virophages discovered in a freshwater lake in China. Front. Microbiol. *7*, 5. https://doi.org/10.3389/fmicb.2016.00005.

Gonzalez, J.M., and Suttle, C.A. (1993). Grazing by marine nanoflagellates on viruses and virus-sized particles – ingestion and digestion. Mar. Ecol. Prog. Ser. *94*, 1–10.

Greber, U.F., Willetts, M., Webster, P., and Helenius, A. (1993). Stepwise dismantling of adenovirus 2 during entry into cells. Cell *75*, 477–486.

Iyer, L.M., Aravind, L., and Koonin, E.V. (2001). Common origin of four diverse families of large eukaryotic DNA viruses. J. Virol. *75*, 11720–11734. https://doi.org/10.1128/JVI.75.23.11720-11734.2001.

Iyer, L.M., Balaji, S., Koonin, E.V., and Aravind, L. (2006). Evolutionary genomics of nucleo-cytoplasmic large DNA viruses. Virus Res. *117*, 156–184.

Iyer, L.M., Abhiman, S., and Aravind, L. (2008). A new family of polymerases related to superfamily A DNA polymerases and T7-like DNA-dependent RNA polymerases. Biol. Direct *3*, 39. https://doi.org/10.1186/1745-6150-3-39.

Jordan, I.K., Wolf, Y.I., and Koonin, E.V. (2004). Duplicated genes evolve slower than singletons despite the initial rate increase. BMC Evol. Biol. *4*, 22. https://doi.org/10.1186/1471-2148-4-22.

Kapitonov, V.V., and Jurka, J. (2006). Self-synthesizing DNA transposons in eukaryotes. Proc. Natl. Acad. Sci. U.S.A. *103*, 4540–4545.

Kerepesi, C., and Grolmusz, V. (2017). The 'Giant Virus Finder' discovers an abundance of giant viruses in the Antarctic dry valleys. Arch. Virol. *162*, 1671–1676. https://doi.org/10.1007/s00705-017-3286-4.

King, A.M.Q., Adams, M.J., Carstens, E.B., and Lefkowitz, E.J. (2011). Virus Taxonomy. Ninth Report of the International Committee on Taxonomy of Viruses. (Elsevier Academic, London).

Klose, T., Herbst, D.A., Zhu, H., Max, J.P., Kenttämaa, H.I., and Rossmann, M.G. (2015). A mimivirus enzyme that participates in viral entry. Structure *23*, 1058–1065. https://doi.org/10.1016/j.str.2015.03.023.

Koonin, E.V., and Krupovic, M. (2017). Polintons, virophages and transpovirons: a tangled web linking viruses, transposons and immunity. Curr. Opin. Virol. *25*, 7–15.

Krupovic, M., and Cvirkaite-Krupovic, V. (2011). Virophages or satellite viruses? Nat. Rev. Microbiol. *9*, 762–763.

Krupovic, M., and Koonin, E.V. (2016). Self-synthesizing transposons: unexpected key players in the evolution of viruses and defense systems. Curr. Opin. Microbiol. *31*, 25–33.

Krupovic, M., and Koonin, E.V. (2015). Polintons: a hotbed of eukaryotic virus, transposon and plasmid evolution. Nat. Rev. Microbiol. *13*, 105–115. https://doi.org/10.1038/nrmicro3389.

Krupovic, M., Bamford, D.H., and Koonin, E.V. (2014). Conservation of major and minor jelly-roll capsid proteins in Polinton (Maverick) transposons suggests that they are bona fide viruses. Biol. Direct *9*, 6. https://doi.org/10.1186/1745-6150-9-6.

Krupovic, M., Kuhn, J.H., and Fischer, M.G. (2016). A classification system for virophages and satellite viruses. Arch. Virol. *161*, 233–247. https://doi.org/10.1007/s00705-015-2622-9.

Legendre, M., Audic, S., Poirot, O., Hingamp, P., Seltzer, V., Byrne, D., Lartigue, A., Lescot, M., Bernadac, A., Poulain, J., *et al.* (2010). mRNA deep sequencing reveals 75 new genes and a complex transcriptional landscape in Mimivirus. Genome Res. *20*, 664–674. https://doi.org/10.1101/gr.102582.109.

Levasseur, A., Bekliz, M., Chabrière, E., Pontarotti, P., La Scola, B., and Raoult, D. (2016). MIMIVIRE is a defence system in mimivirus that confers resistance to virophage. Nature *531*, 249–252. https://doi.org/10.1038/nature17146.

Mohanraju, P., Makarova, K.S., Zetsche, B., Zhang, F., Koonin, E.V., and van der Oost, J. (2016). Diverse evolutionary roots and mechanistic variations of the CRISPR-Cas systems. Science *353*, aad5147. https://doi.org/10.1126/science.aad5147.

Mutsafi, Y., Zauberman, N., Sabanay, I., and Minsky, A. (2010). Vaccinia-like cytoplasmic replication of the giant Mimivirus. Proc. Natl. Acad. Sci. U.S.A. *107*, 5978–5982. https://doi.org/10.1073/pnas.0912737107.

Nasir, A., Kim, K.M., and Caetano-Anolles, G. (2012). Giant viruses coexisted with the cellular ancestors and represent a distinct supergroup along with superkingdoms Archaea, Bacteria and Eukarya. BMC Evol. Biol. *12*, 156. https://doi.org/10.1186/1471-2148-12-156.

O'Day, D.H., Suhre, K., Myre, M.A., Chatterjee-Chakraborty, M., and Chavez, S.E. (2006). Isolation, characterization, and bioinformatic analysis of calmodulin-binding protein cmbB reveals a novel tandem IP22 repeat common to many Dictyostelium and Mimivirus proteins. Biochem. Biophys. Res. Commun. *346*, 879–888.

Oh, S., Yoo, D., and Liu, W.T. (2016). Metagenomics reveals a novel virophage population in a tibetan mountain lake. Microbes Environ. *31*, 173–177. https://doi.org/10.1264/jsme2.ME16003.

Parola, P., Renvoisé, A., Botelho-Nevers, E., La Scola, B., Desnues, C., and Raoult, D. (2012). *Acanthamoeba polyphaga mimivirus* virophage seroconversion in travelers returning from Laos. Emerg. Infect. Dis. *18*, 1500–1502. https://doi.org/10.3201/eid1809.120099.

Pei, J., Kim, B.H., and Grishin, N.V. (2008). PROMALS3D: a tool for multiple protein sequence and structure alignments. Nuc. Acids Res. *36*, 2295–2300. https://doi.org/10.1093/nar/gkn072.

Pelletier, N., Raoult, D., and La Scola, B. (2009). Specific recognition of the major capsid protein of *Acanthamoeba polyphaga mimivirus* by sera of patients infected by *Francisella tularensis*. FEMS Microbiol. Lett. *297*, 117–123. https://doi.org/10.1111/j.1574-6968.2009.01675.x.

Piacente, F., Marin, M., Molinaro, A., De Castro, C., Seltzer, V., Salis, A., Damonte, G., Bernardi, C., Claverie, J.M., Abergel, C., *et al.* (2012). Giant DNA virus mimivirus encodes pathway for biosynthesis of unusual sugar 4-amino-4,6-dideoxy-D-glucose (Viosamine). J. Biol. Chem. *287*, 3009–3018. https://doi.org/10.1074/jbc.M111.314559.

Pritham, E.J., Putliwala, T., and Feschotte, C. (2007). Mavericks, a novel class of giant transposable elements widespread in eukaryotes and related to DNA viruses. Gene *390*, 3–17.

Raoult, D., Audic, S., Robert, C., Abergel, C., Renesto, P., Ogata, H., La Scola, B., Suzan, M., and Claverie, J.M. (2004). The 1.2-megabase genome sequence of Mimivirus. Science *306*, 1344–1350.

Redrejo-Rodríguez, M., and Salas, M.L. (2014). Repair of base damage and genome maintenance in the nucleo-cytoplasmic large DNA viruses. Virus Res. *179*, 12–25. https://doi.org/10.1016/j.virusres.2013.10.017.

Ronquist, F., and Huelsenbeck, J.P. (2003). MrBayes 3: Bayesian phylogenetic inference under mixed models. Bioinformatics *19*, 1572–1574.

Roux, S., Chan, L-K., Egan, R., Malmstrom, R.R., McMahon, K.D., and Sullivan, M.B. (2017). Ecogenomics of virophages and their giant virus hosts assessed through time series metagenomics. Nat. Commun. *8*, 858.

Santini, S., Jeudy, S., Bartoli, J., Poirot, O., Lescot, M., Abergel, C., Barbe, V., Wommack, K.E., Noordeloos, A.A., Brussaard, C.P., *et al.* (2013). Genome of *Phaeocystis globosa* virus PgV-16T highlights the common ancestry of the largest known DNA viruses infecting eukaryotes. Proc. Natl. Acad. Sci. U.S.A. *110*, 10800–10805. https://doi.org/10.1073/pnas.1303251110.

La Scola, B., Audic, S., Robert, C., Jungang, L., de Lamballerie, X., Drancourt, M., Birtles, R., Claverie, J.-M., and Raoult, D. (2003). A giant virus in amoebae. Science *299*, 2033.

La Scola, B., Desnues, C., Pagnier, I., Robert, C., Barrassi, L., Fournous, G., Merchat, M., Suzan-Monti, M., Forterre, P., Koonin, E., *et al.* (2008). The virophage as a unique parasite of the giant mimivirus. Nature *455*, 100–104. https://doi.org/10.1038/nature07218.

La Scola, B., Campocasso, A., N'Dong, R., Fournous, G., Barrassi, L., Flaudrops, C., and Raoult, D. (2010). Tentative characterization of new environmental giant viruses by MALDI-TOF mass spectrometry. Intervirology *53*, 344–353. https://doi.org/10.1159/000312919.

Suhre, K. (2005). Gene and genome duplication in *Acanthamoeba polyphaga Mimivirus*. J. Virol. *79*, 14095–14101.

Sun, S., La Scola, B., Bowman, V.D., Ryan, C.M., Whitelegge, J.P., Raoult, D., and Rossmann, M.G. (2010). Structural studies of the Sputnik virophage. J. Virol. *84*, 894–897. https://doi.org/10.1128/JVI.01957-09.

Taylor, B.P., Cortez, M.H., and Weitz, J.S. (2014). The virus of my virus is my friend: ecological effects of virophage with alternative modes of coinfection. J. Theor. Biol. *354*, 124–136. https://doi.org/10.1016/j.jtbi.2014.03.008.

Williams, T.A., Embley, T.M., and Heinz, E. (2011). Informational gene phylogenies do not support a fourth domain of life for nucleocytoplasmic large DNA viruses. PLOS ONE *6*, e21080. https://doi.org/10.1371/journal.pone.0021080.

Wodarz, D. (2013). Evolutionary dynamics of giant viruses and their virophages. Evol. Ecol. 3, 2103–2115. https://doi.org/10.1002/ece3.600.

Xiao, C., Kuznetsov, Y.G., Sun, S., Hafenstein, S.L., Kostyuchenko, V.A., Chipman, P.R., Suzan-Monti, M., Raoult, D., McPherson, A., and Rossmann, M.G. (2009). Structural studies of the giant mimivirus. PLOS Biol. 7, e92. https://doi.org/10.1371/journal.pbio.1000092.

Yau, S., Lauro, F.M., DeMaere, M.Z., Brown, M.V., Thomas, T., Raftery, M.J., Andrews-Pfannkoch, C., Lewis, M., Hoffman, J.M., Gibson, J.A., *et al.* (2011). Virophage control of antarctic algal host-virus dynamics. Proc. Natl. Acad. Sci. U.S.A. *108*, 6163–6168. https://doi.org/10.1073/pnas.1018221108.

Yutin, N., Raoult, D., and Koonin, E.V. (2013). Virophages, polintons, and transpovirons: a complex evolutionary network of diverse selfish genetic elements with different reproduction strategies. Virol. J. *10*, 158. https://doi.org/10.1186/1743-422X-10-158.

Yutin, N., Wolf, Y.I., and Koonin, E.V. (2014). Origin of giant viruses from smaller DNA viruses not from a fourth domain of cellular life. Virology *466–467*, 38–52.

Yutin, N., Kapitonov, V.V., and Koonin, E.V. (2015a). A new family of hybrid virophages from an animal gut metagenome. Biol. Direct *10*, 19. https://doi.org/10.1186/s13062-015-0054-9.

Yutin, N., Shevchenko, S., Kapitonov, V., Krupovic, M., and Koonin, E.V. (2015b). A novel group of diverse Polinton-like viruses discovered by metagenome analysis. BMC Biol. *13*, 95. https://doi.org/10.1186/s12915-015-0207-4.

Zablocki, O., van Zyl, L., Adriaenssens, E.M., Rubagotti, E., Tuffin, M., Cary, S.C., and Cowan, D. (2014). High-level diversity of tailed phages, eukaryote-associated viruses, and virophage-like elements in the metaviromes of antarctic soils. Appl. Environ. Microbiol. *80*, 6888–6897. https://doi.org/10.1128/AEM.01525-14.

Zhang, X., Sun, S., Xiang, Y., Wong, J., Klose, T., Raoult, D., and Rossmann, M.G. (2012). Structure of Sputnik, a virophage, at 3.5-Å resolution. Proc. Natl. Acad. Sci. U.S.A. *109*, 18431–18436. https://doi.org/10.1073/pnas.1211702109.

Zhou, J., Zhang, W., Yan, S., Xiao, J., Zhang, Y., Li, B., Pan, Y., and Wang, Y. (2013). Diversity of virophages in metagenomic data sets. J. Virol. *87*, 4225–4236. https://doi.org/10.1128/JVI.03398-12.

Zhou, J., Sun, D., Childers, A., McDermott, T.R., Wang, Y., and Liles, M.R. (2015). Three novel virophage genomes discovered from Yellowstone Lake metagenomes. J. Virol. *89*, 1278–1285. https://doi.org/10.1128/JVI.03039-14

Part IV

Technologies Involving Viruses of Microorganisms

Viruses of Microorganisms and Biotechnology

13

Paul Hyman*

Department of Biology and Toxicology, Ashland University, Ashland, OH, USA.

*Correspondence: phyman@ashland.edu

https://doi.org/10.21775/9781910190852.13

Abstract

A large part of biotechnology in the sense of genetic engineering developed in parallel to increasing knowledge of the molecular biology of bacteria and their viruses, the bacteriophages. Many of the enzymes used were ones evolved by bacteriophages. While in principle, enzymes and other components of other viruses of other microbes could be used, historically, knowledge of bacteriophages predates the viruses of other microorganisms. In this chapter I present an overview of some of the technological methods that have been created using whole bacteriophages or bacteriophage components for: the creation of materials and molecules with novel selected properties (phage display); detection of bacterial pathogens (biosensors); and next-generation vaccines. In considering these, I will both define the technological approaches as well as discuss some of the challenges of taking these technologies from the laboratory to the marketplace.

Introduction

In 1966 a collection of essays dedicated to Max Delbrück, entitled Phage and the Origins of Molecular Biology (Cairns *et al.*, 1966), was published. In this context, 'Molecular Biology' referred to the field of biology studying nucleic acids and gene expression but the title is equally applicable to molecular biology in the biotechnology sense – a set of techniques and tools for cloning genes, protein expression and so on. Many of the earliest enzymes and vectors used in biotechnology were derived from bacteriophages. Restriction enzymes can arguably be considered as having evolved as anti-phage adaptations, although they may also be anti-plasmid agents as well (Bickle and Krüger, 1993). Bacteriophage promoters are used to allow specific expression of cloned genes in part because some have evolved to high specificity with their cognate RNA polymerases (e.g. bacteriophage T7; Tabor and Richardson, 1985; Studier and Moffatt, 1986). Filamentous bacteriophage derived gene sequences were used to produce single-stranded DNA for early methods of DNA sequencing and probe generation, although these methods have largely been supplanted by newer methods (Sambrook *et al.*, 1989). Bacteriophage packaging systems are used to create alternatives to plasmids as gene cloning vectors (Collins and Hohn, 1978; Geider *et al.*, 1985). Overall, because much of the development of gene cloning and molecular biotechnology used *E. coli* for both gene cloning and expression, coliphages were and remain a useful source of genetic elements.

While some methods such as cloning genes for sequencing have been replaced by other methods, bacteriophages and other kinds of viruses still play a vital role in some biotechnology applications. In this chapter I will present a survey of these methods, focusing especially on those that are in development or in use for creating marketable products. For the most part, this chapter will be concerned with bacteriophages since there is very little work in this area being done with viruses of microorganisms other than bacteria, although in principle much of the same work could be done with these viruses as well. The major exception to this is the use of virus capsids as drug carriers in which viruses of both microbes and macroorganisms are used. One application that I will not discuss is phage therapy as this is extensively discussed in Chapter 14 in this same volume.

Phage display

Phage display is a technique for screening a library of protein epitopes or other binding proteins, as can be engineered into a single phage gene, and selecting a small number of those proteins with a specific binding affinity. The technique takes advantage of two properties of a bacteriophage: (1) under most circumstances there is a direct link between the phenotype of a phage (here, its capsid) and its genotype (i.e. its genome); and (2) amplification of a single phage into a large population involves simply growing a stock of that phage. Details of the methods of phage display have been extensively reviewed (Smith and Petrenko, 1997; Barbas *et al.*, 2004; Rakonjac *et al.*, 2011) but briefly, a library of peptides is inserted into a phage protein by inserting a corresponding library of nucleic acid sequences in the protein's gene using directed mutagenesis techniques. The protein must be on the surface of the phage capsid and the insert must be in a location that places it on the capsid surface and must not interfere unduly with the protein's function. A collection of these phages is then exposed to a target material for which binding peptides are sought. Phages with a peptide that binds to the target will do so while the remainder of the phages remain in the media or buffer. The material is then washed to remove non-attaching phages. The attaching phages are then eluted off the target material and used to infect host bacteria to amplify just those attaching phages. The process, which is usually described as 'biopanning', is repeated with the newly amplified phages. Generally 3–5 rounds of biopanning are needed to take a population of 10^8 or more unique phages, with respect to the variable binding peptide, down to sometimes fewer than five different phages although this may not represent the full diversity of the library due to competition of the phages during amplification (Derda *et al.*, 2011). Given that each round of biopanning can be done in a few days with the need to grow a new phage stock being the major time limitation, this method allows for a relatively rapid identification of specific binding peptides from a large collection of potential peptides.

Phage display has been used to identify binding peptides for a wide variety of organic and inorganic molecules for a variety of purposes. Table 13.1 lists some of these. While some uses of phage display-generated peptides have been proposed where the peptide remains on the phage and the intact phage becomes part of the new material, in most cases the peptide, once identified, is synthesized and used separately from the phage. Much of the utility of the

Table 13.1 Applications using binding peptides identified using phage display

Application	Description	Recent review(s)
Cancer chemotherapeutic agents and other disease drug delivery	Phage-derived monoclonal antibodies, peptides for drug and drug carrier targeting	Henry *et al.* (2015); Karimi *et al.* (2016); Nagano and Tsutsumi (2016); Petrenko and Gillespie (2017)
Imaging agents for cancer and other diseases	Phage-derived peptides attached to radioisotopes or other imaging agents	Hu *et al.* (2017)
Pathogen identification and treatment	Phage-derived antibodies (see also 'Phage biosensors')	Henry *et al.* (2015); Kuhn *et al.* (2016); Criscuolo *et al.* (2017)
Tissue regeneration	Phage-derived peptides used to functionalize tissue scaffolds and to promote stem cell growth	Martins *et al.* (2016)
Vaccines	Phage-derived peptides can act as immunogens either as free peptides or as part of the whole phage	Aghebati-Maleki *et al.* (2016)
Mimotyping	Phage-derived peptides that mimic allergens for allergy diagnosis and potential treatment	Chen and Dreskin (2017)

identified peptides arises from the ease of conjugating peptides to other functional elements via simple chemistry. Examples include the use of phage display-derived targeting peptides as drug delivery carriers and for imaging via linkage to radioisotopes and other materials (see the reviews listed in Table 13.1).

Of the uses listed above, the development of monoclonal antibodies (Mabs) using phage display for cancer and pathogen treatments is arguably the most developed with several of these Mabs currently in clinical use. These Mabs are developed using the basic biopanning approach described above except that the library of inserts can be an antibody antigen binding fragment (Fab) heavy chain or light chain fragment which are later combined into a Mab with the selected binding affinity (Frenzel *et al.*, 2017). While not all combinations of selected heavy and light chains will retain strong binding affinity, many do, allowing for the production of Mabs more quickly than those derived by traditional procedures, that is, which use live animals to produce the initial antibody producing cells. Most recently, single chain antibody fragments made by combining the heavy old light chain variable regions into a single polypeptide, or based on camelid (camels, llamas, etc.) antibodies, are being used instead (Ahmad *et al.*, 2012). Termed scFvs (single chain fragment variable), these have the advantage of not having to be combined and rescreened to ensure retention of the desired binding affinity. Furthermore, once the gene for an appropriate scFv is obtained through phage display, it can be put through additional rounds of mutagenesis and rescreening to improve specificity and binding kinetics. Recently, for example, Haylock and colleagues (Haylock *et al.*, 2017) created several scFvs that were specific for a cell marker, CD44v6, that is found on 90% of one class of head and neck squamous cell carcinomas. The binding kinetics and other properties of the scFvs compared favourably to other monoclonal antibodies for use as imaging agents to help identify and locate primary and metastatic tumours. Similarly, antibodies against infectious agents such as viruses and bacteria, have been developed or improved upon using phage display. Wu and colleagues (Wu *et al.*, 2005) used phage display to identify improved binding sequences to create more potent antibodies against respiratory syncytial virus (RSV), a common cause of lower respiratory infections in young children. They did this by creating phage M13 libraries containing mutagenized variants of an Fab fragment of an RSV binding antibody and successfully screening the library for antibodies with stronger binding affinities.

Phage biosensors

The specificity of binding of most bacteriophages to their bacterial hosts was recognized fairly early as a way to identify particular strains of bacteria. This early process, termed phage typing, was first developed in the 1920s (Adams, 1959) and proved useful in epidemiological studies. For phage typing, strains of a particular species of bacteria could be differentiated by the pattern of which phages – often described as typing phages – infected or did not infect each bacterial strain. Phage typing can be considered to be a reflection of bacteriophage host range (Hyman and Abedon, 2010; Williams and LeJeune, 2012). Although today largely supplanted by identification using antibodies, many reports of the isolation of new phages from the 1940s to 1980s were motivated by a search for new typing phages for an increasing variety of bacteria. Today phage typing is only rarely used, especially as a sole method of identification. But for some bacterial species it is still a valuable tool. Ziebell and colleagues, for example, compared phage typing alone and in combination with multiple locus variable number tandem repeat analysis (MLVA) and pulsed field gel electrophoresis (PFGE) (Ziebell *et al.*, 2017). They found that combining phage typing with these other methods could better discriminate among different strains of *Salmonella* than any of the methods alone.

Newer methods using bacteriophages as sensors are directed at detecting the presence of particular bacteria for diagnosis and other detection such as testing foods for bacterial contamination rather than discriminating between strains (Singh *et al.*, 2012; Klumpp and Loessner, 2014). These methods can be divided into a) those that use whole viable phages and b) those that use receptor binding proteins either in capsids or in an isolated form. When whole phages are used, the infection process may occur although only some methods require progeny production and further rounds of infection. Methods that use receptor binding proteins typically attach these to some type of sensor surface using a variety of chemical bonds.

Methods using whole phage and requiring infection (genome entry into host cells)

These methods rely on the specificity of phage infections, that most phages will only infect a limited number of hosts. For some methods, the infection must produce progeny while for others phage gene expression is needed but progeny production is not. The logic of each approach is that if bacteriophages are exposed to a mixture or bacteria, such as from a patient or environmental sample, and the number of phages or the amount of a phage gene product increases, their host cells must be present in the mixture.

An increase in the number of phages was the basis of the FASTPlaque assay (Albert *et al.*, 2004) which used a phage able to infect *Mycobacteria* species to detect *M. tuberculosis.* A patient sample was mixed with the phage and incubated briefly to allow phage adsorption. The cells were then separated from remaining free phage and plated using *M. smegmatus* as a plating indicator, allowing lawn formation much more quickly than if *M. tuberculosis* were used for phage plaquing. An increased plaque count over background indicated that *M. tuberculosis* was present in the original sample. This test could also be adapted to detecting antibiotic-resistant strains of *M. tuberculosis* since adding antibiotics to the original sample would prevent phage growth unless the host bacteria were resistant. This method was comparable in sensitivity to direct staining of samples and direct culture. In addition, it had the advantage of being much more rapid than direct culture of sputum samples and required little technical training, although it was not as sensitive or as rapid as PCR based detection of *M. tuberculosis* which can also be done directly on samples (Singh *et al.*, 2008). At the time of this writing, it appears that these assay kits are no longer commercially available as far as an internet search can ascertain.

A similar approach was used to detect drug-resistant *Staphylococcus aureus* in the Keypath detection system (Bhowmick *et al.*, 2013). In this system a blood sample from a patient is mixed with media along with a phage that infects *S. aureus.* If present, the bacteria are infected and phage numbers increase. Drug resistance status can be determined by splitting cultures into sub-cultures with and without the antibiotic. After a relatively short (5-hour) incubation period, the culture can be analysed using antibodies in a lateral flow device similar to that used for home pregnancy testing (Cox, 2012). An increase in phage numbers to detectable levels indicates the presence of the susceptible bacteria. A device based on this system was approved by the US FDA but after several years the development company, Microphage, Inc., declared bankruptcy (Avery, 2013).

As an alternative to waiting for phage growth, some phages have been genetically modified to emit a signal after their genomes enter the target host cell and begin gene expression. The reporter for these phages is usually either firefly or bacterial luciferase although green fluorescent protein has been used as well. In all cases, the same general strategy is used. A bacteriophage specific for the target bacteria is genetically modified to insert the marker gene or genes in a phage genome site where expression is driven by promoter that expresses early to mid-infection. For example, Loessner and colleagues (Loessner *et al.*, 1996) inserted a fused *Vibrio harveyi luxAB* gene next to a capsid protein gene in the genome of Listeria phage A511. This capsid gene expresses about 15 minutes post-infection and the modified phage was shown to begin producing luciferase mRNA at the same time. After 2 hours of incubation, luciferase bio-luminescence could be detected from levels of as few as 500 cells using a luminometer. At least one company, Sample6 (Woburn, MA, USA) has made a similar system commercially available (http://www.sample6.com/food-pathogen-testing). Their system can reliably detect *Listeria* in contaminated food after a six hour incubation of a wash from the food with the phage (Cappillino *et al.*, 2015). Phages with fluorescent or bio-luminescent markers have been developed and tested in research laboratories for *E. coli*, *Bacillus anthracis*, and *Salmonella* among others (Bárdy *et al.*, 2016).

Methods using receptor binding proteins (free or in capsids)

Some detection methods rely on the specificity of the phage receptor binding proteins to only bind to a particular set of hosts, usually all of the same species. The receptor binding proteins may be in whole capsids or used as isolated proteins. In all cases, the logic of this application is that the binding element (capsid or receptor binding protein) is attached to some sensor material. Target bacterial binding changes some physical or electrochemical property of the material. Detecting that change is interpreted as detecting the target bacteria. This approach has only been demonstrated in laboratory studies including some that used phage display derived binding peptides instead of receptor binding proteins. I will briefly present some information on these demonstrations but more detailed information can be found in several recent reviews including (Singh *et al.*, 2012, 2013; Billington *et al.*, 2014; van der Merwe *et al.*, 2014).

The development of these methods relies on two major components: binding the phage or protein to a sensor material and the detection property used. The latter is often called the signal transduction stage as it converts (transduces) the bacterial binding to a detectable signal.

There are three basic approaches to attaching phages or phage proteins to surfaces to create a biosensor material. First, it can be done using simple electrostatic attractions. For example, Johnson and colleagues (Johnson *et al.*, 2008) simply dipped their transducing material in a solution of phages that had been genetically engineered to bind to *Bacillus anthracis* spores to coat that material with phages. Other examples can be found in the work of Wan and colleagues (Wan *et al.*, 2007) and Lone and colleagues (Lone *et al.*, 2016). Even although these are non-covalent attachments, in the relatively controlled chemical conditions of a detection device, the phage attachment is quite stable. Other researchers have used general protein binding chemicals such as glutaraldehyde, silanes or thiol esters to attach phage (Yang *et al.*, 2006; Handa *et al.*, 2008; Wang *et al.*, 2016). These have an advantage in working with a variety of phages but bind the phages in a random orientation so that not all the receptor binding proteins may be in an optimal orientation for binding the target bacteria. Anany and coworkers (Anany *et al.*, 2011; Lone *et al.*, 2016) attached T4 and T7 phages to a charged cellulose surface in a head-down, tail-out orientation by taking advantage of the relatively negative charge of the phage heads vs. tails under the binding conditions. It is not clear if all phages will be amenable to this method. Finally, come researchers have covalently modified a phage capsid protein such as via a biotinylation site or adding an exposed cysteine to create a specific site for chemical attachment (Dultsev *et al.*, 2001; Tolba *et al.*, 2010). This allows the orientation of tailed phages by using a head protein for the modification site so that the tail fibres or other receptor binding proteins are oriented away from the support material.

The use of isolated phage receptor binding proteins (RBPs) is different from whole phages in several ways. First, the attachment is specific and oriented since the isolated proteins can be genetically engineered to improve isolation, purification and attachment to sensor surfaces. Second, they can be produced using protein expression systems that do not need phage infection, which usually kills the infected host. This allows for an extended period of protein production before the cells are lysed to release the binding protein. Third, not all of the RBP needs to be expressed, only the domain that contains the receptor binding sequences and enough of the protein to hold these amino acid residues in the proper structure. For example, the end of the long tail fibres of bacteriophage T4 contains the RBP peptide segments. When Bartual and colleagues obtained the structure of this RBP domain they only needed to obtain the structure for the assembled C-terminal 241 amino acids of the entire 1026 amino acids long protein (Bartual *et al.*, 2010a,b).

A few studies have demonstrated the utility of this approach. These are listed in Table 13.2. Some of these attach the binding protein or protein fragment to surfaces using a binding element which is added to the protein. Singh and colleagues (Singh *et al.*, 2010) attached phage P22 tailspike proteins to a gold surface by adding a cysteine residue to the N-terminal end of the protein so that the cysteine's affinity for gold both oriented and attached the tailspike protein. Tay and colleagues (Tay *et al.*, 2012) used the same protein to test several different attachment methods to functionalize surfaces for direct detection of *Salmonella* cells which altered the optical properties of tailspike coated beads. Other workers have attached the RBPs to a tag such as GFP to fluorescently 'stain' target cells (Javed *et al.*, 2013). Denyes and colleagues (Denyes *et al.*, 2017) attached the enzyme horseradish peroxidase to the S16 receptor binding protein to create an ELISA-like sandwich assay for detecting *Salmonella* cells.

Table 13.2 Representative studies using RBPs to detect specific bacteria. This table includes examples of proteins used to add specific detection function to biosensor surfaces and for other targeting methods including fluorescent tagging and target cell collection

Target bacteria	RBP and use	Reference
E. coli O157:H7	Unspecified phage tail fibre protein surface functionalization	Rozand and Feng (2009), Savoye *et al.* (2011)
E. coli and *Hafnia alvei*	Phage T4 long tail fibre receptor binding protein fragment surface functionalization	Brzozowska *et al.* (2015)
Campylobacter jejuni	Phage NCTC 12673 receptor binding protein surface functionalization	Singh *et al.* (2011), Poshtiban *et al.* (2013)
C. jejuni and *Campylobacter coli*	Phage NCTC 12673 receptor binding protein fragment alone and fused to GFP	Javed *et al.* (2013)
Salmonella	Phage S16 tail fibre receptor binding proteins coupled to GFP or to horse radish peroxidase	Marti *et al.* (2013), Denyes *et al.* (2017)
Salmonella	Phage P22 tailspike protein surface functionalization	Singh *et al.* (2010), Tay *et al.* (2012)

A second type of bacteriophage protein that has been used in host detection is the cell-wall binding domain (CBD) of phage endolysins. Endolysins are the proteins that effect the breakdown of the host cell well by breaking one of the bonds in peptidoglycan. Other proteins disrupt the cell membrane and the cell consequently lyses, releasing progeny phages (Fischetti, 2010; Shen *et al.*, 2012; Roach and Donovan, 2015). Most endolysins from phages of Gram positive hosts are composed of two domains – the CBD and the enzymatic domain that cleaves peptidoglycan. Because endolysins can break down peptidoglycan from outside of the cell as well as from inside, they are being developed as an antibiotic alternative as described in Chapter 14. Here I will only discus uses relying on the specificity of the CBD.

While peptidoglycan is often discussed as if it is identical in all bacteria, there are sufficient structural variations either in the peptidoglycan or other molecules in the cell wall that allow some CBDs to bind only to the cell wall of specific species while other CBDs have a broader range of bacteria they will bind to (Roach and Donovan, 2015). Because some CBDs function independently of the enzymatic component of the endolysin, some researchers have studied the use CBDs in an analogous manner to the RBPs discussed above. Table 13.3 lists some of the bacterial species that have been targeted by CBD based identification studies as well as the methods of using the CBDs. In all cases, since there is no gene for the native CBD protein alone in a phage, the coding sequence for the CBD domain must be cloned allowing for added modifications, for example to attach the CBD to a surface via a His-tag (Kong *et al.*, 2015) or GFP which can be fused to the CBD creating a chimeric protein (Schmelcher *et al.*, 2010). This tagging can be and has occasionally been done with the entire endolysin protein (Gomez-Torres *et al.*, 2017). In this case, however, binding of the fluorescent tag bound to the CBD had a somewhat broader host range and usually stronger binding (when there was a difference) to various *Clostridia* strains and species (Table 13.3; see also Gomez-Torres *et al.*, 2018). Overall, CBDs should provide a good alternative to RBPs for some groups of bacteria, especially as they appear to be much easier to produce via standard biotechnological approaches as they function as monomers, unlike many RBPs which instead are trimeric proteins.

In all of the studies that attached whole phages, receptor binding proteins, or CBDs to surfaces, the second step in detection of the target bacteria is the transduction of cell binding into a detectable signal. This relies on three general approaches: (a) those that depend on the mass of the target cell(s) changing the mass of the surface material; (b) those that depend on cell presence changing some electrochemical property of the surface; or (c) those that depend on cell presence changing some optical property of the surface. The sensitivity of the detection depends on how few cells are needed to cause a measurable change in the property being monitored but can be as low as hundreds to thousands of cells. In most cases the change can be measured

Table 13.3 Studies using CBDs for identification of pathogens. Note that all the *L. monocytogenes* were identified using one of two of the same CBDs

Target bacteria	CBD and use	Reference
Listeria monocytogenes, *Bacillus cereus*, *Clostridium perfringens*	Cloned CBDs from various phage endolysins attached to paramagnetic beads for separation of target cells from cell mixtures or food samples for identification	Kretzer *et al.* (2007), Walcher *et al.* (2010)
L. monocytogenes	Cloned CBD of phage Ply500 endolysin attached to sensor material for detection of bacteria in contaminated milk samples	Tolba *et al.* (2012)
L. monocytogenes	Cloned CBD of multiple *Listeria* phages on magnetic beads for separation or attached to various fluorescent proteins for identification	Schmelcher *et al.* (2010); Schmelcher and Loessner, (2014)
L. monocytogenes	Chimeric CBD (multiple CBD from different phages combined into a single protein) tagged with fluorescent proteins	Schmelcher *et al.* (2011)
B. cereus	Cloned CBD of phage PBC1 endolysin tagged with GFP or His-tagged for attachment to surface for detecting cell binding by surface plasmon resonance (see Table 13.4)	Kong *et al.* (2015), Kong *et al.* (2017)
B. cereus	Cloned CBD of phage PBC4 endolysin tagged with GFP for identification of *B.cereus*	Na *et al.* (2016)
Multiple *Clostridial* species	Cloned CBD of phage ΦCTP1 endolysin tagged with GFP for detection of *Clostridial* species in cheese	Gomez-Torres *et al.* (2018)

in real-time in contrast to the previously discussed methods of bacterial detection that required hours to allow the phages to infect cells and replicate or express proteins. Table 13.4 lists some examples of the properties that have been used. More detailed descriptions can be found in (Hyman, 2012; Singh *et al.*, 2012; Billington *et al.*, 2014; Schmelcher and Loessner, 2014). The variety of methods available is useful in developing different types of biosensors that might be used for testing patient samples, food samples or other environmental samples. Because the sample may be liquid or a swab of a solid, different types of transducers might be needed, each functionalized with the same phage component.

Overall, the combination of phage-derived recognition/binding proteins with highly sensitive transducing methods holds great promise in the development of novel pathogen detection devices. Technologically, the major competition is the use of monoclonal antibodies as recognition elements, which can be used with the same transducing methods or in other methods such as ELISA assays or lateral flow technologies (as was the Keypath method using bacteriophage). Currently phage-based methods lack the sensitivity to be used reliably without enrichment or culturing of samples (Narsaiah *et al.*, 2012; Billington *et al.*, 2014). If phage molecule-based methods can be shown to provide greater sensitivity, then these methods may yet enter the marketplace.

Phage based vaccines

Within a few years of the discovery of phages, it was shown that bacteriophages could act as antigens and promote the production of phage specific antiserum (Bordet and Ciuca, 1921, as cited in Adams, 1959). This is not surprising; bacteriophages, like other viruses, have a capsid of proteins and other components that are immunologically foreign to animals and this is the basis of the animal immune system's protection against reinfection. Phage-specific antiserum has been used for both identification and experimentation with phages (Adams, 1959; Waddell *et al.*, 2009).

All vaccines are based on the introduction of pathogen antigens to trigger a primary immune response without actually infecting the body with the disease-causing pathogen. While this is most commonly accomplished with attenuated or killed pathogens, some research has been done introducing small numbers of antigenic proteins or genes for antigenic proteins into cells (and allowing the cells to express just the key proteins)(Hokama and Nakamura, 1982). The goal of these alternatives is to increase vaccine safety as attenuated vaccines

Table 13.4 Transduction methods that have been used with phage or phage derived biosensors to produce a detectable signal when target bacteria or other molecules are bound to the biosensor

Property type	Description	References
Mass dependent	Quartz crystal microbalances, acoustic wave and microcantilever detection – methods depend on the oscillation frequency of a sensor that changes when cells bind to the surface of the oscillator	Dultsev *et al.* (2001), Olsen *et al.* (2006); Nanduri *et al.* (2007b)
	Magnetoelastic resonance or restriction – magnetic particles in an oscillating magnetic field change their oscillation frequency when cells bind to the particles	Lakshmanan *et al.* (2007), Wan *et al.* (2007), Johnson *et al.* (2008), Huang *et al.* (2009)
Electrochemical dependent	Impedance of surface coated with phage is altered by cell binding	Yang *et al.* (2006), Shabani *et al.* (2008); Mejri *et al.* (2010), Tolba *et al.* (2012), Tlili *et al.* (2013)
Optical change dependent	Surface plasmon resonance (SPR) – detected as changes in light frequencies interacting with surface electrons caused by cell binding to surfaces that affect the electron charge distribution	Balasubramanian *et al.* (2007), Nanduri *et al.* (2007a,b), Arya *et al.* (2011)
	Long-period fibre gratings or optical fibre gratings – optical fibres vary in efficiency of transmission of light of different frequencies, the transmission spectrum is altered as the fibre binds cells on its surface	Smietana *et al.* (2011), Tripathi *et al.* (2012), Brzozowska *et al.* (2015)
	Opto-fluidic ring resonance – the optical resonance of a fluid-filled capillary tube is altered when cells bind to phage attached to the tube wall	Zhu *et al.* (2008)

occasionally contain virulent mutants and killed pathogen vaccines may be improperly prepared and contain live pathogens. The use of bacteriophages as vaccines for non-phage proteins is a relatively recent addition to these efforts. In this context, bacteriophages are considered, and often described as, viral nanoparticles (VNPs) or virus-like particles (VLPs) since they are no longer functioning viruses (Czapar and Steinmetz, 2017) There are two fundamentally different modalities for bacteriophages as vaccines – DNA vaccines (also called genetic vaccines) and protein vaccines (Clark *et al.*, 2012; Jafari and Abediankenari, 2015). For phage-based DNA vaccines, the phage capsid acts as a carrier of the vaccine DNA, protecting that DNA until the phage is taken up by a cell, with the DNA then released as the capsid is broken down intracellularly, allowing expression of antigen genes in the DNA. In contrast, phage protein vaccines are a variation of phage display (Table 13.1). In this case, the displayed proteins or peptides are the antigen, here a protein fragment, that is being vaccinated against. Whole phages are taken up by phagocytic cells and processed as any other foreign organism is, with antigenic information extracted and used in an immune response. In principle it should also be possible to combine phage DNA and protein vaccines into a single phage although little work has been done in this (Jafari and Abediankenari, 2015).

Phage-based DNA vaccines

One of the earliest demonstrations of a phage based DNA vaccine was accomplished by Clark and March (Clark and March, 2004) who inserted a mammalian expression cassette for the hepatitis B small surface antigen into a phage λ cloning vector. They found that mice receiving two injections of these phage generated antibodies specific for the hepatitis B antigen. In some animals, antibody titres approached those of a standard vaccine response. Furthermore, using λ phages that contained a gene for expressing green fluorescent protein, they were able to detect GFP expression in macrophages suggesting that the phages were reaching immune system cells involved in antigen presentation. Later work with the same hepatitis B antigen encoding phage included a direct comparison of immune system responses to the phage with responses to a standard hepatitis B protein vaccine which found equivalent or better responses to the phage vaccine when tested in rabbits (Clark *et al.*, 2011). Table 13.5 lists a representative sample of other antigens that phage-based DNA vaccines have been tested against. The major challenge of this method appears

Table 13.5 Examples of pathogenic viruses and bacteria that have been targeted by phage DNA vaccines

Pathogen target	Phage system	Reference
Human papilloma virus	λ phage vector	Ghaemi *et al.* (2011)
Chlamydophila abortus outer membrane protein	λ phage vector	Ling *et al.* (2011), Ou *et al.* (2013)
Herpes simplex virus	M13 phage vector	Hashemi *et al.* (2010)
Hepatitis C	λ phage vector	Saeedi *et al.* (2014)
Yersinaa pestis capsule protein	T4 prohead vector	Tao *et al.* (2013b)
HIV gag protein (mRNA vaccine rather than DNA)	MS2 phage capsids	Sun *et al.* (2011)

to be a limitation of DNA vaccines in general rather than of phage as delivery vehicles. Whether injected directly into tissues (naked DNA injection) or via chemical or other carriers, DNA vaccines have inconsistent results apparently due to differences in the DNA being taken up by the appropriate cells and achieving sufficiently high expression levels of the antigenic protein (Hasson *et al.*, 2017).

Phage protein vaccines

The development of these vaccines uses techniques developed for phage display (see 'Phage display' above for details) and they are sometimes referred to as phage display vaccines. The vaccine antigen may be inserted into the same phage surface protein sites that are used for phage display but when combined with selection allows for the development of improved vaccine antigens (Aghebati-Maleki *et al.*, 2016). Just as for phage display, both filamentous and non-filamentous phages have been used for vaccines including phages M13 and fd (filamentous phages), phage T4, phage T7, phage λ, and RNA phages MS2, PP7 and Qβ (Clark *et al.*, 2012). A wide variety of studies have used these viruses to develop potential vaccines against both viral and bacterial pathogens. Table 13.6 provides some examples of these but for a more complete overview several recent reviews are (Henry *et al.*, 2015; Aghebati-Maleki *et al.*, 2016; Criscuolo *et al.*, 2017; Czapar and Steinmetz, 2017).

A related use of phage capsid proteins is to create virus-like particles (VLPs) to carry proteins just as the VLPs can be used to carry DNA. The antigen proteins can be fused to the capsid proteins but in this case, the antigen protein does not need to be on the surface of the VLP. For example, Patterson and colleagues (Patterson *et al.*, 2013) used the phage P22 capsid protein to create VLPs that had an influenza antigen in the interior of the VLP. These were able to provide mice with protective immunity to the influenza virus.

Phage protein vaccines would seem to offer significant advantages over traditional vaccines in safe production since phages are unable to infect human cells, are easy to modify to 'fine-tune' protein antigen sequences and do not need animal cells, which may contain other viruses, to be produced. However, in spite of being studied in research laboratories for over 20 years, no vaccines are being made using this technology. Elsewhere (Hyman, 2012), I have discussed reasons that phage display based materials remain as laboratory demonstrations and many of the same reasons seem to apply to phage-based vaccines. While they do have some

Table 13.6 Examples of phage display vaccines

Pathogen target	Phage system	Reference(s)
HIV	Filamentous phages	Loktev *et al.* (1996)
Yersinia pestis	Phage T4	Tao *et al.* (2013a)
Influenza A	Phage T7	Hashemi *et al.* (2012)
Foot-and-mouth disease	M13 phage	Kim *et al.* (2004)
Candida albicans	fd phage	Yang *et al.* (2005)
Trichinella spiralis	M13 phage	Gu *et al.* (2008)

advantages, they lack the proven record of safety and efficacy seen in conventional vaccines. They also do not seem to offer a pathway to a vaccine that cannot also potentially be developed by more standardized means as both approaches depend on providing the same antigens to an immune system. Their biological nature also adds a level of additional regulatory complexity as an injectable component that lacks a well-established safety record.

Much of this same analysis applies to phage DNA vaccines which are mainly also studied in research labs only. At least one company, BigDNA Ltd., was formed to develop phage DNA vaccines but sold its intellectual property to Eli Lilly several years ago and moved on to work to develop cancer pharmaceuticals as ICENI Pharma (http://www.icenipharma.com/). Subsequently, Eli Lilly's animal health division, Elanco Animal Health, announced a joint venture with EnBiotix (www.enbiotix.com), an engineered phage company focusing on antimicrobials, to develop phage DNA vaccines for use in animals (Business Wire, 2016).

Conclusions

Bacteriophages had a major role in the beginnings of biotechnology and new applications continue to be studied and developed. In this chapter I have highlighted some of these although there are other areas that remain less developed such as phages as immune system modulators (Nicastro *et al.*, 2016; Górski *et al.*, 2017); phages as delivery vehicles for therapeutic molecules (Clark and March, 2006; Haq *et al.*, 2012; Bárdy *et al.*, 2016); polyvalent (broad host range) phages as delivery vehicles to uncharacterized environmental bacteria (Yu *et al.*, 2015); among others. Pires and colleagues (Pires *et al.*, 2016) have recently published a broad overview of the methods used to modify bacteriophages for various methods as well as many of the potential uses. This and the other review articles referenced in this chapter provide much more detail on the rich possibilities for phage to continue to be an important source of new technologies. No doubt they will continue to be used in research and industry for decades to come. Further, as we learn more about viruses of microorganisms other than bacteria, they too will likely become the bases of biotechnology applications as well.

References

Adams, M.H. (1959). Bacteriophages. (New York: InterScience).

Aghebati-Maleki, L., Bakhshinejad, B., Baradaran, B., Motallebnezhad, M., Aghebati-Maleki, A., Nickho, H., Yousefi, M., and Majidi, J. (2016). Phage display as a promising approach for vaccine development. J. Biomed. Sci. *23*, 66.

Ahmad, Z.A., Yeap, S.K., Ali, A.M., Ho, W.Y., Alitheen, N.B., and Hamid, M. (2012). scFv antibody: principles and clinical application. Clin. Dev. Immunol. *2012*, 980250. https://doi.org/10.1155/2012/980250.

Albert, H., Trollip, A., Seaman, T., and Mole, R.J. (2004). Simple, phage-based (FASTPplaque) technology to determine rifampicin resistance of *Mycobacterium tuberculosis* directly from sputum. Int. J. Tuberc. Lung Dis. *8*, 1114–1119.

Anany, H., Chen, W., Pelton, R., and Griffiths, M.W. (2011). Biocontrol of *Listeria monocytogenes* and *Escherichia coli* O157:H7 in meat by using phages immobilized on modified cellulose membranes. Appl. Environ. Microbiol. *77*, 6379–6387. https://doi.org/10.1128/AEM.05493-11.

Arya, S.K., Singh, A., Naidoo, R., Wu, P., McDermott, M.T., and Evoy, S. (2011). Chemically immobilized T4-bacteriophage for specific *Escherichia coli* detection using surface plasmon resonance. Analyst *136*, 486–492. https://doi.org/10.1039/c0an00697a.

Avery, G. (2013). MicroPhage Files for Bankruptcy Protection. Denver Business Journal. https://www.bizjournals.com/denver/news/2013/01/02/microphage-files-for-bankruptcy.html, Last accessed 12-31-2017.

Balasubramanian, S., Sorokulova, I.B., Vodyanoy, V.J., and Simonian, A.L. (2007). Lytic phage as a specific and selective probe for detection of *Staphylococcus aureus* – A surface plasmon resonance spectroscopic study. Biosens Bioelectron 22, 948–955.

Barbas, C.F., Burton, D.R., and Silverman, G.J. (2004). Phage Display: A Laboratory Manual. (NY: Cold Spring Harbor Laboratory Press, Cold Spring Harbor, NY).

Bárdy, P., Pantůček, R., Benešík, M., and Doškař, J. (2016). Genetically modified bacteriophages in applied microbiology. J. Appl. Microbiol. *121*, 618–633. https://doi.org/10.1111/jam.13207.

Bartual, S.G., Garcia-Doval, C., Alonso, J., Schoehn, G., and van Raaij, M.J. (2010a). Two-chaperone assisted soluble expression and purification of the bacteriophage T4 long tail fibre protein gp37. Protein Expr. Purif. *70*, 116–121. https://doi.org/10.1016/j.pep.2009.11.005.

Bartual, S.G., Otero, J.M., Garcia-Doval, C., Llamas-Saiz, A.L., Kahn, R., Fox, G.C., and van Raaij, M.J. (2010b). Structure of the bacteriophage T4 long tail fiber receptor-binding tip. Proc. Natl. Acad. Sci. U.S.A. *107*, 20287–20292. https://doi.org/10.1073/pnas.1011218107.

Bhowmick, T., Mirrett, S., Reller, L.B., Price, C., Qi, C., Weinstein, M.P., and Kirn, T.J. (2013). Controlled multicenter evaluation of a bacteriophage-based method for rapid detection of *Staphylococcus aureus* in positive blood cultures. J. Clin. Microbiol. *51*, 1226–1230. https://doi.org/10.1128/JCM.02967-12.

Bickle, T.A., and Krüger, D.H. (1993). Biology of DNA restriction. Microbiol. Rev. *57*, 434–450.

Billington, C., Hudson, J.A., and D'Sa, E. (2014). Prevention of bacterial foodborne disease using nanobiotechnology. Nanotechnol. Sci. Appl. *7*, 73–83. https://doi.org/10.2147/NSA.S51101.

Bordet, J.,and Ciuca, M. (1921). Determinisme de l'autolyse microbienne transmissible. Compt. Rend. Soc. Biol. *84*, 276–278.

Brzozowska, E., Śmietana, M., Koba, M., Górska, S., Pawlik, K., Gamian, A., and Bock, W.J. (2015). Recognition of bacterial lipopolysaccharide using bacteriophage-adhesin-coated long-period gratings. Biosens. Bioelectron. *67*, 93–99. https://doi.org/10.1016/j.bios.2014.07.027.

Business Wire (2016). Elanco, EnBiotix Establish Collaboration Exploring Engineered Phage Technology. Business Wire. https://www.businesswire.com/news/home/20160616005221/en/Elanco-EnBiotix-Establish-Collaboration-Exploring-Engineered-Phage, Last accessed 12-31-2017.

Cairns, J., Stent, G., and Watson, J.D. (1966). Phage and the Origins of Molecular Biology. (Cold Spring Harbor Laboratory Press, Cold Spring Harbor).

Cappillino, M., Shivers, R.P., Brownell, D.R., Jacobson, B., King, J., Kocjan, P., Koeris, M., Tekeian, E., Tempesta, A., Bowers, J., *et al.* (2015). Sample6 DETECT/L: an in-plant, in-shift, enrichment-free *Listeria* environmental assay. J AOAC Int *98*, 436–444. https://doi.org/10.5740/jaoacint.14-213.

Chen, X., and Dreskin, S.C. (2017). Application of phage peptide display technology for the study of food allergen epitopes. Mol. Nutr. Food Res. *61*, 1600568. https://doi.org/10.1002/mnfr.201600568.

Clark, J., Abedon, S.T., and Hyman, P. (2012). Phages as therapeutic delivery vehicles. In Bacteriophages in Health and Disease, Hyman, P., and Abedon, S.T. eds. (CABI Press, UK), pp. 86–100.

Clark, J.R., Bartley, K., Jepson, C.D., Craik, V., and March, J.B. (2011). Comparison of a bacteriophage-delivered DNA vaccine and a commercially available recombinant protein vaccine against hepatitis B. FEMS Immunol. Med. Microbiol. *61*, 197–204. https://doi.org/10.1111/j.1574-695X.2010.00763.x.

Clark, J.R., and March, J.B. (2004). Bacteriophage-mediated nucleic acid immunisation. FEMS Immunol. Med. Microbiol. *40*, 21–26.

Clark, J.R., and March, J.B. (2006). Bacteriophages and biotechnology: vaccines, gene therapy and antibacterials. Trends Biotechnol. *24*, 212–218.

Collins, J., and Hohn, B. (1978). Cosmids: a type of plasmid gene-cloning vector that is packageable in vitro in bacteriophage lambda heads. Proc. Natl. Acad. Sci. U.S.A. *75*, 4242–4246.

Cox, C.R. (2012). Bacteriophage-based methods of baterial detection and identification. In Bacteriophages in Health and Disease, Hyman, P., and Abedon, S.T. eds. (CABI Press, UK).

Criscuolo, E., Spadini, S., Lamanna, J., Ferro, M., and Burioni, R. (2017). Bacteriophages and their immunological applications against infectious threats. J. Immunol. Res. *2017*, 3780697. https://doi.org/10.1155/2017/3780697.

Czapar, A.E., and Steinmetz, N.F. (2017). Plant viruses and bacteriophages for drug delivery in medicine and biotechnology. Curr. Opin. Chem. Biol. *38*, 108–116.

Denyes, J.M., Dunne, M., Steiner, S., Mittelviefhaus, M., Weiss, A., Schmidt, H., Klumpp, J., and Loessner, M.J. (2017). Modified bacteriophage S16 long tail fiber proteins for rapid and specific immobilization and detection of *Salmonella* cells. Appl. Environ. Microbiol. *83*, e00277–17.

Derda, R., Tang, S.K., Li, S.C., Ng, S., Matochko, W., and Jafari, M.R. (2011). Diversity of phage-displayed libraries of peptides during panning and amplification. Molecules *16*, 1776–1803. https://doi.org/10.3390/molecules16021776.

Dultsev, F.N., Speight, R.E., Fiorini, M.T., Blackburn, J.M., Abell, C., Ostanin, V.P., and Klenerman, D. (2001). Direct and quantitative detection of bacteriophage by 'hearing' surface detachment using a quartz crystal microbalance. Anal. Chem. *73*, 3935–3939.

Fischetti, V.A. (2010). Bacteriophage endolysins: a novel anti-infective to control Gram-positive pathogens. Int. J. Med. Microbiol. *300*, 357–362. https://doi.org/10.1016/j.ijmm.2010.04.002.

Frenzel, A., Kügler, J., Helmsing, S., Meier, D., Schirrmann, T., Hust, M., and Dübel, S. (2017). Designing Human Antibodies by Phage Display. Transfus Med Hemother *44*, 312–318. https://doi.org/10.1159/000479633.

Geider, K., Hohmeyer, C., Haas, R., and Meyer, T.F. (1985). A plasmid cloning system utilizing replication and packaging functions of the filamentous bacteriophage fd. Gene 33, 341–349.

Ghaemi, A., Soleimanjahi, H., Gill, P., Hassan, Z.M., Razeghi, S., Fazeli, M., and Razavinikoo, S.M. (2011). Protection of mice by a λ-based therapeutic vaccine against cancer associated with human papillomavirus type 16. Intervirology *54*, 105–112. https://doi.org/10.1159/000320197.

Gomez-Torres,N., Dunne,M., Garde,S., Meijers,R., Narbad,A., Avila,M., and Mayer,M.J. (2018). Development of a specific fluorescent phage endolysin for in situ detection of *Clostridium* species associated with cheese spoilage. Microb. Biotechnol. 11. 332–345. https://doi.org/10.1111/1751-7915.12883.

Górski, A., Dąbrowska, K., Międzybrodzki, R., Weber-Dąbrowska, B., Łusiak-Szelachowska, M., Jończyk-Matysiak, E., and Borysowski, J. (2017). Phages and immunomodulation. Future Microbiol. *12*, 905–914. https://doi.org/10.2217/fmb-2017-0049.

Gu, Y., Li, J., Zhu, X., Yang, J., Li, Q., Liu, Z., Yu, S., and Li, Y. (2008). Trichinella spiralis: characterization of phage-displayed specific epitopes and their protective immunity in BALB/c mice. Exp. Parasitol. *118*, 66–74.

Handa, H., Gurczynski, S., Jackson, M.P., Auner, G., and Mao, G. (2008). Recognition of *Salmonella typhimurium* by immobilized phage P22 monolayers. Surf. Sci. *602*, 1392–1400. https://doi.org/10.1016/j.susc.2008.01.036.

Haq, I.U., Chaudhry, W.N., Akhtar, M.N., Andleeb, S., and Qadri, I. (2012). Bacteriophages and their implications on future biotechnology: a review. Virol. J. *9*, 9. https://doi.org/10.1186/1743-422X-9-9.

Hashemi, H., Bamdad, T., Jamali, A., Pouyanfard, S., and Mohammadi, M.G. (2010). Evaluation of humoral and cellular immune responses against HSV-1 using genetic immunization by filamentous phage particles: a comparative approach to conventional DNA vaccine. J. Virol. Methods *163*, 440–444. https://doi.org/10.1016/j.jviromet.2009.11.008.

Hashemi, H., Pouyanfard, S., Bandehpour, M., Noroozbabaei, Z., Kazemi, B., Saelens, X., and Mokhtari-Azad, T. (2012). Immunization with M2e-displaying T7 bacteriophage nanoparticles protects against influenza A virus challenge. PLOS ONE *7*, e45765. https://doi.org/10.1371/journal.pone.0045765.

Hasson, S.S.A.A., Al-Busaidi, J.K.Z., and Sallam,T.A. (2017). The past, current and future trends in DNA vaccine immunisations. Asian Pacific J. Trop. Biomed. *5*, 344–353.

Haylock, A.K., Nilvebrant, J., Mortensen, A., Velikyan, I., Nestor, M., and Falk, R. (2017). Generation and evaluation of antibody agents for molecular imaging of CD44v6-expressing cancers. Oncotarget *8*, 65152–65170. https://doi.org/10.18632/oncotarget.17996.

Henry, K.A., Arbabi-Ghahroudi, M., and Scott, J.K. (2015). Beyond phage display: non-traditional applications of the filamentous bacteriophage as a vaccine carrier, therapeutic biologic, and bioconjugation scaffold. Front. Microbiol. *6*, 755. https://doi.org/10.3389/fmicb.2015.00755.

Hokama,Y., and Nakamura,R.M. (1982). Immunology and Immunopathology: Basic Concepts. (Little, Brown and Company, Boston).

Hu, L.Y., Kelly, K.A., and Sutcliffe, J.L. (2017). High-throughput approaches to the development of molecular imaging agents. Mol. Imag. Biol. *19*, 163–182. https://doi.org/10.1007/s11307-016-1016-z.

Huang, S., Yang, H., Lakshmanan, R.S., Johnson, M.L., Wan, J., Chen, I.H., Wikle, H.C., Petrenko, V.A., Barbaree, J.M., and Chin, B.A. (2009). Sequential detection of *Salmonella typhimurium* and *Bacillus anthracis* spores using magnetoelastic biosensors. Biosens Bioelectron *24*, 1730–1736. https://doi.org/10.1016/j.bios.2008.09.006.

Hyman, P. (2012). Bacteriophages and nanostructured materials. Adv. Appl. Microbiol. *78*, 55–73. https://doi.org/10.1016/B978-0-12-394805-2.00003-8.

Hyman, P., and Abedon, S.T. (2010). Bacteriophage host range and bacterial resistance. Adv. Appl. Microbiol. *70*, 217–248. https://doi.org/10.1016/S0065-2164(10)70007-1.

Jafari, N., and Abediankenari, S. (2015). Phage particles as vaccine delivery vehicles: concepts, applications and prospects. Asian Pac. J. Cancer Prev. *16*, 8019–8029.

Javed, M.A., Poshtiban, S., Arutyunov, D., Evoy, S., and Szymanski, C.M. (2013). Bacteriophage receptor binding protein based assays for the simultaneous detection of *Campylobacter jejuni* and *Campylobacter coli*. PLOS ONE *8*, e69770. https://doi.org/10.1371/journal.pone.0069770.

Johnson, M.L., Wan, J., Huang, S., Cheng, Z., Petrenko, V.A., Kim, D.J., Chen, I.H., Barbaree, J.M., Hong, J.W., and Chin, B.A. (2008). A wireless biosensor using microfabricated phage-interfaced magnetoelastic particles. Sens. Actuators A *144*, 38–47.

Karimi, M., Mirshekari, H., Moosavi Basri, S.M., Bahrami, S., Moghoofei, M., and Hamblin, M.R. (2016). Bacteriophages and phage-inspired nanocarriers for targeted delivery of therapeutic cargos. Adv. Drug Deliv. Rev. *106*, 45–62.

Kim, Y.J., Lebreton, F., Kaiser, C., Crucière, C., and Rémond, M. (2004). Expression of a foot-and-mouth disease virus immunodominant epitope by a filamentous bacteriophage vector. Arch. Virol. *149*, 365–377. https://doi.org/10.1007/s00705-003-0192-8.

Klumpp, J., and Loessner, M.J. (2014). Detection of bacteria with bioluminescent reporter bacteriophage. Adv. Biochem. Eng. Biotechnol. *144*, 155–171. https://doi.org/10.1007/978-3-662-43385-0_5.

Kong, M., Shin, J.H., Heu, S., Park, J.K., and Ryu, S. (2017). Lateral flow assay-based bacterial detection using engineered cell wall binding domains of a phage endolysin. Biosens. Bioelectron. *96*, 173–177.

Kong, M., Sim, J., Kang, T., Nguyen, H.H., Park, H.K., Chung, B.H., and Ryu, S. (2015). A novel and highly specific phage endolysin cell wall binding domain for detection of *Bacillus cereus*. Eur. Biophys. J. *44*, 437–446. https://doi.org/10.1007/s00249-015-1044-7.

Kretzer, J.W., Lehmann, R., Schmelcher, M., Banz, M., Kim, K.P., Korn, C., and Loessner, M.J. (2007). Use of high-affinity cell wall-binding domains of bacteriophage endolysins for immobilization and separation of bacterial cells. Appl. Environ. Microbiol. 73, 1992–2000.

Kuhn, P., Fühner, V., Unkauf, T., Moreira, G.M., Frenzel, A., Miethe, S., and Hust, M. (2016). Recombinant antibodies for diagnostics and therapy against pathogens and toxins generated by phage display. Proteomics Clin Appl *10*, 922–948. https://doi.org/10.1002/prca.201600002.

Lakshmanan, R.S., Guntupalli, R., Hu, J., Kim, D.J., Petrenko, V.A., Barbaree, J.M., and Chin, B.A. (2007). Phage immobilized magnetoelastic sensor for the detection of *Salmonella typhimurium*. J. Microbiol. Methods *71*, 55–60.

Ling, Y., Liu, W., Clark, J.R., March, J.B., Yang, J., and He, C. (2011). Protection of mice against *Chlamydophila abortus* infection with a bacteriophage-mediated DNA vaccine expressing the major outer membrane protein. Vet. Immunol. Immunopathol. *144*, 389–395. https://doi.org/10.1016/j.vetimm.2011.08.003.

Loessner, M.J., Rees, C.E., Stewart, G.S., and Scherer, S. (1996). Construction of luciferase reporter bacteriophage A511::luxAB for rapid and sensitive detection of viable *Listeria* cells. Appl. Environ. Microbiol. *62*, 1133–1140.

Loktev, V.B., Ilyichev, A.A., Eroshkin, A.M., Karpenko, L.I., Pokrovsky, A.G., Pereboev, A.V., Svyatchenko, V.A., Ignat'ev, G.M., Smolina, M.I., Melamed, N.V., *et al.* (1996). Design of immunogens as components of a new generation of molecular vaccines. J. Biotechnol. *44*, 129–137.

Lone, A., Anany, H., Hakeem, M., Aguis, L., Avdjian, A.C., Bouget, M., Atashi, A., Brovko, L., Rochefort, D., and Griffiths, M.W. (2016). Development of prototypes of bioactive packaging materials based on immobilized bacteriophages for control of growth of bacterial pathogens in foods. Int. J. Food Microbiol. *217*, 49–58. https://doi.org/10.1016/j.ijfoodmicro.2015.10.011.

Marti, R., Zurfluh, K., Hagens, S., Pianezzi, J., Klumpp, J., and Loessner, M.J. (2013). Long tail fibres of the novel broad-host-range T-even bacteriophage S16 specifically recognize Salmonella OmpC. Mol. Microbiol. *87*, 818–834. https://doi.org/10.1111/mmi.12134.

Martins, I.M., Reis, R.L., and Azevedo, H.S. (2016). Phage display technology in biomaterials engineering: progress and opportunities for applications in regenerative medicine. ACS Chem. Biol. *11*, 2962–2980. https://doi.org/10.1021/acschembio.5b00717.

Mejri, M.B., Baccar, H., Baldrich, E., Del Campo, F.J., Helali, S., Ktari, T., Simonian, A., Aouni, M., and Abdelghani, A. (2010). Impedance biosensing using phages for bacteria detection: generation of dual signals as the clue for in-chip assay confirmation. Biosens. Bioelectron. *26*, 1261–1267. https://doi.org/10.1016/j.bios.2010.06.054.

Na, H., Kong, M., and Ryu, S. (2016). Characterization of LysPBC4, a novel *Bacillus cereus*-specific endolysin of bacteriophage PBC4. FEMS Microbiol. Lett. *363*, fnw092. https://doi.org/10.1093/femsle/fnw092.

Nagano, K., and Tsutsumi, Y. (2016). Development of novel drug delivery systems using phage display technology for clinical application of protein drugs. Proc. Jpn. Acad., Ser. B, Phys. Biol. Sci. *92*, 156–166. https://doi.org/10.2183/pjab.92.156.

Nanduri, V., Balasubramanian, S., Sista, S., Vodyanoy, V.J., and Simonian, A.L. (2007a). Highly sensitive phage-based biosensor for the detection of beta-galactosidase. Anal. Chim. Acta *589*, 166–172.

Nanduri, V., Sorokulova, I.B., Samoylov, A.M., Simonian, A.L., Petrenko, V.A., and Vodyanoy, V. (2007b). Phage as a molecular recognition element in biosensors immobilized by physical adsorption. Biosens. Bioelectron. *22*, 986–992.

Narsaiah, K., Jha, S.N., Bhardwaj, R., Sharma, R., and Kumar, R. (2012). Optical biosensors for food quality and safety assurance-a review. J. Food. Sci. Technol. *49*, 383–406. https://doi.org/10.1007/s13197-011-0437-6.

Nicastro, J., Wong, S., Khazaei, Z., Lam, P., Blay, J., and Slavcev, R.A. (2016). Bacteriophage Applications – Historical Perspective and Future Potential. (Springer International Publishing, Switzerland).

Olsen, E.V., Sorokulova, I.B., Petrenko, V.A., Chen, I.H., Barbaree, J.M., and Vodyanoy, V.J. (2006). Affinity-selected filamentous bacteriophage as a probe for acoustic wave biodetectors of *Salmonella typhimurium*. Biosens. Bioelectron. *21*, 1434–1442.

Ou, C., Tian, D., Ling, Y., Pan, Q., He, Q., Eko, F.O., and He, C. (2013). Evaluation of an ompA-based phage-mediated DNA vaccine against *Chlamydia abortus* in piglets. Int. Immunopharmacol. *16*, 505–510. https://doi.org/10.1016/j.intimp.2013.04.027.

Patterson, D.P., Rynda-Apple, A., Harmsen, A.L., Harmsen, A.G., and Douglas, T. (2013). Biomimetic antigenic nanoparticles elicit controlled protective immune response to influenza. ACS Nano 7, 3036–3044. https://doi.org/10.1021/nn4006544.

Petrenko, V.A., and Gillespie, J.W. (2017). Paradigm shift in bacteriophage-mediated delivery of anticancer drugs: from targeted 'magic bullets' to self-navigated 'magic missiles'. Expert. Opin. Drug Deliv. *14*, 373–384. https://doi.org/10.1080/17425247.2016.1218463.

Pires, D.P., Cleto, S., Sillankorva, S., Azeredo, J., and Lu, T.K. (2016). Genetically engineered phages: a review of advances over the last decade. Microbiol. Mol. Biol. Rev. *80*, 523–543. https://doi.org/10.1128/MMBR.00069-15.

Poshtiban, S., Javed, M.A., Arutyunov, D., Singh, A., Banting, G., Szymanski, C.M., and Evoy, S. (2013). Phage receptor binding protein-based magnetic enrichment method as an aid for real time PCR detection of foodborne bacteria. Analyst *138*, 5619–5626. https://doi.org/10.1039/c3an01100c.

Rakonjac, J., Bennett, N.J., Spagnuolo, J., Gagic, D., and Russel, M. (2011). Filamentous bacteriophage: biology, phage display and nanotechnology applications. Curr Issues Mol Biol *13*, 51–76.

Roach, D.R., and Donovan, D.M. (2015). Antimicrobial bacteriophage-derived proteins and therapeutic applications. Bacteriophage *5*, e1062590. https://doi.org/10.1080/21597081.2015.1062590.

Rozand, C., and Feng, P.C. (2009). Specificity analysis of a novel phage-derived ligand in an enzyme-linked fluorescent assay for the detection of *Escherichia coli* O157:H7. J. Food Prot. *72*, 1078–1081.

Saeedi, A., Ghaemi, A., Tabarraei, A., Moradi, A., Gorji, A., Semnani, S., Soleimanjahi, H., Adli, A.H., Hosseini, S.Y., and Vakili, M.A. (2014). Enhanced cell immune responses to hepatitis C virus core by novel heterologous DNA prime/lambda nanoparticles boost in mice. Virus Genes *49*, 11–21. https://doi.org/10.1007/s11262-014-1070-z.

Sambrook, J., Frisch, E.F., and Maniatis, T. (1989). Molecular Cloning: A Laboratory Manual. (Cold Spring Harbor Laboratory Press, Cold Spring Harbor).

Savoye, F., Feng, P., Rozand, C., Bouvier, M., Gleizal, A., and Thevenot, D. (2011). Comparative evaluation of a phage protein ligand assay with real-time PCR and a reference method for the detection of *Escherichia coli* O157:H7 in raw ground beef and trimmings. J. Food Prot. *74*, 6–12. https://doi.org/10.4315/0362-028X.JFP-10-271.

Schmelcher, M., and Loessner, M.J. (2014). Application of bacteriophages for detection of foodborne pathogens. Bacteriophage *4*, e28137. https://doi.org/10.4161/bact.28137.

Schmelcher, M., Shabarova, T., Eugster, M.R., Eichenseher, F., Tchang, V.S., Banz, M., and Loessner, M.J. (2010). Rapid multiplex detection and differentiation of Listeria cells by use of fluorescent phage endolysin cell wall binding domains. Appl. Environ. Microbiol. *76*, 5745–5756. https://doi.org/10.1128/AEM.00801-10.

Schmelcher, M., Tchang, V.S., and Loessner, M.J. (2011). Domain shuffling and module engineering of *Listeria* phage endolysins for enhanced lytic activity and binding affinity. Microb. Biotechnol. *4*, 651–662. https://doi.org/10.1111/j.1751-7915.2011.00263.x.

Shabani, A., Zourob, M., Allain, B., Marquette, C.A., Lawrence, M.F., and Mandeville, R. (2008). Bacteriophage-modified microarrays for the direct impedimetric detection of bacteria. Anal. Chem. *80*, 9475–9482. https://doi.org/10.1021/ac801607w.

Shen, Y., Mitchell, M.S., Donovan, D.M., and Nelson, D.C. (2012). Phage-based enzybiotics. In Bacteriophages in Health and Disease, Hyman, P., and Abedon, S.T. eds. (CABI Press, UK).

Singh, A., Arutyunov, D., McDermott, M.T., Szymanski, C.M., and Evoy, S. (2011). Specific detection of

Campylobacter jejuni using the bacteriophage NCTC 12673 receptor binding protein as a probe. Analyst *136*, 4780–4786. https://doi.org/10.1039/c1an15547d.

Singh, A., Arutyunov, D., Szymanski, C.M., and Evoy, S. (2012). Bacteriophage based probes for pathogen detection. Analyst *137*, 3405–3421. https://doi.org/10.1039/c2an35371g.

Singh, A., Arya, S.K., Glass, N., Hanifi-Moghaddam, P., Naidoo, R., Szymanski, C.M., Tanha, J., and Evoy, S. (2010). Bacteriophage tailspike proteins as molecular probes for sensitive and selective bacterial detection. Biosens. Bioelectron. *26*, 131–138. https://doi.org/10.1016/j.bios.2010.05.024.

Singh, A., Poshtiban, S., and Evoy, S. (2013). Recent advances in bacteriophage based biosensors for food-borne pathogen detection. Sensors *13*, 1763–1786. https://doi.org/10.3390/s130201763.

Singh, S., Saluja, T.P., Kaur, M., and Khilnani, G.C. (2008). Comparative evaluation of FASTPlaque assay with PCR and other conventional in vitro diagnostic methods for the early detection of pulmonary tuberculosis. J. Clin. Lab. Anal. *22*, 367–374. https://doi.org/10.1002/jcla.20264.

Smietana, M., Bock, W.J., Mikulic, P., Ng, A., Chinnappan, R., and Zourob, M. (2011). Detection of bacteria using bacteriophages as recognition elements immobilized on long-period fiber gratings. Opt. Express. *19*, 7971–7978. https://doi.org/10.1364/OE.19.007971.

Smith, G.P., and Petrenko, V.A. (1997). Phage display. Chem. Rev. *97*, 391–410.

Studier, F.W., and Moffatt, B.A. (1986). Use of bacteriophage T7 RNA polymerase to direct selective high-level expression of cloned genes. J. Mol. Biol. *189*, 113–130.

Sun, S., Li, W., Sun, Y., Pan, Y., and Li, J. (2011). A new RNA vaccine platform based on MS2 virus-like particles produced in *Saccharomyces cerevisiae*. Biochem. Biophys. Res. Commun. *407*, 124–128. https://doi.org/10.1016/j.bbrc.2011.02.122.

Tabor, S., and Richardson, C.C. (1985). A bacteriophage T7 RNA polymerase/promoter system for controlled exclusive expression of specific genes. Proc. Natl. Acad. Sci. U.S.A. *82*, 1074–1078.

Tao, P., Mahalingam, M., Kirtley, M.L., van Lier, C.J., Sha, J., Yeager, L.A., Chopra, A.K., and Rao, V.B. (2013a). Mutated and bacteriophage T4 nanoparticle arrayed F1-V immunogens from *Yersinia pestis* as next generation plague vaccines. PLOS Pathog. *9*, e1003495. https://doi.org/10.1371/journal.ppat.1003495.

Tao, P., Mahalingam, M., Marasa, B.S., Zhang, Z., Chopra, A.K., and Rao, V.B. (2013b). In vitro and in vivo delivery of genes and proteins using the bacteriophage T4 DNA packaging machine. Proc. Natl. Acad. Sci. U.S.A. *110*, 5846–5851. https://doi.org/10.1073/pnas.1300867110.

Tay, L.L., Huang, P.J., Tanha, J., Ryan, S., Wu, X., Hulse, J., and Chau, L.K. (2012). Silica encapsulated SERS nanoprobe conjugated to the bacteriophage tailspike protein for targeted detection of *Salmonella*. Chem. Commun. *48*, 1024–1026. https://doi.org/10.1039/c1cc16325f.

Tlili, C., Sokullu, E., Safavieh, M., Tolba, M., Ahmed, M.U., and Zourob, M. (2013). Bacteria screening, viability, and confirmation assays using bacteriophage-impedimetric/loop-mediated isothermal amplification dual-response biosensors. Anal. Chem. *85*, 4893–4901. https://doi.org/10.1021/ac302699x.

Tolba, M., Ahmed, M.U., Tlili, C., Eichenseher, F., Loessner, M.J., and Zourob, M. (2012). A bacteriophage endolysin-based electrochemical impedance biosensor for the rapid detection of Listeria cells. Analyst *137*, 5749–5756. https://doi.org/10.1039/c2an35988j.

Tolba, M., Minikh, O., Brovko, L.Y., Evoy, S., and Griffiths, M.W. (2010). Oriented immobilization of bacteriophages for biosensor applications. Appl. Environ. Microbiol. *76*, 528–535. https://doi.org/10.1128/AEM.02294-09.

Tripathi, S.M., Bock, W.J., Mikulic, P., Chinnappan, R., Ng, A., Tolba, M., and Zourob, M. (2012). Long period grating based biosensor for the detection of *Escherichia coli* bacteria. Biosens. Bioelectron. *35*, 308–312. https://doi.org/10.1016/j.bios.2012.03.006.

van der Merwe, R.G., van Helden, P.D., Warren, R.M., Sampson, S.L., and Gey van Pittius, N.C. (2014). Phage-based detection of bacterial pathogens. Analyst *139*, 2617–2626. https://doi.org/10.1039/c4an00208c.

Waddell, T.E., Franklin, K., Mazzocco, A., and Johnson, R.P. (2009). Preparation and characterization of anti-phage serum. Methods Mol. Biol. *501*, 287–292. https://doi.org/10.1007/978-1-60327-164-6_24.

Walcher, G., Stessl, B., Wagner, M., Eichenseher, F., Loessner, M.J., and Hein, I. (2010). Evaluation of paramagnetic beads coated with recombinant Listeria phage endolysin-derived cell-wall-binding domain proteins for separation of *Listeria monocytogenes* from raw milk in combination with culture-based and real-time polymerase chain reaction-based quantification. Foodborne Pathog. Dis. *7*, 1019–1024. https://doi.org/10.1089/fpd.2009.0475.

Wan, J., Johnson, M.L., Guntupalli, R., Petrenko, V.A., and Chin, B.A. (2007). Detection of *Bacillus anthracis* spores in liquid using phage-based magnetoelastic micro-resonators. Sens. Actuators B *127*, 559–566.

Wang, C., Sauvageau, D., and Elias, A. (2016). Immobilization of active bacteriophages on polyhydroxyalkanoate surfaces. ACS Appl. Mater. Interfaces *8*, 1128–1138. https://doi.org/10.1021/acsami.5b08664.

Williams, M.L., and LeJeune, J.T. (2012). Phages and bacterial epidemiology. In Bacteriophages in Health and Disease, Hyman, P., and Abedon, S.T. eds. (CABI Press, UK).

Wu, H., Pfarr, D.S., Tang, Y., An, L.L., Patel, N.K., Watkins, J.D., Huse, W.D., Kiener, P.A., and Young, J.F. (2005). Ultra-potent antibodies against respiratory syncytial virus: effects of binding kinetics and binding valence on viral neutralization. J. Mol. Biol. *350*, 126–144.

Yang, L.M., Tam, P.Y., Murray, B.J., McIntire, T.M., Overstreet, C.M., Weiss, G.A., and Penner, R.M. (2006). Virus electrodes for universal biodetection. Anal. Chem. *78*, 3265–3270. https://doi.org/10.1021/ac052287u.

Yang, Q., Wang, L., Lu, D.N., Gao, R.J., Song, J.N., Hua, P.Y., and Yuan, D.W. (2005). Prophylactic vaccination with phage-displayed epitope of *C. albicans* elicits protective immune responses against systemic candidiasis in C57BL/6 mice. Vaccine *23*, 4088–4096.

Yu, P., Mathieu, J., Li, M., Dai, Z., and Alvarez, P.J. (2015). Isolation of polyvalent bacteriophages by sequential multiple-host approaches. Appl. Environ. Microbiol. *82*, 808–815. https://doi.org/10.1128/AEM.02382-15.

Zhu, H., White, I.M., Suter, J.D., and Fan, X. (2008). Phage-based label-free biomolecule detection in an opto-fluidic ring resonator. Biosens. Bioelectron. *24*, 461–466. https://doi.org/10.1016/j.bios.2008.04.028.

Ziebell, K., Chui, L., King, R., Johnson, S., Boerlin, P., and Johnson, R.P. (2017). Subtyping of Canadian isolates of *Salmonella* Enteritidis using Multiple Locus Variable Number Tandem Repeat Analysis (MLVA) alone and in combination with Pulsed-Field Gel Electrophoresis (PFGE) and phage typing. J. Microbiol. Methods *139*, 29–36.

Viruses as Biocontrol Agents of Microorganisms

14

Diana R. Alves[1], Jason R. Clark[2] and Stephen T. Abedon[1*]

[1]Department of Microbiology, The Ohio State University, Mansfield, OH, USA.
[2]FixedPhage Ltd, Glasgow, UK.

*Correspondence: abedon.1@osu.edu

https://doi.org/10.21775/9781910190852.14

Abstract

Biocontrol is the application of organisms, or substances that organisms produce, to reduce or eliminate populations of unwanted species. Within the context of environments, this application is equivalent to the use of pesticides, with the unwanted organism serving as the pest. Within the context of medicine, this application is equivalent to that of, especially, the use of antimicrobial drugs. Typically, for any such agents, the application of greater amounts will result in a greater negative impact on target organisms, but levels of application will be limited by costs, convenience, or potential to cause harm to the environment or to the patient being treated. Biocontrol agents thus may be deemed superior, for example versus the use of synthetic pesticides or antimicrobials, should biocontrol agents supply some combination of greater efficacy potential, lower cost, greater convenience, or lower potential to cause various forms of harm. In terms of viruses as biocontrol agents, we can distinguish between whole viruses versus their products or derivatives. Potential targets can be further differentiated in terms of what can be described as the six kingdoms: Bacteria, Archaea, Fungi, Protista, Animalia, and Plantae. In this chapter we touch upon all of these kingdoms as possessing potential targets for virus-mediated biocontrol; nevertheless, consistent with the theme of this monograph, we focus particularly on the virus-mediated biocontrol of microorganisms. Given its substantial domination of the overall area of study, much of the review is derived from consideration of bacteriophage use against members of domain Bacteria. This is seen within the context of either phage-mediated biocontrol or instead the sub-category of such biocontrol known as phage therapy.

Introduction

The use of organisms or their products to combat other organisms, that is, as biological control or biocontrol agents, allows for exploration of a greater range of potential inhibitory agents and can also result in the use of agents that display lower toxicities to environments, to patients, or to both. Ideal control agents are relatively easily discovered, convenient to use, otherwise economical, and display selective toxicities. The latter describes especially an ability to interfere with the functioning of target organisms but little or at least somewhat reduced potential to interfere with the functioning of basically everything else. Among biocontrol agents there are predatory insects, various cellular microbes whose growth or presence can inhibit the functioning of other organisms, and viruses. Viruses, due to their ubiquity in environments along with their often narrow host ranges, and resulting selective toxicity, have been employed particularly within the context of medicine as antibacterial agents, although also against various pest animals. Within this context as biocontrol agents, viruses can be similar to chemical control agents in terms of a combination of their handling and use as direct antagonists to target organisms.

Here we consider the use of viruses as antimicrobial biological control agents. We divide up the discussion into three parts, first consideration of general principles of virus use as biocontrol agents such as their pharmacology or equivalent concepts; their use or potential use against plants, animals, protists, fungi, Archaea, and Bacteria; and then issues of commercialization. Lastly, we provide a discussion of the use of virus products rather than whole viruses as biocontrol agents, particularly the use of bacteriophage-derived proteins. See Harper (2006, 2013) for earlier and broader overviews of the general topic of the use of microorganisms as biocontrol agents. An overview of the process of virus-mediated biocontrol is presented in Fig. 14.1.

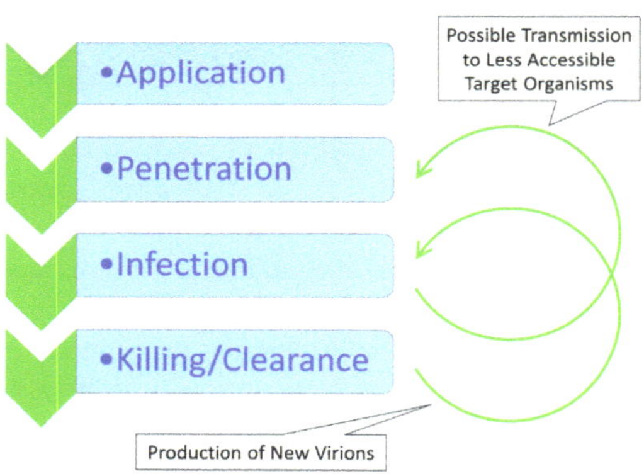

Figure 14.1 Standard virus infection cycle as viewed from the perspective of virus-mediated biocontrol. As with any control agent, the first step of use is some form of application to the to-be-treated individual or environment. The control agent must then reach the target organism (following 'Penetration'). With viruses, biocontrol is mediated via the infection of target organisms. This infection can lead to the production of new virus particles (indicated feedback loops) along with the killing/clearance of target organisms. These new virus particles may then be available for further infection as well as further physical penetration to otherwise less accessible target organism populations, in both cases potentially resulting in greater impact by the biocontrolling viruses than would be the case were killing/clearance not accompanied by the generation of new virion particles.

The 'pharmacology' of virus-mediated biocontrol

The pharmacology of a 'drug', such as a drug that is employed to control microbial agents, typically is distinguished into pharmacodynamic versus pharmacokinetic aspects (Fig. 14.2). Pharmacodynamics refers to both intended and unintended 'big picture' consequences of treatment. For a standard drug, these will include efficacy – does the drug 'work'? – versus side effects (does the drug 'harm'?). In both cases these represent a drug's impact on the body. Pharmacokinetics, by contrast, represents a body's impact on a drug, often in ways that impact the potential for a drug to display efficacy but also the potential for the drug to display side effects. These in a sense are 'little picture' aspects of drug action, that is, important underlying bases of subsequent desired efficacy or undesired side effects. In addition to pharmacodynamic and pharmacokinetic aspects of drugs, there also exist issues of the ease of drug development, the overall costs of producing the drug, and various costs associated with drug use. The latter can refer to the convenience as well as the overall pleasantness of a drug's use, but also downstream aspects such as a drug's impact on environments into which it may be released such as following excretion from the body. A general scheme of the concepts of pharmacodynamics and pharmacokinetics of a biological control agent is presented in Fig. 14.2.

The question of potential efficacy of viruses as 'drugs' (Abedon, 2014a, 2017b), or more broadly as biocontrol agents, we leave to subsequent sections. Focusing on side effects, we can generalize to include not just the negative impact of agents on the bodies of patients but in fact also the negative impact of the application of these agents on treated or downstream environments, e.g. Meaden and Koskella (2013). Such negative impacts typically although not necessarily exclusively will involve damage to species other than target organisms. In strictly pharmacological terms, this would be damage to the body of the treated individual since 'body' in a pharmacological sense can be considered to also include the associated microbiota. Damage to body-associated microorganisms, such as can result in a dysbiosis, therefore also can either represent or give rise to side effects. The issue of damage to non-target species is equivalently applicable to the treatment of environments. As a rule,

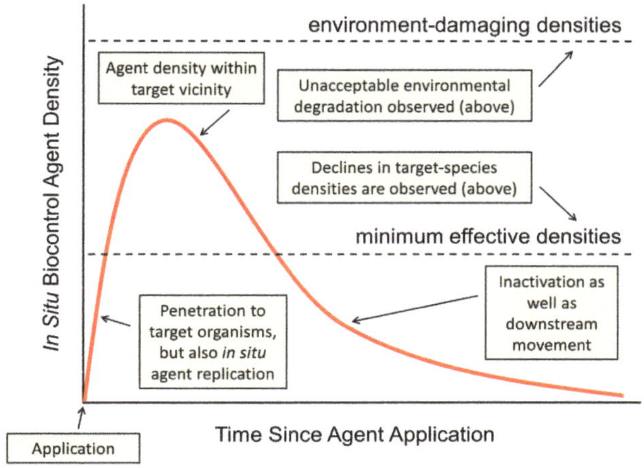

Figure 14.2 Summary of the 'pharmacokinetics' and 'pharmacodynamics' as applied to biocontrol agents. These processes involve the bringing of biocontrol agents to target species and their reaching sufficient agent densities ('little picture' issues). This is application via a combination of agent dispensing or spreading into an environment, agent penetration within that environment to the target species, and, for biocontrol agents as whole organisms, a potential to increase in density in the course of agent reproduction. The result is some agent density within the vicinity of the target species that may or may not – as a 'big picture' issue – be sufficient to control the target species (may or may not, that is, be in excess of some minimum effective density) and may or may not be sufficient to result in environmental harm (may or may not reach environment-damaging densities). Ideally, for biological control agents, their use will be less likely to reach densities that result in environmental harm than will be the case for more traditional, particularly chemical-based control agents. Lastly, agent densities will be expected to eventually decline within environments, either due to inactivation or instead due to agent movement out of the environment being treated. If populations of target species persist, then this tendency for biocontrol agents to decline in number or density may be countered via further agent application. This figure has been adapted from those presented in Abedon (2014a,b).

therefore, it can be preferable to employ biocontrol agents that inherently impact only target organisms, i.e. as can be useful for antimicrobial drugs generally (Curtright and Abedon, 2011), and this can be accomplished in the case of viruses by employing strains that display relatively narrow host ranges (Hyman and Abedon, 2010, Koskella and Meaden, 2013). See Fig. 14.3 for discussion of relationships between the range of species an agent impacts and its economic viability, ease of use, and safety/environmental benignness.

While pharmacodynamics is a description of a drug's impact on bodies, here generalized to biocontrol agent impact on environments, pharmacokinetics instead, following that same theme, is a description of an environment's impact on a drug. So too, then, we may assume that, to at least a first approximation, this may be generalized to include an environment's impact on a biocontrol agent. What do such impacts consist of? They include the ability to move into an environment, the ability to move within an environment, modification of the agent by the environment, and the ability to move out of an environment. In strictly pharmacokinetic terms, these are described as absorption (as in absorption into the blood such as from the small intestine), distribution (as in distribution to non-blood tissues), metabolism (as in the chemical modification of a drug), and excretion (that is, out of the body). For a biocontrol agent being used in a human or other animal, productive movement therefore is particularly towards the target species and may or may not include, for a drug, some combination of movement into and then out of the blood. To describe this movement towards the target organism, we will use the more general concept of 'penetration' rather than absorption and distribution (i.e. Fig. 14.2).

Within the context of treatment efficacy, movement out of the treated environment can be less of a concern for biocontrol agents. This is because, unlike drugs, active processes – particularly excretion via the kidneys – tend to be less effective against whole-organism biological control agents. As less

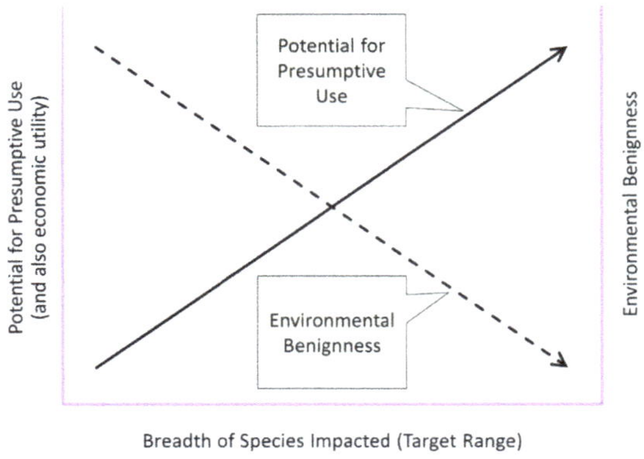

Figure 14.3 The selective toxicity of 'drugs' varies and can be viewed, in part, in terms of the range of species affected. An antimicrobial drug that negatively affects a patient's tissues thus is displaying a toxicity against a species other than the target microorganism, in this case towards the patient. Alternatively, an antimicrobial agent can affect normal microbiota species that are not that target species, resulting also in side effects (i.e. toxicities as more generally considered). Here we have graphed the breadth of species impacted as the independent variable, i.e., as increasing going from left to right. On the dependent axis to the left, we consider antimicrobial agent utility, either overall economically or in terms ease of use. Specifically, agents that display broader target ranges tend to have the potential to make manufacturers more money (in part because the agent may be used against a greater breadth of targets) and often may be used with relatively little prior characterization of the target organism (presumptive use). That relationship is illustrated with the solid-line. On the dependent access to the right we consider agent safety or, more generally, its benignness to environments. The tendency for agents to be more harmful to either patients or environments given a greater target range breadth – that is, to display lower environmental benignness – is illustrated with the dashed line.

frequently considered, however, such movement can be relevant from an downstream environment perspective, even for the use of drugs, as they too can have negative impacts on these subsequently encountered environments, such as following the discharge of antibiotics within sewage effluent into rivers (Ding and He, 2010; Arnold *et al.*, 2014). For biocontrol agents generally, the equivalent concern is one of movement out of the area being treated. That can be an issue, of course, for chemical pesticide use as well, although generally purely chemical agents at least are unable to subsequently reproduce.

That last point brings us to the fourth key aspect of pharmacokinetics, that of metabolism. Metabolism strictly is a description of chemical changes that occur to drugs such as in the course of their modification especially within the liver. These changes can result in drug inactivation, less commonly in drug activation, but also in some cases in conversion of a drug into a more dangerous form. For biocontrol agents that are intact organisms, activation can be viewed as equivalent to reproduction, and can result in an increase in numbers of agents either immediately locally, across entire treated areas, or indeed within environments that are found beyond that which is specifically being treated. For viruses, there is yet another and somewhat important issue that may be discussed under the heading of 'metabolism' and that is a virus' conversion from a relatively inert as well as somewhat benign entity into one that is capable of killing a target organism (Abedon, 2014b). Viruses, that is, exist as either virions or infections (Forterre, 2011). Virions tend to *not* be harmful on their own but instead must being converted into an infecting virus prior to effecting their harm, and in particularly prior to effecting their biocontrol activity.

That conversion from virion to infecting virus generally can occur only under specific circumstances, explicitly those occurring in the course of interaction with target organisms. As a consequence, appropriately chosen viruses can be relatively lacking in negative impacts on environments, or on patients, while still displaying potentially high levels of anti-target organism activity. Indeed, it can be argued that because of this specificity of their anti-biological activities, the pharmacologically 'emergent properties' associated with many viruses, that is, their potential to be unexpectedly harmful to treated environments – such as may be discovered for body environments in the course of clinical trials – can in many cases be less of a concern than the potential for unexpected physiological interactions to derail otherwise promising small-molecule candidate drugs (Curtright and Abedon, 2011). Particularly this is the case when employing viruses whose host ranges do not come close to overlapping valuable non-target species. For example, and as something of an extreme, the host ranges of bacterial viruses are not known to include that most cherished of biocontrol non-target species, *Homo sapiens*.

The means of virus discovery often is different from that of chemical drug discovery, or for that matter the discovery of chemical pesticides. Although chemical control agents certainly can be and are isolated from nature, as well as synthesized *de novo*, the isolation of viruses is often simpler, more rapid and results in greater numbers of potentially useful antimicrobials. In the case of viruses of microorganisms, it is the use of microbiological enrichment techniques (Clokie and Kropinski, 2009; Gill and Hyman, 2010), that is, the selection for those viruses that possess the property of being able to replicate in the presence of a specific target organism, that facilitates this ease of isolation. While the power of this technique can be hampered somewhat given high costs associated with culturing viruses on specific target organisms, for many viruses of microorganisms these techniques are fairly straightforward. So too are means by which virus-containing samples may be concentrated such that heterogeneous mixtures containing large numbers of potentially useful viruses may be applied to large numbers of target organisms, from which enriched viruses may then be isolated and characterized. In addition, biocontrol agents in many cases can, at least in principle, more easily obtain regulatory approval than otherwise equivalent synthetic pesticides (Chan *et al.*, 2013).

Though not necessarily impacting the costs of subsequent trials, this virus potential for enrichment in many cases can result in reasonably facile isolation of new agents possessing relatively novel mechanisms of target-organism killing action, i.e. viruses collectively display a huge diversity of proteins which possess anti-host activity, e.g. Wagemans and Lavigne (2012). In combination with the potential for viruses to be less likely to display pharmacologically or environmentally negative emergent properties, this means that in many cases deep reserves of selectively toxic biocontrol agents may be relatively easily tapped for subsequent application. Alternatively, for living organisms there should always be some concern for the potential for evolution to result in broader and therefore potentially more harmful target specificity. The more similar target organisms are to desirable species, then the greater the potential for such evolution to result in such unintended negative impacts.

A final general pharmacological consideration is the contrasting philosophies of, on the one hand, seeking to match biocontrol agents with target organisms prior to the initiation of treatment versus, on the other hand, the employment of more generally acting agents that may be employed without first putting substantial effort into identifying the target organism or its susceptibility to proposed antagonistic agents. This distinction is relevant because matching biocontrol agents to target organisms generally requires greater handler expertise than the use of agents not requiring substantial target organism characterization prior to control agent application. For viruses, this can be the distinction between the employment of virus panels from which specific viruses can be drawn, as individual viruses in the laboratory are shown to be effective in their infection of targets, or alternatively the development of what are described as 'cocktails' of viruses which collectively can possess host ranges which include multiple desired targets (Pirnay *et al.*, 2011; Chan and Abedon, 2012; Chan *et al.*, 2013, Nilsson, 2014). The former can be more time-consuming immediately prior to the point of biocontrol agent application, as testing along with some expertise can be required to identify the proper, specific viruses to employ. For cocktails of viruses, the concern instead can be that the overall breadth of agents may be sufficient that non-target organisms – either directly or following virus evolution – also might be susceptible to virus action. Alternatively, the breadth of coverage may not be sufficient and some pathogenic strains could evade treatment.

Use of whole viruses as biocontrol agents

Whole viruses, as the phrase is used here, generally refers to infectious virions. That is, the acellular forms of viruses that are responsible for at least localized virus dissemination, protection of virus nucleic acid during this extracellular phase, and acquisition of target cells. Viruses also can be delivered as virus-infected organisms, e.g. virus-infected animals or virus-infected bacteria. These can then supply environments or specific populations of target species with infectious virions, such as *Pantoea agglomerans* as a carrier for bacteriophages targeting the plant pathogen, *Erwinia amylovora* (Boulé *et al.*, 2011). 'Whole virus' thus does not refer to virus products that are isolated from the virus, such as in the course of genetic engineering, and then employed as relatively pure products, including in certain cases as biocontrol agents. In this section we consider the use of whole viruses as biocontrol agents whereas in a subsequent section we consider instead the use of virus-derived products as biocontrol agents. In this section, in particular, we distinguish among whole virus biocontrol agents in terms of the taxonomy of target organisms, i.e. plants, animals, fungi, protists, archaea, and bacteria. In addition, we consider the use of viruses as biocontrol agents of other viruses.

Virus-mediated biocontrol of plants

Numerous plant viruses exist, many of which are economically costly as they can affect the productivity of important crop plants. Use of viruses as biocontrol agents of plants would, by contrast, be expected to target weed species, that is, to serve as biological alternatives, particularly to the use of chemical herbicides. An example of such use is the application Tobacco Mild Green Mosaic Virus to control populations of the weed species, *Solanum viarum*, or tropical soda apple, which is invasive within cattle pastures. Application is achieved via a combination of spraying to apply the virus and abrading the plant to allow virus entry, or by high-pressure virus application to ensure virus penetration directly into the target species (Yandoc-Ables *et al.*, 2006). In a review of the use of plant pathogens to control weeds (Wilson, 2014), the possible natural control of the invasive aquatic plant, *Myriophyllum spicatum*, the Eurasian watermilfoil, is cited. See also Schuytema (1977), for a discussion of the potential to control alligator weed and water hyacinth using viruses.

Virus-mediated biocontrol of animals

The primary use of virus-mediated biocontrol of animals is as non-toxic, high specificity alternatives to standard pesticides. Baculoviruses, for example, are used to control caterpillar populations, such as those of the infamous gypsy moth (Whittle *et al.*, 2008). Impediments to greater use of virus-based insecticides includes relatively high production costs in comparison with standard pesticides and relatively long delays between application and successful reductions in population sizes of target insects. See Harper (2006, 2013) for additional references and further discussion.

Also discussed by Harper is the use of myxoma virus, the viral cause of rabbit haemorrhagic disease, as a biological control agent of rabbits in Australia and, then, New Zealand. Rabbits are not native to Australia but instead were imported by European settlers resulting, upon their escape, in environmental disaster. Virus-mediated biological control, although resulting in substantial rabbit die offs, has not resulted, ultimately, in sufficient control of these feral rabbit populations (Elsworth *et al.*, 2012). While this viral control of rabbits was initially considered successful in Australia, the spread of the disease in Europe, after its introduction by Dr Paul Armand-Delille, to his private estate in France, highlights the need for caution. The disease rapidly spread unchecked from this initial introduction throughout the entire continent (Arthur and Louzis, 1988). In the UK, the public worried about animal cruelty, loss of a valuable food source, danger to pets and human susceptibility, and while the disease was allowed to run its course, deliberate spread (for example by farmers trying to reduce rabbit numbers) was banned (Bartrip, 2008). At present, myxoma virus is sustained as a chronic enzootic in rabbit populations in America, Southern Europe, New Zealand and Australia (Fenner and Fantini, 1999). For another example of virus-mediated biocontrol of animals in Australia, see Pallister *et al.* (2011), who discuss unsuccessful efforts to develop a recombinant virus-based cane toad biocontrol agent. The use of viruses to control the populations of certain animal pests thus is promising, and in some cases somewhat effective (i.e. gypsy moth control), but nonetheless is not yet always effective, particularly over the long term, and not always an economically viable technology.

An alternative approach to species management is to inhibit the growth of target organism populations without necessarily outright killing individual organisms. Under certain circumstances these can be more socially desirable approaches to dealing with species overpopulation as it avoids massive die offs. Towards such ends, from the perspective of virus-based methods of biological control, are efforts that have been put towards the development of virus-based contraceptive agents for the control of vertebrate populations such as mice and rabbits (Hardy *et al.*, 2006). Using recombinant myxoma virus expressing immunocontraceptive antigens,

Mackenzie and colleagues (2006) were able to produce a transient infertility state in 70% of rabbits immunized with the recombinant virus over a period of 70 days.

Mycovirus-mediated biocontrol of fungi

Given the negative impacts of numerous forms of pathogenic fungi and the existence of fungal viruses (mycoviruses), at least some of which have demonstrably negative impacts on their host fungi, it is perhaps only natural that mycoviruses have been proposed as biological control agents of fungi (see Chapter 9 for further discussion of fungal viruses). Most notable among potential targets for such biological control is as a treatment for chestnut blight as caused by the fungus, *Cryphonectria parasitica,* although mycoviruses-associated interference of chestnut blight was in fact observed before the responsible mycoviruses themselves were discovered (Anagnostakis, 1982).

Though promising as an anti-fungal strategy, the use of mycoviruses as biological control agents nevertheless is not as straightforward as biocontrol achieved by other virus types, owing to a combination of mycoviruses typically infecting neither lytically nor lethally, and mostly not appearing to display extracellularly transmissible states. Given a means of mycovirus delivery to target fungi – keeping in mind that it can be technically demanding to introduce a mycovirus into a carrier fungal strain even in the laboratory – then it is particularly the ability of some of these viruses to decrease their host fungus' virulence, that is, induce hypovirulence (Nuss, 2005), that underlies the mycovirus potential to serve as biological control agents. The result can be a blocking of fungus-caused diseases particularly of plants, which has been dubbed 'virocontrol' (Ghabrial and Suzuki, 2009).

The virus Cryphonectria hypovirus 1 (CHV1) has been employed as a biocontrol agent against the chestnut blight pathogen, *C. parasitica.* This mycovirus is transmitted horizontally only during intermingling between cytoplasms (called anastomosis), the result of hyphal fusion between associated individual fungi that display vegetative compatibility. CHV1's transmission between fungi is therefore limited by incompatibilities among fungal strains that limit these associations. The means by which biocontrol is effected, once infection is initiated, is due to the negative consequences of these infections on the fungal host, resulting in reduced virulence against chestnut trees (i.e. hypovirulence). CHV1 thus has successfully demonstrated an ability to control chestnut blight outbreaks in Europe (Anagnostakis, 1982; Nuss, 1992; Heiniger and Rigling, 1994; Nuss, 2005). It has been less successful in the U.S., however, potentially due to transmission difficulties stemming from larger numbers of vegetative fungal compatibility types, or possibly because the American chestnut is a different species from that found in Europe. This relative lack of success of hypovirulence-mediated biocontrol in North America is observed despite the presence on this continent of strains of Cryphonectria hypovirus in addition to CHV1 (Milgroom and Cortesi, 2004).

Jiang *et al.* (2013) review mycoviruses of the fungal pathogen *Sclerotinia sclerotiorum* (affecting, e.g. soybeans), many of which were isolated from hypovirulent strains and which therefore may potentially be employed as biological control agents. In this publication, the authors provide an overview of previous work in the field as well as the obstacles that have so far hindered the use of mycovirus-based biocontrol of this fungal species. More recently, the same group described a mycovirus, SsHADV-1 of *S. sclerotiorum,* that is unusual, among mycoviruses, in that it can be transmitted as purified virions applied extracellularly (Yu *et al.,* 2013). This mycovirus is also unusual among mycoviruses in that it displays a single-stranded DNA genome rather than a typical RNA genome. In field trials it provided promising results, with SsHADV-1-virion treated plants by-and-large protected and mock-treated *S. sclerotiorum*-challenged plants succumbing to this fungal infection.

In addition to these two examples, there is the potential for 'virocontrol' of the plant pathogenic fungus, *Rosellinia necatrix,* cause of white root rot, for which over 400 host species have been identified, including fruit trees (Kondo *et al.,* 2013). Lee *et al.* (2011), furthermore, discuss the potential for the introduction of FgV1-DK21 virus-associated hypovirulence into different species of the genus of pathogenic fungi, *Fusarium.* They accomplish this via protoplast fusion, a means by which vegetative compatibility barriers may be overcome. This has also been accomplished in other mycovirus-fungal pathogen systems. Somewhat more theoretically, van de Sande *et al.* (2010) discuss the potential for mycoviruses to be employed to treat invasive fungal infections of humans. Therefore, perhaps surprisingly, among the viruses used as biocontrol agents of microorganisms, mycoviruses as anti-fungal agents have an impressively long and robust history.

Virus-mediated biocontrol of protists

As both parasitic protozoa and protozoal viruses exist, it would seem natural that the use of the latter to control the former would at least have been proposed. This in fact is the case, as has been covered by two relatively recently published opinion pieces on the subject (Hyman *et al.*, 2013; Keen, 2013), one of which (Keen, 2013) describes the technique as 'Protozoal virotherapy'. Little progress has been made towards the desired goal, however. Emphasis in fact is still very much on identifying appropriate viruses to carry out such biocontrol (see Chapter 11), either on their own or following genetic modification. It is perhaps telling, however, how infrequently these two articles have so far been cited: three for the first, including by the second and also by one of the coauthors of this chapter, and, academically, only once for the second. In addition to the two cited opinion pieces, Klive and Hegarty (1999) less recently speculated that protozoal viruses could be used to reduce population densities of rumen methanogen-supporting protozoa.

An additional approach to virus-mediated control of protists is indirect. This involves the use not of viruses that attack the protozoa themselves but instead the use of viruses which attack vectors involved in transmitting protozoal parasites. For example, these could be the mosquitoes responsible for transmitting *Plasmodium*, the malaria parasite. For malaria this involves the application of Densonucleosis (Denso) viruses which can be genetically modified to carry either anti-mosquito or anti-*Plasmodium* genes (Kamareddine, 2012). The use of such viruses is particularly intriguing since not only can they be transmitted to mosquitoes but they can also be vertically transmitted from parent mosquitoes to offspring. The potential therefore exists to genetically modify mosquitoes, via an infectious agent, so that their capacity to carry the *Plasmodium* parasite is reduced.

Virus-mediated biocontrol of members of domain Archaea

There is little work, at least of which we are aware, in which archaeal viruses (see Chapter 8) have been used as biocontrol agents. Part of the reason for this is that among pathogens, Archaea are at best extremely rare (Abedon, 2017c). Another reason is that within less extreme environments, archaea tend to be found in relatively low densities and are therefore less likely to serve as nuisance organisms. Lastly, our exploitation of more extreme environments has been less extensive than that of less extreme environments, so for the most part emphases on biocontrol tends to overwhelmingly favour their application within less extreme environments. Nonetheless, biofouling particularly in more extreme environments including anaerobic ones, reductions in efficiencies of heterogeneous biotechnological processes such as anaerobic biofuel production, losses of other valuable substances to microbe-mediated degradation, and greenhouse gas emissions, particularly methane, all to a degree and in principle could be at least partially addressed via archaeal virus-mediated biocontrol.

Methanogenic archaea are found as components of gastrointestinal flora, and methane represents a potent if relatively short-lived (in comparison to carbon dioxide) greenhouse gas. Among domesticated animals it is particularly ruminants, and especially cows, that are numerous, large, methanogen harbourers and as a consequence important methane producers. In fact, ruminants emit almost one third of methane from the agriculture sector. One strategy by which bovine methane emissions may be reduced is via selective biocontrol of ruminant microbiota populations. Consequently, the use of methanogen-targeting archaeal viruses has been proposed (Klieve and Hegarty, 1999). Efforts to isolate relevant viruses for this project against *Methanobrevibacter* sp. and *Methanobacterium* spp. have, however, not been successful (Klieve, 2010). Archaeal viruses could be applied to ruminants as a component of their diets, or instead directly to manure. For more discussion, see Jassim and Limoges (2017). It is of related interest that the one area where the most evidence exists for a role of archaea in human disease also involves methanogens, in this case as seen with periodontal disease-associated mixed infections (Nguyen-Hieu *et al.*, 2013).

In addition to biocontrol used to remove unwanted organisms from environments, or virus-mediated therapies to remove unwanted organisms from patients, a third category of virus-mediated selective removal of organisms exists, and that is as a laboratory technique. Specifically towards more efficient enrichment of rarer microorganisms from heterogeneous mixtures, it is possible to selectively eliminate dominant organisms. To our knowledge, this approach has not been employed against archaea, although in principle it could be. It could also be used to isolate rare archaea from complex samples in which members of domain Bacteria greatly dominate. See various publications by Kurtböke and colleagues for more on the use of this technique (Kurtböke *et al.*, 1992, 1993; Kurtböke, 1994; Kurtböke and French, 2007). See also DeMonte and Gupta (1941). In addition, note the use of an equivalent approach to reduce bacterial overgrowth in sputum samples destined for subsequent testing for presence of *Mycobacterium tuberculosis* (Kumar *et al.*, 2007).

Virus-mediated biocontrol of members of domain Bacteria

While the use of bacteriophages – a.k.a. bacterial viruses or phages (see Chapter 7) – as antibacterial agents fits well under the heading of virus-mediated biological control, in fact this use is more widely known instead as phage or bacteriophage therapy. Notwithstanding that popularity, not all phage-mediated biocontrol ought to be described as phage therapy, with the latter phrase best reserved for describing circumstances in which phages are being substituted for antibacterial drugs. That is, phage *therapy* specifically is when phages are being used to treat bacterial infections associated with individual organisms, whether in the guise of medicine or instead veterinary medicine, or indeed even of plants, if it is individual plants versus a plant-containing *environment* that is being treated (Abedon, 2009b). For a major recent review on phage therapy by one of the authors of this chapter, see Abedon (2017b). For a fairly comprehensive list of phage therapy and related articles, see Alves and Abedon (2017a,b).

Among targets of virus-mediated biocontrol, it is pathogenic bacteria that are the most numerous. Indeed, many companies exist or have existed that are involved in the research and development as well as sales of phages for use in therapy or biocontrol (see companies.phage.org). Further, a number of clinics exist and even whole countries or, formerly, even a whole empire, that have commercially developed cultures of phages for use to treat bacterial infections; these most notably are the former Soviet Union and particularly the former Soviet republic of Georgia (Abedon *et al.*, 2011). The Dutch company Micreos Food Safety (formerly EBI Food Safety) markets a phage-based anti-*Listeria* food additive as too does the Baltimore company, Intralytix. The Utah company Omnilytics, by contrast, has been somewhat unique in selling phages targeted to bacterial plant pathogens for sale to farmers. Numerous other companies (see 'Commercial development of bacteriophage-mediated biocontrol' below) along with academic research groups are involved in the development of phages for use against human pathogens whether as potentially found in the food supply (a form of biocontrol) or instead as clinically manifest (a form of therapy).

Numerous general reviews have been published espousing the virtues of phages as biocontrol agents, examining details of the techniques involved, exploring the pharmacology of phage therapy, and otherwise publicizing phage therapeutic use to wider audiences (just from us, for example, see Abedon and Thomas-Abedon, 2010; Kutter *et al.*, 2010; Abedon *et al.*, 2011; Chan *et al.*, 2013; Clark *et al.*, 2012; Clark, 2015; Pirnay *et al.*, 2015; Abedon, 2017b). In addition, there exists a literature in which phages were explored as possible control agents of cyanobacterial blooms within aquatic systems; see Abedon (2009a) for references. An approximation of the total number of experimental or clinical, English-language publications, from 2000 to 2016 involving phage-mediated biocontrol of bacteria (and/or phage therapy) is presented as Fig. 14.4. A more complete listing of the phage therapy or phage-mediated biocontrol literature can be found at publications.phage-therapy.org (Alves and Abedon, 2017b).

Commercial development of bacteriophage-mediated biocontrol

As only bacteriophages among VoMs have so far been commercially developed as anti-microorganism biocontrol agents, in this section we consider exclusively phage commercial development.

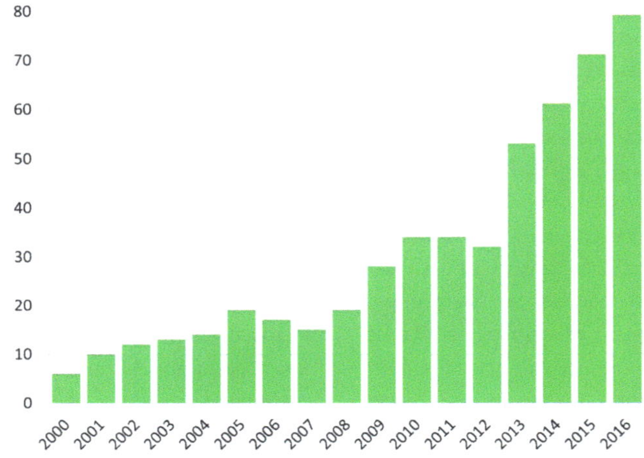

Figure 14.4 Publications were gleaned from the website publications.phage-therapy.org. Summarized in the figure are only experimental studies – not reviews, commentaries, theory, nor relatively trivial antibacterial activity such as broth killing or plaque formation – and these totals should be considered to represent only approximations since it is likely that not all publications were successfully identified. The correlation coefficient (*r*) of the obvious rise, for years 2000 through 2016, is 0.92, with a slope of 4.1, that is, an increase on average of roughly four publications per year. Alternatively, it would appear that there was a substantial jump in productivity, roughly a doubling, going from prior to the year 2009 to following the year 2009. Thus, from 2008 to 2016 the slope instead is 7.4 while *r* is 0.96. Consideration of increases in overall usage of the phrase 'phage therapy' in the literature is presented in Abedon (2017a).

Overview

Historically, the overwhelming success of small-molecule antibiotics, such as penicillin, has undoubtedly played a significant role in limiting the commercial exploitation of bacteriophages, and by extension of other virus-mediated biocontrol strategies (e.g. Clark, 2015). The ability to use these chemical therapies with a relatively crude understanding of the infecting organism, combined with ease of commercial production and affordability, lead to an industry perception that there was little need for alternatives. While biological control strategies do have some advantages, as discussed previously in this chapter, from a commercial perspective the availability of therapies that are cheap, easy to use, and hugely effective, i.e. antibiotics, substantially limits market opportunities for competing technologies. As our understanding of antibiotic use has increased, however, it has become clear that there are issues with their widespread use; this is mainly due to the increasing prevalence and spread of antibiotic resistance among crucial target species but it also reflects the desire for more targeted antibacterial agents, particularly as we have come to appreciate the importance towards health of maintaining diverse microbiomes. The issue with antibiotic resistance results from the spread of resistance genes, exacerbated by overuse of antibiotics, and from a declining pipeline of new antibiotics, while the desire for more targeted antibiotics results from a desire to eliminate pathogens but leave harmless (or helpful) bacteria intact. These factors, combined with a better understanding of the nature of phages and their requirements for effective application, now result in substantial commercial opportunities for virus-mediated biocontrol.

Regulation

At the current time, commercial development of virus-mediated biocontrol is still restricted by several factors. Most notably, the regulatory pathways for antibacterial substances are heavily biased towards chemical antibiotics, particularly in the field of human therapies (Parracho *et al.*, 2012). This results in high levels of regulatory uncertainty, which increases risks for investors and companies in the field. Verbeken *et al.* (2012) produced a thorough discussion of how phages fit into the European regulatory framework, particularly in the UK and Belgium, but while this is detailed and extremely useful, it is still theoretical in nature.

Several companies, however, have made significant regulatory progress, particularly with control of bacterial pathogens with bacteriophages in non-human indications. Omnilytics in the U.S. has two products available for use in agriculture, both under the trade name Agriphage: one targets bacterial tomato pathogens and the other targets bacterial pepper pathogens. Both of these products have been granted regulatory approval through the EPA in America. The most common mechanism for allowing the use of phage products, however, is through the Generally Recognized as Safe (GRAS) pathway in the US, which allows their use under legally defined conditions (Anonymous, 2018): 'Under sections 201(s) and 409 of the Federal Food, Drug, and Cosmetic Act (the Act), any substance that is intentionally added to food is a food additive, that is subject to premarket review and approval by FDA, unless the substance is generally recognized, among qualified experts, as having been adequately shown to be safe under the conditions of its intended use, or unless the use of the substance is otherwise excepted from the definition of a food additive.' That is, GRAS designation allows the use of additives in foods which are recognized by qualified experts to be safe. In total, there have been ten phage-related GRAS applications from four companies [Micreos (formerly EBI), Intralytix, Phagelux and FINK TEC GmbH], two of which have been withdrawn, one of which is pending and seven of which have had no objections raised by the FDA. Those are GRAS applications 198, 298, 435, 468, 528, 603 and 672 and all are available for view on the FDA web site.

Overall, while the regulatory pathway for viruses used in biocontrol applications is still considerably less well defined than that for non-biologicals, in recent years several important strides have been man in non-therapeutic indications with viruses of bacteria. Additionally, while no phage products have been approved for use as human therapies, in the past decade there have been several early-stage clinical trials performed with phage products. Rhoades *et al.* (2009) performed a phase I, i.e. safety clinical trial. No adverse effects were observed but efficacy was not tested. Also, Merabishvili *et al.* (2009) tested a phage cocktail in a pilot clinical study with eight patients, without any adverse effects. This was also a pure safety study and no efficacy data were presented. In 2014, a follow-up paper presented the results of this study in more detail but unfortunately issues with the experimental design 'did not allow for an adequate evaluation of the efficacy of the phage cocktail' (Rose *et al.*, 2014). Exponential Biotherapies is also reported to have performed a phase I clinical trial, although the results have never been published (Kutter *et al.*, 2010). Recent information, however, suggests that there were no issues with the results of the trial but rather that the commercial direction of the company changed (Harper *et al.*, 2014).

Following good safety data in a pre-clinical study in humans (Rainey *et al.*, 2000), where no adverse events were reported, Nestlé undertook a randomized, double blind placebo-controlled trial of orally delivered phages to manage diarrhoea in children in Bangladesh (Nestlé, 2013; Sarker *et al.*, 2016). The study, however, was terminated before the groups were un-blinded and the results known.

The British company Biocontrol (now Ampliphi) has performed what is to date the only phase II clinical trial with a phage therapeutic. A human trial for otitis was performed under the auspices of the Medicines and Healthcare Products Regulatory Agency (MRHA) in the UK (Wright *et al.*, 2009). Differences between the phage and control group were such that the trial was stopped early, so the control group could receive treatment. This trial was the first demonstration of efficacy in a trial performed to Western regulatory standards and is an important step towards the development of regulated phage therapeutics. The phages that were used, however, were not manufactured to GMP (Good Manufacturing Practice), limiting the potential to perform further trials needed for product development. Ampliphi has also recently performed a further phase I trial using phages produced to GMP, which has demonstrated the safety of their AB-SA01 phage product (active against *S. aureus*) in humans (AmpliPhi Biosciences Corporation, 2016). Two further clinical trials are currently open for recruitment: one to allow the use of AB-SA01 in patients with life-threatening conditions (AmpliPhi Biosciences Corporation, 2018b) and a similar study to allow use of a cocktail against *P. aeruginosa* (AB-PA01), also in patients with life-threatening conditions (AmpliPhi Biosciences Corporation, 2018a). Finally, Pherecydes Pharma in France is close to concluding a Phase I/II multicentre clinical trial with 11 clinical partners, to study the effectiveness of phages in the treatment of burn wounds (Anonymous, 2017b).

While regulatory approval is still an issue, albeit one that is now being rapidly addressed, several other unique factors also need to be considered in the commercial development of viruses as biocontrol agents. These are discussed briefly below.

Intellectual property (IP)

The concept of phage therapy was first suggested almost as soon as the nature of phages had been identified (Abedon, 2017b). It is therefore not possible to patent the core concept of phage or virus biocontrol. Additionally, as natural products, it is unlikely that individual phages can be patented, particularly given the 2013 ruling of the USPTO in relation to Myriad genetics (US Court document number 09-cv-4515, 94 USPQ2d 1683). Strictly speaking, this ruling meant that single genes could not be patented but has generally been expanded to limit patenting of naturally occurring genetic sequences. This potential inability to patent a key component of a pharmaceutical product presents a unique challenge in the field. Several companies, however, do have diverse and valuable patent portfolios (and other know-how/trade-secrets), with new patents being regularly filed. These are many and varied but cover areas such as cocktail development, isolation methods, formulation, process development, production, and route/method of application. A ripe area for development of IP is in the development of genetically modified phages, which for example possess extended host ranges or improved growth characteristics, but these advantages are tempered by the fact that as GMOs (Genetically Modified Organisms), the regulatory pathway would almost certainly be more onerous. A more detailed discussion of the issues relating to the patentability of phages is given in De Vos *et al.* (2014).

Product protection

As discussed previously in this chapter, viruses, in ideal circumstances, are readily isolated in the presence of their hosts. Compared with small molecule antibiotics, this process of identifying new antibacterials is relatively quick and inexpensive. This ease of isolation and their self-replicating nature, however, means that it can be difficult to protect commercial phage products from theft. If a company has spent years taking a product through the regulatory process, they do not want that product to be easily available to competitors. With natural virus products generally, although, it is relatively easy to isolate the active ingredient from the product itself and to some extent reverse-engineer the pharmaceutical. This makes know-how less easily protected and increases the commercial risks for companies and investors. The manufacture and formulation of these products, however, is generally non-trivial.

Production

Until relatively recently, manufacture of phages was routine at small scales but relatively unproven at commercial scales, at least to the standards currently required for therapeutic products in Europe and North America. These issues are exacerbated in human therapy where products need to be produced to the standards of Good Manufacturing Practice (GMP) (Anonymous, 2017a). GMP covers a set of standards which almost all products used therapeutically in humans must follow. While GMP and non-GMP processes can be identical, the level of rigour applied to GMP manufacture is substantially greater in terms of quality control, validation, documentation, and record keeping.

Many of the processes used in phage purification are very basic (Vandamme, 2014), are difficult to patent and would not have progressed greatly since phages were first produced commercially. Traditionally, phages have been produced in shaking culture and purification has relied on precipitation, centrifugation, and gradient ultracentrifugation. Gradient ultracentrifugation in particular is considered difficult and extremely expensive to scale up, particularly in the background of GMP manufacturing. To a lesser extent, however, phages have also been grown in fermenters for many years (Siquet-Descans and Calberg-Bacq, 1973) and for some phages processes compatible with industrial production have been developed, e.g. Merabishvili *et al.* (2009). While larger-scale phage production is relatively rare, a number of examples are available. Historically, in the former Soviet republic of Georgia, the Eliava institute at the height of its production was manufacturing tons of phage product on a routine basis. Furthermore, a number of companies that now manufacture their own phage products, most notably Micreos and Phagelux, routinely manufacture phage products on a large-scale, and Ampliphi owns what is currently the world's only GMP phage manufacturing facility. Moreover, a number of companies also exist which specialize in contract phage manufacturer, something that was almost unheard of a decade ago.

Summary

It is fair to say that while there are still several unique challenges that need to be addressed during the commercialization of natural biocontrol agents, the last two decades has seen substantial progress and many of the key questions that have hampered development in the past have been, or are in the process of being addressed. Given the declining pipeline of new antibiotics, combined with the increasing prevalence of antibiotic resistant infections, it seems the time is now ripe for virus-mediated biocontrol to become a commercial reality.

Use of virus products as biocontrol agents

There are a number of general ways that one can distinguish among biocontrol agents. Particularly, these agents may be capable of reproducing, including in terms of using that reproduction towards enhancing their biocontrol abilities. Alternatively, they may be incapable of reproducing, at least *in situ*. An inability to replicate can be viewed as a means of eliminating the potential of biocontrol agents to either move substantially beyond their point of application to an environment, or to otherwise evolve. This inability, however, comes at the expense of just that reproduction, which can serve as a supplementary means of agent application as well as a penetration aid (Fig. 14.1). Indeed, for biocontrol agents an ability to reproduce *in situ* can increase the convenience of use by allowing biocontrol agents to persist longer within the environments to which they have been applied (Fig. 14.2).

Among biocontrol agents that are incapable of reproducing, there are control agents that represent only component parts of larger organisms. Perhaps the most recognizable such agent, although one that typically is not considered within the context of biocontrol, is the production of small-chemical antibiotics by bacteria or fungi, which can then be purified and co-opted to control unwanted bacteria. It is perhaps due to the history of antibiotics, with their widespread adoption having occurred well before it might have been considered to describe them as biocontrol agents, that they are lumped together with chemical rather than biological agents. Alternatively, antibiotics – at least as historically as well as still typically employed – are relatively broadly acting. They consequently are less like biological control agents, which instead tend to be more narrowly acting in terms of what species are affected. Antibiotics thus can perhaps be described as less-than-ideal biocontrol agents, i.e. organism-produced agents that in many cases can have a greater than typically desired impact on non-target species. In addition, in this modern era the concept of antibiotics has been extended to beyond microorganism-produced antibacterial agents to include purely synthetic compounds as well, which certainly are not biological control agents.

Among less-than-whole viruses and their products, we can distinguish at least two basic biocontrol structures. These are viruses that either naturally or as a consequent of artificial manipulation have lost their reproductive abilities but not their target-organism killing abilities. The bacteriophage tail-like colicins produced by certain bacteria, such as R-type pyocins (Scholl *et al.*, 2009), appear to fit in the naturally occurring category. Whole killed or subunit vaccines arguably also can be included among no longer replicating viruses or virus parts that may be employed as control agents, particularly in this case the immunological control of still intact viruses. Alternatively, there are virus products that can be used to kill target organisms or instead disrupt those organisms in a manner such that they either are no longer a threat or instead that they become susceptible to further killing agents, the latter, e.g. such as in terms of biofilm disruption. Two prime examples of such agents are the endolysins and extracellular polymeric substance (EPS) hydrolases, both of which are associated with and can be isolated from numerous bacteriophages. These we consider in greater detail. See also Drulis-Kawa *et al.* (2015).

Endolysins are employed by phages in the course of effecting bacterial lysis. The standard phage-productive life cycle involves the acquisition of a bacterium, infection, associated production of phage progeny virions, and subsequent release of these virions from the infected cell. With tailed phages, this process, although still not fully understood as novel mechanisms keep being discovered (Catalao *et al.*, 2013), generally involves two steps (Young, 2014). The first step is effected by proteins which are known as holins. These proteins enter into the bacterial plasma membrane following their production and, after a delay that is specific for the phage, generate holes in the plasma membrane. These holes have the effect of terminating the infected bacterium's metabolism. Immediately after, a second protein, the endolysin, digests the bacterial cell wall. This results in the cell wall being damaged/degraded to an extent that allows phage virion release.

These endolysins, with for example glycosidase, amidase or endopeptidase activities, can be isolated (Nelson *et al.*, 2012; Shen *et al.*, 2012). This typically is done via cloning and expression of endolysin genes. The endolysin genes also can be modified genetically. Typically, the endolysins display two activities, one of which is responsible for binding to the cell wall of a bacterium and the other which cleaves the cell wall in some manner. This binding and cleavage often occurs with high specificity, to the extent that the proteins are not thought to display inherent toxicities against non-target species, particularly such as ourselves. The major drawback of endolysins is that they are limited by their ability to reach the peptidoglycan layer of their bacterial targets, which in the course of phage infections occurs from within (Schmelcher *et al.*, 2012). By contrast, biocontrol (as mediated by endolysins) requires that endolysins reach the peptidoglycan layer from without. This is far more easily accomplished against the cell walls of Gram-positive bacteria than those of Gram-negative bacteria, which have a resistant outer membrane. Some groups, however, are developing novel strategies which could also allow effective endolysin use against Gram-negative bacteria (Lim *et al.*, 2014; Briers and Lavigne, 2015). This approach of using virus-derived substances, although lacking in agent reproduction *in situ*, nevertheless could substantially broaden our arsenal of relatively small-molecule antibacterials and do so under the legitimate guise of biocontrol.

EPS depolymerases are enzymes that, in various ways, are capable of breaking down bacterial capsules and other extracellular polymeric substances (Shen *et al.*, 2012; Chan and Abedon, 2015; Pires *et al.*, 2016; Gutiérrez *et al.*, 2016). While this action does not directly kill bacteria, these enzymes nevertheless are capable of breaking up aggregations of bacteria, most notably by disrupting biofilms. This may not directly cause the death of target bacteria, but can contribute to their removal from surfaces and can expose bacteria to other antibacterial actions e.g. chemical antibiotics or an infected organism's immune system. As with endolysins, EPS depolymerases can be highly specific for their targets; in this case bacterially produced exopolysaccharide. This specificity can contribute substantially to the safety of these agents, both pharmacologically and also environmentally. Indeed, the fact that both EPS depolymerases and lysins are proteinaceous enzymes should serve to limit their potential to persist in environments. The specificity of EPS depolymerases, however, can be so substantial that this can potentially serve as a hindrance to their utility, necessitating careful matching between enzyme and target.

Conclusion

The utility of biocontrol comes substantially from its potential to broaden our arsenal against pest organisms, particularly as that may be achieved without negatively impacting non-target species. With viruses as biocontrol agents, and perhaps particularly bacteriophages, there also is a utility of substantial quantities in nature of what, in many cases, are relatively easily isolated agents. Thus, the arsenal that may be employed against pest organisms in fact may be substantially broadened via the use of biocontrolling viruses. In other cases, such as in the treatment of chestnut blight or antibiotic resistant bacterial infections, no alternative control agents may be available. In combination with the ability of whole viruses to amplify their numbers *in situ*, spread to not-yet infected target organisms, and otherwise persist for relatively long periods within environments, they have the potential to supplement or replace less versatile or more toxic strategies, particularly chemical control agents, where the use of more environmentally friendly agents is quite welcome. This is particularly significant under circumstances where costs of standard or alternative methods are especially high and where resistance evolution by target species can be effectively mitigated.

References

Abedon, S.T. (2009a). Impact of phage properties on bacterial survival. In Contemporary Trends in Bacteriophage Research, Adams, H.T., ed. (Nova Science Publishers, Hauppauge, New York), pp. 217–235.

Abedon, S.T. (2009b). Kinetics of phage-mediated biocontrol of bacteria. Foodborne Pathog. Dis. *6*, 807–815. https://doi.org/10.1089/fpd.2008.0242.

Abedon, S.T. (2014a). Bacteriophages as drugs: the pharmacology of phage therapy. In: Phage Therapy: Current Research and Applications, Borysowski, J., Międzybrodzki, R., and Górski, A., eds. (Caister Academic Press, UK), pp. 69–100.

Abedon, S.T. (2014b). Phage therapy: eco-physiological pharmacology. Scientifica *2014*, 581639. https://doi.org/10.1155/2014/581639.

Abedon, S.T. (2017a). Bacteriophage clinical use as antibactertial "drugs": utility precedent. In Bugs as Drugs. Cani, P., and Britton, R., eds. (ASM Press, Washington).

Abedon, S.T. (2017b). Bacteriophage Clinical Use as Antibacterial 'Drugs': Utility and Precedent. Microbiol. Spectr. *5*, https://doi.org/10.1128/microbiolspec.BAD-0003-2016.

Abedon, S.T. (2017c). Why archaea are limited in their exploitation of other, living organisms. In Biocommunication of Archaea, Witzany, G. ed. pp. 41–66.

Abedon, S.T., Kuhl, S.J., Blasdel, B.G., and Kutter, E.M. (2011). Phage treatment of human infections. Bacteriophage *1*, 66–85. https://doi.org/10.4161/bact.1.2.15845.

Abedon, S.T., and Thomas-Abedon, C. (2010). Phage therapy pharmacology. Curr. Pharm. Biotechnol. *11*, 28–47.

Alves, D.R., and Abedon, S.T., (2017a). Phage therapy bibliography. Available online: http://publications.phage-therapy.org.

Alves, D.R., and Abedon, S.T. (2017b). An online phage therapy bibliography: separating under-indexed wheat from overly indexed chaff. AIMS Microbiol. 3, 525–528.

AmpliPhi Biosciences Corporation (2016). Ascending dose study of the safety of AB-SA01 when topically applied to intact skin of healthy adults. Avaiable online: https://clinicaltrials.gov/ct2/show/NCT02757755.

AmpliPhi Biosciences Corporation (2018a). Individual patient expanded access for AB-PA01, an investigational anti-Pseudomonas aeruginosa bacteriophage therapeutic. Avaiable online: https://clinicaltrials.gov/ct2/show/NCT03395743.

AmpliPhi Biosciences Corporation (2018b). Individual patient expanded access for AB-SA01, an investigational anti-Staphylococcus aureus bacteriophage therapeutic. Avaiable online: https://clinicaltrials.gov/ct2/show/NCT03395769.

Anagnostakis, S.L. (1982). Biological control of chestnut blight. Science *215*, 466–471.

Anonymous (2017a). Good manufacturing practice and good distribution practice. Avaiable online: http://www.mhra.gov.uk/Howweregulate/Medicines/Inspectionandstandards/GoodManufacturingPractice/

Anonymous (2017b). PhagoBurn. Avaiable online: http://www.phagoburn.eu/

Anonymous (2018). Generally recognized as safe (GRAS). Avaiable online: https://www.fda.gov/Food/IngredientsPackagingLabeling/GRAS/

Arnold, K.E., Brown, A.R., Ankley, G.T., and Sumpter, J.P. (2014). Medicating the environment: assessing risks of pharmaceuticals to wildlife and ecosystems. Philos. Trans. R. Soc. Lond., B, Biol. Sci. *369*, 20130569. https://doi.org/10.1098/rstb.2013.0569.

Arthur, C.P., and Louzis, C.A. (1988). A review of myxomatosis among rabbits in France. Rev. Sci. Tech. Off. Int. Epiz. 7, 959–976.

Bartrip, P.W. (2008). Myxomatosis in 1950s Britain. 20 Century Br. Hist. *19*, 83–105.

Boulé, J., Sholberg, P.L., Lehman, S.M., O'Gorman, D.T., and Svircev, A.M. (2011). Isolation and characterization of eight bacteriophages infecting *Erwinia amylovora* and their potential as biological control agents in British Columbia, Canada. Can J. Plant Pathol. 33, 308–317.

Briers, Y., and Lavigne, R. (2015). Breaking barriers: expansion of the use of endolysins as novel antibacterials against Gram-negative bacteria. Future Microbiol. *10*, 377–390.

Chan, B.K., and Abedon, S.T. (2012). Phage therapy pharmacology phage cocktails. Adv. Appl. Microbiol. *78*, 1–23. https://doi.org/10.1016/B978-0-12-394805-2.00001-4.

Chan, B.K., and Abedon, S.T. (2015). Bacteriophages and their enzymes in biofilm control. Curr. Pharm. Des. *21*, 85–99.

Chan, B.K., Abedon, S.T., and Loc-Carrillo, C. (2013). Phage cocktails and the future of phage therapy. Future Microbiol. *8*, 769–783. https://doi.org/10.2217/fmb.13.47.

Clark, J., Abedon, S.T., and Hyman, P. (2012). Phages as therapeutic delivery vehicles. In Bacteriophages in Health and Disease, Hyman, P., Abedon, S.T., eds. (CABI Press, UK), pp. 86–100.

Clark, J.R. (2015). Bacteriophage therapy: history and future prospects. Future Virol. *10*, 449–461.

Clokie, M.R.J., and Kropinski, A.M. (2009). Bacteriophages. Methods and Protocols. Volume 1: Isolation, Characterization, and Interactions (Humana Press, New York).

Curtright, A.J., and Abedon, S.T. (2011). Phage therapy: emergent property pharmacology. J. Bioanalyt. Biomed. S6:002. https://doi.org/10.4172/1948-593X.S6-002.

De Vos, D., Verbeken, G., Ceulemans, C., Huys, I., and Pirnay, J.-P. (2014). Reintroducing phage therapy in modern medicine: the regulatory and intellectual property hurdles. In Phage Therapy: Current Research and Applications, Borysowski, J., Międzybrodzki, R., and Górski, A., eds. (Caister Academic Press, UK), pp. 289–307.

DeMonte, A.J.H., and Gupta, S.K. (1941). Use of bacteriophage for freeing protozoal cultures of contaminating bacteria (isolation of *Leptospira icterohaemorrhagiae* from mixed infections in guinea-pigs). Indian Med. Gaz. *76*, 154–155.

Ding, C., and He, J. (2010). Effect of antibiotics in the environment on microbial populations. Appl. Microbiol. Biotechnol. *87*, 925–941. https://doi.org/10.1007/s00253-010-2649-5.

Drulis-Kawa, Z., Majkowska-Skrobek, G., and Maciejewska, B. (2015). Bacteriophages and phage-derived proteins–application approaches. Curr. Med. Chem. *22*:1757–1773.

Elsworth, P.G., Kovaliski, J., and Cooke, B.D. (2012). Rabbit haemorrhagic disease: are Australian rabbits (*Oryctolagus cuniculus*) evolving resistance to infection with Czech CAPM 351 RHDV? Epidem. Inf. *140*, 1972–1981.

Fenner, F., and Fantini, B. (1999). Biological Control of Vertebrate Pests: the History of Myxomatosis – an Experiment in Evolution (CABI Publishing, Oxford).

Forterre, P. (2011). Manipulation of cellular syntheses and the nature of viruses: the virocell concept. Comptes Rendus Chimie *14*, 392–399.

Ghabrial, S.A., and Suzuki, N. (2009). Viruses of plant pathogenic fungi. Annu. Rev. Phytopathol. *47*, 353–384. https://doi.org/10.1146/annurev-phyto-080508-081932.

Gill, J.J., and Hyman, P. (2010). Phage choice, isolation, and preparation for phage therapy. Curr. Pharm. Biotechnol. *11*, 2–14.

Gutiérrez, D., Rodríguez-Rubio, L., Martínez, B., Rodríguez, A., and García, P. (2016). Bacteriophages as weapons against bacterial biofilms in the food industry. Front. Microbiol. *7*, 825. https://doi.org/10.3389/fmicb.2016.00825.

Hardy, C.M., Hinds, L.A., Kerr, P.J., Lloyd, M.L., Redwood, A.J., Shellam, G.R., and Strive, T. (2006). Biological control of vertebrate pests using virally vectored immunocontraception. J. Reprod. Immunol. *71*, 102–111.

Harper, D.R. (2006). Biological control by microorganisms. In The Encyclopedia of Life Sciences (John Wiley & Sons), pp. 1–10.

Harper, D.R. (2013). Biological Control By Microorganisms. eLS. (John Wiley & Sons, UK).https://doi.org/10.1002/9780470015902.a0000344.pub3.

Harper, D.R., Burrowes, B.H., and Kutter, E.M. (2014). Bacteriophage: therapeutic uses. eLS. (John Wiley & Sons, UK) https://doi.org/10.1002/9780470015902.a0020000.pub2.

Heiniger, U., and Rigling, D. (1994). Biological control of chestnut blight in Europe. Ann. Rev. Phytopathol. *32*, 581–599.

Hyman, P., and Abedon, S.T. (2010). Bacteriophage host range and bacterial resistance. Adv. Appl. Microbiol. *70*, 217–248. https://doi.org/10.1016/S0065-2164(10)70007-1.

Hyman, P., Atterbury, R., and Barrow, P. (2013). Fleas and smaller fleas: virotherapy for parasite infections. Trends Microbiol. *21*, 215–220. https://doi.org/10.1016/j.tim.2013.02.006.

Jiang, D., Fu, Y., Guoqing, L., and Ghabrial, S.A. (2013). Viruses of the plant pathogenic fungus *Sclerotinia sclerotiorum*. Adv. Virus Res. *86* 215–248.

Kamareddine, L. (2012). The biological control of the malaria vector. Toxins *4*, 748–767. https://doi.org/10.3390/toxins4090748.

Keen, E.C. (2013). Beyond phage therapy: virotherapy of protozoal diseases. Future Microbiol. *8*, 821–823. https://doi.org/10.2217/fmb.13.48.

Klieve, A. (2010). Reducing emissions from livestock research program: archaeaphage therapy to control rumen methanogens (Meat and Livestock Australia, North Sydney).

Klieve, A.V., and Hegarty, R.S. (1999). Opportunities for biological control of ruminal methanogenesis. Australian J. Agric. Res. *50*, 1315–1319.

Kondo, H., Kanematsu, S., and Suzuki, N. (2013). Viruses of the white root rot fungus, *Rosellinia necatrix*. Adv. Virus Res. *86*, 177–214. https://doi.org/10.1016/B978-0-12-394315-6.00007-6.

Koskella, B., and Meaden, S. (2013). Understanding bacteriophage specificity in natural microbial communities. Viruses *5*, 806–823. https://doi.org/10.3390/v5030806.

Kumar, V., Balaji, S., Gomathi, N.S., Venkatesan, P., Sekar, G., Jayasankar, K., and Narayanan, P.R. (2007). Phage cocktail to control the exponential growth of normal flora in processed sputum specimens grown overnight in liquid medium for rapid TB diagnosis. J. Microbiol. Meth. *68* 536–542.

Kurtböke, D.I. (1994). Rapid detection of *Clavibacter toxicus* and of its bacteriophage responsible for annual ryegrass toxicity in Australia and the effect of selected herbicides on toxin production. Actinomycetes *5*, 31–39.

Kurtböke, D.I., Chen, C.F., and Williams, S.T. (1992). Use of polyvalent phage for reduction of streptomycetes on soil dilution plates. J. Appl. Bacteriol. *72*, 103–111.

Kurtböke, D.I., and French, J.R. (2007). Use of phage battery to investigate the actinofloral layers of termite gut microflora. J. Appl. Microbiol. *103*, 722–734.

Kurtböke, D.I., Murphy, N.E., and Sivasithamparam, K. (1993). Use of bacteriophage for the selective isolation of thermophilic actinomycetes from composted eucalyptus bark. Can. J. Microbiol. *39*, 46–51.

Kutter, E., De Vos, D., Gvasalia, G., Alavidze, Z., Gogokhia, L., Kuhl, S., and Abedon, S.T. (2010). Phage therapy in clinical practice: treatment of human infections. Curr. Pharm. Biotechnol. *11*, 69–86.

Lee, K.M., Yu, J., Son, M., Lee, Y.W., and Kim, K.H. (2011). Transmission of *Fusarium boothii* mycovirus via protoplast fusion causes hypovirulence in other phytopathogenic fungi. PLOS ONE 6:e21629.

Lim, J.A., Shin, H., Heu, S., and Ryu, S. (2014). Exogenous lytic activity of SPN9CC endolysin against gram-negative bacteria. J. Microbiol. Biotechnol. *24*, 803–811.

Mackenzie, S.M., McLaughlin, E.A., Perkins, H.D., French, N., Sutherland, T., Jackson, R.J., Inglis, B., Muller, W.J., van Leeuwen, B.H., Robinson, A.J., and Kerr, P.J. (2006). Immunocontraceptive effects on female rabbits infected with recombinant myxoma virus expressing rabbit ZP2 or ZP3. Biol. Reprod. *74*, 511–521.

Meaden, S., and Koskella, B. (2013). Exploring the risks of phage application in the environment. Front. Microbiol. *4*, 358. https://doi.org/10.3389/fmicb.2013.00358.

Merabishvili, M., Pirnay, J.-P., Verbeken, G., Chanishvili, N., Tediashvili, M., Lashkhi, N., Glonti, T., Krylov, V.,

Mast, J., Van Parys, L., *et al.* (2009). Quality-controlled small-scale production of a well-defined bacteriophage cocktail for use in human clinical trials. PLOS ONE *4*:e4944.

Milgroom, M.G., and Cortesi, P. (2004). Biological control of chestnut blight with hypovirulence: a critical analysis. Annu. Rev. Phytopathol. *42*, 311–338. https://doi.org/10.1146/annurev.phyto.42.040803.140325.

Nelson, D.C., Schmelcher, M., Rodriguez-Rubio, L., Klumpp, J., Pritchard, D.G., Dong, S., and Donovan, D.M. (2012). Endolysins as antimicrobials. Adv. Virus Res. *83*, 299–365. https://doi.org/10.1016/B978-0-12-394438-2.00007-4.

Nestlé (2013). Antibacterial treatment against diarrhea in oral rehydration solution. Available online: https://clinicaltrials.gov/ct2/show/NCT00937274.

Nguyen-Hieu, T., Khelaifia, S., Aboudharam, G., and Drancourt, M. (2013). Methanogenic archaea in subgingival sites: a review. APMIS *121*, 467–477. https://doi.org/10.1111/apm.12015.

Nilsson, A.S. (2014). Phage therapy – constraints and possibilities. Ups. J. Med. Sci. *119*, 192–198. https://doi.org/10.3109/03009734.2014.902878.

Nuss, D.L. (1992). Biological control of chestnut blight: an example of virus-mediated attenuation of fungal pathogenesis. Microbiol. Rev. *56*, 561–576.

Nuss, D.L. (2005). Hypovirulence: mycoviruses at the fungal-plant interface. Nat. Rev. Microbiol. *3*, 632–642.

Pallister, J.A., Halliday, D.C., Robinson, A.J., Venables, D., Voysey, R.D., Boyle, D.G., Shanmuganathan, T., Hardy, C.M., Siddon, N.A., and Hyatt, A.D. (2011). Assessment of virally vectored autoimmunity as a biocontrol strategy for cane toads. PLOS ONE *6*, e14576. https://doi.org/10.1371/journal.pone.0014576.

Parracho, H.M., Burrowes, B.H., Enright, M.C., McConville, M.L., and Harper, D.R. (2012). The role of regulated clinical trials in the development of bacteriophage therapeutics. J. Mol. Genet. Med. *6*, 279–286.

Pires, D.P., Oliveira, H., Melo, L.D., Sillankorva, S., and Azeredo, J. (2016). Bacteriophage-encoded depolymerases: their diversity and biotechnological applications. Appl. Microbiol. Biotechnol. *100*, 2141–2151.

Pirnay, J.P., Blasdel, B.G., Bretaudeau, L., Buckling, A., Chanishvili, N., Clark, J.R., Corte-Real, S., Debarbieux, L., Dublanchet, A., De, V.D. *et al.*, (2015). Quality and safety requirements for sustainable phage therapy products. Pharm. Res. *32*, 2173–2179.

Pirnay, J.P., De Vos, D., Verbeken, G., Merabishvili, M., Chanishvili, N., Vaneechoutte, M., Zizi, M., Laire, G., Lavigne, R., Huys, I., *et al.* (2011). The phage therapy paradigm: prêt-à-porter or sur-mesure? Pharm. Res. *28*, 934–937. https://doi.org/10.1007/s11095-010-0313-5.

Rainey, P.B., Buckling, A., Kassen, R., and Travisano, M. (2000). The emergence and maintenance of diversity: insights from experimental bacterial populations. Trends Ecol. Evol. *15*, 243–247.

Rhoads, D.D., Wolcott, R.D., Kuskowski, M.A., Wolcott, B.M., Ward, L.S., and Sulakvelidze, A. (2009). Bacteriophage therapy of venous leg ulcers in humans: results of a phase I safety trial. J. Wound Care *18*, 237–238, 240–243. https://doi.org/10.12968/jowc.2009.18.6.42801.

Rose, T., Verbeken, G., Vos, D.D., Merabishvili, M., Vaneechoutte, M., Lavigne, R., Jennes, S., Zizi, M., and Pirnay, J.P. (2014). Experimental phage therapy of burn wound infection: difficult first steps. Int. J. Burns Trauma *4*, 66–73.

Sarker, S.A., Sultana, S., Reuteler, G., Moine, D., Descombes, P., Charton, F., Bourdin, G., McCallin, S., Ngom-Bru, C., Neville, T., *et al.* (2016). Oral phage therapy of acute bacterial diarrhea with two coliphage preparations: a randomized trial in children from Bangladesh. EBioMedicine *4*, 124–137.

Schmelcher, M., Donovan, D.M., and Loessner, M.J. (2012). Bacteriophage endolysins as novel antimicrobials. Future Microbiol. *7*, 1147–1171. https://doi.org/10.2217/fmb.12.97.

Scholl, D., Cooley, M., Williams, S.R., Gebhart, D., Martin, D., Bates, A., and Mandrell, R. (2009). An engineered R-type pyocin is a highly specific and sensitive bactericidal agent for the food-borne pathogen *Escherichia coli* O157:H7. Antimicrob. Agents Chemother. *53*, 3074–3080.

Schuytema, G.S. (1977). Biological Control of Aquatic Nuisances: A Review. U.S. Environmental Protection Agency, Corvalis, Oregon.

Shen, Y., Mitchell, M.S., Donovan, D.M., and Nelson, D.C. (2012). Phage-based enzybiotics. In Bacteriophages in Health and Disease, Hyman, P., and Abedon, S.T., eds. (CABI Press, UK), pp. 217–239.

Siquet-Descans, F., and Calberg-Bacq, C.M. (1973). Large scale production of bacteriophage Φ x 174. Biotechnol. Bioeng. *15*, 927–932. https://doi.org/10.1002/bit.260150510.

van de Sande, W.W., Lo-Ten-Foe, J.R., van Belkum, A., Netea, M.G., Kullberg, B.J., and Vonk, A.G. (2010). Mycoviruses: future therapeutic agents of invasive fungal infections in humans? Eur. J. Clin. Microbiol. Infect. Dis. *29*, 755–763.

Vandamme, E.J. (2014). Phage therapy and phage control: to be revisited urgently!! J. Chem. Technol. Biotechnol. *89*, 329–333.

Verbeken, G., Pirnay, J.P., De, V.D., Jennes, S., Zizi, M., Lavigne, R., Casteels, M., and Huys, I. (2012). Optimizing the European regulatory framework for sustainable bacteriophage therapy in human medicine. Arch. Immunol. Ther. Exp. *60*, 161–172.

Wagemans, J., and Lavigne, R. (2012). Phages and their hosts, a web of interactions – applications to drug design. In Bacteriophages in Health and Disease, Hyman, P., and Abedon, S.T. eds. (CABI Press, UK), pp. 119–133.

Whittle, A., Lenhart, S., and White, K.A. (2008). Optimal control of gypsy moth populations. Bull. Math. Biol. *70*, 398–411. https://doi.org/10.1007/s11538-007-9260-7.

Wilson, C.L. (2014). Use of plant pathogens in weed control. Ann. Rev. Phytopathol. 7, 411–434.

Wright, A., Hawkins, C.H., Anggård, E.E., and Harper, D.R. (2009). A controlled clinical trial of a therapeutic bacteriophage preparation in chronic otitis due to antibiotic-resistant *Pseudomonas aeruginosa*; a preliminary report of efficacy. Clin. Otolaryng. *34*, 349–357.

Yandoc-Ables, C.B., Rosskopf, E.N., and Charudattan, R. (2006). Plant pathogens at work: progress and possibilities for weed biocontrol part 2: improving weed control efficacy. APSnet Features. https://www.apsnet.org/publications/apsnetfeatures/Pages/WeedBiocontrolPart2.aspx.

Young, R. (2014). Phage lysis: three steps, three choices, one outcome. J. Microbiol. *52*, 243–258. https://doi.org/10.1007/s12275-014-4087-z.

Yu, X., Li, B., Fu, Y., Xie, J., Cheng, J., Ghabrial, S.A., Li, G., Yi, X., and Jiang, D. (2013). Extracellular transmission of a DNA mycovirus and its use as a natural fungicide. Proc. Natl. Acad. Sci. U.S.A. *110*, 1452–1457. https://doi.org/10.1073/pnas.1213755110

Methods and Technologies to Assess Viral Interactions in the Aquatic World

15

Stéphan Jacquet[1], Xu Zhong[2], Peter Peduzzi[3], T. Frede Thingstad[4], Kaarle J. Parikka[5] and Markus G. Weinbauer[6]*

[1]INRA, UMR CARRTEL, Thonon les Bains, France.
[2]Department of Earth, Ocean and Atmospheric Sciences, University of British Columbia, Vancouver, BC, Canada.
[3]Department of Limnology and Bio-Oceanography, University of Vienna, Vienna, Austria.
[4]Department of Biology and Hjort Centre for Marine Ecosystem Dynamics, University of Bergen, Bergen, Norway.
[5]LabMCT, Belgian Department of Defence, Queen Astrid Military Hospital, Brussels, Belgium.
[6]Ocean Observatory of Villefranche-sur-Mer, Sorbonne University, Villefranche-sur-Mer, France

*Correspondence: wein@obs-vlfr.fr

https://doi.org/10.21775/9781910190852.15

Abstract

In this chapter, we summarize and discuss methods used to study viral interactions in aquatic environments. These methods to assess interactions between viruses, hosts and the environment (including other viruses) are operationally separated into interactions at the molecular, single-cell, and community levels. Many of these methods benefit from the methodological advancements in the field of molecular biology including the 'Omics area'. There are, however, numerous methods from other fields. We also present examples of unanswered questions along with problems solved by these questions. Finally, we conclude that viral interactions with the abiotic environment have received surprisingly little attention, so far, from a methodological perspective.

Introduction

This chapter focuses on methodological developments dealing with aquatic viral ecology, viral interactions with hosts, and virus diversity. Although the chapter discusses techniques which are particularly applicable to the study of aquatic viral ecology, many of these techniques can also be applied to viruses found in terrestrial environments (see Chapter 4). In addition, rather than being comprehensive, the chapter focusses on key concepts along with some under- or not fully considered research avenues.

Techniques to study virus and virus–host interactions

In the past, studies on viruses as well as virus–host interactions in aquatic systems have suffered from technical limitations. Today, however, there is a large toolbox available for a variety of analyses in viral ecology, in particular in the probing of virus–host interactions. Elsewhere in this volume we consider insights into aquatic viral ecology (see Chapter 6) which have benefited from this toolbox, but here we summarize existing and potentially new techniques for the study of these interactions.

This section is separated into three parts: (1) methods detecting organic molecules and the interactions of such molecules with viral particles (methods at the molecular level), (2) methods focusing on the localization of viruses and viral gene products or their trafficking within a cell (methods at the single-cell level) and (3) methods used at the community level, that is, as consisting of multiple species living within the same environment. A summary of the three types of techniques including potential applications, advantages, and disadvantages is presented in Tables 15.1, 15.2, and 15.3.

Methods at the molecular level (detection of organic molecules)

Spectroscopy

Mass spectrometry (MS) measures the mass-to-charge ratio of charged molecules or molecule fragments (e.g. peptides, lipids, etc.). From this feature, it is possible to differentiate and determine a molecular mass as a means of molecule identification and quantification. MS is widely used in proteomics (Domon and Aebersold, 2006; Keller and Hettich, 2009; VerBerkmoes *et al.*, 2009; Zhou *et al.*, 2011) for peptide/protein identification, in peptide sequencing (Syka *et al.*, 2004), in relative and absolute quantification of proteins, and in the determination of post-translational protein modification. Zhou *et al.* (2011) have provided detailed information about MS-based methods. Significant progress in sensitivity and accuracy of mass spectrometers (LC MS/MS, FT-ICR MS, LTQ-Orbitrap MS, TOF MS, etc.) now enables high-throughput protein identification and quantification in the low-ppm to sub-ppm range. As well, MS has emerged as a tool used in environmental proteomics (Schneider and Riedel, 2010). Banfield *et al.* (2005) used MS for the first time to profile the proteins of a microbial community (metaproteomics) of an acid mine drainage, but no study has been performed so far for viruses.

Viruses encode multifunctional proteins that interact with host-cell proteins involved in constantly adapting to and modulating the host environment. Viswanathan and Fruh (2007) provided insights via a global monitoring on the impact of viral infection on the proteome of the host cell, highlighting interactions between viral and host proteins, in particularly using quantitative or semi-quantitative mass spectrometry-based analysis of viral and cellular proteomes. MS can also be used to identify and quantify small molecules (peptides, microRNA, RNAi, crRNA). Recently, Ankrah *et al.* (2014) employed LC MS/MS to monitor the changes of dissolved organic matter (DOM) concentration and composition of a lysate originating from the activity of a lytic phage infecting a *Sulfitobacter* strain. Their main findings were that (i) viral infection can redirect 75% of nutrients into virions (viral particles) and (ii) metabolic activity differs between virus-infected cells and uninfected cells, suggesting that viral infection alters host physiology.

Infrared (IR) spectroscopy deals with light with a longer wavelength and lower frequency than visible light. It covers a range of techniques, mostly based on absorption spectroscopy and consists of a simultaneous infrared spectroscopic measurement and an electrodeposition to collect viruses from aqueous samples. These techniques can be used to identify and study chemicals and are widely

Table 15.1 Summary of methods or approaches to study virus diversity and abundance

Methods	Applications	Advantages	Disadvantages
Methods to study virus abundance			
Direct counting			
Epifluorescence microscopy (EFM)	Enumeration of total viral abundance	Sensitive detection and accurate quantification of viruses	No reliable detection and counts of RNA and ssDNA viruses; possibility to confound small bacteria and big viruses
Flow cytometry (FCM)	Enumeration of total viral abundance and several viral subgroups	High throughput analysis, sensitive detection, and accurate quantification of viruses; better discrimination between viruses and prokaryotes compared to EFM	No reliable detection and counts of RNA and ssDNA viruses; very small viruses could fall below the detection limit
Transmission electron microscopy (TEM)	Enumeration of total viral abundance and specific morphotypes	Specific information about the morphology and the size of viral particles	Low throughput, low precision, low detection limit, time consumption and costly

Table 15.1 Continued

Methods	Applications	Advantages	Disadvantages
Indirect counting			
Plaque assays (PAs)	Enumeration of viruses specific for a selected host	Quantification of the abundance of infectious units that cause the lysis of a particular host	Host cells and viruses must be cultivable
Most-probable-number assays (MPNs)	Enumeration of viruses specific for a selected host	Quantification of the abundance of infectious units that cause the lysis of a particular host	Host cells and viruses must be cultivable
Quantitative PCR (qPCR)	Enumeration of specific viral types	Sensitive and accurate quantification of specific viral groups for which direct counting methods are limited	Knowledge of the number of copies of marker genes per genome necessary, since targeted genes can occur in several copies per genome
Fluorescence *in situ* hybridization (FISH)	Enumeration of specific viral types	Rapid and high throughput analysis	Probes are limited to specific groups; drawbacks are possible if the probe is not fully specific
Microarray	Assessment of both abundance and richness of specific viral communities	Rapid and high throughput analysis	Probes are limited to specific groups; drawbacks are possible if the probe is not fully specific
Methods to study virus diversity			
Culture-independent diversity studies			
PCR-based approaches			
Fingerprinting methods	Assessment of both abundance and richness of specific viral communities	Cost-effective, rapid	Specific for groups of viruses depending on primer design; low throughput
Cloning–sequencing	Assessment of both abundance and richness of specific viral communities	Cost-effective	Specific for groups of viruses depending on the primer design
Deep amplicon sequencing	Assessment of both abundance and richness of specific viral communities	High throughput	Specific for groups of viruses depending on the primer design
Metagenomics	Assessment of both abundance and richness estimation of total viral community	No need for prior knowledge of viruses or viral sequences present in the samples;	Possible low sequencing depth of coverage, short read length, sequencing error, microbial contamination, majority of sequences in virome has no affiliation
Pulsed field gel electrophoresis (PFGE)	Assessment of genome size distribution of viruses (isolated and of communities)	No PCR steps involved	Limited to dsDNA viruses
Culture-dependent diversity studies			
Above methods in culture-independent studies	Confident and robust results on viral microdiversity of specific hosts	Circumvention of the potential problem of non-specificity of primers	Host cells and viruses must be cultivable

Table 15.2 Summary of methods used to study virus morphology, molecular structure and some virus–host interactions at molecular level

Methods	Applications	Advantages	Disadvantages
Electron microscopy			
Cryo-electron microscopy (cryo-EM)	Structural details of viral particles (e.g. topology and accessory arrangement); resolution 0.5–2 nm	No crystallisation of samples needed	Description of only the 'average' structure of entire virus population; not applicable to pleiomorphic, architecturally uniform, and virus >100 nm
Transmission electron microscopy (TEM)	Particle and cell 2D morphology imaging, e.g. used to assess virus morphology and estimate the viral burst size; also used to estimate viral infection of bacteria; resolution 0.1 nm	Classical method to visualize a virus or viral particle and provide size and morphological details	Complete sample dehydration usually required
Atomic force microscopy (AFM)	Nanoimaging 3D structure of protein and nucleic acid (i.e. secondary and tertiary structure of large RNA molecules); resolution < nm	Mechanically and electronically more straightforward than electron microscopy; inexpensive and easy to install; assessment of ultrastructure at nanoscale; analysis asymmetrical and non-uniform objects and viruses larger than 100 nm	
Crystallography			
X-ray crystallography	Protein 3D structure e.g. unveiling the crystal structure of lysin PlyC from the streptococcal bacteriophage or the structural basis for CRISPR RNA-guided DNA recognition by Cascade; resolution 2.3 nm		Crystallization of sample needed before analysis potentially affecting the state of protein/RNA structural conformation; limited to particles larger than 100 nm
Spectrometry			
Infrared spectrometry	Identification of types of virus, based on distinct vibrational fingerprints of viral components (nucleic acids, proteins, phospholipids, and other small molecules); resolution 20–30 nm	Fast and easy	Low resolution
Nuclear magnetic resonance (NMR) spectrometry	Assessment of structural details of the coat proteins and DNA of phages at the atomic resolution		Restricted to a molecular mass of up to ~100 kDa and RNAs of up to 35 kDa; gram quantities of material required; model building required for interpretation
Mass spectrometry (MS)	Used for peptide/protein identification, peptide sequencing, relative and absolute quantification of proteins, and the determination of post-translational modifications		

Table 15.3 Summary of methods or approaches to study virus–host interactions

Methods	Applications	Advantages	Disadvantages
Single-virus tracking/ imaging	Developed to follow the fate of individual viral particles, monitor dynamic interactions between virus and cellular structure/ component, study the viral entry (virus–receptor interaction, penetration, internalization), intracellular transportation, genome release, nuclear transport, and cell-to-cell transmission		Only applicable for virus cultures and not for environmental samples
PhageFISH	Identification of specific viruses infecting specific hosts; identification of modes of infection (lytic, lysogenic or chronic infection)	Assessment of viruses and hosts in a culture-independent way	Specific probes for each virus–host pair required
Viral-tagging	Assessment of infection of specific hosts by specific viruses	High-throughput counting of fluorescently labelled viruses	Only applicable for isolated viruses–host systems
Microfluidic digital PCR	Assessment of co-occurrence of phages and hosts	Fast screening of virus–host pairs from natural samples; independent of culturing	Possible false positive signals can occur when 1) multiple phage genes are released from a lysed cell, 2) multiple bacterial 16S rRNA genes from cells adhered to the same chamber, or 2) fluorescence signal spilled over from neighbouring chambers
Single-cell amplified genomics	Identification of individual cells and infected viruses	Viable approach to obtain genomes of unculturable viruses; assessment of viruses and hosts in a culture independent way	Low throughput, costly and time consuming

used tools for material identification and secondary structure analysis in chemistry, biology, and biochemistry. Therefore, IR spectroscopy can also be used to identify and quantify different types of viruses based on distinct vibrational fingerprints for viral components (nucleic acids, proteins, phospholipids, and other small molecules) (Vargas *et al.*, 2009). IR spectroscopy can provide information on the composition of a material within minutes. Applying IR to drinking water, which is an oligotrophic (low-nutrient) environment, low-concentrations of viral particles could be detected with good sensitivity within a few minutes. Until recently, diffraction limitations prevented nanoscale protein studies. Amenabar *et al.* (2013), however, developed a method, referred to as the Fourier transform infrared nanospectroscopy (nano-FTIR), to overcome this limitation and to map protein structure at the nanoscale with a 30-nm lateral resolution and sensitivity for individual protein complexes.

Nuclear magnetic resonance (NMR) spectroscopy exploits the magnetic properties of atomic nuclei. It determines the physical and chemical properties of atoms or the molecules in which they are contained. The intramolecular magnetic field around an atom in a molecule changes the resonance frequency, thus giving access to details of the electronic structure of a molecule. NMR spectroscopy is a powerful tool to study and model the three-dimensional structure of proteins and nucleic acids at an atomic resolution (Tzakos *et al.*, 2006; Dingley *et al.*, 2008). The use of NMR as a multiprobe method enables detailed structural studies on conformational equilibria, molecular dynamics, and the relative orientation of protein domains or proteins in a macromolecular complex as well as the identification of intermolecular interactions at protein-protein, protein-ligand, or protein–nucleic acid interfaces (Tzakos *et al.*, 2006). An example for NMR application to viruses at the atomic resolution has been the description of the structural details of the proteins and DNA of a filamentous bacteriophage (Opella *et al.*, 2008). NMR is suited to analyse binding epitopes of ligands bound to receptor proteins. It has been used to identify and characterize the binding of an antiviral compound to native human rhinovirus serotype 2 (Benie *et al.*, 2003). NMR is also suitable for RNA structural determination and thus for binding affinity interaction between microRNA and protein, or microRNA and microRNA (Scott and Hennig, 2008). Current NMR and labelling techniques for robust structure determination are now feasible for proteins with a molecular mass of up to ~100 kDa and RNAs of up to 35 kDa (Tzakos *et al.*, 2006).

X-ray crystallography

X-ray crystallography (XRC) is a tool used for identifying the atomic and molecular structure of a crystal, in which the crystalline atoms cause a beam of incident X-rays to diffract in many specific directions. By measuring the angles and intensities of these diffracted beams, a three-dimensional picture of the density of electrons within the crystal can be produced. From this electron density, the mean positions of the atoms in the crystal can be determined, as well as their chemical bonds, their disorder and various other information. XRC is a fundamental research tool that has shaped our understanding of biological structures and functions at the molecular level (Yonath, 2011). It provides the information required to create accurate three-dimensional models of organic molecules, namely methodological mapping the position of each and every atom that constitutes the studied object.

The biggest problem when using X-ray crystallography for environmental applications is that the sample needs to be crystallized prior to analysis. Crystallization may affect the native state of the analysed protein or the structural conformation of RNA. Nevertheless, X-ray crystallography remains useful in virology as indicated by the two following examples. McGowan *et al.* (2012) described the crystal structure of lysin PlyC from a streptococcal bacteriophage. They found that PlyC contain two components (PlyCA and PlyCB) which play different roles in phage-induced bacterial lysis. PlyCA contains a glycoside hydrolase domain in addition to the previously recognized cysteine, histidine-dependent amidohydrolases/peptidases (CHAP) catalytic domain. The presence of eight cell-wall-binding domains together with two catalytic domains may explain the extraordinary potential of the PlyC holoenzyme to target bacteria. X-ray crystallography has also revealed the composition and structure of what is known as Cascade, consisting of CRISPR-associated Cas proteins and crRNA, and the structural basis for CRISPR RNA-guided DNA recognition by Cascade (Jore *et al.*, 2011).

Electron microscopy

An electron microscope uses a beam of accelerated electrons as a source of illumination. Since the wavelength of an electron can be up to 100,000 times shorter than that of visible light photons, the electron microscope has a higher resolving power than a light microscope and can reveal the structure of objects or features of objects which are smaller than a wavelength of light. Electron microscopy (EM) is a classical method used for the visualization of viruses and virus-like particles and provides morphological details (Ackermann, 2012). There are many other applications of EM at the molecular level. For example, immuno-EM allows the detection of specific proteins in ultrathin tissue sections. Antibodies labelled with heavy metal particles (e.g. gold) can be directly visualized using transmission electron microscopy (TEM). Viruses and viral proteins can be localized within cells using such techniques.

Transmission EM is the traditional method used to describe morphological features of isolated viruses. It offers also the possibility to describe viruses on organic particles as long as these particles are not electron dense. Upon specific collection of organic particles and embedding them into a resin, ultrathin sections can be made. Subsequently the viruses can be enumerated (Peduzzi and Weinbauer, 1993). Also, the fraction of host cells containing mature virions can be quantified (Proctor and Fuhrman, 1991). Using conversion factors and models, an estimation of the percentage of infected cells and viruses mediated mortality can be made. Scanning electron microscopy (SEM) in principle is also applicable to the study of viral particle interactions. So far, however, only specific and potentially large viruses can be detected by SEM.

In Cryo-EM, also known as electron cryotomography, multiple tilted 2D EM images made at cryogenic temperatures are combined to obtain a 3D reconstruction of a sample. It is based on automated collection, cataloguing and the elaboration of mathematical techniques in order to analyse images of viral particles. These are then used for the reconstruction of viral particle structures from a collection of particle observations, taken from multiple orientations/angles. Cryo-EM can thus be used to obtain structural details of viral particles (e.g. topology and accessory arrangement) at sub-nanometre scale. This method has been used for example in the characterization the marine siphovirus Vibrio phage SIO-2 (Lander *et al.*, 2012).

Atomic force microscopy

Atomic force microscopy (AFM) is a very high-resolution type of scanning probe microscopy that forms images of surfaces using a mechanical probe on a cantilever. When the tip of the probe is brought into the proximity of a sample surface, the forces between the tip and the sample lead to a deflection of the cantilever, which can be measured. Resolution on the scale of fractions of a nanometre has been obtained. This technique has been reviewed by Kuznetsov *et al.* (2011) for the application of imaging viruses and virus-infected cells. Atomic force microscopy is also a promising technique for the nano-imaging and the determination of the overall 3D structure of large RNA molecules. Its recent application on a hepatitis C virus (Gilmore *et al.*, 2014) opened up a whole new avenue for the study of the secondary and tertiary structure of long RNA molecules. Like cryo-EM, AFM does not require the crystallization of viral particles and has a similar level of resolution. It can be used to analyse viruses that do not have symmetrical or uniform architectures, while this is not the case for cryo-EM and X-ray crystallography (which are both based on an average estimation of symmetrical viral structures). Moreover, AFM can also be used for the analysis of giant viruses with sizes larger than 1000 nm, whereas both cryo-EM and X-ray chromatography are limited to smaller particles (Kuznetsov *et al.*, 2011).

Summary and outlook

Only NMR and AFM allow for structural investigations of nucleic acids, as they provide high resolution. It is noteworthy that only MS and EM have been employed for environmental samples so far; other techniques require large amounts of material for robust analysis or detection, and have therefore been only used in laboratory studies with cultured viruses. An improvement of these techniques is required for environmental applications within complex communities, to allow *in-situ* and reproductive analysis from as little as, for example, 1 ml of raw water.

Methods at the single-cell level

Electron microscopy

Transmission EM reveals 2D images and requires staining with heavy metal dyes or shadowing for contrast enhancement. By examining and quantifying viruses in cells, TEM is also used for the estimation of virus-mediated bacterial mortality (Proctor and Fuhrman, 1990; Weinbauer *et al.*, 1993, Binder, 1999). Modification and improvement of TEM resolution have been developed to meet different needs of application, such as seen with what is described as high-resolution TEM (HRTEM).

Fluorescence *in situ* hybridization-based approaches

Fluorescence *in situ* hybridization (FISH) is a powerful tool for the single cell detection of different bacterial types and has also been applied to viruses. Allers *et al.* (2013) developed the phageFISH method, consisting of an improved version of the geneFISH protocol (Moraru *et al.*, 2010) that increases gene detection efficiency from 40% to >92% at the single-cell level. This method was optimized for the detection and visualization of intracellular as well as extracellular phage DNA and allows a simultaneous identification and quantification of host cells during all stages of infection. Briefly, it uses a catalysed reporter deposition (CARD)-FISH protocol (Pernthaler *et al.*, 2002; Amann and Fuchs, 2008) to link, on the one hand, the florescence-labelled rRNA detection and, on the other hand, the detection of viral DNA. Visualization is carried out using epifluorescence microscopy. This approach offers the opportunity to link viruses and hosts in a culture-independent manner and to study the virus–host interaction at the single-cell level. It is based on an appropriate probe design, which is achieved given *a prior* knowledge of phage and host genes. This method has been validated using a marine virus–host model (e.g. *Pseudoalteomonas* sp. strain H100 and its phage PSA-HP1, Allers *et al.*, 2013), but has not yet been applied to field samples. It could be potentially useful for environmental surveys and lineage-specific population ecology of free phages (such as for well documented viruses, like cyanomyoviruses, T4-like myophages, etc.).

Kenzaka *et al.* (2010) developed the 'cycling primed *in situ* amplification fluorescence *in situ* hybridization' (CPRINS-FISH) method for studying virus-mediated gene transfer in freshwater ecosystems. In this method, viruses were labelled via the *gfp* or *bla* gene cloned into the genome and detected by using epifluorescence microscopy. Direct viable counting combined with CPRINS-FISH revealed that more than 20% of the cells carrying the transferred gene retained their viability. This brought into question the previous assumptions that a proportion of transferred genes inside recipient cells may be destroyed. These results, however, suggested that DNA exchange among bacteria via phages in natural aquatic environments may be more frequent than previously thought.

Single-virus tracking

Single-virus tracking (Brandenburg and Zhuang, 2007) is a FISH method in living cells and offers the possibility to detect individual viruses in living cells. Using this method, both viral and cellular components (e.g. protein, genome, or membrane, etc.) are labelled using specific fluorescent probes. The labelled viruses then can be visualized in live cells by using the fluorescence microscopy (e.g. epifluorescence microscopy, confocal laser scanning microscopy, or total internal reflection fluorescence microscopy). This single-virus imaging method allows following of the fate of individual particles, monitoring dynamic interactions between viruses and cellular structures/components, and studying viral entry (virus–receptor interaction, penetration, internalization), intracellular transportation, genome release, nuclear transport, and cell-to-cell transmission (Marsh and Helenius, 2006; Brandenburg and Zhuang, 2007; Sun *et al.*, 2013). For example, Hübner *et al.* (2009) monitored the movement of Gag-GFP in HIV-infected cells and found that the Gag proteins migrate and assemble into button-like structures adjacent to neighbouring cells. The authors further showed that the button-like Gag structures could enable cell-to-cell transmission and infection of HIV.

Flow cytometry-based approach

In addition to phageFISH, Deng *et al.* (2012) also developed a viral-tagging (VT) method that uses flow cytometry (FCM) to count fluorescently labelled viruses. This FCM-based approach allows high-throughput detection and sorting of co-occurring viruses and hosts. This method was tested and validated using two cultivated hosts (the cyanobacterium *Synechococcus* and the gammaproteobacterium *Pseudoalteromonas*) and their phages (podo-, myo-, and siphoviruses) by comparing results with a conventional method (liquid and plaque assay) (Deng *et al.*, 2012).

Microfluidic digital PCR

The microfluidic digital polymerase chain reaction (PCR) technique (Tadmor *et al.*, 2011) serves to link single bacterial cells harvested from a natural environment to a viral marker gene, i.e. to detect an infected cell. Briefly, both a viral gene and the bacterial 16S rRNA gene are amplified using specific primers in each microfluidic PCR chamber that ideally contains only a single bacterial cell. They can be thereafter labelled using two fluorescence-labelled probes, one for the virus and another for the host. The co-localization of virus and host can thus be visualized and virus–host pairs can be identified by sequencing (Tadmor *et al.*, 2011).

Single-cell/virus-genomic approach

Using single-cell genomics, viral DNA can also be sequenced (Ishoey *et al.*, 2008; Woyke *et al.*, 2010; Stepanauskas, 2012), thus providing a means to reveal organismal interactions in uncultivated microbes. Yoon *et al.* (2011) used fluorescence-activated cell-sorting to obtain individual heterotrophic protist cells from a 50-ml seawater sample, and then applied whole-genome-sequencing of three sorted and single uncultivated marine protist cells (Picobiliphyta). They showed that genome data from one cell were dominated by sequences from a widespread single-stranded DNA virus, MS584–5, which was closely related to *Nanoviridae*, using the Rep gene as the phylogenetic proxy. Results showed a lytic interaction with the host. This virus was absent from the other two cells, however. Both cells also contained non-eukaryote DNA derived from marine Bacteroidetes and large DNA viruses. This single-cell-genomic approach (i) revealed potentially complex biotic interactions among previously uncharacterized marine microorganisms, and (ii) provides a new approach to study, without cultivation artefacts, the viral interactions with their hosts (protists and prokaryotes) *in situ*.

Summary and outlook

PhageFISH allows *in situ* tracking of viruses in the environment. Viral-tagging (VT) and microfluidic PCR do not allow *in situ* analysis but can instead be used to track viruses at different time points in virus–host systems. Additionally, the VT method enables high-throughput analysis as it employs sorting by FCM. Using VT, PhageFISH or microfluidic PCR requires targeted virus and host sequences to be known for probe or primer design. In contrast, the single-cell/virus genomics approach is not limited by this constraint.

Interactions at the community level

Quantifying the virosphere

The focuses of viral ecology have evolved during the past decade. After a bloom of viral abundance studies lasting about two decades, the interests of viral ecology have shifted from the quantitative to the qualitative exploration of the viriosphere. Consequently, there has been little development of the quantification techniques. Within the reports of viral and host abundance, inference of interactions was made using the virus-to-prokaryote ratio (VPR), using the numerical dominance (or lack thereof) of viral particles over hosts as indicative of high/low viral dynamics. This has since been shown to be overly simplistic (Parikka *et al.*, 2016). Before the use of direct counting methods, viral abundance in aquatic ecosystems was investigated using indirect methods by plaque-assay on agar plates (yielding plaque-forming units (PFU)) or by the most-probable-number assays (MPN) in liquid medium. The discovery of the high incidence of viral particles in various marine waters by Bergh *et al.* (1989) launched numerous investigations of viral abundance in a variety of aquatic ecosystems using different direct counting techniques. The methods commonly used are TEM, epifluorescence microscopy (EFM) and FCM, which were developed for viral enumeration roughly in that order. Since it is not always possible to prove a viral origin by using these methods, one finds also the term virus-like particle (VLP) in the literature.

While the majority of current studies presenting data on viral numbers used EFM and FCM, due to methodological problems with the TEM approach a new technique also using TEM has been recently developed. Quantitative TEM (qTEM) (Brum *et al.*, 2013) is based on the use of an air-driven ultracentrifugation of viral samples onto hydrophilic-rendered Formvar copper grids (Brum and Steward, 2010). The method, described by Hammond *et al.* (1981) for animal viruses and as subsequently adapted to phage enumeration (Maranger *et al.*, 1994), has been redescribed for prokaryotic virus quantification and evaluated by Brum and co-workers (Brum and Steward, 2010; Brum *et al.*, 2013). The authors report no significant differences between the morphological data of viruses between absorption of marine samples onto grids and the deposition of viruses directly onto grids by qTEM hence, no biases are expected. Viral enumeration using qTEM thus allows morphological data analysis in addition to providing quantitative data on viruses, which represents an advantage of this direct counting method compared to EFM and FCM.

Viral enumeration using EFM is essentially a variation of the methodology applied originally to prokaryotic enumeration, where cells are captured on filters, stained with fluorescent dyes, and counted using a microscope (Francisco *et al.*, 1973; Daley and Hobbie, 1975). As this can be laborious, Cunningham *et al.* (2015) developed the 'wet mount' method, where stained samples are mounted directly onto slides without first capturing viral particles onto a filter. In this method, the sample is first concentrated (if necessary), stained with a fluorescent dye and then a known concentration of silica microsphere beads is added prior to observation. The sample is then wet mounted to a slide and the viral numbers are counted directly. Their concentration in the sample is finally inferred from their ratio in relation to added microsphere beads. This method is rapid, accurate and ca. 500-fold less expensive in material costs compared to the classical filter-mounting method.

Requiring more equipment and expertise than EFM, flow cytometry is also based on the staining of the nucleic acids of viral particles using fluorescent dyes. FCM has the distinct advantage of a rapid and high-throughput analysis of samples over TEM and EFM, processing at a rate of thousands of events per second (Brussaard *et al.*, 2010; Wang *et al.*, 2010) and the possibility to distinguish different viral populations (Marie *et al.*, 1999; Brussaard *et al.*, 2000). As virtually all types of samples can be eluted into water after sample treatment, EFM and FCM are applicable to all types of samples. For viral enumeration in aggregates (e.g. aquatic snow), another technique was proposed by Luef *et al.* (2009) and validated by Peduzzi *et al.* (2013) as an alternative to the three mentioned counting methods: confocal laser scanning microscopy (CLSM). Preceding enumeration, the matrix material of aggregates is stained using lectin and prokaryotic cells and viral particles are dyed using SYBR Green I. Cryosections are also performed on larger aggregates, allowing better detection and distribution of viruses, bacteria and aggregate constituents. Although enumeration seems to be somewhat difficult due to problems with digital image analysis of the CLSM pictures, confocal laser microscopy offers the benefit of the visualization of viral particle and prokaryotic cellular distribution within aggregates, and thus is a promising avenue for the ecological studies of aquatic aggregates. Allen and colleagues (2011) combined genomics with CLSM to sort and sequence single viral particles, yielding 'Single Virus Genomics'. Using a mixed viral assemblage of fluorescently dyed *Escherichia coli* phages λ and T4, they sorted viral particles, using a forward scatter photomultiplier tube for more sensitivity, into agarose beads applied to 'multi-well' microscope slides. The nanolitre droplets containing the sorted virions were overlaid with additional agarose for embedding and stabilizing the particles for CLSM. The individual virions were then visualized with CLSM and beads, with only a single virion chosen for further analysis. Whole genome amplification via multiple displacement amplification (MDA) was performed *in situ* and subsequently sequenced.

Most of the development of methods related to EFM and FCM has been in sample preparation in order to better distinguish viral particles from background noise. Little has been done, however, to enhance the detectability of viruses with genomes other than double-stranded DNA. Regardless of the stainability of single- and double-stranded DNA and RNA (Tuma *et al.*, 1999; Shibata *et al.*, 2006), ssDNA and RNA viruses (and viruses of small genomes in general) are either scarcely detected or undetectable (Brussaard *et al.*, 2000; Holmfeldt *et al.*, 2012). Although it has been presumed that dsDNA phages are the prevailing viruses in aquatic ecosystems (Steward *et al.*, 1992; Breitbart *et al.*, 2002; Weinbauer, 2004; Comeau *et al.*, 2010), more recent studies suggest an unexpected importance of ssDNA and RNA viruses (Angly *et al.*, 2006; Lang *et al.*, 2009; Rosarioa and Breitbart, 2011; Roux *et al.*, 2012). Therefore, the future of the improvement in FCM probably lies in the development of methods to detect ssDNA and RNA viruses.

In medicine, developments for the detection of RNA viruses, such as HIV, have been made by adding markers, such as combinations of beads and antibodies. In this technique either several (Kim *et al.*, 2009) or a single (Arakelyan *et al.*, 2013) virion is attach to magnetic beads and then revealed by specific antiviral fluorescent antibodies, giving the name 'Flow Virometry'. Gaudin and Barteneva (2015) developed the 'Flow Virometry Assay' in which they detected and sorted ssRNA Junin-viruses (a causative agent of Argentine haemorrhagic fever), while being able to retain virion infectivity in the process. The authors used a flow cytometer equipped with a powerful laser (300 mW and 488 nm), a digital focusing system (DFS), and using the forward scatter channel option for the multiplier tube (FSC-PMT). By optimizing the parameters of the DFS, a picomotor-driven focusing device that adjusts the beam in order to obtain a smaller focal spot, they were able to detect small lipid microvesicles. The application of this technology to viral ecology could enable the exploration of the ssDNA, dsRNA and ssRNA viruses, as well as other elements of horizontal gene transfer, such as gene transfer agents (GTA), membrane vesicles (MVs) and VLPs. As pointed out by Forterre *et al.* (2013), the presence of the aforementioned elements of horizontal gene transfer elements are likely to interfere with the counts of *bona fide* viral particles in a given sample. There is currently few data on the abundance of MVs, GTAs, BLPs and free nucleic acids, although their ubiquitous production in aquatic ecosystems has been established (Mashburn-Warren and Whiteley, 2006; McDaniel *et al.*, 2010; Chiura *et al.*, 2011; Gaudin *et al.*, 2013; Lang *et al.*, 2012).

Fingerprints

DNA fingerprinting describes a collection of techniques originally developed to identify individual organisms but later adapted for species identification and studies of diversity. Originally accomplished with restriction analysis and Southern blotting techniques, PCR has become a core technique. DNA sequencing technologies especially those used in metagenomics analysis are also used.

Although several biases can be associated to the use of fingerprint methods, these techniques remain very useful, as they are cost-effective and reproducible (e.g. denaturing gradient gel electrophoresis, DGGE; Zhong and Jacquet, 2013; Zhong *et al.*, 2014). In addition to fingerprinting methods, sequencing the amplicons (using next generation sequencing (NGS), i.e. deep-sequencing or ultra-deep sequencing) of the cloned amplicons (using traditional Sanger sequencing method, i.e. cloning–sequencing) have been widely used. The classical cloning–sequencing method is based on the construction of a clone library for amplicon DNA fragments with a plasmid vector for its expression in a host cell (e.g. *E. coli*, etc.) and clones are then sequenced by using the Sanger method (Zhong *et al.*, 2015). With the improvements in sequencing technologies in recent years, the high throughput 'next generation sequencers' (e.g. 454 pyrosequencing, Illumina, SOLiD, Helixo, PacificBio), it is now possible sequence millions/billions of DNA fragments per day, with costs that are increasingly affordable. Such an approach has the advantage of revealing what is rare, i.e. the 'rare biosphere'(Sogin, 2006), as it allows deep sequencing (Simon and Daniel, 2011).

Randomly amplified polymorphic DNA (RAPD)-PCR and separation by gel electrophoresis of viral concentrates from isolates or natural communities yield a simple fingerprint based on size patterns (Winget and Wommack, 2008). These fragments can be used as probes or sequenced. The same can be done for the host. A high seasonal variability of deep-sea viruses as well as the co-variation with host community structure has been observed using such a technique (Winter and Weinbauer, 2010).

Pulsed field gel electrophoresis (PFGE) allows fingerprinting virioplankton based on size fractionation of intact genomic DNA. This technique has proved on many occasions, since the pioneer study of Wommack *et al.* (1999), to be very powerful in giving a first insight on genome size distribution in a variety of aquatic environments. It provided the first data on viral community dynamics related to their genome sizes in an aquatic environment. PFGE, however, can only provide a rough minimum estimate of the virioplankton diversity (Parada *et al.*, 2008). Indeed, it can only be used for dsDNA viruses. Also, PFGE can only reveal abundant groups and hence is limited for the detection of minor groups (Filippini and Middelboe, 2007, Colombet *et al.*, 2006, Zhong *et al.*, 2014). Because a good PFGE performance requires a large amount of DNA (corresponding to ca. 10^9 viruses per plug), insufficient DNA loads may make bands and thus viruses invisible. To obtain a sufficient concentration of VLPs, additional steps are required to concentrate viruses, which can also produce biases. Moreover, one PFGE band, representing a subpopulation of dsDNA viruses (characterized by having the same genome size), may contain several genetically and morphologically different viruses or viral groups, and the composition of this subpopulation (band) may vary with time (Jamindar *et al.*, 2012). It is noteworthy that PFGE can be very powerful if combined with PCR when investigating the diversity of specific viral groups, based on the analysis of the DNA bands isolated from PFGE gel slices. This type of analysis has been useful in detecting connections between virus phylogenetic affiliation and genome size (Sandaa and Larsen, 2006; Sandaa *et al.*, 2008; Jamindar *et al.*, 2013; Zhong *et al.*, 2014).

PCR is a sensitive tool that has been successfully used to explore viral diversity using degenerate primers targeting phylogenetic markers. Unlike prokaryotes and eukaryotes, for which universal genes and primers exist, the application of PCR to viruses is trickier. Although no common genes can be found for all viruses, some conserved core genes exist in different viral groups that can be used as genetic markers for these viral groups (Rohwer and Edwards, 2002). The differences in sequences found for these genes correspond to potentially different genotypes, and the number of genotypes can be used as a proxy of viral species richness (Weinbauer and Rassoulzadegan, 2004). These group-specific gene markers are, for instance, *g20*, *g23*, and *g43* for T4-like *Myoviridae* including cyanophages (Fuller *et al.*, 1998; Zhong *et al.*, 2002; Desplats *et al.*, 2003; Filée *et al.*, 2005; Comeau *et al.*, 2008; Sullivan *et al.*, 2008; Marston *et al.*, 2013); *polA* for T7-like *Podoviridae* (Breitbart *et al.*, 2004; Labonté *et al.*, 2009; Chen *et al.*, 2009; Dekel-Bird *et al.*, 2013); *psbA* for most isolated cyanophages (Suvillan *et al.*, 2006); *polB* and *mcp* for nucleocytoplasmic large DNA viruses (NCLDV) (Chen *et al.*, 1996; Larsen *et al.*, 2008). To analyse PCR amplicons and assess viral diversity, fingerprinting methods (see above), cloning–sequencing (clone library construction followed by Sanger sequencing), or deep-sequencing (amplicons tagged plus massive sequencing by NGS can be used (for instance Zhong *et al.*, 2013, 2014, 2015).

Metagenomics, metatranscriptomics, metaproteomics, metabolomics

Metagenomics is the study of genetic material recovered directly from environmental samples. The broad field may also be referred to as environmental genomics, ecogenomics, or community genomics. Likewise, studies at the community levels are called metatranscriptomics when gene expression is studied based on RNA sequence and metaproteomics when gene expression is studied for the proteins. Metatranscriptomics and metaproteomics studies have not been yet performed at the community levels. Metabolomics is the scientific study of chemical processes involving metabolites. More specifically, metabolomics is the study of the unique chemical fingerprints that specific cellular processes leave behind. The metabolome represents the collection of all metabolites, which are the end products of cellular processes. Gene expression data based on mRNA along with proteomic analyses reveal the set of gene products being produced within a cell at a given moment or under a given set of environmental conditions, data that represents one aspect of cellular functions. Conversely, metabolomic profiling can give an instantaneous snapshot of the physiology of the cell. Metabolomics at the community level (community metabolomics or 'metametabolomics') remains unapplied to viruses.

Metagenomics is a powerful technology, developed during the last two decades, which allows both phylogenetic and/or functional studies on microbial/viral communities using total DNA/RNA, extracted from the environment (see Chapter 5). One of the approaches is the whole metagenome sequencing using the Sanger method to sequence the shot-gun libraries with the goal of studying microbial/viral phylogenic and functional diversity. Using this approach, it was shown that marine viruses are extremely diverse and more than half of the sequences obtained were unknown (Breitbart *et al.*, 2002; 2004; Bench *et al.*, 2007; Williamson *et al.*, 2008). These studies thus indicated the great force and potential of sequencing for examining microbial community diversity. Recent high-throughput sequencing technologies have provided a rapid and robust resolution for exploring the vast taxonomic diversity of viruses as they enable the direct sequencing of DNA without clone library construction and associated biases. The first application of metagenomics for the study of aquatic viruses was carried out on seawater-samples by Angly *et al.* (2006). Compared to other studies (Breitbart *et al.*, 2002, 2004; Bench *et al.*, 2007), which used the traditional sequencing method, this approach lead to the discovery of a large and unknown viral diversity. Since then, it has been applied for studying viral communities in oceans (Angly *et al.*, 2006), lakes (López-Bueno *et al.*, 2009; Lauro *et al.*, 2011; Roux *et al.*, 2012; Brum *et al.*, 2015a,b; Skvortsov *et al.*, 2016; Aguirre de Cárcer *et al.*, 2016), reclaimed water (Rosario *et al.*, 2009) and desert ponds (Fancello *et al.*, 2013), as well as for revealing viral–host interactions (Reyes *et al.*, 2010; Rodriguez-Brito *et al.*, 2010).

Unlike PCR-based methods, the metagenomic approach does not rely on prior knowledge of viruses or viral sequences present in the samples (Mokili *et al.*, 2012). Such an approach is therefore well adapted for the investigation of the diversity of viruses, for which no universal phylogenetic marker exists. The total available virome sequences make the study of viral taxonomic diversity easier and more comprehensive (Rosario and Breitbart, 2011). Hence, viral metagenomes allow the examination of the whole virioplankton diversity, providing both (i) functional and phylogenetic diversity analysis as well as (ii) abundance and richness at the same time. Although the application of NGS in metagenomic has extended our ability to assess multiple metagenomes and increased our knowledge on environmental microbial diversity, it currently also suffers from problems such as those associated to the low sequencing depth of coverage, short read length, sequencing error, microbial contamination, and from the fact that the majority of sequences in virome typically are found to have no known affiliation (Mokili *et al.*, 2012). Despite all drawbacks mentioned above, the current high-throughput sequencers are powerful and especially so as the stability, fidelity, high-throughput characteristics, read-lengths, and single-molecule sequencing technologies improve.

It is also noteworthy that methods exist to assess the diversity of RNA viruses (Culley *et al.*, 2006), by which the RNA virome is firstly reverse-transcribed to complimentary DNA prior to sequencing. Recent studies have revealed that, in some situations, virioplankton could be largely represented by RNA viruses (Culley *et al.*, 2006; Lang *et al.*, 2009; Steward *et al.*, 2013). Steward *et al.* (2013) investigated both RNA and DNA viromes of the California coastal seawater and demonstrated that RNA viruses are as numerous as DNA viruses, accounting for 38–60% of total viral abundance. Molecular surveys using degenerate primers to target the RNA-dependent RNA polymerase gene of picorna-like viruses have shown that, in addition to the handful of isolates, a very diverse pool of uncultivated picornaviruses exist in seawater (Culley *et al.*, 2003; Culley and Steward, 2007; Gustavsen *et al.*, 2014). The quantification of these RNA viruses, however, has remained unachieved so far.

While viral metagenomics have been applied more and more during the last decade, only a few studies have been made available using transcriptomics (Brum and Sullivan, 2015). To give a recent example, Lin *et al.* (2016) studied the transcriptomic responses of *Prochlorococcus* infected by a cyanomyovirus under phosphorus limitation, a strong selective force in the ocean. They could reveal that transcripts of the phosphorus acquisition genes such as *pstS* in the uninfected cells were enriched after phosphorus limitation but also in the infected cells. By contrast, other genes, such as ATP synthetase and ribosomal protein genes, were depleted in uninfected cells after phosphorus limitation but were enriched in infected cells. Their study also revealed that phage *pstS* transcript number per cell was almost 20 times higher than the host copy, suggesting this may help to maintain the host phosphate uptake rate during viral infection.

Summary and outlook

Fingerprints allow for rapid detection of changes in the viral groups making up communities found in experimental set-ups and systems, and along environmental gradients such as temperature, light, water depth (pressure), salinity, pCO2 levels or trophic status of the system. Some fingerprints can also provide sequence information (e.g. from excised DGGE or RAPD-PCR bands) or can be combined with subsequent approaches to obtain sequence information (e.g. excised PFGE bands + fingerprinting + sequencing). There are no common genes for viruses, not even for the monophyletic group of the *Caudovirales* (Ackermann, 1998). Hence, unless genomic DNA (or RNA) is the target, fingerprints are only targeted against specific viral groups, and this limits the information obtainable for entire communities. Also, the majority of environmental sequences, such as those obtained by metagenomic approaches, have no homologues in databases of isolated viruses. Hence, we do not know what most of the sequences encode.

Conclusions

Overall, the number of methods used to analyse virus interactions at the molecular, cellular and community level have increased tremendously during the last decades. Recently, new analyses to deal with large-scale data sets have allowed assessment of the potential for co-occurrence of organisms and non-living environmental conditions including the role of viruses in the ocean. For example, the implication of viruses associated with carbon export in nutrient-poor regions of the ocean has been described (Chow *et al.*, 2013).

The toolbox for the more direct study of the interactions between viruses and the abiotic environment nevertheless is poorly filled, especially in non-aquatic systems. Epifluorescence microscopy has been used extensively to quantify viruses in the water column and in the sediments of aquatic systems and represents a golden standard, of sorts, against which other methods such as flow cytometry (Brussaard *et al.*, 2010) have been tested. Epifluorescence microscopic enumeration has also been used to roughly quantify the proportion of 'free-living' viruses and the proportion of viruses attached to particles such as debris (marine and lake snow) or suspended sediment material (Weinbauer *et al.*, 2009). Upon selective collection of organic particles and separation of viruses from the organic matrix, viruses can be counted more specifically. Recently, epifluorescence microscopy based approaches to assess viral production and virus-mediated mortality of prokaryotes (virus reduction approach) have been applied to organic particles (Bettarel *et al.*, 2016). Further improvements will be required for the detection of ssDNA, dsRNA and ssRNA viruses as well as elements of horizontal gene transfer. For these, some methodologies exist in the medical field.

Laser scanning microscopy (LSM) has been used to detect viruses on and within particles such as marine, lake and river snow, soot or desert dust and obtain digitally created high-quality images. This method is useful to estimate viral abundance on particulate material, such as transparent exopolymeric particles (TEP) and other types of aquatic aggregates (e.g. Weinbauer *et al.*, 2009; Peduzzi *et al.*, 2013), however it also allows assessment of the distribution of viruses on particles. In combination with specific markers, the distribution of specific viruses or viral groups on and within particles thus could be assessed in the future. An LSM picture of a river aggregates in combination with lectin detection of the aggregate matrix, with viruses, is shown in Fig. 15.1. It can be anticipated that such methods, and other techniques such as AFM, will increase our knowledge how viruses interact e.g. with organic and inorganic particles.

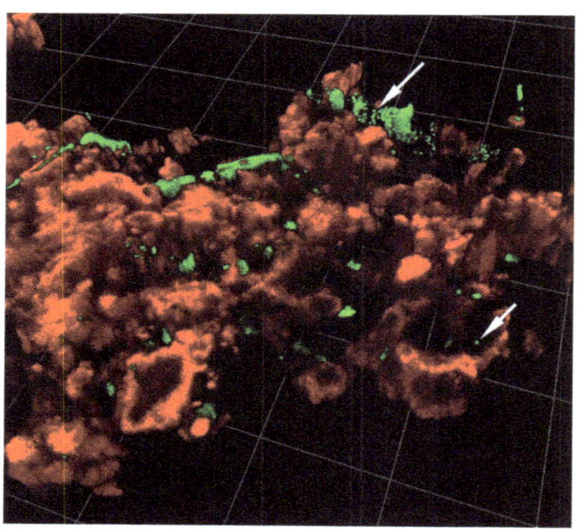

Figure 15.1 3D volume reconstruction of a river aggregate by confocal laser scanning microscopy. Arrows point to viruses. Green: Nucleic acid (staining by SYBRGreen I); red: glycoconjugates (detection by lectin).

Acknowledgements

This work was financed by the ANR-project ANCESTRAM (French Science Ministry).

References

Ackermann, H.W. (1998). Tailed bacteriophages: the order Caudovirales. Adv. Virus Res. *51*, 135–201.

Ackermann, H.W. (2012). Bacteriophage electron microscopy. Adv. Virus Res. *82*, 1–32. https://doi.org/10.1016/B978-0-12-394621-8.00017-0.

Aguirre de Cárcer, D., López-Bueno, A., Alonso-Lobo, J.M., Quesada, A., and Alcamí, A. (2016). Metagenomic analysis of lacustrine viral diversity along a latitudinal transect of the Antarctic Peninsula. FEMS Microbiol. Ecol. *92*, fiw074. https://doi.org/10.1093/femsec/fiw074.

Allen, L.Z., Ishoey, T., Novotny, M.A., McLean, J.S., Lasken, R.S., and Williamson, S.J. (2011). Single virus genomics: a new tool for virus discovery. PLOS ONE *6*, e17722. https://doi.org/10.1371/journal.pone.0017722.

Allers, E., Moraru, C., Duhaime, M.B., Beneze, E., Solonenko, N., Barrero-Canosa, J., Amann, R., and Sullivan, M.B. (2013). Single-cell and population level viral infection dynamics revealed by phageFISH, a method to visualize intracellular and free viruses. Environ. Microbiol. *15*, 2306–2318. https://doi.org/10.1111/1462-2920.12100.

Amann, R., and Fuchs, B.M. (2008). Single-cell identification in microbial communities by improved fluorescence *in situ* hybridization techniques. Nat. Rev. Microbiol. *6*, 339–348. https://doi.org/10.1038/nrmicro1888.

Amenabar, I., Poly, S., Nuansing, W., Hubrich, E.H., Govyadinov, A.A., Huth, F., Krutokhvostov, R., Zhang, L., Knez, M., Heberle, J., *et al.* (2014). Structural analysis and mapping of individual protein complexes by infrared nanospectroscopy. Nat. Commun. *4*, 2890.

Angly, F.E., Felts, B., Breitbart, M., Salamon, P., Edwards, R.A., Carlson, C., Chan, A.M., Haynes, M., Kelley, S., Liu, H., *et al.* (2006). The marine viromes of four oceanic regions. PLOS Biol. *4*, e368.

Ankrah, N.Y.D., May, A.L., Middleton, J.L., Jones, D.R., Hadden, M.K., Gooding, J.R., LeCleir, G.R., Wilhelm, S.W., Campagna, S.R., and Buchan, A. (2014). Phage infection of an environmentally relevant marine bacterium alters host metabolism and lysate composition. ISME J. *8*, 1089–1100.

Arakelyan, A., Fitzgerald, W., Margolis, L., and Grivel, J.-C. (2013). Nanoparticle-based flow virometry for the analysis of individual virions. J. Clin. Invest. *123*, 3716–3727.

Banfield, J.F., Verberkmoes, N.C., Hettich, R.L., and Thelen, M.P. (2005). Proteogenomic approaches for the molecular characterization of natural microbial communities. OMICS *9*, 301–333. https://doi.org/10.1089/omi.2005.9.301.

Bench, S.R., Hanson, T.E., Williamson, K.E., Ghosh, D., Radosovich, M., Wang, K., and Wommack, K.E. (2007). Metagenomic characterization of Chesapeake Bay virioplankton. Appl. Environ. Microbiol. *73*, 7629–7641.

Benie, A.J., Moser, R., Bäuml, E., Blaas, D., and Peters, T. (2003). Virus-ligand interactions: identification and characterization of ligand binding by NMR spectroscopy. J. Am. Chem. Soc. *125*, 14–15. https://doi.org/10.1021/ja027691e.

Bergh, O., Børsheim, K.Y., Bratbak, G., and Heldal, M. (1989). High abundance of viruses found in aquatic environments. Nature *340*, 467–468. https://doi.org/10.1038/340467a0.

Bettarel, Y., Motegi, C., Weinbauer, M.G., and Mari, X. (2016). Colonization and release processes of viruses and prokaryotes on artificial marine macroaggregates. FEMS Microbiol. Lett. *363*, fnv216. https://doi.org/10.1093/femsle/fnv216.

Binder, B. (1999). Reconsidering the relationship between virally induced bacterial mortality and frequency of infected cells. Aquat. Microb. Ecol. *18*, 207–215.

Brandenburg, B., and Zhuang, X. (2007). Virus trafficking – learning from single-virus tracking. Nat. Rev. Microbiol. *5*, 197–208.

Breitbart, M., Salamon, P., Andresen, B., Mahaffy, J.M., Segall, A.M., Mead, D., Azam, F., and Rohwer, F. (2002). Genomic analysis of uncultured marine viral communities. Proc. Natl. Acad. Sci. U.S.A. *99*, 14250–14255. https://doi.org/10.1073/pnas.202488399.

Breitbart, M., Felts, B., Kelley, S., Mahaffy, J.M., Nulton, J., Salamon, P., and Rohwer, F. (2004). Diversity and population structure of a near-shore marine-sediment viral community. Proc. Biol. Sci. *271*, 565–574. https://doi.org/10.1098/rspb.2003.2628.

Brum, J.R., Ignacio-Espinoza, J.C., Roux, S., Doulcier, G., Acinas, S.G., Alberti, A., Chaffron, S., Cruaud, C., de Vargas, C., Gasol, J.M., *et al.* (2015a). Ocean plankton. Patterns and ecological drivers of ocean viral communities. Science *348*, 1261498. https://doi.org/10.1126/science.1261498.

Brum, J.R., and Sullivan, M.B. (2015b). Rising to the challenge: accelerated pace of discovery transforms marine virology. Nat. Rev. Microbiol. *13*, 147–159. https://doi.org/10.1038/nrmicro3404.

Brum, J.R., Schenck, R.O., and Sullivan, M.B. (2013). Global morphological analysis of marine viruses shows minimal regional variation and dominance of non-tailed viruses. ISME J. *7*, 1738–1751. https://doi.org/10.1038/ismej.2013.67.

Brum, J.R., and Steward, G.F. (2010). Morphological characterization of viruses in the stratified water column of alkaline, hypersaline Mono Lake. Microb. Ecol. *60*, 636–643. https://doi.org/10.1007/s00248-010-9688-4.

Brussaard, C.P., Marie, D., and Bratbak, G. (2000). Flow cytometric detection of viruses. J. Virol. Methods *85*, 175–182.

Brussaard, C.P.D., Payet, J.P., Winter, C., and Weinbauer, M.G. (2010). Quantification of aquatic viruses by flow cytometry. In Manual of Aquatic Viral Ecology. Weinbauer, M.G., Rowe, J.M., and Wilhelm, S.W. eds. (American Society of Limnology and Oceanography, Waco, Texas), pp 102–109.

Chen, F., Suttle, C.A., and Short, S.M. (1996). Genetic diversity in marine algal virus communities as revealed by sequence analysis of DNA polymerase genes. Appl. Environ. Microbiol. *62*, 2869–2874.

Chen, F., Wang, K., Huang, S., Cai, H., Zhao, M., Jiao, N., and Wommack, K.E. (2009). Diverse and dynamic populations of cyanobacterial podoviruses in the Chesapeake Bay unveiled through DNA polymerase gene sequences. Environ. Microbiol. *11*, 2884–2892. https://doi.org/10.1111/j.1462-2920.2009.02033.x.

Chiura, H.X., Kogure, K., Hagemann, S., Ellinger, A., and Velimirov, B. (2011). Evidence for particle-induced horizontal gene transfer and serial transduction between bacteria. FEMS Microbiol. Ecol. *76*, 576–591. https://doi.org/10.1111/j.1574-6941.2011.01077.x.

Chow, C.E., Kim, D.Y., Sachdeva, R., Caron, D.A., and Fuhrman, J.A. (2014). Top-down controls on bacterial community structure: microbial network analysis of bacteria, T4-like viruses and protists. ISME J. *8*, 816–829. https://doi.org/10.1038/ismej.2013.199.

Colombet, J., Sime-Ngando, T., Cauchie, H.M., Fonty, G., Hoffmann, L., and Demeure, G. (2006). Depth-related gradients of viral activity in Lake Pavin. Appl. Environ. Microbiol. *72*, 4440–4445.

Comeau, A.M., and Krisch, H.M. (2008). The capsid of the T4 phage superfamily: the evolution, diversity, and structure of some of the most prevalent proteins in the biosphere. Mol. Biol. Evol. *25*, 1321–1332. https://doi.org/10.1093/molbev/msn080.

Comeau, A.M., Arbiol, C., and Krisch, H.M. (2010). Gene network visualization and quantitative synteny analysis of more than 300 marine T4-like phage scaffolds from the GOS metagenome. Mol. Biol. Evol. *27*, 1935–1944. https://doi.org/10.1093/molbev/msq076.

Culley, A.I., Lang, A.S., and Suttle, C.A. (2003). High diversity of unknown picorna-like viruses in the sea. Nature *424*, 1054–1057. https://doi.org/10.1038/nature01886.

Culley, A.I., Lang, A.S., and Suttle, C.A. (2007). The complete genomes of three viruses assembled from shotgun libraries of marine RNA virus communities. Virol. J. *4*, 69.

Culley, A.I., Lang, A.S., and Suttle, C.A. (2006). Metagenomic analysis of coastal RNA virus communities. Science *312*, 1795–1798.

Cunningham, B.R., Brum, J.R., Schwenck, S.M., Sullivan, M.B., and John, S.G. (2015). An inexpensive, accurate, and precise wet-mount method for enumerating aquatic viruses. Appl. Environ. Microbiol. *81*, 2995–3000. https://doi.org/10.1128/AEM.03642-14.

Daley, R.J., and Hobbie, J.E. (1975). Direct counts of aquatic bacteria by a modified epifluorescence technique. Limnol. Oceanogr. *20*, 875–882.

Dekel-Bird, N.P., Avrani, S., Sabehi, G., Pekarsky, I., Marston, M.F., Kirzner, S., and Lindell, D. (2013). Diversity and evolutionary relationships of T7-like podoviruses infecting marine cyanobacteria. Environ. Microbiol. *15*, 1476–1491. https://doi.org/10.1111/1462-2920.12103.

Deng, L., Gregory, A., Yilmaz, S., Poulos, B.T., Hugenholtz, P., and Sullivan, M.B. (2012). Contrasting life strategies of viruses that infect photo- and heterotrophic bacteria, as revealed by viral tagging. MBio *3*, e00373–12. https://doi.org/10.1128/mBio.00373-12.

Desplats, C., and Krisch, H.M. (2003). The diversity and evolution of the T4-type bacteriophages. Res. Microbiol. *154*, 259–267.

Dingley, A.J., Lorenzen, I., and Grötzinger, J. (2008). NMR analysis of viral protein structures. Methods Mol. Biol. *451*, 441–462. https://doi.org/10.1007/978-1-59745-102-4_30.

Domon, B., and Aebersold, R. (2006). Mass spectrometry and protein analysis. Science *312*, 212–217.

Fancello, L., Trape, S., Robert, C., Boyer, M., Popgeorgiev, N., Raoult, D., and Desnues, C. (2013). Viruses in the desert: a metagenomic survey of viral communities in four perennial ponds of the Mauritanian Sahara. ISME J. *7*, 359–369. https://doi.org/10.1038/ismej.2012.101.

Filée, J., Tétart, F., Suttle, C.A., and Krisch, H.M. (2005). Marine T4-type bacteriophages, a ubiquitous component of the dark matter of the biosphere. Proc. Natl. Acad. Sci. U.S.A. *102*, 12471–12476.

Filippini, M., and Middelboe, M. (2007). Viral abundance and genome size distribution in the sediment and water column of marine and freshwater ecosystems. FEMS Microbiol. Ecol. *60*, 397–410.

Forterre, P., Soler, N., Krupovic, M., Marguet, E., and Ackermann, H.W. (2013). Fake virus particles generated by fluorescence microscopy. Trends Microbiol. *21*, 1–5. https://doi.org/10.1016/j.tim.2012.10.005.

Francisco, D.E., Mah, R.A., and Rabin, A.C. (1973). Acridine orange-epifluorescence technique for counting bacteria in natural waters. Trans. Am. Microsc. Soc. *92*, 416–421.

Fuller, N.J., Wilson, W.H., Joint, I.R., and Mann, N.H. (1998). Occurrence of a sequence in marine cyanophages similar to that of T4 g20 and its application to PCR-based detection and quantification techniques. Appl. Environ. Microbiol. *64*, 2051–2060.

Gaudin, M., Gauliard, E., Schouten, S., Houel-Renault, L., Lenormand, P., Marguet, E., and Forterre, P. (2013). Hyperthermophilic archaea produce membrane vesicles that can transfer DNA. Environ. Microbiol. Rep. *5*, 109–116. https://doi.org/10.1111/j.1758-2229.2012.00348.x.

Gaudin, R., and Barteneva, N.S. (2015). Sorting of small infectious virus particles by flow virometry reveals distinct infectivity profiles. Nat. Commun. *6*, 6022. https://doi.org/10.1038/ncomms7022.

Gilmore, J.L., Aizaki, H., Yoshida, A., Deguchi, K., Kumeta, M., Junghof, J., Wakita, T., and Takeyasu, K. (2014). Nanoimaging of ssRNA: genome architecture of the Hepatitis C Virus revealed by atomic force microscopy. J. Nanomed. Nanotechol. S5 https://doi.org/10.4172/2157-7439.S5-010.

Gustavsen, J.A., Winget, D.M., Tian, X., and Suttle, C.A. (2014). High temporal and spatial diversity in marine RNA viruses implies that they have an important role in mortality and structuring plankton communities. Front. Microbiol. *5*, 703. https://doi.org/10.3389/fmicb.2014.00703.

Hammond, G.W., Hazelton, P.R., Chuang, I., and Klisko, B. (1981). Improved detection of viruses by electron microscopy after direct ultracentrifuge preparation of specimens. J. Clin. Microbiol. *14*, 210–221.

Holmfeldt, K., Odić, D., Sullivan, M.B., Middelboe, M., and Riemann, L. (2012). Cultivated single-stranded DNA phages that infect marine Bacteroidetes prove difficult to detect with DNA-binding stains. Appl. Environ. Microbiol. *78*, 892–894. https://doi.org/10.1128/AEM.06580-11.

Hübner, W., McNerney, G.P., Chen, P., Dale, B.M., Gordon, R.E., Chuang, F.Y., Li, X.D., Asmuth, D.M., Huser, T., and Chen, B.K. (2009). Quantitative 3D video microscopy of HIV transfer across T cell virological synapses. Science *323*, 1743–1747. https://doi.org/10.1126/science.1167525.

Ishoey, T., Woyke, T., Stepanauskas, R., Novotny, M., and Lasken, R.S. (2008). Genomic sequencing of single microbial cells from environmental samples. Curr. Opin. Microbiol. *11*, 198–204. https://doi.org/10.1016/j.mib.2008.05.006.

Jamindar, S., Polson, S.W., Srinivasiah, S., Waidner, L., and Wommack, K.E. (2012). Evaluation of two approaches for assessing the genetic similarity of virioplankton populations as defined by genome size. Appl. Environ. Microbiol. *78*, 8773–8783. https://doi.org/10.1128/AEM.02432-12.

Jore, M.M., Lundgren, M., van Duijn, E., Bultema, J.B., Westra, E.R., Waghmare, S.P., Wiedenheft, B., Pul, U., Wurm, R., Wagner, R., *et al.* (2011). Structural basis for CRISPR RNA-guided DNA recognition by Cascade. Nat. Struct. Mol. Biol. *18*, 529–536. https://doi.org/10.1038/nsmb.2019.

Keller, M., and Hettich, R. (2009). Environmental proteomics: a paradigm shift in characterizing microbial activities at the molecular level. Microbiol. Mol. Biol. Rev. *73*, 62–70. https://doi.org/10.1128/MMBR.00028-08.

Kenzaka, T., Tani, K., and Nasu, M. (2010). High-frequency phage-mediated gene transfer in freshwater environments determined at single-cell level. ISME J. *4*, 648–659. https://doi.org/10.1038/ismej.2009.145.

Kim, B.C., Ju, M.K., Dan-Chin-Yu, A., and Sommer, P. (2009). Quantitative detection of HIV-1 particles using HIV-1 neutralizing antibody-conjugated beads. Anal. Chem. *81*, 2388–2393. https://doi.org/10.1021/ac802267u.

Kuznetsov, Y.G., and McPherson, A. (2011). Atomic force microscopy in imaging of viruses and virus-infected cells. Microbiol. Mol. Biol. Rev. *75*, 268–285. https://doi.org/10.1128/MMBR.00041-10.

Lander, G.C., Baudoux, A.C., Azam, F., Potter, C.S., Carragher, B., and Johnson, J.E. (2012). Capsomer dynamics and stabilization in the T = 12 marine bacteriophage SIO-2 and its procapsid studied by CryoEM. Structure *20*, 498–503. https://doi.org/10.1016/j.str.2012.01.007.

Labonté, J.M., Reid, K.E., and Suttle, C.A. (2009). Phylogenetic analysis indicates evolutionary diversity and environmental segregation of marine podovirus DNA polymerase gene sequences. Appl. Environ. Microbiol. *75*, 3634–3640. https://doi.org/10.1128/AEM.02317-08.

Lang, A.S., Rise, M.L., Culley, A.I., and Steward, G.F. (2009). RNA viruses in the sea. FEMS Microbiol. Rev. *33*, 295–323. https://doi.org/10.1111/j.1574-6976.2008.00132.x.

Lang, A.S., Zhaxybayeva, O., and Beatty, J.T. (2012). Gene transfer agents: phage-like elements of genetic exchange. Nat. Rev. Microbiol. *10*, 472–482. https://doi.org/10.1038/nrmicro2802.

Larsen, J.B., Larsen, A., Bratbak, G., and Sandaa, R.A. (2008). Phylogenetic analysis of members of the Phycodnaviridae virus family, using amplified fragments of the major capsid protein gene. Appl. Environ. Microbiol. *74*, 3048–3057. https://doi.org/10.1128/AEM.02548-07.

Lauro, F.M., DeMaere, M.Z., Yau, S., Brown, M.V., Ng, C., Wilkins, D., Raftery, M.J., Gibson, J.A., Andrews-Pfannkoch, C., Lewis, M., *et al.* (2011). An integrative study of a meromictic lake ecosystem in Antarctica. ISME J. *5*, 879–895. https://doi.org/10.1038/ismej.2010.185.

Lin, X., Ding, H., and Zeng, Q. (2016). Transcriptomic response during phage infection of a marine cyanobacterium under phosphorus-limited conditions. Environ. Microbiol. *18*, 450–460. https://doi.org/10.1111/1462-2920.13104.

López-Bueno, A., Tamames, J., Velázquez, D., Moya, A., Quesada, A., and Alcamí, A. (2009). High diversity of the viral community from an Antarctic lake. Science *326*, 858–861. https://doi.org/10.1126/science.1179287.

Luef, B., Neu, T.R., and Peduzzi, P. (2009). Imaging and quantifying virus fluorescence signals on aquatic aggregates: a new method and its implication for aquatic microbial ecology. FEMS Microbiol. Ecol. *68*, 372–380. https://doi.org/10.1111/j.1574-6941.2009.00675.x.

Maranger, R., Bird, D.F., and Juniper, S.K. (1994). Viral and bacterial dynamics in Artic sea ice during the spring algal bloom near Resolute, N.W.T., Canada. Mar. Ecol. Prog. Ser. *111*, 121–127.

Marie, D., Brussaard, C.P.D., Thyrhaug, R., Bratbak, G., and Vaulot, D. (1999). Enumeration of marine viruses in culture and natural samples by flow cytometry Appl. Environ. Microbiol. *65*, 45–52.

Marsh, M., and Helenius, A. (2006). Virus entry: open sesame. Cell *124*, 729–740.

Marston, M.F., Taylor, S., Sme, N., Parsons, R.J., Noyes, T.J.E., and Martiny, J.B.H. (2013). Marine cyanophages exhibit local and regional biogeography. Environ Microbiol, *15*, 1452–1463.

Mashburn-Warren, L.M., and Whiteley, M. (2006). Special delivery: vesicle trafficking in prokaryotes. Mol. Microbiol. *61*, 839–846.

McDaniel, L.D., Young, E., Delaney, J., Ruhnau, F., Ritchie, K.B., and Paul, J.H. (2010). High frequency of horizontal gene transfer in the oceans. Science *330*, 50. https://doi.org/10.1126/science.1192243.

McGowan, S., Buckle, A.M., Mitchell, M.S., Hoopes, J.T., Gallagher, D.T., Heselpoth, R.D., Shen, Y., Reboul, C.F., Law, R.H., Fischetti, V.A., *et al.* (2012). X-ray crystal structure of the streptococcal specific phage lysin PlyC. Proc. Natl. Acad. Sci. U.S.A. *109*, 12752–12757. https://doi.org/10.1073/pnas.1208424109.

Mokili, J.L., Rohwer, F., and Dutilh, B.E. (2012). Metagenomics and future perspectives in virus discovery. Curr. Opin. Virol. *2*, 63–77. https://doi.org/10.1016/j.coviro.2011.12.004.

Moraru, C., Lam, P., Fuchs, B.M., Kuypers, M.M., and Amann, R. (2010). GeneFISH – an *in situ* technique for linking gene presence and cell identity in environmental microorganisms. Environ. Microbiol. *12*, 3057–3073. https://doi.org/10.1111/j.1462-2920.2010.02281.x.

Opella, S.J., Zeri, A.C., and Park, S.H. (2008). Structure, dynamics, and assembly of filamentous bacteriophages by nuclear magnetic resonance spectroscopy. Annu. Rev. Phys. Chem. *59*, 635–657. https://doi.org/10.1146/annurev.physchem.58.032806.104640.

Parada, V., Baudoux, A.C., Sintes, E., Weinbauer, M.G., and Herndl, G.J. (2008). Dynamics and diversity of newly produced virioplankton in the North Sea. ISME J. *2*, 924–936. https://doi.org/10.1038/ismej.2008.57.

Parikka, K.J., Le Romancer, M., Wauters, N., and Jacquet, S. (2016). Deciphering the virus-to-prokaryote ratio (VPR): insights into virus-host relationships in a variety of ecosystems. Biol. Rev. Camb. Philos. Soc. *92*, 1081–1100. doi: 10.1111/brv.12271.

Peduzzi, P., Agis, M., and Luef, B. (2013). Evaluation of confocal laser scanning microscopy for enumeration of virus-like particles in aquatic systems. Environ. Monit. Assess. *185*, 5411–5418. https://doi.org/10.1007/s10661-012-2955-8.

Peduzzi, P., and Weinbauer, M.G. 1993. Effect of concentrating the virus-rich 2-200 nm size fraction of seawater on the formation of algal flocs (marine snow). Limol. Oceanogr. *38*, 1562–1565.

Pernthaler, A., Pernthaler, J., and Amann, R. (2002). Fluorescence in situ hybridization and catalyzed reporter deposition for the identification of marine bacteria. Appl. Environ. Microbiol. *68*, 3094–3101.

Proctor, L.M., and Fuhrman, J.A. (1990). Viral mortality of marine bacteria and cyanobacteria. Nature *343*, 60–62.

Proctor, L.M., and Fuhrman, J.A. (1991). Roles of viral infection in organic particle flux. Mar. Ecol. Prog. Ser. *69*, 133–142.

Reyes, A., Haynes, M., Hanson, N., Angly, F.E., Heath, A.C., Rohwer, F., and Gordon, J.I. (2010). Viruses in the faecal microbiota of monozygotic twins and their mothers. Nature *466*, 334–338. https://doi.org/10.1038/nature09199.

Rodriguez-Brito, B., Li, L., Wegley, L., Furlan, M., Angly, F., Breitbart, M., Buchanan, J., Desnues, C., Dinsdale, E., Edwards, R., *et al.* (2010). Viral and microbial community dynamics in four aquatic environments. ISME J. *4*, 739–751. https://doi.org/10.1038/ismej.2010.1.

Rohwer, F., and Edwards, R. (2002). The Phage Proteomic Tree: a genome-based taxonomy for phage. J. Bacteriol. *184*, 4529–4535.

Rosario, K., Nilsson, C., Lim, Y.W., Ruan, Y., and Breitbart, M. (2009). Metagenomic analysis of viruses in reclaimed water. Environ. Microbiol. *11*, 2806–2820. https://doi.org/10.1111/j.1462-2920.2009.01964.x.

Rosario, K., and Breitbart, M. (2011). Exploring the viral world through metagenomics. Curr. Opin. Virol. *1*, 289–297. https://doi.org/10.1016/j.coviro.2011.06.004.

Roux, S., Krupovic, M., Poulet, A., Debroas, D., and Enault, F. (2012). Evolution and diversity of the Microviridae viral family through a collection of 81 new complete genomes assembled from virome reads. PLOS ONE *7*, e40418. https://doi.org/10.1371/journal.pone.0040418.

Sandaa, R.A., and Larsen, A. (2006). Seasonal variations in virus-host populations in Norwegian coastal waters: focusing on the cyanophage community infecting marine *Synechococcus* spp. Appl. Environ. Microbiol. *72*, 4610–4618.

Sandaa, R.A., Clokie, M., and Mann, N.H. (2008). Photosynthetic genes in viral populations with a large genomic size range from Norwegian coastal waters. FEMS Microbiol. Ecol. *63*, 2–11.

Schneider, T., and Riedel, K. (2010). Environmental proteomics: analysis of structure and function of microbial communities. Proteomics *10*, 785–798. https://doi.org/10.1002/pmic.200900450.

Scott, L.G., and Hennig, M. (2008). RNA structure determination by NMR. Methods Mol. Biol. *452*, 29–61. https://doi.org/10.1007/978-1-60327-159-2_2.

Shibata, A., Goto, Y., Saito, H., Kikuchi, T., Toda, T., and Taguchi, S. (2006). Comparison of SYBR Green I and SYBR Gold strains for enumerating bacteria and viruses by epifluorescence microscopy. Aquat. Microb. Ecol. *43*, 223–231.

Simon, C., and Daniel, R. (2011). Metagenomic analyses: past and future trends. Appl. Environ. Microbiol. *77*, 1153–1161. https://doi.org/10.1128/AEM.02345-10.

Skvortsov, T., de Leeuwe, C., Quinn, J.P., McGrath, J.W., Allen, C.C., McElarney, Y., Watson, C., Arkhipova, K., Lavigne, R., and Kulakov, L.A. (2016). Metagenomic Characterisation of the Viral Community of Lough Neagh, the Largest Freshwater Lake in Ireland. PLOS ONE *11*, e0150361. https://doi.org/10.1371/journal.pone.0150361.

Sogin, M.L., Morrison, H.G., Huber, J.A., Mark Welch, D., Huse, S.M., Neal, P.R., Arrieta, J.M., and Herndl, G.J. (2006). Microbial diversity in the deep sea and the underexplored 'rare biosphere'. Proc. Natl. Acad. Sci. U.S.A. *103*, 12115–12120.

Stepanauskas, R. (2012). Single cell genomics: an individual look at microbes. Curr. Opin. Microbiol. *15*, 613–620. https://doi.org/10.1016/j.mib.2012.09.001.

Steward, G.F., Culley, A.I., Mueller, J.A., Wood-Charlson, E.M., Belcaid, M., and Poisson, G. (2013). Are we missing half of the viruses in the ocean? ISME J. *7*, 672–679. https://doi.org/10.1038/ismej.2012.121.

Steward, G.F., Winker, J., Cochlan, W.P., Smith, D.C., and Azam, F. (1992). Estimation of virus production in the sea: II field results. Mar. Microb. Food Webs *6*, 79–90.

Sullivan, M.B., Lindell, D., Lee, J.A., Thompson, L.R., Bielawski, J.P., and Chisholm, S.W. (2006). Prevalence and evolution of core photosystem II genes in marine cyanobacterial viruses and their hosts. PLOS Biol. *4*, e234.

Sullivan, M.B., Coleman, M.L., Quinlivan, V., Rosenkrantz, J.E., Defrancesco, A.S., Tan, G., Fu, R., Lee, J.A., Waterbury, J.B., Bielawski, J.P., *et al.* (2008). Portal protein diversity and phage ecology. Environ. Microbiol. *10*, 2810–2823. https://doi.org/10.1111/j.1462-2920.2008.01702.x.

Sun, E., He, J., and Zhuang, X. (2013). Live cell imaging of viral entry. Curr. Opin. Virol. *3*, 34–43. https://doi.org/10.1016/j.coviro.2013.01.005.

Syka, J.E., Coon, J.J., Schroeder, M.J., Shabanowitz, J., and Hunt, D.F. (2004). Peptide and protein sequence analysis by electron transfer dissociation mass spectrometry. Proc. Natl. Acad. Sci. U.S.A. *101*, 9528–9533. https://doi.org/10.1073/pnas.0402700101.

Tadmor, A.D., Ottesen, E.A., Leadbetter, J.R., and Phillips, R. (2011). Probing individual environmental bacteria for viruses by using microfluidic digital PCR. Science *333*, 58–62. https://doi.org/10.1126/science.1200758.

Tuma, R.S., Beaudet, M.P., Jin, X., Jones, L.J., Cheung, C.Y., Yue, S., and Singer, V.L. (1999). Characterization of SYBR Gold nucleic acid gel stain: a dye optimized for use with 300-nm ultraviolet transilluminators. Anal. Biochem. *268*, 278–288.

Tzakos, A.G., Grace, C.R., Lukavsky, P.J., and Riek, R. (2006). NMR techniques for very large proteins and rnas in solution. Annu. Rev. Biophys. Biomol. Struct. *35*, 319–342. https://doi.org/10.1146/annurev.biophys.35.040405.102034.

Vargas, C.A., Wilhelm, A.A., Williams, J., Lucas, P., Reynolds, K.A., and Riley, M.R. (2009). Integrated capture and spectroscopic detection of viruses. Appl. Environ. Microbiol. *75*, 6431–6440. https://doi.org/10.1128/AEM.02036-08.

VerBerkmoes, N.C., Denef, V.J., Hettich, R.L., and Banfield, J.F. (2009). Systems biology: Functional analysis of natural microbial consortia using community proteomics. Nat. Rev. Microbiol. *7*, 196–205. https://doi.org/10.1038/nrmicro2080.

Viswanathan, K., and Früh, K. (2007). Viral proteomics: global evaluation of viruses and their interaction with the host. Expert Rev. Proteomics *4*, 815–829. https://doi.org/10.1586/14789450.4.6.815.

Wang, Y., Hammes, F., De Roy, K., Verstraete, W., and Boon, N. (2010). Past, present and future applications of flow cytometry in aquatic microbiology. Trends Biotechnol. *28*, 416–424. https://doi.org/10.1016/j.tibtech.2010.04.006.

Weinbauer, M.G. (2004). Ecology of prokaryotic viruses. FEMS Microbiol. Rev. *28*, 127–181. https://doi.org/10.1016/j.femsre.2003.08.001.

Weinbauer, M.G., Fuks, D., and Peduzzi, P. (1993). Distribution of Viruses and Dissolved DNA along a Coastal Trophic Gradient in the Northern Adriatic Sea. Appl. Environ. Microbiol. *59*, 4074–4082.

Weinbauer, M.G., and Rassoulzadegan, F. (2004). Are viruses driving microbial diversification and diversity? Environ. Microbiol. *6*, 1–11.

Weinbauer, M.G., Bettarel, Y., Cattaneo, R., Luef, B., Maier, C., Motegi, C., Peduzzi, P., and Mari, X. (2009). Viral ecology of organic and inorganic particles in aquatic systems: avenues for further research. Aquat. Microb. Ecol. *57*, 321–341. https://doi.org/10.3354/ame01363.

Williamson, S.J., Cary, S.C., Williamson, K.E., Helton, R.R., Bench, S.R., Winget, D., and Wommack, K.E. (2008). Lysogenic virus-host interactions predominate at deep-sea diffuse-flow hydrothermal vents. ISME J. *2*, 1112–1121. https://doi.org/10.1038/ismej.2008.73.

Winget, D.M., and Wommack, K.E. (2008). Randomly amplified polymorphic DNA PCR as a tool for assessment of marine viral richness. Appl. Environ. Microbiol. *74*, 2612–2618. https://doi.org/10.1128/AEM.02829-07.

Winter, C., and Weinbauer, M.G. (2010). Randomly amplified polymorphic DNA reveals tight links between viruses and microbes in the bathypelagic zone of the Northwestern Mediterranean Sea. Appl. Environ. Microbiol. *76*, 6724–6732. https://doi.org/10.1128/AEM.00531-10.

Wommack, K.E., Ravel, J., Hill, R.T., Chun, J., and Colwell, R.R. (1999). Population dynamics of chesapeake bay virioplankton: total-community analysis by pulsed-field gel electrophoresis Appl. Environ. Microbiol. *65*, 231–240.

Woyke, T., Tighe, D., Mavromatis, K., Clum, A., Copeland, A., Schackwitz, W., Lapidus, A., Wu, D., McCutcheon, J.P., McDonald, B.R., *et al.* (2010). One bacterial cell, one complete genome. PLOS ONE *5*, e10314. https://doi.org/10.1371/journal.pone.0010314.

Yonath, A. (2011). X-ray crystallography at the heart of life science. Curr. Opin. Struct. Biol. *21*, 622–626. https://doi.org/10.1016/j.sbi.2011.07.005.

Yoon, H.S., Price, D.C., Stepanauskas, R., Rajah, V.D., Sieracki, M.E., Wilson, W.H., Yang, E.C., Duffy, S., and Bhattacharya, D. (2011). Single-cell genomics reveals organismal interactions in uncultivated marine protists. Science *332*, 714–717. https://doi.org/10.1126/science.1203163.

Zhong, Y., Chen, F., Wilhelm, S.W., Poorvin, L., and Hodson, R.E. (2002). Phylogenetic diversity of marine cyanophage isolates and natural virus communities as revealed by sequences of viral capsid assembly protein gene g20. Appl. Environ. Microbiol. *68*, 1576–1584.

Zhong, X., and Jacquet, S. (2013). Prevalence of viral photosynthetic and capsid protein genes from cyanophages in two large and deep peri-alpine lakes. Appl. Environ. Microbiol. *79*, 7169–7178.

Zhong, X., and Jacquet, S. (2014). Contrasting diversity of phycodnavirus signature genes in two large and deep western European lakes. Environ. Microbiol. *16*, 759–773. https://doi.org/10.1111/1462-2920.12201.

Zhong, X., Colombet, J., Pradeep-Ram, A.S., and Jacquet, S. (2014). Diversity of genome size, morphology, abundance and functional role of the virioplankton in Lakes Annecy and Bourget over a one-year period. Microb. Ecol. *67*, 66–82.

Zhong, X., Guidoni, B., Jacas, L., and Jacquet, S. (2015). Structure and diversity of ssDNA Microviridae viruses in two peri-alpine lakes (Annecy and Bourget, France). Res. Microbiol. *166*, 644–654. https://doi.org/10.1016/j.resmic.2015.07.003.

Zhou, S., Liu, R., Zhao, X., Huang, C., and Wei, Y. (2011). Viral proteomics: the emerging cutting-edge of virus research. Sci. China Life Sci *54*, 502–512. https://doi.org/10.1007/s11427-011-4177-7

Current Books of Interest

Plant-Microbe Interactions in the Rhizosphere	2019
Porcine Viruses: From Pathogenesis to Strategies for Control	2019
Lactobacillus Genomics and Metabolic Engineering	2019
Cyanobacteria: Signaling and Regulation Systems	2018
Viruses of Microorganisms	2018
Genes, Genetics and Transgenics for Virus Resistance in Plants	2018
DNA Tumour Viruses: Virology, Pathogenesis and Vaccines	2018
Pathogenic *Escherichia coli*: Evolution, Omics, Detection and Control	2018
Postgraduate Handbook: A Comprehensive Guide for PhD and Master's Students and their Supervisors	2018
Enteroviruses: Omics, Molecular Biology, and Control	2018
Molecular Biology of Kinetoplastid Parasites	2018
Bacterial Evasion of the Host Immune System	2017
Illustrated Dictionary of Parasitology in the Post-genomic Era	2017
Next-generation Sequencing and Bioinformatics for Plant Science	2017
The CRISPR/Cas System: Emerging Technology and Application	2017
Brewing Microbiology: Current Research, Omics and Microbial Ecology	2017
Metagenomics: Current Advances and Emerging Concepts	2017
Bacillus: Cellular and Molecular Biology (Third Edition)	2017
Cyanobacteria: Omics and Manipulation	2017
Foot-and-Mouth Disease Virus: Current Research and Emerging Trends	2017
Brain-eating Amoebae: Biology and Pathogenesis of *Naegleria fowleri*	2016
Staphylococcus: Genetics and Physiology	2016
Chloroplasts: Current Research and Future Trends	2016
Microbial Biodegradation: From Omics to Function and Application	2016
Influenza: Current Research	2016
MALDI-TOF Mass Spectrometry in Microbiology	2016
Aspergillus and *Penicillium* in the Post-genomic Era	2016
The Bacteriocins: Current Knowledge and Future Prospects	2016
Omics in Plant Disease Resistance	2016
Acidophiles: Life in Extremely Acidic Environments	2016
Climate Change and Microbial Ecology: Current Research and Future Trends	2016
Biofilms in Bioremediation: Current Research and Emerging Technologies	2016
Microalgae: Current Research and Applications	2016
Gas Plasma Sterilization in Microbiology: Theory, Applications, Pitfalls and New Perspectives	2016

Full details at www.caister.com

Index

Short virus names (including abbeviations)

A

B

C

Long virus names (not abbreviated)

Other virus names and descriptors

Non-virus taxa

Additional terms

C

D

Q

R

S

T

www.ingramcontent.com/pod-product-compliance
Lightning Source LLC
LaVergne TN
LVHW070847210926
848482LV00016B/884